BEYOND THE ENERGY CRISIS

Opportunity and Challenge

(in Four Volumes)

Proceedings of the Third International Conference on Energy Use Management
Berlin (West), October 26 - 30, 1981

Organized by:

The Interdisciplinary Group for Ecology, Development and Energy (EDEN)
Los Angeles, California, USA
Paris, France
Berlin (West), Federal Republic of Germany

In Cooperation with:

The Commission of the European Communities, Brussels
The Senate of Berlin
Bundesanstalt für Materialprüfung Berlin (BAM)
Technical University of Berlin
Organization for Economic Cooperation and Development (OECD)
U.S. Department of Energy, Washington
Alliance to Save Energy
University of Arizona

PREFACE

The last eight years have seen much discussion of energy throughout the world and some increasingly important new policies, technologies, economics, and political attitudes. Although the changes have been small and to date have not had the impact that many of us wish they would have, they signal the coming of a new revolution. It may be unequivocally stated that the world as we have known it is changing dramatically. The past few decades, which were characterized by abundant supplies of cheap energy, are no longer with us. In the future, energy will be more costly and less available and at the same time in greater demand due to a growing world population and the growing aspirations of that population.

To avoid world conflicts and economic disruptions as cheap fossil fuels become more scarce and their maldistribution becomes more acute, new perspectives are required leading to long-term planning within nations and between them. A new era is emerging in which there exist opportunities and challenges in social, political, and economic terms for interdependence and partnership between the developed and developing world.

Fortunately, we are also at a time when there is a growing awareness of the importance and vital necessity of energy. There are many economic, social and technical measures being developed to help increase the efficiency with which we use our precious energy resources.

ICEUM-III is dedicated to examining approaches and means to achieve and maintain economic well being in an energy efficient society. The emphasis in the organization of the program is international and interdisciplinary. The expression of diverse philosophical views is encouraged. Four broad mechanisms of change and modification provide the underlying themes of the presentations and discussion:
- Rational Economic and Energy Policy   - Technology and Innovation
- Institutional Organization   - Social and Cultural Values
As a principal goal, the Conference will provide insight into the extraordinary energy problems faced by the world during the next two decades, and will suggest new solutions and approaches.

This conference in Berlin assembles many of the world's leading experts in the field of efficient energy use, more appropriately referred to as Energy Use Management. It is our earnest hope that the Conference will provide a forum for exchanging ideas and will help stimulate innovations and concepts for a future energy efficient society.

      Rocco Fazzolare   Craig B. Smith

ORGANIZATION AND PLANNING COMMITTEES

Conference Chairmen:

Craig B. Smith
Anco Engineers, Inc.
9937 Jefferson Blvd.
Culver City, California, USA

Rocco Fazzolare
Dept. of Nuclear and Energy
Engineering
University of Arizona
Tucson, Arizona 85721, USA

Executive Advisory Board:

U. Businaro, Director of Research,
FIAT, Orbassano, Italy

Tina Hobson, Director,
Office of Consumer Affairs, U.S.
DOE, Washington, D.C., USA

Gerald Leighton, U.S.
Dept. of Energy, Washington,
D.C., USA

Robert Maxwell, Publisher,
Pergamon Press,
Oxford, England

G. Mélèse D'Hospital, Consultant,
General Atomic Company,
San Diego, California, USA

Christof Rohrbach, Vice-President,
Bundesanstalt für Materialprüfung
(BAM), Berlin (West),
Federal Republic of Germany

Chauncey Starr, Vice Chairman,
Electric Power Research Institute
Palo Alto, California, USA

Organization Board:

J. M. Gibb, Commission of the
European Communities, Luxembourg

Bob Jackson, International Affairs
U.S. Department of Energy
Washington D.C., USA

Klaus Künkel
Technical University of Berlin,
Berlin (West), Federal Republic of
Germany

Jean Mascarello
Direction des Etudes et
Recherches d'Electricité de France,
Paris, France

Maxine Savitz,
U.S. Dept. of Energy,
Washington, D.C., USA

Hasso Schreck
Technical University of Berlin
Berlin (West), Federal Republic
of Germany

Congress Organizing Bureau:

DER-Congress, German Convention
Service, Joachimstater Strasse 19,
D-1000 Berlin 15

Program Committee:

K. K. Murthy, UNESCO, Jamaica
David Pilati, Esalen Institute, California
Jeffrey M. Klopatek, Arizona State University, Arizona
Fritz Kalhammer, Electric Power Research Institute, California
Thomas V. Long, University of Chicago, Illinois
Volker Hartkopf, Carnegie-Mellon University, Pennsylvania
Bill S. Stout, Texas A and M, Texas
Margaret Fels, Princeton University, New Jersey
Rosalyn Barbieri, Jet Propulsion Laboratory, California
James Dyer, Long Beach State University, California

Staff/Publicity/Student Program:

Katherine Little, University of Arizona, Arizona
Muriel Willens, Anco Engineers, Inc., California
Marcia Untracht, Anco Engineers, Inc., California
Brunhilt Jaeger, Berlin (West), Federal Republic of Germany
Dietmar Winje, Technical University, Berlin (West), Federal Republic of Germany

TECHNICAL PROGRAM OUTLINE AND LIST OF SESSIONS

TOPIC A INTERNATIONAL POLICIES AND STRATEGIES
Sessions

A-1   The Oil Transition
A-2   Progress in Improving Energy Utilization 1970-1980
A-3   Energy Paths for Industrializing Economies
A-4   Energy Constraints and Strategies for Development
A-5   The Oil Exporting Developing World
A-6   Policy Formulation for Energy Saving and Substitution

TOPIC B MODELS, ANALYSIS, ASSESSMENT

B-1   End Use Models
B-2   Modeling Energy Use of Buildings
B-3   National Energy Assessment
B-4   Global Energy Supply Assessments
B-5   Energy Information Systems
B-6   Use of Energy Policy Models: Case Histories
B-7   The Limits of Analysis

TOPIC C ECOLOGY AND ENVIRONMENT

C-1   Energy and Water Conflicts
C-2   Health and Environmental Implications of World Energy Policy
C-3   Environmental Significance of Utilizing Biomass for Energy
C-4   Ecological Implications of Alternative Energy Sources
C-5   Oil Shale and Tar Sands: Ecological Consequences of their Development
C-6   Environmental Consequences of Energy Development: Global CO_2
C-7   Environmental Consequences of Energy Development: Acid Precipitation
C-8   Energy, Environmental Quality, and Human Ecology

TOPIC D NEW ENERGY TECHNOLOGIES

D-1   Advanced Energy Conversion Technologies
D-2   Technologies and Energy Flow for Dual Energy Use Systems
D-3   Energy Storage and Management
D-4   Utilization of Environmental and Waste Heat
D-5   Utilization of Solar Energy
D-6   Control of Energy Systems

TOPIC E   INDUSTRIAL PRODUCTIVITY AND DEVELOPMENT

E-1   Energy Management Via Material Recovery
E-2   Industrial Energy Use in Developing Countries
E-3   The Rational Use of Energy in Major Industrial Sectors
E-4   Energy Management Through New Generic Technologies
E-5   Thermodynamic Analysis of Industrial Production

TOPIC F   COMMUNITY SYSTEMS AND BUILDINGS

F-1   Demographics, Living Patterns, and Energy Use
F-2   Energy Management for Building Occupant Productivity and
           Well-being
F-3   Commercial Buildings
F-4   Monitoring and Auditing of Energy Performance
F-5   Institutions, Incentives, and Energy Use in Buildings
F-6   Integrated Urban Developments
F-7   Material Selection and Component Design
F-8   Residential Buildings

TOPIC G   AGRICULTURE AND RURAL DEVELOPMENT

G-1   Energy Management/Conservation in Production Agriculture
G-2   Energy Management/Conservation in Food Processing
G-3   Solar Applications in Agriculture
G-4   Biomass Production
G-5   Biomass Conversion: Direct Combustion/Gasifiers/Pyrolysis
G-6   Alcohol Fermentation
G-7   Biogas Production
G-8   Vegetable Oil for Diesel Fuel

TOPIC H   TRANSPORTATION

H-1   Automotive Fuel Economy: Policy
H-2   Automotive Fuel Economy: Technology
H-3   New Fuels for Transportation
H-4   Transportation and World Energy: Current Demands, Future
           Trends and Implications
H-5   Transportation and World Energy: Contingency Planning and
           Urbanization Issues
H-6   Intercity Transportation Systems

TOPIC I   ECONOMICS AND CHANGING LIFESTYLES

I-1   Strategies for Education
I-2   Interaction of Government Institutions
I-3   Promotion of New Energy Technologies
I-4   Trade-off Analysis and Energy Options Decision Making
I-5   The Press and Information Transfer
I-6   Energy Consumption and the Quality of Life
I-7   Commmunity Energy Planning and Participation

TOPICAL COORDINATORS AND SESSION LEADERS

TOPIC A
INTERNATIONAL POLICIES AND STRATEGIES

Topical Coordinator: K. K. Murthy
UNESCO, Kingston, Jamaica

A-1 THE OIL TRANSITION

Session Co-Leaders: Mans Lonnroth
Secretariat for Futures Studies,
Stockholm, Sweden

M. Fred Gorbet
Organization for Economic Cooperation
and Development, International Energy
Agency, Paris, France

A-2 PROGRESS IN IMPROVING ENERGY UTILIZATION 1970-1980

Session Leader: Lee Schipper
Lawrence Berkeley Laboratory,
Berkeley, California, USA

A-3 ENERGY PATHS FOR INDUSTRIALIZING ECONOMIES

Session Co-Leaders: J. W. Dias
Energetica, Rio de Janeiro, Brazil

Robert Jackson
U.S. Department of Energy,
Washington, D.C., USA

A-4 ENERGY CONSTRAINTS AND STRATEGIES FOR DEVELOPMENT

Session Leader: Emanuel Glakpe
Howard University, Washington D.C., USA

A-5 THE OIL EXPORTING DEVELOPING WORLD

Session Co-Leaders: Ali Kettani
University of Petroleum and Minerals,
Dhahran, Saudi Arabia

Nelson Ngoka
University of Ife, Ile, Ife
Nigeria

A-6   POLICY FORMULATION FOR ENERGY SAVING AND
      SUBSTITUTION

      Session Co-Leaders: Christian Waeterloos
                              Commission of the European Communities,
                              Brussels, Belgium

                              Gregory Daneke
                              University of Arizona, Tucson, Arizona, USA

TOPIC B
MODELS, ANALYSIS, ASSESSMENT

Topical Coordinator: David Pilati
Esalen Institute, California, USA

B-1   END USE MODELS

      Session Leader: Jerry Jackson
                          Georgia Institute of Technology, Atlanta,
                          Georgia, USA

B-2   MODELING ENERGY USE OF BUILDINGS

      Session Leader: Metin Lokmanhekim
                          Lawrence Berkeley Laboratory, California, USA

B-3   NATIONAL ENERGY ASSESSMENT

      Session Leader: J. E. Samouilidis
                          National Technical University, Athens, Greece

B-4   GLOBAL ENERGY SUPPLY ASSESSMENTS

      Session Leader: J. Michael Gallagher
                          Bechtel National, Inc., San Francisco,
                          California, USA

B-5   ENERGY INFORMATION SYSTEMS

      Session Leader: Wesley Foell
                          University of Wisconsin, Madison, Wisconsin,
                          USA

B-6   USE OF ENERGY POLICY MODELS: CASE HISTORIES

      Session Leader: John Weyant
                          Stanford University, Palo Alto, California, USA

B-7   THE LIMITS OF ANALYSIS

   Session Co-Leaders: Larry Mayer
                  University of Pennsylvania, Philadelphia,
                  Pennsylvania, USA

                  Harry Otway
                  Joint Research Centre, Ispra (Varese), Italy

TOPIC C
ECOLOGY AND ENVIRONMENT

Topical Coordinator: Jeffrey M. Klopatek
Arizona State University, Tempe, Arizona, USA

C-1   ENERGY AND WATER CONFLICTS

   Session Co-Leaders: Martha Gilliland
                  Energy Policy Studies, Inc., Omaha, Nebraska,
                  USA

                  Dietmar Winje
                  Technische Universitat Berlin, Berlin,
                  Federal Republic of Germany

C-2   HEALTH AND ENVIRONMENTAL IMPLICATIONS OF WORLD ENERGY
      POLICY

   Session Co-Leaders: Robert A. Lewis
                  US Department of Energy, Washington, D.C.
                  USA

                  Paul Muller
                  University of Saarland, Saarbrucken,
                  Federal Republic of Germany

C-3   ENVIRONMENTAL SIGNIFICANCE OF UTILIZING BIOMASS FOR ENERGY

   Session Leader: Valcav Smil
                  University of Manitoba, Winnipeg, Canada

C-4   ECOLOGICAL IMPLICATIONS OF ALTERNATIVE ENERGY SOURCES

   Session Co-Leaders: Robert W. Brocksen
                  Electric Power Research Institute, Palo Alto,
                  California, USA

                  Jorge Vieira da Silva
                  Universite de Paris, Paris, France

C-5   OIL SHALE AND TAR SANDS: ECOLOGICAL CONSEQUENCES OF
      THEIR DEVELOPMENT

    Session Co-Leaders: Willard R. Chappel
                                University of Colorado-Denver, Denver,
                                Colorado, USA

                                Trevor E. Brown
                                ESSO - Australia, Ltd., Sydney, New South
                                Wales, Australia

C-6   ENVIRONMENTAL CONSEQUENCES OF ENERGY DEVELOPMENT:
      GLOBAL CO_2

    Session Co-Leaders: Jill Jäger
                                  Kernforschungzentrum Karlsruhe, Karlsruhe,
                                Federal Republic of Germany

                                Jerry S. Olson
                                Oak Ridge National Laboratory, Oak Ridge,
                                Tennessee, USA

C-7   ENVIRONMENTAL CONSEQUENCES OF ENERGY DEVELOPMENT: ACID
      PRECIPITATION

    Session Co-Leaders: Gunnar Abrahamsen
                                  Norsk Institutt for Skogforskning
                                  AS, Norway

                                Hans Seip
                                Central Institute for Industrial Research,
                                Oslo, Norway

C-8   ENERGY, ENVIRONMENTAL QUALITY, AND HUMAN ECOLOGY

    Session Co-Leaders: Helmut Knötig
                                  Technische Universität Wien, Vienna, Austria

                                Erich Panhauser
                                Technische Universitat Wien, Vienna, Austria

TOPIC D
NEW ENERGY TECHNOLOGIES

Topical Coordinator: Fritz Kalhammer
Electric Power Research Institute, Palo Alto,
California, USA

D-1 ADVANCED ENERGY CONVERSION TECHNOLOGIES

    Session Co-Leaders: Anthony Liccardi and/or John Neal
                         U.S. Department of Energy, Germantown,
                         Maryland, USA

                         R. Holighaus and/or H. J. Stoecker
                         KFA Jülich, Julich, Federal Republic of Germany

D-2 TECHNOLOGIES AND ENERGY FLOW FOR DUAL ENERGY USE
SYSTEMS

    Session Co-Leaders: D. Limaye
                         Synergetic Resources Inc.
                         Philadelphia, Pennsylvania

                         P. Fegers
                         Commission of the European Commmunities,
                         Brussels, Belgium

D-3 ENERGY STORAGE AND MANAGEMENT

    Session Co-Leaders: W. Fischer
                         Brown, Boveri & Cie, Heidelberg,
                         Federal Republic of Germany

                         M. Aureille
                         Electricité de France, Moret-sur-Loing,
                         France

D-4 UTILIZATION OF ENVIRONMENTAL AND WASTE HEAT

    Session Co-Leaders: K. F. Ebersbach
                         Forschungsstelle für Energiewirtschaft,
                         Munich, Federal Republic of Germany

                         M. Heurtin
                         Electricite de France, Moret-sur-Loing,
                         France

                         PHilip S. Schmidt
                         University of Texas, Austin Texas

D-5 UTILIZATION OF SOLAR ENERGY

Session Co-Leaders: David Feasby
Solar Energy Research Institute, Golden,
Colorado, USA

Wolfgang Palz
Commission of the European Communities
Brussels, Belgium

D-6 CONTROL OF ENERGY SYSTEMS

Session Leader: Russell B. Spencer
ANCO Engineers, Inc., Culver City,
California, USA

TOPIC E
INDUSTRIAL PRODUCTIVITY AND DEVELOPMENT

Topical Coordinator: Thomas Long
University of Chicago, Chicago, Illinois, USA

E-1 ENERGY MANAGEMENT VIA MATERIAL RECOVERY

Session Co-Leaders: Michael Henstock
University of Nottingham, Nottingham, England

Staf Spaepen
Centre d'Energie Nucleaire, Brussels, Belgium

E-2 INDUSTRIAL ENERGY USE IN DEVELOPING COUNTRIES

Session Leader: Leon de Rosen
Consultant, Versailles, France

E-3 THE RATIONAL USE OF ENERGY IN MAJOR INDUSTRIAL SECTORS

Session Leader: Rick Moll
Statistics Canada, Ottawa, Canada

E-4 ENERGY MANAGEMENT THROUGH NEW GENERIC TECHNOLOGIES

Session Co-Leaders: A. J. Streb
U.S. Department of Energy, Washington, D.C.
USA

C. C. Bradley
U.K. Department of Industry, London, England

E-5   THERMODYNAMIC ANALYSIS OF INDUSTRIAL PRODUCTION

   Session Co-Leaders: R. Stephen Berry
                  University of Chicago, Chicago, Illinois, USA

                  Bjarne Andresen
                  University of Copenhagen, Copenhagen,
                  Denmark

TOPIC F
COMMUNITY SYSTEMS AND BUILDINGS

Topical Coordinators: Volker Hartkopf
Carnegie-Mellon University, Pittsburgh,
Pennsylvania, USA

Bernd Seidel
Dipl. Ing. Architect, Berlin,
Federal Republic of Germany

F-1   DEMOGRAPHICS, LIVING PATTERNS, AND ENERGY USE

   Session Leader: Ira Robinson
               Calgary University, Alberta, Canada

F-2   ENERGY MANAGEMENT FOR BUILDING OCCUPANT PRODUCTIVITY AND
      WELL-BEING

   Session Co-Leaders: Michael K. J. Anderson
                  ANCO Engineers, Inc., Culver City, California
                  USA

                  P. O. Fanger
                  Technical University of Denmark, Lyngby,
                  Denmark

F-3   COMMERCIAL BUILDINGS

   Session Leader: Michael Holtz
               Solar Energy Research Institute, Golden,
               Colorado, USA

F-4   MONITORING AND AUDITING OF ENERGY PERFORMANCE

   Session Co-Leaders: Arthur Rosenfeld
                  Lawrence Berkeley Laboratory, Berkeley,
                  California, USA

                  Albert Dupagne
                  Universite de Liège, Liège, Belgium

F-5 INSTITUTIONS, INCENTIVES, AND ENERGY USE IN BUILDINGS

Session Leader: John Cable
Buildings and Community Systems,
U.S. Department of Energy, Washington, D.C.,
USA

F-6 INTEGRATED URBAN DEVELOPMENTS

Session Co-Leaders: Vivian Loftness
Interatom
Bergisch Gladbach, Federal Republic of
Germany

H. H. Leijendeckers
Van Heugten, Nijmegen, Netherlands

F-7 MATERIAL SELECTION AND COMPONENT DESIGN

Session Leaders: Peter Mill
Department of Public Works, Ottawa, Canada

Gerd Hauser
University of Essen, Essen,
Federal Republic of Germany

F-8 RESIDENTIAL BUILDINGS

Session Leaders: Enzio Guisti
University of Florence, Florence, Italy

Jacques Michel
Architect
Neuilly Sur Seine, France

TOPIC G
AGRICULTURE AND RURAL DEVELOPMENT

Topical Coordinators: William S. Stout
Texas A & M University, College Station,
Texas, USA

Horst Gohlich
Technische Universität Berlin, Berlin,
Federal Republic of Germany

G-1 ENERGY MANAGEMENT/CONSERVATION IN PRODUCTION
AGRICULTURE

Session Leader: David C. White
Ministry of Agriculture, Fisheries & Food,
London, United Kingdom

G-2   ENERGY MANAGEMENT/CONSERVATION IN FOOD PROCESSING

Session Leader: Larry Kelso
U.S. Department of Energy, Washington, D.C.
USA

G-3   SOLAR APPLICATIONS IN AGRICULTURE

Session Leader: F. Wieneke
Universität Göttingen, Göttingen,
Federal Republic of Germany

G-4   BIOMASS PRODUCTION

Session Leader: Kurt B. Heden
National Swedish Board for Energy Source
Development, Spanga, Sweden

G-5   BIOMASS CONVERSION: DIRECT COMBUSTION/GASIFIERS/
PYROLYSIS

Session Leader: A. Strehler
Technische Universitat Munich, Munich,
Federal Republic of Germany

G-6   ALCOHOL FERMENTATION

Session Leader: James Coombs
Consultant, Reading, United Kingdom

G-7   BIOGAS PRODUCTION

Session Leader: W. Baader
Bundesforschungsanstalt für Landwirtschaft (FAL)
Braunschweig-Völkenrode,
Federal Republic of Germany

G-8   VEGETABLE OIL FOR DIESEL FUEL

Session Leader: J. J. Bruwer
Division of Agricultural Technical Services
Silverton, Republic of South Africa

TOPIC H
TRANSPORTATION

Topical Coordinator: Margaret Fels
Princeton University, Princeton, New Jersey,
USA

H-1 AUTOMOTIVE FUEL ECONOMY: POLICY

Session Co-Leaders: Barry McNutt
U.S. Department of Energy, Washington, D.C.
USA

Frank von Hippel
Princeton University, Princeton,
New Jersey, USA

H-2 AUTOMOTIVE FUEL ECONOMY: TECHNOLOGY

Session Leader: Ulrich Seiffert
Volkswagenwerk AG, Wolfsburg,
Federal Republic of Germany

H-3 NEW FUELS FOR TRANSPORTATION

Session Leader: Charles Gray
U.S. Environmental Protection Agency,
Ann Arbor, Michigan, USA

H-4 TRANSPORTATION AND WORLD ENERGY: CURRENT DEMANDS,
FUTURE TRENDS AND IMPLICATIONS

Session Leader: William Porter
U.S. Department of Energy, Washington,
D.C., USA

H-5 TRANSPORTATION AND WORLD ENERGY: CONTINGENCY PLANNING
AND URBANIZATION ISSUES

Session Leader: William Porter
U.S. Department of Energy, Washington D.C.,
USA

H-6 INTERCITY PASSENGER AND FREIGHT SYSTEMS

Session Leader: Umberto Montalenti
National Research Council of Italy, Turin, Italy

TOPIC I
ECONOMICS AND CHANGING LIFESTYLES

Topical Coordinator: Rosalyn Barbieri
Jet Propulsion Laboratory, Pasadena, California,
USA

I-1 STRATEGIES FOR EDUCATION

Session Co-Leaders: Kathleen Wulf
University of Southern California
Los Angeles, California, USA

Debra Langford
U.S. Department of Energy, Washington, D.C.,
USA

I-2 INTERACTION OF GOVERNMENT INSTITUTIONS

Session Co-Leaders: Eugene Frankel
U.S. House of Representatives
Health Science and Technology Committee
Washington, D.C., USA

Mans Lonnroth
Secretariat for Futures Studies, Stockholm,
Sweden

I-3 PROMOTION OF NEW ENERGY TECHNOLOGIES

Session Co-Leaders: Alan S. Hirshberg
Booz-Allen, Hamilton, Inc., Bethesda,
Maryland, USA

Gunther Schafer
FhG, Karlsruhe, Federal Republic of Germany

I-4 TRADE-OFF ANALYSIS AND ENERGY OPTIONS DECISION MAKING

Session Leader: Climis A. Davos
University of California, Los Angeles,
California, USA

I-5 THE PRESS AND INFORMATION TRANSFER

Session Leader: Joel Strasser
Hill and Knowlton, New York, New York, USA

I-6   ENERGY CONSUMPTION AND THE QUALITY OF LIFE

   Session Leader: Peter de Leon
                         Science Applications, Inc., Englewood,
                         Colorado, USA

I-7   COMMUNITY ENERGY PLANNING AND PARTICIPATION

   Session Leader: Tina Hobson
                         U.S. Department of Energy, Washington, D.C.
                         USA

CONTENTS
Volume I

TOPIC A - INTERNATIONAL POLICIES AND STRATEGIES

An Energy Strategy For a Developing Country
J. R. Acosta — 3

Energy Conservation in the Household Sector: Economic Implications for the European Community
F. Conti, G. Graziani, G. Helcké, M. Maineri, M. Paruccini, R. Peckham and C. Zanantoni — 11

Energy as a Constraint on Development: The Case of Brazil
J. W. C. Dias — 19

Sweden in the Year 2020 - A Human Ecology Approach
B. Eriksson — 27

Government Financing of Oil and Gas Related Research and Development
D. Fee — 35

Regional Energy Policy Analysis: The Wisconsin Experience
W. Foell and M. Hanson — 43

Human Labor and Machine Inputs: An Energy Analysis for Agricultural Production in the Less Developed World
S. Freedman — 55

Energy Management and Strategies in a Developing Country with a Centrally Planned Economy - Romanian Issues and Solutions
A. V. Gheorghe — 63

Energy Resources and Economic Development in West Africa
E. K. Glakpe and R. Fazzolare — 73

A Discussion on the Methods of Solving China's Rural Energy Problem
D. Keyun and Z. Qingche — 85

Solving the Energy Problem in Latin America
J. R. Moreira and J. Goldemberg — 93

Energy Policy-International Strategies and Alternatives
K. K. Murthy — 103

Assessment of Nigeria's Energy Policy
N. I. Ngoka — 109

The Role of Public Participation in Energy End Use Management
R. W. Rycroft and J. L. Regens — 127

Ideas on Economic and Energy Growth and on the Goal of the Economy in Industrialized Countries
W. Schönherr   133

Energy Planning in Brazil
A. C. M. Sousa and R. G. Esteves   141

International Technology Assessment with the Markal Energy Model
St. Rath-Nagel, G. Giesen, A. Hymmen, K. Maher, M. Müller and H. Vos   149

Improvements in Industrial Efficiency as a Final Benefit of an Energy Conservation Program in Argentina
M. S. Ussher   161

TOPIC B - MODELS, ANALYSIS, ASSESSMENT

Residential End-use Energy and Load Forecasting; A Microsimulation Approach
S. D. Braithwait and A. Goett   181

Multiobjective Power Systems Generation Planning
J. C. N. Clímaco and A. Traca-Almeida   189

Impact of Residential Load Management on Generation Planning Reliability
S. B. Dhar and W. R. Puntel   197

The Statistics of Energy Conservation
R. Everett   207

Proposed: A Fundamental Energy Data System for States
M. F. Fels   215

Nuclear Power - View From France
A. Ferrari   223

The Impacts of Higher Energy Import Prices on European Economies
Y. Guillaume, M. Konings, O. Rouland and F. Thys-Clément   229

A Dynamic Linear Programming Model for Gas Supply Planning
H. Gündogdu and I. Kavrakoglu   239

Energy for the Rest of the Century
J. F. Gustaferro   247

Potential Energy Savings in the Residential Sector of the United States
J. Ingersoll   263

A Commercial End Use Model of Energy Use and Peak Demand
J. Jackson and R. Lann   275

Towards the Quantitative Assessment of Energy Data
G. E. Liepins   285

Residential End Use Demand Modeling: Improvements to the ORNL Model
J. E. McMahon   297

Alternative Off-Oil Scenarios for Ontario
R. H. H. Moll, K. H. Dickinson and R. B. Hoffman   305

A Master Plan for the Use of Coal Resources in Turkey
A. G. Pasamehmetoglu, M. Oskay, I. Nisanci, C. Dagli and K. Yurtseven   317

Towards a Universal Model of Energy Demand
P. C. Roberts   325

Application of an LP Model to Strategic Planning of Multinational Cooperative
R&D Programs
V. L. Sailor   335

GREPOM: A Greek Energy Policy Model Preliminary Description
J.-E. Samouilidis   345

Methods for Evaluating Energy Conservation
J.-E. Samouilidis, G. Kontoroupis, N. Koumoutsos   351

Scenario 2000 - Energy for the FRG
T. Schott, J. Nitsch and H. Klaiß   367

Alternative Means of Coping with National Energy Emergencies
J. H. Sorensen   385

A Regional Time-and Space Dependent Energy Model
E. Thöne, R. Friedrich and K. H. Höcker   395

Energy Supply Options for the 1980's. A Japanese Perspective
T. Tomitate   403

Residential Hourly and Peak Demand Model
G. Verzhbinsky and M. D. Levine   411

The Thermal Zone Module: An Approach to Integrating Energy Decisions into
the Building Design Process
N. Yoran, A. Yoran and V. Hartkopf   421

CONTENTS
Volume II

TOPIC C - ECOLOGY AND ENVIRONMENT

Effects of Air Pollution on Forests
G. Abrahamsen
433

Environmental Considerations in the Site Finding for Thermal Power Plants
U. Bernard and R. Friedrick
447

Australian Oil Shale Development
T. Brown
455

The Scottish Shale Oil Industry
F. M. Cook
463

Policy Options for Energy/Water Issues in the Western U.S.
M. D. Devine
477

Water Conservation with Energy Development in the San Juan River Basin,
New Mexico
M. W. Gilliland and L. B. Fenner
485

Fuel From Biomass
N. E. Good
491

Assessing Health and Environmental Effects of a Developing Fuel Technology
R. H. Gray and H. Drucker
499

Soil Loss and Leaching, Habitat Destruction, Land and Water Demand in
Energy-Crop Monoculture: Some Quantitative Limits
V. P. Gutschick
509

Prediction, Risk and Uncertainty in Environmental Assessment of Energy
Development
A. Hirsch
519

A Critical Review of Recent Studies of the Impact on Climate of an
Increased Atmospheric Carbon Dioxide Concentration
J. Jäger
529

Effects of Acid Precipitation on Elemental Transport from Terrestrial to
Aquatic Ecosystems
D. W. Johnson
539

Energy and Information, the Two Lone Common Denominators of all
Systems: Steps Toward the "Integrative Approach" of Human Ecology
H. Knötig
547

Effect of Acid Precipitation on Soil
E. Matzner and B. Ulrich
555

Atmospheric Deposition of Acidifying Substances
R. Mayer
565

Health Impacts of U.S. Energy Policy Alternatives
R. O. McClellan, R. G. Cuddihy, F. A. Seiler and W. C. Griffith
573

Energy Policy and Green Environment on the Basis of Ecology
A. Miyawaki
581

Possibilities of Surveying Synecological Effects (Summary)
P. Müller
589

Impact Assessment - A Contribution to Human Ecology
I. Paul
599

Acid Rain, Eastern U.S.A. - Problems of Control
R. M. Perhac
607

Some Environmental Considerations in Siting a Solar/Coal Hybrid Power
Plant
R. L. Perrine
623

Environmental Quality Effects of Alternative Energy Futures: A Caveat on the
Use of Macro-Modelling
J. L. Regens and D. A. Bennett
631

Energy Policy, Air Pollution, and Community Health
H. W. Schlipköter and U. Ewers
643

Mechanisms of Surface Water Acidification
H. M. Seip
651

Ecological Implications of Air Pollutants from Synthetic Fuels Processing
D. S. Shriner, S. B. McLaughlin and G. E. Taylor
669

Can the Poor Countries Afford Biomass Energies?
V. Smil
677

Environmental Aspects of Alberta Oil Sands Development
K. R. Smith
687

Conflicts of Water Use and Energy Development in the Ohio River Basin
H. T. Spencer and C. A. Leuthart
693

Estimating the Future Input of Fossil Fuel CO_2 into the Atmosphere by
Simulation Gaming
I. Stahl and J. Ausubel
699

Is the Large-Scale Destruction of Tropical Rain Forests Necessarily Crucial
for the Global Carbon Cycle?
N. Stein
715

Environment and Health Implications of Canadian Energy Policy
P. M. Stokes and J. B. Robinson
727

Environmental Problems by Energy Production in Denmark
P. B. Suhr
737

The Lurgi Ruhrgas (LR) Process. An Environmentally and Economically Sound
Oil Shale Retort
H. Weiss and F. Gonnert
745

Interdependence Between Electricity Production and Water Use
D. Winje   753

TOPIC D - NEW ENERGY TECHNOLOGIES

Récupération de Chaleur par Thermofrigopompe dans l'Industrie Laitiere
Y. Almin et G. Laroche   765

Cogeneration Applications in Commercial Facilities
M. K. J. Anderson   775

Large Demonstration Solar Project in Italy
P. Baronti and G. Benevolo   785

Control Systems for District Heating Subscriber Stations, and Heating
Installations
J. Boel   791

The Role of the Packaged ORC System in Industrial Waste Heat Recovery
L. Y. Bronicki and W. E. Rushton   801

Chaffage de Serres a l'Aide des Rejets Thermiques de Centrales en Circuit
Ferme
Y. Cormary et C. Nicolas   809

Hoffmann-La Roche Slow-Speed Diesel Cogeneration Project
J. P. Davis and S. E. Nydick   819

Preliminary Design Study of Underground Pumped Hydro and Compressed Air
Storage in Hard Rock
A. Ferreira and P. E. Schaub   839

Fuel Cell Power Plants for Electric Utilities
A. P. Fickett, E. A. Gillis and F. R. Kalhammer   855

The Reality of Onsite Fuel Cell Energy Systems in the 1980's
V. B. Fiore and R. T. Sperberg   865

Eurelios, The World's First Operating Solar Power Tower Plant (1MWe1)
J. Gretz   873

Batteries for Electric Vehicles
B. Hartmann   881

Industrial Process Heat Applications for Solar Thermal Technologies
D. W. Kearney and D. Feasby   891

Trends in District Heating
M. Larsen   901

Thermodynamic Characteristics of Endogenous Fluids in Relation to their
Utilization on a Wider Scale in Electricity Generation
C. Latino and O. Sammarco   913

Air Storage Gas Turbine Power Stations an Alternative for Energy Storage
J. Lehman   923

The U.S. Department of Energy High Temperature Turbine Technology
Program
G. Manning and J. Neal   939

U.S. Gasification Technology
C. L. Miller   947

Installation Demands and Techniques for District Heating Pipes
H. C. Mortensen and J. Christiansen   959

Pressurized Fluidized Bed Combustion
S. Moskowitz   971

District Heating Metering Systems in Denmark
N. Nedergaard   979

Fluidized Bed Combustion and Its Potential Use for a Variety of Solid Fuels
J. N. Nikolchev   989

100MW Coal Gasification-Combined Cycle Cool Water Project
D. R. Plumley   999

EPRI Residential Solar Demonstration
G. G. Purcell   1007

Multicomponent Renewable Energy Supply for Low Temperature District
Heating Systems
H. C. Rasmussen and J. Jennsen   1017

Energy and Microcomputers: An Overview
R. B. Spencer   1029

The German Development Program in Coal Gasification and Liquefaction
H.-J. Stöcker, R. Holighaus and J. Batsch   1041

A Microcomputer-Based System for Power Demand Monitoring and Control
A. Traca de Almeida and A. Gomes Martins   1049

Project - Process Surplus Heat
M. Uhrskov   1057

TOPIC E - INDUSTRIAL PRODUCTIVITY AND DEVELOPMENT

Energy Analysis in Scrap-Based Mini-Mills: An Evaluation of Different
Methods of Electric Arc Furnace Operation
A. Borroni, C. M. Joppolo, B. Mazza, G. Nano and D. Sinigaglia   1061

Conversion of Refuse to Storable Fuel
G. P. Bracker and H. Sonnenschein   1071

Development of a Commercial Fluid Bed Package Boiler
L. Brealey and J. H. Wilson   1079

Energy Conservation Through the Use of Gas Technology
I. M. Coult   1087

An Energy Analysis of the Reconstruction of a Car Starter Motor
T. R. Cox and M. E. Henstock   1095

The Composition of Thermoeconomic Flow Diagrams
R. A. Gaggioli and W. J. Wepfer   1107

Integrated Multi-Task Energy Systems
K. Illum   1115

Energy Saving in Continuous Casting Process by New Concepts in the
Design of Equipment
B. Indyk and R. Wilson 1127

Answers: A Comprehensive Solution to the Economic, Technical and
Institutional Problems of Solid Waste Reuse
P. F. Mahoney and G. L. Sutin 1139

Energy-Conserving Industrial Plant Design - An Energy Efficient Brewery
J. M. Newcomb 1151

Empirical Energy Requirements for Several Ethanol-From-Grain
Operations
R. A. Plant and R. A. Herendeen 1159

Energy Recovery from Low-Grade Fuels and Wastes
M. Rasmussen 1165

Finite Time Constraints and Availability
M. H. Rubin, B. Andresen and R. S. Berry 1177

The Science of Energetics in the Exergy Crisis Or How is Thermodynamics
made Really Useful?
T. S. Sorensen 1185

Production with Minimal Energy Use
M. Splinter and W. Willeboer 1201

Energy Management - A Methodology for Project Evaluation
T. J. Stenlake 1209

Development Trends in the Field of Waste and Residue Incineration
K. J. Thomé-Kozmiensky 1221

Industrial Energy Conservation as an Investment Allocation Problem
B. de Vries, D. Dijk and E. Nieuwlaar 1229

CONTENTS
Volume III

TOPIC F - COMMUNITY SYSTEMS AND BUILDINGS

Power Impact Assessment - Case Study: Pasadena, California, Redevelopment
Project
J. I. Baum   1241

Erfahrungen einer Eigentümergemeinschaft bei der Planung ihrer Wohnanlage,
unter Berücksichtigung Alternativer Energiesysteme
E. Beinroth   1251

Energy Conservation in the Transportation Sector Through Spatial Planning
U. Bernard, R. Friedrich and S. Gepp   1259

Impact of Energy Management Programs on Hospital Energy Bills
L. Corum   1265

Enhancement of Built-Up Cities as a Conservation Strategy
D. A. Dove   1273

Waste Energy for Space Heating in Residential Buildings
E. Giusti   1283

Urban Sunspaces: Urban Energy Conservation with Atria and Arcades
J. W. Glassel   1293

Evaluation of Housing Heat Gains Due to Metabolism, Artificial
Lighting, Appliances and Domestic Hot Water Use
J. M. Hauglustaine   1297

Effects of Wintry and Summery Heat Protection Measures on the Energy-
Balance of Dwelling Houses
G. Hauser   1305

Energy Conscious Building Design: American Telephone and Telegraph
Company's New World Headquarters
E. P. Hodges   1317

Instrumentation and Monitoring of South Oakland Solar House
L. R. Hoffman and V. Hartkopf   1323

Constraints and Potential for Energy Conservation in the Small-Consumer
Sector
E. Jochem, G. Angerer, E. Gruber, U. Hauser and T. Mentzel   1327

Energiesparen im Mietwohnungsbau
D. A. Kolb   1337

Energy Concepts of Urban Renewal in Historic Areas. Berlin Chamisso Platz
Project
D. A. Kolb   1341

Study on Ecology and Urban Renewal for Inner City Blocks
M. Küenzlen   1345

District Heating Management in Denmark
L. Larson   1349

The Combination of Total-Energy and Solar Assisted Heat Pumps at the
Project "De Achtse Barrier" at Eindhoven
P. H. H. Leijendeckers   1357

Some Economic Welfare Thermal Energy Storage Technology: A Preliminary
Regional Analysis for the U.S. Residential Sector
B.-C. Liu, R. F. Giese and J. Stavrou   1371

Greece/Germany's New Energy Efficient Communities Lykovrissi Solar
Village
V. Loftness and F. Boese   1383

In Energy Conservation and Passive Solar Design, Materials and
Components are Critical
V. Loftness and F. Boese   1391

Passive Measures of Energy Conservation Applied to Inner City Buildings
V. Nikolic   1399

Analyse de Differents Types d'Architecture Solaire
J. Michel   1403

Analysis of Architectural Shapes and Climatic Envelopes
J. Michel   1413

Energy Savings by Heating Control
D. Oswald   1421

Changing Energy Efficiency in Buildings - Institutional Implications
A. J. Penz   1427

Indoor Air Pollution: Conservation and Health Implication
R. M. Perhac and D. J. Moschandreas   1435

District Heating via Combined Heat and Power - A Past and Future
Comparison of Europe to the U.S.A.
D. J. Santini   1449

Energy Considerations in Local Planning
G. F. Schaefer and U. Gundrum   1465

Indicators of Residential Energy Use and Conservation - An International
Study
L. Schipper, S. Meyers and A. Ketoff   1473

Energy Saving By Daylight?
K. Stolzenberg   1483

The Potential for Energy Conservation Through the Controlled Ventilation
of Domestic Buildings
L. Trepte and V. Meyringer   1495

Energy and Life-Cycle Cost Analysis of a Six-Story Office Building
I. Turiel   1503

Housing Rehabilitation and Construction: An Analysis of Energy Use
R. Woodbury, V. Hartkopf and J. L. Onaka   1513

TOPIC G - AGRICULTURE AND RURAL DEVELOPMENT

The Use of Vegetable Oils in Straight and Modified Form as Diesel Engine
Fuels
D. M. Bacon, F. Brear, I. D. Moncrieff and K. L. Walker   1525

Energy Use in Small Farm High Yielding Variety Rice Production in Low
Income Countries
D. Boughton and J. Wicks   1535

Energy Consumption and Efficiency in the United States Food Processing
and Marketing Sector
J. Broder, J. Beierlein, K. Schneeberger and D. Van Dyne   1545

Biogas - An Alternative Fuel for Automotive Application
S. Büttner   1553

On-Farm Demonstration Program Using Solar Energy for Heating of
Livestock Shelters and Drying of Crops
W. T. Cox   1561

Wirkungsgrade von Solardachsystemen bei Alternativer Nutzung
W. Dernedde   1567

Trocknung von Trauben mit Solarenergie (Drying of Grapes by Using Solar
Energy)
W. Eißen   1573

Drying of Cassava with Solar Heated Dehumidified Air
G. O. I. Ezeike   1579

Integrating Biomass Production Activities with Operating Farms. The Case of
Grain Residue Harvesting
S. J. Flaim, B. F. Neenan and H. O. Mason   1587

On-Farm Preparation of Sunflower Oil Esters for Fuel
J. Fuls and F. J. C. Hugo   1595

Alcohol as an Alternative Fuel in Agriculture
M. Graef   1603

Laboratory Procedures for Investigating Some Fuel Properties of
Sunflower Oil Esters in Diesel Engines
C. S. Hawkins, J. Fuls and F. J. C. Hugo   1611

Effect of Some Important Parameters on the Performance of Simple Flat
Plate Collectors
S. M. Ilyas, W. Grimm and F. Wienke   1619

Economic Feasibility of Alternative Fuels for Food Processing Plants in the
United States
H. B. Jones, Jr. and W. K. Whitehead   1627

Sunflower Oil and Methyl Ester as Fuels for Diesel Engines
K. R. Kaufman, M. Ziejewski, M. Marohl and A. E. Jones   1635

The Use of Agricultural Energy-Sources - An Economic Evaluation
W. Kleinhanss   1645

Protein and Energy from Wet Green Crop Fractionation
A. A. Lepidi, R. Fiorentini, M. P. Nuti and C. Galoppini   1657

Macro-level Implications of Using Soybean Oil as a Diesel Fuel
W. Lockeretz   1667

Vegetable Oil as a Diesel Fuel - Soybean Oil
R. McCutchen   1679

Gas as an Alternative Fuel for Agriculture
G. J. Mejer   1687

On-Farm Production of Fuel From Vegetable Oil
E. J. Merrikin and J. A. Ward   1697

Driver Information Displays - A Step to Optimum Tractor Operation
K. H. Mertins and H. Göhlich   1707

Solar Water Heating for Food Processing Plants
P. Z. Mintzias   1715

Marine Biomass as a Renewable Energy Source
J. G. Morley   1727

The Utilization of Wood-Products and Straw for Combustion
Ch. Nilsson   1731

Minimizing Fuel Energy Use in Fruit and Vegetable Food Production and
Transportation
M. O'Brien and R. P. Singh   1741

Landwirtschaftlicher Energiebedarf und Substitutionsmöglichkeiten durch
Biogas
H. W. Orth   1745

Energy Cost Management in Food Processing
L. D. Pedersen and W. W. Rose   1753

The Potential for Biomass Energy and Programs for its Development in the
Caribbean Basin
C. Peterson and K. Farrell   1761

On Farm Production of Sunflower Oil for Fuel
G. Pratt, L. Backer, K. Kaufman, L. Jacobsen, C. Olson, P. Ramdeen,
W. Dinusson, D. Helgeson, L. Schaffner and H. Klosterman   1767

On-Farm Sunflower Oil Extraction for Fuel Purposes
M. Prinsloo and F. J. C. Hugo   1775

Economic and Energetic Aspects of Using Solar Energy Water Heaters in
Polish Agriculture
J. Pyrko   1783

Biomass Fuel as an Oil Substitute for Residential and Commercial Space-
Heating
J. G. Riley and N. Smith   1791

Biomass Energy Conversion in Hawaii
R. Ritschard and A. Ghirardi   1799

Energy Analysis of an Agriculture Alcohol Fuel System
E. D. Rodda and M. P. Steinberg   1807

Rape Seed Oil as an Alternative Fuel for Agriculture
F. Schoedder   1815

Energiebilanz der Biogasproduktion aus Pflanzen und Flüssigmist
F. Schuchardt   1823

Conservation of Thermal Energy Use in Fruit and Vegetable Processing
R. P. Singh and M. O'Brien   1829

Energy Use in Irrigation
E. T. Smerdon and E. A. Hiler   1837

Development and Demonstration of Solar Malt Kilning
C. C. Smith   1847

Heat From Straw and Wood
A. Strehler   1855

Bioenergetic Trend - A Key to Solving Energy, Food, and Ecological
Problems
I. I. Sventitsky   1863

Pyrolysis of Waste Biomass in Developing Countries
J. W. Tatom, H. W. Welborn and R. D. Hardy   1871

Permanently Stratified Lakes and Ocean Basins as Possible Source of Methane
for Developing Countries
K. Tietze   1889

Energy Use for Ethanol Production
W. Vergara and J. R. Castello Branco   1901

Surface Heating Greenhouses with Waste Heated Water
P. N. Walker   1911

Cyclic Load Testing of Agricultural Tractors in Fuel Research -
Instrumentation and Equipment for a Test Facility
A. N. v. d. Walt and F. J. C. Hugo   1919

Diesel Engine Tests with Sunflower Oil as an Alternative Fuel
A. N. v. d. Walt and F. J. C. Hugo   1927

Agriculturally Based Oils as Engine Fuels for the Australian and South
Pacific Region
J. F. Ward   1935

Experiences on Farm-Scale Digestion with Piggery Waste
R. Wenzlaff   1939

Reducing the Energy Required for Mechanised Cultivations in Developing
Countries
T. J. Willcocks   1945

Results of a Long Term Engine Test Based on Rape Seed Oil Fuel
M. Wörgetter   1955

Improved Anaerobic Digestion by Application of Constant Concentration of
Digested Sludge Recycling
P. Y. Yang, S. Y. Nagano, J. K. Lin and Y. T. Wong   1963

Batch and Continuous Ethanol Fermentation Process with Constant Cell
Recycling
P. Y. Yang and J. K. Lin   1971

CONTENTS
Volume IV

TOPIC H - TRANSPORTATION

Modal Shift: How Much a Policy May Affect and Reduce Fuel Consumption.
What Are the Perspectives?
C. Abacoumkin   1981

Energy and Economic Benefits of National Railroad Electrification in the
United States
H. Cooper, Jr. and R. Buck   1991

Transportation Systems Planning and Energy Management: An International
Perspective
M. Hanson and W. Foell   2003

Energy Requirements for the Motor Car - Analysis and Opportunities for
Conservation
A. Herham and M. Jacobson   2015

Fuel Economy Improvements for Road Transport in France. Achievements and
Prospects
C. Lamure   2029

Competitivity of Non-Contact Guided Transport System Compared to High
Speed Railway
F. Di Majo   2037

The Impact of the TGV on Total Energy Consumption on the Paris-South
East Route
R. Monnet   2045

Der Spezifische Energieverbrauch Verschiedener Schneller
Intercityverbindungen (Auto, Bus, Eisenbahn, Flugzeug, Magnetschwebebahn)
W. Schwanhäußer   2053

Energy Considerations in the Analysis of Alternative Urban Transportation
Investments
P. W. Shuldiner, M. Jacobs and J. M. Ryan   2057

Intermodal Freight Transportation: a Contribution to Energy Management
R. W. Sparrow   2065

Future Propulsion Systems for Cars
P. Walzer   2077

TOPIC I - ECONOMICS AND CHANGING LIFE STYLES

Facing the Energy Challenge. The Limitations and Possibilities of
Administrative Decision-Making
M. Arnestad and T. R. Burns 2095

Energy Management is Not a Technological Problem
H. L. Breckenridge 2107

Research on Final Use of Energy
F. Cabrini, P. Pascoli and T. Sinibaldi 2121

Changing Lifestyles: Economic Stagnation, Postmaterialistic Values and the
Role of the Energy Debate
J. Conrad 2127

Energy Use Management: The Critical Issues of Interdependency and
Indeterminateness
C. A. Davos 2135

Industrial Marketing Research for an Emerging Technology: A Case Study
P. J. Grogan 2149

How the European Press Communicates News on Energy Alternatives
P. Hoffmann 2159

Influence of Consumer Behaviour on Energy Demand of Households
P. Iblher and W. Brög 2163

Solar Energy Education from a Federal Perspective
D. D. Langford 2175

An Application-Oriented Method and a Successful National Education
Program for Energy Savings in Existing Buildings
K. Meier and P. Schlegel 2179

Energy Saving in Transportation: The French Policy
G. Monot 2185

The Esalen Energy Program: On the Road to Self-Sufficiency
D. A. Pilati and R. Mozic 2189

A Quantitative Evaluation of Government Promoted Energy Saving Projects
N. Scheirle and R. Wartmann 2201

Comparative Assessment of Energy Systems
B. Sorensen 2209

The Press and Information Transfer: Communicating Energy Alternatives
Enhancing the Energy Communications Pie
J. A. Strasser 2217

How Industry Communicates News on Energy Alternatives
J. B. Wright 2225

Strategies for Infusing a Solar Energy Curriculum into Schools Around the
World
K. M. Wulf, S. Lampert and G. Yanow 2231

Essential Ingredients in the Development and Implementation of Energy
Curriculum for Elementary Schools
K. Wulf, A. Brown, V. Johnson and E. Walton 2235

APPENDIX - LATE SUBMISSIONS

TOPIC A

Energy Consumption and Economic Growth in Israel - Trend Analysis
(1960-1979)
J. Bargur and A. Mandel   A1

Ill-Educable Nature of U.S. Energy Policy
G. A. Daneke   A19

TOPIC B

Building Energy Use Modeling in Sweden by Julotta
K. Källblad and F. Higgs   A37

Analysis of Canadian Energy Policy
D. Quon, S. Wong and S. Singh   A51

The Experiences of the Energy Modeling Forum
J. P. Weyant   A59

TOPIC C

Freshwater Acidification in Scandinavia and Europe - An Overview
W. Dickson   A69

The U.S. Department of Energy's Oil Shale and Tar Sands Environmental
Mitigation Program
A. M. Hartstein   A77

Ecological Planning in High Density Urban Areas
P. Krusche and M. Weig-Krusche   A83

A Regional Environmental Approach to Eastern Devonian Oil Shale
Development
W. J. Mitsch and C. G. Lind   A91

Acidification - Effects on Aquatic Organisms
I. P. Muniz   A101

The Role of the Biosphere in the Carbon Cycle
J. S. Olson   A125

TOPIC D

TOTEM: A Modular System for Distributed Cogeneration
F. P. Ausiello   A141

Artificial Drying of Wood Chips for Energy Purposes
G. Gustafsson   A151

District Heat from Modular Cogeneration Plants
K. Hein   A173

TOPIC E

Energy Efficiency in the Chemical Industry
R. U. Ayres, K. Subrahmanian and A. Werner   A181

TOPIC F

Gesundes Energie- und Anlagensparendes Raumklima Grundlegende
einleitende Betrachtungen
R. Ayoub   A193

Passive Solar Heating, Passive Cooling and Ventilation in the Same
Building without Equipment
R. Ayoub   A203

Basic Principles and Applications of Isolated Facades with
Selfcontrolled Transparency for Solar Energy with the means of Optical
Heatcontrol
H. Köster   A225

Multi Storey Housing Energy as a Determinant in Planning
S. Los   A237

Building Codes and Energy Conservation Standards for Austria
E. Panzhauser   A245

Likely Impacts of the Energy "Crisis" on Recent Population
Redistribution Trends in Canada
I. M. Robinson   A251

Energy and Urban Futures
J. Van Til   A265

Self-Sufficient Energy Systems for a Small Community
R. A. von Oheimb   A273

TOPIC G

World Overview of Plant Oils for Fuel: Status of Research and
Applications Technology
J. J. Bruwer and F. J. C. Hugo   A287

Biomass Energy Conversion using Fluidized-Bed Technology
W. A. LePori, R. G. Anthony, R. B. Griffin, T. C. Pollock, A. R. McFarland
and C. B. Parnell, Jr.   A297

TOPIC H

The U.S. and the IEA: Can We Live Up to Our Commitments?
W. J. Kruvant   A305

Negotiating Transportation Fuel Purchase Contracts
A. H. Levine   A311

TOPIC I

The Solar Energy/Utility Interface — Workshops in Conflict Resolution
R. H. Bezdek   A317

Human Values, Energy Options and Cost Effectiveness
J. J. Califano   A327

Evaluating the National Passive Promotion Program, Plans and Progress
S. Heffernan and A. S. Hirshberg   A335

Early Market Experience of Solar Energy in the United States
A. S. Hirshberg   A345

The Nature and Scope of Instructional Modules under Development by
Climate, Ocean, Land and Discovery - A Scientific Study of the Polar
Regions (COLD)
W. C. Kazanjian   A359

Attitudes Towards Using Solar Equipment in Germany - A Preliminary Report
H. J. Klein   A365

Methodological Approaches to the Assessment of Social and Societal Risks
O. Renn   A375

Results of a Comparative Survey on the Psychological Perception of
Technology and Risk
O. Renn and C. Schlupp   A395

Attitudes Toward Different Kinds of Energy Use
Part 1: - A Theoretical Approach on a Macro Social Level
W. Ruppert   A409

Passive Economics and Market Analysis Model
M. R. Sedmak and A. S. Hirshberg   A415

AUTHOR INDEX   A425

TOPIC C

Ecology and Environment

ENVIRONMENTAL CONSIDERATIONS IN THE SITE FINDING FOR THERMAL POWER PLANTS

U. Bernard* and R. Friedrich**

*Institut für Landschaftsplanung, First Dept. of Architecture and Regional
and Municipal Planning, Universität Stuttgart, Keplerstrasse 11,
D-7000 Stuttgart 1, Federal Republic of Germany
**Institut für Kernenegetik und Energiesysteme, Fifth Dept., Universität
Stuttgart, Pfaffenwaldring 31, D-7000 Stuttgart-80,
Federal Republic of Germany*

ABSTRACT

This abstract gives an introduction to the site dependant model of the allocation of power plants. The model considers aspects such as regional planning, town planning, ecology and environment, economies, technology, and others simultanously. The methodology applied in this model permits its application in large regions such as a state as well as in smaller areas such as municipalities or communities.

KEYWORDS

Energy utilities; power generation; environmental impact assessment; regional and community planning; power plant site finding (allocation); electronic data processing.

INTRODUCTION

The demands and aims in the field of power economy decisions are linked with many fields such as economy, engineering, regional policy, and ecology. Therefore, it is important to consider and to investigate all fields mentioned, when choosing sites for power plants.

The knowledge about and the assessment of the numerous interactions within the system 'Energy-Environment-Economy' is conducive in achieving correct and acceptable decisions and measures. Today it practically is impossible to find an ideal site for a large-scale project such as a Thermal Power Plant in the state of Baden-Württemberg. Because of the high density of the population (255 inhabitants/square kilometer) and the actual land use pattern, a great deal of conflict, criticism, and much distrust of people will arise.

In the FRG, regional planning policy is determined to allocate all human activities in the region optimally and to minimize their environmental impacts. In order to avoid difficult and dangerous locations, the complex procedure of finding Thermal Power Plant sites is to be discussed as early as in the preliminary planning phases.

The objective is a solution which considers all spatial regards: ecology and environmental impacts, regional planning objectives, cost, and acceptance by the people.

This paper will discuss some of partial aspects which have to be considered in the finding of sites, particularly the environmental aspects.

GENERAL METHODOLOGICAL CONSIDERATIONS

It is close to impossible to find sites in this country which completely fulfill all criteria mentioned. Therefore, all advantages and disadvantages of possible locations have to be taken into consideration and need to be assessed. Only the optimal alternative, with reference to the goals of society or its elected representatives, can be chosen.

It appears to be important to process the site selection on the grounds of quantitative, and scientifically developed methods. Firstly, the large number of criteria and alternatives make it very difficult for the decision maker to consider all aspects of the problem in a reasonable and consistent manner. Secondly, the lack of transparency of a merely qualitative method may lead to criticism and distrust in the decision by the concerned population.

The purpose of this report is to show how, based on a concrete calculation, a transparent and consistent selection of site zones can be made on the grounds of a quantitative method. The region of investigation is Baden-Württemberg, a country located in the South-West of the Federal Republic of Germany. 9,2 million people live here on an area of about 35.000 km².

Many of the relevant criteria which are presently available in the first step of investigation refer directly to the site, not to the environment of the site. Therefore, the smallest spatial classification unit must be at least as small as the area of a single site. Hence, a square grid with 500 m x 500 m elements is used. The entire area investigated is divided into 245.000 grid cells.

The large quantity of data needed for the evaluation was caused by this high number of cells and requires the use of electronic data processing. The modular computer program system COPLAN has been developed to perform the necessary calculations with a reasonable amount of computer time.

At first the requirements to be met for power plant sites are listed in a hierarchical order. The last level of the hierarchy is occupied by operational site criteria. For every criterion an indicator has to be computed. The indicator shows for every possible site, how well the criterion belonging to the indicator is being met. There are two types of criteria:

i) By means of excluding criteria a zone is excluded from further consideration, when one of the values of at least one indicator exceeds a certain limit.

ii) For the remaining zones which still cover a large area, every indicator belonging to a non-excluding criterion is converted into a component utility value which states in a cardinal scale the degree of fulfillement of every criterion.

In the next step, the component utilitiy values are aggregated for each zone into a single utility value which gives the qualification for the zone as possible power plant site.

ENVIRONMENTAL IMPACTS OF THERMAL POWER PLANTS

In this context, the detrimental impacts relate to the following areas:
. contamination of ground-water
. contamination of fresh-water
. pollution of atmosphere
. contamination of soil
. space exhaustion.

The following main land uses will be influenced:

Residential Areas: Problems and conflicts resulting from construction and operation of power plants can hardly be avoided in areas with medium to high density.
The quality of living within those residential areas adjacent to the plant will
be disturbed by noise, exhaust fumes, waste heat, and visual pollution. The
disturbances will differ depending on the location of the plant as related to the
residential area.

Agriculture: Agriculture will be influenced negatively as follows
. directly through loss of land,
. indirectly through profit losses as a result of additional immissions.
The loss of mostly high productive agricultural land (especially fertile riparian
soils) amounts to 50 - 200 ha, depending on size and make of plant. The area lost
can be even larger, if additional industries, new living quarters, and their infrastructure (as roads) become necessary (regional and local secondary effects).

Forestry: The construction of a power plant within forests does not only occupy
forest land, but also has an impact on the surrounding forest which is detrimental
to its wellknown functions and qualities. This is true for the immediate
surroundings as well as for the greater area, the impact itself depending on the
quantity and quality of the forest.

Recreation: High concentrations of immissions, noise within larger areas, and
structures dominating the appearance of the landscape, are capable of reducing
recreational qualities of a region. Particularly recreation areas of greater than
regional importance will be affected in their perceived image by highly visible
power plant structures, such as large chimneys and cooling towers of 200 m hight
and 100 m diameter.

STARTING TO SOLVE THE PROBLEMS IN THE FIELD OF LANDSCAPE ECOLOGY

For planning and assessment purposes, the problem solving process has to start
out with the relationship between cause (= origin of impact), impact, and object
of impact (affect). This chain of cause and effects is based in the idea that
using the land(scape) as a living environment for man does not only mean using and
accepting what Nature offers (soils, water, air), but also affecting the potential
of the land materialisticly and influencing the image of appearance of the landscape. Thus, the natural basis of land use will be changed which means that
within the control system "environment" man introduces feed backs into the partial
systems which cannot be buffered anymore, and which will lead to the destruction
of natural resources in the end.

The fundamental and life relevant functions of an area are being affected greately,
or limited, by negative feed backs or destruction. Such functions are living,
working, providing goods and services, education, recreation, foodproduction, etc.
A possible strategy for solving the problems around power plants is the definition
of areas with similar ecological and functional character. Such a definition is the
ecological zones system of the FRG (published by Bundesanstalt für Raumforschung).

Special characteristics of sensitivity are being used to apprehend landscape units that can be assessed together according to their ecological affect. Such characteristics are, for example, density of waterway systems, groundwater system, morphology, vegetation structues of greater importance, and the mesoklimatic lability. Together, they are being expressed by the Translation factor $\propto$. The higher the $\propto$-factor is, the more likely and the greater are long distance affects. With plant sites in a valley, for example ($\propto$ = 10.0), long distance affects (such as quick transport of large amounts of radioactive seepage, or the potential danger of accumulation of redaiactive gases in the lower valley; so called "aquiferouses") are more likely as with plants sited on high plateaus without depressions and with good ventilation ($\propto$ = 2.0).

RESEARCH RESULTS OF THE FIRST STEP OF INVESTIGATION

Figure 1 shows $\propto$-factor values in staggered shades of grey. The darker shades indicate higher $\propto$-factor assessments and higher landscape sensitivity.

A good scale for the mathematical solution of landscape features is the 500 x 500 m grid. On the one hand, this grid allows calculation of conditions in larger regions (such as earth quake zones or bioclimatic zones). The processing of data classes with relatively undefined margins is possible. On the other hand, this size of grid also permits the assessment of ecological interrelationships which take place in rather small areas (i.e. forest edges, adjacent open field, or settlement edges). These small areas then will be seen in a scale that suits the scale of the project.

Figure 2 shows the result of some calculations in the target field "Landscape Ecology", according to the first evaluations: The darker shaded areas show the higher multiattribute utility data and the lower site qualification. The criteria are (numbers in brackets show the weight coefficients): ground water (34), soil potential (8), climatic potential (8), biotic resources (31), and ecological sensitivity of natural environment (19). Fresh water resources which quantity of flow remain below a certain standard, have been excluded. Figure 3 shows the ecological result for coal fired power plant.

Figure 4 shows the results of a calculation which gives the qualification for each zone as a possible site for a nuclear power plant with circuit cooling. The darker shaded areas show the higher aggregated multiattribute utility data and the lower site qualification.

The calculation results depend on the evaluation and assessment parameters fed into the computer up to this point, i.e. the results depend on the approved preference structure that has been chosen by the scientists.

REGIONAL AND LOCAL INVESTIGATIONS (SECOND STEP)

Starting out with some of the zones for power plant sites as mentioned above, this chapter shall discuss research that is intended to define single spots within these zones with the assistance of ecological criteria. The most important of these criteria are:

i)   Natural Factors: morphology, soils, geology, hydrology, waters (surface), climate, vegetation, animal life.

ii)   Nature and Landscape Protection: protected areas, negatively affected areas.

iii) Land Use, Goals for Regional Development: settlement, transportation, agriculture, forestry, regional planning, groundwater, protection zones.

iiii) Socio-Cultural Importance of the Area (Landscape): image and appearance of
      landscape, recreation potential, cultural importance.

For every single criterion, special investigations will produce indicating values
of their features. Some of these might be indicators for exclusion of site.
Ground water protection zones would be such an indicator. Indicators which do not
exclude the site serve to assess and evaluate the site. Such an indicator would be
the land use, present and as planned.

The same is true for the trend in regional development which indicates general
regional planning objectives with reference to population data, structure of
settlement, and the limits of environmental impacts (pollution). Finally, a
criterion shall be mentioned which is capable of showing the present and future
impacts certain well defined areas suffer, if they have been assigned a site for
a nuclear power station. It becomes apparent that a considerable present impact
in the environment already exists: especially noise and air pollution.

SUMMARY

At Stuttgart University, a large-scale computer model has been designed. This
model is capable of simulating the interactions between plants and the environment.
One part of this model is the so called area model which is dictated by the site,
and which works with quantitative calculations. It is based on the computer program
"COPLAN". A staggered procedure starting out with zones for sites, and ending with
close up views of areas within these zones, is used to define possible locations
for thermal power plants.

The combination of this computer aided method with the classic methods of planning
will produce a site finding model which can be applied universally. The values
used in the assessment depend on the preferences of the individual user of the
model. Using the simulation model with varied preferences may produce a much more
"objective" basis for site finding than has been possible up to now.

REFERENCES

Bernard, U., Friedrich, R. und Kaule, G. (1978). Ecological Aspects of Landscape
   in the System "Energy-Environment-Economy" when chossing the site of Thermal
   Power Plants. Landschaft + Stadt 10, (3), 125-136.

Bernard, U., Friedrich, R. und Kaule G. (1979). Indikatoren der Umweltqualität
   als Steuerungsmittel in der Landschaftsentwicklung. Arbeitsbericht 12, Institut
   für Landschaftsplanung, Universität Stuttgart.

Bernard, U., et al. (1978). Landscape - Ecological Study for the Siting of a
   Nuclear Power Plant in Baden-Württemberg. Unveröffentlichtes Manuskript.

Höcker, K.-H., Unger, H. (1979). Simulation des Systems Energie-Wirtschaft-Umwelt
   für begrenzte Wirtschaftsräume am Beispiel Baden-Württembergs. Zusammenfassender
   Schlussbericht. Institut für Kernenergetik und Energiesysteme, IKE K-54-20.

Friedrich, R. (1979). A Computer-aided Method for Evaluation of Site Zones for
   Power Plants. Transactions of the American Nuclear Society, Volume 31, p. 468-
   472.

Bernard, U. (1979). Regional Policy and Ecologial Aspects of Landscape when
   choosing Site Areas for Thermal Power Plants. Transactions of the American
   Nuclear Society, Volume 31, p 472-475.

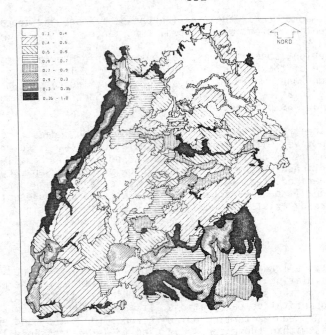

Fig. 1. Sensitivity of ecosystems with respect to affection of
landscape factors caused by nuclear power plants.

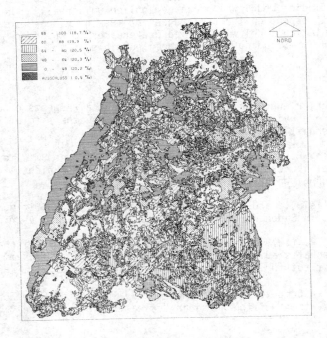

Fig. 2. Utility values in the target field 'ecology' for
nuclear power plants with closed-circuit-cooling.

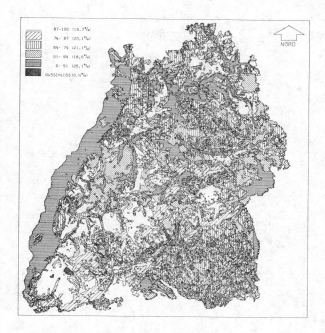

Fig. 3. Utility values in zhe target field 'ecology' for coal
fired power plants with closed-circuit-cooling.

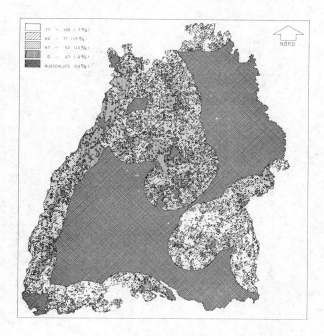

Fig. 4. Aggregated multiattribute utility data for a nuclear
power plant with closed-circuit-cooling.

AUSTRALIAN OIL SHALE DEVELOPMENT

T. Brown

Rundle Project, P.O. Box 1189, Gladstone, Qld, Australia

ABSTRACT

The exploitation of oil shale in Australia began in 1865 and these early mines produced commercial products until 1952. The recent revived interest in syncrude products due to increased oil demands has renewed exploration and development feasibility of oil shale deposits in Australia. Reference to environmental research and government regulations are discussed for a Queensland deposit.

KEYWORDS

Oil shale; environmental baseline; Australia; energy demand.

INTRODUCTION

The history of oil shale in Australia began in the mid nineteenth century and provided oil, kerosene, and other tar products to the Australian and world markets until 1952. The renewed interest in synthetic crudes with the predicted oil supply and demand to the end of this century has stimulated the exploration for oil shale with significant deposits being found in the last 5-10 years. Environmental impacts associated with development of the resource are being addressed in view of the proposed projects, with appropriate investigatory programmes for each site, process and mining method.

HISTORICAL BACKGROUND

In 1802 the first recorded discovery of oil shale in Australia was made in the Blue Mountains in New South Wales. It was described as a "bituminous schist" by the Frenchmen who were part of a scientific expedition exploring the foothills of the coastal mountains.

The first commercial oil shale plant was built at Mount Kembla in 1865, and between 1865 and 1952 sixteen mines achieved commercial production of oil from shale (Table 1). All these ventures were based in New South Wales and Tasmania, and the average oil yield from these small deposits was high. The shale seams were thin, usually less than 1 metre thick, and were removed by underground mining.

The retorts used at many of the sites were vertical Scottish Pumpherston retorts which were modified to suit the Australian shale.

Table 1. History of Commercial Oil Shale in Australia

Date	Company	Location	No. of Retorts	Av. Oil Production
1865	Pioneer Kerosene Co.	Mt. Kembla N.S.W.		
1866-1911	N.S.W. Shale Oil Co.	Sydney works NSW Hartley Vale NSW	100 47	800 1/tonne 800 1/tonne
1873-1903	Aust. Kerosene Oil and Mineral Co.	Joadja NSW		450 1/tonne
1910-1934	Tasmanian Shale & Oil Co.	Latrobe District Mersey River Tas.	4	180 1/tonne
1911-1912	Commonwealth Oil Corp.	Newnes NSW	64	495 1/tonne
1914-1922	Comm. Oil Corp. and John Fell	Newnes NSW	32	495 1/tonne
1939-	National Oil Co.	Glen Davis NSW	100	260 1/tonne
1939-1945	Lithgow Oil Pty. Ltd.	Marangaroo NSW		

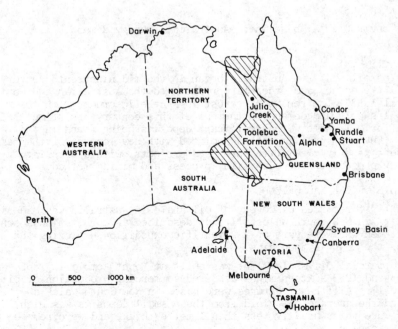

Fig.1 : Location of oil shale resource

AUSTRALIAN OIL SHALE RESOURCE

Oil shale occurs in three states of Australia, (Fig. 1);
.New South Wales deposits which were exploited between 1865-1952;
.Tasmania where small deposits were mined 1910-1934; and
.Queensland where exploration has been centred in recent years.

The Australian National Energy Advisory Committee (1978) reported that inferred reserves of in situ oil are 3×10^{12} barrels, and at present the proven reserves in the well defined deposits is 12×10^9 barrels. (Table 2). Exploratory drilling is still underway at several other sites in Queensland (e.g. Duaringa, Pluto, Byfield, Alpha, Nargoorin, Yamba), and this will increase the proven reserves significantly in the next 12 months.

As a comparison to the Australian Reserves, the Green River formation in the Rocky Mountains, U.S.A. has an estimated reserve of 8×10^{12} barrels of oil.

Table 2   Australian Major Oil Shale Deposits

Deposit	Age	Average Grade Litres/tonne	Reserves Shale 10^6 tonne	Oil 10^6 barrels
Condor	Tertiary	60-70	18000	6000
Rundle	Tertiary	100 approx.	5000	2600
Stuart	Tertiary	100 approx.	5000	2000
Yamba	Tertiary	60		1000
Julia Creek	Cretaceous	50-80	4000	1500
Alpha	Permian	350	24	19
Sydney Basin	Permian	up to 1000	50	200
Tasmania	Permian	up to 150	25	20

AUSTRALIAN ENERGY - SUPPLY AND DEMAND

Australia's known energy reserves would exceed the predicted national demand over the next 20 years, except that the available energy sources do not match the pattern of demand, (Table 3). The requirement for oil exceeds the supply and this imbalance must be supplemented by new discoveries of crude oil, development of synthetic fuels, changes to alternative transportation which will utilize alternative fuels, or an increasing dependance on imported crude oils.

In an Esso Report "Australian Energy Outlook" (1980) they state "....Australia is forecast to consume, in the next two decades, more than twice as much oil as will be available from idigenous reserves. About 2.5 billion barrels of known recoverable oil including condensate from natural gas fields, remain to be produced, while the country is forecast to use 5.3 billion barrels from 1980-2000. This means that Australia will need a further 2.8 billion barrels of oil between 1980 and 2000 ....
...On the basis of any sensible probability of discoveries, the gap between oil demand and supply will remain large through to 2000 if snythetic liquid fuel sources are not developed. Production from existing wells is beginning to decline sharply. But fortunately the area comprising development potential made up by recent discoveries and re-evaluated areas will allow production to be maintained at or above 400,000 barrels per day until the mid 80s before production from existing discoveries will decline rapidly."

Table 3.   Australia's Known Reserves of Primary Energy

Produce	Billion of barrels oil equivalent	
	Reserves	Consumption (1980-2000)
Oil	2.5	5.3
LPG	0.9	0.5
Gas	4.8	2.3
Uranium	26.0	-
Coal	170	7.4

As there can be no guarantee of finding more conventional oil reserves in quantities sufficient to prevent loss of Australia's present level of self sufficiency, unconventional sources of oil must be developed. Figure 2 illustrates that the projected development of synthetic crude can account for nearly half of Australia's total indigenous oil production by 2000, and that oil from shale could supply better than 30% of that demand.

It has been projected that oil from shale could provide 250,000 - 300,000 barrels per day of syncrude by the year 2000, and to achieve this production will require large amounts of capital, and extensive development of new technology. The dedication of areas of land and its subsequent restoration to an acceptable land use will demand extensive measures of appropriate environmental management throughout the life of the projects.

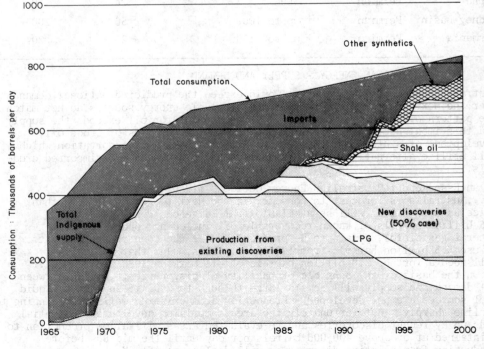

Fig. 2 : Australian oil supply and demand

GOVERNMENT POLICIES

Federal Government policy of pricing indigenous crude oil at import parity has encouraged the exploration for oil shale, and the crude oil allocation scheme ensures that locally produced crude oil is taken by Australian oil refineries. The extension of this policy to synthetic crude oils is likely and would thus ensure the marketability of shale oil until supply exceeded internal demand.

Australia has a basic policy that exploitation of energy resources must involve more than 50% Australian ownership, and extension of this policy to include oil shale retorting and upgrading plants is likely. The Federal Government does not offer specific development grants or taxation relief to synfuel projects, but policies do exist for large projects such as oil shale processing to be afforded special taxation and depreciation allowances in the early years of production.

Prior to Government approvals being granted for any of the oil shale developments it will be necessary for the individual projects to present environmental impact statements (EIS) which satisfy both Federal and State Government legislation requirements. The Federal Government requires an EIS for Foreign Investment Review Board (FIRB) approval which enables 50% or less of overseas equity in a project, and this EIS is subject to public review. In the case of Queensland projects approval of an EIS is necessary, and the standards to be met are covered in environmental regulations contained in the Queensland Legislative framework which provides extensive coverage for control of pollution, and regulation and conservation of resources, in approximately 40 State Acts and Regulations. More project specific controls are contained in an individual development Franchise Agreement which can be negotiated for each project (e.g. Rundle Oil Shale Agreement Act 1980).

GENERAL OIL SHALE PROJECT DESCRIPTION

Mining - Many of the Australian oil shale deposits have a low overburden to shale stripping ratio (approx. 1:1) and the soft nature of the shales makes them particularly suitable to open cut mining. The thickness of the oil bearing shales (cutoff yield 50 litres/tonne) in the coastal Queensland deposits of Condor, Rundle and Stuart vary between 300 to 350 metres. The shale will be mined using bucketwheels or draglines and conveyors and the raw shale will be retorted above ground.

Retorting - The retort units heat the crushed raw shale feedstock to around 500°C and this effects pyrolysis of the kerogen bonded within the shale to produce gaseous hydrocarbons which are condensed, and upgraded by hydrogenation to provide shale oil suitable as conventional refinery feedstock.

There are several retorting methods available overseas which are in a pilot plant stage and are being considered for use in Australia. These are basically either heat transfer retorts (e.g. Union A, Lurgi-Ruhrgas, TOSCO Paraho), hot gas retorts (e.g. Union B, Petrosix) or circular grate retorts (e.g. Superior, Dravo).

Solid Waste Disposal - Following mining and processing the disposal of over-burden, waste shale (subeconomic grade), spent shale and fines constitute a major part of the solid handling of the project. Management strategies for the stabilization of the wastes require early planning due to the bulk of the materials which will be involved (planned 500,000 to 1,000,000 tonnes/day), and the suitability of each waste for revegetation will affect the mode and site of disposal. The resulting areas of stabilized, revegetated dumps will be planned to provide areas suitable for a proposed future land use (e.g. possible forrestry, grazing, cultivation).

ENVIRONMENTAL IMPLICATIONS OF OIL SHALE DEVELOPMENT

The nature and extent of environmental impacts arising from oil shale development may be project, process and/or site specific and variables will include:

- mining method - opencut or underground
- geotechnical and geochemical nature of the shale
- retorting site - above ground or in situ
- retort technology - direct heat or heat transfer retorts
- solid and liquid waste composition
- waste disposal strategies
- local environmental sensitivities
- local infrastructure and social structure.

Impacts associated with Australian development must be investigated on a site and regional area basis due to the nature of the projects and their possible widespread and long term effects on industrial and population growth, recreation resources, hydrology, and ecology. Environmental studies must therefore provide sufficient baseline data in the following categories to enable future monitoring programmes to detect any change which is approaching unacceptable levels, and thus enable mitigation strategies to be implemented.

. Emission impact causing deterioration in general air quality due to retort and fugitive dust emission from the plant, mine, and waste disposal.

. Impact on water quality and hydrology due to the mining activity, waste water disposal, runoff water and leachate control.

. Land disturbance from mining activity will destroy the natural features of the area and adequate restoration management needs to be undertaken throughout the life of the project.

. Habitat alteration of both terrestrial and aquatic communities in the disturbed areas require quantification

. Occupational health - workers protection and general safety aspects investigated before process startup.

. Socioeconomic impact caused by the enormity of the project and the associated large workforce requires long term development plans for the area.

The location of the Australian oil shale deposits vary considerably with respect to geographical meteorological, biological and sociological environments, and the sensitivity of each location to alteration or disruption of one or all of these areas requires the collection of adequate baseline data to predict the capacity of the existing environment to resist detrimental impact. The coastal deposits require considerable effort on baseline biological studies due to their close proximity to mangrove areas, and estuarine fisheries.

ENVIRONMENTAL BASELINE STUDIES

The Rundle Oil Shale Project on the central Queensland coast which has been the subject of a detailed feasibility study, has had a concerted effort placed on the establishment of an environmental baseline programme to enable data collected to be used in the early planning stage of the mine and process plant. This will ensure the minimal impact will occur from the beginning of the project, and the ongoing monitoring programme will be designed using the baseline data to define the most practical and sensitive monitoring studies. The initiated baseline studies cover the following major programmes:

Air quality - meteorological data is being collected for:

rainfall	-	duration, intensity and total
temperature	-	maximum and minimum
humidity	-	continuous recording
evaporation	-	net
sunshine	-	duration and intensity
wind	-	speed and direction
barometric pressure		

A network of wind speed and direction monitors has been designed to provide data for the air quality model, and the resulting predictions will enable a regional understanding to be obtained on the dispersion and ground level concentrations expected from the mine and process emissions.
Air quality measurements for SO_2, NO_x, CO, ozone, and dust are at present being made and the regional levels which constitute the background will be used to verify the air quality model when sufficient data is collected.

Water hydrology - The freshwater streams demonstrate mainly seasonal flow, with only one major creek which has measurable flow beyond the summer rains. The hydrology of these creeks is being monitored to enable water control strategies to be designed for the mine area, with minimal impact on the environment, and maximum use of available fresh water on site for use in the process plant, revegetation, or spent shale moistening. The construction of a retention dam and diversion channel on the major creek is planned to control the water which would otherwise flow across the resource area, and this will provide a source of fresh water of approximately 30,000 megalitre storage.

Water Quality - The establishment of a water quality baseline for the streams, estuarine, marine, and groundwater regimes has been planned to enable a valid indication of seasonal variability in the parameters selected over the period prior to project startup. The parameters selected for measurement cover a wide range of routine cations and anions, plus organic analyses. This data will provide a base for detection of any fugitive leakage or seepage of liquid wastes.

Mangrove ecology - These studies are part of the environmental baseline programme due to the location of the mining operation beside a subtropical marine coastal area. The Narrows is a tidal marine channel which separates the mainland resource area from Curtis Island. The tidal flats are heavily wooded with mangrove stands that are host to complex ecological communities which includes a commercial mudcrab (Scyalla serrata) fishery.
The mapping and community structure of the mangrove areas, plus population density and species diversity of the associated fauna is an important part of the data base which will be used for comparative monitoring during the life of the project.

Waste Management - The disposal of the various solid and liquid wastes from the mining and process plant, and the subsequent stabilization and rehabilitation programme is planned to investigate the suitability of solid wastes for revegetation. A waste water management investigation has been designed to determine the treatment facilities required to allow reuse of waste waters where possible, or their disposal in an environmentally safe manner following treatment. Characterization of the wastes is being performed on available samples as they are collected from exploratory drilling and process plant trials. More representative waste samples will be tested as larger trial batches are extracted and processed.

Occupational Health - The programme to investigate the toxicological properties of the various oil products and wastes will be carried out as representative samples are collected from process trials. The results of these tests will enable the formulation of a safe working environment for the operation staff.

Socioeconomic Impact - The magnitude of oil shale mining and processing projects demand a large workforce, which in itself generates considerable social and economic pressures on the areas that will service the projects. For example, the Rundle Project, located near the city of Gladstone (pop 22,000) could increase the city's population threefold during the construction period of 8-10 years. Adequate forward planning is needed to enable housing and associated services to be available when required.

REFERENCES

Bell, A. (1981) Ecos, 27, 11-15

Cane, R.F. (1979) Proceedings of the 12th Oil Shale Symposium, 17-25

Esso Australia (1980) Australian Energy Outlook

Ferguson, P. (1980) History of the Development of the Oil Shale Industry in Australia. Southern Pacific Petroleum N.L. and Central Pacific Minerals N.L.

Saxby, J. (1980a) ERT, July, 30-32
        (1980b) ERT, Sept. 24-27

THE SCOTTISH SHALE OIL INDUSTRY

F. M. Cook

Littleton Cottage, Cultoquhey, By Crieff, Perthshire PH7 3NF, UK

ABSTRACT

One hundred and thirty one years ago a Scottish chemist, Dr. James Young, registered a patent describing a method of producing and refining oil from bituminous coal. This "cannel" coal, as it was called, occurred near the town of Bathgate in Central Scotland, and it was there that Young erected the first Oil Works for the production and refining of mineral oil. Young's patent and the erection of the Bathgate Works heralded the advent of a completely new industry in Scotland, the Shale Oil Industry, because the limited supply of "cannel" coal diverted Young's activities to the processing of oil shale, which existed in the Lothian Region of Scotland.

When Young's patent expired in 1864 there was a rapid expansion in shale oil production, due to a large number of oil shale companies starting up in business. By the beginning of this century the throughput of oil shale per annum had reached a peak of three million tons. The output of oil was about two million barrels and the Industry gave employment to 10,000 people. However, this throughput was not enough to make operations economical and in 1919 the separate Shale Oil Companies amalgamated to form one company, Scottish Oils Ltd. (a BP subsidiary). This made for greater efficiency, and constant attention to improvements in the design and operations of the retorts for processing the oil shale helped to keep production costs down. A tax or duty concession on indigenous oil products, operated by the U.K. Govt., was also a factor in keeping the Industry alive. The withdrawal of the concession in 1964 was anticipated by the Industry and it ceased operations in 1962.

KEYWORDS

Oil Shale Industry; Scottish Shale Oil; Scottish Shale Oil History; retorting; refining; environmental effects; health effects.

INTRODUCTION

Exploiting sources of energy, hitherto neglected because of high cost or lack of technology, is currently one of the World's most urgent tasks, because one of our main sources of energy, conventional crude oil, is expected to be used up early in the next century. One energy source which is at present under-exploited is oil shale.

Oil shale, of course, is not a new source of energy. As early as 1838 there was small scale production of oil from oil shale in France. However, shale oil operations in France were rapidly surpassed in magnitude and technology by the inception, a decade or so later, of a shale oil industry in Scotland which operated for more than a century. Its origin and early development are usually associated with James Young, a Glasgow chemist. It was he, who in 1850, registered patent No. 13292 for "Treating bituminous coals to obtain paraffine and oil containing paraffine therefrom." It specified the distillation of bituminous coal at low red heat to produce crude oil and the treatment of the distillates from that crude oil with acid and alkali. The patent also detailed the extraction and refining of paraffin wax.

Coal of the type referred to in Young's patent was known as "cannel" or Boghead coal and was mined at Boghead near Bathgate, which is a small town 23 miles (37kms) west of Edinburgh. The name "cannel" was applied to the coal because it burned with a long luminous flame like a candle and "cannel" is the old Scots word for candle.

It was on the outskirts of Bathgate that Young in 1851 built the first oil works to process "cannel" coal. At first he concentrated on the production of lubricating oil, paraffin (kerosine) for lamps, and naphtha for use as a solvent in the rubber and paint industries. In the first few years of its existence the Bathgate Works processed a few hundred tons of "cannel" coal per annum, but by the year 1860 the throughput had risen to over 10,000 tons. However, the amount of coal was limited. It was estimated that the coalfield originally contained approximately 200, 000 tons, and this was not all available to Young. Some was being exported to the U.S.A. and some to Mandal in Norway where the Salvesen brothers built an oil works.

Young realized therefore that he had to find another raw material and he turned his attention to oil shale, which existed in the Lothian region of Scotland, as well as in certain other parts of the United Kingdom. He already had competitors in Scotland who were using oil shale, but, of course, they had to pay him a royalty until the expiry of his patent in 1864. He also obtained royalties from oil operators in the U.S.A., but it was from the U.S.A. that his fiercest competition was to come. The discovery of natural petroleum in Pennsylvania in 1859 heralded the advent of cheap oil products into the United Kingdom.

However, before this happened, Young had commenced processing oil shale at his Bathgate Works. The yield of oil from oil shale was relatively low compared with "cannel" coal. The best quality oil shale produced about 35 to 40 gallons per ton (156 to 180 litres per ton) of oil, whereas the yield from the Boghead coal was 120 gallons per ton (537 litres per ton). Consequently the ash content from shale processing was much higher than from "cannel" coal and this has resulted in the landscape in the oil shale region of Scotland being dotted with "bings" i.e., huge piles of spent shale.

In 1865 Young built a new works solely to process oil shale. These works were reputed to be the largest in the world at that time and were located at a place called Addiewell, near West Calder, a small town five miles (8 kms) south of Bathgate. Like the Bathgate Works, the Addiewell Works produced naphtha, paraffin for lamps, wax and lubricating oil. Young concentrated on the quality and the output of paraffin (kerosine) from both his Works and it was his concentration on paraffin that resulted in his friend, Dr. Livingstone, the explorer, naming him "Paraffin" Young.

The expiry of Young's patent in 1864 was the signal for a number of shale oil companies to start up in business in Scotland. In all, it is recorded that there were 120 Oil Works built, but they were not all built at the same time. In fact many of the Companies, who built these Works were not in business long enough to earn a dividend.

By 1905 the following shale oil companies were the only ones still in existence:

	Year of Registration
The Broxburn Oil Company Limited	1877
The Dalmeny Oil Company Limited	1871
The Oakbank Oil Company Limited	1885
The Pumpherston Oil Company Limited	1883
James Ross and Company Limited	1883
Young's Paraffin Light and Mineral Oil Company Limited	1886

The throughput of oil shale from these six Companies reached a peak of just over 3 million tons per annum at the beginning of this century, but the yield of oil per ton of shale had dropped from 34 gallons (154 litres) in 1879 to 27 gallons (122 litres) in 1910. This drop in yield, coupled with competition from imported oil products, threatened the extinction of the Scottish Shale Oil Companies. In 1918 they formed the Scottish Shale Oil Agency to streamline their marketing operations, but this was not enough in itself to keep them in business. In 1919 they approached the Anglo-Persian Oil Company Limited (now the British Petroleum Company Limited) for assistance. The result was the formation of a new company, Scottish Oils Limited, a subsidiary of BP, which controlled the operation of the six separate companies and centralized common services. Its headquarters were at Uphall, which is 14 miles (22 kms) due west of Edinburgh.

The area of oil shale operations covered by the new Company is shown on Fig. 1. It was approximately triangular in shape. The base of the triangle, approximately 8 or 10 miles long (13 or 16 kms), extended along the line of the River Forth, west of Edinburgh. The apex lay some 6 miles (9 kms) south, in the moorland district of Cobbinshaw and Tarbrax. (See Fig. 1)

In this area of oil shale seams occur in the lower part of the carboniferous system. There are about seven different seams or groups of seams distributed among beds of sandstone, limestone, marl and poor quality coal. They are not regular in formation and exist in separate basins which are severely faulted and in some places disrupted by volcanic intrusions. The higher seams gave a better yield of oil than those lower down.

In general Scottish oil shale can be divided into two types, plain and curly. The latter can be cut by a pocket knife into curly shavings - hence the name. It probably evolved from the plain type which is more common. The color of the oil shale is dark brown or black, and is of a laminated structure. Its organic content, from which the oil is produced, is called "Kerogen" (named by Prof. Crum-Brown F.R.S.). To decompose the kerogen to produce oil requires the oil shale to be heated over a range of temperature from $350^{\circ}C$ to $500^{\circ}C$. A typical analysis of Scottish oil shale is given in Table 1.

There were three main operations involved in the production of oil from oil shale in Scotland:

(1) Procuring the shale by mining or opencast methods.
(2) Retorting the shale to produce gas, ammonia, crude naphtha and crude oil.
(3) Refining the crude oil and naphtha to produce marketable products.

(1) Procuring the raw material. There were two mining methods for procuring the shale from underground. The first was the stoop and room or pillar system, in which a network of roads or rooms were driven through the seams, leaving pillars

FIG. 1a. Scotland.

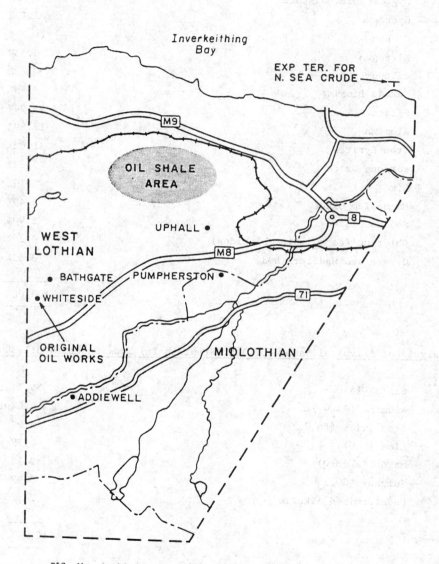

FIG. 1b. Lothian region of Scotland.

TABLE 1 Analysis of a Typical Sample of Scottish Oil Shale

Hygroscopic Moisture	3.25%
Hydrogen	2.17%
Carbon	14.27%
Nitrogen	0.53%
Sulphur	0.23%
Carbon Dioxide	3.95%
Silica	41.94%
Alumina	17.89%
Iron Pyrites	1.02%
Ferrous Oxide	3.82%
Lime	3.19%
Magnesium Oxide	1.49%
Sodium Oxide	2.40%
Chlorine (as soluble chlorides)	0.20%
Oxygen and undetermined	3.65%
	100.00%

Analysis of A Typical Sample of Ash (Spent Shale) from Scottish Oil Shale

Silica (SiO_2)	55.7%wt
Alumina (Al_2O_3)	25.0%wt
Iron Oxide (Fe_2O_3)	9.9%
Lime (CaO)	2.7%
Magnesia (MgO)	3.1%
Sulphur (SO_3)	0.9%
Undetermined Alkalis	2.7%
	100.0%

to support the roof. The pillars were afterwards removed, and if the workings were near the surface, ground subsidence could result. The second method was the longwall system which involved removing the whole working face of the seam in one operation. Many of the underground seams were reached by inclined mines starting from a surface outcrop and going downwards along the dip of the strata to the trough of the oil shale basin. In some cases vertical shafts were used to reach the oil shale seam. The seams were four to twelve feet (1.2m to 3.66m) thick, and in some instances mining operations were carried out a depth of more than 1000 feet (305 metres). The miners used explosives to dislodge the shale, which was removed from the working face in hutches, each capable of carrying one ton (Long ton 2240 lbs or 1016 kgs). The output from a mine could be about 1000 tons per eight-hour shift.

A limited amount of shale was obtained from opencast mining and an attempt was made just before the industry closed down in the 1960's to produce oil from shale by heating it in situ underground, but the project was abandoned because of the poor yield.

(2) Retorting. An oil shale retorting Works usually obtained its supply of shale from an adjacent mine, but this could be augmented by supplies from one or two other mines. The shale was conveyed in hutches or by conveyor belt from the adjacent mine and fed into a shale breaker or crusher, which had a toothed drum capable of breaking the shale into pieces of 3" or 4" cube size (7 to 10 cms). In the case of mines which were distant from the retorting works, the shale was conveyed in ten ton rail wagons to the Works.

The main feature of the Works were the lines of retorts built adjacent to one another in rows called "benches".

Throughout the century of Scottish shale oil operations there were a number of retort designs. Young started with horizontal retorts, similar to those which were being used in the gas industry at the time. Then he discarded these in favor of a vertical type which consisted simply of vertical cast iron pipes 12 ft in length (3.65m) by 14 inches diameter (36 cms) set in a brick furnace. The retorts were grouped in sets of four. They were charged hourly and the oil shale was fed in at the top of the retorts. The bottom of the retorts protruded through the furnace into a water trough.

The ultimate in retort design were vertical retorts erected by Scottish Oils Limited in 1942 at Westwood, a site not far from Young's Addiewell Works. A feature of the design was the way in which steam and air injection was used to improve the overall heat distribution throughout the oil shale on its downwards passage through the retort. The performance was vastly superior to previous designs, the throughput being trebled, and the retorts were thermally self-sufficient. In other words, no other fuel was required to augment shale gas for heating the oil shale. Producer gas was used initially for bringing the retorts on stream.

Westwood Works consisted of two benches of 52 retorts with a capacity of about 1200 tons of oil shale per day, to give an output of about 600/700 barrels of oil per day. The retorts were of the vertical continuous type, 34 ft high (10 m), the upper 14 ft (4 m) being cast iron, and the lower portion being brickwork. The cross section was rectangular and at the top $2'-9\frac{1}{2}"$ x $1'-2\frac{3}{8}"$ (84 cms by 36 cms), expanding to $4'-8"$ by $1'-10"$ (142 cms by 56 cms) at the bottom. The increasing cross sectional area from top to bottom allowed the oil shale to expand as it was heated on its passage down through the retort. A cross section of this retort is shown in Fig. 2.

The shale broken into pieces by the shale crusher was fed onto a conveyor belt which delivered the material to the main conveyor belt running from ground level

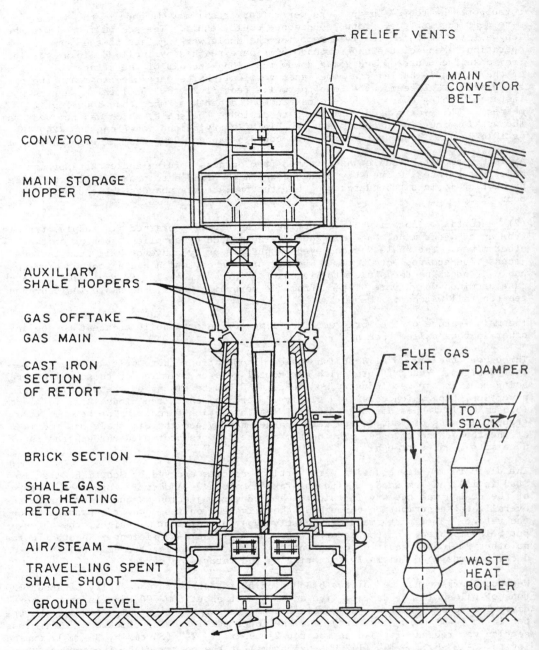

RELIEF VENTS

MAIN CONVEYOR BELT

CONVEYOR

MAIN STORAGE HOPPER

AUXILIARY SHALE HOPPERS

GAS OFFTAKE

GAS MAIN

CAST IRON SECTION OF RETORT

FLUE GAS EXIT

DAMPER

TO STACK

BRICK SECTION

SHALE GAS FOR HEATING RETORT

AIR/STEAM

TRAVELLING SPENT SHALE SHOOT

GROUND LEVEL

WASTE HEAT BOILER

FIG. 2. Scottish oil shale retort type erected at Westwood Works

to the top of the retorts. A reversible conveyor system on the top was then used
to supply shale to any of the storage bunkers. There was one storage bunker for
every four retorts with a capacity of 2,000 tons. From the main storage bunker the
shale was fed to an auxiliary hopper and from there the gravity flow through the
retort was regulated by adjusting the discharge mechanism at the bottom. Air and
steam were injected into the bottom of the retorts counter current to the passage
of the shale, which at the hottest zone in the retort reached a temperature of
785^0C (1450^0F). The spent shale, which was discharged continuously into hoppers,
was quenched by water sprays, thus producing some of the steam for the retorting
process. The steam which was injected into the base of the retort served the purpose of removing the oil vapor from the hot zone to a cooler one to avoid cracking.
It also improved heat distribution throughout the retort and it reacted with the
residual carbon in the shale to form gas, and with the nitrogen in the shale to
form ammonia.

Air injection, which had to be carefully regulated, assisted the heating of the
shale by combining with part of the carbon in the shale. The gaseous products from
the retorts, consisting of crude oil vapors, steam and ammonia and permanent gas
were drawn off by fans or exhausters, capable of handling one million cubic feet
of gas per hour (28,320 cu m/hr). Some of the flue gas heat was recovered in waste
heat boilers. Crude oil and ammonia liquor were collected from condensers. Un-
condensed gas containing appreciable quantities of ammonia and light hydrocarbons
was scrubbed with water and gas oil. The water absorbed the ammonia and the gas
oil, the light hydrocarbons. A steam stripper was used to recover the light oil
or scrubber naphtha as it was called. The scrubbed gas was returned to the retorts
as fuel. A typical yield from a retort with a throughput of 12 tons of shale per
day was:

Crude Oil	19.5 gals per ton	(87 litres per ton)
Crude Naphtha	3.7 gals per ton	(17 litres per ton)
Sulphate of Ammonia	26 lbs per ton	(11.6 kgs per ton)

(3) Refining. Scottish crude shale oil is generally classified as paraffinic,
but differs from conventional crude petroleum in that it contains a greater proportion of unsaturated hydrocarbons, large quantities of nitrogen bases of the
pyridine and quinoline type, and an appreciable amount of phenols. These constituents were responsible for the discoloration, unpleasant smells and gumming ten-
dencies of the distillates if they were not properly treated. Treatment of distillates with sulphuric acid and caustic soda was a fundamental part of the refining process. The low aromatic content of the oil resulted in the motor gasoline
having a low octane rating. However, the sulphur content was less than ½%.

In the early days of the Scottish shale oil industry the refining scheme included
the production of gasoline, solvent naphtha, kerosene, gas oil, lubricating oil,
wax, fuel oil, resin and coke. (See Fig. 3.) Later lubricating oil production
was cut out and diesel oil production was maximized. (See Fig. 4.)

Both schemes involved heavy chemical treatment, with consequent losses, although
the tar from the treatments was used for refinery fuel, mostly on the coking plant.

During the last 20 or 30 years of oil shale operations all the crude shale oil
from the retorting Works was processed in Pumpherston Refinery. Distillation of
the crude shale oil was carried out in a conventional tube still of about 4,000
barrels/day capacity, which was linked to a battery of coking stills for handling
the residue. The products from the still were:

		S.G. at 60^0F (15.6^0C)
Crude Gasoline	4.6%	.766
Wax Free Fraction	22.0%	.837

BEC 2 - F

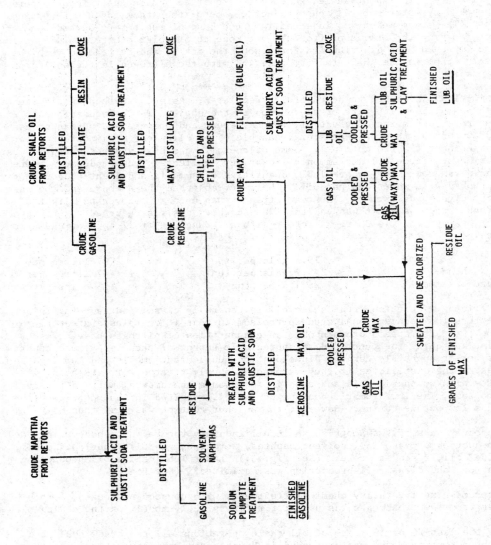

FIG. 3 Early shale oil refining scheme which includes lubricating oil

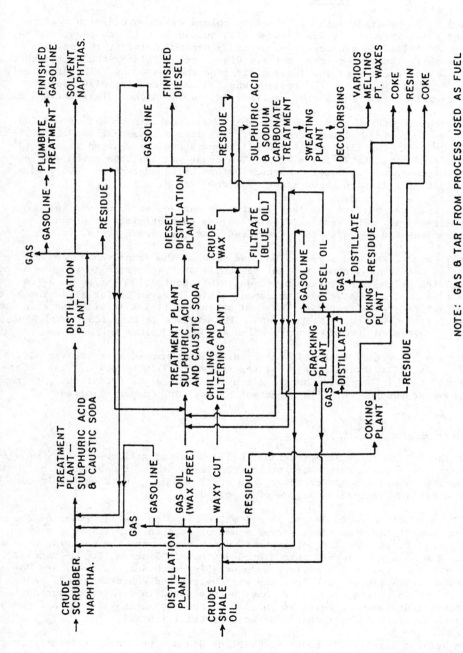

FIG. 4   Scottish shale oil refining scheme for Max$^{\underline{m}}$ Diesel Oil Production

NOTE: GAS & TAR FROM PROCESS USED AS FUEL

Wax Bearing Fraction	52.0%	.878
Residue	21.4%	.954
	100.0%	

The overhead from the crude unit distillation column was combined with crude naphtha from the retorting works, then treated with sulphuric acid and caustic soda, before being distilled to produce gasoline and a naphtha solvent, the residue after treatment being routed to a re-run unit for diesel oil. This re-run unit also handled the wax free cut plus the dewaxed oil from the wax plant. The diesel oil was taken off as a tray cut and the residue was fed to a cracking plant to produce gasoline and diesel oil. The latter was recycled to the diesel distillation plant.

Wax was removed from the wax-bearing cut, which had a certain amount of wax free oil blended with it, in a filter press, operating in two stages at 500 lbs/sq inch (35 kgms/sq cm). In the first stage the oil temperature was reduced to 38^0F (3^0C) and in the second stage to 20^0F (-6.7^0C). The crude wax was then given acid and soda treatment to remove unstable compounds before being segregated into grades and decolorized by filtering through the "Fullers Earth".

The acid tar from the various chemical treatments was hot water washed to remove most of the free acid. The weak acid solution was then utilized to produce ammonium sulphate from the ammonia liquor received from the retort Works.

In the early days of the Industry, no use was made of the spent shale from the retorting works, but in 1934 a brick plant was built at Pumpherston to produce a good quality building brick with a crushing strength of about 3,500 lbs/sq inch (245 kgms/sq cm). A typical analysis of the spent shale is given in Table 1. The brick-making process involves pulverizing the spent shale and mixing it with hydrated lime and lime slurry in the correct proportions. The mixture is then fed to brick moulding machines to produce "green bricks", which are steam heated in autoclaves.

In 1948 the manufacture of synthetic detergents was commenced by using the olefin content of a gas oil fraction in the boiling range of 199^0C to 309^0C. The detergent plant is still in operation but is not using a shale oil feedstock. It trades under the name of Young's Paraffin Light and Mineral Oil Company Limited.

Environmental and Health Effects

Throughout the 100 years of Scottish Shale Oil operations millions of tons of oil shale were processed, and today some of the results of this, the bings or heaps of spent shale from above ground retorting, are still visible, marring the landscape in certain areas in the Lothian Region of Scotland.

However, it is possible to landscape an oil shale bing, so that it enhances the environment, and at the same time, serves a useful purpose. This is the case at Dalmeny, a small village on the outskirts of Edinburgh, where the material from a spent shale bing has been formed into a bunded area of 90 acres (36½ hectares), where storage tanks for half a million tons of BP's North Sea crude oil are completely hidden from view (except from the air) by grassed embankments of spent shale, on which small trees and shrubs have been planted in a layer of top soil. This landscaping achievement resulted in 1976 in the site being given the European Architectural Heritage Premier Award for an industrial installation.

Apart from being unsightly the spent shale bings did not present the Industry with a problem. They were not a health hazard, in that materials leached out of the spent shale were not detrimental to agricultural crops or to nearby rivers.

475

Because of the nature of oil shale, miners found underground conditions better than those prevailing in coal mines, and fires or explosions in shale mines were very rare. The main hazard to health lay in the refining of Scottish Shale Oil because of its carcinogenic characteristics. The point of contact for employees was the wax extraction process, where contact with crude wax from the filter presses could result in some of them contracting skin cancer. The filter press type of process has now, of course, been superceded by the solvent extraction process, where it is unnecessary for employees to come in contact with wax.

Because of the high chemical treatment associated with shale oil refining, the effluent had to be very carefully controlled. In many cases it was pumped down a disused mine without affecting rivers or agricultural land. The low sulphur content of Scottish shale oil precluded any serious health hazard from the emission of sulphur compounds into the atmosphere. However, burning acid tar, arising from the chemical treatment of shale oil, was a problem, in that it produced acid smuts from the chimney stacks. Eventually improvements to the treatment of the acid tar and an improved type of tar burner did much to eliminate atmospheric pollution.

CONCLUSION

For the last 50 years of its existence the shale oil industry in Scotland owed its survival in part to its concentration on improving retort design, so as to attain greater operating efficiency. The other factor which allowed it to survive so long was the fact that the British Government did not demand the full rate of duty or tax on it products. It was only when the Government gave notice in 1962 or thereabouts that it intended to remove the tax concession in 1964 that the Industry was obliged to close down. By that time the number of employees had dropped to about 2,000 or 3,000 from a peak figure of about 10,000.

Now the present price of crude oil makes shale oil production economically feasible, especially in the U.S.A and Australia, where the deposits are so enormous. And although history is supposed to repeat itself, oil shale history in Scotland is unlikely to do so. The remaining oil shale deposits are too low in "kerogen" and too small in quantity.

POLICY OPTIONS FOR ENERGY/WATER ISSUES IN THE WESTERN U.S.

M. D. Devine

Science and Public Policy Program, University of Oklahoma, Norman, Oklahoma 73019, USA

ABSTRACT

Water availability and water quality problems are critical factors in shaping the development of energy resources in the Western U.S. This paper identifies the potential impacts of energy development on water resources, discusses the issues raised by these impacts, and evaluates several policy options for addressing these issues. The findings presented in this paper are based on the results of a multi-year study, funded by the U.S. Environmental Protection Agency, on the environmental, social, and economic issues associated with western energy development. The study covered the development of six energy resources (coal, oil shale, crude oil, natural gas, uranium, and geothermal) in eight Rocky Mountain/Northern Great Plains states.

KEYWORDS

Energy development, water resources, energy/water requirements, water quality, water policy, water management.

INTRODUCTION

Given the substantial energy resources in the western U.S. (see Fig. 1), this region is expected to be a major contributor to increased domestic supplies. The production of uranium and coal has increased greatly in the past five years; and pressures for expanded development of western energy resources will increase dramatically in response to recent legislation calling for a major synthetic fuel program, primarily from coal and oil shale. If such a program does take place, literally all of the oil shale development and a substantial part of coal synfuel development will occur in the western states.

Three factors regarding water resources are critical to western energy development. First a large percentage of surface water is already committed to use, particularly in the Colorado River Basin. Thus, new water-right applications, even if they represent only a very small part of a state's total supply, may be competing for a very small quantity of uncommitted supplies. A second factor is that energy resources are often located disadvantageously with respect to water supplies. For example, approximately 85 percent of the nation's high-grade recoverable oil shale reserves are located in a two-county area of western Colorado. This concentration of a resource occurs in an area where most streams are fully used and where flowing

Fig. 1. Eight-State Study Area

water is particularly valued for environmental purposes. The third, and perhaps most important, factor is that a new group of interests is aggressively pursuing and acquiring rights for water, for example, to develop energy resources, to meet the expanding needs of Indians for irrigation, to protect water quality, and to preserve instream values such as recreation and aquatic life. Based on this, it is easy to understand why water availability and quality questions generate intense conflicts in the West--resolution of the issues will have a profound effect on the basic economic, cultural, and social character of the region (White and others, 1979a, 1979b; Science and Public Policy, 1981a, 1981b).

IMPACTS AND ISSUES

Western energy development decisions will be entangled in a number of water use and water quality protection issues, some of which are briefly summarized below. Many of these conflicts will exist even without energy development because, as described above, the region is generally water-short and because a variety of demands are being made for the use and protection of the resource. Not only will energy development exacerbate these issues but they, in turn, can directly influence energy resource development decisions.

Water requirements for energy development. Table 1 presents water requirements for the energy facilities considered in this study. In estimating future water requirements for energy development in the region, considerable uncertainty exists in several areas:

- There is a lack of experience with commerical-size facilities for new technologies such as oil shale and coal synfuel processes;

- The type of cooling process chosen can create significant differences in water requirements; and

- There are inherent difficulties in estimating future energy development levels in the region over the next twenty years and beyond.

TABLE 1   Water Requirements For Energy Facilities

Facility and "Standard" Size	Water Requirements per Facility (acre-ft/year)
Coal-fired power plant (3,000 MWe)	23,900 - 29,800
Coal gasification (250 MMscfd)	4,890 - 8,670
Coal liquefaction (100,000 bbl/day)	9,230 - 11,750
TOSCO II oil shale (100,000 bbl/day)	12,900 - 18,600
Modified _in situ_ oil shale (100,000 bbl/day)	7,600
Uranium mine and mills (1,000 mtpy)	270 - 300
Slurry pipelines (25 MMtpy)	18,400
Surface coal mines (25 MMtpy)	Neg. - 1,240
Geothermal (100 MWe)	12,700 - 13,700

MMscfd = million standard cubic feet per day   MMtpy = million tons per year
mtpy = metric tons per year   Neg. = negligible

Thus, precise estimates of the extent to which water availability will affect energy development are not possible. _However, it is clear that enough water physically exists on a region-wide basis to support very large levels of energy development._ For example, the total water requirements for our Nominal Demand scenario (which represents a very high level of development by the year 2000) would consume, at most, 22 percent of the remaining unused water supplies in the Upper Colorado and Upper Missouri river basins. Nevertheless, water availability and quality issues are certain to affect the overall level, location, and types of energy development that will occur. Even if water shortages are not a concern region-wide, water is relatively scarce in many energy-rich regions, including the Powder River Basin in Wyoming and Montana, the oil shale area of western Colorado and eastern Utah, and the Four Corners area of New Mexico. In addition, a variety of institutional and political questions will determine the amount of water made available to supply new energy projects as discussed below.

Increasing demands for water use. Although agriculture is the dominant water user in the study area, there is a rapidly growing demand for water from other users, as indicated previously. These demands can influence the level of agricultural production in the region and, thus, threaten many existing interests and values which have developed within a predominantly agricultural economy. In addition, disputes among energy, environmental, Indian, agricultural, municipal, and other interests set an increasingly complex context for water policymaking.

Uncertainty and complexity of the water policy system. Managing water resources is made more difficult by the complex combination of state water law, federal water policies, court cases, interstate agreements, and international treaties which determine how water will be used. One of the most serious deficiencies of this system is that a state appropriation system typically settles disputes over water use only after damage of a water right has occurred. Thus, the current system, as practiced in most states, is designed primarily to react to questions about the legality of rights and uses and not to manage the resource on a day-to-day basis. A second major deficiency is that the water allocation system in most western states is not tied directly to water quality protection. A third deficiency is that the system offers few incentives for conserving water as exemplified by the "use or lose" and "nonimpairment" doctrines.

Reserved water rights. Reserved rights recognize that when the U.S. establishes a federal reservation such as a national park, military installation, or Indian reservation, a sufficient quantity of water is "reserved" to accomplish the purposes for that particular land. This reserved rights doctrine is significant because the federal government and Indian tribes own large amounts of western land--about 70 percent of the land in the CRB--and thus large quantities of water are at stake. At the present time relatively little water has been put to use under this doctrine; however, since Indians and federal agencies could exercise these rights for large quantities of water, this situation adds uncertainty as to the amount of water available to existing and future users under state appropriation systems.

Pollution from energy facilities. Current federal and state regulations are designed to strictly limit the direct discharge of pollutants from energy facilities into surface streams. Nevertheless, even if these standards are met, energy production and conversion processes can pollute both surface and groundwater either through the disposal of waste products or through the disruption and contamination of aquifers during mining and in situ recovery operations. For aquifer contamination from mining and in situ recovery, information on the extent of the potential impacts and possible control technologies is inadequate. Control of seepage and runoff from waste disposal sites constitutes a potentially serious problem, especially after a facility has shut down. Given current federal and state regulations and the level of uncertainty, existing environmental regulations are inadequate to ensure that long-term or irreversible damage to surface and groundwater quality does not occur.

Salinity control. Salinity has already been singled out for regulatory control by the federal government and by each of the states in the CRB and is of increasing concern in the Yellowstone River Basin. The major sources of salinity currently are natural salt flows and runoff from irrigated agriculture. Energy development can exacerbate salinity problems by increasing salt loading in runoff from mining areas and due to the concentrating effect of consumptive water use. Thus, energy development is likely to intensify conflicts over salinity control.

POLICY OPTIONS

While the impacts and issues outlined above represent a more pessimistic perspective than other studies similar in scope, a variety of technological and institutional alternatives exists for avoiding or mitigating many of these problems. Table 2 lists a selected set of the options in five categories which were included in the study. However, given the brevity of this paper, only the findings for a few of these options will be summarized below.

Dry and Wet/Dry Cooling

One approach examined for conserving water in energy facilities was through the use of some degree of dry cooling. Since water for cooling is the largest single water requirement for electric power plants and synthetic fuel facilities, the choice of cooling technology is crucial. Table 3 summarizes the findings on water savings. Wet/dry cooling of power plants can reduce water requirements by about 70 percent, while intermediate wet cooling of coal synfuel plants can reduce water requirements by about 20 to 30 percent. These water savings can increase economic costs. In the case of synfuel plants, however, the economic penalty for intermediate wet cooling is expected to be less than 0.5 percent. For power plants the economic penalty would typically be from 3 to 5 percent. In addition, with all dry cooling of power plants, plant efficiency and capacity could be lowered due to larger energy requirements to run the fans and due to higher condenser temperatures on hot

TABLE 2 Selected Policy Options for Water
Quantity/Quality Issues

Approach	Examples of Specific Alternatives
Conservation of water	o Modified process designs in synfuel plants o Utilization of dry or wet/dry cooling o Improved irrigation efficiency
Augmentation of supply	o Utilization of saline water in energy conversion facilities o Weather modification o Interbasin diversion projects
Water quality protection	o Water quality control plans for all energy projects o Salinity "offset" policy
State water administration and management	o Water rights exchange systems (e.g., "water banks") o Coordinated administrative structures (e.g., "Utah Plan")
Regional and federal roles	o Federal technical assistance (e.g., grants to develop information systems) o Federal-Interstate Compact Commissions

summer days. For this reason, wet/dry cooling is generally much more practical than all dry cooling for power plants in the western U.S.

Although the economics are not always favorable, conservation techniques in energy facilities may be attractive for other reasons. In some water-short areas, water supplies may be physically or institutionally limited; thus, water conserving technologies may be necessary. And, even if sufficient water is available, a facility with smaller water requirements will be less threatening to other water users. This factor is likely to become increasingly important in the siting process. Water conservation for energy resource development may be the easiest set of alternatives to implement. Conservation for energy resource development can save large percentages of water and has less uncertainty associated with its effectiveness and

TABLE 3 Water Savings for Coal Conversion Facilities[a] with
Alternative Cooling Systems

| Plant (size) | Water Requirements | | |
	High Wet Cooling	Intermediate Wet Cooling	Minimum Wet Cooling
Power Plant (3,000 MWe)	23,900-29,800	5,500-9,500	NC
Lurgi Gasification (250 MMscfd)	4,900-7,100	3,300-5,600	2,900-5,200
Synthane Gasification (250 MMscfd)	7,700-8,700	5,900-6,700	5,500-6,300
Liquefaction (100,000 bbl/day)	9,200-11,800	7,500-9,700	7,000-9,100

NC = not considered   [a]Range is across six sites considered.

cost. It appears easier to implement new regulations on energy resource development than on agriculture, and energy industries are probably better able than farmers to afford the cost increases associated with water conservation efforts.

Saline Water Use by Energy Facilities

Saline water (containing more than 1,500 parts per million of TDS) is a resource not suitable for use by many of the competitors for western water, particularly agriculture and municipalities. However, energy developers can afford to treat saline water at only modest price increases. Several sources exist: (1) surface waters such as the Powder River of the Yellowstone River Basin, (2) aquifers such as the Madison or Dakota, (3) irrigation return flows from irrigation projects, and (4) municipal wastewater.

The economic costs of using saline water include supply, treatment, and disposal of brines. As shown in Table 4, total cost increases from using saline water will be modest: a maximum of 6 percent for gasification, 3 percent for electric power generation, and 2 percent for liquefaction. Since in the calculations of Table 4 no credit has been given for the nominal treatment, supply, and disposal costs for facilities handling fresh water, these numbers are worst case estimates.

Several incentives may exist for energy developers to use saline water. These include the possibility that fresh water supplies will be unavailable or limited; the possibility that political conflicts, lengthy adjudication proceedings, and other siting delays could be reduced by using this "unwanted" water source; and the need to have alternative water sources as a hedge against the uncertainties of the water situation in the West. States may also encourage saline water use; for example, as part of a salinity control program or to facilitate energy development. State policies in this regard could include allowance of rate adjustments (so that increased costs are borne by consumers) and financial subsidies such as deferred or reduced taxes.

TABLE 4 Estimated Costs of Using Saline Water in Energy Conversion Facilities

Category	Gasification ($/Mcf)	Liquefaction ($/bbl)	Electric Power Plant (¢/kWh)
Boiler	0.01	0.10	0.007
Process	0.009	0.06	Neg.
Cooling Towers	0.03	0.03	0.05
Disposal	0.16	0.52	0.014
Water Supply	0.06	0.22	0.044
Total Costs	0.27	0.93	0.11
Approximate Product Price	4.50	40.0	4.0
Cost as Percent of Price	6%	2%	3%

Mcf = thousand cubic feet   kWh = kilowatt-hour

Salinity Offset Policy

The salinity offset policy would establish an administrative mechanism to prevent increased salinity levels due to energy development. Under this policy, proposed energy projects which increase salinity (either through salt-loading or through concentrating effects) would only be allowed to proceed if some "offsetting" activity were implemented so that downstream salinity concentrations would not be worsened. Offsets could be obtained, for example, by energy developers paying for a proportion of improved agricultural irrigation projects or for a proportion of desalination plants. The primary advantage to this approach is economic efficiency, since energy developers could use the lowest cost "offsets" available. Increased energy costs are expected to be small. For example, if offsets were obtained by paying a proportion of the costs for salinity control projects already being undertaken energy costs for three facilities analyzed are estimated to increase by 0.02 to 0.5 percent

Two important disadvantages exist to this policy. First, the salinity models required to predict salinity impacts are not well developed. Second, by adding another requirement to energy facility siting permits, energy development could be slowed or constrained. However, states could help to alleviate this problem, for example, by banking offsets. The legal basis and practical feasibility of such a policy is uncertain. One type of salinity offset policy has been attempted in Colorado by a local government agency, but the plan was vetoed by the governor. This plan would have affected a number of future water users in the state and was therefore strongly opposed by several groups. By limiting the offset policy to energy development activities the opposition would be reduced, although important equity questions would remain.

Coordinated State Administrative Structures

The purpose of this option is to increase the responsibility for day-to-day management of water and to explicitly integrate water quality and quantity concerns. The following briefly summarizes the findings for one particular administrative system which would combine elements of bottom-up planning with centralized agency control, which is termed the "Utah Plan" (Dewsnup and Jensen, 1975). Although never implemented, the basic idea of this plan was to organize water resource planning around local areas or "hydrological units." Within each unit, water planning staffs, in cooperation with local interests, would collect data and formulate the basics of a tentative water plan. This plan would identify the current water supplies, usage, and alternatives for conserving and augmenting that supply. These plans would be reviewed for comment at several levels before being sent to the agency having overall authority over the state's water resources. After holding public hearings on the plan, the agency would finalize and adopt the unit plan. The plan would serve as a legal guide to state and local decisions concerning water use, and would also serve to advise federal agencies. State and local water management agencies would be legally required to follow the unit water plans in allocating water.

The primary benefits of such coordinated water administration would be increased capacity to integrate water quality and water quantity concerns and increased flexibility--because diverse interests are represented at the level of hydrological units, and thus multiple demands on water can be accommodated. For example, in Utah energy development could receive high priority in a unit representing the east-central part of the state while environmental or other values may take precedence in other units. This idea contrasts significantly with current state systems which typically place priorities for water use on a state-wide basis. Finally, such an approach is explicitly designed to improve the quality of information about water resources by developing detailed information about hydrological units. It could also reduce much of the uncertainty faced by prospective users under the appropriation system. For example, a prospective appropriator would know from a unit plan if unappropriated water existed, if any existing rights were available for purchase and transfer, and if a specific use of water is compatible with the water use plan.

There are serious impediments to establishing coordinated water administration in the western U.S. The "Utah Plan" directly challenged agricultural interests, and the plan was rejected by the Utah legislature. Whether integrated planning and management would promote an acceptable allocation of water would largely depend on the degree of consensus among the various competing interests in a unit: agricultural producers, environmental groups, industrial development interests, energy producers, and Indian tribes. The failure of this idea to gain acceptance in Utah exemplifies the typical problems facing attempts to implement change toward more comprehensive water management planning. The current system seems quite adequate from the perspective of the traditional users who are generally satisfied with their allocations. This position, however, may contrast with the position of some state-level decisionmakers interested in promoting municipal and industrial water uses which might yield greater opportunity for employment and income for the state's residents.

CONCLUSION

A central conclusion from our study is that not all of the legitimate demands for water can be met. Thus, conflicts among agricultural, municipal, energy, environmental, Indian, and other users are likely to increase. Our analysis of policy options does not identify any one alternative as the best choice. This is largely because of the nature of water problems and issues, which incorporate a wide range of values and change from location to location. Thus, a mix of policy responses--both technological and legal/institutional--will be needed. However, attempts to change the current system should be based on three basic principles: (1) Water availability and water quality are inherently related; future water policies should consider them together. (2) Although states are primarily responsible for water resource policy, future water policies should encourage basin-wide approaches to water management. (3) The causes of water problems and the implications of policy alternatives vary considerably across the West. Future water policies, even if established regionally, should be implemented in ways that allow for local differences.

REFERENCES

Dewsnup, Richard L., and Dallin W. Jensen (1975). Proposed Procedures for Planning, Allocating and Regulating Use of Water Resources in Utah, 2 vols. Utah Division of Water Resources, Salt Lake City.

Science and Public Policy Program, University of Oklahoma (1981a). Energy From the West: A Technology Assessment of Western Energy Resource Development. University of Oklahoma Press, Norman.

Science and Public Policy Program, University of Oklahoma (1981b). Energy From the West: Water Policy Report. U.S. Environmental Protection Agency, Washington, D.C.

White, Irvin L., and others (1979a). Energy From the West: Policy Analysis Report. U.S. Environmental Protection Agency, Washington, D.C.

White, Irvin L., and others (1979b.) Energy From the West: Impact Analysis Report, 2 vols. U.S. Environmental Protection Agency, Washington, D.C.

WATER CONSERVATION WITH ENERGY DEVELOPMENT IN THE SAN JUAN RIVER BASIN, NEW MEXICO

M. W. Gilliland and L. B. Fenner

Energy Policy Studies Inc., Omaha, Nebraska, USA

ABSTRACT

Energy resource development (coal mining, synthetic fuel production, and electric power generation) is expanding in the Colorado River Basin of the Western U.S. where water is already scarce and where the traditional economy and culture are built around irrigated agriculture. Conflicts over who will receive how much water are significant and environmental impacts are expected to be major. The use of water conservation techniques, such as dry cooling in energy facilities, is often proposed as a mechanism to alleviate conflicts and mitigate impacts. In this study, the effects of the best available water conservation techniques on projected impacts in one sub-basin of the Colorado River Basin were analyzed. Techniques included lining existing irrigation canals, conversion of existing flood and furrow irrigation to sprinkler irrigation, the use of dry cooling in existing and new electric power and synthetic fuels facilities, and the use of groundwater pumped from new uranium mines for irrigation. The effects of these were analyzed in toto for the San Juan River Basin in New Mexico. Results are somewhat dualistic. On the other hand, future energy development is expected to be so massive that even this high degree of water conservation can do little to change the water quantity, water quality, and instream impacts expected over the next 25 years. On the other hand, water conservation may do just enough to alleviate potential conflicts among states over water allocations and salinity standards and to ameliorate confrontations among parties-at-interest over instream resources and instream needs.

KEYWORDS

Water and energy; the Colorado River Basin; salinity impacts; instream impacts; water conservation.

INTRODUCTION

Numerous water conservation and management strategies have been proposed as a means to ameliorate the water quantity and quality impacts associated with energy

485

development in the Western U.S. This paper describes the effects of some of these water conservation strategies on impacts in the San Juan River Basin of New Mexico (one of the sub-basins of the Upper Colorado River Basin). In the San Juan Basin, conflicts over water use exist primarily between existing agricultural users and emerging energy developers. Obvious water quantity conflicts are exacerbated by water quality problems and, more recently, by environmental interest in the protection of instream resources.

Under the Upper Colorado River Compact, New Mexico has been alloted between 798 and 867 million cubic meters of water.[1] Depletions (the amount of water consumed and no longer available within the Basin) now total about 390 million cubic meters or about one-half of the allotment. Development (including agricultural, energy and municipal) that is already planned, however, will use the remaining half of New Mexico's allotment. This development is expected to occur within the next 25 years. We evaluated the impacts of this future development on water quantity, water quality, and instream resources assuming, first, that development occured without serious attempts to manage or conserve water and, secondly, that the best available water conservation strategies were employed by agricultural water users and within the energy industry. This paper: (i) defines future water development, (ii) summarizes our methods of analysis, and (iii) describes the results.

DEFINITION OF FUTURE WATER DEVELOPMENT

The development projected for San Juan County over the next 25 years is summarized in Table 1. As indicated, the amount of land irrigated for crop production is expected to increase 250 percent from 19,500 to 67,500 hectares. Electric power generation capacity is to reach 8000 Mwe, a 200 percent increase, and the production of synthetic gas from coal could reach 42.5 million cubic meters per day. In order to provide sufficient coal to the mine-mouth electric power and synthetic fuel plants as well as to provide for coal exports, the amount of coal mined is expected to increase by an order of magnitude (Table 1).

TABLE 1 Development in San Juan County, New Mexico: Current and Projected

	1974	Projected
Agriculture (hectares irrigated)	19,425	67,570
Energy Development		
Coal Mining (10^6 tonnes/yr)	9.8	92.0
Electric Power (Mwe)	2,660	8,000
Synthetic Gas (10^6 m^3/day)	0	42.5
Uranium Mining-Milling (tonnes/day)	0	2,400
Oil Production (bbls/day)	7,600	100,000

Other water Users: exports from the San Juan Basin, evaporation, municipal and manufacturing.

[1]New Mexico is entitled to 11.25% of the water in the Colorado River to which the Upper Colorado River Basin States are entitled. The U.S. Department of Interior uses a lower estimate of the total Colorado River flow than does New Mexico to calculate the allotment, accounting for the range in values.

As planned, the coal facilities will employ "all wet" cooling systems. Moreover, the 19,500 hectares of agricultural land that are now served with unlined canals and employ the inefficient flood and furrow irrigation method are not expected to convert to the more efficient sprinkler irrigation method with lined canals. The use of the best available water conservation strategies would, of course, call for "all dry" cooling systems for the energy facilities and lined irrigation canals and sprinkler irrigation systems on the agricultural land. Additionally, the groundwater extracted from the underground uranium mines during dewatering could be used for irrigation of agricultural land or to reclaim surface coal mine land.

METHODS OF ANALYSIS

The water impacts of planned development, with and without water conservation, were analyzed in three categories: water quantity, water quality, and instream resources. All analyses were carried out only for an average water supply year; drought years were not considered. Impacts were analyzed first for development as it is now planned and then for the same development but using the best available water conservation strategies (including dry cooling, lined canals, sprinkler irrigation, and reuse of the uranium mine water).

Water quality impacts were evaluated in terms of depletions, as distinguished from diversions and return flows. The diversion is the quantity of water withdrawn from the river for use elsewhere. For some uses, a portion of this returns to the River via percolation through the soils, as runoff, or through an outflow pipe-- the return flow. The remainder of the diversion is consumed and represents the depletion. The diversion, return flows, and depletions assumed in these analyses are given in Table 2 for one hectare of irrigated crop land and for unit size energy facilities.

TABLE 2 Water Use Patterns for Agriculture and Energy: San Juan County, N.M.

	Diversion	Return Flow	Depletion
Agricultural Land[a] (thousand m^3/ha/yr)			
Unlined Canals & Flood/Furrow	15.24	9.11	6.13
Lined Canals & Sprinkler	9.15	3.00	6.16
Electric Power Plant[b] (300 Mwe)			
Wet Cooling (thousand m^3/yr)	36,900	0	36,900
Dry Cooling (thousand m^3/yr)	7,700	0	7,700
Synthetic Gas Plant[b] (7.1 million m^3/day)			
Wet Cooling (thousand m^3/yr)	9,100	0	9,100
Dry Cooling (thousand m^3/yr)	6,800	0	6,800

m^3/ha/yr = cubic meters per hectare per year

[a] New Mexico Water Quality Control Commission, 1976; New Mexico Interstate Stream Commission, 1976; Utah State University, 1975.
[b] Water Purification Associates, 1977 and 1979.

Water quality impacts were evaluated in terms of salinity. Salinity increases are caused both by the salt loads carried from agricultural lands to the River with return flows, and by the salt concentrating effect induced with evaporation and with the consumption of high quality upstream water. Salinity increases were projected using an input-output technique in which salt and water masses were tracked through the San Juan River Basin as well as the Lower Colorado River Basin.

Instream resources are generally defined in terms of values that relate to the flow of the river itself--navigation, hydropower, fisheries, and/or native aquatic and riparian plant and animal communities. Using the method developed by Tennant (1975), instream needs for the San Juan River were defined as a percentage of average flow at various locations. "Excellent" instream conditions are maintained if 50% of the average annual river flow is maintained April through September and 30% of the average annual flow is maintained during March through October. Corresponding percentages for "good" instream conditions are 40% and 20% respectively. Larger flows are recommended for April through September because that period represents the recreation season and the growth stage for most fish.

RESULTS

Water impacts, expressed in terms of depletions, salinity, and instream flow, are summarized in Table 3 for: (i) current conditions, (ii) development as it is now planned, and (iii) development employing the best available water conservation strategies. A "standard" against which to compare the impacts is also included although, in the case of instream flow, these standards have not been legally adopted in New Mexico.

TABLE 3 Water Impacts Currently and As a Result of Future Development With and Without Water Conservation

	"Standard"	Current	Future As Planned	With Water Conservation
Depletions (million m^3/yr)	798-867[a]	386.6	871.0	770.7
Salinity (mg/l)				
At Shiprock, N.M.	None	434	510	472
At Imperial Dam, AZ	879[b]	836	874	865
Instream Flow at Shiprock (m^3/s)				
Average Annual	None	56.2	35.3	40.6
In September	22.5-28.1[c]	28.3	20.1	24.6
In December	11.2-16.8[c]	37.1	15.7	18.5

[a] This is New Mexico's water allotment under the Upper Colorado River Compact.
[b] This is the numeric criterion at Imperial Dam as set forth by the Colorado River Basin Salinity Control Forum, 1978.
[c] The lower value corresponds to the maintenance of "good" instream conditions and the higher value to "excellent" instream conditions.

As indicated, depletions will increase by 125% with development as planned and by 99% with the same development employing water conservation strategies. Water use under planned development (871.0 million m^3/yr) is about equal to the higher estimate of New Mexico's water allotment (867 million m^3/yr). With water conservation, water use will be 770.7 million m^3/yr, less than the lower estimate of New Mexico's water allotment (798 million m^3/yr). Thus, development with water conservation, while it reduces consumption by only 100.3 million m^3/yr, alleviates any conflict over New Mexico's allotment.

Water quality impacts were evaluated as the salinity increase caused by development in the San Juan River at Shiprock, New Mexico and in the Colorado River at Imperial Dam, Arizona. Shiprock was chosen since it is within San Juan County almost at the point where the San Juan leaves New Mexico and enters Utah. Imperial Dam is in southwest Arizona where the Colorado River enters Mexico. As indicated in Table 3, a salinity standard exists at Imperial Dam; the seven states in the Colorado River Basin have agreed to maintain salinity concentrations in the Lower Colorado River at or below 1972 levels (Colorado River Basin Salinity Control Forum, 1978). For Imperial Dam, this is 879 milligrams per liter (mg/l).

As indicated in Table 3, salinity at Shiprock is now 434 mg/l; with planned development it would increase to 510 mg/l. With the same development employing water conservation strategies, the salinity increase would be only half as great (472 mg/l). At Imperial Dam, planned development in San Juan County would increase salinity from 836 to 874 mg/l which is just under the standard of 879 mg/l. With water conservation, the increase would still be substantial (to 865 mg/l). Not only does development risk a conflict among states over who is causing a standards violation, but economic losses associated with the use of more saline water can also be heavy. In 1972, the U.S. Department of Interior (1974) estimated that each milligram per liter salinity increase at Imperial Dam has an environmental and lost agricultural production cost of about $230,000.

Finally, instream impacts are also listed in Table 3. Under planned development, the annual average instream flow of the San Juan River will decline 37 percent from 56.2 to 35.3 cubic meters per second (m^3/s). Development with water conservation would make the decline only 28 percent to 40.6 m^3/s. On a monthly basis, low flows occur in September and December. In September, a flow of 22.5 m^3/s is required to maintain good instream conditions and 28.1 m^3/s is required for excellent conditions. Currently, excellent conditions are maintained (Table 1); but under planned development the September flow of 20.1 m^3/s will be less than that required for good conditions. With water conservation, good instream conditions but not excellent conditions can be maintained in September.

If development were to proceed with the use of water conservation strategies instead of as planned, the water impacts of development would obviously be reduced. But these improvements would accrue at a dollar cost. The use of dry cooling in electric power plants would increase the cost of electricity by about 13%. However, dry cooling in synthetic gas plants would increase costs less than one percent. Costs of farming in San Juan County would increase by at least 50% if canals were lined and sprinklers installed (Gilliland and Fenner, 1981).

SUMMARY

Water conservation in agriculture and energy facilities can mitigate some of the water quantity, water quality, and instream impacts that will occur as a result of planned development in San Juan County over the next 25 years. But, water conservation will not "come close" to maintaining current conditions. Water conservation would reduce the depletions associated with planned development by 11.5%; depletions would still be double their current value. Similarly, water conservation would reduce the salinity effect at Imperial Dam of development in San Juan County by only 1 percent; the salinity at Imperial Dam would still be 3.5% greater than currently. And, the annual average flow of the San Juan River will be 37% less than it is now without water conservation and 28% less with water conservation. On the other hand, although the improvements induced by these water conservation strategies seem small and will be expensive, they could ameliorate serious conflicts among states and among parties-at-interest. Specifically, water conservation means that full development can occur: (i) with depletions remaining below the lowest estimate of New Mexico's water allotment, (ii) without violation of salinity standards in the Lower Colorado River Basin, and (iii) with the maintenance of "excellent" instream conditions in 11 out of 12 months of the year and "good" conditions in the 12th month (September).

ACKNOWLEDGEMENT

This work was carried out under a subcontract with Science and Public Policy Program, the University of Oklahoma, Norman, who was contracted to the Office of Energy, Minerals, and Industry; Office of Research and Development, U.S. Environmental Protection Agency (EPA Contract No. 68-01-1916).

REFERENCES

Colorado River Basin Salinity Control Forum (1978). "Proposed 1978 Revision Water Quality Standards for Salinity Including Numeric Criteria and Plan of Implementation for Salinity Control: Colorado River System". Salt Lake City, Utah.

Gilliland, M.W., and L.B. Fenner (1981). Resources and Conservation (forthcoming).

New Mexico Interstate Stream Commission (1975). San Juan County Water Resources Assessment for Planning Purposes: County Profile. Santa Fe, N.M.

New Mexico Water Quality Control Commission (1976). San Juan River Basin Plan. Santa Fe, N.M.

Tennant, D.L. (1975). Instream Flow Regimes for Fish, Wildlife, Recreation, and Related Environmental Resources, U.S. Fish and Wildlife Service, Billings, Montana.

U.S. Department of Interior (1974). Colorado River Water Quality Improvement Program: Status Report. Bureau of Reclamation, Washington, D.C.

Utah State University (1975). Colorado River Regional Assessment Study. Utah Water Research Laboratory, Logan, Utah.

Water Purification Associates (1977). Water Requirements for Steam Electric Power Generation and Synthetic Fuel Plants in the Western U.S. U.S. Environmental Protection Agency, Washington, D.C.

Water Purification Associates (1979). Wet/Dry Cooling and Cooling Tower Blowdown Disposal in Synthetic Fuel and Steam-Electric Power Plants. U.S. Environmental Protection Agency, Washington, D.C.

FUEL FROM BIOMASS

N. E. Good

*Department of Botany and Plant Pathology, Michigan State University,
East Lansing, Michigan, USA*

ABSTRACT

The net energy gain which can be realized from the harvesting of vegetation must be calculated from the total photosynthesis (primary production) after subtracting that part which is inaccessible and that part which is needed for purposes other than fuel, then multiplying by a fraction which represents the proportion of the energy in the raw biomass which is conserved in the fuel form ultimately used. From the amount thus determined must be subtracted all the energy costs of growing, harvesting, and processing the vegetation. Since the entire primary production in the United States is already considerably below our energy consumption, it is obvious that only a very small proportion of our industrial fuel could ever come from biomass.

KEYWORDS

Photosynthesis, water use, biomass, energy, fuel.

INTRODUCTION

Modern society is based on the ready accessibility and lavish use of transportable fuel. But accessible fossil fuels, especially fluid fossil fuels, are being depleted at an alarming rate. A serious search for alternative energy sources is long overdue. We must look objectively at all potential energy sources, eschewing alike reflex optimism and reflex pessimism. The purpose of this paper is to explore the implications of a return to the use of vegetation as a renewable replacement for fossil fuels. In the paper we will discuss only the supply of energy obtainable from biomass, leaving to others a consideration of the social implications of this falling back on the ways of our ancestors.

The potential supply of energy from biomass depends on plant growth. Therefore we will discuss the upper limits of plant growth. The supply also depends on the accessibility of the produced biomass, that is on how much can be harvested with a net energy gain. It depends too on the availability of the accessible part, since we can only use that which is still left after we have removed the amount needed for food, lumber, and fiber. It depends, in addition, on the efficiency with which the energy in the raw biomass can be retained in the products of the conversion into usable fuel. Finally, it depends on how much energy has to be expended in growing, harvesting, and processing the biomass; all of this energy cost must be subtracted from the energy in the accessible, available, processed fuel before we can estimate the net energy gain to be had.

These considerations can be summarized in an equation,

$$F_B = E (P-I-U) - F_C$$

where F_B is the net energy gain from biomass, P is the primary productivity of the photosynthetic process, I is the amount of biomass which is inaccessible as defined, U is the amount of biomass which is unavailable because is is needed for other purposes, E is the proportion of the energy of the biomass still in the processed fuel, and F_C is the energy required for growing, harvesting and processing the biomass. Precise values for these quantities, applicable on a global or even on a national scale, are hard to come by, but precise principles can be stated with confidence. These principles are too often overlooked.

I. PLANT PRODUCTIVITY

The amount of biomass produced is very simply the total amount of carbon dioxide reduced by photosynthesis less the amount of reduced product used by the respiration of the plant. Photosynthesis in land plants is restricted by the availability of water (1), or if it is not actually so restricted, the availability of water sets an inescapable upper limit. This dependence on water in crops and natural vegetation has been noted by all of us. Indeed it is an inevitable consequence of the laws of physics, since leaves cannot take up carbon dioxide without losing a large amount of water.

Let us think for a moment about the quantitative aspects of the exchange of water for carbon dioxide. The air inside a leaf is virtually saturated with water vapor but the carbon dioxide outside a leaf is in a relatively dry medium, the atmosphere. Thus if carbon dioxide is to diffuse into a leaf, there must be a diffusion pathway and water vapor must be able to diffuse out along the same pathway. Now it is a well-known fact that diffusion rates depend on the concentration differences at the two ends of the diffusion pathway and on the relative mobilities of the diffusing substances (water and carbon dioxide):

$$\frac{\text{moles of } H_2O \text{ lost}}{\text{moles of } CO_2 \text{ gained}} = \frac{([H_2O]\text{inside} - [H_2O]\text{outside}) \times \text{mobility } H_2O}{([CO_2]\text{outside} - [CO_2]\text{inside}) \times \text{mobility of } CO_2}$$

where square brackets mean concentrations in the gas phase. The mobilities of gaseous H_2O and CO_2 are inversely proportional to the square roots of the molecular masses. Therefore the mobility of H_2O/mobility of CO_2 = the square root of 44/18 or 1.56. The concentration of CO_2 inside a leaf depends very much on the rate of photosynthesis and on the stomatal opening and, indeed, it is only by varying the internal CO_2 concentration that a plant has any control over the efficiency with which it can use water to buy carbon dioxide. Some reasonable values which have been measured and give an idea of the problem are as follows:

Water vapor inside (saturated air at 35°C) = 5.6×10^{-2} atm.
Water vapor outside (40% relative humidity at 30°C) = 1.7×10^{-2} atm.
CO_2 concentration outside (300 ppm)   = 3×10^{-4} atm.
CO_2 concentration inside:
    for a C_4 plant (1/2 outside)   = 1.5×10^{-4} atm.
    for a C_3 plant (3/4 outside)   = 2.25×10^{-4} atm.

Putting these values into equation 2 we find that the C_4 plant must use 406 moles of water for every mole of carbon dioxide it reduces by photosynthesis and that the C_3 plant must use 812 moles of water.

These calculations are not idle exercises. Actual water loss measurements bear them out (2). In fact, most field water losses are larger (1) since the calculation presuppose constant high rates of photosynthesis with optimum adjustment of stomatal apertures and make no allowance for the fact that plants use water for evaporative cooling even if photosynthesis is for some reason slow. Moreover, plants do not all seem to have evolved to use water with maximum efficiency. After all, a plant is saving water for its competitors when it conserves soil moisture. Many plants have therefore opted for a prodigal use of water in a race with their neighbors.

If we assume a year-round, country-wide ratio of one molecule of carbon dioxide fixed per 1000 molecules of water transpired, we can compute annual photosynthesis from annual precipitation. The average rainfall over the United States is about 75 cm and the total area is about 8×10^{12} m^2. Thus the volume of rain is about 6×10^{12} m^3 (3.3×10^{17} moles). The run-off plus evaporation from bare rock and water surfaces probably amount to about one third of the total, which leaves 2.2×10^{17} moles of water available for evapotranspiration. This then suffices for the fixation of 2.2×10^{14} moles of carbon dioxide (1). However, each mole of carbon dioxide, when reduced to the carbohydrate level, conserves 115 Kcal in the biomass produced. Therefore we can expect an upper limit of annual photosynthetic conservation of solar energy of $2.2 \times 10^{14} \times 115 \times 10^3$ calories or 1.06×10^{20} Joules. (That unfortunate unit, the "quad", is 1.05×10^{18} Joules).

It should be emphasized that this is not an estimate of current photosynthesis but rather an estimate of the highest yield of total biomass obtainable with forseeable technology, given the rainfall patterns prevailing in the recent past. More to the point may be best estimates of the actual productivity of crops and natural vegetation at these prevailing levels of precipitation. Burwell (3) has estimated that in 1974 the total gross energy yield of agriculture, silviculture, pasture, and natural vegetation in the 48 contiguous states was 27.7 quads, just under 3×10^{19} Joules. This value is only 30% of the theoretical maximum arrived at by assuming even distribution of water and uniformly efficient use of water. It may be a little too low since it probably does not include all underground parts. However, it is obviously a more realistic estimate of current biomass production then is the inflated theoretical value.

Before leaving this matter of maximum plant productivity it would be well to dispel some particularly pervasive false hopes. Modern agriculture has increased crop yield very greatly, partly by the use of improved plants and partly by the use of fertilizers. Three pertinent points need to be made about these increases. First, a large part of the gain in yield has been made by increasing the proportion of the plant which is of economic value and only a smaller part of the gain is due to increased overall plant productivity. Second, such improvements as there have been in primary productivity have been made from a very low base, a base where primary productivity was being sadly depressed by poor agriculture. Natural vegetation tends to make good use of water and it is rare indeed to find crops, even the best well-fertilized crops, exceeding the productivity of the natural vegetation they replace. Hence, we cannot rightfully expect to increase productivity very much above the values given in the preceding paragraph, no matter how heavily we fertilize. Third, we cannot fertilize heavily if the yield is to be energy, since any increase in the energy in the biomass must exceed the energy cost of the fertilizer several-fold if a net energy gain from fertilization is to be realized (see below). For all of these reasons, we are probably not going to be able to increase primary productivity in the United States much beyond 5×10^{19} Joules per annum, although for the argument I will assume some much higher rates close to the theoretical maximum.

II. ACCESSIBILITY OF BIOMASS

It is axiomatic that no energy source can be considered accessible if the energy cost of collecting and processing the fuel equals or exceeds the energy content of the fuel acquired. In fact, the energy cost of collecting fuel must be substantially less than its energy content or the fuel will be prodigiously expensive. In the case of biomass there is another constraint on accessibility, the need to maintain the productivity of the system if we are to look on biomass as a renewable resource. This need precludes many inexpensive systems of harvesting.

It is one of the principles of thermodynamics that concentrating substances in itself requires energy. For this reason, a great deal of the photosynthetic productivity of arid regions and mountains must remain forever inaccessible, except perhaps to the few residents such areas can support. Thus sparse and remote vegetation cannot be used at all as fuel for an industrial, urban economy. A more serious limitation on accessibility has to do with the fact that no harvest ever obtains all of the plant. It probably will not be possible to harvest plant roots to any significant extent, not only because of the cost in energy of the operation, but also because root-harvesting would in most cases interfere with the continuity of productivity. Below-ground parts of plants vary from a low of about 20% in some forests and crops in humid regions to a high of close to 70% in unfavorable conditions. Compare, for instance, the accumulation of biomass above ground in the boles of forest trees and below ground in the sod of a grassland. But also in forests the boles of trees represent only a portion of the produced biomass. The leaves and self-pruned branches, even in forests being clear cut repeatedly, must represent in the order of 30% of the products of photosynthesis.

Clearly the amount of biomass which can be considered accessible depends on the use to which the biomass is to be put. If we live in woods and use woman-power liberally to collect twigs, a great deal more can be used than if we want to make alcohol to fuel cars or to generate electricity to heat water. In the context of our present industrial society, less than half of the total products of photosynthesis can be considered accessible, probably very much less than half. Only that part which is above ground and has accumualted to a considerable density can possibly qualify as a source of industrial fuel.

III. AVAILABILITY OF BIOMASS

Most of the productivity of other lands is already committed to feeding, clothing, and housing humanity. Because of the still modest population of the United States relative to the very great agricultural productivity, we do not here sense the degree of this commitment fully. But even in the United States a very large portion of annual biomass production is so destined. Furthermore, as fossil fuels become scarcer, there will be an increased demand on biomass, not as fuel but as a source of chemicals.

In trying to estimate the actual proportion of accessible biomass already committed to the production of food and fiber, we encounter again the problem of maintenance of productivity. How much of farm "residues" can be removed safely without damage to soils, especially if rising fuel costs limit the availability of fertilizers. In the long run, it may be fortunate that many farm residues are not accessible, i.e. their collection and conversion would not yield as much fuel as would be used in collecting and converting them. But, this being so, the productivity of agricultural land must remain out of all proportion to the actual energy content of our food. The much larger gross productivity is nearly all unavailable as an industrial fuel. Furthermore, food being by far our most important export, the proportion of plant production which must be set aside for food is likely to increase rather than decrease. When we realize that the most productive land all over the world is already devoted to agriculture and that it is on this good land that plant production is most accessible, it becomes obvious that a very high proportion of accessible biomass is not to be had for fuel.

A different set of considerations determines the availability of forest biomass. It is true that the energy content of the lumber from a forest is only a small part of the energy originally conserved by photosynthesis in that forest. However, the growing of trees large enough to use for boards or plywood entails a great deal of loss of biomass back to the system. Trees only become large over years in which they have shed innumerable crops of leaves, killed off their competitors, and dropped their own lower branches. This loss of biomass is a price which must be paid for lumber since, as we have already pointed out, the shed leaves and dead branches are not accessible for industrial fuel. The same principle applies, albeit with less force, in the growing of pulpwood.

The availability of fuels from our farms and forests naturally depends on our techniques of agriculture and foresty and on our eating habits. Farming, lumbering and eating as we do, we "use" all of the productivity of our farms and forests already. No doubt we can grow more food on the same land, we can put less food into domestic animals, we can cultivate a little more land, we can grow better forests, and we can harvest them more efficiently. Perhaps with all of these changes we can make half of the accessible biomass available for fuel where now there is only a little available. Let us make this optimistic assumption for purposes of discussion. But, at the same time, we must realize that this half cannot be obtained without major changes in our way of life, not all of which will be enthusiastically accepted.

IV. FUEL CONVERSION EFFICIENCY

In assessing the amount of energy to be had from biomass, we must not overlook the need to convert crude biomass into usable fuels. The energy content of standing trees is often very much larger than the energy content of the various forms of processed fuel obtained therefrom. The difference depends critically on the form of the fuel. Rural, decentralized economies can make effective use of wood for part of their energy needs, at least for heating houses, and in this case the potential energy yield of burning wood is almost as great as the energy content of the wood itself (less only the significant loss associated with the evaporation of residual water). The energy contents of derived "fuels", for example, ethanol, methanol, methane, and electricity, may be quite small when compared to the energy content of the vegetation.

Let us consider why the conversion of biomass into these convenience fuels must always involve large losses. The conversions suggested involve biological, chemical or physical processes. The biological and chemical processes on the one hand, and the physical processes on the other hand involve quite different kinds of losses which must be discussed separately:

The biological and chemical reactions responsible for fuel production founder on the inconvenient fact that biomass is chemically diverse. This chemical diversity, which makes biomass a potential treasure-house of prefabricated substances only waiting to be exploited, nevertheless poses great problems in the production of fluid fuels. For instance, the conversion of wood into alcohol is complicated by the fact that the polymers of glucose (such as celluose), which can be fermented easily, are mixed with polymers of mannose which cannot be fermented quite so easily, polymers of pentoses which are much more difficult to ferment, and many other substances such as lignins and proteins which can hardly be fermented at all. Then we must consider the fact that biological and chemical reactions only occur if they are thermodynamically down-hill and the products contain less energy than the starting materials. Thus if one makes ethanol from maize grain (a concentration of highly fermentable starch), the energy content of the alcohol is only a little over one half of the energy content of the grain. If we were to consider the whole maize plant, the yield would be less, and it would be still worse if trees were used. For these two reasons: the chemical diversity of biomass and the adverse thermodynamic consequences of spontaneous interconversions, it would be prudent to expect average conversion efficiencies below 50%, even if we exercise great wisdom in the selection and integration of conversion processes.

The physical process used for the conversion of biomass energy into electrical energy via the usual heat engine is dreadfully inefficient unless the low grade residual heat is used. Proponents of mid-forest generating plants seem to have not given this disadvantage the attention it deserves. If wood chippers are used to harvest trees and the resulting choppped whole trees are promptly burned, it is probable that a good deal less than 25% of the energy in the trees will appear in the electricity produced (this without making any allowance for the energy costs of harvesting). Heat engines have a maximum thermodynamic efficiency which is quite low. Without considering any practical losses due to friction in the turbines and resistance in the generators and transformers, the efficiency is only $(T_2 - T_1)/T_2$, where T_1 is the absolute temperature of the heat source, the boiler, and T_2 is the absolute temperature of the cooling tower. As a result, the efficiency of the best oil or coal-fired generating plant is only about 35%. However, chopped wet wood could not begin to match this value, if only because so much energy must be expended in heating and evaporating the large amount of water in the wood.

V. ENERGY COSTS OF GROWING, HARVESTING, AND PROCESSING BIOMASS

These costs are of two kinds, direct energy inputs and and energy inputs associated with building and maintaining the equipment used. Because tractors, trucks, wood chippers, and fertilizer plants are unlikely to be fueled by logs, the greater part of these costs will have to be paid with already converted convenience fuels.

Many exhaustive studies of the energy inputs associated with various biomass production systems have been undertaken but only two will be mentioned here. These two have been singled out, not because of the plausibility of the schemes analyzed but because they illustrate the problem so dramatically. Pimental et al. (4) have attempted to assess the energy costs of producing maize grain. They have concluded that the energy expended amounts to about one third of the energy content of the grain. Others have been so optimistic as to claim an energy cost of only one fifth. A first impression might be, therefore, that the growing of maize is a reasonable way to augment fuel supplies. However, we must first estimate the direct and indirect costs of converting the grain into alcohol. We must subtract these costs from the energy content of the alcohol produced and not from the energy content of the grain itself. To illustrate, let us make some plausible but optimistic guesses as to the values involved. Let us suppose that the direct and indirect costs of growing and harvesting the grain are 25% of the energy content of the grain, that the direct and indirect costs of converting the grain into alcohol are another 25% and that 50% of the energy in the grain remains in the alcohol. With these assumptions, there will be no net energy gain at all. Plotkin (5) has made a careful analysis of the energy costs of making alcohol for "gasohol", even allowing for the food value of the grain residues and the beneficial effects of alcohol on the combustion of gasoline in engines. From his study he has concluded that there is a net energy loss unless fossil fuels are used for the conversion process. Thus making alcohol from grain is only viable as a very indirect way of transferring energy from fossil fuels to alcohol, not as a way of extending our supply of fossil fuels.

Similar calculations could be made with regard to the production of other kinds of fuel from other kinds of biomass. Some such processes would yield a net energy gain and some would not. It is not our purpose here to deny the possibility of obtaining an energy gain from biomass but rather to point out how small the gain will be unless circumstances are very favorable and a minimum of conversion losses are incurred. Biomass is already being used under some of these favorable circumstances and only a limited amount of expansion of biomass use will yield significant gains. In other words, the amount of biomass which is accessible (as defined in section II) is much more limited than one might have thought.

VI. CONCLUSIONS

I do not denigrate efforts to expand and improve the use of biomass for fuel. As we are continually reminded, every little bit counts. But I do deplore the suggestion that biomass can replace much of our fossil fuel and allow us to continue business as usual, our business of profligate energy consumption. We can do immense damage if we imply answers where none exist, since by so doing we foster complacency and inhibit the search for real solutions. For instance, Johansson and Steen (6) plan to fill 70% of Sweden's energy needs from forest products within 35 years, and our own Congressional Office of Technology Assessment would supply 15 to 20% of our much larger energy budget from biomass (7). Neither report has given sufficient weight to factors of accessibility and availability, neither has given sufficient thought to conversion losses, and both have overlooked the fact that very large production costs must be paid in the coinage of expensive processed fuel.

Present energy consumption in the United States exceeds the TOTAL energy now conserved by photosynthesis by a sizable margin and probably exceeds any plausible future photosynthesis. When we consider how little of the primary productivity remains after allowing for inaccessibility, unavailability, conversion losses, and production costs, this startling statistic should give us pause. Let us assume that the solar energy annually conserved by photosynthesis is about 8×10^{19} Joules, an unrealistically high value approximately equal to our annual energy consumption. Let us further assume that 40% of the biomass produced is accessible by the criteria stated, and that 50% of this accessible biomass might be available for fuel. The entire supply of fuel biomass would then contain 1.6×10^{19} Joules or about 16 quads, close to the OTA estimate. If now we want to make the biomass into something industrially useful, we must expect average converson losses of the order of 50%. The residual energy is now down to 8×10^{18} Joules. From this must be subtracted all of the energy expended in growing, harvesting and processing. The more biomass we try to get, the larger will be these costs as a proportion of the total. An improbably low cost, applicable only for easily obtained biomass undergoing minimal conversion, would be of the order of 50%. Therefore the net energy gain from biomass would be 4.0×10^{18} Joules or 5% of our current energy use. This is only a little over twice the energy already being used from the direct burning of wood and it may be that we are already approaching a practical limit, the local use of unconverted biomass. For the reasons stated, further expansion of the use of biomass would rapidly get us into the area of high transportation costs or high conversion costs, where net gains become so problematic.

REFERENCES

1) Rosenzweig, M. L. (1968). Net primary productivity of terrestrial communities: Prediction from climatological data. American Naturalist, 102, 67.

2) Shantz, A. L., and L. N. Piemeisel (1927). The water requirement of plants at Akron, Colorado. J. Agric. Res., 34, 1093.

3) Burwell, C. C. (1978). Solar biomass energy: An overview of the U.S. potential. Science, 199, 1041.

4) Pimentel, D., L. E. Hurd, A. C. Bellotti, M. J. Forster, I. N. Oka, O. D. Sholes, R. J. Whitman (1973). Food production and the energy crisis. Science, 182, 443.

5) Plotkin, S. E. (1980). Energy from biomass: the environmental effects. Environment, 22, 6.

6) Johansson, T. B., and P. Steen, Solar Sweden. Secretariat for Futures Studies. Stockholm.

7) Office of Technology Assessment, U.S. Congress, Washington, D.C. (1980). Energy for biological processes.

ASSESSING HEALTH AND ENVIRONMENTAL EFFECTS OF A DEVELOPING FUEL TECHNOLOGY

R. H. Gray and H. Drucker

Coal Liquefaction Environmental Research Program, Pacific Northwest Laboratory, P.O. Box 999, Richland, Washington 99352, USA

ABSTRACT

Increasing energy demands, coupled with rising prices and an unstable world oil market, have stimulated international interest in developing alternative sources of fuel. Because of vast national coal reserves, the U.S. Department of Energy (DOE) is currently evaluating development of several coal liquefaction processes. Direct liquefaction [solvent refined coal (SRC-I and -II), donor solvent, and H-coal] processes hydrogenate pulverized coal under high temperature and pressure to produce gaseous, liquid or solid products. Although these processes are promising options for producing fuel, potential health and environmental effects from exposure to coal liquefaction products, process streams or wastes must be carefully assessed.

We describe, here, an integrated approach which couples engineering development with health and environmental assessment to aid design of environmentally acceptable coal liquefaction processes. The research is organized in four phases which correlate with and provide information early in process development.

Phase I involves short-term, less expensive screening of coal-derived materials for potential biomedical and ecological effects. Phase II involves longer-term, more expensive biological and ecological assays to evaluate materials considered most representative of potential commercial practice. Phase III evaluates effects of process modification, control technologies and changing operational conditions on potential health and ecological effects. Chemical fractionation and analyses are performed on materials studied in Phases I, II and III to determine compounds or compound classes responsible for effects. Phase IV develops and applies methods to monitor for site-specific effects at potential demonstration and/or commercial facility sites.

Results to date suggest that although coal liquids are mutagenic in microbial test systems, genetically active chemical constituents, such as primary aromatic amines (PAA), occur primarily in those portions of the wide-boiling-range materials that distill above 700°F. Hydrotreating the higher-boiling-range materials reduces PAA concentration and lowers mutagenic potency by up to 99%. Hydrotreating also lowers the concentration of phenolic compounds, which are major determinants of acute and chronic ecological toxicity. Use of these and

500

other data in the design and management of environmentally acceptable coal conversion processes is discussed.

KEYWORDS

Coal liquefaction; solvent refined coal; biomedical effects; ecological fate and effects; chemical characterization; control technology; health and environmental assessment; hydrotreatment; fractional distillation.

INTRODUCTION

Increasing energy demands, coupled with a shortage of natural gas, a desire to reduce dependence on foreign oil, and constraints on combustion of high-sulfur coal, have stimulated development of alternative energy sources in the U.S. Hydrogenation of coal under high temperature and pressure results in a variety of gaseous, liquid and solid products that may serve as substitutes for coal or oil.

Four direct coal liquefaction processes are currently under development, and may be ready for commercialization by the 1990s. These include two SRC (SRC-I and -II) processes, Exxon donor solvent (EDS) and H-coal. The SRC-I process produces a low-sulfur, low-ash, solid fuel for use in industry and in electric power production. The other three processes produce primarily liquid fuel products.

Both SRC process options are being studied at a 30- to 50-ton/day pilot plant at Ft. Lewis, Washington. The SRC-I process is also being evaluated at a 6-ton/day pilot plant at Wilsonville, Alabama. The H-coal process is being studied at a 200- to 600-ton/day facility at Catlettsburg, Kentucky; and a 250-ton/day EDS pilot plant is operational at Baytown, Texas.

In parallel with engineering development, the U.S. Department of Energy (DOE) has established a program to evaluate the environmental acceptability of these processes. Epidemiological studies and toxicological research on coal liquefaction, gasification and coking process materials suggest that constituents of coal tars and heavy coal liquids may have carcinogenic properties (Hueper, 1956a, 1956b; Mazumdar and co-workers, 1975; Weil and Condra, 1960). Research on direct coal liquefaction processes support this conclusion (PNL, 1979a, 1979b). Additionally, the high phenol content and complex nature of coal liquids may imply greater acute and chronic toxicity, if these materials are released to the environment, than that observed for crude petroleums (Strand and Vaughan, 1981).

Several organizations were selected by DOE to study the biomedical, environmental and safety aspects of direct coal liquefaction processes. The Pacific Northwest Laboratory (PNL), initially assigned responsibility for evaluating SRC materials, prepared comprehensive health and environmental effects research program plans (PNL, 1980, 1981). Similarly, a program plan for H-coal was prepared by the Oak Ridge National Laboratory (ORNL,1980). Objectives of DOE's Coal Liquefaction Environmental Research Program, implemented by PNL, are to: 1) identify and assess potential biomedical and environmental effects of coal liquefaction materials (products and byproducts); 2) identify compounds or compound classes responsible for effects; and 3) provide biomedical and environ-

mental data for use by process engineers in the design and development of environmentally more-acceptable commercial coal liquefaction processes.

RESEARCH STRATEGY

Although objectives of PNL's research focus on assessing potential health and
environmental effects of commercialization, there are currently no large-scale
(i.e., demonstration- or commercial-scale) coal liquefaction facilites in existence. Therefore, research and assessement must utilize materials produced at
existing subcommercial-scale facilities, which are generically representative of
possible demonstration or commercial facility operations. In this effort,
environmental research accompanies engineering development (Fig. 1) so that
research results may influence process design. For comparative purposes, we
have also evaluated other, more-familiar materials. These materials, which
include shale oil, petroleum crudes, other fossil-derived materials, and pure
forms of known chemical mutagens and carcinogens, help to place in clearer
perspective results obtained with SRC and other coal liquefaction materials.

The research program is being performed in four phases which chronologically
overlap. Phase I research (screening) evaluates materials from existing coal
liquefaction facilities (bench, process demonstration unit, pilot) which may or
may not be representative of physical and chemical properties of materials from
potential demonstration or commercial facilities. These materials are subjected
to a battery of short-term biomedical and ecological assays. Chemical fractionation and analysis are also performed to determine compounds and compound
classes of greatest potential concern. In some cases, chemical fractionation is
required to provide materials amenable to biomedical and ecological testing.
Phase I research provides a first estimate of the environmental properties and
potential effects of coal liquefaction materials.

Phase II research evaluates materials that are considered most representative of
potential demonstration or commercial practice. These materials are subjected
to longer-term, more expensive biomedical assays and ecological analyses relative to effects and environmental fate. Again, chemical fractionation and
analyses will be used to provide materials amenable to bioassay and to determine
compounds and compound classes of greatest concern.

Phase III research evaluates the influence of process or operational modifications and control technology options on potential health and environmental effects of coal liquefaction materials. Short-term biomedical and ecological assays used in Phase I, which correlate with longer-term assays used in Phase II,
are used in Phase III. As in Phases I and II, integrated chemical fractionation
and analytical methods are used to determine compounds and compound classes of
concern.

Phase IV research will identify and develop methods and initiate monitoring for
potential health and environmental effects at actual demonstration and/or commercial facility sites. Data from Phases I, II and III will indicate potential
biomedical and ecological effects and identify chemicals of concern, and will be
used to select methods with which to monitor the workplace and local environment. Phase IV will also include studies of the fate and transport of coal
liquefaction materials in the environment.

502

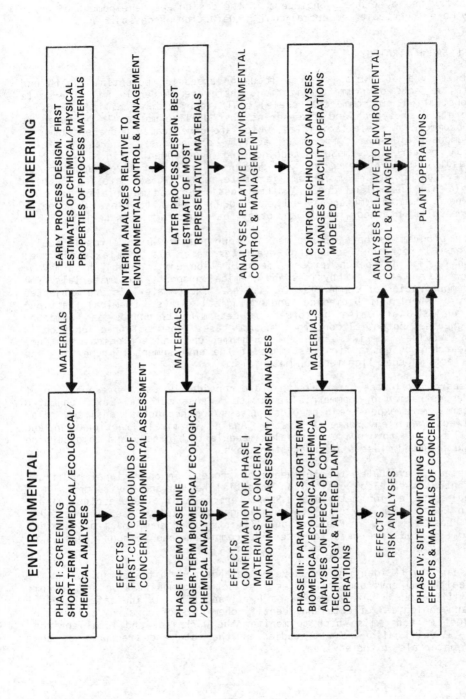

Fig. 1. Strategy for conducting biomedical and ecological research program to aid design and development of environmentally acceptable coal liquefaction processes.

Phase IV research may be initiated before construction of SRC demonstration facilities (at Ft. Martin, West Virginia and/or Newman, Kentucky). Since no EDS or H-coal demonstration facilities are planned, and existing pilot plants may not be representative of commercial development, little or no field monitoring is anticipated for these process options during the pilot or demonstration phase. However, methods relevant to monitoring commercial EDS and H-coal facilities will be developed.

SIGNIFICANT RESULTS

Results of PNL efforts to evaluate potential biomedical and ecological effects of various SRC-I and -II materials are documented in numerous reports (Becker, Woodfield and Strand, 1978; Mahlum, 1981; Pelroy, 1981; Pelroy and Wilson, 1981; PNL, 1979a, 1979b, 1981; Strand and Vaughan, 1981; Weimer and co-workers, 1980; and others) and in symposia and open literature publications (Mahlum, Felix and Gray, in press; Pelroy and Gandolfi, 1980; Pelroy and Petersen, in press; Pelroy and Stewart, in press; Pelroy and Wilson, in press; States and co-workers, 1979; Wilson and co-workers, 1980; Wilson and Pelroy, in press; Wilson, Pelroy and Cresto, 1980, in press; and others).

Potential microbial mutagenesis and chemical characteristics, in vitro mammalian-cell toxicity and transformation, as well as acute, subchronic, and developmental toxicity in rats, have been evaluated for SRC-II light, middle and heavy distillates (LD, MD and HD, respectively), and for SRC-I light oil (LO), wash solvent (WS) and process solvent (PS). Mutagenic potency and chemical characteristics have also been studied for distillation cuts representing different boiling-point ranges of SRC-II PS. Epidermal carcinogenesis (skin-painting) studies have been conducted with SRC-II LD and HD and their solvent extracted fractions. Similar studies are currently underway with SRC-I PS. Tissue distribution and metabolism of SRC-I PS after oral administration to rats have also been evaluated; studies are underway, in rats, on effects of inhaling PS aerosols.

Samples of an SRC-II distillate blend (2.9:1, MD:HD), a potential commercial product, and its solvent-extracted fractions, have been evaluated for mutagenic activity before and after hydrotreating (a process which catalytically stabilizes liquid hydrocarbon products and/or removes objectionable elements by reacting them with hydrogen). Hydrotreating is one potential method of refining or upgrading the raw product prior to marketing. Acute and chronic toxicities of water-soluble fractions of the SRC-II 2.9:1 distillate blend to various aquatic organisms (e.g., algae, zooplankton, insect larvae and fish) have also been determined. Studies to evaluate the behavioral response of fish to water-soluble fractions of the 2.9:1 blend are underway. Additionally, effects of the SRC-II blend and SRC-I solid product on growth of barley have been studied.

Significant results from the efforts described above include the following: SRC-I PS and SRC-II HD, both high-boiling-point [>260°C (>500°F)] materials, were mutagenic in microbial test systems; lower-boiling-range SRC-I LO and WS, and SRC-II LD and MD, were not. Studies with distillation cuts showed that the genetically active constituents of SRC-II PS occurred in those liquids boiling above 371°C (700°F), which comprised only 20% of the wide-boiling-range material. Solvent extraction, coupled with in vitro assay, showed that the basic and tar fractions were responsible for the mutagenic activity of high-boiling-range SRC materials. Additional chemical fractionation and characterization efforts suggested that mutagenic acitivity was due to 3- and 4-ring primary aromatic amines (PAA).

Hydrotreating reduced the mutagenicity of the SRC-II 2.9:1 MD:HD blend: Mild hydrotreating lowered mutagenic potential by 10%; severe hydrotreating reduced total mutagenic response to about 1% of that obtained for the untreated blend. Hydrotreating also reduced concentrations of two genetically active chemical classes, the polynuclear aromatic hydrocarbons (PAH) and PAAs. For example, concentrations of benzo[a]pyrene, a known mutagen/carcinogen, were reduced at least 75% by hydrotreatment. Furthermore, PAAs, detected in the distillate blend, were not found in hydrotreated materials.

Other in vitro studies indicated that materials showing mutagenic activity in microbial assays also caused transformation in cultured mammalian cells. Cellular transformation involves morphological and biochemical changes which permit unrestricted growth and is a further indication of potential carcinogenicity. Results of skin-painting studies with mice suggest good correlation with results of cellular assays: SRC-II HD, shale oil and benzo[a]pyrene all resulted in high incidences of skin tumors.

SRC-II LD, MD and HD, and SRC-I LO and PS, were moderately toxic (the LD_{50} for an acute, 3-day study ranged from 2.3 to 3.8 g/kg body weight) after oral administration to rats. SRC-I WS was the most toxic SRC material tested (acute $LD_{50} = 0.56$ g/kg), probably reflecting its high phenolic content.

Effects of SRC-II LD, MD and HD, and SRC-I LO, WS and PS, on fetal development were determined after oral administration of SRC materials to pregnant rats at 7 to 11 and 12 to 16 days gestation (dg). Fetal growth and survival were decreased by all materials administered at either gestation period. Doses producing fetal effects also produced indications of maternal toxicity. Administration of SRC-I PS and SRC-II HD at 12 to 16 dg increased the incidence of fetal malformations -- primarily, cleft palate. Neoplastic (immature) lungs and herniated diaphragms were also observed in fetuses after administration of HD.

Ecological studies showed that water-soluble fractions of the SRC-II 2.9:1 MD:HD blend produced both acute (mortality) and chronic (inhibition of growth, reproduction) effects in aquatic organisms representing various trophic levels. Furthermore, the hazard posed to aquatic organisms from this material was greater than that posed by a No. 6 or No. 2 fuel oil or a Prudhoe Bay crude oil. Acute toxicity of the SRC-II blend appears to result from high concentrations of low-molecular-weight, easily degradable phenolic compounds. Chronic toxicity appears to be due to high-molecular-weight, more-persistent compounds (including phenols). Although fish avoid acutely lethal concentrations of the SRC-II blend in water, they may not avoid concentrations causing chronic effects. Algal populations exposed to SRC-II concentrations that cause either acute or chronic effects may recover if the toxic material is removed.

The SRC-II 2.9:1 blend mixed in soil was toxic to barley; however, toxic effects were reduced when the blend overwintered in soil. Similar studies showed that the SRC-I solid product was nontoxic; although layering the SRC-I solid product under a soil overburden retarded root growth.

IMPLICATIONS

The results described above have significant implications for coal liquefaction process designers and developers. Some coal-derived materials have mutagenic, teratogenic and carcinogenic properties in laboratory test systems. Additionally, they cause acute and chronic effects in organisms representing various ecological trophic levels. Knowledge of potential detrimental biomedical and ecological effects, and their causative agents, can be used to design more environmentally acceptable coal conversion processes. For example, if mutagenic activity of coal-derived materials is due primarily to PAAs that are found only in higher-boiling-range liquids, distillation cuts might be adjusted so that the commercial product contains little mutagenic potential. Alternatively, distillation cuts could be selected to concentrate mutagenically active compounds in a relatively small process stream, which might receive special treatment (e.g., hydrotreating). In addition to lowering mutagenicity, hydrotreating also lowers concentrations of SRC constituents (e.g., phenols) responsible for acute and chronic ecological toxicity. Although engineering and economic factors must also be considered, the growing biomedical and ecological data base can now be incorporated into the decision-making process. If adjusting distillation cuts, and/or hydrotreating, are not feasible, process designers can begin, now, to develop special handling and accident prevention procedures.

Accidental spills of coal liquids are a potential threat to ecological systems. Low-molecular-weight compounds, such as phenols, which are responsible for acute effects, degrade rapidly. However, high-molecular-weight compounds are more persistent, and may bind to soils and sediments, causing chronic effects at very low concentrations. Although mobile organisms, such as fish, avoid acutely toxic concentrations of SRC liquids, they may be unable to detect and avoid concentrations causing chronic effects. Thus, spill prevention becomes a major consideration. Assuming that accidental spills of coal liquids will occur, cleanup procedures should be developed to minimize potential chronic effects in aquatic and/or terrestrial systems. At least some aquatic populations (e.g., algae) can recover following exposure to toxic concentrations of a coal liquid after the material is removed.

CONCLUSIONS

Although the success of U.S. efforts to develop environmentally acceptable coal liquefaction processes remains to be demonstrated, chemists, biologists, ecologists and engineers are involved in a concerted effort to solve mutual problems early in technology development. Health and environmental problems are being considered during the design stages -- before construction rather than after regulatory agencies require studies. These efforts provide guidance for appropriate process and product modifications, control technology options, mitigative strategies, accident prevention, spill cleanup and solid-waste disposal procedures to minimize any potential adverse human health and ecological effects. Only through a cooperative, interdisciplinary effort will this goal be achieved.

ACKNOWLEDGMENTS

We thank W. J. Bair, S. Marks, L. D. Perrigo, R. C. Thompson, and D. G. Watson, who critically reviewed the manuscript. These studies reflect the cooperative efforts of numerous individuals representing the staffs of the Pacific Northwest

Laboratory, the U.S. Department of Energy, the Pittsburg & Midway Coal Mining Co., Gulf Mineral Resources Co., and the International Coal Refining Co. Health and environmental studies conducted by PNL are supported by the U.S. Department of Energy under Contract DE-AC06-76RLO 1830 and by the U.S. Environmental Protection Agency under Contract TD1123, with the Pacific Northwest Laboratory, operated by Battelle Memorial Institute.

REFERENCES

Becker, C. D., W. C. Woodfield, and J. A. Strand (1978). Solvent Refined Coal Studies: Effects and Characterization of Treated Solvent Refined Coal Effluent, PNL-2608. Pacific Northwest Laboratory. NTIS, Springfield, VA.

Hueper, W. C. (1956a). Experimental carcinogenic studies on hydrogenated coal oils: I. Bergius Oils. Ind. Med. Surg., 25, 51-55.

Hueper, W. C. (1956b). Experimental carcinogenic studies on hydrogenated oils: II. Fisher-Tropsch Oils. Ind. Med. Surg., 25, 459-462.

Mahlum, D. D. (Ed.) (1981). Chemical, Biomedical and Ecological Studies of SRC-I Materials from the Ft. Lewis Pilot Plant: A Status Report, PNL-3474. Pacific Northwest Laboratory. NTIS, Springfield, VA.

Mahlum, D. D., D. W. Felix, and R. H. Gray (Eds.). Coal Conversion and the Environment: Chemical, Biomedical and Ecological Considerations. DOE Symposium Series (in press).

Mazumdar, S., C. Redmond, W. Sollecito, and N. Sussman (1975). An epidemiological study of exposure to coal tar pitch volatiles among coke oven workers. J. Air Pollut. Control Assoc., 25, 382-389.

ORNL (1980). Environmental and Health Program for H-Coal Pilot Plant. Oak Ridge National Laboratory, Oak Ridge, TN.

Pelroy, R. A., and A. J. Gandolfi (1980). Use of a mixed-function amine oxidase for metabolic activation in the Ames/Salmonella assay system. Mutat. Res., 72, 329-334.

Pelroy, R. A. (Ed.) (1981). The Mutagenic and Chemical Properties of SRC-I Materials: A Status Report, PNL-3604. Pacific Northwest Laboratory. NTIS, Springfield, VA.

Pelroy, R. A., and M. R. Petersen. Mutagenic characterization of synthetic fuel materials by the Ames/Salmonella assay system. Mutat Res. (in press).

Pelroy, R. A., and D. L. Stewart. The effect of nitrous acid on the mutagenicity of two coal liquids and their genetically active chemical fractions. Mutat. Res. (in press).

Pelroy, R. A., and B. W. Wilson. Relative concentrations of polyaromatic primary amines and azaarenes in mutagenicially active nitrogen fractions from a coal liquid. Mutat. Res. (in press).

Pelroy, R. A., and B. W. Wilson (Eds.) (1981) Fractional Distillation as a Stragegy for Reducing the Genotoxic Potential of SRC-II Coal Liquids: A Status Report, PNL-3787 (Draft). Pacific Northwest Laboratory, Richland, WA.

PNL (1979a). Appendix to Biomedical Studies on Solvent Refined Coal (SRC-II) Liquefaction Materials: A Status Report, PNL-3189. Pacific Northwest Laboratory. NTIS, Springfield, VA.

PNL (1979b). Biomedical Studies on Solvent Refined Coal (SRC-II) Liquefaction Materials: A Status Report, PNL-3189. Pacific Northwest Laboratory. NTIS, Springfield, VA.

PNL (1980). Solvent Refined Coal-II (SRC-II) Detailed Environmental Plan, PNL-3517. Pacific Northwest Laboratory. NTIS, Springfield, VA.

PNL (1981) Volume I, Solvent Refined Coal-I (SRC-I) Joint Environmental Plan: Health and Ecological Effects (Draft). Pacific Northwest Laboratory, Richland, WA.

States, J. B., C. D. Becker, M. J. Schneider, and J. A. Strand (1979). Response of aquatic biota to effluents from the Ft. Lewis, Washington Solvent Refined Coal Pilot Plant, pp. 212-220. In Proceedings of the Symposium on Potential Health and Environmental Effects of Synthetic Fossil Fuel Technologies. CONF-780903, NTIS, Springfield, VA.

Strand, J. A. and B. E. Vaughan (Eds.) (1981) Ecological Fate and Effects of Solvent Refined Coal (SRC) Materials: A Status Report (Draft), PNL-3119. Pacific Northwest Laboratory, Richland, WA.

Weil, C. S., and N. I. Condra (1960). The hazards to health in the hydrogenation of Coal-II. Carcinogenic effect of materials on the skin of mice. Arch. Environ. Health, 2, 187.

Weimer, W. C., B. W. Wilson, R. A. Pelroy, and J. C. Craun (1980). Initial Chemical and Biological Characterization of Hydrotreated Solvent Refined Coal (SRC-II) Liquids: A Status Report, PNL-3464. Pacific Norhtwest Laboratory. NTIS, Springfield, VA.

Wilson, B. W., J. C. Craun, R. A. Pelroy, M. R. Peterson, and D. W. Felix (1980). Strategies for characterizing the fate of bioactive materials during coal liquefaction product upgrading. In Proceedings, Second U.S. Department of Energy Environmental Control Symposium, Volume 1, Fossil Energy. CONF-800334/1, NTIS, Springfield, VA.

Wilson, B. W., and R. A. Pelroy. Effects of catalytic hydrogenation on coal liquids which exhibit microbial activity. Toxicol. Environ. Chem. Rev. (in press).

Wilson, B. W., R. A. Pelroy, and J. T. Cresto (1980). Identification of primary aromatic amines in mutagenically active subfractions from coal liquefaction materials. Mutat. Res., 79, 190-202.

Wilson, B. W., R. A. Pelroy, and J. T. Cresto. In vitro assay for mutagenic activity and gas chromatographic mass spectral analysis of coal-liquefaction material and the products resulting from its hydrogenation. Fuel (in press).

SOIL LOSS AND LEACHING, HABITAT DESTRUCTION, LAND AND WATER DEMAND IN ENERGY-CROP MONOCULTURE: SOME QUANTITATIVE LIMITS

V. P. Gutschick

*Environ. Sci. Group LS-6, Los Alamos Nat. Lab.,
Los Alamos, NM 87545, USA*

ABSTRACT

I resolve three classes of environmental impacts: essentially irreversible impacts such as gene-pool loss, often comprising unavoidable costs per unit power; and incremental uses of either renewable resources/environmental tolerances (consider water, soil) or nonrenewable resources (fossil fuels, phosphorus reserves) needed to sustain or restore the environment. Impacts per unit power produced can be large, and consequent limits to power production can be low for biomass systems, because of low power production per area and fair-to-poor net energy balance. I quantify these limits and impacts for a selected number of biomass energy systems. I then draw preliminary conclusions about preferred systems (by type of national resource base) and about institutional barriers to reducing impacts and improving acceptable levels of power production.

KEYWORDS

Biomass energy; environmental impacts; net energy balance; limitations; alcohols; jojoba; gene pool.

INTRODUCTION

The environmental impacts of growing biomass for energy, especially for liquid automotive fuels, are potentially large. They are sensitive to the low power production per unit area (high land requirement) and to net energy balances. In this paper, I assemble initial quantitative estimates for impacts per unit power, within several classes of impacts...and conversely, for limits to power produced, if one avoids irreversible impacts. Four major biomass technologies are considered.

METHOD

Three classes of environmental impacts may be distinguished: (1) Irreversible but often avoidable environmental degradations leading to losses in productivity bases, when steady-state renewal rates are exceeded. Examples include simple soil erosion and massive ecosystematic failures (clearing tropical wet forests). By "irreversibility," I mean loss for at least one human generation. Losses usually extend beyond potential energy productivity to include local food productivity and even global genetic resources. (2) Increments in use of renewable environmental resources, requiring tradeoffs against current uses of these resources (land, water, etc.) or commitment of future reserve capacities. The finitude of renewal capacities, as of soil erosion tolerance (Larson, 1979; Soil Conservation Service, 1971, 1977), limits power production from biomass. All resource uses also bear costs, which are either internalized by producers, to be added to products costs, or externalized, to be borne by society as a whole (such as soil erosive losses). Externalized costs can be insidious and large, and they merit intensive scrutiny. In any event, one must evaluate resource uses per unit of power produced. One must therefore know (a) yield of raw biomass energy per area (for land use) and per input (of water, etc.), and (b) net energy balance in total fuel cycle. The latter has been a source of controversy in ethanol production for "gasohol" (Anon., 1980; Berger, 1979; Chambers and others, 1979; Cheremisinoff, 1979; da Silva and others, 1978; Krochta, 1980; Ladisch and Dyck, 1979), and it incorporates several subtleties. (3) Increments in use of nonrenewable resources, mostly fossil fuels and phosphorus reserves. These are not directly environmental impacts, but "in second order" their depletion affects our ability to restore environments damaged by biomass production, such as P-depleted leached areas.

I do not discuss the more abstract environmental values--aesthetics, recreation, and the like--which are hard to quantify but very important; I leave these for later work.

FOCUS ON SOME MAJOR IMPACTS: CLASSIFICATION AND INITIAL QUANTIFICATION

Class 1: "Irreversible" Impacts

1. Permanent agricultural/silvicultural productivity losses. In temperate zones, where most industrial nations lie, the major threats are soil erosion and loss of tilth (organic matter important for aeration, water retention, cation exchange capacity). Erosion is predictable quantitatively (as by the Universal Soil Loss Equation; Wischmeier and Smith, 1965) and is a graded, noncatastrophic process. It can be tolerated in a steady state (Gupta, Onstad, and Larson, 1979; Soil Conservation Service, 1977), at a rate dependent upon local soil nature and climate, but recovery from time-integrated excesses is at a slow, geologic rate, perhaps 100 to 10,000 years for full depth (Biswas, 1979). In the U.S., even our "advanced" agricultural practices exceed erosion tolerance limits over perhaps 2/3 of our food cropland! (Gupta, Onstad, and Larson, 1979; Soil Conservation Service, 1967) Some choices of energy crops are superior for minimizing erosion potential; the continuous cover provided by grasses and trees is preferred (Plotkin, 1980) over bare-soil periods attained with most annual crops.

In the wet tropics, clearing the forest for cropping herbaceous or non-native woody crops often leads to irreversible failures. In recent times, even enthusiasts for such schemes are re-evaluating the ecological price (Dickson, 1978). The failure, a loss of soil and nutrients and of regenerative capacity of the original ecosystem for up to several hundred years (Bennema, 1977; Brünig, 1977; Farnsworth and Golley, 1974; May, 1975), is rapid and hard to reverse, once started. Contributing are several factors: (1) Fast erosion of bared soil in high-rainfall areas (Bennema, 1977; Brünig, 1977); (2) Fast leaching of mineral nutrients to deep zones or groundwater, ordinarily prevented by tight biological recycling achieved by plants' woodiness and by dense root mats (Herrera and others, 1978; Stark and Jordan, 1978; Went and Stark, 1968); (3) Exacerbating the latter, a very low reserve of mineral nutrients (Bennema, 1977); (4) Low resilience of animal and plant species diversity, perhaps because slowly-reproducing "K-selected" species evolve in the stable tropical forests (May, 1975); (5) Loss of recolonization ability (Gomez-Pompa, Vazquez-Yanes, and Guevara, 1972; Kozlowski, 1979), including germination areas for seeds because bare soils are hotter; (6) Reduction of special microflora needed for nutrient recycling...microfloral species that are

not supported in the monocultures replacing forests. In general, we lack knowledge of these component aspects (Mooney and others, 1980).

Loss of land productivity for many years after a few years' extraction of biomass
energy gives an even poorer power production per area of land used. Consider
temperate crops yielding 20 tonnes per hectare of dry matter per year, as raw
energy (before processing energy losses and debits for auxiliary energy inputs).
Assume that productivity is lost for 100 years after 30 years' use. The power
production is then, on the average,

$$P = \frac{energy}{time} = \frac{(30 \text{ yr}) (20 \times 10^3 \text{ kg ha}^{-1}\text{yr}^{-1}) (18 \times 10^6 \text{ J kg}^{-1})}{(130 \text{ yr}) (3.1 \times 10^7 \text{ s yr}^{-1})} = 2.7 \text{ kW ha}^{-1} \ !$$

For a tropical crop yielding 50 tonnes /ha for 6 years with 100 years subsequent
loss, the rate is only 1.6 kW/ha. Compare these figures with those for one of the
most land-intensive fossil-fuel production methods, the strip mining of coal from
a modest seam (5m thick, 80% recoverable), with any given part of the mine land
tied up for 1 to 10 years. The power production here is

$$P = \frac{(0.80 \text{ recovery}) (5 \times 10^4 \text{ m}^3\text{ha}^{-1}) (1.4 \times 10^3\text{kg coal m}^{-3}) (26 \times 10^6 \text{ J kg}^{-1})}{(1 \text{ to } 10 \text{ yr}) (3.1 \times 10^7 \text{ s yr}^{-1})}$$

$$= 4.7 \text{ to } 47 \text{ MW ha}^{-1}$$

2. Gene-pool loss. Clearing the whole ecosystem of native vegetation (and animals)
contributes to extinction of wild plant species at a rate that is hard to quantify. Even in a narrow, economic sense, this is a loss of (1) potential new
economic plants not yet identified (Schultes, 1979), and (2) genetic diversity in
relatives of existing crops (Day, 1977), for interbreeding that helps the latter
retain pest resistances as pests co-adapt. Such breeding is necessary on a 10-to-
20 year rotation, and many plant scientists see the need for expanding the pool of
genes beyond our limited breeding stock. Once lost, the wild gene-pool is eter-
nally lost. This may be the most severe impact of increased land use for biomass
energy, yet the least understood and quantified.

3. Carbon dioxide addition to the atmosphere? Replacement of the native plant
communities by energy-crop or other monocultures (except perhaps by trees) may
substantially decrease both standing biomass and soil organic matter (Bolin, 1977;
Botkin, 1977; Broecker and others, 1979; Moore, 1977; Revelle and Munk, 1977;
Siegenthaler and Oeschger, 1978; Wong, 1979; Woodwell and Houghton, 1977). The
carbon content of the latter two reservoirs currently totals about twice the
atmosphere's carbon contained as carbon dioxide (Botkin, 1977; Revelle and Munk,
1977; Woodwell, 1978). Consider a conversion of x percent of average biomass on
earth to energy crops with half the original biomass' density. If half the injected carbon dioxide remains in the air, as apparently occurs for fossil-fuel
derived carbon dioxide (Siegenthaler and Oeschger, 1978), then this might contribute an increase in atmospheric CO_2 content of x/2 percent.

The nature, degree, and negative implications of climatic changes that might be
induced by significantly increased levels of carbon dioxide are being avidly
researched and hotly debated today (Bach, 1978; Baes and others, 1977; National
Academy of Sciences, U.S.A., 1977). Agriculture itself, including silviculture,
might suffer crucial negative impacts. The slight increase in growth rate from
better availability of carbon dioxide (Kramer, 1981; Wong, 1980) might not occur
in open fields over whole seasons (Botkin, 1977; Hicklenton and Jolliffe, 1980;
Kramer, 1981), and it may be more than cancelled by shifts in rainfall zones
(Bach, 1978; National Academy of Sciences, U.S.A., 1977). In any event, carbon
dioxide injection and subsequent climatic changes are reversible only by oceanic
uptake over long times, 500 to 5000 years (Lamb, 1972: National Academy of

Sciences, 1977; Siegenthaler and Oeschger, 1978). In our own interest, we might need to stay within a tolerance of a factor-of-two increase.

In contrast to biomass use, fossil fuel use contributes to atmospheric carbon dioxide increase continuously, in proportion to integrated energy rather than power. With one-half retention in the air, the increase is about 5 parts per million (320 ppm is the present level) per 10^{21} J produced (world annual production is about 0.3×10^{21} J). Biomass use contributes only during initial land conversion and not during steady-state use--perhaps 26 ppm increase per 10^{21} J per year of added power.[1] However, it is much debated (see references at start of section) if land conversion has clearly contributed to CO_2 increases historically.

Class 2: Effects of Using Renewable Environmental Resources

Water use. Biomass energy production requires far more water per unit of energy produced than any other technology. Even coal liquefaction, criticized for a high water use that approximates two volumes of water per volume of liquid fuel produced (0.08 cubic meters per gigajoule), is dwarfed by biomass production that required 600 to 2000 times greater use per unit energy. Much of this is truly an added transpirational demand over that of the natural ground cover. Water use should be charged against biomass production whether water is drawn directly from managed surface supplies for irrigation, or only diverted from recharging aquifers. The fractional commitment of water resources for raw (not even net) biomass energy far exceeds the fractional contribution to energy supply, by perhaps a factor of four in the U.S.[2] Povich (1978) has noted the strong constraint on biomass power capacity enforced by finite water availability. Note that there are two additional consequences of water use for environmental quality: drying up large wild areas, as now in the Owens Valley because of California water projects, and flooding large catchment/impoundment areas to bring more water under management.

2. Land use, within soil erosion tolerances. The calculational method used for permanent productivity losses earlier can be adapted here, dropping the extra span of time lost for regeneration. In the temperate-zone example, the power production becomes 11.5 kW/ha. Thus, if arable land constitutes 20% of total land area, the rate of committing our arable land area exceeds by a factor of 1.3 the rate of meeting our (U.S.) energy demand, assuming the latter is 2.6×10^{12} W. This factor is inflated by low net energy balances in biomass energy production (at least 2 to 10 J in as raw energy per J out) and by prior commitment of a large fraction of land to raising food. In the tropics, the factor is more favorable because power demand is lower in tropical nations, but the sustained-yield area may also be a notably lower fraction of total area.

The erosion tolerance is a crucial factor limiting the effective area for biomass production. The tolerance on marginal land, which land is often touted as a solution for areal limitations, is low. Larson (1979), Roller and others (1975), Tyner (1980), and others have discussed the limitation, to perhaps one or two percent of total power demand in the U.S.

Land use, considered as a resource denial as its primary environmental effect,

---

[1] I assume that the same primary productivity is retained, and that the average distribution of biomass parallels that of productivity, which totals 3.1×10^{21} J/yr.
[2] I assume a low water use of 80 $m^3 GJ^{-1}$, appropriate for only 5% of area irrigated.

also has other impacts, such as reversible habitat destruction (of non-endangered species).

3. Air and water quality. In comparison to conventional energy technologies, biomass energy production rates rather well regarding air-quality impacts, from production through end use (Berger, 1979; Cheremisinoff, 1979). Nitrogen oxide and particulate emissions are comparable in both technologies, sulfur oxide emissions are lower with biomass. Biomass impacts water quality more, mostly because of salinization of surface water by irrigation demands and because incremental use of nitrogenous fertilizers leads to nitrate loading of surface waters and aquifers (Alexander, 1972). Degradation of water quality effects both human economic and ecosystematic losses. For lack of space here, I leave the topic for future work.

Class 3: Effects of Using Nonrenewable Resource Reserves

Drawdown of soil nutrient reserves by harvesting plant parts for processing off-site provides the main occasion for using nonrenewable reserves, as of phosphorus (Holt, 1979). The redistribution of plant ash after fuel conversion is generally uneconomic, because the ash nutrients have low value compared to the tractor fuel required for redistributing them. Phosphorus use is an essentially irreversible commitment without building up soil reserves; only the soluble forms are in short supply in soil, and applied phosphorus is soon transformed to insoluble forms (Olsen and Fried, 1957; Povich, 1978). Phosphorus demand places a limitation on total biomass energy (not power) producible (Povich, 1978). Nitrogen demand has similar consequences, though it acts by demanding fossil fuels for producing fertilizer; I give no quantitative discussion here (see my report, 1977).

PRODUCTIVITY AND NET ENERGY BALANCE: SCALING FACTORS FOR IMPACTS

Productivity per unit of land is low; only 0.5% to 3% of solar energy, itself a diffuse source, is fixed. Productivity per unit of water used is also low. Both are variable, however, and selection of best species and cultivars is merited (Roller and others, 1975; Tyner, 1980). Perennials are desired, especially those that regrow unattended after harvest (e.g., trees that coppice). The productivity of trees is high, while among herbaceous species the C-4 photosynthetic species are often superior in both land and water uses, under select conditions (Bishop and Reed, 1976; Loomis and Gerakis, 1975; Mooney, 1972; Zelitch, 1975).

In net energy balance, the energy value per unit weight of biomass is primary. As dry weight, it is nearly invariably 18 kJ/g. Costs of water removal can debit a modest fraction of this value. The energy debits for inputs of chemical fertilizers and pesticides, for tractor fuels and irrigation, and for amortized energy of manufacturing farm equipment are generally near 20% of raw biomass energy (Heichel, 1976; Pimentel and Terhune, 1977, Roller and others, 1975). These debits are less in nonintensive agriculture (Pimentel and Terhune, 1977). Post-harvest energy costs for transportation can be substantial, and they worsen with increased size of operation. Overall, because fossil-fuel inputs are a fair fraction of energy output, Roller and coworkers (1975) declare that the prospect of inexpensive raw biomass and expensive fossil fuel to be replaced is illusory.

Between raw biomass input and fuel output lie many processing energy losses and energy debits for processing-heat inputs. These are most severe in ethanol production. The numerous authors cited under "Method" debate the precise energy balance; at best, it is marginal. However, some researchers claim, with some justification, that not all energy inputs should be weighed equally--that solid fuels which are abundant, such as coal, should hardly be debited, i.e., that only liquid fuel balances matter. (This is only appropriate for nations rich in solid fuels.) I have discussed energy balance on a liquids-only basis recently (1980) for coal liquefaction. Regarding process-heat inputs, the gasification of biomass, with or without subsequent conversion to liquid fuels such as methanol, is most favorable. Median thermal efficiencies near 50% are likely (Reed, 1976).

A few subtleties remain in evaluating energy value of the final product (Plotkin, 1980). Alcohols as motor fuels deserve an octane credit against crude petroleum, because one can forgo some refining steps that consume a significant energy fraction. Some people (Anon., 1980; Stumpf, 1980) claim additional credit is due for raised efficiency in combustion with alcohols, but the magnitude of the claim is dubious (Berger, 1979). Non-energy byproducts merit some energy credits, also. In ethanolic fermentations, dried distillers' grain has an energy equivalent about equal to the cost of alternatively supplying cattle with other protein-rich feed (perhaps 20% of combustive value, therefore). However, the market for it is quite saturable, and credits will not persist at high biomass production rates (Plotkin, 1980). One last subtlety is that some primary products are non-energy, e.g., jojoba waxes, but they substitute for petroleum-derived products. Their energy equivalent is several times larger (three?) than their raw combustive value, because much processing energy is required to make petroleum feedstocks to the same products. These are quite favorable--perhaps most favorable (Budiansky, 1980)--products of biomass.

INITIAL ASSESSMENTS OF FOUR BIOMASS ENERGY TECHNOLOGIES

Definitions: Y_r = yield, raw energy (good crop, not best), in GJ ha^{-1}yr^{-1}

W = water use, from managed supplies, in m^3 GJ^{-1}

e = net energy ratio, (fuel out minus process inputs)/(raw input)

L = land use, in 10^6 ha required to produce 10^{18} J yr^{-1}, net energy

P = power producible, in 10^{18} J yr^{-1}, in specified region, within environmental constraints (avoiding irreversible direct impacts)

Ethanol--from Food Grains; Temperate Zone

Y_r = 54 (wheat, kernels only) to 180 (corn, ears plus one-third residue) (Roller and others, 1975)

W = 0 (dryland wheat) to 1400 (irrigated corn; Pimentel and Terhune, 1977)

e = -0.5 to +0.5 (see Refs. at "Methods"); likely +0.1 with very good technology; routinely +0.5 on liquids-only basis with very good technology (my estimates)

$P \leq 1$ in U.S.; limits are erosion tolerances, then water availability--to about 50×10^6 ha? (See Larson, 1979; Povich, 1978; Roller and others, 1975; Tyner, 1980)

Special impacts: increased carbon dioxide in air is likely very small; most land to be used has low cover or has been cleared. Gene-pool loss probably small; mostly lost already? Water-quality impacts fair; depend on wheat/corn mix.

Ethanol--from Cellulosic Residues; Temperate Zone

Y_r: about 1.5 to 2 times larger than above; most other factors scale down by inverse of this factor. See Larson (1979), for example, for details.

Ethanol--from Sugar Cane and Root Crops; Tropics

Y_r = 250 to 900 (from 50 tonnes ha^{-1}yr^{-1}, estimated; see Zelitch, 1971; lower figure accounts for low sugar/high cellulose in non-maturing Amazonian plantings
e = o.3?; L = 4
Limitations: in new land clearing, by danger of irreversible ecosystematic and
soil destruction; by erosion tolerance also? (Needs study.) Power limits need
study
Special impacts: Irreversible gene-pool loss currently in progress; perhaps a significant increase in atmospheric CO_2 level

Methanol--from Grains and Herbaceous Crops; Temperate Zone

(Via gasification and catalytic reaction of resulting CO and H_2.)
e = +0.5; all impacts about 20% as large as with ethanol from food grains
P: about 5 times larger than latter case

Jojoba Wax, Energy Equivalent; Semi-Tropics

Y_r= 10? (about 0.5 tonne way per hectare per year); W = 0? (dryland)
e = 2 = 3(feedstock credit ratio) x 0.7 (Extraction debits); L approx. 50
P $\leq$ 0.25 in U.S.; limitation is primarily by area of suitable climate and soil
Special impact: selectively high loss of arid-area gene-pool

Wood, for Synthesis Gas/Methanol

Y_r = 200 (pine, in temperate zone: Roller and others, 1975) to 600 (eucalyptus)
W: approximately zero?; e = 0.4 (more processing debits than for herbaceous crops)
L: about 4 to 12
Limitations: need much study

CONCLUSIONS AND FURTHER OBSERVATIONS

(1) Environmental constraints on net power production are strong..presuming that
nations do want to preserve agricultural bases (I have some doubts). The prospects are best in the tropics, relative to demand, but much care is needed to
avoid catastrophic impacts. (2) Preferences: methanol is better than ethanol;
residues are better than total energy crops, at least in temperate zone. (3) Erosion tolerant arable land is quite limited; marginal land gives little hope
(Plotkin, 1980; Roller and others, 1975), and will likely not favor energy production over food production. (4) Impacts to watch most closely are permanent losses
of productivity and of gene pools; the latter are historically high in temperate
zones and increasing in the tropics (with little effective concern by temperate-
zone nations; Janzen, 1973). (5) Externalization of environmental costs by
producers must be countered; the largest of these costs potentially exceed by far
the short-term benefits provided. The modern economic panacea, "the marketplace,"
is part of the problem rather than of the solution here, especially in global impacts such as carbon dioxide, in that it rewards externalization and promotes the
"tragedy of the commons." (6) Relative to coal liquefaction (Gutschick, 1980) as
an alternative liquid-fuel source, biomass energy technologies promise to impact
the environment much more overall, and should be preferred only by coal-poor
nations of limited power demand. For coal-rich nations, liquid-fuel conservation
by increasing task efficiencies (Ford and others, 1975) almost surely deserves

first consideration, both environmentally and economically.

REFERENCES

Alexander, M. (1972). Environmental impact of nitrate, nitrite, and nitrosamines. In L. J. Guarraia and R. K. Ballentine (Eds.), The Aquatic Environment: Microbial Transformations and Water Management Implications. U.S. Environmental Protection Agency, Washington, D.C. Pp. 135-146.

Anon. (1980). Gasohol: does it save energy? Environ. Sci. Technol. 14, 140-141.

Bach, W. (1978). The potential consequences of increasing CO_2 levels in the atmosphere. In J. Williams (Ed.), Carbon Dioxide, Climate, and Society. Pergamon Press, Oxford. Pp. 141-167.

Baes, C. F., H. E. Goeller, J. S. Olson, and R. M. Rotty (1977). Carbon dioxide and climate: the uncontrolled experiment. Am. Sci. 65, 310-320.

Bennema, J. (1977). Soils. In P. T. de T. Alvim and T. T. Kozlowski (Eds.), Ecophysiology of Tropical Crops. Academic Press, New York. Pp. 29-55.

Berger, J. E. (1979). The use of gasohol. Ecolibrium (Shell Oil Co.) 8, 10-13.

Bishop, D. G., and M. L. Reed (1976). The C_4 pathway of photosynthesis: ein Kranz-Typ Wirtschaftswunder? Photochem. Photobiol. Rev. 1, 1-69.

Biswas, M. R. (1979). Environment and food production. In M. R. Biswas and A. K. Biswas (Eds.), Food, Climate, and Man. Wiley, New York. Pp. 125-158.

Bolin, B. (1977). Changes of land biota and their importance for the carbon cycle. Science 196, 613-615.

Botkin, D. B. (1977). Forests, lakes, and the anthropogenic production of carbon dioxide. BioScience 27, 325-331.

Broecker, W. S., T. Takahashi, H. J. Simpson, and T.-H. Peng (1979). Fate of fossil fuel carbon dioxide and the global carbon budget. Science 206, 409-418.

Brünig, E. F. (1977). The tropical rain forest--a wasted asset or an essential biospheric resource? Ambio 6, 187-191.

Budiansky, S. (1980). Chemical feedstocks from biomass. Environ. Sci. Technol. 14, 642.

Chambers, R. S., R. A. Herendeen, J. J. Joyce, and P. S. Penner (1979). Gasohol: does it or doesn't it produce positive net energy? Science 206, 789-795.

Cheremisinoff, N. P. (1979). Gasohol for Energy Production. Ann Arbor Science Publishers, Ann Arbor, Michigan.

da Silva, J. G., G. E. Serra, J. R. Moreira, J. C. Concalves, and J. Goldemberg (1978). Energy balance for ethyl alcohol production from crops. Science 201, 903-906.

Day, P. R. (Ed.) (1977). The Genetic Basis of Epidemics in Agriculture. Ann. N. Y. Acad. Sci. 287.

Dickson, D. (1978). Brazil learns its ecological lessons--the hard way. Nature 275, 684-685.

Farnsworth, H. C., and F. B. Golley (Eds.) (1974). Fragile Ecosystems: Evaluation of Research and Applications in the Neotropics. Springer, Berlin.

Ford, K. W., G. I. Rochlin, R. H. Socolow, D. L. Hartley, D. R. Hardesty, M. Lapp, J. Dooher, F. Dryer, S. M. Berman, and S. D. Silverstein (Eds.) (1975). Efficient Use of Energy. American Institute of Physics, New York.

Gomez-Pompa, A., C. Vazquez-Yanes, and S. Guevara (1972). The tropical rain forest: a nonrenewable resource. Science 177, 762-765.

Gupta, S. C., C. A. Onstad, and W. E. Larson (1979). Predicting the effects of tillage and crop residue management on soil erosion. J. Soil Water Conserv. 34, 77-79.

Gutschick, V. P. (1977). Long-term strategies for supplying nitrogen to crops. Los Alamos National Laboratory, Los Alamos, New Mexico. Report LA-6700-MS.

Gutschick, V. P. (1980). A preliminary assessment of the environmental, health, and safety issues in coal liquefaction. Los Alamos National Laboratory, Los Alamos, New Mexico. Report LA-8578-MS.

Heichel, G. H. (1976). Agricultural production and energy resources. Am. Sci. 64, 64-72.

Herrera, R., T. Merida, N. Stark, and C. F. Jordan (1978). Direct phosphorus transfer from leaf litter to roots. Naturwissenschaften 65, 208-209.

Hicklenton, P. R., and P. A. Jolliffe (1980). Alterations in the physiology of CO_2 exchange in tomato plants grown in CO_2-enriched atmospheres. Can. J. Bot. 58, 2181-2189.

Holt, R. F. (1979). Crop residues, soil erosion, and plant nutrient relationships. J. Soil Water Conserv. 34, 96-98.

Janzen, D. H. (1973). Tropical agroecosystems. Science 182, 1212-1219.

Kozlowski, T. T. (1979) Tree Growth and Environmental Stresses. University of Washington Press, Seattle.

Kramer, P. J. (1981). Carbon dioxide concentration, photosynthesis, and dry matter production. BioScience 31, 29-33.

Krochta, J. M. (1980). Energy analysis for ethanol. Calif. Agric. (June), 9-11.

Ladisch, M. R., and K. Dyck (1979). Dehydration of ethanol: New approach gives positive energy balance. Science 205, 898-900.

Lamb, H. H. (1972). Climate: Present, Past, and Future. Methuen, New York.

Larson, W. E. (1979). Crop residues: energy production or erosion control? J. Soil Water COnserv. 34, 74-76.

Loomis, R. S., and P. A. Gerakis (1975). Productivity of agricultural ecosystems. In J. P. Cooper (Ed.), Photosynthesis and Productivity in Different Environments. Cambridge University Press, Cambridge. Pp. 145-172.

May, R. M. (1975). The tropical rainforest. Nature 257, 737-738.

Mooney, H. A. (1972). The carbon balance of plants. Annu. Rev. Ecol. Syst. 3, 315-346.

Mooney, H. A., O. Björkmann, A. E. Hall, E. Medina, and P. B. Tomlinson (1980). The study of the physiological ecology of tropical plants--current status and needs. BioScience 30, 22-26.

Moore, P. D. (1977). Atmospheric carbon dioxide and forest clearance. Nature 268, 296-297.

National Academy of Sciences, U.S.A. (1977). Energy and Climate. Washington, D.C.

Olsen, S. R., and M. Fried (1957) Soil phosphorus and fertility. In The 1957 Yearbook of Agriculture. U.S. Department of Agriculture, Washington, D.C. Pp. 94-100.

Pimentel, D., and E. C. Terhune (1977). Energy and food. Annu. Rev. Energy 2, 171-195.

Plotkin, S. (1980). Energy from biomass. Environment 22, 6-13, 37-40.

Povich, M. J. (1978). Fuel farming--water and nutrient limitations. AIChE Symp. Ser. 181 No. 74, 1-5.

Reed, T. B. (1976). When the oil runs out: a survey of our primary energy sources and the fuels we can make from them. In Proc. Conf. on Capturing the Sun Through Bioconversion. Washington Center for Metropolitan Studies, Washington, D. C. Pp. 366-388.

Revelle, R., and W. Munk (1977). The carbon dioxide cycles and the biosphere. In Energy and Climate. National Academy of Sciences, U.S.A., Washington, D.C. Pp. 140-158.

Roller, W. L., H. M. Keener, R. D. Kline, H. J. Mederski, and R. B. Curry (1975). Grown organic matter as a fuel raw material resource. National Aeronautics and Space Administration report NASA-CR-2608. Ohio Agricultural Research and Development Center, Wooster, Ohio.

Schultes, R. E. (1979). The Amazonia as a source of new economic plants. Econ. Bot. 33, 259-266.

Siegenthaler, U., and H. Oeschger (1978). Predicting future atmospheric carbon dioxide levels. Science 199, 388-395.

Soil Conservation Service (1967). Conservation Needs Inventory. U.S. Department of Agriculture, Washington, D.C.

Soil Conservation Service (1971). Basic statistics--national inventory of soil and water conservation needs. Statistical Bulletin No. 461. U.S. Department of

Agriculture, Washington, D.C.

Soil Conservation Service (1977). Cropland erosion. In Second National Water Assessment. Water Resources Council, Washington, D.C.

Stark, N. M., and C. F. Jordan (1978). Nutrient retention by the root mat of an Amazonian rain forest. Ecology 59, 434-437.

Stumpf, U. (1980). The CTA alcohol program. In Proc. Bio-energy '80; World Congress and Exposition. Bio-energy Council, Washington, D.C. P. 417.

Tyner, W. E. (1980). Agricultural energy production potential. In C. D. Scott (Ed.), Second Symp. Biotechnology in Energy Production and Conservation. Wiley-Interscience, New York. Pp. 81-89.

Went, F. W., and N. Stark (1968). The biological and mechanical role of soil fungi. Proc. Nat. Acad. Sci. U.S.A. 60, 497-504.

Wischmeier, W. H., and D. D. Smith (1965). Predicting rainfall-erosion losses from cropland east of the Rocky Mountains. Agriculture Handbook No. 282. U.S. Department of Agriculture, Washington, D.C.

Wong, C. S. (1979). Carbon input to the atmosphere from forest fires. Science 204, 209-210.

Wong, S. C. (1980). Elevated atmospheric partial pressure of CO_2 and plant growth. I. Interactions of nitrogen nutrition and photosynthetic capacity in C_3 and C_4 plants. Oecologia 44, 68-74.

Woodwell, G. M. (1978). The carbon dioxide question. Sci. Am. 238 No. 1, 34-43.

Woodwell, G. M., and R. A. Houghton (1977). Biotic influences on the world carbon budget. Phys. Chem. Sci. Res. Rep. 2, 61-72.

Zelitch, I. (1971). Photosynthesis, Photorespiration, and Plant Productivity. Academic, New York.

Zelitch, I. (1975). Photosynthesis and plant productivity. Science 188, 626-633.

A CRITICAL REVIEW OF RECENT STUDIES OF THE IMPACT ON CLIMATE OF AN INCREASED ATMOSPHERIC CARBON DIOXIDE CONCENTRATION

J. Jäger

Fridtjof-Nansen-Str. 1, 7500 Karlsruhe 41, Federal Republic of Germany

ABSTRACT

The atmospheric carbon dioxide concentration is observed to be increasing. Concern about this increase arises because of the possibility that substantial increases could lead to possibly irreversible
and undesirable climatic changes. This paper reviews the methods
and results of studies of these potential climatic impacts. It is
concluded that no method (modeling or analogue studies) can be used
presently to predict how the climate system will respond to a
doubling (or more) of the atmospheric carbon dioxide concentration.
Climate models give acceptable estimates of the response of global
surface temperature but not of regional temperature and rainfall.
The study of past climates and recent warm periods can be used as a
complement of the modeling approach but not as a replacement.

KEYWORDS

climate, carbon dioxide, climate models, greenhouse effect.

INTRODUCTION

Observations of the amount of carbon dioxide in the atmosphere have
indicated that the carbon dioxide concentration is increasing (e.g.
Keeling and coworkers, 1976). For instance, at Mauna Loa in Hawaii
the annual average concentration of carbon dioxide has increased fr
from just over 315 ppm in 1958 to almost 336 ppm in 1978. This increase is confirmed by measurements made elsewhere and the long-
term increase is clearly global. It is believed that part or all of
this increase is due to the release of carbon dioxide into the
atmosphere by the burning of fossil fuels (coal, oil, gas). It has
been argued recently that the release of carbon dioxide due to deforestation and land use changes, especially in the tropics, has
also contributed to the observed carbon dioxide increase.

Concern about the increase in atmospheric carbon dioxide concentration arises because of the possibility that substantial increases
could lead to possibly irreversible and undesirable climatic changes.
As will be shown in this paper, a global increase in the atmospheric

carbon dioxide concentration can be expected to cause an increase in the globally-averaged surface temperature. However, the climate system is complex and the temperature change would vary regionally. It is even possible that some areas would cool even if there is a globally averaged warming. Along with changes of temperature, one can expect changes in the precipitation distribution and in other climatic elements. In order to evaluate the potential consequences of climatic changes for economic and social sectors, such as agriculture or water resources, it is necessary to be able to forecast the potential climatic changes for a given increase in the global carbon dioxide concentration. This is not an easy task and many questions remain to be answered. While climatologists have considerable confidence in predictions about changes in globally-averaged, annually-averaged, equilibrium temperatures, the uncertainties increase when attempts are made to consider a less-idealized version of the climate system. This paper shows what methods are available for predicting the climatic effects of an increase in the atmospheric carbon dioxide concentration and indicates the problems and uncertainties inherent in such evaluations. The survey starts with the simplest types of prediction and proceeds to consider increasingly complex versions of the climate system.

THE 'GREENHOUSE EFFECT'

Many studies made with models of varying complexity have shown that an increased atmospheric carbon dioxide concentration produces a warming of the earth's surface and of the lower atmosphere. This warming is due to the fact that carbon dioxide is a good absorber and emitter of long-wave radiation. The increased carbon dioxide in the atmosphere therefore leads to increased downward emission of long-wave radiation and a surface warming. This effect has often been referred to as the "Greenhouse Effect", although the analogy is not perfect. The warming of the lower atmosphere is due to the increased downward emission from the carbon dioxide in the atmosphere.

FEEDBACK PROCESSES IN THE CLIMATE SYSTEM

Calculations show that a doubling of the atmospheric carbon dioxide concentration would give rise to a net heating of the lower atmosphere, oceans, and land by a global average of about 4 Wm^{-2} (Ramanathan, Lian and Cess, 1979; NAS, 1979). It is estimated by NAS (1979) that this value is correct within $\pm$ 25%. However, greater uncertainties arise in estimating the change of globally-averaged surface temperature which would result from the above change in heating rate. This is because a number of feedback processes can increase or decrease the heating rate.

The most important feedback process is due to the fact that the higher surface temperature leads to an increased water vapour content in the atmosphere as a result of increased evaporation. The increase of absolute humidity increases the absorption of long-wave radiation thereby adding to the effect of the carbon dioxide (positive feedback). Another feedback process is the temperature-ice albedo feedback. In this case the increased surface temperature causes increased melting of snow and ice. The melting means that the reflectivity (albedo) of the surface is decreased from that of bright snow or ice to that of ground or water or some mixture. The decreased reflectivity means that more solar radiation can be absorbed and the surface temperature can further increase (positive feedback).

Taking known climatic feedback processes into account, NAS (1979) estimated that the globally-averaged surface temperature increase, if the atmospheric carbon dioxide concentration were doubled, would be between 1.6 and 4.5 K with 2.4 K a likely value.

LATITUDINAL AND SEASONAL VARIATIONS

Ramanathan, Lian and Cess (1979) have pointed out that the radiative heating processes due to increased carbon dioxide are extremely dependent upon the atmospheric temperature and humidity distributions both of which are seasonally and latitudinally dependent. The latitudinal and seasonal variations of the impact of increased atmospheric carbon dioxide concentrations were investigated by Ramanathan, Lian and Cess (1979) using two relativaly simple numer- ical models of the climate system. It was found, for instance, that although a doubling of the atmospheric carbon dioxide concentration leads to a global average change in the heating rate of 4 Wm^{-2}, in winter the change in heating rate of the surface and lower atmosphere decreased from 4.6 Wm^{-2} at the equator to 2.2 Wm^{-2} at 80°N. This illustrates the latitudinal differences. It was also found that northward of 20°N the heating is larger in the summer than in the winter and the difference between the summer and winter heating in- creases with increasing latitude. In addition to latitudinal and seasonal differences in heating due to increased atmospheric carbon dioxide levels, Ramanathan, Lian and Cess found that the heating effects differ significantly between clear sky and overcast condit- ions.

For a doubling of the atmospheric carbon dioxide concentration, Ramanathan, Lian and Cess calculated that the hemispherically- averaged annual surface temperature increase would be 3.3 K. The annual surface temperature increase was found to vary from just over 3 K in low latitudes to more than 5.5 K in high latitudes. In the low latitudes virtually no seasonal variability in the temperature increase was found. In high latitudes, the temperature increase was found to be greatest in spring at 65°N and spring and summer at 85°N. This spring/summer amplification is due to the temperature-ice albedo effect discussed earlier.

GENERAL CIRCULATION MODEL STUDIES

In atmospheric general circulation models (GCMs) the three-dimension- al behaviour of the atmosphere is simulated by solving time-dependent dynamic equations governing atmospheric motion. Other processes are included in GCMs, by including equations for radiation and for processes which occur on a scale which is smaller than the resolution of the model, such as cloud formation. GCMs are generally considered to be more accurate than the more simplified climate models but improvements are still required, especially in the treatment of sub- grid scale processes such as cloud formation. In the earliest versions, the atmospheric circulation was simulated with the assump- tion that the ocean surface temperature remains constant at predeter- mined values during the course of the simulation. In the case of experiments to determine the impacts of a doubling of the carbon dioxide concentration this assumption is not necessarily valid.

The first study made with a GCM was that of Manabe and Wetherald (1975). The numerical model was run with an atmospheric carbon di- oxide concentration of 300 ppm in the first simulation and 600 ppm in the second simulation. The equations are solved for grid points but not for a realistic geography. The global average surface temperature increase was calculated to be 2.9 K, which was in line with estimates from simpler models of a 1.5 - 3 K increase for a doubling of the carbon dioxide concentration (Schneider, 1975). The temperature increase was considerably amplified in the polar region, where it was as much as 10 K. A further model result was that the global precipitation increased by 7%. There were, however, several shortcomings as far as the model was concerned and these have been discussed by Manabe and Wetherald (1975) and Smagorinsky (1977). In particular, there was no heat storage by continents or oceans and no transport of heat by the ocean. The ocean was basically treated as a "swamp" from which water could freely evaporate. In addition, the cloud distribution was fixed at present-day observed values in both simulations.

Manabe and Wetherald (1980) used a similar model to that above in which a simplified computation of cloud cover was included. In the experiment with a doubled atmospheric carbon dioxide concentration the average surface warming was found to be 3.0 K, which agrees closely with the previous result. The temperature change was also amplified in the polar area with a maximum increase of almost 8 K at 80° latitude.

Another version of the same GCM was used by Manabe and Stouffer (1979) to investigate the effects of a carbon dioxide concentration doubling. In this case the atmospheric model was coupled to an ocean layer which was 68 m deep. The ocean layer, the equivalent of a mixed layer, was assumed to be static (no horizontal or vertical currents). In this version of the model, the cloud cover was fixed and seasonal insolation values were used. The geography used in this model was realistic. The average surface temperature increase for a doubling of the atmospheric carbon dioxide concentration was about 2 K, which is significantly less than the 3 K found by Manabe and Wetherald (1975, 1980). Gates (1980) suggests that this result shows that the combined effects of a realistic geography and a seasonally interacting ocean serve to place the GCM's response close to that found with the best of simplified climate models.

Gates and Cook (1980) used a GCM in which the boundary conditions were seasonally varying and the ocean surface temperatures were kept equal to those of the present day. This assumes infinite ocean heat capacity with no feedback: as the atmospheric temperature changes, there is no response in the ocean surface temperatures. The model's response is therefore confined to the direct radiative effects of increased carbon dioxide, a situation which might arise during the 20 years after the carbon dioxide doubling as the oceans are slow to respond. The result of a doubling of the atmospheric carbon dioxide concentration was a global average surface temperature increase of 0.3 K in January. This warming is smaller than in other GCM experiments because in this case the oceans are serving as a sink for much of the extra long-wave radiation caused by the increased carbon dioxide, since the ocean surface temperature was not permitted to increase as the anomalous radiation is absorbed. In July the globally averaged tropospheric temperature increased by 0.33 K for a doubled atmospheric carbon dioxide concentration. An interesting result of the simulations made by Gates and Cook is that there was no significant change in precipitation, which is in contrast to the results of GCMs with a "swamp" ocean.

The results of GCM experiments with a doubled atmospheric carbon dioxide concentration have been reviewed in a number of recent papers including NAS (1979), MacCracken (1980), Washington and Ramanathan (1980) and Gates (1980). It is pointed out by NAS (1979) that the presently available GCMs produce time-averaged mean values of meteorological variables, such as temperature and rainfall, that correspond quite well with those observed when global or latitudinal averages are considered. However, NAS (1979) concludes that regional distributions are not well simulated due to model shortcomings, including the poor treatment of cloud, precipitation and orographic effects. This leads NAS (1979) to conclude that no prediction of regional climatic changes could be made on the basis of available GCM results.

The numerical models which have been used so far to investigate the potential climatic impacts of an increased atmospheric carbon dioxide concentration have considered very simplified versions of the oceanic component of the climate system. It has been shown, however, in a number of studies that it is necessary to consider both the atmospheric and oceanic circulation as a coupled system in order that a realistic evaluation of climatic impacts can be made.

The climate modeling studies that have been discussed so far in this paper have all considered the equilibrium climatic response to a step function in the carbon dioxide concentration. For instance, the atmospheric carbon dioxide concentration in a model has been doubled and the difference between two equilibrium states of the model (with and without the carbon dioxide doubling) has been investigated. In reality the atmospheric carbon dioxide concentration is increasing year by year and not as a step function. Also, the time lags introduced by the heat capacity of the ocean suggest that the equilibrium temperature change would be reached some time after the carbon dioxide doubling. Schneider and Thompson (1980) have reviewed the question of the transient response to a carbon dioxide concentration increase in detail. Their study suggests that the time evolution of regional climatic anomalies as a result of an increase in the carbon dioxide concentration could well be different from that suggested by equilibrium climate modeling experiments.

SCENARIOS OF PAST CLIMATE

Since there are many uncertainties in the results of climate model experiments to investigate the impact of increasing atmospheric carbon dioxide concentration, particularly in terms of the regional climatic anomalies and the precipitation changes as a result of a carbon dioxide increase, other approaches to answering the questions have been taken.

One approach is to look at evidence of past climates. Kellogg (1978) suggested that one way to find out what a warmer earth might be like is to study climatic evidence from periods when the earth was warmer than it is now. A period chosen by Kellogg for such a study is that of 4000 - 8000 years ago, known as the "Altithermal" or "Hypsithermal", when evidence from the northern hemisphere suggests that the climate was warmer than now. The evidence suggests that during this warmer period North Africa was generally more favourable for agriculture than it is now, that Europe was wetter, Scandinavia was drier and a belt of grassland extended across North America in what subsequently became forest land. Kellogg rightly cautions the reader that this evidence must not be taken as a literal representation of what might occur if the earth were to become warmer again, since the causes of the previous warming could have been quite different from those of the potential warming due to anthropogenic changes. The causes of the Altithermal warming are not clear. Another problem is that the Altithermal was time-transgressive. That is, the peak warming occured at different times in various places.

An extension of the approach taken by Kellogg (1978) has been made by Flohn (1980). Flohn suggests that a warming of 1 K would be equivalent to the early Middle Age warming of 900-1100 AD. A warming of 1.5 K would be equivalent, according to Flohn, to the postglacial warm period, referred to as the Altithermal above, which Flohn dates at about 5500 to 6500 years before present. A warming of 2-2.5 K would be equivalent to the last interglacial period of about 125,000 years ago. Lastly, Flohn considers a scenario of a 4 K warming, in which case the Arctic Ocean is assumed to be ice-free. Recent data cited by Flohn has suggested that the Arctic Ocean has not been ice-free in the last 2.5 million years. Flohn (1980) has made a detailed analysis of the climatic conditions during each of these periods, but does also ask the question: Can climatic history repeat itself ? This question is important if the palaeoclimatological evidence is to be used for the construction of climate scenarios for a potential carbon dioxide-induced warming. Flohn points out that two boundary conditions which existed during the Altithermal period have basically changed since then. Firstly, the presence of permanent ice sheets in Eastern Canada influenced the atmospheric circulation at that time and these ice sheets are no longer in existence. Secondly, since the Altithermal man has considerably altered the climatic boundary conditions.

In contrast to studies of climates of the geological past, a number of recent studies have used instrumental observations of temperature rainfall and pressure during the present century as a basis for discussing the response of the climate system to a warming. Williams (1980) has looked at regional rainfall, pressure and temperature anomalies in the Northern Hemisphere for seasons within the last 70 years when the Arctic was warm. The reason for choosing the Arctic as an indicator was a result of the model and observational studies that have indicated that the Arctic is more sensitive to climatic changes. Clearly, the warm Arctic seasons in the last 70 years were not a result of a carbon dioxide increase, rather they were the result of non-linear interactions between components of the climate system.

Neither the study of Williams (1980) nor the similar study of Wigley, Jones and Kelly (1980) claims that the distributions of temperature anomalies can be taken as predictions of the changes to be expected as a result of a doubling of the carbon dioxide concentration of the atmosphere. However, the results of such studies should be useful in guiding the development of scenarios of potential changes. An advantage of this type of analysis is that it illustrates clearly that large coherent anomalies are a basic response to climatic forcing and that seasons respond differently because of the different climatic processes that dominate in each season.

A similar study is reported by Namias (1980), who used 1000- to 700-mbar thickness data from the Northern Hemisphere from the period 1951-1978. In one case Namias looked at the 9 warmest winters during this period. In the second case, the 9 warmest winters north of 60°N were studied. Namias concludes, in agreement with the two studies mentioned above, that a moderate carbon dioxide-induced warming would not be uniform but rather would be regional in character.

The studies of Williams (1980), Wigley, Jones and Kelly (1980), and Namias (1980) have been reviewed and compared recently by Sear and Lough (1981). These authors conclude that since the temperature results of Wigley, Jones and Kelly and Williams show marked similarities, a certain degree of confidence might be placed in this type of scenario development. They suggest that it would be useful to consider runs of warm years and cold years in order to construct patterns of changes with some element of air-sea-ice interaction included. They also point out that the work on observational data might indicate possible "key" regions, where changes in the recent past have occured earliest or have been largest. These key regions could be monitored so that it might be possible to recognize as soon as possible any changes which could be precursors of anomalies resulting from a carbon dioxide-induced warming.

Pittock and Salinger (1981) have reviewed the various approaches to the problem of estimating the regional climatic impacts of a carbon dioxide-induced warming, with application of these approaches to Australia and New Zealand. The authors point out that none of the methods is perfect and none of the results can be given a great degree of confidence but that each individual method has its merits and a combination of the approaches could give useful results, especially if there is agreement between the results of different methods.

SUMMARY AND CONCLUSIONS

Observations of the amount of carbon dioxide in the atmosphere indicate that the carbon dioxide concentration is increasing. It is believed that part or all of this increase is due to the release of carbon dioxide into the atmosphere by the burning of fossil fuels (coal, oil and gas). It is also argued that the release of carbon dioxide due to deforestation and soil destruction has also contributed to the observed carbon dioxide increase.

A global increase in the atmospheric carbon dioxide concentration can be expected to cause an increase in the globally-averaged surface temperature. However, the climatic system is complex and non-linear and the temperature change would vary regionally. Along with changes in temperature, one can expect changes in the precipitation distribution and in other climatic elements.

Taking known climatic feedback processes into account, NAS (1979) estimated that the global average surface temperature increase, if the atmospheric carbon dioxide concentration were doubled, would be between 1.6 and 4.5 K, with 2.4 K a likely value.

A number of approaches are being taken towards predicting what the regional changes of temperature, rainfall and other climatic variables would be if there were a significant carbon dioxide increase.

The first approach is through the use of numerical models of the atmospheric circulation. Presently available models produce time-averaged mean values of meteorological variables that correspond reasonably well to those observed when global or latitudinal averages

are considered. However, NAS (1979) concluded that regional distributions are not so well simulated because of model shortcomings, including the poor treatment of cloud, precipitation and orographic
effects.

The numerical climate models that have been used so far to investigate the potential impacts of an increased carbon dioxide concentration have considered very simplified versions of the oceanic
component of the climate system. There is a strong need to consider
both the atmospheric and oceanic circulations as a coupled system,
in order that a realistic evaluation of the climatic impacts of an
increased atmospheric carbon dioxide concentration can be made.

The majority of climate model studies carried out so far have considered the equilibrium climate response to a step function in the
atmospheric carbon dioxide concentration. Recent model results have
shown, however, that the time evolution of regional climatic
anomalies as a result of a continuous increase in the carbon dioxide
concentration could well be different from that suggested by equilibrium climate modeling experiments.

Another approach to predicting climate anomalies resulting from a
carbon dioxide increase is to study climatic evidence from geological periods when the earth was warmer than it is now. This approach
is not without its shortcomings, especially since the causes of
the previous warming could have been quite different from those of
the potential warming due to anthropogenic impacts. Nevertheless,
the approach is useful in illustrating actual patterns of climatic
change that have been observed to occur in the past.

A similar approach uses instrumental observation of temperature and
rainfall and pressure during the present century as a basis for
discussing the response of the climate system to a warming. Several
studies have looked at the regional anomalies in meteorological
variables during warm seasons or years in this century. Again, such
studies cannot serve as predictions but do show where the climate
system responds more or less to forcing.

In conclusion, it is clear that there is no single, reliable method
available for predicting how the climate system will respond to a
doubling (or more) of the atmospheric carbon dioxide concentration.
Climate models give fairly accurate estimates of the response of the
globally-averaged surface temperature. However, the environmental
and social scientists need information on regional changes in seasonal temperature and rainfall, for example, in their impact analyses.
The study of past climates and recent warm periods can be useful in
illustrating the climatically sensitive areas and the potential
range of climatic responses. As such, it could be used as a complement of the modeling approach but not as a complete replacement.
Finally, it will be necessary, sooner or later, to consider the
future evolution of climate not only under the influence of carbon
dioxide increases but also due to simultaneous man-made changes of
the climatic boundary conditions.

REFERENCES

Flohn, H. (1980). Possible climatic consequences of a man-made global
    warming. RR-80-30. International Institute for Applied Systems
    Analysis, Laxenburg, Austria.

Gates, W. L. (1980). Modeling the surface temperature changes due
    to increased atmospheric CO_2. In W. Bach, J. Pankrath, and J.
    Williams (Eds.), _Interactions of Energy and Climate_. D. Reidel,
    Dordrecht, Holland.

Gates, W. L. and K. H. Cook (1980) Preliminary analysis of experim-
    ents on the climatic effects of increased CO_2 with the OSU
    atmospheric general circulation model. Climatic Research Instit-
    ute, Oregon State University, Corvallis.

Keeling, C.D., R. B. Bacastow, A. E. Bainbridge, C. A. Ekdahl, P. R.
    Guenther, and L. S. Waterman (1976). Atmospheric carbon dioxide
    variations at Mauna Loa Observatory, Hawaii. _Tellus_, _28_, 538-
    551.

Kellogg, W. W.,(1978). Global influences of mankind on the climate.
    In J. Gribbin (Ed.), _Climatic Change_. Cambridge University
    Press, Cambridge.

MacCracken, M. C. (1980). Climate research. Carbon Dioxide Research
    Progress Report, DOE-EV-0071. U. S. Department of Energy,
    Washington D. C.

Manabe, S. and R. T. Wetherald (1975). The effects of doubling the
    CO_2 concentration on the climate of a general circulation model.
    J. Atmos. Sci., _32_, 3-15.

Manabe, S. and R. T. Wetherald (1980). On the distribution of climat-
    ic change resulting from an increase in CO_2 content of the
    atmosphere. _J. Atmos. Sci._, _37_, 99-118.

Manabe, S. and R. J. Stouffer (1979). A CO_2 climate sensitivity
    study with a mathematical model of the global climate. _Nature_,
    _282_, 491-492.

Namias, J. (1980). Some concomitant regional anomalies associated
    with hemispherically averaged temperature variations. _J. Geo.
    Res._, _85_, 1585-1590.

NAS (1979). Carbon dioxide and climate:A scientific assessment.
    National Academy of Sciences, Washington, D.C.

Pittock, A. B. and M. J. Salinger (1981). Towards regional scenarios
    for a CO_2-warmed earth. Submitted for publication.

Ramanathan, V., M. S. Lian, and R. D. Cess (1979) Increased atmos-
    pheric CO_2: Zonal and seasonal estimates of the effect on the
    radiation balance and surface temperature. _J. Geophys. Res._,
    _84_, 4949-4958.

Schneider, S. H. (1975). On the carbon dioxide-climate confusion.
    J. Atmos. Sci., _32_, 2060-2066.

Schneider, S. H. and S. L. Thompson (1980) Atmospheric CO_2 and
    climate: Impact of the transient response. Submitted to _J.
    Geophys. Res._

Sear, C. B. and J. M. Lough (1981) Three scenarios for a warm, high
    CO_2 world. Presented at the American Meteorological Society
    First Conference on Climatic Variations, San Diego, California.

Smagorinsky, J. (1977). Modeling and predictability. In _Energy and
    Climate_, National Academy of Sciences, Washington, D.C.

Washington, W. M. and V. Ramanathan (1980) Climatic response due to
    increased CO_2 : Status of model experiments and the possible ro
    role of the oceans. Carbon Dioxide and Climate Research Program
    Progress and Planning Meeting, CONF-8004110, U.S. Department
    of Energy, Washington D.C.

Wigley, T.M.L., p. D. Jones, and P. M. Kelly (1980). Scenario for a
    warm high-CO_2 world. _Nature_, _283_, 17-21.

Williams, J. (1980). Anomalies in temperature and rainfall during
    warm Arctic seasons as a guide to the formulation of climate
    scenarios. _Climatic Change_, _2_, 249-266.

EFFECTS OF ACID PRECIPITATION ON ELEMENTAL TRANSPORT FROM TERRESTRIAL TO AQUATIC ECOSYSTEMS

D. W. Johnson

Environmental Sciences Division, Oak Ridge National Laboratory, P.O. Box X, Oak Ridge, TN 37830, USA

ABSTRACT

Significant progress has been made in terrestrial-aquatic transport methodology. Several techniques and conceptual frameworks are available for assessment of acid rain effects on these transport processes. Using the anion mobility model, for instance, it is possible to assess the relative effects of acid rain vs natural, internal acid production on elemental transfer rates. However, further research is needed in order to quantify these effects on a regional scale for sensitivity assessments. Sensitivity must be defined as to what ecosystem and what effects are being considered. Criteria have been developed, but there is a major need for more information on natural acid production.

KEY WORDS

Leaching, H^+, Al^{3+}, cations, runoff, anions, carbonic acids, organic acids, site sensitivity.

INTRODUCTION

Processes governing elemental transport from soils to surface waters are of crucial importance in determining the effects of acid precipitation on both terrestrial and aquatic ecosystems; such processes govern the elemental losses from terrestrial ecosystems and contribute to inputs to aquatic ecosystems. Due to the numerous deleterious effects of acidification on aquatic ecosystems, the transport of H^+ and Al^{3+} from terrestrial to aquatic ecosystems has received a great deal of attention (Cronan and Schofield, 1979; Seip, 1980; Abrahamsen and Stuanes, 1980). Seip discusses many aspects of this particular type of elemental transport in the following paper. This paper provides a general review of elemental transport from terrestrial ecosystems as affected by acid rain, including effects on nutrient ions as well as H^+ and Al^{3+}.

Pathways of Elemental Transfer

Elements may be transported from terrestrial to aquatic ecosystems by soil leaching or storm runoff. In the case of leaching, solutions can be substantially altered by a variety of chemically active sites within the soil profile and

underlying bedrock prior to leaving the terrestrial and entering the aquatic ecosystem. In the case of runoff, soil-water interactions are much reduced and solutions may enter aquatic ecosystems in a much less altered state than is the case
with leachate. In some regions runoff during snowmelt can be a significant epi-
sodic pathway for H^+ and $A\ell^{3+}$ transport to aquatic ecosystems, creating very
deleterious effects on aquatic life (Schofield, 1980). The relative occurances of
soil vs bedrock and the depths of soils on a given area will strongly affect the
relative importance of runoff vs leaching in elemental transport (Seip, 1980;
Galloway and others, 1980), although even barren rock can alter to some extent the
chemical composition of solutions flowing over it (Abrahamsen and others, 1979).
There is also evidence that subsurface runoff (i.e., runoff within surface soils,
immediately above less permeable soil horizons) may transport solutions directly
from surface soils to aquatic ecosystems without being altered further by contact
with subsurface horizons (Johnson and Henderson, 1979; Christopherson and Wright,
1980).

Leaching processes have been studied in great depth by soil scientists (e.g.,
Biggar and others, 1966; Dutt and Tanji, 1962; Cole and others, 1975), but a
review of them would be too lengthy to include here. Certain aspects of soil
leaching warrant some discussion in the context of acid rain effects, however, and
these are briefly reviewed in the following section.

Anion Mobility and Soil Leaching

The concept of anion mobility is gaining recognition as a useful tool for both
understanding and assessing the effects of acid precipitation on elemental leaching (Johnson and Cole, 1977; Cronan and others, 1978; Johnson, 1980; Cole and
Johnson, 1977; Reuss, 1978; Seip, 1980). This concept, first introduced by Gorham
(1958) and Nye and Greenland (1960), revolves around the fact that total anions
must balance total cations in solution. Since soils typically have more cation
than anion exchange or adsorption capacity, it is useful to visualize total ionic
leaching as a result of the introduction of mobile anions to soil solution
(Johnson and Cole, 1980).

With respect to acid precipitation, it has been repeatedly demonstrated that soil
sulfate adsorption prevents the leaching of either nutrient cations or H^+ and
$A\ell^{3+}$ (Johnson and Cole, 1977; Singh, 1980; Singh and others, 1980; Khanna and
Beese, 1978; Farrell and others, 1980), whereas the reverse is true when sulfate
is not adsorbed (Cronan and others, 1978; Abrahamsen and Stuanes, 1980). Soils
rich in amorphous iron and aluminum oxides are effective sulfate adsorbers
(Johnson and others, 1979; Johnson and Henderson, 1979), whereas soils low in
these constituents or high in organic matter are generally less effective sulfate
adsorbers (Johnson and others, 1980).

Nitrate is sometimes associated with acid precipitation and differs considerably
from sulfate in that it is very poorly adsorbed to most soils (Johnson and Cole,
1980). However, nitrate is quickly immobilized by biological processes in
N-limited ecosystems, and since N-limitations are common in forested regions of
the world, nitrate is rarely mobile (Abrahamsen, 1980).

Seip (1980) examines the potential roles of direct input (i.e., precipitation with
little or no alteration), introduction of mobile anions, and changes in soil pH as
mechanisms of surface water acidification in Norway. He concludes that the contributions of direct input and soil pH change are small, and that the introduction
of sulfate (which appears to be relatively mobile in these ecosystems) is the most
likely explanation for surface water acidification. This is because soils in
these regions are acidic with low base contents, and therefore a significant
portion of the increased cation concentration required to balance the increased

sulfate concentration must be H^+ and Al^{3+}. Emperical studies support this theory (Abrahamsen and Stuanes, 1980), but there remains some controversy about the magnitude of pH change this mechanism could produce relative to observed changes in surface water acidity (Abrahamsen and Stuanes, in prep.). Mechanisms of surface water acidification are discussed in detail by Seip in the following paper in this volume.

Site Sensitivity to Acid Precipitation

Various schemes for assessing site sensitivity to acid rain effects have been proposed. Those directed toward aquatic effects have emphasized bedrock geology (Norton, 1980; Hendry and others, 1980), while those concerned with terrestrial effects have emphasized cation exchange capacity and base saturation (McFee, 1980; Klopatek and others, 1980). For the reasons previously discussed, sulfate adsorption capacity ought to be included in the sensitivity criteria for both aquatic and terrestrial impacts (Johnson, 1980), but unfortunately the data base for the latter is much more limited than that for the criteria used until now. Nonetheless, some general, theoretical predictions can be made based upon sulfate adsorption and base content criteria.

If base content and sulfate adsorption are low, soils are only moderately sensitive to cation leaching and acidification (Table 1). This is due to the reduced efficiency of metal cation displacement by H^+ in acid soils (Wiklander, 1980). However, aquatic ecosystems surrounded by such soils would be sensitive to acidification according to the mobile anion mechanism described by Seip (1980), assuming bedrock geology is rated sensitive, also. If a soil has low base content and high sulfate adsorption, the soil is more sensitive to acidification than in the above case because both incoming SO_4^{2-} and H^+ are retained in the soil (Johnson and Cole, 1977), and the displacing efficiency of H^+ is not a factor. If a soil has a medium base content (i.e., slightly acid), it is relatively susceptible to cation loss and further acidification according to Wiklander (1980). However, since almost all incoming H^+ displaces soil cations, aquatic ecosystems are not acidified and receive base cations leached from the soil in the short term. In the long term, waters entering aquatic ecosystems may become more acid as soils become more acid; therefore, a "moderate" rating was (arbitrarily) assigned to sensitivity to aquatic ecosystem acidification. Again, if a soil with medium base content has high sulfate adsorption capacity, terrestrial cation losses are low, aquatic sensitivity is low, and only sensitivity to soil acidification is high (since H^+ and SO_4^{2-} are retained). Soils with high base content are generally regarded as insensitive (McFee, 1980; Klopatek and others, 1980). However, it is important to define sensitivity since such soils may be very "sensitive" with respect to the rate of base cation loss, and aquatic ecosystems may be sensitive with respect to increased cation inputs if sulfate adsorption is low. Only if sulfate adsorption is high are these soils insensitive with respect to all effects listed in Table 1.

Natural Acid Production

While considering the sensitivity of terrestrial and aquatic ecosystems to acid precipitation effects, it is important to realize that acids are produced naturally within these systems (Rosenqvist, 1977; Rosenqvist and others, 1980; Reuss, 1977). The effects of atmospheric acid inputs must be viewed as an addition to natural, ongoing acidification and leaching processes within soils due to carbonic acid formation, organic acid formation, tree cation uptake and a variety of other processes (Johnson and others, 1977; Sollins and others, 1980; Andersson and others, 1980). Unfortunately, the data base for including natural acid formation criteria into regional sensitivity assessments is extremely limited. Thus,

TABLE 1 Theoretical Sensitivities of Terrestrial and Aquatic Ecosystems
to Acid Precipitation Effects Given Soil Base Content (eq/ha)
and Soil Sulfate Adsorption Capacity (eq/ha)

Soil Properties		Terrestrial Sensitivity		Aquatic Sensitivity*	
Base content	Sulfate adsorption	Cation loss	Acidification	Cation input	Acidification
low	low	moderate	moderate	moderate	high
low	high	low	high	low	low
medium	low	high	high	high	moderate
medium	high	low	high	low	Low
high	low	high	low	high	low
high	high	low	low	low	Low

*Assuming bedrock falls into sensitive catagories according to Norton (1980) and Hendry and others (1980).

current sensitivity rating schemes, by default, assume that atmospheric inputs add significantly to internal acid production, an assumption that is by no means universally accepted (e.g., Rosenqvist, 1977; Rosenqvist and others, 1980).

It is important to distinguish between acidification and elemental leaching when considering the role of natural acid formation, also. Carbonic acid is a major leaching agent in some forest soils (McColl and Cole, 1968; Nye and Greenland, 1960), yet it does not produce low pH (i.e., < 5.0) solutions under normal conditions (Johnson and others, 1975, 1977; McColl and Cole, 1968). Organic acids may contribute substantially to elemental leaching in forest soils undergoing podzolization (Johnson and others, 1977), and they can produce low pH (i.e., < 5.0) in unpolluted natural waters as well (Rosenqvist, 1977; Johnson and others, 1977; Johnson, 1981). Also, since leaching is only one of several processes that affect soil acidity (other major factors being humus buildup, plant cation uptake, and mineral weathering; Ulrich, 1980), the relative contribution of atmospheric acid deposition to elemental leaching may be quite different from the relative contribution of atmospheric deposition to soil acidification.

CONCLUSIONS

Much has been learned about processes of elemental transport from terrestrial to aquatic ecosystems and the effects of acid precipitation upon them. It is possible to infer the relative roles of internal vs atmospheric acid inputs in soil leaching by means of lysimeter studies. Significant progress has been made in assessing the relative contributions of internal vs atmospheric inputs to the overall process of soil acidification, using some recently developed conceptual frameworks for exactly that purpose. Similarly, new theories for surface water acidification by acid precipitation have been developed using concepts originally applied to soil leaching processes.

Further research is needed on the effects of acid precipitation on terrestrial-aquatic transport in several areas, however. As always, a better understanding of mechanisms is needed, especially with regard to acid production and transport through soils. Perhaps the most pressing need is to provide decision-makers with quantitative information on the effects of acid precipitation. There is currently considerable pressure in the U.S. and Canada to define sensitive areas so that the problem can be quantified on a geographical basis, at least. Sensitivity" needs to be clearly defined in relation to the target ecosystem, the ecological effects being discussed, and the magnitudes of effects considered detrimental. The problem of quantifying acid rain effects may prove quite large in view of the variability in natural ecosystems and in atmospheric inputs, but quantification is an essential element to the ultimate resolution of the question of acid rain effects.

ACKNOWLEDGMENTS

Research sponsored jointly by the National Science Foundation's Ecosystem Studies Program (DEB-7824395) grant to the University of Washington, and the Office of Health and Environmental Research, U.S. Department of Energy, under contract W-7405-eng-26 with Union Carbide Corporation. Publication No. 1779, Environmental Sciences Division, ORNL.

REFERENCES

Abrahamsen, G. (1980). Acid precipitation, plant nutrients, and forest growth. In D. Drabløs and A. Tollan (Eds.), Ecological Impact of Acid Precipitation. Johs. Grefslie Trykkeri A/S, Mysen, Norway. pp. 58-63.

Abrahamsen, G., and A. O. Stuanes. (1980). Effects of simulated rain on the effluent from lysimeters with acid, shallow soil rich in organic matter. In D. Drabløs and A. Tollan (Eds.), Ecological Impact of Acid Precipitation. Johs.Grefslie Tyrkkeri, A/S, Mysen, Norway. pp. 152-153.

Abrahamsen, G., and A. O. Stuanes. Acid precipitation, sulfur cycle, and lake acidification. In preparation.

Abrahamsen, G., A. O. Stuanes, and K. Bjor. (1979). Interaction between simulated rain and barren rock surface. Water, Air, Soil Pollut., 11, 191-200.

Andersson, F., T. Fagerstrom, and S. I. Nilsson. (1980). Forest ecosystem responses to acid deposition - hydrogen ion budget and nitrogen/tree growth model approaches. In T. C. Hutchinson and M. Havas (Eds.), Effects of Acid Precipitation on Terrestrial Ecosystems. Plenum Press, New York. pp. 319-334.

Biggar, J. W., D. R. Nielson, and K. K. Tanji. (1966). Comparison of computed and experimentally measured ion concentrations in soil column effluents. Trans., ASAE. pp. 784-787.

Christopherson, N., and R. F. Wright. (1980). Sulfate at Birkenes, a small forested catchment at Storgma, Southern Norway. In D. Drabløs and A. Tollan (Eds.), Ecological Impact of Acid Precipitation. Johs. Grefslie Trykkeri A/S, Mysen, Norway. pp. 286-287.

Cole, D. W., W. J. B. Crane, and C. C. Grier. (1975). The effect of forest management practices on water chemistry in a second-growth Douglas-fir ecosystem. In B. Bernier and C. H. Winget (Eds.), Forest Soils and Land Management. Les Presses de l' Universite Laval, Quebec. pp. 195-208.

Cole, D. W., and D. W. Johnson. (1977). Atmospheric sulfate additions and cation leaching in a Douglas-fir ecosystem. Water Resour. Res., 13, 313-317.

Cronan, C. S., W. A. Reiners, R. L. Reynolds, and G. E. Lang. (1978). Forest floor leaching: Contributions from mineral, organic, and carbonic acids in New Hampshire subalpine forests. Science, 200, 309-311.

544

Cronan, C. S., and C. L. Schofield. (1979). Aluminum leaching response to acid precipitation: Effects on high elevation watersheds in the Northeast. Science, 204, 304-306.

Dutt, G. F., and K. K. Tanji. (1962). Predicting concentrations of solutes in water percolated through a column of soil. J. Geophys. Res., 67, 3437-3439.

Farrel, E. P., I. Nilsson, C. O. Tamm, and G. Wiklander. (1980). Effects of artifical acidification with sulfuric acid on soil chemistry in a Scots pine forest. In D. Drabløs and A. Tollan (Eds.), Ecological Impacts of Acid Precipitation, Johs. Grefslie Trykkeri A/S, Mysen, Norway. pp. 186-187.

Galloway, J. N., C. L. Schofield, G. R. Hendry, E. A. Altwicker, and D. E. Troutman. (1980). An analysis of lake acidification using annual budgets. In D. Drabløs and A. Tollan (Eds.), Ecological Impacts of Acid Precipitation. Johs. Grefslie Trykkeri A/S, Mysen, Norway. pp. 254-255.

Gorham, E. (1958). Free acids in British soils. Nature, 181, 106.

Hendry, G. R., J. N. Galloway, S. A. Norton, C. L. Schofield, D. A. Burns, and P. W. Schaffer. (1980). Sensitivity of the eastern United States to acid precipitation impacts on surface waters. In D. Drabløs and A. Tollan (Eds.), Ecological Impact of Acid Precipitation. Johs. Grefslie Trykkeri A/S, Mysen, Norway. pp. 216-217.

Johnson, D. W. (1980). Site susceptibility to leaching by H_2SO_4 in acid rainfall. In T. C. Hutchinson and M. Havas (Eds.), Effects of Acid Precipitation on Terrestrial Ecosystem. Plenum Press, New York. pp. 525-536.

Johnson, D. W. (1981). The natural acidity of some unpolluted waters in southeastern Alaska and potential impacts of acid rain. Water, Air, Soil Pollut. (in press).

Johnson, D. W., and D. W. Cole. (1977). Sulfate mobility in an outwash soil in western Washington. Water, Air, Soil Pollut., 7, 489-495.

Johnson, D. W., and D. W. Cole. (1980). Anion mobility in soils: Relevance to nutrient transport from forest ecosystems. Environ. Internat., 3, 79-80.

Johnson, D. W., D. W. Cole, and S. P. Gessel. (1975). Processes of nutrient transfer in a tropical rain forest. Biotropica, 7, 208-215.

Johnson, D. W., D. W. Cole, and S. P. Gessel. (1979). Acid precipitation and soil sulfate adsorption in a tropical and in a temperate forest soil. Biotropica, 11, 38-42.

Johnson, D. W., D. W. Cole, S. P. Gessel, M. J. Singer, and R. V. Minden. (1977). Carbonic acid leaching in a tropical, temperate, subalpine and northern forest soil. Arct. Alp. Res., 9, 329-343.

Johnson, D. W., and G. S. Henderson. (1979). Sulfate adsorption and sulfur fractions in a highly-weathered soil under a mixed deciduous forest. Soil Sci., 128, 34-40.

Johnson, D. W., J. W. Hornbeck, J. M. Kelly, W. T. Swank, and D. E. Todd. (1980). Regional patterns of soil sulfate accumulation: Relevance to ecosystem sulfur budgets. In D. S. Shriner, C. R. Richmond and S. E. Lindberg (Eds.), Atmospheric Sulfur Deposition: Environmental Impact and Health Effects. Ann Arbor Science, Ann Arbor, Michigan. pp. 507-520.

Khanna, P. K., and F. Beese. (1978). The behavior of sulfate salt input in podzolic brown earth. Soil Sci., 125, 16-22.

Klopatek, J. M., W. F. Harris, and R. J. Olson. (1980). A regional ecological assessment approach to atmospheric deposition: Effects on soil systems. In D. S. Shriner, C. R. Richmond, and S. E. Lindberg (Eds.), Atmospheric Sulfur Deposition: Environmental Impact and Health Effects. Ann Arbor Science, Ann Arbor, Michigan. pp. 559-554.

McColl, J. G., and D. W. Cole. (1968). A mechanism of cation transport in a forest soil. Northwest Sci., 42, 134-140.

McFee, W. D. (1980). Sensitivity of soils to long-term acid precipitation. In D. S. Shriner, C. R. Richmond, and S. E. Lindberg (Eds.), Atmospheric Sulfur Deposition: Environmental Impact and Health Effects. Ann Arbor Science, Ann Arbor, Michigan. pp. 495-506.

Norton, S. A. (1980). Geologic factors controlling the sensitivity of aquatic ecosystems to acidic precipitation. In D. S. Shriner, C. R. Richmond, and S. E. Lindberg (Eds.), Atmospheric Sulfur Deposition: Environmental Impact and Health Effects. Ann Arbor Science, Ann Arbor, Michigan. pp. 521–532.

Nye, P. H., and D. J. Greenland. (1960). The soil under shifting cultivation. Commonwealth Bureau of Soils Tech. Comm. No. 51, Commonwealth Agricultural Bureaux, Farnham Royal, Bucks.

Reuss, J. O. (1977). Chemical and biological relationships relevant to the effect of acid rainfall on the soil-plant system. Water, Air, Soil Pollut., 7, 461–478.

Reuss, J. O. (1978). Simulation of nutrient loss from soils due to rainfall acidity. Research Report EPA-600/3-78-053, U.S. Environmental Protection Agency, Corvallis, Oregon.

Rosenqvist, I. Th. (1977). Sur Jord/Surt Vann., Ingeniorforlaget, Oslo, Norway. 123 p.

Rosenzvist, I. Th., P. Jørgensen, and H. Rueslatten. (1980). The importance of natural H^+ production for acidity in soil and water. In D. Drabløs and A. Tollan (Eds.), Ecological Impact of Acid Precipitation. Johs. Grefslie Trykkeri A/S, Mysen, Norway. pp. 240–241.

Schofield, C. L. (1980). Processes limiting fish populations in acidified lakes. In D. S. Shriner, C. R. Richmond, and S. E. Lindberg (Eds.), Atmospheric Sulfur Deposition: Environmental Impact and Health Effects. Ann Arbor Science, Ann Arbor, Michigan. pp. 345–356.

Seip, H. M. (1980). Acidification of freshwater - Sources and Mechanisms. In D. Drabløs and A. Tollan (Eds.). Ecological Impacts of Acid Precipitation, Johs. Grefslie Trykkeri A/S, Mysen, Norway. pp. 358–366.

Singh, B. R. (1980). Sulfate sorption by acid forest soils. In D. Drabløs and A. Tollan (Eds.), Ecological Impact of Acid Precipitation. Johs. Grefslie Trykkeri A/S, Mysen, Norway.

Singh, B. R., G. Abrahamsen, and A. Stuanes. (1980). Effect of simulated acid rain on sulfate movement in acid forest soils. Soil Sci. Soc. Am. J., 44, 75–80.

Sollins, P., C. C. Grier, F. M. McCorison, K. Cromack, R. Fogel, and R. L. Fredriksen. (1980). The internal element cycles of an old-growth Douglas-fir ecosystem in western Oregon. Ecol. Mon. 50, 261–285.

Ulrich, B. (1980). Production and consumption of hydrogen ions in the ecosphere. In T. C. Hutchinson and M. Havas (Eds.), Effects of Acid Precipitation on Terrestrial Ecosystems. Plenum Press, New York. pp. 255–282.

Wiklander, L. (1980). The sensitivity of soils to acid precipitation. In T. C. Hutchinson and M. Havas (Eds.), Effects of Acid Precipitation on Terrestrial Ecosystems. Plenum Press, New York. pp. 553–568.

ENERGY AND INFORMATION, THE TWO LONE COMMON DENOMINATORS OF ALL SYSTEMS: STEPS TOWARD THE "INTEGRATIVE APPROACH" OF HUMAN ECOLOGY

H. Knötig

International Organization for Human Ecology, Karlsplatz 13, A-1040 Vienna, Austria

ABSTRACT

Mankind has gained immense knowledge about an enormous number of problems through-out the whole universe by consistent progress of specialisation in the realm of all natural and social sciences. The reverse side of this bright medal is the inability of solving a number of urgent problems; this inability is related to the decisive feature all these problems have in common, i.e. the diversity of their roots which are subject of very different scientific disciplines. Specific obstacles prevent an easy way of integration; considerations about ways of re-integrating scientific re-sults from different fields show opportunities on the one hand and the fact that all these ways are long ones on the other. "Energy" and "information" are seen as helpful concepts because these two are common to all subjects of natural and social scientific work. Therefore they are part of the basic conceptual framework of human ecology which aims at the requested re-integration under the aspect "interrelation-ships between human beings and the outside world surrounding them". The way from a multidisciplinary via an interdisciplinary to an integrative approach is outlined in short. The considerations are related to the problems of "energy use management" since these are understood to be typical for the category of "problems unsolvable without an integrative approach".

KEYWORDS

Human ecology; integrative approach; material-energetic aspect; informatory aspect; ecological potency; ecological valency.

A very interesting sentence in a circular (dated 1980-03-16) of Dr. KLOPATEK, Coor-dinator of Topic "C" of this Conference reads:

"Let me just remind you that
energy is the lone common
denominator to all systems

and that it requires assistance from all disci-plines - engineers, ecologists, architects, socio-logists, etc. to solve the local as well as inter-national problems related to energy needs consump-tions and shortages."

It is the answer to the report given in the preceding sentence of this circular: "The overwhelming response that we have received from many Europeans, as well as non-Europeans, is that the conference is too broad, too diverse and therefore not able to address the critical problems in their field or specialty."

By formulating this sentence Dr. KLOPATEK probably has touched upon t h e critical problem of the contemporary attempts of science to contribute to solving the most urgent problems of developed (maybe "over-developed") and developing nations. Mankind today knows so much and therefore can do so much - but this knowledge is distributed to a great variety of scientists, technologists, medical doctors, ... with a decisive lack of cooperation or even coordination.

This lack of coordination is not the result of any ill will - it is simply the result of the lack of sufficient opportunity for cooperation. One must not forget that all scientific progress was gained by progress in specialisation of scientific work. By necessity this will be the case also in future; there is no other way of gaining additional knowledge in the different sciences but by progress in specialisation.

However, the progress in the different specialties means progress in consistently developing the different bodies of concepts, ideas and terms. And more and more they differ one from another to such an extent that they have nearly nothing in common. This is shocking but we can't help it - except we would be ready to stop further specialisation and thus further progress in science.

Therefore, when trying to overcome the difficulties arising from progress in specialisation, we have to look for another way. To speak more precisely, we have to look for means of re-integrating scientific findings. Human Ecology i s such an attempt at re-integration (at least in the opinion of I.O.H.E., the International Organization for Human Ecology). This attempt - understood as o n e among others - is based on the interrelationships between human beings and the outside world surrounding them (formulated in accordance with the definition of "ecology" in HAECKEL, 1868).

Let's now look whether it seems to be meaningful to cope with problems of "energy use management" under a human ecological perspective. Human ecological efforts pay primarily in cases where it is important to consider natural and social scientific problems together in an integrating manner. There is probably no doubt that "energy use management" relates to both scientific realms - to the natural scientific one as well as to the social scientific; and therefore it seems appropriate to treat such problems from a human ecological point of view. In order to see this more clearly, it may be helpful to formulate some important problems of "energy use management" before presenting some tools provided by human ecology.

Probably one should start with the question "what is the reason why human beings wish to command energy?" (assuming for the moment we have adequate understanding with regard to what is meant by "energy"). The next question would be "how could the demand be met?" and "should this demand be met in any case?" This latter one is already related to a certain degree to the second major group of questions, the first one relating to the "acceptance" of the different kinds of energy production and energy supply.

In general this second group of questions belongs to the concept "impact": "what is the impact of the different kinds of energy production and supply on nature (recreational sites, ...) and residential habitats?", "what is the impact of certain kinds and amounts of energy supply on employment and on the availability of the different items of the 'civilized' world?"

Of course, one could formulate a lot of other such questions but these surely are typical ones when dealing with energy use management. In every one of the problems implied by these questions the amount of energy in question and the specific form of availability of energy are as essential as the nature of human wishes and needs, as the specific ways of social cooperation, ...

No doubt, these terms belong to different disciplines widely ranging from physics to psychology and sociology. This is not the place to deal with any one of the above problems in exact detail but in spite of this it is clear that all the different parts of such a problem are of equal importance - may they belong to physics, to technology, to physiology, to psychology, to economics, to sociology or to any other scientific field.

Of course, it would be a good idea to start with treating every one of the parts implied above within the respective disciplines and then to coordinate the different findings. In reality, this "multidisciplinary approach" is the usual one. However, this proves to be insufficient since the "coordination" - by necessity - remains always an unfulfilled wish. "By necessity" since every scientific field in order to succeed in gaining knowledge characteristic for its field and in solving the problems specific to its realm has built up its own framework of ideas and concepts based on the way of thinking most appropriate and best adapted to its work.

The most essential feature of any of these bodies of concepts is the concentration on the questions understood to be the most important ones for the respective science. By way of complementary formulating this means to exclude all problems which do not directly belong to these "important questions". No doubt, this is the only way to deal successfully with special problems - and therefore we initially stated that there will be no further progress in gaining knowledge except by further scientific specialisation. In our days information theory of the human nervous system makes quite clear why there is no other way of using the human brain for the above mentioned purposes: its information processing capacity, of course, is limited and thus the more complex the specific way of processing turns out to be (i.e. the more difficult the solving of a problem becomes) the more limited the range of inputs has to be. KEIDEL (1963) has published specific figures about the capacity of human information processing - it is awfully limited (as everyone of us surely experiences himself/herself every day).

Anyway, the above required re-integration of knowledge (for specific problems) is not to be achieved by attempts of "coordination" in a mere multidisciplinary way because a body or framework of concepts of which the conceptual bodies of the different disciplines in question are part of, would be an indispensible prerequisite. Just this is lacking - by necessity as we understood above.

Attempts which are sometimes made on the next highest level regarding the coordination of the findings of the different disciplines - in order to give a coherent answer to politicians and other persons asking for it - have a high reputation. It is the "interdisciplinary approach" which implies that the results of the different natural and social sciences are not merely delivered one by one but that an attempt is made by scientists of these different fields to dialogue with one another.

However, in this case everyone speaks in his own scientific language and t r i e s to understand the languages of his various colleagues. To be honest, such a trial is sometimes really fruitful but at any rate it is a very rare event - therefore the high reputation. Nevertheless, even this is not sufficient for a real re-integration aimed at coherent answers as mentioned above. To make meaningful attempts to this end, it is necessary to use an "integrative approach" (cf. KNÖTIG, 1979).

Of course, this "integrative approach" can be no miracle that abolishes all difficulties we have experienced as yet. It is no magic wand that may produce such a body of concepts that embraces all the conceptual bodies of the specific scientific disciplines as mentioned above.

On the other hand, there is the indispensible need for a "guiding model" when attempting to use in a simultaneous and integrating manner the results of the different disciplines. This "guiding model", however has probably to be formulated in terms of a terminology newly designed for this purpose. At any rate, this new "language" needs at least a few unequivocal contacts to the "languages" of the different special (natural and social) sciences. Starting from there, attempts might be made to set up a terminology for such "guiding models" - always having in mind that this model can never have the same degree of accuracy as one in the different special disciplines and that there is always a danger that this model is not as appropriate as would be possible at this time.

With respect to the kind of such a model in the case of human ecology the important role of the characteristics of human beings has to be taken into account; since human beings are living beings the concept "system" is decisive for the structure of the desired "guiding model". (By the way, it is the same reason why contemporary biological - or general - ecology to a large extent is formulated in terms of the "ecosystems theory".)

Regarding the "unequivocal contacts" with the other scientific languages as requested above, the following three can at present be enumerated:
- series of natural numbers and logical thinking
- energy
- information

Of course, it is possible that may be detected more concepts that are common to all natural and social sciences; of course, there are different concepts not common to a l l sciences but common to a number of them.

When discussing these three "common denominators", common to all subjects of all disciplines or the scientific statements upon them, one is primarily convinced that the first of them does not show specific problems - all that is included here seems to be a self-understood prerequisite for any effort that truly applies to scientific work. Indeed, reflections about these items are a considerable part of the work of metamathematics and epistemology - but this should not be our problem here.

Our problem, on the contrary, should be to consider the two other "common denominators" and the conclusions one may draw in trying to use an "integrative approach" in order to treat problems of energy use management in a human ecological manner. The title of this contribution states that they are the lone two, common to all systems; we will become aware that they are common to all subjects of scientific work. How does this become apparent in the course of human ecological considerations?

Human ecological work should rest, at least in the opinion of I.O.H.E., on two pillars:
- integrative approach
- case studies
The latter ones should provide the opportunity to consistently examine the integrative approach's concrete manifestations as to their usefulness, appropriateness, and shortcomings.

In the course of such human ecological work that is based on the "integrative approach", the two above-mentioned "common denominators" mostly are introduced by the pair of concepts
- "material-energetic aspect",
- "informatory aspect".

The first one, the "material-energetic aspect", refers to the fact that nothing is recognized by any science as "existing" if there is not any amount (> 0) of matter and/or energy present. Thus, one may state there is no subject for any science if there is no amount (> 0) of matter and/or energy. In terms of a complementary formulation:

any scientific subject (entity, process, ...) shows a material-energetic aspect.

These quantities of matter and/or energy, characterizing the material-energetic aspect, are distributed in space and time and thus constitute a certain space-time structure; one always can correlate an "information content" (measured by "bits") with this structure - thus characterizing the "informatory aspect". One should note

the use of the word "correlate" in the above sentence; there is n o statement like
this: "the respective structure may be fully described by a certain information
content". "Information" has a broader meaning than "information content"; measurable
and countable, at any rate, is only the latter one. As a result, however, one can
formulate:

   any scientific subject shows an informatory aspect.

Once more referring to the two "common denominators" stated above, the correlation
between one of them, namely "energy", and the "material-energetic aspect" should be
discussed a little more in detail. "Energy" may be considered as the truly scientific nucleus of the "material-energetic aspect". Present-day physicists understand
"matter" just as a specific form of appearance of "energy" (like light, vibration,
...). Moreover, one can see the close connection even between those items which
are implied in the everyday terms "matter" and "energy" in the most common events,
e.g. nutrition: food uptake means incorporating specific matter but is mostly seen
under the aspect of how much energy ("Jouls") may be released in the body and used
for running all the physiological processes of a human body and - the surplus - for
being stored again in the form of specific kinds of matter; or heating: one buys
oil, coal, wood, ... per kilogram but calculates how much thermic energy could be
gained by oxidizing the carbon contained in the fuel. The ultimate basis, of course,
would be the Einstein equation $E = m \cdot c^2$ which gives the exact equivalents of amount
of energy and amounts of mass, the essential characteristic of matter. (A hint for
scientists especially interested in these problems: the paper PANZHAUSER - KNÖTIG,
1980 shows that even fairy tales, mathematical sentences, ... only exist with the
help of a material energetic aspect.)

Starting from these considerations one may, in building a model of the interrelationships between human beings and the outside world surrounding them, understand
the human being(s), the respective outside world (= "HAECKELian environment" in human ecological language), and the "pattern of environmental relationships" existing
between them - f o r t h e p u r p o s e of the investigation - as systems
which are matter/energy processing and information processing. A contribution on
"stress" (KNÖTIG - MAYER, 1981) just in preparation for a volume devoted to this
problem field shows that on this basis the different circumstances - belonging to
different scientific disciplines - may be described in one, consistent language to
such a degree of precision that drawing of meaningful conclusions is possible.

To state it more in detail, in the terminology of I.O.H.E. (cf. KNÖTIG, 1976a)
the one human being (in case of an autecological or individual ecological investigation) as well as those human beings, called "ensemble" (in case of a demecological or social ecological investigation) who is/are chosen for the purpose of the
respective human ecological consideration is designated as the "System S 2" (or
simply "S 2") and the remaining part of the world (i.e. the totality of existing
entities minus "S 2") is called "System S 3" (or "S 3"). The interrelationships between an "S 2" and an "S 3" are called the "environmental relationships" (of this
specific "S 2"). The totality of these "environmental relationships" ("e.r."s), the
"pattern of environmental relationships", is called "System S 1" (or "S 1").

Of course, every "e.r." distinctively depends on both the relevant characteristics
of "S 2" and of "S 3" (cf. KNÖTIG, 1980):

$$\text{e.r.} = F([ch_{S2}]_i, [ch_{S3}]_j)$$

$[ch_{S2}]_i$ ...... relevant characteristic of "S 2"

$[ch_{S3}]_j$ ...... relevant characteristic of "S 3"

$i \quad = \quad 1, ..., m$

$j \quad = \quad 1, ..., n$

(In case "S 2" as well as "S 3", i.e. the whole world [= "S 2" + "S 3"] would be
completely known in all details all the different "e.r."s could be just calculated;

since this is not only far from reality but impossible in principle, it seems reasonable to investigate the individual "e.r."s according to the above function [more precisely: functional]).

However, the "e.r."s not only depend on "S 2" and "S 3" but they influence these two systems as well. This fact is important in a twofold way:

- the influence on "S 2", i.e. the influence on a human being or a number of human beings is important to him/her or them,

- the influence on "S 3" is an influence partly on other human beings (which is of direct importance to them) and partly on the other parts of the "world" which constitute the "S 3" of these other human beings as well (and therefore is of indirect importance to them).

The details of the sources of these influences of the different "e.r."s as well as the details of the impacts on the different human beings on the one hand and on the different other parts of the world on the other are subjects of a great number of individual natural and social sciences.

The task of human ecology is to put all the different statements of these sciences into one single statement referring to the decisive problems. - As stated above, this task is to be fulfilled by necessity with less precision than that typical for the statements of the individual sciences and even this reduced goal is often only a hope for the future as the desired result of hard work seriously initiated just in the last one or two decades.

However, there are already initial outlines how to cope with this task. One of the most important is the stating of "energy" and "information" as common denominators to the subjects of all natural and social sciences - as explained above. Within the framework of human ecological concepts another pair of concepts, namely "ecological potency" and "ecological valency" (cf. KNÖTIG - MAYER, 1981) is probably very helpful in many cases as well.

"Ecological potency" designates the meaning of the totality of an "S 2"'s characteristics for the coping of this "S 2" with its "S 3". - "Ecological valency" designates the meaning of the totality of the "environmental factors", i.e. the characteristics of an "S 3" for the "S 2" which determines this "S 3".

In general one could say that the basic consideration of human ecology (and of the aut- and demecology of any other species) is the ratio "ecological potency" :" ecological valency".

Of course, when it comes to the distinct "case study", it is not as easy to use this pair of concepts in a meaningful way as it may look at first glance. At any rate, one has to consider in what way any change in "S 3" influences "S 2" and vice versa. However, this "vice versa" does not mean any symmetry: the way from "S 3" to "S 2" is always dominated to a high degree by the organizational structure of the respective "S 2", i.e. a human individual or a number of them. Of course, even in the latter case the organizational characteristics of human individuals are of decisive importance - the structure of the ensemble is to be taken into account i n   a d-
d i t i o n.   In order to characterize the way from "S 3" to an "S 2" that consists of a single human being, it is best to start with UEXKÜLL (e.g. UEXKÜLL, 1928; 1957 [1934]). He states that every living being (and thus every human being) c r e-
a t e s its "Umwelt" (environment; exactly: "UEXKÜLLian environment") which means the internal presentation (or model, maybe "map") of the outside world surrounding it ("HAECKELian environment"). Since this "UEXKÜLLian environment" is the basis for feeling, experiencing, acting, ... of the respective human individual it is probably the most important part within the whole "General Inter-Action Scheme" (G.I.A.S.) of human ecology (cf. KNÖTIG - MAYER, 1981). It is based on
- the "HAECKELian environment" ("S 3") of this human being
                as well as on
- the kind of transformation that "creates", produces the "UEXKÜLLian envi-

ronment" out of parts of the "HAECKELian environment".

The specific kind of this creational process ("UEXKÜLL transformation") is determined - in the terms of the Stockholm Stress Research School (cf. KAGAN - LEVI,
1974) - by the "psycho-biological programme" of the human being. UEXKÜLL himself
explained the determination of this process - both for human beings and for other
living beings as well - by the "Bauplan", i.e. an old biological concept that for
our purposes should be replaced by the concept introduced above.

When considering these problems, one must never forget that the "afferent" way (in
the language of physiology) from "S 3" to "S 2" represents only one half of the UEX-
KÜLL transformation. The way in the opposite direction, the "efferent" one, is also
determined by the above-mentioned "psycho-biological programme" (or the "Bauplan").

Nevertheless, there is a decisive difference between the two ways, expressed above
by denying a symmetry, since the kind of matter/energy and information processing
on the way "S 3" → "S 2" → "S 3" is always shaped by the "psycho-biological programme". (In case of a demecological investigation, i.e. taking an ensemble of human
beings as "S 2", the processing is shaped by the different "psycho-biological programmes" of the individual members of the ensemble; they determine the realization
of the interrelationships between any of these individuals and the "S 3" of the
whole ensemble in a direct way on the one hand and they determine the realization of
the interrelationships between the whole "S 2" [ensemble] and its "S 3" in an indirect way by shaping the interrelationships among the individuals, and thus the
structure of the ensemble, on the other. - The way "S 2" → "S 3" → "S 2" is not as
"easy" to be characterized. There are no specific features of the respective processing - except those governing the "efferences" of the "S 2" and its "afferences";
there is, first of all, no general regularity with respect to the kind of parts of
the "S 3" which are predominantly addressed by "S 2"'s actions ("efferences"). The
next stratum of uncertainties refers to the vast multitude of different processes
initiated or influenced by those actions. It is the general task of (natural and social) sciences to clarify to the highest possible degree the concrete structure of
these processes. (The expected results are to be used under the headings "technology", "politics", "economics", ... in order to cope with the "world" - or the parts
of it of interest at this time).

A last shock is experienced when considering this high complexity of the nature of
human environmental relationships (i.e. the life of human beings in the world) and
becoming aware that all the above explanations do not refer to human beings exclusively; on the contrary, they basically refer to the ecology of every species of
living beings. The characteristic that separates human beings from all other living
beings (the "specificum humanum") is the faculty of "mapping the primary mapping
process" ("meta-mapping" in accordance with an idea of BATESON; cf. RÖSSLER, 1977).
Therefore a human being can anticipate different "models" ("UEXKÜLLian environments") not realized as yet - and can choose one of them in order to try to realize
it.

When considering these problems of the "specificum humanum" it is easy to see that
"information" in the case of h u m a n ecology is of special importance. The great
problem of every individual human being, of the different ensembles (families, peer
groups, communities, staffs, classes, tribes, nations, ...) and of mankind as a
whole is the c h o i c e. This choice is the earmark of Homo sapiens. The decisive question is the guidelines for the choice. There one is confronted with the
well-known problems of values, wishes, norms, ... Unfortunately, this is an unli-
mited field of problems; fortunately, even here one can take the first steps with
the help of the human ecological tools presented above (and some others not presented here).

Of course, when Human Ecology is understood as a science (or as a collection of
scientific efforts) it can n o t provide any norms, values or anything like that.
The utmost it can do is to give hints what the results probably will be when apply-

ing the different norms or values. Maybe it is justified to give one more hint: it seems meaningful to further the "Weiterschreiten in der Mensch-Werdung" (approximately: "progress in the realization of human capacities"; cf. KNÖTIG, 1976b) - and to lift all the barriers on the way.

As a result one may state that there is surely no short cut to reasonably deal with problems like "energy use management"; human ecology as well as any other (natural or social) science has no magic faculties. On the contrary, to elaborate meaningful results (instead of decision-making on the basis of "feelings") will be a long and arduous task. Nevertheless, there is a possibility to realize this task; it will be a long way to go yet but there is hope for fulfillment.

REFERENCES

HAECKEL, E. (1868). Generelle Morphologie der Organismen, vol.2: Allgemeine Entwickelungsgeschichte der Organismen. Reimer, Berlin.

KAGAN, A.R. - LEVI, L. (1974). Health and Environment - Psychosocial Stimuli: A Review. Soc.Sci.and Med., 8, 225 ... 241.

KEIDEL, W.-D. (1963). Beispiele und Probleme einer kybernetischen Physiologie des ZNS und der Sinne. Bericht über den 23. Kongreß der Dtsch.Ges.f.Psych., Sept.1963 in Würzburg (Vlg.f.Psych./Dr.C.J.Hogrefe, Göttingen), 103 ... 123.

KNÖTIG, H. (1976a). Terminology, Session Report. Coll.intern., 1, 119 ... 144.

KNÖTIG, H. (1976b). Umweltgestaltung und Umweltbeeinflussung durch Juristen: Unreflektierte und wissenschaftlich erarbeitete Leitvorstellungen. KNÖTIG (ed.) 1976, 835 ... 849.

KNÖTIG, H. (1979). From a Multidisciplinary via an Interdisciplinary to an Integrative Approach. Coll.intern., 4, 231 ... 236.

KNÖTIG, H. (1980). Development and Environmental Protection: the Integrative Approach of Human Ecology. Invited Lecture at the Regional Symposium on Prospects of Development and Environmental Protection in the Arab Gulf Countries. Preprint I.O.H.E., Vienna.

KNÖTIG, H. - MAYER, H. (1981). Menschliche Umweltbeziehungen in der Sichtweise von Streßforschung und Humanökologie. In preparation.

KNÖTIG, H. (ed.) (1976). Proc.of the Intern.Meet.on Hum.Ecol., 1975 in Vienna. Georgi Publ.Comp., St. Saphorin.

PANZHAUSER, E. - KNÖTIG, H. (1980). Complexity of Models: Can the Development of Biophysics Serve as an Example for Establishing Human Ecology? Coll.intern., 5, 156 ... 169.

RÖSSLER, O. (1977). Deductive Biology and General Interaction Scheme. Coll.intern., 2, 16 ... 24.

SCHILLER, C.H. (ed.) (1957). Instinctive Behavior. The Development of a Modern Concept. Methuen & Co. Ltd., London.

UEXKÜLL, J.v. (1928). Theoretische Biologie. 2nd ed., Springer, Berlin.

UEXKÜLL, J.v. (1957 [1934]). A Stroll through the Worlds of Animals and Men. A Picture Book of Invisible Worlds. SCHILLER (ed.) 1957, 5 ... 80.

EFFECT OF ACID PRECIPITATION ON SOIL

E. Matzner and B. Ulrich

*Institut für Bodenkunde und Waldernährung, Büsgenweg 2,
D-3400 Göttingen, Federal Republic of Germany*

ABSTRACT

The effect of acid precipitation on soil depends on the rate of
hydrogen load (sum of input and internal production) and the rate of
hydrogen consumption within the soil. Considering the buffer systems
acting in the soil and their buffer rates and capacities, it is
evident that under the present load of acidity soil acidification
occures and the aluminum buffer range will be reached in soils free
of $CaCO_3$. The effect of acid precipitation is not limited to buffering reactions but influences also decomposer and root activity.
The changes in decomposition generate ecosystem internal H^+-produc-
tion which accelerate the process of soil acidification. In analogy
to the mobilization of aluminum as a consequence of soil acidification, heavy metals like Mn, Fe, Co, Cu, Zn, and Cd are mobilized
in the upper soil and reach high concentrations in soil solution to
a level, where toxic effects on soil microorganisms and vegetation
may occur.

KEYWORDS

Acid rain, buffer systems of soils, ecosystem internal H^+-production,
soil acidification, heavy metals.

INTRODUCTION

In Central Europe, acid precipitation became a widespread phenomenon
around 1960. As a consequence of the increasing combustion of fossil
fuels the connected sulfur emissions have lead to a heavy load of

acidity on limnic and terrestrial ecosystems. Having high filtering
action upon air pollutants ,i.e.SO_2,forest ecosystems and forest
soils in particular are considerably affected by acid rain.

During the evaluation of an IBP study of forest ecosystems in the
Solling region the effect of acid precipitation on the soil was
measured since 1966 under a beech and a spruce stand. (For site
description see Ellenberg, 1971) The annual rates of total atmospheric hydrogen ion input to the soils investigated are given in
Tab. 1.

TABLE 1   Hydrogen Ion Input to the Soils of the Beech
and the Spruce Stand ($keq.ha^{-1}.a^{-1}$)

	1969	1970	1971	1972	1973	1974	1975	1976
Beech	1.56	1.67	1.41	1.10	1.22	1.28	1.38	1.11
Spruce	2.92	3.20	3.30	3.00	2.94	3.20	3.30	2.50

The hydrogen ion input from wet and dry deposition varies under
beech from 1.1 to 1.67 $keq.ha^{-1}.a^{-1}$ with a mean total deposition
rate of about 1.3 $keq.ha^{-1}.a^{-1}$. Under spruce the mean hydrogen ion
input to the soil is about 3.0 $keq.ha^{-1}.a^{-1}$ and exceeds that under
beech more than twice.
The effect of acid precipitation on soil depends, in principle, upon
the rate of hydrogen load of the soil and on the rate of hydrogen
consumption within the soil. If the rate of hydrogen consumption is
exceeded by the rate of hydrogen load, acidification results. Calculating physico-chemical equilibria (Ulrich et al., 1979), one can
distinguish different buffer ranges of soils.
The first buffer range which should be specified here is the Carbonate buffer range assuming that $CaCO_3$ reacts according to
equation (1).

(1) $CaCO_3 + H_2SO_4 \longrightarrow Ca^{2+} + SO_4^{2-} + CO_2 + H_2O$

The rate of hydrogen consumption in this buffer range is 2 keq
hydrogen per kmole Ca, while the soil solution is buffered at a pH
value between 6.2 and > 8. A lime content of 1 % within the soil
corresponds to a buffer capacity of 300 $keq.ha^{-1}.dm^{-1}$ which is quite
high in comparison to the rates of hydrogen-ion input from acid
precipitation as given above. But such a high buffer rate is only

reached if the lime distribution is homogeneous throughout the soil
profile. In most soils developed on limestone the lime content of
the uppermost horizons is represented by stones while the fine
material is almost free of lime. Under these conditions as well as
in soils having no lime content, the weathering of silicates consumes H^+ ions by liberating alcali and earth alcali ions; the proton
is converted to undissociated silica acid and finally to H_2O. In
this Silicate buffer range pH is kept at or above pH 5 as long as
their are no net acid formations besides the formation of carbonic
acid. The buffer rate of the silicates depends upon the content of
weatherable minerals and amounts to 0.2 to 2.0 $keq.ha^{-1}.a^{-1}$ (Ulrich
et al., 1979). It seems that in most soils of Central Europe the
rate of H^+ ion load exceeds under the influence of acid rain the
buffer rate by silicate weathering. Thus the system switches over to
the next buffer system following at lower pH.
If the pH value decreases below 5.0 liberation of Al from Aluminum-
-silicates occurs. This process is also connected with an equivalent
H^+ consumption, as it is shown in simplified form in equation 2.

(2) $\equiv Si - O - Al(OH) - Si \equiv + H_3O \rightarrow \equiv SiOH + Al(OH)_2^+$

Dependent upon pH and anions present the Al ions are liberated in
the form of various monomeric and polymeric ion species with charges
between +0.5 (at pH close to 5) and +3 (at pH close to 4) per Al
atom (Nair, 1978). Between pH 5 and 4.2 the Al ions are mainly
bound by the clay minerals and do not appear in the soil solution.
Their binding leads to a change in the composition of the exchangeable cations by displacement of Ca and Mg and finally results
in an almost complete loss of the exchangeable Ca and Mg. Because of
the change in composition of the exchangeable cations, this buffer
range is called the Exchange buffer range. The buffer capacity is
equal to the total cation exchange capacity (CEC), which can be
roughly calculated as C % . 7 $(keq.ha^{-1}.dm^{-1})$, where C % is the
clay content in percent of weight, assuming illite as dominating
clay mineral.
If the pH in the soil solution falls below 4.2, which is the case in
most forest soils in Central Europe, aluminum ions will reach
remarkable concentrations within the soil solution and may cause
toxic effects to bacteria and plant roots. Below pH 3 Aluminum will
be rapidly lost from the soil. The buffer range between pH 4.2 and

3.0 is therefore called Aluminum buffer range. It has a buffer capacity of about 100 to 150 keg . C % .ha^{-1}.dm^{-1}. The buffering capacity is very high and would only be exhausted after weathering of the total clay content of the soil. But pH values below 3.0 which can be found in the uppermost horizons of strongly acidified soils indicate that the rate of silicate weathering and clay destruction does not keep up with the rate of proton load. In this case iron oxides come into play and react as buffer. Depending upon the solubility of the iron oxides this can start already at pH 3.8 and may be described by equation (3).

(3) $$Fe(OH)_3 + 3 H^+ \longrightarrow Fe^{3+} + 3 H_2O$$

The processes occuring in this Iron buffer range are known as podzolization. When the aluminum and iron buffer ranges are reached during acidification of the soil, most heavy metals become mobilized and their concentration in soil solution increases. This mobilization takes place in analogy to the liberation of Al from silicates by weathering. In addition to this, the heavy metals bound to soil organic matter are equally mobilized, for heavy metal organic complexes are, in general, less stable under low pH conditions. This is of primary importance for elements like Pb, Cd or Hg, which normally are present in rocks and soils in very low concentrations, but are strongly accumulated in top soils due to deposition of atmospheric pollutants (Heinrich and Mayer, 1977). These heavy metal pollutants are preferably bound to soil organic matter.

Summarizing these considerations it is obvious that, under the present load of acidity, soils which are free of lime will reach the Exchange and Aluminum buffer range throughout the whole soil profile, losing all exchangeable Ca and Mg. As a consequence the calcium and magnesium supply of the stands becomes insufficient while bacteria and fine roots may be damaged by increasing aluminum and heavy metal concentrations in the soil solution.

Changes in chemical soil status due to acid precipitation have been followed during the OBP study within the beech and the spruce stand. Some results are given in Fig. 1 and Fig. 2. The pH (CaCl$_2$) lies between 3 and 4 with only slight changes. The concentration of Al and Fe ions in the equilibrium soil solution (ESS) substantially increased between 1966 and 1973 and has remained at this level until

1979. From this it is evident that the soil shifted in the deeper
layers from the Exchange buffer range to the Aluminum buffer range
between 1966 and 1973.

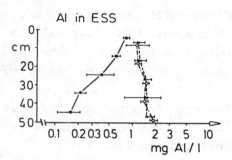
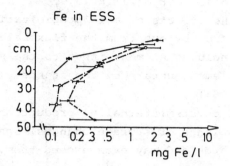

Fig. 1   Mean values and value ranges of concentrations
         of Al and Fe in the equilibrium soil solution
         (ESS) in 1966 (●), 1973 (+) and 1979 (O) under
         beech.

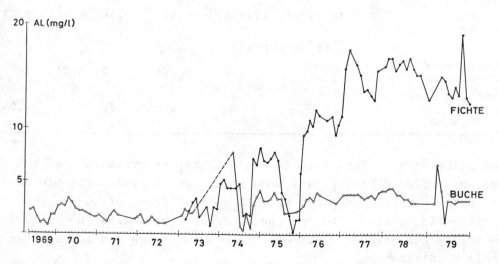

Fig. 2   Mean values of Aluminum concentration within
         the seepage water under beech and spruce.

The Al concentration in seepage water (at a depth of 90 cm) under
beech increased from 1969 to 1979 about two fold, a more drastic
developement is evident for the spruce stand having a higher load of
acidity. Under spruce the Al concentration increased from 1 - 2 mg/l
to 15 - 18 mg/l in 1979.

The effects of acid precipitation on the soils investigated becomes
also obvious from the flux balance of the soil. The flux balance
includes element input to the soil, output with seepage water and
element uptake by the stands. (For detail see Ulrich and Matzner,
1981).

Ecosysteminternal H^+ production (calculated from the cation/anion
balance of storage changes) reaches rates equal to H^+ deposition
rates and may even exceed them, as indicated in Tab. 2.

TABLE 2   Ecosystem internal H^+ Production and total
H^+ Load of the Beech and the Spruce Stand
$(keq.ha^{-1}.a^{-1})$

| | B e e c h | | | | | | | $\bar{x}$ 1969 -1975 |
	1969	1970	1971	1972	1973	1974	1975	
Internal H^+ production	2.85	3.01	0.99	1.99	2.39	1.60	2.32	2.16
total H^+ load of the ecosystem	4.48	5.04	2.28	3.69	4.19	3.18	4.01	3.84
S p r u c e								
Internal H^+ production					3.46	4.12	2.81	3.46
total H^+ load of the ecosystem					6.86	7.78	7.11	7.25

From the data in Tab. 2 it is evident that an evaluation of the
effects of acid precipitation on soils must include possible changes
of soil internal processes as well as the direct effect on the
soil buffer system. The opening of ecosystem internal H^+ sources can
multiply the direct effect of acid precipitation and accelerates
soil acidification.

Probably via Aluminum toxicity acid precipitation leads to changes
in decomposition conditions and to the accumulation of organic
matter at the soil surface (Bååth et al., 1980). This effect was

measured for the Solling stands, comparing the element stores
within the humus layer over an extended period of time (Tab. 3).

TABLE 3 Stores of Carbon and Nitrogen within the Humus
Layer of the Beech and the Spruce Stand ($kg.ha^{-1}$)

B e e c h

	C	s_x	N	s_x
1966 (n=2)	14800	(430)	809	(40)
1973 (n=4)	20900	3120	953	220
1979 (n=3)	22300	1320	1010	60

S p r u c e

	C	s_x	N	s_x
1968 (n=4)	24500	2160	960	83
1973 (n=3)	28200	4840	1130	300
1979 (n=3)	34300	2680	1250	16

Accumulation of C, N and other elements within the humus layer,
which is supported by measurements of humus mineralisation in
lysimeter studies, results in a reduced N-supply of the forest
vegetation. It seems also to be the cause for the soil internal H^+
production by discoupling the biogeochemical cycle between mineralisation and ion uptake (Ulrich et al., 1979).
Table 4 shows annual changes of element storage within the mineral
soil. Losses of Ca, Mg and Al from the soil occur under beech as
well as under spruce while for H^+ and for sulfate the soil reacts
as a sink. Both processes, the source function for Ca, Mg and Al and
the sink function for H^+ and SO_4 reach higher rates under spruce
than under beech according to the rate of H^+ and SO_4 input. The
retention of SO_4 may take place as $AlOHSO_4$ (v.Breemen, 1973; Meiwes,
1979; Fassbender und Matzner, 1977), which can be considered as
another effect of acid precipitation on soils with low pH status.

The flux balance of the Solling forest soils for heavy metals
(Mayer, 1981) shows considerable losses of Mn, Co, and Cd and gains
of Pb, Cr, and Cu. Iron, Ni, and Zn fluxes are almost balanced.
Closer inspection of heavy metal distribution within the ecosystem
shows that all heavy metals investigated, with the exception of Mn,
are accumulated in the humus layer covering the soils while they
are leached from the mineral soil. High concentrations in soil

TABLE 4 Annual Changes of Element Stores within the
Mineral Soil of the Beech and the Spruce Stand
$(kg.ha^{-1}.a^{-1})$.

	1969	1970	1971	1972	1973	1974	1975	$\sum$	$\bar{x}$ 1969- 1975
			B	e	e	c	h		
H	+1.79	+2.27	+1.56	+1.15	+1.18	+0.78	+0.44	+9.17	+ 1.31
Ca	-20.6	-25.6	-1.8	-6.8	-17.4	-7.2	-2.6	-81.6	-11.6
Mg	- 2.7	- 5.1	-0.2	-1.9	-2.7	-1.6	-0.8	-15.0	- 2.14
Al	- 8.4	-23.4	-3.1	-4.4	-7.3	-18.6	-10.5	-75.7	-10.8
S	+26.8	- 4.6	+20.4	+13.8	+9.8	- 9.5	-14.4	+71.2	+10.2
			S	p	r	u	c	e	
H					+2.66	+2.59	+2.49	+7.74	+ 2.58
Ca					-15.1	-16.6	-12.8	-44.5	-14.6
Mg					- 0.9	- 2.0	- 0.6	- 3.5	- 1.17
Al					- 7.9	-32.5	-14.4	-54.8	-18.3
S					+64.2	+37.1	+55.7	+157.	+52.0

solution within upper mineral soil are, therefore, found in the
case of Mn, Co, and Cr. There is little known on possible toxic
effects of these concentrations on soil microorganisms and higher
plant roots. Experimental evidence shows that decomposer microorganisms are most sensible in this respect, and disturbance of
nitrogen mineralisation and nutrient supply of vegetation is to be
expected (Tyler, 1975).

REFERENCES

Abrahamsen, G., K. Bjor, R. Horntvedt, and B. Tveite (1976). Effects
of acid precipitation on coniferous forest. In: Impact of Acid
Precipitation on Forest and Freshwater Ecosystems in Norway.
Ed. F. H. Braekke. pp. 38-63 SNSF project, Aas-NLR, Norway.

Bååth, E., B. Berg, U. Lohm, B. Lundgren, H. Lundkvist, T. Rosswall,
B. Söderström and A. Wiren (1980). Soil organisms and litter
decomposition in a Scots Pine Forest-effects of experimental
acidification. In: Effects of Acid Precipitation on Terrestrical
Ecosystems. Ed. T.C. Hutchinson and M. Havas, Plenum, New York,
pp. 375-381.

Benecke, P. (1979). Der Wasserhaushalt eines Buchen- und eines Fichtenwaldökosystems im Hochsolling (Methoden und Ergebnisse).
Habil. Univ. Göttingen.

Van Breemen, N. (1973). Soil forming processes in acid sulphate soils. In: Acid sulphate soils. Ed. H. Dost, pp. 66-130, Wageningen

Cronan, C. S. (1980). Controls on leaching from coniferous forest floor microcosms. Plant and Soil 56 No.2, 301-322.

Eaton, J. S., G. E. Likens and F. H. Borman (1973). Throughfall and stemflow in a northern Hardwood forest. J. Ecol. 61, 495-508.

Ellenberg, H. (ed.) (1971). Integrated experimental ecology, methods and results of ecosystem research in the German Solling Projekt. Ecological Studies 2.

Fassbender, H. W. und E. Matzner (1977). Zur Bildung von basischen Aluminiumsulfaten im Boden. Mitt.Dtsch.Bodenkundl.Gesellsch. 25, 175-182.

Heinrichs, H. und R. Mayer (1977). Distribution and cycling of mayor and trace elements in two Central European forest ecosystems. J. Environ. Qual. 6, pp. 402-407.

Hutchinson, T.C. and M. Havas (Ed.) (1980). Effects of acid precipitation on terrestrical ecosystems. Plenum. New York, pp

Likens, G. E., F. H. Borman, R.S. Pierce, J. S. Eaton and N. M. Johnson (1977). Biogeochemistry of a forested ecosystem. Springer Verlag pp. 146.

Matzner, E. und B. Ulrich (1980). The transfer of chemical elements within a heath ecosystem (Calluna vulg.) in Northwest Germany. Z. Pflanzenern. Bodenkd.

Matzner, E. und W. Hetsch (1981). Beitrag zum Elementaustrag mit dem Sickerwasser unter verschiedenen Ökosystemen im nordwest-deutschen Flachland. Z.Pflanzenern.Bodenkd. 144, 64-73.

Matzner, E., D. Hübner und W. Thomas (1981). Content and storage of polycyclic aromatic hydrocarbons in two forested ecpsystems in Northern Germany. Z.Pflanzenern.Bodenkd. im Druck.

Matzner, E. und J. Prenzel (1981). in Vorbereitung

Mayer, R. und H. Heinrichs (1980). Flüssebilanzen und aktuelle Änderungsraten der Schwermetall-Vorräte in Wald-Ökosystemen im Solling. Z. Pflanzenern.Bodenkd. 143, 232-246.

Mayer, R. (1981). Natürliche und anthropogene Komponenten des Schwermetallhaushalts von Waldökosystemen. Habilitationsschrift Universität Göttingen.

Meiwes, K.J. (1979). Der Schwefelhaushalt eines Buchenwald- und eines Fichtenwaldökosystems im Solling. Göttinger Bodenkundl. Ber. 60, pp. 107.

Nair, V. D. (1978). Aluminium species in soil solutions. Göttinger Bodenkundl. Berichte 52. 1-122.

Ulrich, B., R. Mayer, P. K. Khanna (1979). Deposition von Luftver-unreinigungen und ihre Auswirkungen in Waldökosystemen im Solling. Schriften Forstl.Fak.Univ.Göttingen, Bd. 58, Sauer-länder's Verlag, Frankfurt, Main, pp. 291.

Ulrich, B., R. Mayer und P.K. Khanna (1980). Chemical changes due to acid precipitation in a loess-derived soil in central europe. Soil Science 130, No.4, 193-199.

Ulrich, B. und E. Matzner (1981). Bilanzierung jährlicher Element-flüsse in Waldökosystemen im Solling (in preparation).

Tyler, G. (1975). Effect of heavy metal pollution on decomposition rates and mineralisation rates in forest soils. Int.Conf. Heavy Met. in the Environment. Toronto, Canada, Oct. 1975, 217-226.

ATMOSPHERIC DEPOSITION OF ACIDIFYING SUBSTANCES

R. Mayer

*Institut für Bodenkunde und Waldernährung der Universität Göttingen,
Büsgenweg 2, D-3400 Göttingen, Federal Republic of Germany*

ABSTRACT

Deposition processes for acidifying substances in the atmosphere,
finally leading to the appearance of acid precipitation, and their
controlling factors are reviewed. Knowledge of these is insufficient,
at present, for precise predictions of wet and dry deposition. Also
direct measurement of acid input to terrestrial surfaces is complicated, therefore indirect methods are employed. Data are presented
for the deposition rates of acidifying substances to terrestrial
ecosystems, measured over an extended period of time.

INTRODUCTION

Sulfur dioxide, SO_2, and oxides of nitrogen, NO_x, released into the
atmosphere mainly by combustion of fossil fuels, are recognized as
main causes for the existence of an acid precipitation problem. Increasing use of fossil fuels connected with growing industrial production and automobile traffic has led to an increasing acidity in
precipitation. Between 1967 and 1972, the mean pH in precipitation
at a remote station in the Black Forest of Southern Germany (Schau-
insland) decreased continuously for almost one unit from around pH
5.9 to 4.2, with considerable seasonal variations (Kayser et al.
1974), i.e. the hydrogen ion concentration increased at the factor
10 almost.

Ecological impact and damage is closely linked to acidity in rain or
snow melt water, i.e. to their hydrogen ion concentration. Hydrogen
ions may be present in the incoming rain, or be generated at the
interface between atmosphere and land or sea surface by chemical reaction and/or incorporation of pollutants into wet precipitation.
Therefore, mechanisms of wet and dry deposition of pollutants have to
be considered to understand the phenomenon of acid precipitation.

Wet deposition of pollutants is almost exclusively controlled by processes in the atmosphere, while dry deposition (including deposition
of fog) can either be controlled by processes in the atmosphere or
at the accepting surface.

It is not the intention here to discuss the nature and efficiency of

565

single deposition mechanisms, as it has been done by several authors in recent papers (e.g. Hidy 1973, Rasmussen et al. 1975, Slinn 1976, Schwela 1977, Fowler 1980) but rather to give a short survey on the processes involved in the removal of sulfur and nitrogen compounds from the atmosphere.

WET DEPOSITION

This term includes deposition of pollutants by rain or snow. There is a number of individual mechanisms by which pollutants may enter cloud or rain droplets. These processes are often called rainout when occuring within the cloud, and washout when the pollutant is taken up by the falling raindrop. Rainout and washout processes may well be identical.

When SO_2 gas is released into the atmosphere it may be dissolved in water droplets where it may further be oxidized to form sulfuric acid in an ionized form. A great number of interacting chemical reactions including photochemical and catalytic reactions may be involved in this process when other pollutants (as e.g. NO_x) are present in the atmosphere. In the same way oxides of nitrogen may be converted into nitric acid, HNO_3.

Sulfate and nitrate-containing aerosols may be captured by cloud and rain droplets due to inertial impaction, interception, Brownian diffusion, thermo- and diffusiophoresis. More significant for their removal is the fact, that these aerosols act as very effective condensation nuclei, thus giving rise to the formation of larger droplets which are subsequently rained out, together with their pollutant load. In the atmosphere of industrial countries this type of aerosols is probably the most important component of cloud condensation nuclei.

During their travel through the atmosphere the droplets may pick up more pollutants. Their final deposition to the surface is governed by air flow and gravitational forces. Surface has influence on the wet deposition rate as far as it influences the wind field. Therefore, in a given region wet deposition of pollutants is proportional to the degree of atmospheric pollution and to the amount of precipitation. Modification of this relationship by form, physical state and chemical reactivity of the accepting surface (i.e vegetation or cultivation type, moisture conditions etc.) will, under most conditions, be insignificant.

DRY DEPOSITION

This term includes
- sedimentation of large solid particles ($> 10 \mu m$, ice crystals excluded) due to gravitational forces
- the impaction of aerosols ($< 10 \mu m$) onto surfaces
- the absorption of gases.

In the case of SO_2 and NO_x, only the latter two mechanisms are of importance, aerosols being formed by conversion from gaseous compounds in the described way.

Dry deposition is controlled by atmospheric factors as well as by surface-linked factors. To make this clear, it is necessary to examine the structure of the lower atmosphere:
Free atmosphere is separated from the surface by a turbulent boundary layer (thickness from 10^1 to 10^2 m) where molecular viscosity of the air is controlling air movement. A very small laminar sublayer

of a few mm thickness, adjacent to the surface itself, is separating surface from boundary layer.

Gases and aerosols are transported through the atmosphere and across the viscous boundary layer by turbulent diffusion, and across the laminar sublayer by molecular diffusion in the case of gases and particles < 1 μm, by impaction and gravitational forces in the case of particles > 1 μm diameter.

It is convenient to describe the transfer to a surface by defining a deposition velocity and its reciprocal, the total resistance to the transfer between free atmosphere and accepting surface. The total resistance, r_t, can be subdivided into single resistances:

1. atmospheric resistance
   which is the sum of

   a. resistance to turbulent transport in the boundary layer, r_a, depending upon structure of the air flow above the accepting surface,

   b. resistance to the transfer across the laminar sub - layer by molecular diffusion, r_b, depending upon the molecular properties of the gas/aerosol particle,

2. surface resistance, r_s,
   which is the sum of resistances on or within the acceptor surface as , e.g. , stomatal resistance, cuticular resistance, soil resistance etc.

If surface resistance becomes very small

$$r_s \ll r_a + r_b$$

then total resistance approaches atmospheric resistance

$$r_t \approx (r_a + r_b)$$

and the surface behaves as a perfect sink.

Numerous investigations reported in literature have added to our knowledge of deposition processes and velocities to natural surfaces and have shown the relative significance of single components of the total transport resistance. A few results are summarized:

1. Deposition of gases SO_2, NO_x, and HNO_3

   - Water bodies (oceans, inland water surfaces) are perfect sinks, therefore deposition is only controlled by atmo - spheric resistance.

   - Soils with high pH and high water content have a small surface resistance compared to dry or low pH soils.

   - In vegetation-covered surfaces the stomatal resistance is small when stomata are open. When the vegetation canopy is wet, surface resistance approaches zero.

2. Deposition of aerosols

   Atmospheric transport is almost always rate-limiting. From theoretical considerations and experimental evidence the following general trends can be deduced (cf. Slinn 1976):

   - Deposition velocity is increasing with roughness height in the atmospheric boundary layer which, in turn, is de - pendent upon the shape and form of the surface.

- Deposition velocity to a vegetation canopy is increasing with wind speed in canopy and canopy height up to a saturation value.

ASSESSMENT OF DEPOSITION RATES

Although a lot of experimental and theoretical work has been done to investigate and explain atmospheric deposition, present knowledge is not sufficient to predict deposition rates pecisely in any individual situation occuring in the field. This shortcoming is due to the insufficient understanding of transformation and removal processes as well as to the lack of information on all boundary conditions in any given situation over an extended period of time. Grennfelt et al. (1980) have given a rough estimate of the total input of acidifying substances to a coniferous forest ecosystem in southern Sweden based upon literature data on deposition velocities for individual compounds. They found that dry deposition of SO_2 gas is the dominant process, followed by wet deposition of sulfate, dry deposition of particulate sulfate (aerosols) and deposition of mist and fog.

The experimental verification of such estimates would be very desirable. Unfortunately, the task to determine dry and wet deposition rates to natural sufaces directly encounters large experimental difficulties (cf. Galloway and Parker,1980). These difficulties are due to the impossibility to construct an acceptor surface which is able to duplicate the acceptor properties of a natural surface , especially when complicated surfaces like forest canopies or natural vegetation are concerned. After what has been said on the different deposition mechanisms, this experimental drawback applies primarily to the assessment of dry deposition.

Lack of experimental control of models describing transport and removal of atmospheric pollutants is most detrimental when environmental effects of acid precipitation are to be evaluated in remote, rural areas where it is extremely difficult to relate sources of pollutants to effects or damages. Therefore, indirect methods of determining deposition rates of acidifying substances to natural surfaces have been developed. One method is the use of the total element balance of a watershed in which all inputs and outputs can be controlled. This method yields mean deposition rates, integrating over a large area (watershed) and a long period of time (> 1 year). Eaton et al. (1980) found the following deposition rates of sulfur for a forest covered watershed (Hubbard Brook, New Hampshire, USA):

Total annual input 18.8 $kg.ha^{-1}.a^{-1}$

- Wet deposition (with
  bulk precipitation) 12.7 $kg.ha^{-1}.a^{-1}$

- Dry deposition (gaseous
  uptake and impaction) 6.1 $kg.ha^{-1}.a^{-1}$

Here dry deposition has been estimated from total budget by making assumptions for severel unknown source and sink functions like gaseous losses, weathering rates and accumulation rates for sulfur in the mineral fraction of the soil.

In another approach deposition of atmospheric pollutants to forests has been evaluated by using the forest canopy itself as acceptor surface. Total deposition is measured below the tree canopy giving allowance for substances originating from internal leaching. This method

allows to evaluate deposition rates integrated over a period of
months or years and over a homogeneous sector of a forest landscape.
It is thus possible to compare deposition rates to different types
of vegetation under the same geographical and meteorological conditions.

Deposition rates measured and calculated in this way for a beech
and a spruce forest in the Solling are given in Table 1:

TABLE 1 Deposition rates of sulfur to a beech and spruce forest ecosystem in the Solling

Age of the stands: Beech (Fagus sylvatica) 130 yrs.
Spruce (Picea abies) 90 yrs.

Year	Wet Deposition Beech or Spruce	Dry Deposition		Total Deposition	
		Beech	Spruce	Beech	Spruce
		$kg.ha^{-1}.a^{-1}$			
1969	24.0	32.6	56.0	56.6	80.0
1970	27.2	23.1	61.0	50.3	88.2
1971	22.2	18.3	54.3	40.5	76.5
1972	25.2	21.2	43.8	46.4	69.0
1973	21.1	16.7	60.1	37.8	81.2
1974	23.2	32.2	70.0	55.4	93.2
1975	25.3	26.2	77.6	51.5	102.9
1976	23.0	25.1	50.6	48.1	73.6

These forests are located in a remote mountain area, about 150 km
east of the main industrial center of northern Germany, the Ruhr
district. The meteorological conditions are characterized by high
precipitation (1070 mm per year) quite evenly distributed over the
year, and high frequency of fog. A detailed description of the mode
of evaluating deposition rates for this area is given by Ulrich et
al. (1979). It is repeated here briefly:

Since sedimentation of large particles containing sulfur is negligible in this area, and dry deposition of aerosols to the smooth surface of a collector above the ground is small, bulk precipitation
collectors give good approximations for wet deposition rates. Internal leaching of S (originating from root uptake, not from atmospheric deposition)from the beech canopy is an important process only in
senescent leaves at the end of the vegetation period (september/octo-
ber). It can be estimated by comparison of the element concentration
of sulfur in precipitation below the canopy. These estimates indicate that internal leaching is small compared to the rates of wet deposition measured below the canopy (<5 %). Therefore, even with
quite rough estimates of internal leaching rates dry deposition can
be calculated in good approximation as difference between

    (a) wet deposition measured below the forest canopy, Dw_b , and

    (b) sum of wet deposition measured above the forest canopy, Dw_a , and internal leaching, Li

$$Dd = Dw_b - (Dw_a + Li) \qquad eqn.(1)$$

where Dd is the dry deposition rate.

The difference Dd, calculated according to eqn.(1), is very large in the Solling forest ecosystems during the winter months (november to april), in the order of 10 to 20 kg sulfur per hectare under beech, and 30 to 50 kg sulfur per hectare under spruce (Mayer and Ulrich, 1978). It is obvious that this difference must be due primarily to atmospheric deposition, remembering that total flux of sulfur with annual litterfall is only in the order of 6 kg sulfur per hectare and year (Meiwes 1979) both in beech and in spruce forest. Therefore, the doubts expressed by Galloway and Parker (1980) wether this method is applicable can be resolved here. Under different circumstances and for different elements internal leaching may be of much more importance and uncertainty in determining leaching rates may render the calculation of dry deposition from equation (1) very unreliable.

From Table 1 it becomes obvious that dry deposition of aerosols and/ or gases may be the dominant depositional process. Within a given area the relative importance of single processes depends largely upon the quality of the accepting surface. There are considerable differences between the years, more so in dry deposition rates than in wet deposition rates.

It seems that deposition of SO_2 and sulfur compounds derived from SO_2 is most sensitive to the conditions at the surface, due to its emission as gas and its chemical reactivity and solubility. The same is probably true for nitrogen oxides and, in consequence, for the hydrogen ion input with acid precipitation.

FUTURE NEEDS

At present it appears that research along the following lines is primarily needed:

1. Dry deposition of small particles and aerosols to natural surfaces, especially to vegetation.

2. Simultaneous assessment of deposition rates by micrometeorological methods and budget methods.

3. Buffering processes taking place at vegetation surfaces, influencing acidity of precipitation reaching the ground.

REFERENCES

Eaton, J.S., G. Likens, and F.H. Borman (1980). Wet and dry deposition of sulfur at Hubbard Brook. In Th.C. Hutchinson and M.Havas (Eds.), Effects of Acid Precipitation on Terrestrial Ecosystems, Plenum Press, New York and London, pp. 69 - 76.
Fowler, D. (1980). Removal of sulfur and nitrogen compounds from the atmosphere in rain and by dry deposition. In D. Drabløs and A. Tollan (Eds.), Ecological impact of acid precipitation, Proc. of an Int. Conf., Sandefjord, Norway, March 11-14, 1980, Oslo-Ås, pp. 22-32.
Galloway, J.N., and G.G. Parker (1980). Difficulties in measuring wet and dry deposition on forest canopies and soil surfaces. In Th.C. Hutchinson and M. Havas (Eds.), Effects of Acid Precipitation on Terrestrial Ecosystems, Plenum Press, New York and London, pp.57 -68.
Grennfelt, P., C. Bengtson, and L. Skarby (1980). An Estimation of Atmospheric Input of Acidifying Substances to a Forest Ecosystem. In Th. C. Hutchinson and M. Havas (Eds.), Effects of Acid Preci-

tation on Terrestrial Ecosystems, Plenum Press, New York and London, pp. 29-40.

Hidy, G.M. (1973). Removal processes of gaseous and particulate pollutants. In S.I. Rasoul (Ed.), Chemistry of the lower atmosphere, Plenum Press, New York and London, pp. 121-176.

Kayser, K., U. Jessel, A. Köhler und G. Rönicke (1974). Die pH-Werte des Niederschlags in der Bundesrepublik Deutschland 1967-1972. DFG-Komm. z. Erforschung der Luftverunreinigungen, Mitt. IX, Verlag H. Boldt, Boppard.

Mayer, R.,and B. Ulrich (1978). Input of Atmospheric Sulfur by Dry and Wet Deposition to Two Central European Forest Ecosystems. Atmospheric Environment, 12, pp. 375-377.

Meiwes, K.J. (1979). Der Schwefelhaushalt eines Buchenwald- und eines Fichtenwaldökosystems im Solling. Gött. Bodenkundl. Ber.,60,pp. 1-108.

Rasmussen, K.H., M. Taheri, and R.L. Kabel (1975). Global emissions and natural processes for removal of gaseous pollutants. Water, Air and Soil Pollution, 4, pp. 33-64.

Schwela, D. (1977). Die trockene Deposition gasförmiger Luftverunreinigungen. Schriftenreihe der Landesanstalt für Immissionsschutz, Heft 42, Verlag W. Girardet, Essen, pp. 46-85.

Slinn, W.G. N. (1976). Some approximations for the wet and dry removal of particles and gases from the atmosphere. USDA Forest Service General Technical Report NE-23, pp. 857-894.

Ulrich, B., R. Mayer,and P.K. Khanna (1979). Deposition von Luftverunreinigungen und ihre Auswirkungen in Waldökosystemen im Solling. Schriften aus der Forstlichen Fak. Univ. Göttingen und der Nieders. Forstl. Versuchsanstalt, Bd. 58, J.D. Sauerländer's Verlag, Frankfurt am Main, 291 p.

HEALTH IMPACTS OF U.S. ENERGY
POLICY ALTERNATIVES

R. O. McClellan, R. G. Cuddihy, F. A. Seiler and W. C. Griffith

Lovelace Inhalation Toxicology Research Institute,
P.O. Box 5890, Albuquerque, NM 87185, USA

ABSTRACT

Formalized cost-benefit analyses that include health risk assessments provide a logical framework for bringing together the various inputs needed to choose between energy technologies. This paper briefly reviews cost-benefit analysis methodology using as an example an assessment of health risks of diesel exhaust particles. The assessment used a series of interrelated models to project population exposures and health effects. In the absence of a demonstrated cause-effect relationship for exposure to diesel exhaust and lung cancer in man, models based on coke oven workers and cigarette smokers were used. This assessment provides a model for conducting assessments of other energy technologies.

KEYWORDS

Risk-benefit analysis; health risk assessment; diesel; lung cancer; energy technologies; nuclear power; coal combustion.

INTRODUCTION

Events in recent years have served to focus public attention on how the availability and cost of energy affects individuals, communities, states, nations, and the world. In addition to concern for energy costs, it is now recognized that reduced energy supplies can seriously disrupt society. This has stimulated increased efforts to develop strategies to assure that future energy needs are met.

Fortunately, most countries have a range of available energy alternatives. These include different resource options, such as oil, coal, uranium, solar, wind and water. Also, there are options as to how the resource may be used, such as combustion, gasification or liquefaction of coal. Finally, choices exist as to end use; industrial processes, home heating or as diesel fuel for vehicles. In addition, conservation is an option. Selection of any of these alternatives will result in different benefits and associated costs, including environmental and health costs.

Development of national energy strategies is complex and includes consideration of numerous political, social, economic and resource availability factors and health and environmental impacts. Some factors are highly intangible and basically represent value judgments. Others are subject to measurement or estimation using scientific techniques. In view of the complexity, it is not surprising that considerable effort has been expended in developing and using analytical tools such as cost-benefit analyses as an aid in decision making.

In this paper, we discuss cost-benefit analyses, give an example of one phase of a specific analysis, and offer comments on the usefulness and limitations of cost-benefit analyses in establishing energy strategies and policy.

COST-BENEFIT ANALYSES

Cost-benefit analysis is a tool that provides a logical framework for assembling, examining and interrelating all relevant inputs to reach a decision that balances costs and benefits. This process (Fig. 1) consists of two major elements, an assessment phase and an evaluation phase. Each phase is a critical part of the total process; and neither stands alone as a cost-benefit analysis.

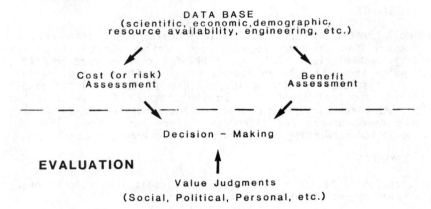

ASSESSMENT

DATA BASE
(scientific, economic, demographic, resource availability, engineering, etc.)

Cost (or risk) Assessment

Benefit Assessment

Decision – Making

EVALUATION

Value Judgments
(Social, Political, Personal, etc.)

Fig. 1. Schematic representation of the cost-benefit analysis program.

The assessment phase assembles all available data and interrelates it in a manner that yields quantitative outputs either as benefits or costs. It would be ideal if the outputs could be expressed in equivalent monetary units. Unfortunately, this cannot always be done without introducing value judgments. For example, in considering health costs, one may estimate the number of excess deaths or days of illness attributable to a particular technology. In most cases, conversion of these outputs into monetary units requires judgments about the value of a human life or a day of illness. These value judgments are appropriately included in the second phase of the analyses.

Two features of the assessment phase are worthy of note. First, it should represent the best estimate of cost or benefit at a particular time. This requires that all relevant data be considered. When multiple pieces of data are available, the basis for including or excluding data should be explicitly stated. If the desired information is not available, a best estimate must be used with an explicit statement as to how the estimate was developed. Second, to the maximum extent possible, a measure of uncertainty should be provided for all intermediate values and final cost (or risk) value(s).

The second element of the cost-benefit analyses is the evaluation phase. This phase uses the quantitative outputs of the assessment phase along with other non-quantitative inputs. The latter include societal and personal values, public opinion, political judgments and other factors needed to reach an informed decision. Clearly, the evaluation phase is subjective and contrasts with the objective nature of the assessment phase. It can be noted in Fig. 1 that the decision making process involves value judgments.

A final point regarding cost-benefit analyses is that there may be instances in which the relative benefits of two or more alternatives are set aside and only the costs are considered. This is the case in the health risk assessment example that follows in which an a priori assumption was made that the U.S. populace will utilize light duty motor vehicles in 1995 and the choice is between diesel and spark ignition engines.

HEALTH RISK ASSESSMENT FOR DIESEL EXHAUST PARTICLES

Having considered the broad aspects of cost-benefit analyses, we will now consider a detailed example of the assessment phase applied to an energy issue. Specifically, we will summarize one aspect of a recent assessment of the health impacts of increased use of diesel engines in light duty vehicles (Cuddihy and colleagues, 1980). An important aspect of this study was the potential health risks of diesel exhaust particles. Concern for the health risks of diesel exhaust particles was motivated by recognition that diesel engines emit substantially greater amounts of particulate material than do gasoline-fueled engines. Further, these particles are readily respirable and a large portion of their mass is readily extracted with organic solvents and has been shown to be mutagenic to bacteria and mammalian cells and carcinogenic when applied to the skin of rodents.

The overall assessment was multi-faceted and considered impacts of manufacturing and servicing diesel vehicles and industries required to support diesel or gasoline spark ignition vehicles. Thus, inherent in the study was the assumption that society would expect the benefits of using vehicles and the choice was only between vehicle options. That portion of the assessment dealing specifically with diesel exhaust particle emissions and their potential health risks is described here.

The approach used in making the assessment is shown schematically in Fig. 2. The analysis moves stepwise from the pollutant source to an estimate of health effects by using both models and direct observations. As will be emphasized later, it is important that each model be realistic and, if possible, verifiable by direct observation or the use of surrogates. In this context, a surrogate is viewed as a material that behaves in a manner analogous to the material of interest within the pathway being evaluated.

Assumptions used in estimating annual diesel particle emissions in 1995 in the United States are shown in Table 1. The quantities of exhaust particles and particle associated benzo(a)pyrene were projected in the atmosphere using a dispersion model that took account of wind speed and direction, atmospheric mixing, diffusion, settling and resuspension of particles. The dispersion modeling provided estimates of average air concentration of diesel exhaust particles throughout the United States. For urban areas, an average air concentration of 0.1 µg per m^3 weighted by population density was projected. For rural areas the value was 0.01 µg per m^3. For major population centers with densities of 2,000 to 16,000 people per km^2, the projected ground air concentrations of diesel exhaust particles were between 0.3 and 2 µg/m^3. Air concentrations of diesel vehicle particles were projected to be 2 to 3 times higher in urban street canyons and, depending on the traffic density, building dimensions and meterological conditions, may reach 20 µg/m^3.

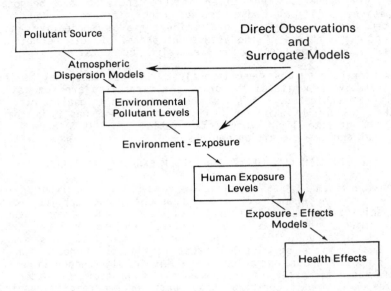

Fig. 2. Approach used to assess the health risks of diesel exhaust particles.

TABLE 1 Summary of Assumed Parameters for Estimating Diesel Exhaust
Particle Emissions in 1995

Total Light Duty Vehicles[a]	1.7×10^8
Total Light Duty Diesel Vehicles[a]	3.4×10^7
Total Light Duty Diesel Travel[a]	6.0×10^{11} km
Particle Emissions Rate[a]	0.1 g/km
Particle Associated Benzo(a)pyrene[b]	12 mg/kg
Total Diesel Exhaust Particle Emissions[c]	6.0×10^7 kg
Total Diesel Particle Associated Benzo(a)pyrene[c]	700 kg

[a]Estimated.
[b]Based on measurements.
[c]Calculated from estimates and measurements.

Fortunately, considerable data is available on atmospheric lead that originates from vehicle exhaust in small particles. Thus, lead was used as a surrogate for diesel exhaust particles to verify the appropriateness of the dispersion modeling. The lead data also were used to project levels of diesel exhaust particles in dust in residential areas after long periods of time, 0.2 to 2 mg per gram of dust. The potential significance of these projections will be discussed later.

Current use of light duty diesel vehicles is low and it has not been feasible to actually measure diesel exhaust particle concentrations at sufficient locations to verify the dispersion models. Further, diesel exhaust particles and their chemical constituents are not sufficiently unique that they can readily be distinguished from the substantially greater amounts of these constituents emitted from other sources.

Based on recent work with particles similar in size to diesel exhaust particles (Wolff and colleagues, 1981), it was assumed that 25 percent of the diesel exhaust particles that are inhaled deposit in the lower respiratory tract. This leads to projected annual depositions in the United States population of 60 kg of diesel exhaust particles. Using data available on lead, it also was estimated that approximately 1000 kg of diesel exhaust particles would be ingested each year as a result of their presence in dust.

Next, attention was directed to predicting health effects. Fortunately, from the viewpoint of public and occupational health, there are no data establishing a causal relationship between exposure to diesel exhaust and lung cancer (National Academy of Science, 1980). Also, studies completed to date with laboratory animals exposed to diesel exhaust have failed to demonstrate late-occurring health effects even at exposure levels in excess of 1 mg/m^3.

In the absence of direct observations, alternative approaches were needed. The alternative approaches made use of extensive short-term test data (obtained from studies in bacteria and mammalian cell and tissue systems) and epidemiological data obtained from cigarette smokers and coke oven workers. The short-term test results collectively suggested that, on a per unit particle mass basis, diesel exhaust particles are no more mutagenic than cigarette smoke or coke oven emissions (National Academy of Science, 1980). The same statement can be made for carcinogenic potency, although based on less data. With these findings, the approach used assumed that an upper-bound estimate of diesel exhaust particle carcinogenicity could be made from the human data available on cigarette smokers and coke oven workers. The lower boundary was considered to be zero health effects.

We derived annual lung cancer risk factors of 5000 per 100,000 people per µg of benzo(a)pyrene per m^3 of air and 150 per 100,000 people per mg of particle per m^3 of air. These factors were used in combination with exposure concentration data developed earlier to project the lung cancer incidence. Using the risk factor derived for benzo(a)pyrene, the annual excess lung cancer incidence attributed to diesel exhaust particles for urban residents and rural residents were 10 and 0.4 cases per year, respectively. The corresponding values using the total particle mass risk factor were 26 and 1 cases per year.

These risks are placed in perspective in Table 2. The lung cancer risks attributable to diesel exhaust particles are substantially less than those of smokers. The highest exposure case for diesel exhaust particles, the city street canyon, has an associated risk about equal to that of the rural non-smoker.

Some general comments concerning the diesel health risk assessment illustrate the utility of the approach and its possible application to other technologies. The process involved a mixture of data and estimated values. That more actual data was not available is surprising since diesel-powered vehicles have been used for many years. This may reflect society's acceptance of the familiar. Although the amount of actual data was less than desired, it was more than is available today for assessing many other technologies, for example, health risks to workers in a coal liquifaction plant.

An additional benefit of the assessment process lies in identifying new information and research needs. The calculation of the relative amounts of diesel exhaust particles entering the body via ingestion relative to inhalation (1000 kg vs 60 kg) emphasizes the need to obtain information on the biological fate of organic compounds associated with the ingested particles. It is likely that current scoping risk assessments of emerging technologies can identify critical information needs that will facilitate more definitive assessments.

Table 2 Cancer (Cancers/Year) for Smokers, Non-Smokers and Coke Oven Workers and Estimated Additional Cases Attributable to Diesel Exhaust Particle Exposure

Population	Probability
Coke Oven Workers[a]	4×10^{-3}
Average U.S. Smoker[a]	8×10^{-4}
Average Urban Non-Smoker[a]	7×10^{-5}
Average Rural Non-Smoker[a]	3×10^{-5}
Diesel Exposure	
Street Canyon Resident[b]	$1-3 \times 10^{-5}$
High Density Urban Resident[b]	$1-3 \times 10^{-6}$
Low Density Urban Resident[b]	$1-3 \times 10^{-7}$
Rural Resident[b]	$1-3 \times 10^{-8}$

[a]measured
[b]estimated

Finally, it is appropriate to emphasize that the risk estimates are tentative and subject to revision based on the availability of new data. Despite their tentative nature, it is encouraging that the upper boundary estimates using the two exposure-effect estimators agreed within a factor of three. Further, a similar low estimate of lung cancer risk for diesel exhaust particles was obtained by Harris (1981). In considering the risk values, their magnitude relative to the risk of lung cancer in non-smokers and its induction by cigarette smoke and other agents is more important than the absolute numbers. The decision-maker faced with questions of regulation of diesel particle emissions can be confident that even for the highest exposures the lung cancer risk is substantially less than that of cigarette smokers or coke oven workers and about equivalent to the spontaneous risk in rural non-smokers.

It must be recognized that the foregoing health risk assessment was narrow in focus. In future assessments, other variables also should be considered; such as high exposure individuals, injury and fatality rates for occupants of diesel vehicles. Beyond these obvious considerations, it also may be expected that industries required to produce and service light duty diesel vehicles might have associated costs that are different than those encountered in producing and servicing a comparable fleet of gasoline spark ignition vehicles. Examination at this level of detail could be accomplished with a risk accounting method of analysis of health impacts based on input-output models of the U.S. economy (Hamilton, 1980).

CONSIDERATIONS USING COST-BENEFIT ANALYSES

Returning to a broader consideration of cost-benefit analyses, it is important to keep in mind that these involve both assessment and evaluation phases. In many cases, it is difficult to sort out the two phases and identify the role of value judgments. This is illustrated by considering the results of an analysis conducted by Inhaber (1979) of the health impacts of both conventional and non-conventional energy technologies. His results are compared in Table 3 with a subsequent analysis by Holdren and associates (1979).

In comparing the results, one is struck by the close agreement of the estimates of the two studies for the conventional technologies and by the greater
disagreement for the non-conventional technologies. For the latter, Holdren's
estimates are all lower, largely due to their individual perspectives on the
use of different energy technologies. Inhaber assumed that the non-conventional technologies required backup provided by using coal-fired plants.
Holdren assumed that the new technologies would reduce demand for energy from
conventional technologies and not require new backup, thereby reducing risks
because less conventional generated energy was needed. Thus, subtle differences in cost accounting perspectives can have a tremendous impact on the
assessment even though careful review reveals how the costs arise.

An additional point can be made by considering the health risk estimates for
two conventional technologies: coal and nuclear. These data suggest a 10 to
200 fold greater health impact from the use of coal than from nuclear power.
Restated, these results suggest that a health risk penalty of a factor of 10
to 200 is being incurred for each megawatt-year of electrical output being

TABLE 3 Health Risks of Alternative Energy Technologies per Megawatt-Year
(Expressed as WDL - Worker Days Lost, PDL - Public Days Lost)

	Inhaber			Holdren and Associates		
CONVENTIONAL						
Nuclear						
WDL	1.7	-	8.7	3.1	-	12
PDL	.3	-	1.5	0.3	-	70
Oil						
WDL	2	-	18	3	-	19
PDL	9	-	1900	9	-	1000
Coal						
WDL	18	-	73	19	-	43
PDL	20	-	2000	20	-	1500
NON-CONVENTIONAL						
Solar Heating						
WDL	91	-	100	11	-	17
PDL	4.6	-	9.5	2.1	-	5.5
Wind						
WDL	220	-	290	9.7	-	10
PDL	22	-	540	.2	-	.5
Ocean Thermal						
WDL	23	-	30	2.2	-	4.6
PDL	.8	-	1.4	.4	-	.9
Photovoltaic						
WDL	140	-	190	5.0	-	14
PDL	10	-	510	.9	-	2.2

installed today with coal as the energy source rather than nuclear power.
This apparently irrational choice may be due to several factors. One possible
explanation is that other offsetting costs favor installation of coal-fired
generating facilities over nuclear facilities. This explanation is question-
able since comparisons that included construction and operating costs have
usually favored nuclear over coal-fired plants. A second possible explanation
is uncertainty over future operating costs; for example, unknown costs related
to storage or reprocessing of spent nuclear fuel or disposal of nuclear waste
may be viewed as outweighing the apparent health risk advantages. A third
possible explanation is that the perceived health risks of nuclear power are
greater than the actual risks, contrary to the assessments of Inhaber and

Holdren. This may be related to concern for accidents, despite rigorous analyses conducted both in the U.S. (Nuclear Regulatory Commission, 1975) and Germany (Birkhofer, 1980) which indicate that the risk of such accidents is small when viewed relative to other risks. A fourth possible explanation is that government regulations, or other policy actions may favor one alternative. During the Carter administration, a number of actions, taken in part because of concern for proliferation of nuclear weapons, tended to discourage the construction of nuclear power plants relative to coal-fired plants. This course may be reversed under the Reagan administration.

Increased use of cost-benefit analyses can assist in formulating national strategies for energy production and usage. In performing such analyses, it is important to distinguish between the assessment and evaluation phases and to identify the contributions of actual data, numerical estimates and value judgments in the analyses.

ACKNOWLEDGEMENT

Research was conducted under U. S. Department of Energy Contract No. DE-AC04-76EV01013.

REFERENCES

Birkhofer, A. (1980). The German risk study for nuclear power plants. IAEA Bulletin 22, 23-33.

Cuddihy, R. G., F. A. Seiler, W. C. Griffith, B. R. Scott and R. O. McClellan (1980). Potential health and environmental effects of diesel light duty vehicles. DOE Research and Development Report, LMF-82, National Technical Information Service, Springfield, VA 22161.

Hamilton, L. D. (1980). Comparative risks from different energy systems: evolution of the methods of studies. IAEA Bulletin 22, 35-72.

Harris, J. E. (1981). Potential risk of lung cancer from diesel engine emissions. Report to the Diesel Impacts Study Committee, National Research Council, National Academy Press, Washington, DC.

Holdren, J. P., K. Anderson, P. H. Glenck, I. Mintzer, G. Morris and K. R. Smith (1979). Risk of renewable energy sources: a critique of the Inhaber report. Energy and Resources Group, University of California, Berkeley, CA, ERG 79-3.

Inhaber, H. (1979). Risk with energy from conventional and nonconventional sources. Science 203, 718-723.

National Academy of Sciences (1980). Health effects of exposure to diesel exhaust. The Report of the Health Effects Panel of the Diesel Impacts Study Committee, National Research Council, Washington, DC.

U. S. Nuclear Regulatory Commission (1975). Reactor safety study. An assessment of accident risks in U.S. commercial nuclear power plants. Report WASH-1400. (NUREG-75/014).

Wolff, R. K., G. M. Kanapilly, P. B. DeNee and R. O. McClellan (1981). Deposition of 0.1 μm chain aggregate aerosols in beagle dogs. J. Aerosol Sci. 12: 119-129.

ENERGY POLICY AND GREEN ENVIRONMENT ON THE BASIS OF ECOLOGY

A. Miyawaki

*Dep. Vegetation Science, Inst. Environmental Sci. and Techn.,
Yokohama National University, Yokohama, Japan*

ABSTRACT

Energy problem and environmental problem are two sides of a single
issue. Whenever a new energy base on an industrial complex is to be
built or expanded it is strongly recommended that an environmental
protection forest(EPF) be also created in and around its grounds.
The EPF has been developed on the besis of numerous ecological
studies. This paper presents basic ecological researches which
prwceded the formation of the living filter in Japan, and some case
examples of EPFs.

KEYWORDS

Living filter; environmental protection forest(EPF); potental natural
vegetation map; actual vegetation map; phytosociological releve;
system of energy-living filter-man; green alarm device.

PROBLEM OF ENERGY AND ENVIRONMENT

Energy problem and environmenatl problem are two sides of a single
issue. Constraction and expansion of a new energy base should be
allowed only when the living environment of its vicinity is to be
better re-created and perenially guaranteed. In more specific terms,
formation of environmental protection forest is recommended. Environmental Protection Forest is diverse and stable in nature, and biolo-
gically represents the integral of a given region's indigenous natural
environment , upon which human lives are profoundly dependent.
   Since the climate of the Japanese Archipelago is temperate and
rainy, more than 99% of the land was coverd by woods and forests. The
woods and forests which mainly occupied the southern half of the
country were evergreen broad-leaved forests dominated by such evergreen species as Castanopsis,Persea and Quercus(Cyclobalanopsis).
The woods and forests in the northern half and those in the zones at
the altitude of 800 - 1000 m and above in the southern Japan were
summergreen broad-leaved forests with species such as Fagus and
Quercus.

Human beings migrated to the Archipelago 2000 years ago, and started to slowly cultivate land and develop industries. In the past 100 years the pace of land development and industrizytion has been radically accerated. The new civilization created by the populationincrease and rapid urbanization has inevitably resulted in nature destruction, such as deterioration and demolishment of vegetation.

Although continuously destroying nature represented by the natural forest on one hand, the Japanese, like many other peoples in the rest of the world, have created, cherished and preserved nature in certain locations, after many trials and errors over a long period of time. This is Native Forest as symbolized by the "Chinju-no-mori" forest in and around towns villages. "Chinju-no-mori" has been under the protection of communities' residents. The Native Forest also includes natural growth, which is vulnerable to human intervention, along steep slopes and seashores.

Rapid growth of industries, urban centers and traffic facilities in the recent years has been sustained by the increasing number of energy bases. The pace and scale of the energy base constraction have escalated to such a degree that the Native Forest, the symbol of indigenous natural environment prserved and protected from the olden times, has begun to be destroyed in many parts of the country.

Land development and construction of indstrial complexes such as energy bases are being undertaken on a large scale, in response to the needs(energy, resources and environment) of a new era. However, not only physical & chemical surveys but also thorough investigation of biocommunity should be required before commencing such development and construction, for bio-community relates to human existance at a profound level and embodies,in its own life form, the integral of natural environment in a given location. Vegetational survey is particularly important since vegetation is the producer in the ecosystem, and basic constituent of the biosphere of which man is a part.

Implementation of energy policies entails the construction of energy bases and industrial facilities in various places. Levelling of hills and reclamation of the sea would inescapably accelerate the deterioration of nature, namely destruction of vegetation and subsequent loss of nature's diversity, and impoverishment of the biosphere. In the light of such alarming prospect, environmental assessment alone would not suffice; not only restroration of natire, but also creation of diverse and stable biological environment in and around new energy and industrial sites are in need.

In Japan, our recommendation has been widely accepted. Natural environment survey(vegetationalsurvey) is now a prerequisite to the construction of hydraulic, thermal, atomic and geothermal power plants, factories, coastal industrial sites, residential complexes and traffiic facilities. Also, time is ripe for making it a premise to disgnose current conditions of the greens in the vicinity concerned and compile an actual vegetation map, a base map for nature restoration and a potential natural vegetation map, i.e. a qualitative illustration of the habitat in question. Realistic planning and implementation of nature preservation, restoration and creation as represented by the Environmental Protection Forest are necessary.

METHODS AND PROCESS OF VEGETATIONAL SURVEY AND
CREATION OF ECOLOGICAL ENVIRONMENTAL PROTECTION
FORESTS

The native forest -environmental protection forest(EPF)- in and

around an energy base is deeply rooted in the indiginous potentiality
of the site,and its environmental protection functions are diverse.
Therefore, vegetational survey both in terms of ecology and phytosociology is an important step to be taken prior to the formulation
of the scheme of the EPF. At the same time, community units are
phytosociologically determines. Then actual and potential natural
vegetation mapping is conducted using these community units.

Only on the basis of careful phytosociological studies both in
the field and in the laboratory, creation of an ecologically
feasible EPF can be realized.

An EPF in the site of an energy base, for instance, can be
created by following the steps in the flow chart in fig. 1 (Ecological field survey: releve: Vegetationsaufnahme).

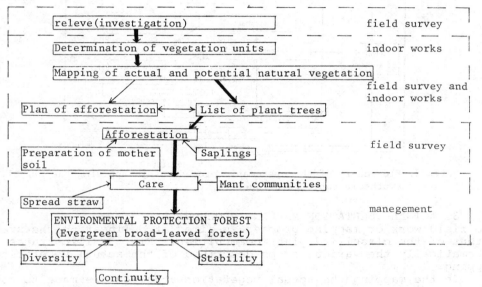

Fig. 1. Flow chart from the field survey until creation of the ecoligical
        protection forests.

1.Releve(field investigation)

The first and most important step for the creation of the environmental protection forest is to collect reliable and quantitative
field data. This is done in the field by adopting the Releve Technique of Braun-Blanquet(1928, 1951, 1954). The releve is equivalent
to a sample plot in vegetation analysis. All the existing plant
species have to be properly identified regardless of their vegetative
state being a seedling or a mature plant individual.

It is of further importance to sample as many local plant
communities as possible to facilitate later the distinction of vegetation units like association, alliance, oder and class. Of great
significance to the classification of plant communities is to recognize the relationship and patterns of plant communities with micro-
relief, landform, climate and other environmental factors.

2. SYNTHESIS AND CLASSIFICATION OF PLANT COMMUNITIES
   BY SYNTHESIS TABLES

A synthesis table is a tabular arrangement and compilation of raw field data in a more appropriate format for synthesis. Community units such as local community and association are determined by repeating the tablework process as illustrated in Fig. 2.

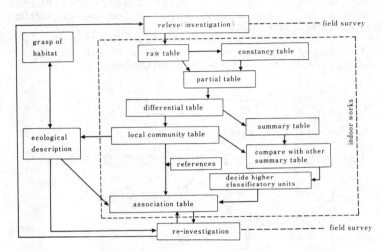

Fig. 2. Flow chart of the systematic procedures in applying the Br.-Bl. synthesis table technique.

3. ACTUAL VEGETATION MAPPING PROCEDURES

The field work or mapping process is aimed at delineating the areal extent of the classified plant communities. Fig. 3. illustrates schematically the various steps described of the actual vegetation mapping.

In the mapping the actual vegetation attempts are made to assess the degree of human. Upon human impact certain sensitive floristic elements of the natural association have been replaced by other, often weedy species. Often, the total natural species composition has been eliminated as in the case of drastic conversion from natural forest to farmland. By mapping the potential natural vegetation on a map base one can make inferences as to the substitute vegetation types.

Releve
↓
Classification based on floristic criteria
↓
Establishing the mapping key for the actual vegetation map (legend)
↓
Field mapping of the actual vegetation map
↓
Mapping correct with the aid of aerial photographs & topographic maps
↓
The final map preparation of the actual vegetation

Fig. 3. Process of the actual vegetation mapping

4. PREPARATION OF THE POTENTIAL NATURAL VEGETATION MAP

It is very important to prepere an additional map from the above
phytosociological map which delineates pristine or primeaval vegetation types, not influenced by man. The philosophy behind such a map
is that the pristine natural vegetation would be most insicative of
its environment or habitat and illustrate best the land productivity
on a long term basis.

The following schematic representation illustrates the various
steps in the preparation of the potential natural vegetation map.

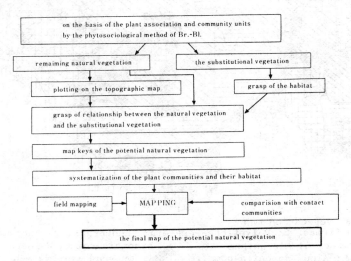

Fig. 4. Flow chart of the potential natural vegetation mapping.

Before constructing or expanding a new energy base, an Actual
Vegetation Map of the site should be drawn up. This could be used
as a basic diagnosis map, by which ecologically valuable vegetation
and biocenoces are identified and necessary protective measures can
be chosen.   In addition, a potential natural vegetation map is
also useful as a diagnosis map in creating an EPF in the site of the
new energy base.

On the basis of these ecological studies, EPFs, have been
initiated in the site of thermal, atomic and geothermal power plants
in Fukushima, Sodegaura, Tsuruga, Tanagawa, Genkai and Hachobaru in
Japan. These young forests are growing steadily to prove, hopefully, to be the forerunners of global development of the EPF.

CONCLUSION AND SUMMERY

These diagnosis and proposals from the piont of view of ecology
and phytosociology are being actively adopted and utilized in todays
Japanese energy strategies. Gradually, Environmental Protection
Forest, a creative ecological response to an emerging new era, is
taking root in various energy and industrial bases throughout the
country.

The significance of the System of Energy-Living Filter-Man,

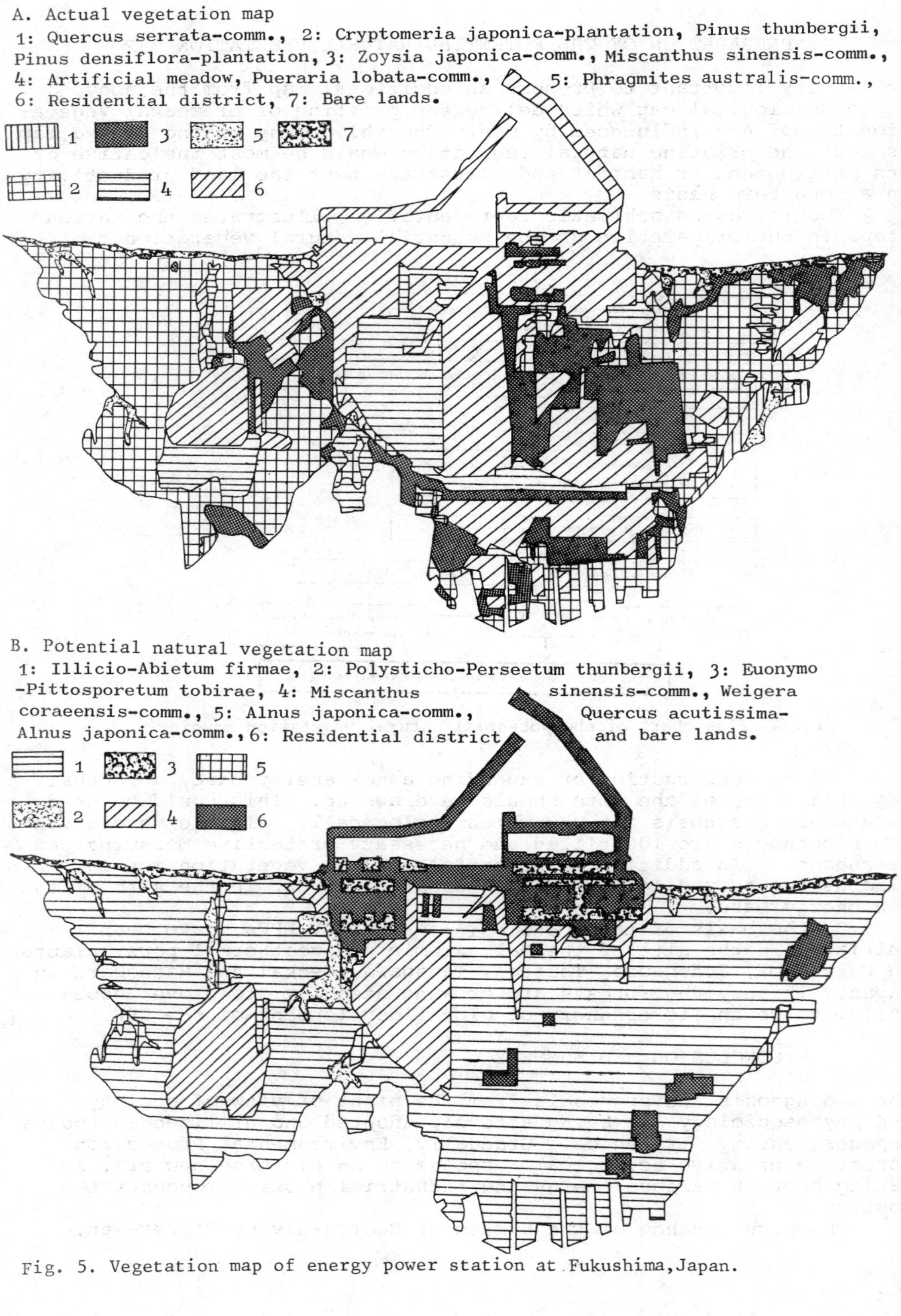

A. Actual vegetation map
1: Quercus serrata-comm., 2: Cryptomeria japonica-plantation, Pinus thunbergii,
Pinus densiflora-plantation, 3: Zoysia japonica-comm., Miscanthus sinensis-comm.,
4: Artificial meadow, Pueraria lobata-comm., 5: Phragmites australis-comm.,
6: Residential district, 7: Bare lands.

B. Potential natural vegetation map
1: Illicio-Abietum firmae, 2: Polysticho-Perseetum thunbergii, 3: Euonymo
-Pittosporetum tobirae, 4: Miscanthus sinensis-comm., Weigera
coraeensis-comm., 5: Alnus japonica-comm., Quercus acutissima-
Alnus japonica-comm., 6: Residential district and bare lands.

Fig. 5. Vegetation map of energy power station at Fukushima, Japan.

in which energy bases and local residents coexists through the utilization of Green Filter, is certain to increase year by year.
This system provides a crucial ecological and environmental approach
to solve energy problems with the participation of local residents.
The forthcoming presentation shall deal with the current conditions
and some examples of how certain sapects of the global energy problem
are being slowly but steadily solved in Japan. It would be appre-
ciated if participants of the conference would be interested in
watching how our approach will further develop and become increasingly effective in creating healthy living environment without unnecessarily hampering industrial development.

REFERENCES

Miyawaki, A.(1977). Encyclopedia of Sci. & Techn.,3,1-533. Gakken,
   Japan.
Miyawaki, A.(edt)(1980). Vegetation of Japan,1. Shibundo-Japan.
Miyawaki, A.(edt)(1981). Vegetation of Japan,2. Shibundo-Japan.
Miyawaki, A, and S. Okuda (1977). Vegetation and Environmental Pro-
   tection, 361.367. Maruzen-Tokyo.
Miyawaki, A. and K. Suzuki (1980). Phytocoenologia 7, 492-506.
Miyawaki, A., K. Suzuki and K. Fujuwara (1977). Naturaliste Can. 104,
   97-107.
Miyawaki et al. (1977). Bull. Envir. Sci. Techn. Yokohama Nat. Univ.2,
   95-114.
Miyawaki et al. (1978). Bull. Envir. Sci. Techn. Yokohama Nat. Univ.4,
   113-148.

POSSIBILITIES OF SURVEYING
SYNECOLOGICAL EFFECTS
(SUMMARY)

P. Müller

Department of Biogeography, University of Saarland,
6600 Saarbrücken, Federal Republic of Germany

ABSTRACT

The ecological effects and consequences of decisions of energy and
economic policies can by no means be adequately assessed with the
legal, medical and toxicological means and measures so far in use.
After testing different ecological control methods, a screening
system is suggested which consists of an experimental monitoring
programme (including effect and trend monitoring) and the analyses
of food chains of selected organisms at the same site. In this
context, the monitoring programmes serve to standardize individual
components of the ecosystem (e.g. exposure of genetically comparable Hypogymnia physodes, Lolium multiflorum or Populus nigra for
the assessment of air pollution), whereas food chain analyses are
used to elucidate the accumulation, degradation, metabolisation
and effects of certain environmental chemicals in significant
free-living organisms of the ecosystems. Experiences gathered in
screening immissions from different sources in Saarbrücken (FRG),
Porto Alegre (Brazil) and North Cameroun are presented.

KEYWORDS

Synecology; ecosystems; experimental monitoring; food chains.

ENERGY POLICY AND ENVIRONMENT ASSESSMENT

Decisions of energy and economic policies decide on the pollution
of ecological systems and man's health. The scope of action is
being more and more restricted by the lack of (immediately available
and cost-saving) alternatives, and prognostic statements as to the
consequences of man's activities are becoming more and more problematic. For this reason, ecosystem research must clarify

- whether and if so, why our 'life form' makes sick;

- which are the effects and consequences of man's activities on the
  environmental qualities of urban and other ecosystems;

- which repercussions on human populations must be expected and

- whether it is possible to draw timely conclusions from the reactions of the ecosystems in which we live on the hazards and threats to human life (Müller, 1980).

It is, of course, obvious that we may come up with an environment-friendly 'intelligent' technology against nature-changing techniques, particularly as the former enables the export-dependent industrialized countries to follow up competitive advantages on a long-term basis. Of course, it is meaningful to transfer dose-effect-relations established in the laboratory to an innovative policy of critical values. Naturally, the best structural policy at the moment is the one which tries to realize both an optimum technology and an optimum environment at the same location.

However, the decisive questions are quite different in that they read as follows:

1. Do we have sufficient knowledge to correctly value chemico-physical processes world-wide?

2. Are autecological and toxicological analyses a sound basis of hazard evaluations e.g. in location decisions?

The discussion of the climatic changes as a result of increasing CO_2-concentrations or the importance of fluoromethanes illustrates how far we are from being able to clearly answer the first question in the affirmative. Yet, the situation is not bad at all in terms of available data. What is lacking in my opinion is a group of scientists who have a chance to think of and develop conceptions and schemes rather than being forced to rush from one research project to the other. This is the reason why environmental protection programmes in Europe, the U.S.A. and Japan are wanting in clearly defined priorities. It is the hazardous political everyday business rather than ecological comprehension which determines the materials to be 'in' next.

Autecological and toxicological basic research has supplied valuable information . Particularly, testing procedures such as those defined in the 'Chemikaliengesetz' (Chemicals Act) of the Federal Republic of Germany (cf. also Rule 79/831 of the EC Council of 18 September 1979) are based on the results of investigations carried out in ecophysiological and toxicological laboratories. In conformity with the Chemicals Testing Programme of the O.E.C.D. (14 March 1980), all new materials entering the market are tested for their physico-chemical properties (such as water-octanol distribution coefficient, complexing capacity etc.) as well as their biological effects (such as acute and subacute toxicity, mutagenicity, cancerogenicity, teratogenicity etc.). From 1982, every new substance must pass through the following toxicological examinations:

1. Acute toxicity (uptake per os; uptake by inhalation; cutaneous uptake; skin irritation test; eye irritation test; dermal sensibilisation);

2. Toxicity after repeated application (subacute toxicity; rat as experimental animal);

3. Other effects (screening tests for mutagenicity and cancerogenicity);

4. Effects on organisms (toxicity in fishes, Daphniae and bacteria);

5. Degradation (biotic degradation in the water; abiotic degradation).

It is appalling for an ecologist to admit that there are, at present, no better tests for evaluating the effects of decisions of energy and economic policies on the environment. Our German legislation, which is certainly exemplary, cannot fill the gaps in the basic evaluation criteria either. The continued discussions on the amendment of the 'Bundesimmissionsschutzgesetz' (Federal Immission Protection Act) and the 'TA-Luft' (technical instructions for keeping the air clean) are a striking evidence of this observation.

SYNECOLOGICAL EVALUATION PROCEDURES

What has been lacking so far are standardized procedures for the evaluation of the effects of environmental chemicals on the ecosystem. In contrast to autecological analyses (analyses of individual components), synecological analyses try to elucidate the effects and consequences of individual substances or complex mixtures of substances on the ecosystems. The degradation or accumulation of the majority of environmental chemicals are not only influenced by abiotic factors but, naturally, also by the biological matrix (cf. among others Clemmesen and Hjalgrim-Jensen, 1981, Morton and Lewis, 1980, Müller 1979, 1980, 1981). Differences of age, sex or the allele type influence the accumulation rates and their effects just as much as do the organ-specific deposits.

TABLE 1   Concentrations of Pb and Cd in different organs of
          juvenile and adult females of Accipiter gentilis in
          Saarland (1981) (mg/kg/dry matter)

muscle tissue			brain
juvenil	Pb	0,10	0,11
	Cd	0,007	-
adult	Pb	0,43	0,40
	Cd	0,014	0,002

The more diverse the biocenosis (for diversity problems cf. Müller,
1981), the greater the uncertainty with which predictions can be
made on the consequently effects of the application of an insecticide (cf. Kohli et al., 1973, Müller et al., 1979, Müller, Nagel and
Flacke, 1980). This uncertainty is even increased by the complex
nature of individual components of the ecosystem (such as the soil),
so that it must be borne in mind that a simple transference of laboratory results to field conditions will always lead to misconcep-
tions. To fill these gaps of knowledge, synecological studies in the
U.S.A., in Japan, Europe and the U.S.S.R. (cf. also ORD programmes;
EPA studies; MAB projects) concentrate especially on the simulation
of the circulation of materials and energy in natural ecosystems
and the analysis of artificial ecosystems which can be manipulated
by man. In the latter attempts are made to imitate the complexity
of field conditions without always realizing that even with ecosystem models it is not possible to eliminate the basic problems
of transferring information from the laboratory to the field.

Normally, short food chains (3 - 7 species), sediment biota or sedi-
ment- water body correlations can be reproduced without major
difficulties in model systems (cf. among others Po-Young et al.,
1975, Metcalf, 1975, 1977, 1979, Rao, 1978, Khan et al., 1976, 1978,
Hill and Wright, 1978, Brown, 1978, Ratclife, 1978). Comprehensive
experiences are available from Great Britain, the U.S.A., Japan,
the U.S.S.R. and Australia. In the Federal Republic of Germany,
similar approaches have been started in which model programmes are
run with 'comparative chambers' and complex biocenoses.

However, even these testing procedures cannot cover up the fact that
the assessment of environmental noxious agents can only be carried
ahead

- by the development of standardized control procedures of
  biosystems in the field;

- by the development of tests which are reproducible in
  natural ecosystems after the application of defined substances;

- by the development and control of effect and trend monitoring systems (experimental monitoring) and

- by food chain and population analyses in the field.

Controls of the reactions of organisms, populations, area systems
(Müller 1979) and biocenoses can cover the whole area under investi-

gation (cf. among others 'Kartierung der Pflanzen Mitteleuropas'
(Plant Mapping in Central Europe); 'Erfassung der westpaläarktischen
Organismen' (Survey of the West Palaearctic Organisms)). In ecological long-term programmes, even the behaviour of complex ecosystems
is taken into account as an indicator of changes in the human
environment (cf. among others the MAB study by Ellenberg, Fränzle
and Müller 1978 for the Federal Republic of Germany; the NSF
study 1979 for the U.S.A.). Residue analyses of free-living populations of numerous plant and animal species are available (such as
the O.E.C.D. programmes; cf. Blau and Neely, 1976, Brown, 1978,
Gish and Christensen, 1973, Holden, 1970, 1973, Khan, 1977, Murphy
and Strassman, 1978, Matsumura, 1977, Metcalf, 1977, Müller, 1979,
Po-Young et al., 1975, Rao, 1978). Without the knowledge of the
ecological data (such as prey animals, size of the habitat, food
chains etc.) their interpretation is, in many cases, of no avail
whatsoever.

EXPERIMENTAL MONITORING

As a hazard indication, the combination of food chain analyses and
experimental monitoring seems to be particularly promising. It is
the task of experimental monitoring to put the spatio-temporal behaviour of ecologically effective factors and ecological elements to
the uses of the assessment of the environment. Within the experimental monitoring, effect monitoring shall permit conclusions to be
drawn from the reactions of ensitive organisms which are used for
this purpose to the site-related relative toxicity of a noxious
substance or a mixture of noxious substances. Trend monitoring is
used to register the accumulation of noxious substances in exposed
or free-living biosystems in terms of time. In trend monitoring,
it is, therefore, best to use less sensitive organisms, which, on
the other hand, are able, as collecting substrates, to accumulate
a representative share of the substances existing in the area.
Effect monitoring of the type of lichen monitoring only permits
statements to be made on the ecotoxicological effect of, for instance, a complex immission type. Normally, a specific noxious agent
can only be eliminated by a simultaneous combination of effect and
trend monitoring at a given site. Only accumulation indicators are
suitable for setting up standardized trend monitoring system (cf.,
among others, grass culture procedures with Lolium multiflorum).
Generally, the analysis of the substances contained in the exponates
allows conclusions to be drawn on the substances which also play a
role in the noxious impact on more sensitive species of the same
site. Furthermore, an area-related transference of the results to
human populations becomes feasible (especially with species used
for food); finally, it is possible to relate the findings to the
noxious impact on free-living populations.

Naturally, only substances characterized by high persistency and
bioaccumulation as well as chronically toxic properties are suitable
for trend monitoring, such as:

- aliphatic and alicyclic substances (haloalkanes, haloalkenes,
  anhydrides, ketones, ester etc.);

- aromatics (such as halophenols, nitrophenols, amines,
  polycyclic-aromatic hydrocarbons, PAHs)

- organometallic compounds

- halogenated hydrocarbons
- heterocyclic compounds
- inorganic compounds (such as As, Be, Cd, Co, Cr, F, Hg, Mo, Mn, Ni, Pb, Se, Tl, V, Zn, methyl-Hg).

At the moment, multiple species monitoring systems are being tried out in which animals and plants of partly different ecology are exposed at the same time with the aim to gain deeper insight into more complex immission types via a multiple reaction potential. The uptake of materials (such as heavy metals, DDT, PCB, PAH) which is species- and partly also sex-specific is understood as an expression of the relative pollution of a site at the time of measuring. Practical experience is available, among others, from the following plant and animal species in the Saarbrücken area: Lolium multiflorum, Sambucus nigra, Brassica oleracea, Plantago major, Taraxacum officinale, Picea omorica, Populus nigra, Aesculus, Plantanus Hypnum cupressiforme, Sphagnum ssp., Vulpes vulpes, Turdus merula, Dreissena polymorpha, Physa acuta, Lumbricus ssp., Lacerta agilis, L. vivipara, L. muralis, Apis, Carabiden.

One of my colleagues, Mr. Wagner, has built up a monitoring programme with Populus nigra "Italica", which covers the whole area of the Federal Republic of Germany.

TABLE 2   Heavy metal concentrations in leaves of Populus nigra "Italica" Aug. 1979 (conc. in mg/kg dry matter)

LOCATION	Cd	Pb	Zn
Oker (north. Harz)	37.1	42.9	976
Essen, municipal forest	3.84	18.7	212
Stolberg east	14.5	51.1	812
Saarbrücken-Burbach	0.88	8.5	172
" -Fh. Neuhaus	1.25	5.4	364
Luxembourg west	0.34	3.5	98
Failly (Lorraine, France)	0.59	3.0	119
Braunstein (Palatinate Forest)	0.54	2.6	154
Mössingen (Swabian Alb)	1.00	1.6	199

TABLE 3   Heavy metal contents of plants and soil (Saarbrücken, Ev. Hospital)(conc. in mg/kg dry matter)

SAMPLE	Cd	Pb	Zn
Poplar leaves 1979	1.06	3.18	185
Poplar (act. mon.)'80	0.79	4.60	117
Parsley '80	1.14	2.61	89
Chives '80	1.35	1.36	49
Urtica urens '80	0.50	2.47	101
Soil '79			
A$_{ph}$ 0-20 cm	0.41	1339	185
A$_{h2}$ 20-40 "	0.61	254	346
B$_{v}$ 40-60 "	0.73	284	366

Such monitoring systems are area-related biotests with exposed organisms. They supply information on the relative threat to the exposed substrates in the analysed areas and allow statements to be made on the burdening of individual components of the ecosystem. Effect and trend monitoring systems have been developed in recent years only, and aquatic organisms (cf. Final Draft Document of the 'International Workshop on Monitoring Environmental Materials and Specimens', 23-28 October, 1978, Berlin), useful animals and terrestrial invertebrates (Müller, 1978, 1979) have mainly been investigated. Laboratory tests and exposure in the field of the species used under standardized conditions must be feasible.

FOOD WEB CONTROL AS A MEANS OF ASSESSMENT

The populations and substrates involved in food webs are integrators of the burdening of areas and ecosystems. For 'short' food chains, which must be representative of different areas of the Federal Republic of Germany (i.e. the species involved must occur and be available both in Kiel and Munich), continuous analyses are feasible, and they are, in fact, partly past the trial phase in several towns and their surroundings, where they have been carried out over the complete areas in question; however, where more complex food webs are involved, they can only be successfully carried through via a continuous analyses and monitoring of the ecosystems.

After several years of - partly negative - experiences and detailed discussions with North American colleagues (US Workshop in Washington 1978) my study group compiled the 'food web species' listed in Fig. 1 for Germany and started continuous population and residue analyses in Saarland. It was shown that, with the knowledge of the ecological variables, the analyses of the soil types (different plant species), of earth-worms (different genera), ground beetles (Carabidae), blackbirds (Turdus merula), insectivores (Sorex), honey-bees (Apis) and foxes (Vulpes vulpes) carried out at the same site yielded especially useful area-related results.

The control of the 'food web species' calls for a well coordinated and highly qualified team working perfectly together in a strictly organized system. This is the only way to ensure the reproducibility of the webbing quantities and the burdening parameters in a constantly changing environment. The basic knowledge which is still lacking in this field cannot be magically produced by planning models, however sophisticated they may be. Control of food web species means the analysis of all prey animals, to cite an example, and of all substances they have taken up (TABLE 4).

TABLE 4   Residues of Pb and Cd (mg/kg dry matter) in Lacerta muralis and its prey animals at two sites in Saarbrücken in 1979

Saarbrücken (Mine of Jägersfreude)		Pb	Cd
Lacerta muralis	27 males	8.02 (2.55-31.03	1.73 (0.36-7.73)
	15 females	4.61 (1.89- 7.99)	0.47 (0.25-0.75)
Beutetiere	Coleoptera	2.35	1.25
	Isopoda	43.45	11.87
	Araneida	12.70	1.33
	Myriapoda	20.52	4.24
	Oligochaeta	75.32	13.92

Saarbrücken (central station)

Lacerta muralis	8 males	23.0 (6.39-40.44)	1.52 (0.45-2.72)
	4 females	18.82(9.48-18.67)	0.64 (0.41-0.85)
Coleoptera		2.72	1.33
Isopoda		61.34	14.05
Araneida		17.53	5.50
Myriapoda		40.91	10.52
Oligochaeta		77.56	26.73

Fig. 1   Food web species in the 'Stadtverband Saarbrücken' which
are regularly subjected to population-ecological and
residue analytic investigations with- in the framework
of a BMI project. Species of higher trophic levels
(such as Vulpes vulpes, Accipiter gentilis, Strix aluco)
are controlled telemetrically.

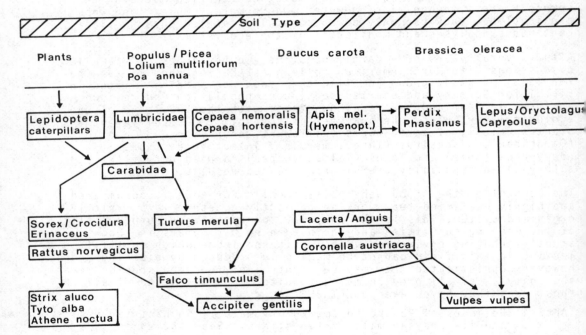

Such analyses, if accompanied by population controls in the field, permit a tight correlation between substrates, prey animals, preditors and other components of the ecosystem (such as soil, water, air). As could be demonstrated by insecticide controls in Tropical Africa, affected human populations at the site must also be included to some extent. This is also true of the Saarbrücken agglomeration area (correlation of PAHs in ground animals and bronchial carcinoma).

In my opinion, the environment-changing effects of our energy and economic policies can only be assessed by the creation of permanent investigation areas (world-wide network), in which experimental monitoring programmes and food web analyses are followed up in a regular and continuous way.

REFERENCES

Blau, H. and W. Neely (1976). Adv.Ecol.Res. 9, 133-163.
Brown, A. (1978). Ecology of Pesticides. John Wiley, New York.
Clemmesen, J. and S. Hjalgrim-Jensen (1981). Ecotoxicology and Environmental Sofety 5, 15-23.
Ellenberg, H., O. FränzTe and P. Müller (1978). Ökosystemforschung im Hinblick auf Umweltpolitik und Entwicklungsplanung. BMI, Bonn.
Gish, C. and R. Christensen (1973). Environ. Sci. Technol. 7, 1060-1062.
Hill, J. and S. Wright (1978). Pesticide Microbiology. Acad. Press, New York.
Kohli, J., I. Weisgerber, W. Klein and F. Korte (1973). Chemosphere 4, 153-156.
Müller, P. (1977). Belastbarkeit von Ökosystemen. Energie und Umwelt, Düsseldorf.
Müller, P. (1978). Urbane Ökosysteme. BMI, 1 - 303, Bonn.
Müller, P. (1979). Basic ecological concepts and urban ecological systems. Biogeographica 15, 209-223.
Müller, P. (1980). Verh.Dtsch.Zool.Ges. 1980, 57-77, Stuttgart.
Müller, P. (1980). Biogeographie. UTB, Stuttgart.
Müller, P. (1981). Arealsysteme und Biogeographie. Ulmer, Stuttgart.
Müller, P., P. Nagel and W. Flacke (1980). Ökosystemare Wirkungen einer Dieldrin-Applikation im Rahmen der Tsetsefliegen-Bekämpfung im Hochland von Adamaoua, Kamerun. GTZ-Mitt. 1-196, Eschborn.
Müller, W., G. Nohynek, F. Korte and F. Coulston (1979). Z. Naturforsch. 34c, 340-345.
Matsumura, F. (1977). Absorption, Accumulation and Elimination of Pesticides by Aquatic Organisms. Plenum, New York.
Metcalf, R. (1977). Model Ecosystem Studies of Bioconcentration and Biodegradation of Pesticides. Plenum, New York.
Morton, M. and R.A. Lewis (1980). Annual changes in blood parameters, organ weights, and total body lipid in Peromyscus maniculatus and Microtus ochrogaster from a Prairie environment. J. interdiscipl. Cycle Res. 11 (3) 209-218.
Po-Young, L., R. Metcalf, L. Furman, R. Vogel and J. Hasset (1975). J. Environm. Qual. 4, 147-163.
Rao, K.R. (1978). Pentachlorphenol. Plenum, New York.
Wagner, G. (1981). Monitoring of airborne Heavy Metal pollution using selected accumulation indicators. Intern. Conf. Heavy Metals in the Environment, Amsterdam.

ACID RAIN, EASTERN U.S.A. - PROBLEMS OF CONTROL

R. M. Perhac

Environmental Assessment Department, Electric Power Research Institute, Palo Alto, California, USA

ABSTRACT

Extensive measurements in the eastern U.S.A show that precipitation is averaging about pH 4.2 on a yearly basis. Subtle north-south trends are apparent, and pH tends to be lowest in summer. No evidence of change in acidity is apparent over the past few years. The extent to which acidity has changed over the past few decades is uncertain. Data are conflicting. Some data indicate that pH has dropped; others indicate no change. Data from stations which have been in continuous operation for a number of years in New Hampshire, New York, Pennsylvania and Tennessee give no indication of change.

The quantitative role of anthropogenic activity, particularly coal burning by power plants, is uncertain. Atmospheric sulfates often show changes in concentrations that do not reflect changes in SO_2 emissions. Meteorological variables, not just SO_2 emissions, play a significant role in atmospheric sulfate levels. Variability in rain pH does not correlate simply with changes in air quality. Because of the many uncertainties, we cannot judge what effect increasing or decreasing emissions will have on the acidity or distribution of precipitation. Reliable models which relate quantitatively emissions of precursors to acidity of rain have yet to be developed.

KEYWORDS

Acid rain; acid precipitation; air quality; air quality control.

INTRODUCTION

Whether or not acid rain occurs is not an issue. Such rains are falling over, at least, much of the eastern third of the United States. The acid rain issue focuses on three hypotheses: (1) that acid rain has been increasing (in acidity and geographic extent) in the United States over the past few decades; (2) that acid rain is causing severe damage to ecosystems; and (3) that coal burning by electric power plants is a major contributor to acid precipitation. The present controversy in America centers on the extent to which people feel that the hypotheses are true.

This paper will focus only on the first and third hypotheses. For the purpose of discussion, let us accept that acidic deposition has the potential for causing severe ecological damage or that it is, in fact, causing extensive damage. What, then, do we do to reduce the acidity of precipitation? What control measures will be effective in lessening the potential impact of acidic deposition? The first and third hypotheses bear on these questions of control.

ACID PRECIPITATION IN THE EASTERN U.S.A.

Throughout most of the eastern third of the United States, precipitation with an annual mean pH of about 4.2 is falling (Table 1). This value is based on nearly two years of data from nine stations supported by the Electric Power Research Institute (EPRI) and from data from eight stations (the MAP3S Program) maintained by the U.S. Department of Energy and the U.S. Environmental Protection Agency (Dana, 1980). Other measurements throughout the East support the observation that the present pH of precipitation is generally between 4.0 and 4.5.

TABLE 1 AVERAGE pH (AUGUST 1978 – JUNE 1980)

	pH	Uncertainty
Duncan Falls, Ohio	4.13	0.12
Giles County Tennessee	4.12	0.09
Indian River, Delaware	4.22	0.11
Lewisburg, West Virginia	4.31	0.19
Montague, Massachusetts	3.96	0.15
Raleigh, North Carolina	4.50	0.13
Roanoke, Indiana	4.19	0.11
Rockport, Indiana	4.10	0.07
Scranton, Pennsylvania	4.11	0.08

The acids in eastern rain are principally the strong ones, sulfuric and nitric in a ratio of about 6:4, respectively. (Strong acids, in fact, comprise over 80 percent of the acids in the rain.) The acidity shows a subtle seasonal pattern, being somewhat greater in summer than in winter (Fig. 1). Geographic trends, likewise, are subtle. We see little change in an east-west direction over a distance of about 1500 km; however, slightly less acidic conditions prevail as one goes from north to south. Perhaps the most notable feature of the precipitation is the striking differences in pH from one rain event to the next (Fig. 2).

That acid precipitation is now falling is unquestioned. We are not so certain, however, about what changes may have occurred over the past decades. Some data seem to show an increase in acidity (both in terms of decreasing pH and increase in geographic area being affected) over the past three or four decades. Other data suggest no change.

Most of the claims for increasing acidity are based on studies by Likens (1976) and Cogbill and Likens (1974). Using data from either rain chemistry or rain acidity, they constructed a series of maps for three time periods (1955/56, 1965/66, and 1972/73) which, by use of pH contours, show both an increase in rain acidity and in the area being affected by such rains (Fig. 3). Using the numbers on the maps, the conclusion must be drawn that precipitation has become more acid since the mid 1950s. The value of the data, however, can be questioned for three reasons. Firstly, the different networks for the three time periods involve different sets of monitoring stations, and we cannot be certain of the reliability of comparing data from one time period to another if different sets of stations are involved. Secondly, different methods of collecting precipitation were used for the different time periods, and different collection procedures have a bearing on the pH that would be measured (Miller and Everett, 1979). Thirdly, different techniques were used for arriving at acidity (pH) values. For example, for the 1955/56 map, pH values were calculated from chemical analyses of the precipitation. For the 1972/73 map, pH values were obtained by direct measurement with a pH meter. Tyree (1981) has shown that measured and calculated pH values from 32 samples collected by the World Meteorological Organization differ by an average of approximately 0.9 ±0.6 pH units (range = 0.02 to >2.7) with no consistency in direction, hence no simply correction factor can be applied to one set of data to bring it in line with another. This uncertainty in measurement (0.9 pH units) is greater than the increases in acidity shown by the maps for the two time periods 1955/56 and 1972/73. But let us accept that the data are valid. We must, then, conclude that rain in the eastern United States is becoming more acidic.

Other studies, however, suggest that no change in acidity has occurred. Likens and others (1977) have maintained a station at Hubbard Brook, New Hampshire (U.S.A.). For at least the 10-year period 1964-1974 (Fig. 4), no overall trend in rain acidity was statistically significant (Likens et al., 1977, p. 36). Similarly, the U.S. Geological Survey has measured total deposition (wet and dry) at nine stations in New York State (U.S.A.) and Pennsylvania (U.S.A.) since 1965 (Fig. 5). The data from the stations (Fig. 6,7) show no significant change in acidity (Hansen and Hidy, 1980, p. 3-47). And a number of monitoring projects run by the U.S. Tennessee Valley Authority (some starting as early as 1971) show no consistent change in acidity in part of the southeastern U.S.A. (TVA, 1980).

We find, therefore, that data on changing acidity are both confusing and conflicting. The U.S. Environmental Protection Agency points out that "The absence of a precipitation monitoring network throughout the United States in the past makes determination of trends in pH an extremely difficult and controversial topic." (EPA, 1980, Chapt. 8).

THE ROLE OF COAL BURNING

One hypothesis of the acid rain issue is that coal burning by electric utility companies is a major cause of acid rain in America. If so, then a control strategy could focus on reduction of emissions from power plants. In order to judge the effectiveness of reducing emissions, we should quantify the following functional relation:

$$acid\ rain = f\ (emissions)$$

Unfortunately, the relation is not simple. It is a complex one comprising at least six components: (1) emissions, (2) chemical conversions of primary emitted compounds (e.g., SO_2) into secondary products (e.g., sulfates or sulfuric acid), (3) atmospheric transport, (4) scavenging efficiency of clouds, (5) in-cloud processes which result in the acidity of the rain drop, and (6) changes in rain drop chemistry during fall through the atmosphere. We must be able to quantify these components if we are to put the above functional relation into a usable equation--

one which defines the utility role in acid rain. And we must have a quantitative relation if we are to judge the efficacy of any control strategy which is to be promulgated.

Unfortunately, we can quantify only the first of the six components. We have good data on utility emissions, particularly for SO_2. In the eastern half of the United States, utility operations account for about 60-65 percent of the total SO_2 emissions (Fig. 8). A striking feature of utility emissions is the consistency from day to day and from week to week. Typically, utility SO_2 emissions vary by less than 10-15 percent on a day-by-day basis.

In contrast to SO_2 emissions, ambient sulfate concentrations may vary considerably. Typically, ambient concentrations in the East are 5-10 $\mu g/m^3$; 75 percent of the time the ambient level is less than about 10 $\mu g/m^3$ (Fig. 9). High values, however, do occur (Fig. 9). The cumulative frequency plot (Fig. 9) shows many sulfate values above 25 $\mu g/m^3$. We do not yet understand all the reasons for the occurrence of the high values but we do know that both air temperature and dew point temperature play an important role in increasing the atmospheric content of sulfates (Fig. 10) even though emissions of the precursor SO_2 may be changing little. As a result, we have many observations which show SO_2 emissions changing by only 5, 10 or 15 percent (day-by-day) whereas sulfate concentrations are changing by hundreds of percent (Fig. 11). If we try to relate SO_2 emissions to a specific sulfate, e.g., sulfuric acid, our lack of knowledge becomes even more apparent.

Many cities have experienced marked decreases in SO_2 levels without commensurate decreases in sulfate concentrations (Altshuler, 1976). This discrepancy has been explained by many by recourse to long-range transport. In other words, a city may reduce its SO_2 levels simply by burning cleaner fuels but the sulfate levels might not be affected if the sulfates are a result of material being transported in from a long distance. Similarly, acid rain in the eastern U.S.A. has been attributed to SO_2 emissions from the Midwest coupled with conversion to and long-range transport of sulfates. Unfortunately, our ability to model, or even to recognize, long-range transport is still in a nascent stage. As a result of two years of detailed study in the East (EPRI, 1981), we do not feel that we can distinguish with any sense of certainty transported material from the local background of pollutants. And on the few occasions when we feel that we can recognize transport, we also recognize that explanations other than long-range transport can account for the field observations. For example, during a major sulfate episode (July 20-23, 1978), field measurements showed that a mass of sulfate built up (to levels of 35-40 $\mu g/m^3$) over the Ohio Valley (U.S.A.). Over the next three days (Fig. 12-15), this mass of sulfate "moved" from west to east a distance of about 1000 km. Such movement would represent true long-range transport of a mass of pollutant. The observation can be explained, however, without recourse to long-range transport. Instead it can be explained by the meteorological condition (which started the episode in the first place) moving from west to east and converting local emissions of SO_2 to sulfate during its movement from the Ohio Valley to the Atlantic coast. Our data (or modelling) are not sufficient to distinguish the two cases. Both are permissive. The above observations are based on detailed aerometric measurements at 54 stations (Fig. 16) as part of EPRI's Sulfate Regional Experiment (Perhac, 1978). Because our long-range transport models have a high degree of uncertainty in their estimates, we are in a difficult position in trying to evaluate what sort of change in emissions would effect what sort of change in ambient levels of pollutants.

The problem of relating emissions to acid rain becomes even more difficult when considering in-cloud processes. We cannot quantify the scavenging efficiency of clouds. Identical concentrations of pollutants in the atmosphere can result in markedly different concentrations within the cloud itself. And our knowledge is scanty about the physico-chemical processes that cause cloud water to become acidic. Finally, the scientific community has barely started to investigate changes in rain chemistry during fall of the drops. Even if we could predict what concentration of ambient pollutants (in the vicinity of a cloud) would result from a particular emissions level, we could not predict the acidity of rain that would result from that ambient pollutant concentration. We know for example that the hydrogen ion content of rain can change drastically from one storm to the next--by factors of more than 100 (Fig. 3)--whereas the ambient concentration of pollutants is not changing nearly that much.

Until we understand the chemical conversion of emissions, the role of long-range transport, and cloud processes, we cannot model the relation between emissions and acidity of precipitation. Without that modeling capability, we cannot assess the role of utility operation in the occurrence or distribution of acid rain. We do not know, for example, what effect a 20 percent increase or decrease in emissions will have on the acidity of rain. Too many uncertainties exist.

In view of these uncertainties, regulatory action is difficult. The regulator faces the problem of trying to design a control strategy without knowing quantitatively (or even qualitatively) how effective that strategy will be. For acid rain, it is easy to propose a reduction in emissions from power plants (because it will do no harm and it may do some good). But reducing emissions is not a simple matter. Reducing sulfur dioxide emissions, for example, may involve the installation of scrubbers at a cost of $100 million per plant--a cost which is ultimately borne by the consumer.

The regulator is in a difficult position. Should he propose an emissions control strategy which may be costly to the consumer but without the knowledge of how effective that strategy will be in correcting a problem whose severity is still not well established. The issues are complex and simple solutions are not readily apparent.

WORKS CITED

Altshuler, A. P., 1976, Regional transport and transformation of sulfur dioxide to sulfates in the U.S.: Jour. Air Pollut. Control Assoc., v. 26, p. 318-324.

Cogbill, C. V. and G. E. Likens, 1974, Acid Precipitation in the northeastern United States: Water Resources Res., v. 10, p. 1133-1137.

Dana, M. T. (ed.), 1980, The MAP3S precipitation chemistry network: Battelle Pacific Northwest Laboratory RPT PNL-3400.

EPA, 1980, External review draft No. 1, Air quality criteria document for particulate matter and sulfur oxides: U.S.E.P.A., 1980.

EPRI, 1981, SURE - Sulfate regional experiment: in press.

Hansen, D. A. and G. M. Hidy, 1980, Examination of the basis for trend interpretation of historical rain chemistry in the eastern United States: Environmental Res. & Technology Rpt. P-A097 (Westlake Village, Calif., U.S.A.).

Likens, G. E., 1976, Acid precipitation: Chem. Engr. News, Nov. 22, 1976, p. 29-44.

_____, F. H. Borman, R. S. Pierce, J. S. Eaton, and N. M. Johnson, 1977, Biogeochemistry of a forested ecosystem: Springer-Verlag, New York, 146 p.

Miller, M. L. and A. G. Everett, 1979, History and trends of atmospheric nitrate deposition in the eastern U.S.A.: Am. Chem. Soc., paper presented before Div. Environ. Chem., Washington, DC, Sept. 1979.

Perhac, R. M., 1978, Sulfate regional experiment in northeastern United States: the "SURE" program: Atmos. Environ., v. 12, p. 641-647.

TVA, 1980, What's wrong with rain: Impact, Tenn. Vy. Authority, v. 3, no. 4, p. 2-10.

Tyree, S. Y., 1981, Rainwater acidity measurement problems: Atmos. Environ., v. 15, p. 57-60.

MEAN pH VS TIME

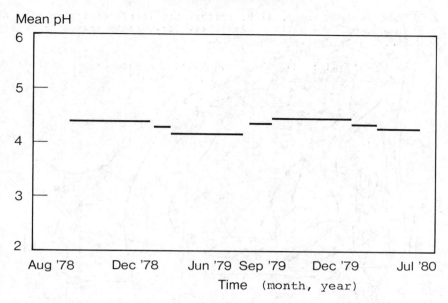

FIGURE 1

pH (event basis) VS TIME
Indian River, Delaware

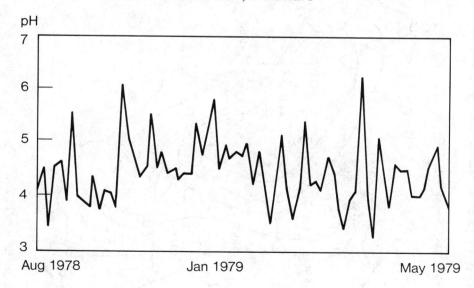

FIGURE 2

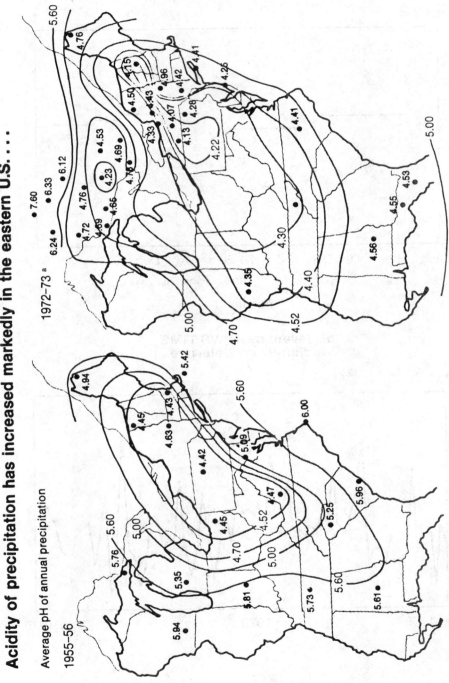

Acidity of precipitation has increased markedly in the eastern U.S....

Average pH of annual precipitation

1955–56

1972–73 [a]

FIGURE 3

After Likens (1976)

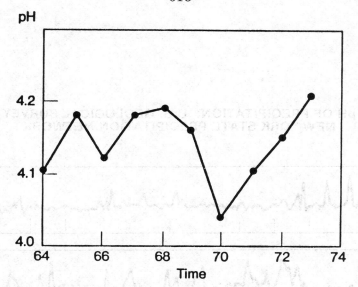

"There has been a slight upward trend in . . . concentration of
hydrogen ion between 1964–1965 and 1970–1971, followed by a
downward trend until 1973–1974; overall (1964–1974), however, no
trend in concentration is statistically significant."

Liken, G. E. et al., 1977, *Biogeochemistry of a Forested Ecosystem*,
Spring-Verlag, New York, p. 36.

FIGURE 4

U.S. GEOLOGICAL SURVEY: PRECIPITATION MONITORING STATIONS

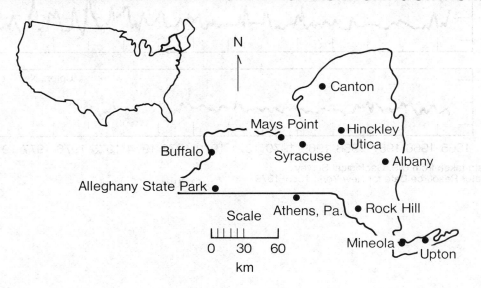

FIGURE 5

**pH OF PRECIPITATION: U.S. GEOLOGICAL SURVEY
NEW YORK STATE PRECIPITATION NETWORK**

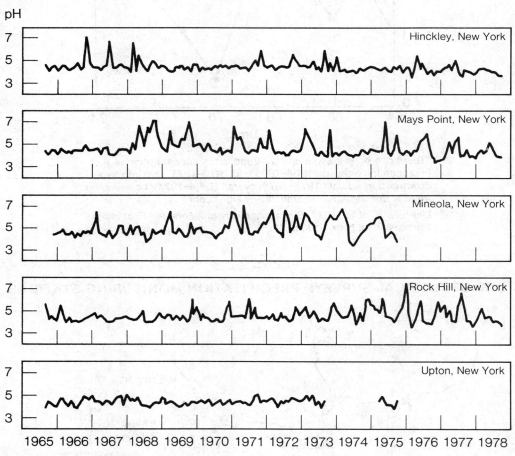

Data taken from U.S. Geological Survey
Water Resource Data for New York, 1965–1979.

FIGURE 6

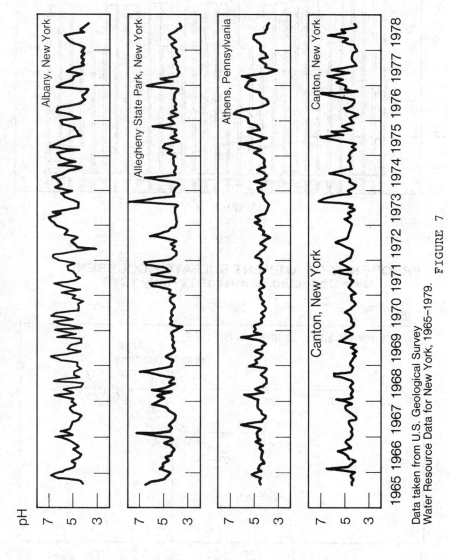

pH OF PRECIPITATION: U.S GEOLOGICAL SURVEY
NEW YORK STATE PRECIPITATION NETWORK

Albany, New York

Allegheny State Park, New York

Athens, Pennsylvania

Canton, New York

Canton, New York

pH

7
5
3

7
5
3

7
5
3

7
5
3

1965 1966 1967 1968 1969 1970 1971 1972 1973 1974 1975 1976 1977 1978

Data taken from U.S. Geological Survey
Water Resource Data for New York, 1965–1979.

FIGURE 7

SO₂ EMISSIONS

Emissions (10⁴ t/d)

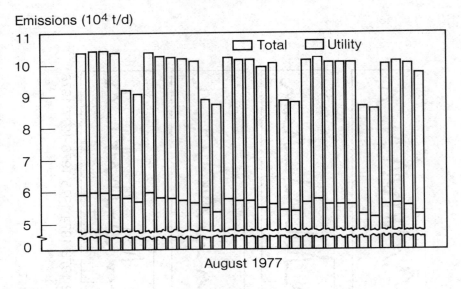

August 1977

FIGURE 8

FREQUENCY OF AMBIENT SULFATE OCCURRENCE
Data Collected August 1977–July 1978

Sulfate (μg/m³)

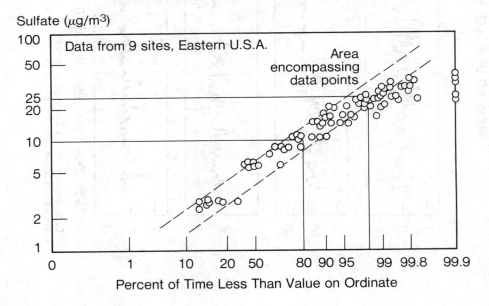

FIGURE 9

SO₄⁼ CONCENTRATION VS DEW POINT

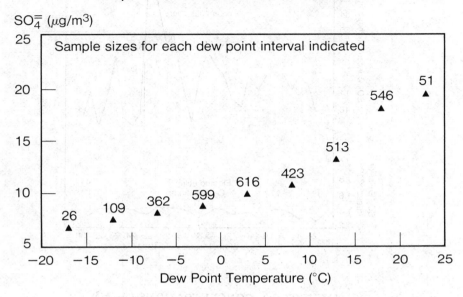

FIGURE 10 a.

SO₄⁼ CONCENTRATION VS TEMPERATURE

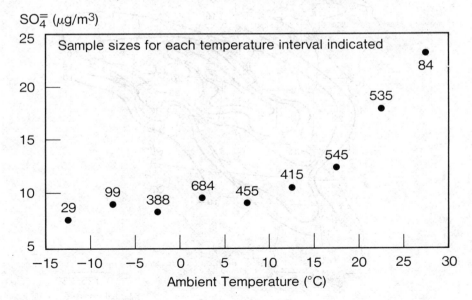

FIGURE 10 b.

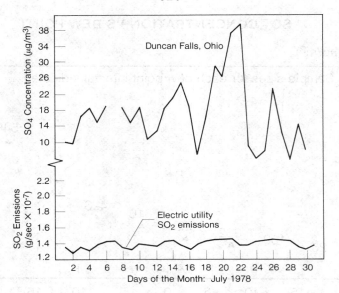

FIGURE 11

AMBIENT SO$_4$ CONCENTRATIONS (μg/m^3)
July 20, 1978

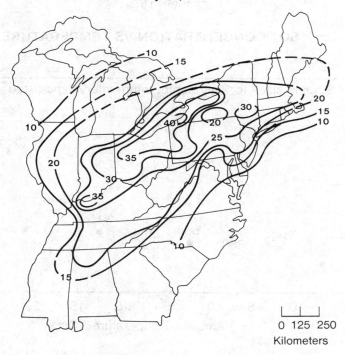

FIGURE 12

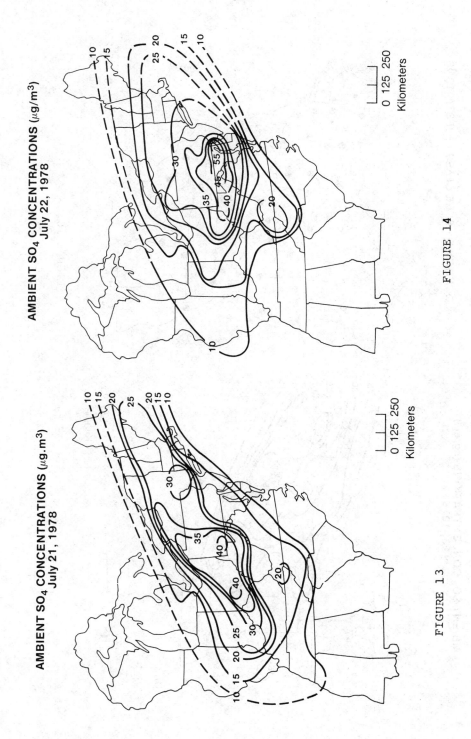

AMBIENT SO₄ CONCENTRATIONS (μg/m³)
July 22, 1978

0 125 250
Kilometers

FIGURE 14

AMBIENT SO₄ CONCENTRATIONS (μg.m³)
July 21, 1978

0 125 250
Kilometers

FIGURE 13

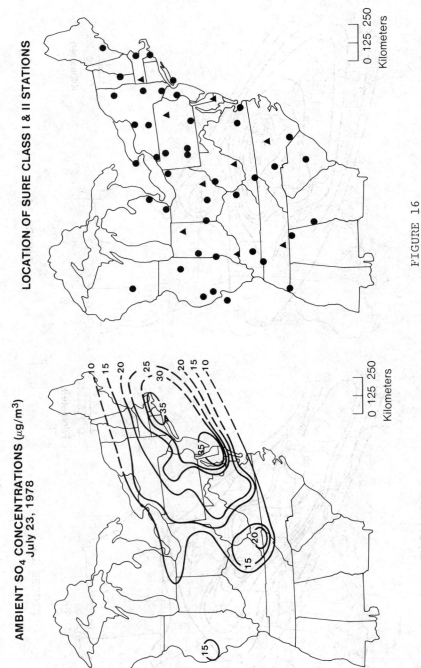

AMBIENT SO₄ CONCENTRATIONS (μg/m³)
July 23, 1978

FIGURE 15

LOCATION OF SURE CLASS I & II STATIONS

FIGURE 16

SOME ENVIRONMENTAL CONSIDERATIONS IN SITING A SOLAR/COAL HYBRID POWER PLANT

R. L. Perrine

*Environmental Science and Engineering, University of California,
Los Angeles, California, USA*

ABSTRACT

A Solar thermal/fossil hybrid power plant could prove useful to meet combined needs such as for repowering "old" fossil systems, reducing fossil dependence, and meanwhile facilitating cogeneration. However, unique environmental problems arise from interactive systems. This study took as a starting point a Rockwell International hybrid design, and a suitable site in California with good available data. Assessment covered topics such as impacts on the solar system of fossil emissions, cooling tower salts, and fugitive sources, and impacts of the combined systems on their surroundings. In several areas useful quantification is possible, though in others data are insufficient. Air quality concerns are paramount.

KEYWORDS

Solar-coal hybrid; solar thermal; repowering; hybrid systems interaction; fugitive emissions; heliostat performance degradation.

INTRODUCTION

Historically, the fuels of choice for the production of electricity have been oil and natural gas. These no longer represent either cheap or reliable energy sources, however. Our long-term future clearly rests in large part on use of renewable resources. In the decade of the 90's it will be technically possible for solar energy to begin to fill the traditional role of fossil fuels.

There are two basic solar electricity options: photovoltaics, and solar thermal systems. For a stand-alone facility including dedicated storage, current projection of technological advances may give photovoltaic systems an apparent edge. However, solar thermal systems retain advantages in several important situations. If there is a need for cogeneration--the shared production of electricity and process heat--solar thermal energy systems are required. There also may exist a currently fossil-fueled plant for which, over an extended lifetime, repowering can be justified. A solar thermal/fossil hybrid system, utilizing fossil input to all but eliminate energy storage needs, could prove useful in many such situations. Environmental protection could be enhanced while (hopefully) reducing costs and assuring reliability through use of an available and relatively cheap fuel such as coal.

There are, however, unique environmental problems associated with any attempt to combine the best of two worlds. The purpose of the study behind this paper (ESE 1980) was to address the constraints unique to siting a solar/coal hybrid system. We anticipate that in the future other hybrid configurations may be thought useful. Thus an added benefit of the present results may be to anticipate the nature of concerns accompanying development of other interactive systems.

System Design

Our study took as a starting point the Rockwell International Energy Systems Group's 430 MWe Solar Central Receiver/Coal Hybrid Power System Design (Rockwell 1980), so that effort could focus promptly on the issues of environmental impacts. This design, selected because of its scale and degree of definition, was the largest for which design data had been developed. It utilizes a field of 61,000 heliostats as its primary energy collection source. Sodium is the energy transfer medium and also provides short-term (three hour) storage. A coal-fired sodium heater operates in parallel; routinely at 20% of capacity and so able to ramp to full power in less than five minutes. A summary of power system characteristics is provided by Table 1 and a sketch of power flow is given in Fig. 1. It should be noted that the system as sketched represents a highly optimistic efficiency near 43%. Using sodium and such approaches as cascading turbines, the optimistic figure might be achieved. Exhaust gases are cleaned using dry fuel gas desulfurization for SO_2 removal, and baghouses to remove particulates. NO_x levels are lowered by burner design and control of combustion air. Wet or wet/dry cooling towers are available options.

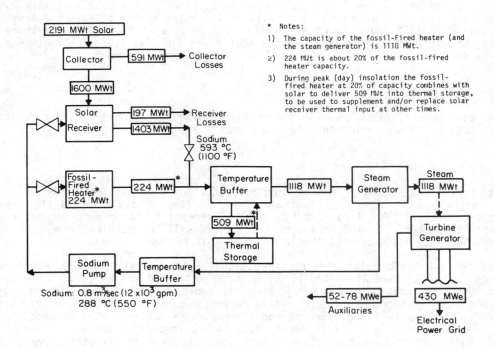

* Notes:

1) The capacity of the fossil-Fired heater (and the steam generator) is 1118 MWt.

2) 224 MWt is about 20% of the fossil-fired heater capacity.

3) During peak (day) insolation the fossil-fired heater at 20% of capacity combines with solar to deliver 509 MWt into thermal storage, to be used to supplement and/or replace solar receiver thermal input at other times.

Fig. 1. Hybrid Plant Power Flow (Rockwell 1980).
(Numbers approximate, scaled from Rockwell's figures.)

TABLE 1   Summary of Solar Receiver-Coal Hybrid
Power System Characteristics (Rockwell 1980)

Design Point Power Levels:
  During receiver operation ....430 Mwe net
  Operation exclusively from thermal storage..........................430 MWe net
Thermal Storage Capacity at 100% Load.................................3 hours
Plant Availability (exclusive of sunshine)...........................$0_{.}9$
Land Area...12.34 km^2 (4.77 mi^2)
Exclusion Area..site dependent
Glass Area...2.98 km^2 (1.15 mi^2)
Number of Heliostats...60,676
Heat Rejection............................2 mechanical draft, wet cooling towers
Tower Height...330 m (1083 ft)
Distance to Cooling Towers.............................1568 m (5140 ft)
Electric Power Generation Subsystem Feedback Conditioning:
  Dissolved solids..20-50 ppb
  pH..9.5
Water Consumption:
  Heliostat cleaning................1.47×10^7 l/yr (3.88×10^6 gal/yr; 12 AF/yr)
  Cooling towers...................7.68×10^9 l/yr (2.03×10^9 gal/yr; 6227 AF/yr)
  Flue Gas Desulfurization........4.04×10^9 l /yr (1.07×10^9 gal/yr; 3282 AF/yr)
Coal Consumption at 100% Load.............................203 MT/hr (223 tons/hr)
Coal Storage Capacity:
  Dead.....................258,000 MT (52 days at capacity or 283,800 tons)
  Live..19,544 MT (21,500 tons)
Transmission Lines (assumed)......................................115 KV
Nominal Design Wind (at reference height
  of 10 m (30 ft))...3.5 m/s (8 mi/hr)
Maximum Operating Wind (at reference height,
  including gusts)...16/m/s (36 hr)
Maximum Survival Wind (at reference height,
  including gusts)...40 m/s (90 mi/hr)
Seismic Environment (Uniform Building Code Zone 3
  survival earthquake, horizontal and vertical)....................0.25 g
Rain Survival:
  Average annual...750 mm (30 in)
  Maximum 24-hr rate.......................................75 mm (3 in)
Operating Lifetime...30 years
Operating Ambient Air temperature.....................-30 to 50°C (-20 to 120° F)

Example Site

As a highly suitable site for which good data were available, the Palo Verde South (Blythe Site) in Riverside County, California, was utilized (Fig. 2). This site had been chosen by San Diego Gas and Electric Company for its once-proposed Sun-desert Nuclear Plant (SDGE 1976). Characterized by its very high insolation, typical of southwestern U.S. deserts, it also shows unusually low seismicity, is remote from population growth yet near transmission corridors, does not consist of unduly sensitive areas, harbor unusual vegetation or endangered species, and though water-short has available agricultural return flows as coolant (SDGE 1977).

POTENTIAL FOR SOLAR/COAL HYBRID SYSTEMS INTERACTION

The primary concern in our assessment has been the potential for interaction and the emergence of synergistic impacts from elements of the hybrid system, which might lead to unique environmental constraints.

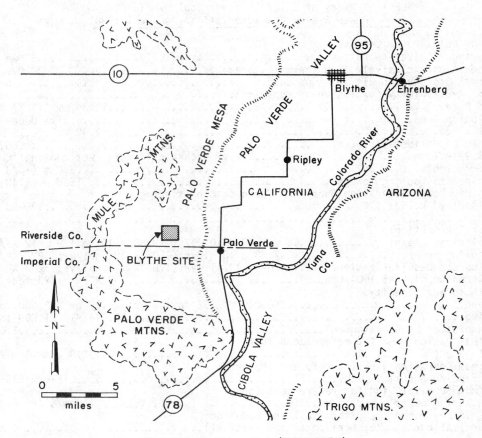

Fig. 2 Location Map (SDGE 1976)

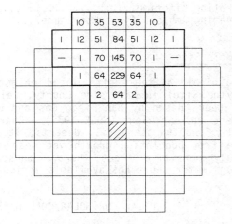

Fig. 3 Number of heliostats per cell, preferred commercial system (Rockwell 1980).

Fig. 4 Predicted particulate mass per heliostat in each 30-day month (in grams).

TABLE 2   Emissions Standards Applicable to 430 MWe Coal Power
Plant in Riverside County, California (SCE 1979)

SCAQMD Rule No.*	Pollutant	Maximum Allowable Emissions
405	Particulate Matter	2.9 g/sec (23 lb/hr)
431.3	Sulfur Dioxide	319 g/sec (2530 lb/hr)
475/1135.1	Nitrogen Oxides	92 g/sec (730 lb/hr)
407	Carbon Monoxide	1484 g/sec (11780 lb/hr)

*SCAQMD South Coast Air Quality Management District, State of California

Coal Combustion Emissions and Heliostat Performance

Atmospheric emissions from coal combustion lead to a first concern: the impact on heliostat performance. Figure 3 shows numbers of heliostats within cells distributed as for the preferred commercial system. Each cell is 369 m on a side. Applicable emissions standards for a 430 MWe plant are given in Table 2 (SCE 1979). Dispersion modeling using established methods (Turner 1969) can be used to predict ambient particulate levels within polluted air parcels reaching the heliostats. Three types of meteorologial conditions can cause worst-case ground-level results from the elevated source: a turbulent and well-mixed unstable atmosphere; trapping of the plume by the base of an inversion above the plume stack; or fumigation when emissions are released into a stable inversion layer, and then entrained into the mixed layer as surface heating breaks the inversion. The first of these was selected as most representative for use in this study. The lowest wind speed observed at the site, about 3.2 km/hr (2 mph), was assumed to be from the south in order to maximize heliostats exposed to the emissions-laden plume.

A simple model was developed to estimate particle impaction and mass deposition rates on heliostat surfaces (ESE 1980). Multiplying the ambient particulate concentration in $\mu g/m^3$ by the ground level wind speed in m/s produced a particle impaction rate in $\mu g/m^2$-sec. This is the rate of impaction on the row of heliostats first reached by the plume. Further downwind results are influenced by upstream removal, and a small correction is needed. Most important, however, is the fraction of impinging particles which actually stick to a heliostat surface. Empirical data to date (for example, King and Meyers 1978) appear to us to not serve as a useful empirical base for our model. Thus we arbitrarily assume, as an approximation and for purposes of calculation, that 10% of the total mass of particles computed as impacting a heliostat surface stick to that surface, and hence are removed from air. In reality, this "sticking coefficient" will most likely not remain constant; rather, it will vary with wind speed and direction, particle size and chemical composition, temperature, humidity, and degree of soiling of the heliostat surface. The model also assumed a uniform 45° angle for heliostats. In reality these will vary both in space and time. Thus results should be viewed as providing first-cut, order-of-magnitude estimates only.

Figure 4 presents predicted mass of coal particulates deposited on a single heliostat in a cell within a 30-day month. The model predicts a worst-case maximum of 229 g deposited, which translates into 6.5 g of particulate matter per m^2 area. A major result is to show that deposition is highly dependent on location in the field, and most likely not all heliostats will require the same washing schedule. A second useful result is that worst-case intra-plant air quality impacts can be mitigated through changes in layout; carefully selecting relative positions of sources and heliostat receptors to generally minimize adverse impact. In fact, a very effective design change would leave an "open" quarter downwind of the coal stack and cooling towers. Solar collection could be somewhat hindered, but there are at least possible tradeoffs to be evaluated. A critical need is development of a sound theoretical and experimental basis from which to model the actual aerosol impaction and sticking on heliostat surfaces.

Impact of Salt Emissions from Cooling Tower Operation

Cooling tower emissions can also impact heliostat performance. Modeling had been done for the Sundesert project (SDGE 1977), thus results were simply scaled to our use. Wet mechanical draft cooling towers represented a worst-case equivalent to the anticipated wet/dry towers from the standpoint of salt deposition. Meterological conditions assumed were not documented in the original reference, though conditions similar to those for coal particulate deposition should apply. Source location and size thus were simply adapted to the current problem.

Maximum salt deposition rate predicted is 50 kg/km^2-month for a July period, equivalent to 0.05 gm/m^2-month and thus much less than the maximum particulate deposition rate from coal burning sources (6 gm/m^2-month). Use of different models invited disagreement. However, an apparently insignificant deposition rate from a larger source (about 11 g/sec compared with 3 g/sec for coal particulates) suggests substantially different model assumptions and need for re-evaluation.

Whereas coal combustion particulates following scrubbing and collection should be very small and readily drift almost unaffected by gravity, salt particles would likely grow larger and settle out readily. If an equivalent impaction model to that for coal were developed, we would thus anticipate larger impacts. Because of different source locations, the pattern of fallout of salt particles differs from coal, leading to some levelling of overall spatial dependence of particle impacts.

Fugitive Emissions Sources

Fugitive dust emissions from outside the plant. Another important concern is fugitive dust. The desert is a "play" area for many recreationists, and a large, highly visible solar facility is certain to become a significant attraction to area visitors. Off-road vehicles, careening across the neighboring terrain in order to get a better look, could cause a significant dust problem and soiling of heliostats. Natural dust storms characteristic of deserts add to the man-made burden.

Studies have shown that fine particles will collect on heliostat surfaces even when they are turned face down (King and Meyers 1978). When face up, larger particles as well are likely to be trapped. The widespread use of off-road vehicles thus is a potentially very important source in sites such as that considered. The only feasible control is to establish an exclusion area; i.e., fence off an area large enough so that dust would settle out to near background levels.

Quantifying impacts of off-road vehicle operation is difficult at best because of many complicating factors; vehicle size and type, speed, location, local meteorology, and desert soils and terrain. The first step in estimating fugitive dust impacts is to compute an emission factor. A formula has been developed (EPA 1978a) in which we use these assumptions: 25% silt content of the surface, 40 mph average vehicle speed; an average vehicle weight of 1 ton, and 345 dry days per calendar year. Substituting these values leads to a presumed worst-case emissions factor of 6.43 lb per vehicle mile travelled. To use the model, off-road vehicles are considered a ground-level line source at the north edge of the heliostat field, with wind from the north. To obtain a line source emissions rate we assume 10 vehicles, each travelling about 1/10 their daily recreational mileage along an effective line source length of 3.2 mi (5.1 km). With 8 hours per day of operation, these values lead to an emission rate of 6.3 x 10^{-3} g/m-sec.

Meteorological conditions expected to produce maximum ground-level concentrations from ground-level sources are a stable atmosphere and low wind speed (Turner 1969); in this case 0.9 m/s. Two exclusion areas were considered, extending 1- and 2-mile distances beyond the plant boundary. Predicted mass deposition of fugitive dust in kg per cell over one month for the 1-mile exclusion area is shown by Fig. 5. Data for a 2-mile exclusion area show only a modest drop-off, with the peak loading falling from 80 to 68 kgm/cell over the 30 day period.

Fugitive emissions from coal handling and in-plant vehicle operation. Coal handling constitutes a source much like vehicular travel. Both ground level (loaders, stockpiles, transfer points) and elevated sources (hopper building, crusher, live storage and coal silos) contribute. In this study those above 100 m are considered elevated sources. Paved roads were proposed by Rockwell to minimize plant fugitive dust sources. Entrained dust from travel on roads should be about 3.2 g/km-vehicle. Impacts can be calculated by the methods illustrated.

Fugitive dust impacts on the environment. Clearing of vegetation and construction will cause a regular loss of about 4 million tons of dust per year, deposited mostly downwind or northeasterly of the site. The Barstow Solar Plant site shows a corona of sand formed from wind erosion (Lindberg 1980). Concern is not for material "lost", but light particles blown back onto heliostats when wind reversals occur. Construction is expected to produce about 1.2 tons of dust per acre per month (EPA 1978b), a total of 216,000 tons over 5 years of construction.

Summary of fugitive source impacts. Figure 6 shows mass deposition rates (g/m^2) for both coal combustion and fugitive dust over a 30-day month, for the northern fraction of the heliostat field most subject to impact. Coal figures are given at the top and dust (in parenthesis) below for each cell. Results suggest that, allowing for other effects also, regular reflectance maintenance will be essential.

A related impact of the plant on the desert environment is expected. Fine desert soils form at the surface into a thin crust which protects underlying fines from erosion. "Desert pavement" is formed by densely packed pebbles and stones, cemented by encrustation with various salts, gypsum, lime and silicates, often coated with a "desert varnish". Damage, as with vegetation in the harsh desert environment, can be long term; unlikely to be repaired within a useful lifetime.

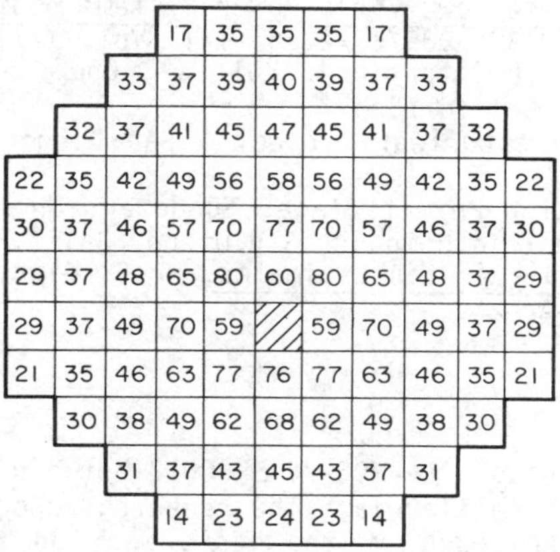

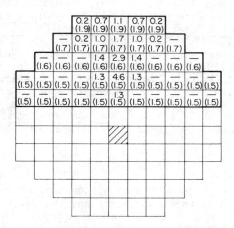

Fig. 5 Predicted fugitive dust deposition rates per cell from a one-mile exclusion zone around the plant (kg/cell-30 day)

Fig. 6 Combined coal and fugitive dust deposited on heliostats over a 30-day month in grams/m^2)

Phytogenic Emissions

Some desert vegetation naturally emits gaseous hydrocarbons. Little data are available on which to base useful emission estimates; none for plants such as the creosote bush. Very crude estimates suggest a source of the order of 1 gm/sec near our site. Of limited importance by itself, such material could add to and provide the cementing agent to tie together other particulates, and thus increase the heliostat-degrading impact of the several sources of concern.

SUMMARY OF IMPACTS

Several environmental impacts have been discussed which are unique to the solar-coal hybrid facility concept. These include impacts of portions of the plant on other portions, impacts of the plant on its surroundings, and impacts of the surroundings on the plant. Broad-ranging impacts in the categories of land use, geology, hydrology, vegetation and wildlife, health and safety, legal and institutional constraints, and esthetics also must be considered for any such facility. However, it appears the concerns of greatest importance lie with air quality. Thus while the development of joint use of solar energy and coal could reduce many adverse health impacts of total dependence on coal, there remain several technical issues in conflict. For these significant future research is essential.

REFERENCES

EPA (1978a). "Emission Factors for Mining Operations". Two-page worksheet, U.S. Environmental Protection Agency, Region VII', Denver , Colorado (March 1978).

EPA (1978b). "Compilation of Air Pollutant Emission Factors", 3rd Ed., Supplement #8, U.S. Environmental Protection Agency, Washington, D.C.

ESE (1980). Environmental Effects of Solar Thermal Power Systems. Vol. I, Environmental Assessment, Vol. II, Air Quality and Meteorological Impacts; Reports UCLA 12/1282 and 12/1283, Environmental Science and Engineering/ Laboratory of Biomedical and Environmental Sciences, Univ. of California, Los Angeles.

King, D.L. and J.E. Meyers (1978). "Environmental Reflectance Degradation of CRTF Heliostats", Sandia Laboratories, Div. 4713, Albuquerque, New Mexico.

Lindberg, Robert G. (1980). Personal Communication.

Rockwell International (1980). Solar Central Receiver Hybrid Power Systems Sodium Cooled Receiver Concept, Final Report. ESG-79-30; DOE/ET/20567-1/1. Vol. I-IV.

SCE (1979). "California Coal Project, Notice of Intention", Vol. II, Chap. XII, Southern California Edison Company, Rosemead, California.

SDGE (1976). "Notice of Intention for Sundesert Nuclear Project", San Diego Gas and Electric Company, San Diego, Calif.

SDGE (1977). "Environmental Report, Construction Permit Stage, Sundesert Nuclear Plants, Units 1 and 2", San Diego Gas and Electric Company, San Diego, Calif.

Turner, D.B. (1969). Workbook of Atmospheric Dispersion Estimates. U.S. Dept. of Health, Education andd Welfare, PHSP No. 999-AP-26.

ACKNOWLEDGEMENT

We are pleased to acknowledge support under contact DE-AM03-76-SF00012 between the U.S. Department of Energy and the University of California. The research reported is from an interdisciplinary group effort, with much of the basic work on this phase contributed by Dr. Carolyn T. Hunsaker, Dr. Donald B. Hunsaker, Jr., and Dr. Robert G. Lindberg. Their contributions were essential and highly valued.

ENVIRONMENTAL QUALITY EFFECTS OF ALTERNATE ENERGY FUTURES: A CAVEAT ON THE USE OF MACRO-MODELING

J. L. Regens and D. A. Bennett

Office of Strategic Assessment and Special Studies, Office of Research and Development, U.S. Environmental Protection Agency, Washington, DC, USA

ABSTRACT

This paper discusses the difficulties in using macro-level, computer-based models as forecasting tools to project environmental quality effects of alternative energy futures. It also addresses the prospects of formulating energy-environment policy on the basis of such macro-models as well as alternate approaches.

KEY WORDS

Macro-modeling; environmental residuals; forecasting; SEAS; pollutant trends.

INTRODUCTION

During the period from 1975 to 1979, total domestic energy consumption in the United States increased from 70.8 quadrillion Btu's (quads) to 78.0 quads which translates into an increase of 2.0 percent on an annual basis. At the same time, total domestic energy production increased from 60.1 quads to 62.8 quads with reliance on imports to cover the shortfall growing from 14.1 to 19.3 quads (U.S. Department of Energy, 1979). Long-term forecasts of energy demand reveal projections for the year 2000 ranging downward from almost 200 quads to less than 100 quads in the most recent forecasts. The differences reflect varying assumptions about such factors as energy costs, availability, economic growth, conservation, and energy mix (see Energy Policy Project of the Ford Foundation, 1974; Adelman and others, 1975; Lovins, 1977; Stobaugh and Yergin, 1979; Schurr and others, 1979; Landsberg and others, 1979; U.S. Department of Energy, 1979). Regardless of which, if any, of the forecasts are isomorphic with future reality, energy production and consumption will result in some environmental quality effects.

The principal factors concerning energy as it affects environmental quality are presented in Fig. 1. The fuel extraction, processing, and consumption stages generate air, water, and solid waste pollutants as byproducts or residuals which are subject to various environmental control measures. Moreover, residual generation is dependent not only upon governmental policies and regulations but a number of other interacting variables including economic growth, energy costs, conservation measures, and market penetration by new technologies. To the extent that the relationships among those variables can be specified, it may be possible to estimate empirically the environmental quality effects of alternative energy futures using macro-level, computer-based models as forecasting tools. The

632

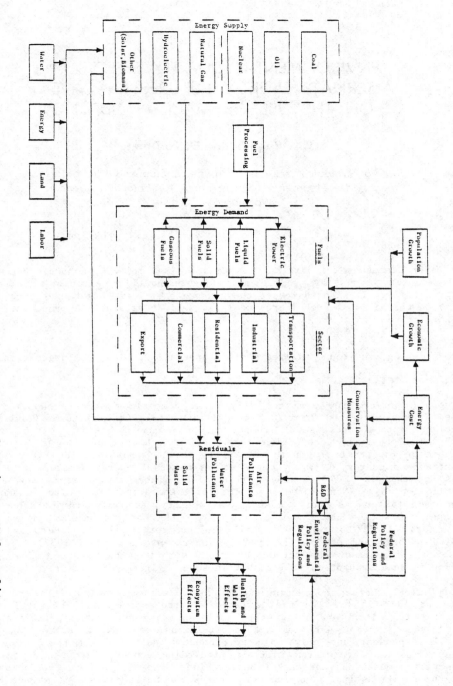

Fig. 1: Factors Influencing Energy and Environmental Quality

desirability of such macro-models stems in large part from the assumption that they not only provide information about aggregate environmental impacts but also permit decision-makers to identify the consequences of changes in environmental policy. Thus, such models can be used to guide policy as well as to identify effects.

Presumably, varying scenario assumptions or models might produce divergent results, while employing the same linear model under similar scenarios ought to result in comparable findings. Interestingly enough, however, such common sense assumptions are not necessarily the case. The primary purpose of this paper is to discuss the difficulties in projecting residuals estimates and formulating energy-environment policy on the basis of such macro-models and to discuss briefly alternative approaches.

SCENARIO ASSUMPTIONS

Recent experience has shown how hard it is to predict future patterns of energy supply and demand (see Fig. 2). This suggests the need to be humble about our forecasting efforts. It also underlines the necessity of clearly delineating the assumptions underlying different scenarios used to represent possible energy futures and their environmental implications. This can be illustrated by the experiences of the Office of Strategic Assessment and Special Studies, U.S. Environmental Protection Agency (EPA) in applying the Strategic Environmental Assessment System (SEAS) model to long-range forecasting.

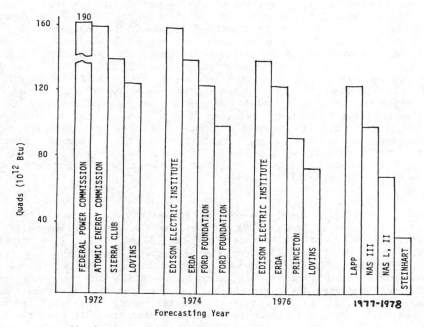

Fig. 2: Forecasts of Energy Demand for 2000 or 2010

The high growth scenario presented in the EPA's Environmental Outlook 1980 report is based upon an assumed growth in Gross National Product (GNP) of 3.5 percent per year and a population increase of 0.8 percent per year from 1975 to 2000

(U.S. Environmental Protection Agency, 1980: 12-17). Similarly, it assumes that total energy supply will expand at an average 2.1 percent per year. Most of this growth results from burning more coal, increasing nuclear powered electricity generation, and obtaining oil from Western oil shales as well as increases in consumption of both domestic and imported oil. New energy technologies for renewable resources are allocated a less significant role; solar, geothermal, and biomass sources providing about 3 percent of the total energy supply in 2000. Overall, the demand for energy use by the transportation, industrial, residential, and commercial sectors as well as export is expected to grow from 53 quads to 85 quads from 1975 to 2000. The commercial and industrial sectors account for the greatest growth because those sectors are projected to experience substantial economic growth.

In evaluating the environmental implications of our energy supply and demand scenario, a regulatory background for emission standards must also be established. Since 1976, five major environmental laws which influence the ways energy is produced and consumed and therefore the impact of energy utilization on environmental quality have been enacted or amended significantly:

o Toxic Substance Control Act of 1976 (PL94-469)
o Resource Conservation and Recovery Act of 1976 (PL94-580)
o Clean Air Act as amended in 1977 (PL95-95)
o Clean Water Act of 1977 (PL95-217)
o Surface Mining Control and Reclamation Act of 1977 (PL95-87)

Recent energy-related legislation also has environmental implications. The Energy Supply and Environmental Coordination Act of 1974 (PL93-319) and the Power Plant and Industrial Fuel Use Act of 1978 (PL95-620) stipulate that coal replace oil and natural gas as fuel in new electric utilities and large industrial boilers. Two basic assumptions are made to incorporate the effects of these regulations into the SEAS model: (1) that all sources will attain full compliance with all applicable regulations by the time specified in the law, and (2) that no additions or changes will be made to the standards promulgated as of July 1, 1978. The effect of employing these assumptions as opposed to alternative ones may be to understate potentially significant emission levels wherever noncompliance problems are expected to be acute (e.g., mobile source controls) and overstate pollution problems in areas where future regulatory initiatives might be anticipated (e.g., controls in combustion-related nitrogen oxide emissions, controls on heavy trucks and buses). This should be kept in mind when one compares the residuals estimates generated using the SEAS model.

OVERVIEW OF SEAS

The SEAS model, originally developed by the EPA and now maintained jointly by EPA and the U.S. Department of Energy (DOE), functions via a set of interacting energy, economic, regional, and environmental modules (House, 1977; MITRE Corporation, 1980a). The model can be illustrated graphically with a flow diagram (Fig. 3). The logic of the SEAS model is relatively straightforward. The energy and economic modules simulate national level energy consumption and economic activity. The regionalizaion modules disaggregate those projections and apportion them among various geographic subdivisions. The environmental modules then estimate the pollution levels and resource requirements associated with those levels of activity.

An important feature of the modules is their interaction, particularly in the estimation of national economic activity. After a preliminary economic projection is made, it is calibrated to the energy strategy, energy investment, and pollution abatement investment estimates. Thus, if those estimates are altered, so are the economic projections and the industry outputs from non-energy sectors of the economy which are "shared" or allocated to geographic units down to the state

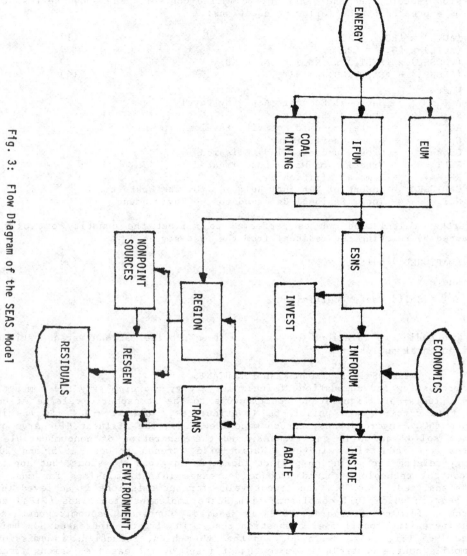

Fig. 3: Flow Diagram of the SEAS Model

level. Given the activity level for each unit, the total volume of each environmental residual is estimated for production processes operating without environmental controls (i.e., gross pollution). The amount of pollution "captured" by control technologies also is estimated to reflect the extent and efficiency of treatment. The amount captured is subtracted from the gross to provide an estimate of "net" pollution (i.e., residual emissions) released to the environment. Thus, residual estimates are sensitive to energy/economic activity levels, gross emission factors, and the degree of pollution control. The above series of steps can be expressed by the following equations:

$$(ACT_n) \times (RS_s) = ACT_s \tag{1}$$
$$(ACT_s) \times (G_s) = GROSS_s \tag{2}$$
$$(GROSS_s) \times (FWLT_s) \times (RE_s) = CAPTURED_s \tag{3}$$
$$(GROSS_s) - (CAPTURED_s) = NET_s \tag{4}$$

Where,

ACT_n = National Economic Activity Level
RS_s = Share of State s
ACT_s = State Economic Activity Level
G_s = Gross Coefficient
$Gross_s$ = Gross (uncontrolled) Emissions
$FWLT_s$ = Fraction of Waste Load Treated
RE_s = Removal Efficiency
$CAPTURED_s$ = Amount of Residual Removed from the Wastestream
NET_s = Total Residual Released to the Environment

A further calculation can be performed to reflect the quantity of solid waste produced by removing the residual from the wastestream.

$$(CAPTURED_s) \times (SC_s) = SW_s \tag{5}$$

Where,

SC_s = Solid Waste Coefficient

The preceeding calculations are aggregated to provide national trend estimates in environmental pollutants.

TRENDS IN ENVIRONMENTAL POLLUTANTS

The processing and combustion of fuels for energy--particularly the combustion of fossil fuels--are significant contributors to the atmospheric release of each of the five criteria air pollutants (particulates, sulfur oxides (SO_x), nitrogen oxides (NO_x), hydrocarbons, carbon monoxide (CO)), the effluent discharge of dissolved solids and oils and greases, and the generation of noncombustible solid wastes and industrial sludges. Nation-wide, trends between 1975 and 2000 in energy-related air pollution project increases in gross emissions, but application of control technology should produce decreases in net emissions for most pollutants analyzed. For SO_x, the lowest value for net emissions is achieved in 1985, the year in which full compliance with State Implementation Plans (SIPs) and New Source Performance Standards (NSPS) is assumed. Continued economic growth coupled with increasing fossil fuel combustion result in slightly increased SO_x emissions after 1985 (Fig. 4A). With NO_x, on the other hand, reductions in emissions due to mobile source controls are more than off-set by increased emissions from point sources, particularly electric utilities and industrial combustion (Fig. 4B).

With the exception of total dissolved solids and oils and greases, the energy technologies included in our definition of energy-related activities will not directly contribute significantly to point source water pollution. However, by emitting quantities of SO_x and NO_x into the atmosphere, combustion sources

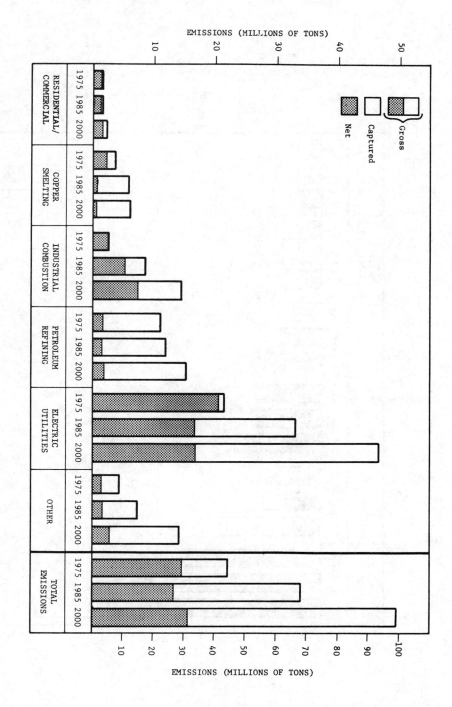

Fig. 4A: Trends in SOx emissions by source, 1975-2000.

638

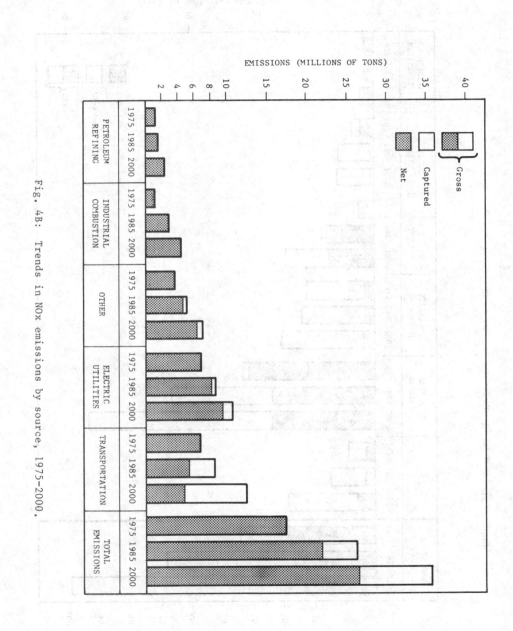

Fig. 4B: Trends in NOx emissions by source, 1975-2000.

contribute to acid rain that may lower pH levels in surface and groundwaters. Further, emissions of trace metals, in particular heavy metals, to the atmosphere may add to the concentration of these metals in water bodies. While the emissions of criteria air pollutants are projected to decline from 1975 to 2000, levels of noncombustible solid wastes and scrubber sludge are expected to increase dramatically due to the reduction of air pollutants. The major cause of the increases will be attributable to the imposition of stringent requirements for the control of criteria air pollutants--SIPs, NSPS, and Best Available Control Technology (BACT) regulations. Air residuals captured thus become solid waste (U.S. Environmental Protection Agency, 1980).

Subsequent to the analysis yielding the above trends, EPA and DOE undertook a comprehensive review of the SEAS model (MITRE Corporation, 1980b). This culminated in an updated version of SEAS (the SEAS Reference System). In order to evaluate the extent to which the updated version produced differing projections of environmental residuals, an analysis comparing 1975 and 1985 estimates was performed using virtually identical national high growth scenario assumptions. Major differences in air pollutant emissions occurred between the two analyses. Examples are shown in Table 1.

TABLE 1   Summary of Differences in Air Pollutant Residuals Estimates (SEAS versus SEAS Reference System Analyses)

Pollutant	Industry	Year	SEAS Value	SEAS Reference System Value
Gross Particulates (TSP)	Pre-1976 Coal Fired Utilities	1975	45.2	2.9
Gross Particulates (TSP)	Pre-1976 Coal Fired Utilities	1985	41.0	0.4
Gross Sulfur Oxides (SO_x)	Petroleum Refining & Storage	1975	11.0	1.3
Gross Sulfur Oxides (SO_x)	Petroleum Refining & Storage	1985	12.0	1.4
Gross Carbon Monoxide (CO)	Petroleum Refining & Storage	1975	9.6	3.6

CAUSES OF DIFFERENCES IN ENVIRONMENTAL TRENDS

Because the scenario assumptions were essentially identical, this raises the question of what factors might account for those differences in environmental trends which do arise between the two runs. With respect to the national level scenario, the only major difference occurs in the energy supply assumptions for solar and biomass. Although the demand assumptions are the same for both analyses, the SEAS Reference System energy technology mix differs reflecting contributions by solar and biomass. As a result, allocations by energy type were changed in the Energy Systems Network Simulator Module (ESNS). These differences, however, cause little or no differences in the estimated residuals output.

The first potentially significant source for differences is the allocation of regional fuel use shares. While the values did not change significantly for the non-energy sectors as well as most of the energy sectors, Federal Region estimates of utility and industrial combustion emissions are very different and account for differences exhibited in regional level estimates. These differences reflect changes in the updated SEAS model's Electric Utility Module (EUM), Industrial Fuel Use Module (IFUM), and the siting files for future synthetic fuel facilities. For the SEAS reference system model, existing (pre-1976) plant capacity factors by fuel type were adjusted to published 1975 data using Oak Ridge National Laboratory's Generating Unit Reference File to include small plants omitted in the previous SEAS model. In addition, the initial run incorporated state level emission factors from the Federal Power Commission's Form 67 instead of unit specific factors. Projected reductions in generation by old plants are accounted for through selective retirement rather than decreasing generation of all plants within a state by a common proportion with remaining capacity adjusted to meet projected demand. The IFUM equation for calculating fuel consumption values for each plant type was revised to reflect coal conversion projections, and the synfuels files were updated to reflect coal availability and environmental considerations.

A second major source of differences in the results is due to changes made in primary residual coefficients. A major portion of the coefficient changes involved replacing old estimates with values from the Department of Energy's then best estimates for energy technologies (U.S. Department of Energy, 1980). Other coefficients were revised when inaccurate (i.e., transposed digits), incomplete or out-of-date data were identified. For example, the CO values for the steel industry were changed as a result of error correction. Similarly, the solid waste portion of the data base was altered to reflect a new residuals taxonomy, changes in solid waste coefficients, and changes in amounts of other pollutants captured. Because the new taxonomy imposed stricter definitions on the residuals, this had a major impact on the estimates of total dissolved solids released from power plants.

The most far reaching source of differences, however, was the base line calibration performed on the 1975 economic activity levels and the incorporation of a new version of INFORUM into the SEAS Reference System model. The 1975 base year estimates of economic activity were adjusted for the updated analysis to reflect actual 1975 conditions. The expansion of the INFORUM module from a 185-sector to a 200-sector national economy is also responsible for some of the significant differences in residual estimates. As a result, the differences we find in the environmental trends are primarily due to altering the economic activity estimates rather than changes in regional shares or primary residual coefficients.

Clearly, the basic source of divergence in the pollutant residuals estimates for the two analyses is estimate variation. While the INFORUM module was expanded, the differences between SEAS and the updated SEAS Reference System version do not constitute a respecification of the model. Instead, because the components of the SEAS model remain the same, the differences are attributable to a transformation of segments of the data in the matrices utilized to estimate the residuals. Thus, assuming adequate theoretical basis exists for specifying properly the system of equations constituting the model (i.e., the validity issue), estimate reliablity is the fundamental constraint on employing macro-models such as SEAS to project aggregate pollutant residuals. Moreover, if the data reliability criterion is not met, the equations cannot be estimated correctly irrespective of the presumed conceptual validity of the model.

CONCLUSIONS

Our discussion raises the following questions: can a macro-model ever achieve the necessary data reliability, and if achievement of reliability is to be seriously attempted, can the cost in resources be afforded? In the SEAS Reference System there are an estimated 5 million data points. As a result, for SEAS, and most other "top down" macro-models, the answers to the questions above would seem to be no, if they are to be used over an extended period of time, unless substantial resources are committed to continual data validation and updating.

A collateral question that needs to be raised by those interested in the environmental or ecological effects of energy futures is: can a macro-model, even if
highly reliable, allow the accurate estimation of effects upon receptors of pollutants? Again the answer would seem to be no, because for most analytical methods
or models to estimate effects requires quite precise location of sources of pollutants. In the SEAS model economic activity and concommitant generation of residuals are "shared" only to the state level. Pollutant discharges to water and
solid waste disposal are usually in the vicinity of the source, and air emissions
that may be transported hundreds of kilometers still need to be located as
aggregated plumes or within grids, typically 20-150 kilometers on a side. In
principle emissions could be shared to individual energy sources found in the
SEAS siting files. However, in practice a much more efficient method is to build
conceptually simple, interactive models from the "bottom up" using appropriate
siting files. Several emissions scenarios may then quickly be compared and
coupled to transport-deposition models. This approach has been used by Fay and
Rosenzweig (1980) for SO_x pollutants in air, and has been adopted for non-criteria
pollutants as well. One of the present authors has recently participated in
applying this approach to demonstrate that levels of lead and cadmium deposition
in the United States can be adequately described by transport from major smelters
and power plants (see Fig. 5). The approach used could be refined considerably
without loss of the virtues of transparency of method and immediate results for
further analysis.

Fig. 5: Lead Deposition (g/ha yr)

In most instances, contingency-type analyses from "bottom-up" models can be run
using a fraction of the analysis time and resources that would be necessary to
obtain comparable results from a macro-model. In addition, to maintain data

reliablity the macro-model also requires resources and manpower levels similar to those required for analyses. Thus, the costs for a comprehensive look into the future usually far outweigh the benefits to policy makers or environmental scientists.

ACKNOWLEDGEMENTS

The contents of this paper do not necessarily reflect the views or policies of the Environmental Protection Agency or the United States Government. We appreciate suggestions and comments from James B. Sullivan, Brian Price, and Thomas Wolfinger.

REFERENCES

Adelman, M. A., A. A. Alchian, J. DeHaven, G. W. Hilton, M. B. Johnson, H. Kahn, W. J. Mead, A. Moore, T. G. Moore, and W. H. Riker (1975). No Time to Confuse. Institute for Contemporary Studies, San Francisco.

Energy Policy Project of the Ford Foundation (1974). A Time to Choose. Ballinger, Cambridge, MA.

Fay, J. A. and J. J. Rosenzweig (1980). An analytical diffusion model for long distance transport of air pollutants. Atmospheric Environment 14: 355-365.

House, P. (1977). Trading Off Environment, Economics, and Energy--A Case Study of EPA's Strategic Environmental Assessment System (SEAS). Lexington Books, Lexington, MA.

Landsberg, H. H., K. J. Arrow, F. M. Bator, K. W. Dam, R. W. Fri, E. R. Fried, R. L. Garwin, S. W. Gouse, W. H. Hogan, H. Perry, G. W. Rathjens, L. E. Ruff, J. C. Sawhill, T. C. Schelling, R. Stobaugh, T. B. Taylor, G. P. Thompson, J. L. Whittenberger, and M. G. Wolman (1979). Energy: The Next Twenty Years. Ballinger, Cambridge, MA.

Lovins, A. B. (1977). Soft Energy Paths. Harper and Row, New York.

MITRE Corporation (1980a). Introduction to the Strategic Environmental Assessment System. MTR-80W115. MITRE Corporation, McLean, VA.

MITRE Corporation (1980b). The Status of the Strategic Environmental Assessment System. MTR-80W206. MITRE Corporation, McLean, VA.

MITRE Corporation (1980c). Energy trends for research outlook 1981. Draft Working Paper (WP-80W00707). MITRE Corporation, McLean, VA.

Schurr, S.H., J. Darmstadter, H. Perry, W. Ramsay, and M. Russell (1979). Energy in America's Future. Johns Hopkins University Press, Baltimore.

Stobaugh and D. Yergin (Ed.) (1979). Energy Future. Random House, New York.

U.S. Department of Energy (1979). Annual Report to Congress, Volume II: Data. DOE/ETA-0173(79)/2. Energy Information Administration, Washington.

U.S. Department of Energy (1980). Technology Characterizations Environmental Information Handbook. DOE/EV-0072.

U.S. Environmental Protection Agency (1980). Environmental Outlook 1980. EPA/600/8-80-003. Office of Research and Development, Washington.

U.S. Environmental Protection Agency (1981). Atmospheric deposition of heavy metals. Draft Working Paper. Office of Research and Development, Washington.

ENERGY POLICY, AIR POLLUTION, AND COMMUNITY HEALTH

H.-W. Schlipköter and U. Ewers

*Med. Institut für Umwelthygiene an der Universität Düsseldorf,
Gurlittstrasse 53, D-4000 Düsseldorf 1, Federal Republic of Germany*

KEYWORDS

Energy consumption; combustion processes; emissions; air pollution; health effects; respiratory cancer.

INTRODUCTION

The combustion of fossil fuels for heating and energy production respresents the commonest and most widespread source of atmospheric pollution. In 1975 about 93% of the world's energy consumption were supplied by the combustion of fossil fuels (Council on Environmental Quality, 1980). Generally it is assumed that the world's energy demand will increase dramatically in the next years. Although nuclear energy, solar energy, and renewable energy sources will contribute in a growing manner, the consumption of fossil fuels and, therefore, the total mass of emissions caused by combustion processes will increase significantly during the next decades (Council on Environmental Quality, 1980).

Combustion processes do emit significant quantities of pollutants, which can be harmful to man and to the environment. The objective of the present paper is to review briefly some experimental and epidemiological aspects of health effects known to be associated with air pollution. Some practical conclusions with respect to the demands to future energy policy will be discussed from the viewpoint of community health at the end of this paper.

EMISSIONS FROM COMBUSTION PROCESSES

The emissions generated by combustion processes can be categorized as a) gases and vapors, and b) particulate matter (solid particles and liquid droplets). The most important gases and vapors are carbon dioxide, sulfur oxides, nitrogen oxides, carbon monoxide and gaseous hydrocarbons. The latter two are only emitted by incomplete combustion processes. Solid particles are generally ash from the fuel and/or carbonaceous material (soot) resulting from incomplete combustion. Liquid droplets generally consist of unburned and/or partially combusted fuel droplets. Sulfur trioxide may be present as droplets, if the flue gas temperature is below about 200°C; however, from technical reasons sufficiently high flue gas temperatures are generally maintained to avoid the formation of these acid droplets.

As shown in TABLE 1 the major part of emissions released into the atmosphere in the FRG are formed by combustion processes and fuel processing. In the last years important progress has been achieved in the reduction of particulate matter due to improvements of gas cleaning equipments and technology, control efforts, process and operational modifications and material substition (particulary the substitution of coal by oil and natural gas). However, a substantial increase of the total emissions of nitrogen oxides has occurred since 1965, which led to a continuous increase of the nitrogen oxide concentration in the ambient urban air (Umweltbundesamt, 1977; Sachverständigenrat für Umweltfragen 1978). No changes or improvements have been achieved with respect to the total emissions of sulfur dioxide, carbon monoxide, and hydrocarbons.

TABLE 1 Emission from combustion processes, fuel processing, and other industrial processes in the Federal Republic of Germany 1975 (Umweltbundesamt, 1977).

	Particulate matter		SO_2		NO_x		Hydrocarbons	
	Mill.t	%	Mill.t	%	Mill.t	%	Mill.t	%
Energy use – combustion processes	416.5	74.8	3471	95.4	1816	98.5	938.o	51.7
Powerplants and heating plants	172.5	3o.7	17o4	46.8	688	37.5	8.1	o.4
Industrial combustion processes	153.5	27.5	1214	33.4	458	24.8	33.2	1.8
House heating	72.o	12.9	473	13.o	115	6.3	134.7	7.4
Traffic	18.5	3.3	8o	2.2	555	3o.1	762.o	42.o
Fuel processing[+)]	39	7.o	–	–	–	–	139.8	7.7
Non-energetic industrial processes	1o3.8	18.6	163	4.4	24	1.3	736.3	4o.6
Total	56o	1oo	363o	1oo	184o	1oo	181o	1oo

[+)] Coal-conversion processes, coal gasification, petroleum refining etc.

Similar to many large cities and industrial areas in Europe there is a general decline in sulfur dioxide and particulate matter concentration in the FRG, which can be attributed in part to declining emissions (particulate matter), improved dispersion from chimneys, the elimination of sources that contribute substantially to local concentrations (domestic fires, particularly coal burning), fuel substitution, and control effort.

TOXIC AND HAZARDOUS POLLUTANTS FROM COMBUSTION PROCESSES

Emissions due to combustion processes contribute substantially to air pollution, particularly in urban and industrialized areas. Many pollutants which are released into the atmosphere by combustion processes are toxic and/or hazardous to man because either (a) they affect the respiratory organs; (b) they enter the body via the lung and may cause toxic effects in various organs and tissues; or (c) they are carcinogenic.

Among the pollutants of major interest are (a) irritative gases such as sulfur oxides, nitrogen dioxide, and ozone; (b) toxic gases such as carbon monoxide; (c) suspended particulate matter containing toxic and carcinogenic agents such as polycyclic aromatic hydrocarbons (PAH), lead, cadmium, arsenic etc.

Metals are present in liquid and solid fuels. Metals in liquid fuels, e.g. lead in gasoline, are generally entrained in the flue gas upon combustion and emitted into the air. For coal burning it is known that most of the trace elements remain associated with the bulk inorganic combustion residues (ash). However, mercury and other volatile trace elements and their compounds, e.g. selenates and arsenates, can escape from a combustion system because of their high volatility. In general trace element emissions depend on the combustion conditions as well as on the effectiveness of the flue gas cleaning equipment.

The class of PAH, which generally are adsorbed on the surface of soot particles from incomplete combustion prosses, include a great variety of polycyclic aromatic and heteroaromatic hydrocarbons (Umweltbundesamt, 1979). This class of material is yet poorly defined both from the viewpoint of collection and analysis, and the effect of combustion conditions on their generation. For power plants, which operate under relatively steady state conditions with high flame temperatures and good air-fuel mixing, PAH emission levels are believed to be low. Smaller combustion systems with inherent quenching by cold surfaces and short residence times in the flame tend to emit larger quantities of PAH. Combustion systems with poor air-fuel mixing (such as small coal-fired systems) are relatively high emitters of PAH and have been the most important sources in the past in the FRG as well as in many other European countries. Systems operating in an on-off mode (with poor mixing at start-up and shutdown) also emit high concentrations of PAH.

EFFECTS OF AIR POLLUTION ON HUMAN HEALTH - GENERAL REMARKS

The effects of air pollution on personal or community health may be summarized as follows (Goldsmith and Friberg, 1977):

a) acute sickness and excess mortality due to an abrupt rise in the concentrations of air pollutants to very high levels;
b) chronic diseases due to prolonged exposure to lower levels of air pollutants;
c) alterations of important physiological functions such as ventilation of the lung, transport of oxygen by hemoglobin, sensory acuity, time interval estimation, or other functions of the nervous system;
d) impairment of performance such as athletic activities, motor vehicle operation, or complex tasks such as learning;
e) untoward symptoms, such as sensory irritation, which in the absence of an obvious cause such as air pollution, might lead a person to seek medical attention and relief;
f) storage of potentially harmful materials in the body;
g) discomfort, odor, impairment of visibility, or other effects of air pollution sufficient to cause annoyance or to lead individuals to change residence or

place of employment.

Furthermore, the possibility that general air pollution is a causal factor in lung cancer has given rise to considerable concern. This issue will be discussed, however, later on.

Another point to be mentioned is the great individual variability in responsiveness to air pollution in all populations. Susceptibility is believed to be great among premature infants, the newborn, the elderly, and the infirm. Those with chronic diseases of the lungs or heart also have to be considered to be at particular risk. Because of the great variability in sensitivity to air pollution in the population, data concerning health effects on healthy persons may not be as important as the responses of those individuals likely to be the most sensitive. Legislative actions and the control of air pollution, to the extent that they are based upon health effects, should be based on the most sensitive groups of persons in the population. This principle requires that these sensitive groups are definable in terms of age and/or medical status.

CONFOUNDING VARIABLES

Many common air pollutants are also substances to which persons are exposed in their occupation or due to their personal lifestyle. With respect to causing lung cancer or chronic pulmonary diseases in the whole population the effects of occupationally related exposures and of smoking, particularly cigarette smoking, are more important than the effects of general air pollution. It must be stressed, however, that for many workers cessation of work does not terminate their exposure, since many substances, to which they are exposed at work, also occur in the ambient air. Other common air pollutants may exert aggravating or synergistic effects in connection with certain occupationally related exposures. Another point of importance is that general air pollution and smoking can have a synergistic effect, when both factors are present. This means that cigarette smokers are at an unusual high risk if they live in highly polluted areas, and that the effects of general air pollution on the respiratory organs are more likely to occur in cigarette smokers.

It should be noted that a number of other confounding variable have to be taken into account when considering health effects associated with air pollution:
a) domestic pollution· home heating, cooking, and other operations can generate air pollutants whose health effects have been overlooked for a long time;
b) meteorological factors such as temperature, ventilation, moisture, solar radiation etc.; c) socioeconomic and sociocultural factors.

ACUTE AIR POLLUTION EFFECTS

There are numerous reports in the literature showing that episodes with very high air pollution are associated with a significant increase of mortality and morbidity. Effects were primarily seen in those already ill, old, or enfeebled. Respiratory diseases and disorders, including bronchitis, respiratory tract irritation, shortness of breath, cyanosis, emphysema, and heart diseases were the symptoms most frequently observed. The episodes have always occurred under extraordinary meteorological conditions that greatly reduced the transport and diffusion of air pollutants. Most episodes have also occurred under circumstances, in which fog and high concentrations of particulate matter were present. It is likely, therefore, that a combination of aerosols and gaseous pollutants was involved. For further details the reader is referred to the literature (Goldsmith and Friberg, 1977, and references cited there-in).

AIR POLLUTION AS A CAUSAL FACTOR IN CHRONIC RESPIRATORY DISEASE

There is considerable epidemiological evidence that sufficient exposure to air pollution of various types is a likely causal factor in chronic bronchitis and a suspected one in emphysema. There seems to be a relationship between the degree of air pollution and the prevalence of chronic bronchitis (TABLE 2). Air pollution certainly is not the only cause, nor perhaps even the most important initiating cause. In any case, however, air pollution can act as a promoting or aggravating factor in the presence of impaired lung function or serious chronic lung diseases. Moreover, there is suggesting evidence of an aggravation of chronic bronchitis in cigarette smokers due to general air pollution (see TABLE 2).

TABLE 2 Prevalence of chronic bronchitis (according to WHO definition) in some areas of Nordrhein-Westfalen. The degree of air pollution is: BDA < Bo $\approx$ Re < Du[a], (Ministerium für Arbeit, Gesundheit und Soziales NW, 1980).

	Prevalence in %[b] (Number of subjects examined)			
	BDA	Bo	Re	Du
Males (Non-smokers)	1o.6 (132)	21.9 (128)	12.6 (95)	14.7 (170)
Males (Smokers)	24.1 (112)	29.9 (97)	25.4 (63)	38.1 (134)
Females (Non-smokers)	6.7 (283)	6.5 (217)	9.o (178)	9.5 (232)
Females (Smokers)	6.7 (3o)	17.6 (34)	21.1 (19)	28.o (25)
Children	1.2 (769)	2.o (514)	3.2 (339)	2.5 (473)

[a] BDA = Borken, Dülmen, Ahlen; Bo = Bochum; Re = Recklinghausen; Du = Duisburg
[b] Corrected for confounding variables

Since normally a great variety of pollutants is present in the air it is difficult to attribute certain effects to the influence of specific air pollutants. As a general rule the water solubility of a gaseous pollutant determines what proportion is deposited in the upper airway, and what proportion reaches the terminal sacs of the lung, the alveoli. For example sulfur dioxide, which has a high water solubility, is absorbed in the upper respiratory tract and has as its primary reaction pattern an irritation of the airways and an increased airway resistance. Nitrogen dioxide and ozone are examples of relatively water insoluble gases. These substances have their major biological reactions at the alveolar level and may interfere with the gas exchange between air and blood. The deposition of particles largely depends on the particle size (aerodynamic diameter). The combined action of sulfur dioxide and suspended particulate matter is believed to have a synergistic, i.e. a more than additive effect (WHO, 1979a).

SUBCLINICAL AIR POLLUTION EFFECTS

Subclinical, i.e. asymptomatic effects of air pollutants include alterations of

physiological and biochemical functions and parameters such as respiratory functions, enzyme activities, sensory and reflex functions or complex neuropsychologi-
cal performances. Some of these effects have been observed under exposure conditions present in many European cities and industrialized areas. However, it is
unclear, to what extent these alterations can be compensated by the healthy organ-
ism and to what extent they are impairments of physiological functions representing
early signs of illness, which, in combination with other factors, could develop
to more serious diseases. For further information the reader is referred to special
monographs and papers (WHO, 1977; WHO, 1978; WHO, 1979a; WHO, 1979b; Needleman,
1980).

AIR POLUTION AND RESPIRATORY CANCER

There are a number of arguments suggesting that air pollution may play a role as
a causal factor of respiratory cancer:

a) Potent carcinogenic agents are found in polluted atmospheres.The most important class of carcinogenic agents is the class of polycyclic aromatic hydrocarbons (PAH), which are formed by incomplete combustion processes. Many of
these compounds have been shown to be carcinogenic in animal experiments and
to be mutagenic in various short-term tests (Umweltbundesamt, 1979).

b) Occupational experience shows that the inhalation of various agents, which
are found in polluted atmosheres, can cause cancer, particularly of the respiratory organs.

c) There is an excess of lung cancer in urban areas, which has been attributed
to urban air pollution. It is difficult, however, to prove a relationship between lung cancer rates and the degree of air pollution. A positive correlation is found with population density (FIGURE 1).

d) Cigarette smoking generally s accepted as the main cause of lung cancer.
Urban air contains similar agents to those found in cigarette smoke. This is
particularly true for the emissions of small coal-fired combustion systems
(house-heating) and motor vehicle exhausts.

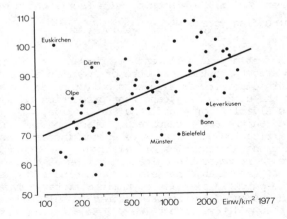

Fig. 1.   Age-adjusted lung cancer death rates for males in relation to the
population densities (inhabitants per km^2) in the cities and counties
of Nordrhein-Westfalen (r = 0.61; p < 0.001). Rates are means from
1975-77, and are given in units per 100 000 male inhabitants (according to Ziem-Hanck and coworkers, 1979).

Available evidence is not yet convincing, however, that general urban air pollution plays a role as a causal factor of lung cancer. One part of uncertainty can be attributed to the dominant role of cigarette smoking as a cause of lung cancer. Smoking, which is believed to account for about 80 % of all lung cancer deaths, is so important that the lesser role of other inhaled carcinogenic agents is hardly to detect. Another important cause of uncertainty are the long time spans involved. Based on our knowledge on the biological effects of smoking and occupational exposures, it usually requires several decades of exposure, before alterations in the occurence of lung cancer are detectable. Residential mobility and, additionally, problems of estimating the exposure levels make it very difficult to study the relationship between urban air pollution and lung cancer.

In conclusion it can be stated that there may be other explanations of the 'urban factor', e.g. domestic and occupational exposures, a greater occurence of infections of the respiratory organs due to a higher population density etc. The evidence presently available does not confirm the conclusion that general urban air pollution per se is the factor. However, it is not possible to reject the possibility that air pollution plays a role as a carcinogenic, cocarcinogenic or promoting factor. Air pollution, therefore, remains a "suggested" explanation for the urban excess of lung cancer.

CONCLUSIONS

There are sufficient experimental data as well as occupational-related experiences showing that common air pollutants are capable to cause respiratory diseases such as bronchitis, emphysema, and cancer of the respiratory organs. Subclinical, i.e. asymptomatic alterations of certain physiologically important functions can be observed under today's exposure conditions present in many urban areas. In many cases such effects may be attributable to the action of air pollutants.

Although there is not always conclusive evidence that urban air pollution is a causal agent for excess mortality or increased numbers of cases with defined diseases the well-documented possibility that air pollutants may act as a causal, promoting or aggravating factor allows to draw some practical conclusions:

a) The control of soot, suspended particles, and products of incomplete combustion are one of the highest and most urgent goals of air pollution control.

b) Whatever the risks may be that are associated with air pollution efficient emission controls and a general reduction of energy consumption and use should have a high priority. Combustion processes represent the most important and most widespread source of atmospheric pollution. Based on the present structure of the use of energy resources, a reduction of energy consumption will - pari passu - substantially decrease air pollution. A reduction of energy consumption without greater restrictions in comfort and services should be attainable by a more efficient use of primary energy. Efficient house insulation, district heating systems by thermal power plants, extension of the public transport systems, particularly in urban areas, offer a great potential for substantial energy savings. A reduced energy input, particularly in house heating and in the traffic sector, both of which substantially contribute to urban air pollution, would - pari passu - help to reduce health risks associated with air pollution.

c) Not only considerations of community health, but also a great variety of political, economic, resource, and environmental problems associated with a continued rapid growth of energy consumption have focussed attention on the benefits of conservation and increased efficiency of energy use. Efforts to energy saving and greater efficiency of energy use as well as the development of so-called "soft-energy-technologies" should, therefore, be intensified in the future.

REFERENCES

Council on Environmental Quality (1980). The Global 2000 Report to the President. German Edition. Verlag Zweitausendeins, Frankfurt. pp. 726-789.

Goldsmith, J.R., and Friberg, L.T. (1977). Effects of ir pollution on human health. In A.C. Stern (Ed.), Air Pollution, Vol. 2, 3rd ed., Academic Press, New York. pp. 458-610.

Ministerium für Arbeit, Gesundheit und Soziales NW (1980). Luftreinhalteplan Ruhrgebiet Mitte 1980-1984. Universitätsdruckerei Bonn. pp. 164-198.

Needleman, H.L., Ed. (1980). Low-level lead exposure: The clinical implications of current research. Rven Press, New York.

Sachverständigenrat für Umweltfragen (1978). Umweltgutachten 1978. Kohlhammer, Stuttgart. pp. 130-175.

Sachverständigenrat für Umweltfragen (1981). Sondergutachten 'Energie und Umwelt'. Manuscript, in press (Kohlhammer, Stuttgart).

Umweltbundesamt (1977). Materialien zum Immissionsschutzbericht der Bundesregierung. E.Schmidt-Verlag, Berlin-W.

Umweltbundesamt (1979). Luftqualitätskriterien für ausgewählte polyzyklische aromatische Kohlenwasserstoffe. Berichte 1/79. E.Schmidt-Verlag, Berlin-W.

WHO (1977). Environmental Health Criteria 4. Oxides of Nitrogen. WHO, Geneva.

WHO (1978). Environmental Health Criteria 7. Photochemical Oxidants. WHO, Geneva.

WHO (1979a). Environmental Health Criteria 8. Sulfur Oxides and suspended particulate matter. WHO, Geneva.

WHO (1979b). Environmental Health Criteria 13. Carbon monoxide. WHO, Geneva.

Ziem-Hanck, U., F.Pott, and U.Krämer (1979). Bösartige Neubildungen der Atmungsorgane nach den amtlichen Sterbedaten in der Bundesrepublik Deutschland unter besonderer Berücksichtigung Nordrhein-Westfalens. Umwelthygiene (= Med. Institut für Umwelthygiene, Jahresbericht 1979, Band 12). Giradet, Essen. pp. 114-126.

MECHANISMS OF SURFACE
WATER ACIDIFICATION

H. M. Seip

*Central Institute for Industrial Research, P.O. Box 350 Blindern,
Oslo 3, Norway*

ABSTRACT

Acid precipitation and other possible causes of acidification of streams and lakes
are discussed. Important processes and possible mechanisms are considered, and special emphasis is given to soil-water interactions. It is concluded that the increased concentrations of sulphate in runoff caused by acid precipitation explain at
least a substantial part of the observed acidification. Changed soil acidity, caused by for example changes in land use, may also play a role in freshwater acidification in some areas.

KEYWORDS

Acid precipitation; acid rain; acidification mechanisms; water acidification; soil-
water interactions.

INTRODUCTION

The recent acidification[1] of freshwater in many parts of the world is well documented (see *e.g.* Conroy and co-workers, 1976; Likens and co-workers, 1979; Overrein,
Seip, and Tollan, 1980; Schofield, 1976), and is discussed in other presentations
at this conference.

The main reason is usually thought to be the increased deposition of acidifying
components, though other causes have been suggested (Rosenqvist, 1978, 1980).

The regional picture seems to be in agreement with the theory that the acidification is dominantly caused by acid precipitation, with the result modified by geological conditions. The time trends seem also to support this assumption. We may
take southern Norway as an example. Though occasional reports of fish kills, pos-

---

[1] The acidity is expressed either by the hydronium-concentration, $[H_3O^+]$ (for simplicity $[H^+]$) or by $pH = -\log [H^+]$. When pH decreases by 1 unit the H^+-concentra-
tion increases by a factor of 10.

sibly caused by acid water, exist from the beginning of this century, the large decline at least in trout population occurred later, say from around 1960 (Muniz and Leivestad, 1980).

However, correlations of this kind may be accidental and do not prove a causal relationship. An understanding of the processes and mechanisms involved in acidification is therefore needed. The discussion presented here is mainly based on results obtained within the Norwegian SNSF-project (Acid Precipitation — Effects on Forest and Fish) (cf. Overrein, Seip, and Tollan, 1980).

PROCESSES IN A CATCHMENT OF POSSIBLE IMPORTANCE FOR ACIDIFICATION

A catchment is a very complex system, and a large number of processes may possibly play a role in connection with acidification. Already as the precipitation passes the canopy of the vegetation the composition is considerably changed. Other important factors are

- CO_2-pressure in soil.

- Uptake and release by vegetation.

- Cation exchange.

- Weathering.

- Oxidation and reduction reactions (sulphur and nitrogen compounds).

- Accumulation and depletion in the catchment, especially of sulphur compounds.

- Dissolution of organic acids and other weak acids.

The CO_2-pressure, which in the atmosphere is about 0.3 mbar, may be more than 100 times greater in the soil. Such high CO_2 pressures in equilibrium with otherwise pure water, result in pH-values less than 4.7 (see *e.g.* Bolt and Bruggenwert, 1976). However, the runoff is not in equilibrium with the CO_2-pressure in the soil, but with a pressure more similar to that of the atmosphere. It is therefore unlikely that carbonic acid will cause a pH much below 5.5 in surface waters.

A few essential points with respect to effects of vegetation and ion exchange are illustrated in Fig. 1. Here it is indicated that the roots take up various cations and release H^+, resulting in a soil acidification if plant matter is accumulated or removed. This is typical for coniferous forest on acid soil. On the other hand, most crop plants take up more anions than cations and consequently have to liberate bicarbonate ions (Nye and Tinker, 1977).

Cation exchange is one of the most important processes we have to consider (Bolt and Bruggenwert, 1976). Soil particles have normally a negatively charged surface and therefore a layer of cations close to the surface (Fig. 1). These cations may be interchanged with those in the solution. Cation exchange may increase or decrease the H^+-concentration in the runoff. Thus when a dilute solution of neutral salts percolates through acid soils, the leachate becomes acid because of exchange of other cations with hydrogen ions (Wiklander, 1975). It follows that the acidity of the runoff depends on the acidity of the soil, the ion content of the precipitation as well as on the degree of contact between water and soil.

Weathering of most minerals leads to consumption of H^+-ions (quartz is an exception). The H^+-ions[2] in solution are replaced by other cations as Ca^{2+}, Mg^{2+}, Na^+ and K^+. Carbonic acid[2] is a very important weathering agent. Thus complete (congruent) dissolution of calcite may occur according to the reaction:

$$CaCO_3 + H_2CO_3^* \rightarrow Ca^{2+} + 2HCO_3^-$$

Cation Exchange

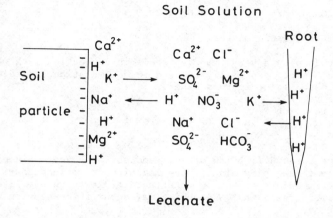

Fig. 1.   Schematic illustration of cation exchange between the soil solution and a soil particle (left) and a root (right).

In most cases other minerals are formed, as shown for the ultimate weathering of K-feldspar below:

$$KAlSi_3O_8 + H^+ + 7H_2O \rightarrow Al(OH)_3 + K^+ + 3H_4SiO_4$$

Since the protons are not necessarily supplied by carbonic acid, this equation has been formulated with H^+ instead of $H_2CO_3^*$.

Oxidation and reduction reactions of sulphur and nitrogen compounds in soil may be purely chemical processes, but usually microorganisms are involved. Acidification following an oxidation of sulphur or sulphide may be exemplified by the reactions:

---

[2] Here denoted $H_2CO_3^*$ which refers to the sum of hydrated and unhydrated dissolved CO_2.

$$4FeS_2 + 2H_2O + 15O_2 \rightarrow 2Fe_2(SO_4)_3 + 4H^+ + 2SO_4^{2-}$$

$$S + 3/2O_2 + H_2O \rightarrow 2H^+ + SO_4^{2-}$$

The first of these reactions will normally be followed by hydrolysis

$$2Fe_2(SO_4)_3 + 12H_2O \rightarrow 4Fe(OH)_3 + 6SO_4^{2-} + 12H^+$$

Occasional serious drops in pH may occur because of oxidation of sulphur compounds after long dry periods or as a result of drainage of bogs (see, for example, Ødelien, Selmer-Olsen, and Haddeland, 1975).

The sulphide in the above equation may be a result of previous reduction of sulphate. If the reducing agent is organic matter, the total reaction may summarily be given as

$$SO_4^{2-} + 2CH_2O + 2H^+ \rightarrow H_2S + 2CO_2 + 2H_2O$$

where CH_2O represents organic matter.

The sulphur compounds play a very important role in connection with water acidification and various aspects especially related to accumulation and release of these compounds are discussed later in this paper.

The nitrification reaction

$$NH_4^+ + 2O_2 \rightarrow 2H^+ + NO_3^- + H_2O$$

may also contribute to acidification of soil and surface water. Since the nitrate produced is quickly taken up by plants, the leaching is not necessarily increased significantly. The immediate effect on water acidity is therefore likely to be small. A short-term increase in NO_3 leaching may occur in connection with fertilizing or clear-cutting.

The concentration of NH_4^+ in the precipitation in southern Norway is generally 35-60% of the H^+-concentration though higher values do occur, probably due to local farming. If all the ammonium ions were oxidized to nitrate, the amount of protons produced would be comparable to that of the precipitation. However, the role of NH_4^+ and NO_3^- in acidification is very complex. Plants take up nitrogen both as NH_4^+ and NO_3^-. If the uptake is in equivalent amounts, no net release of H^+ or bicarbonate occurs. Ammonium ions may also participate in ion-exchange.

Though organic acids may cause low pH-values in soft-water lakes, there is no evidence for a hypothesis that the recent regional acidification is caused by such compounds. Wright and Henriksen (1978) found no significant correlation between pH and total organic carbon in 155 lakes included in a regional survey in southern Norway. Many of the low pH-lakes in southern Norway are exceptionally clear. Furthermore, when a lake becomes more acid, it is often found that the transparency increases and that the content of organic matter decreases (Almer and co-workers, 1978).

It follows from the above discussion that the interactions between precipitation (or meltwater) and soil and vegetation are of utmost importance for the composition of runoff. In the SNSF-project these interactions have been studied in laboratory experiments and in natural catchments (> about 0.1 km^2). As a link between these very different types of studies we have also used mini-catchments, i.e. natural catchments covering only 5-1500 m^2. It is obviously not possible to cover all aspects of freshwater acidification here. In the next section a few results and some general conclusions are presented. More details are given by Overrein, Seip, and Tollan (1980) and in other SNSF-publications (see Tollan, 1980).

ACIDIFICATION MODELS

To clarify the following discussion I will consider three possible conceptual models for freshwater acidification:

1. Model based on a direct effect; $i.e.$ assuming that a substantial fraction of the precipitation reaches rivers and lakes essentially unchanged.

2. Model emphasizing the increased deposition of mobile anions (SO_4^{2-}).

3. Model based on effects on freshwater through a change of soil acidity.
   a. Caused by acid precipitation.
   b. Other causes ($e.g.$ changed vegetation).

Direct Effects

We will start with the simplest model and consider direct effects. The acidification of rivers and lakes is then determined by the pH of the precipitation and the fraction of the precipitation essentially unchanged. At least the precipitation falling directly into rivers and lakes may be considered unchanged. However, this is normally only a small part, often 5-10%.

The importance of cation exchange has been illustrated by many experiments, $e.g.$ by using artificial precipitation containing radioactive calcium, $^{45}Ca^{2+}$ (Dahl and co-workers, 1979a). The studies showed considerable adsorption of calcium even on barren rock covered only by patches of lichens.

Studies with calcium tracer on shallow soil showed usually very effective adsorption of calcium when water percolates through soil. Considerable tracer concentration in the runoff was found only in one of six experiments on shallow soil. The reason for the high concentration in one case may be that channel flow has occurred, reducing the contact between soil and water. Sodium ions ($^{24}Na^+$) move on the other hand rather quickly through soil (Christophersen and co-workers, to be published).

The contact between meltwater from snow and soil is likely to be different from that between rain and soil uncovered by snow or ice. Dahl and co-workers (1979b, 1980) studied this by spraying radiactive tracers on the snow in mini-catchments in Storgama. It was found that 80-90% of the Ca-tracer sprayed on the snow in mini-catchments were adsorbed by soil or vegetation during snowmelt. This indicates some reduction in the contact between soil and meltwater compared to that between rain and soil in the summer. However, the interaction is still important.

From these and other studies it seems likely that the direct effect is small in most cases. It is most important if the catchment has relatively large areas of nearly barren rock or water surface.

Model Emphasizing the Increased Deposition of Mobile Anions

We will now discuss a model for acidification where the total composition of the precipitation is considered; the emphasis is on the increased deposition of mobile anions (SO_4^{2-}). The movement of SO_4^{2-} through a catchment will also be considered. To simplify the discussion we assume here that there is no significant change in the soil acidity.

The importance of sulphate anions for transport of cations through the soil was pointed out more than 20 years ago by Gorham (1958), and has later been discussed by several authors (*e.g.* Cronan and co-workers, 1978; Johnson and Cole, 1977; Overrein, Seip, and Tollan, 1980).

Let me first mention a very simple experiment using artificial precipitation in a mini-catchment (Seip, Gjessing, and Kamben, 1979). A sprinkler was used for spraying artificial precipitation with varying pH. The other ionic concentrations were close to the average values in the precipitation in the area. The runoff followed two tracks, the water in one of them being somewhat more acid than in the other. pH in the runoff varies with that in the input, though the amplitude is considerably smaller, *i.e.* 0.2-0.4 units (Fig. 2). There is a high correlation between the concentrations of H^+ and sulphate in the runoff, though the two tracks give slightly different regression lines. The sum of H^+ and Al^{3+} gives even better correlation (Fig. 3).

Discharge and concentrations of various ions in runoff from a mini-catchment during snowmelt are shown in Fig. 4 (Seip and co-workers, 1980). We see that during snowmelt there is a period with a rapid decrease in the concentrations of ionic components. This is probably related to the well-known fractionation effect observed when snow is melted; *i.e.* most of the ionic impurities are found in the first meltwater (Johannessen and Henriksen, 1978). We may also note that the nitrate concentrations are much greater than normally found in rivers and lakes in Norway. The decrease in the H^+-concentration observed during the melting period is probably mainly due to similar changes in the concentration of mobile anions, since the contact between meltwater and soil is important even during snowmelt.

During the first part of the snowmelt there may be some frost in the soil. At a later stage the discharge is often large. During most of the snowmelting period the runoff is therefore likely to pass parts of the catchment with little neutralizing capacity, and pH in the runoff will decrease rapidly with increasing concentrations of mobile anions.

The movement of sulphate through soil has been studied in the Storgama area using radioactive tracers. Experiments with artificial precipitation indicate that sulphate ($^{35}SO_4^{2-}$) may move as fast as tritiated water through the soil at high rain intensities (Dahl and co-workers, 1979a).

However, this is probably not generally true. Another experiment from the mini-catchments illustrates this point. In June 1978 tracer was sprayed on one of the mini-catchments. Only about 50 l water was used, and no immediate runoff was observed. The accumulated runoff and the accumulated amount of sulphate tracer in runoff are given in Fig. 5. Even in this case where the soil cover is very shallow, only about 40% of the added sulphate has been found in runoff after about 4 months. The remainder of the sulphur tracer was stored both as sulphate and as other sulphur compounds in soil and vegetation (Dahl and co-workers, 1979b, 1980).

Based primarily on the SNSF-results we may say that sulphate shows nearly an input-output balance in many catchments if periods of several years are considered. However, according to our previous discussion this is a result of an approximate steady state; sulphate and other sulphur compounds are involved in complicated processes in the catchment.

Artificial precipitation

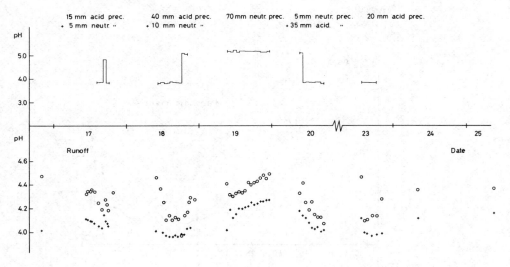

Fig. 2.   The pH in artificial rain (upper part) and in runoff (lower part)
from a mini-catchment. The symbols (0,+) correspond to different
outlets.

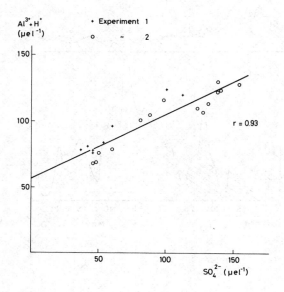

Fig. 3.   Sum of the concentrations of aluminium (considered as Al^{3+}) and
H^+ plotted against concentrations of sulphate. The values are
for the runoff in two experiments with artificial precipitation.

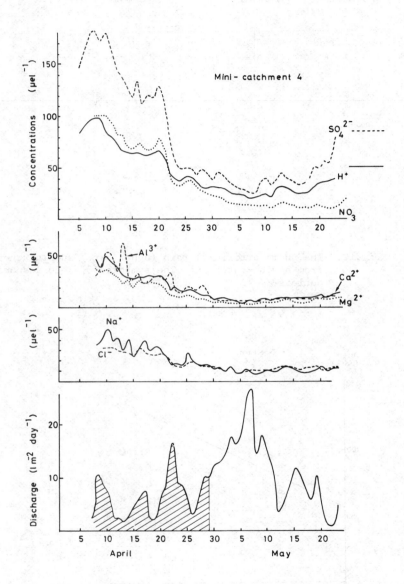

Fig. 4. Discharge and concentrations measured in runoff from a mini-catchment during snowmelt 1978. Average concentrations for SO_4 and H^+ during summer and autumn 1978 are given at the right. Shaded area in the lower figure corrresponds to 30% of the run-off in the period.

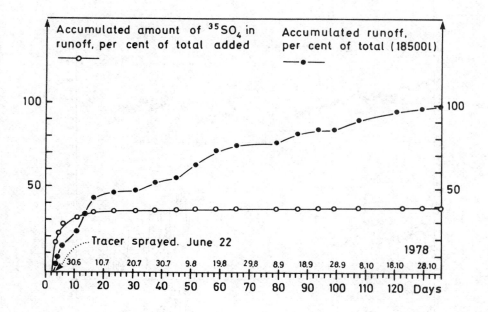

Fig. 5.   Accumulated amounts of runoff and sulphate tracer in runoff in a
mini-catchment experiment.

The retention and release of sulphate are probably of great importance in many
catchments. Thus Likens, Bormann, and Eaton (1980) found an accumulation of sulphate
during summer followed by a later release for forested watershed ecosystems of the
Hubbard Brook Experimental Forest. Christophersen and Wright (1981) found the same
trend in the input-output budget for Birkenes, a 0.4 km^2 catchment in southernmost
Norway. They have also presented a mathematical model for simulating the sulphate
concentrations in streamwater. A main feature is accumulation of sulphate during
dry periods in the summer followed by a wash-out during heavy autumn rain. The
model has later been extended to include major cations, $i.e.$ H$^+$, Al-ions and Ca^{2+} +
Mg^{2+} (treated together as M^{2+}). Na$^+$ and Cl$^-$ occur in nearly equivalent concentrations in streamwater and these ions were not included in the model (Christophersen,
Seip, and Wright, in prep.). The agreement between simulated and measured values was
quite satisfactory as illustrated in Fig. 6 for 1974. Notice that high H$^+$ concentrations occur in streamwater for high discharge and high sulphate concentrations.

Fig. 7, which is modified after an illustration used by Rosenqvist (1978), shows
very schematically the inputs and outputs to a catchment. If the bicarbonate concentration in the runoff is appreciable, there is no great acidification problem.
We will concentrate on catchments with negligible bicarbonate concentrations in
the runoff. As discussed previously, the NO$_3^-$-concentrations in runoff are usually
quite small.

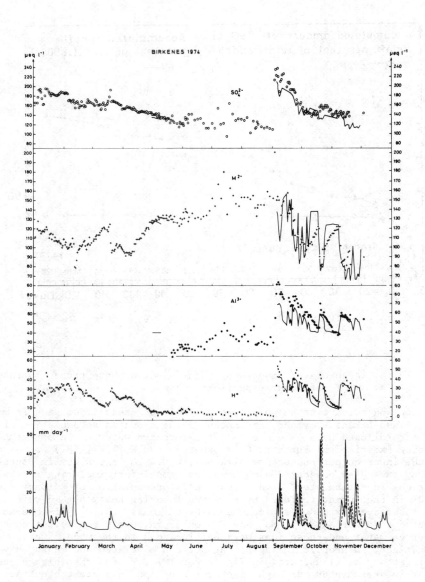

Fig. 6. Results of model calculations compared to measured values for the Birkenes catchment. Simulated (line) and observed concentrations of various ions and (bottom) simulated (dashed) and observed (full line) discharge.

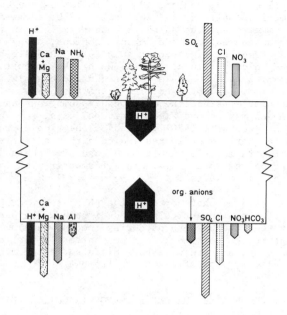

Fig. 7. Schematic illustration of the input and output of ions for a catchment. The broad arrows inside the "box" represent production and consumption of H^+ in the catchment.

The input and output of chloride are usually nearly equal. Chloride in the precipitation is accompanied by a nearly equivalent amount of sodium. Many studies indicate that sodium ions move fairly easily through soil. The long-term acidifying effect of sodium in the precipitation should then be small. The situation may be different for short periods. Skartveit (1980) found that cation exchange involving sodium must be included to explain the seasonal variations in pH in runoff in some catchments on the west coast of Norway, where the precipitation in periods has extreme sodium concentrations of about 1 meql^{-1} or even more. Neglecting sodium and chloride, the importance of sulphate for the transport of cations, including H^+ and Al, becomes evident. To explain acidification it is important to consider the possibilities for leaching as well as the size of the large reservoir and the production and consumption of H^+ indicated by the two broad arrows inside the "catchment" in Fig. 7.

A rough estimate of the effect on pH in stream water caused by an increase in the sulphate concentration may be obtained very simply. If we consider a catchment where most of the sodium in the runoff is of atmospheric origin, an increase in the concentration of mobile anions in the runoff must result in an increase mainly in H^+-, Ca-, Mg- and Al-concentrations. Results presented elsewhere (cf. Overrein, Seip, and Tollan, 1980) indicate that as a rough approximation we may assume a linear relationship between the concentrations of H^+ and sulphate. If the concentrations of bicarbonate, nitrate and organic anions are small, a direct proportionality, $[H^+] = a[SO_4]$, may be valid.

According to the OECD report on long-range transport of air pollutants (OECD, 1977), the natural emissions in Europe are not likely to exceed 10% of the total (cf. Semb, 1978; Cullis and Hirschler, 1980). Similar estimates have been made for regions in North America (Galloway and Whelpdale, 1980; Adams and co-workers, 1980). It seems therefore likely that the sulphate concentrations in runoff today are about ten times greater than the natural one in some regions, though the factor will be smaller where there are other significant sources of sulphate besides atmospheric deposits of anthropogenic or natural origin. Using the proportionality relation given above and assuming as an example a fivefold increase in the sulphate concentration, a pH shift of 0.7 is obtained. The proportionality relation will in most cases give an overestimate, since the mentioned assumptions can only be approximately valid. The model may be modified in several ways, *e.g.* by using Gapon's equation (Bolt and Bruggenwert, 1976) instead of the simple proportionality between H^+ and sulphate. Various estimates indicate that a five to tenfold increase in the sulphate concentrations in runoff may cause pH-changes in the range 0.4-1.0, say from 5.5 to between 5.1 and 4.5, in streamwater with low content of organic matter (Seip, unpubl. results).

The simple model illustrates that not only the pH of the precipitation is important in estimating the effects. In particular, the difference between sulphate and nitrate should be noted. Except perhaps during the snowmelting period, most of the NO_3^- will be taken up in the catchment, and only a small fraction will appear in runoff. Thus sulphate and nitrate will have different effects not only on terrestrial systems, but also for acidification of freshwater.

EFFECTS ON FRESHWATER THROUGH A CHANGE IN SOIL ACIDITY

So far we have largely ignored possible changes in soil acidity. Obviously such changes may occur for example because of increased leaching of base cations. Since most of the precipitation is in contact with the soil, this will also affect the composition of the runoff.

A change in the soil acidity may be caused by

a.   acid precipitation

or

b.   other factors (for example timber production, changed grazing pressure, afforestation, or other vegetation changes).

a)   By Acid Precipitation

Acid precipitation is one of several possible sources of soil acidification. Table 1 is taken from a paper by Wiklander. He predicts acid precipitation to cause the largest soil acidification when the soil pH is around 5. The effect is probably less on more acid soils due to lower replacing power of H^+. However, this leads to a prediction of increasing effect of acid precipitation on water acidity with increasing soil acidity.

Calculations indicate that acid precipitation may, in some areas, change the soil acidity significantly in perhaps one or a few decades (Overrein, Seip, and Tollan, 1980; Reuss, 1980).

TABLE 1 Acidifying effect of acid precipitation on soil and water
        (from Wiklander, 1980)

Soil pH	$\frac{\Delta Me}{\Delta H}$*	Soil	Water
8	1	None - slight	None
7	1	Buffering important	None
6	1	Buffering important	None
5	<1	Strong	Slight
4	<1	Medium - slight	Increasing
3	<<1	Very slight	Increasing

* $\frac{\Delta Me}{\Delta H}$ expresses the efficiency of H^+ to replace base cations. ΔMe represents the amount of base cations replaced by a minute amount of H^+ added (ΔH).

Experiments have shown that application of sulphuric acid tends to decrease pH and the base saturation of the soil. Except close to large emitters, there are few field observations to prove acidification of soil by acid precipitation. However, results reported by Troedsson (1980) and by Linzon and Temple (1980) seem to indicate that such an effect may exist.

Acid precipitation may influence plant growth. Both increase and decrease in growth have been observed depending on the species studied, the actual composition of the precipitation as well as other factors. In most areas acid precipitation contains elevated concentrations of NH_4^+ and NO_3^-. This is likely to contribute to an increased growth, which may result in soil acidification. It is at present not possible to estimate the importance of this mechanisms, but the hypothesis may deserve attention in future studies.

b)   Other Causes

The acidity of the soil and the leaching of cations may also be affected by agricultural and silvicultural activities such as grazing, afforestation, and timber production.

It is clearly difficult to obtain reliable quantitative estimates of these effects. Calculations on cation removal by timbering and accumulation in growing trees indicate that this should not have a great effect in most regions in Norway, especially not in heath and mountain areas. On the other hand, soil acidification may have occurred in catchments with extensive reforestation. Also replacement of deciduous trees by conifers is likely to cause acidification.

Decreased grazing pressure is among the factors mentioned as possible reasons for acidification of soil and water. It has been stated that decreased grazing may change the vegetation and lead to increased accumulation of acid humus (Rosenqvist, 1978, 1980). It is likely that the grazing pressure has decreased in many areas in Norway. However, in many remote outfields and mountain areas, mainly grazed by sheep and reindeer, there has probably been an increase in the grazing pressure (Figs. 8 and 9). Nevertheless, some of these areas are among those with the most severe acidification problems.

Fig. 8. Landscape from Njardarheim, a mountain area with severe
acidification problems.

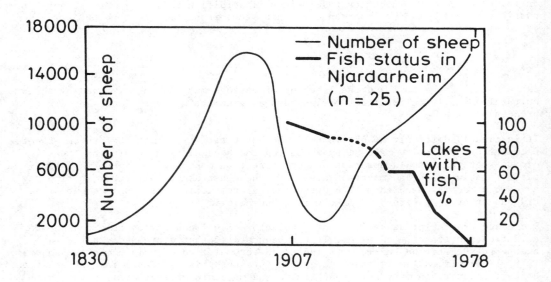

Fig. 9. Fish status development in 25 lakes in Njardarheim, which have
now lost the fish population, compared to number of sheep in
the area.

In more productive areas the situation is extremely complex. It is possible that pH in some streams and lakes has decreased by more than one unit. It seems difficult, with our present knowledge of the acidification mechanisms, to ascribe such changes only to acid precipitation. Though the effect of changes in land use is not easily quantified, the result is probably enhanced acidification in some areas. However, the great variation in vegetation, land use and in changes in land use in areas where acidification is a problem, makes it difficult to explain regional acidification primarily by these factors.

CONCLUSIONS

It is still not possible to quantify precisely the contributions from various processes involved in acidification. The conclusions given below are primarily based on Norwegian studies. Referring to the three conceptual models discussed for acidification, our present knowledge may be summarized as follows:

The fraction of the precipitation reaching rivers and lakes essentially unchanged is normally small. Thus the direct effects are small, but not negligible.

Changes in the soil may occur as a result of acid precipitation. This may contribute to freshwater acidification. It is difficult to quantify the contribution, but it is not likely to account for a major part of the observed pH changes in freshwater.

Though there seems to be no consistent pattern of changes in land use that may explain regional acidifications, these factors are probably contributing in some areas.

The most plausible explanation of the recent acidification of freshwater is provided by the mobile anion concept. While most of the atmospheric nitrate is retained in the catchment, sulphate anions are usually quite mobile. The increased concentrations of cations, including H^+, resulting from an increase in the sulphate concentrations in runoff, explain at least a substantial part of the observed acidification.

Taking all evidence into account, there seems to be no reason to doubt that the change in the composition of the precipitation has played an important role in the acidification of freshwater. Studies on the kinetic aspects of acidification ought to play an important role in future work.

REFERENCES

Adams, D.F., S.O. Farwell, M.R. Pack, and E. Robinson (1980). Estimates of natural sulfur source strengths. In D.S. Shriner, C.R. Richmond, and S.E. Lindberg (Eds.), *Atmospheric Sulphur Deposition. Environmental Impact and Health Effects*. Ann Arbor Science, Ann Arbor, pp. 35-45.

Almer, B., W. Dickson, C. Ekström, and E. Hörnström (1978). Sulfur pollution and the aquatic ecosystem. In J.O. Nriagu (Ed.), *Sulfur in the environment, Part II, Ecological Impacts*, John Wiley & Sons, pp. 271-311.

Bolt, G.H., and M.G.M. Bruggenwert (Eds.) (1976). *Soil Chemistry, A. Basic elements*. Elsevier.

Christophersen, N., and R.F. Wright (1981). Sulfate flux and a model for sulfate concentration in streamwater at Birkenes, a small forested catchment in southernmost Norway. *Water Resour. Res., 17,* 377-389.

Conroy, N., K. Hawley, W. Keller, and C. Lafrance (1976). Influences of the atmosphere on lakes in the Sudbury area. Proc., First Spes. Symp. on Atmosph. Contr. to the Chemistry of Lake Waters, *J. Int. Assoc. Great Lakes Res.*, *2*, 146-165.

Cronan, C.S., W.A. Reiners, R.C. Reynolds Jr., and G.E. Long (1978). Forest floor leaching: Contributions from mineral, organic and carbonic acids in New Hampshire subalpine forests. *Science*, *200*, 309-311.

Cullis, C.F., and M.M. Hirschler (1980). Atmospheric sulphur: natural and man-made sources. *Atmos. Environ.*, *14*, 1263-1278.

Dahl, J.B., C. Qvenild, O. Tollan, N. Christophersen, and H.M. Seip (1979a). Methodology of studies of chemical processes in water runoff from rock and shallow soil using radioactive tracers. *Water Air Soil Poll.*, *11*, 179-190.

Dahl, J.B., C. Qvenild, H.M. Seip, and O. Tollan (1979b). Omsetning av kalsium og sulfat i smeltevann og i regnvann på små felter undersøkt ved hjelp av radioaktive tracere. (Chemical processes involving calcium and sulphate in water runoff from rock and shallow soil studied by using radioactive tracers.) SNSF-project, IR49/79, 65 p.

Dahl, J.B., C. Qvenild, O. Tollan, N. Christophersen and H.M. Seip (1980). Use of radioactive tracers to study runoff and soil-water interactions in natural mini-catchments. In D. Drabløs and A. Tollan (Eds.), *Ecological impact of acid precipitation*, pp. 160-161, SNSF-project.

Galloway, J.N., and D.M. Whelpdale (1980). An atmospheric sulfur budget for eastern North America. *Atmos. Environ.*, *14*, 409-417.

Gorham, F. (1958). Free acids in British soils. *Nature*, *181*, 106.

Johannessen, M., and A. Henriksen (1978). Chemistry of snow meltwater. Changes in concentrations during melting. *Water Res.*, *14*, 615-619.

Johnson, D.W., and D.W. Cole (1977). Sulfate mobility in an outwash soil in Western Washington. *Water Air Soil Poll.*, *7*, 489-495.

Likens, G.E., R.F. Wright, J.N. Galloway, and T.J. Butler (1979). Acid rain. *Sci. Am. 241*, *4*, 43-51.

Likens, G.E., F.H. Bormann, and Eaton (1980). Variations in precipitation and streamwater chemistry at the Hubbard Brook Experimental Forest during 1964 to 1977. In T.C. Hutchinson and M. Havas (Eds.), *Effects of Acid Precipitation on Terrestrial Ecosystems*. Plenum Press, New York, pp. 443-464.

Linzon, S.N., and P.J. Temple (1980). Soil resampling and pH measurements after an 18-year period in Ontario. In D. Drabløs and A. Tollan (Eds.), *Ecological impact of acid precipitation*, pp. 176-177, SNSF-project.

Muniz, I.P., and H. Leivestad (1980). Acidification - effects on freshwater fish. In D. Drabløs and A. Tollan (Eds.), *Ecological Impact of Acid Precipitation*, pp. 84-92, SNSF-project.

Nye, P.H., and P.B. Tinker (1977). *Solute movement in the soil-root system*. Blackwell Sci. Publ., Oxford.

Ødelien, M., A.R. Selmer-Olsen, and I. Haddeland (1975). Investigation of some redox processes in peat and their influence on runoff water. *Acta Agric. Scand.*, *25*, 161-166.

OECD (1977). *The OECD Programme on Long Range Transport of Air Pollutants*. Paris 1977.

Overrein, L.N., H.M. Seip, and A. Tollan (1980). *Acid Precipitation - Effects on Forest and Fish. Final Report of the SNSF-project 1972-1980*. SNSF-project, FR19/80.

Reuss, J.O. (1980). Simulation of soil nutrient losses resulting from rainfall acidity. *Ecol. Model.*, *11*, 15-38.

Rosenqvist, I.Th. (1978). Acid precipitation and other possible sources for acidification of rivers and lakes. *Sci. Total Environ.*, *10*, 39-49.

Rosenqvist, I.Th. (1980). Influence of forest vegetation and agriculture on the acidity of freshwater. In J.R. Pfafflin and E.N. Siegler (Eds.), *Advances in Environmental Science and Engineering*, *3*, Gordon and Breach Science Publ., 56-79.

Schofield, C.L. (1976). Acid precipitation: Effects on fish. *Ambio*, *5*, 228-230.

Seip, H.M., E.T. Gjessing, and H. Kamben (1979). Importance of the composition of the precipitation for the pH in runoff - Experiments with artificial precipitation on partly soil covered "mini-catchments". SNSF-project, IR47/79.

Seip, H.M., G. Abrahamsen, N. Christophersen, E.T. Gjessing, and A.O. Stuanes (1980). Snowmelt and meltwater chemistry in mini-catchments. SNSF-project, IR53/80.

Semb, A. (1978). Sulphur emissions in Europe. *Atmos. Environ.*, *12*, 455-460.

Skartveit, A. (1980). Observed relationships between ionic composition of precipitation and runoff. In D. Drabløs and A. Tollan (Eds.), *Ecological Impact of Acid Precipitation*, pp. 242-243, SNSF-project.

Tollan, A. (1980). Annotated bibliography, SNSF-project, 1980.

Troedsson, I. (1980). Ten years acidification of Swedish forest soils (Abstract). In D. Drabløs and A. Tollan (Eds.), *Ecological Impact of Acid Precipitation*, p. 184, SNSF-project.

Wiklander, L. (1975). The role of neutral salts in the ion exchange between acid precipitation and soil. *Geoderma*, *14*, 93-105.

Wiklander, L. (1980). The sensitivity of soils to acid precipitation. In T.C. Hutchinson and M. Havas (Eds.), *Effects of Acid Precipitation on Terrestrial Ecosystems*, Plenum Press, New York, p. 553-567.

Wright, R.F., and A. Henriksen (1978). Chemistry of small Norwegian lakes with special reference to acid precipitaion. *Limnol. Oceanogr.*, *23*, 487-498.

ECOLOGICAL IMPLICATIONS OF AIR POLLUTANTS FROM SYNTHETIC FUELS PROCESSING

D. S. Shriner, S. B. McLaughlin and G. E. Taylor

Environmental Sciences Division, Oak Ridge National Laboratory, Oak Ridge, Tennessee 37830, USA

ABSTRACT

Determination of potential ecological impacts from air pollutants from synfuels facilities involves four major problem areas: (1) effluent characterization, (2) acute and chronic toxicity of effluents, (3) potential for chemical reaction during atmospheric transport and deposition, and (4) biotransformation and accumulation. Compounds of concern, because of our lack of knowledge about their behavior in the environment, include a variety of organics, reduced sulfur and nitrogen compounds, and certain trace metals. Remarks in this paper are confined to problems related to air pollutants projected to be released from coal conversion facilities and do not address problems related to product use or industrial hygiene within the plant itself, except in cases where ventilation streams are exhausted to the atmosphere. An overview is presented of potential sources of atmospheric releases of pollutants resulting from typical synfuels process operations. Potential pollutant mixtures are discussed, as are the implications of such mixtures for ecological impacts on a regional scale.

KEYWORDS

Air pollution; vegetation; ecological effects; synthetic fuels; coal conversion; pollutant interactions.

INTRODUCTION

The potential problems associated with environmental effects of air pollution from coal conversion processes may be organized into four primary areas. In time sequence, the first of these areas is effluent characterization. Because of the (1) rapid and frequent changes in effluent composition associated with the developing technology, (2) very reactive nature of the effluents, and (3) analytical difficulties involved in their characterization at ambient concentrations, good data on effluent characterization have been difficult to obtain. Compounds such as H_2S, COS, CS_2, thiophene, benzene, methane, ethane, and phenols are frequently identified as principal effluent gases (Shults, 1975; Jahnig, 1975; Forney and others, 1974; Sather and others, 1975; Rubin and McMichael, 1974; O'Hara and others, 1974); however, quantitative estimates of their importance may vary by two orders of magnitude (USERDA, 1975). Order-of-magnitude quantitative estimates are essential for early attempts to evaluate potential environmental impacts from conversion processes. A second area of importance is knowledge of the potential of effluent compounds for chemical reaction during the course of atmospheric transport and deposition. Some of the compounds involved (e.g., hydrocarbons, NO_2) have potential for rapid reaction in the presence of ambient oxidant concentrations and ultraviolet light (Spicer and Miller, 1976; Heicklen, 1976). The nature of these reactions, knowledge of their rates, and identity of their reaction products are essential to interpretation of environmental effects at the eventual deposition surface. A third problem area relates to the potential for biological uptake, transformation, and/or accumulation. Effluents may enter the terrestrial ecosystem through processes of gas exchange, active uptake from soil solution, or adsorption. In cases of metabolic transformation and/or accumulation, the potential consequences (e.g., toxic accumulations in plant tissues to be ingested by man) must be evaluated. If uptake or accumulation reaches toxic levels (the fourth problem), acute or chronic toxicity may assume priority. Because most of the terrestrial landscape is covered by vegetation of some type, there is high probability that the primary deposition surface for most gaseous effluents will be plant surfaces (Saunders, 1971). The role vegetation cover provides as an efficient means of detoxification has been suggested (Martin and Barber, 1971) for more common pollutants, such as SO_2, to be of potentially major significance to the ultimate fate and effects of, for example, sulfur input to the landscape. A similar role may likely be operative for a wide range of pollutant compounds.

Determining the ecological implications of developing technologies requires close cooperation between environmental scientists and technologists at the earliest opportunity during the development process. Such cooperation ensures both timely and environmentally compatible technology development -- timely in the sense that expensive pollution abatement "add-ons" will be avoided, as will effects on the environment (Richmond and others, 1976).

SOURCES OF ENVIRONMENTAL RELEASE

Dry and wet deposition of gaseous and particulate atmospheric contaminants released from coal synfuels technologies can or will occur following release of (1) stack emissions, (2) fugitive emissions, (3) cooling tower drift, and (4) ponding volatiles. Many of these possible contaminants (Table 1) are of potential significance because, although their releases may be quantitatively small, the materials are either known hazardous materials or there is little if any information on their transport behavior, fate, or effects in the environment.

Stack Emissions

Primary point souces of atmospheric emissions from synfuels technologies will be from combustion processes (power plant and process heaters). Of primary concern for ecological effects from point source emissions are sulfur dioxide and oxides of nitrogen. Because these pollutants are frequently common to many individual sources within a region and are capable of medium- to long-range atmospheric transport, they will not be addressed in the process-specific context of this paper. Numerous reviews are available which discuss the ecological impacts of sulfur and nitrogen oxide air pollution (Treshow, 1968; Mudd and Kozlowski, 1975; Heck and Brandt, 1976; McLaughlin, 1981; Shriner, 1981).

Coal Storage and Coal Handling

Coal unloading, storage, and preparation areas are difficult to quantitatively assess as potential sources of fugitive dust because of the fact that emissions

TABLE 1 Potential Source Categories and Classes
of Compounds Released to the Atmosphere
from Operating Synfuels Facilities

Source category	Compound class
(1) Stack emissions	SO_2, NO_x, CO_2, CO, particulates
(2) Fugitive emissions, vents	CO, C_1-C_6, BTX,[*] H_2S, COS, mercaptans, phenols, NH_3, HCN particulates, polynuclear aromatic hydrocarbons
(3) Water treatment or ponding volatiles	H_2S, phenols, NH_3, HCN
(4) Cooling tower drift	Chromium and compounds, chlorinated phenols,[†] zinc and compounds, sulfates

[*]Benzene, xylene, toluene
[†]If process waters were used in cooling loop.
(Source: Bombaugh and others, 1980; Gehrs and others, 1981.)

are intermittent, generally at or near ground level. Because of their relatively large size and consequent rapid deposition rate, ecological effects would be expected to be restricted to the localized area of a facility. Two primary types of effects might be expected, one due to the physical mass loading of particles and the second due to chemically active toxic elements associated with the particles. Coal storage and handling impacts are primarily restricted to effects of the first type. The deposition of particles on aboveground plant parts can reduce net radiation input to the plant surface, increase leaf temperatures, and block stomates (Guderian, 1981). Any of the above effects of particle loading can result in reduced primary productivity of the plants and should be the major potential impact.

Fugitive Emissions, Ponding Volatiles

Characterization and quantification of fugitive emissions from coal conversion processing will be an area of significant environmental concern in the near future (Gehrs and others, 1981). The large number of sources, their low concentrations, and their potential reactivity make both quantification and control of fugitive emissions difficult. The compound classes listed under the fugitive emissions source category in Table 1 represent potential concern in the area of worker exposure-related health effects, and these concerns will be the major factor driving control efforts. As a generic source of fugitive hydrocarbon and reduced sulfur gas emissions, however, these miscellaneous vents, valves, flanges, lock-hoppers, and ponds also represent potential for ecological impact. Because of the generally ground-level release of fugitive emissions, however, most ecological impacts will be localized, rather than regionalized in scope.

Identification of problem effluent streams and components from currently available literature has been of marginal utility up to this point in time in the area of fugitive atmospheric emissions, because characterization for engineering purposes may be expressed quantitatively for major constituents (e.g., total sulfur) and not for individual constituents (e.g., trace sulfur species). Typical vegetation response thresholds for gaseous pollutants are in the microgram per liter to milligram per liter range, and therefore, a difference of even 0.01% in source strength may have environmental significance without being amenable to engineering controls. Principal components are reduced sulfur gases (hydrogen sulfide, carbonyl sulfide, carbon disulfide, and methyl mercaptans) and an array of hydrocarbons including ethylene, ethane, propane, butane and butylene, and a variety of aromatic hydrocarbons.

Cooling Tower Drift

Taylor (1980) recently reviewed implications of ecological impacts to the terrestrial environment from cooling tower drift. Typically, because of the level of release and the size distribution involved, the majority of drift aersols are deposited within the immediate vicinity of the drift basin ($\leq$ 1 km). Potential concerns include direct deposition or runoff of heavy metal or organic-contaminated cooling water to surface streams. Additional chemicals may, at times, be added to the cooling water for wood preservation (e.g., acid copper chromate) and could also impct ecological systems (Taylor, 1980).

ECOLOGICAL IMPLICATION

Based on the types of sources and emissions discussed above, several implications are suggested for ecological systems: (1) localized, acute effects on soils, vegetation; (2) localized but chronic effects on vegetation, soils; (3) some contribution to regional air quality patterns; and (4) potential secondary effects on aquatic ecosystems, water quality.

Localized, Acute Effects

Localized, acute injury to vegetation in the vicinity of an operating synfuels facility would only be expected under some type of upset conditions under which high concentrations of phytotoxic gases were released at ground level during a leak or spill (Shriner, 1976; Shinn and Kercher, 1980). Effluent compounds such as carbon monoxide, ammonia, hydrogen cyanide, and benzene, for example, are all harmful to plants only at levels significantly above those of concern for occupational health and safety, and, as such, control strategy priorities should not involve vegetation effects. Major concern must be for effluents which are phytotoxic at levels below health effects standards and for which welfare effects standards do not exist. Most effluents in that category are virtually unknown in terms of their potential ecological impact (Shinn and Kercher, 1980).

Among the more quantitatively significant of the expected emissions for which ambient air quality standards do not exist, but which may be released from synfuels processing, are hydrogen sulfide (H_2S) and carbonyl sulfide (COS). Research in our laboratory was designed to evaluate the comparative phytotoxic effects of these sulfur gas species relative to sulfur dioxide, the major sulfur gas released from conventional coal combustion or from the combustion of tail gases from a coal conversion facility. Results indicate that these three sulfur gases may be ranked in phytotoxicity in the order SO_2 > H_2S > COS, confirming an environmental advantage in selection of technology options to reduce SO_2 emissions in favor of H_2S, which can be more easily recovered from emission streams as elemental sulfur. Results of acute and chronic exposures of sensitive vegetation species to COS, however, suggest that since that sulfur specie is less efficiently removed by sulfur tail-gas clean-up technologies, additional clean-up measures may be required on large-scale facilities to ensure minimal environmental release of COS as well (Auerbach, 1979).

Chronic Effects

Chronic effects in the vicinity of an operating synfuels facility might be expected as the result of the cumulative effect of low concentrations of a toxicant over long periods of time or as the result of additive or synergistic effects from two or three more pollutants in combination.

In the latter case, operation of synthetic fuels facilities may result in localized or even regionalized shifts in ratios of the major criteria air pollutants: SO_2, NO_2, and O_3. These shifts are anticipated in the United States because of (1) increases in efficiency of sulfur oxide control technology, (2) less immediate improvement in nitrogen oxide control technology, and (3) recent relaxation of the oxidant standard combined with the anticipated presence of precursors (hydrocarbons, nitrogen oxides) for the photochemical formation of ozone and other photochemical oxidant compounds.

Reinert and coworkers (1981) evaluated the significance of mixtures of gaseous pollutants in interpreting plant response patterns which may occur with long-term shifts in ambient $SO_2:NO_2:O_3$ ratios. A 3^3 factorial experimental design was employed using the three pollutants at three levels each (0.10, 0.20, and 0.40 ppm). With the test species used (radish, Raphanus sativus), as ozone concentrations increased relative to concentrations of NO_2 or SO_2, greater yield reduction occurred as a general phenomenon. The importance of the increasing ozone concentration decreased as either SO_2 or NO_2 concentration increased. These results suggest that the siting of additional sources of SO_2 and NO_2 in regions currently experiencing elevated ozone levels may have significant ecological implications.

Contributions to Regional Air Quality

Operating synfuels facilities represent potential sources of alteration of regional air quality as well. The major contributions would include: (1) a contributing source of SO_2 and NO_x, both capable of long-range transport and deposition in precipitation as "acid rain" (Shriner and others, 1980); (2) a contributing source of both hydrocarbons and oxides of nitrogen--precursors for photochemical oxidant formation--to regional background levels; and (3) a contributing factor in shifting regional air quality patterns (e.g., $SO_2:NO_2:O_3$ ratios) as a result of changing fuel-use options (Table 2).

Secondary Effects on Aquatic Systems

The potentically most significant secondary impacts to aquatic systems are those discussed under "Cooling Tower Drift", e.g., direct deposition or surface runoff to water bodies, and regional air quality as "acid rain", where acidification of surface waters may be associated with atmospheric deposition of acidic substances.

TABLE 2 Estimated Annual Emissions to Supply* a 1000 MWe Power Plant

Atmospheric emissions (Mg/year)	Process						
	Low Btu gasification		High Btu gasification		Coal liquefaction		Coal combustion
	Coal Source						
	Western (low S)	Illinois (high S)	Western (low S)	Illinois (high S)	Western (low S)	Illinois (high S)	
SO_2	470	1815	1450	8390	1200	1580	102,000
NO_x	910	910	5470	6260	6860	6860	24,000
HC	26	26	93	100	2100	2100	240

*Annual controlled air emissions from a unit plant (output equal to product required to support a 1000-MW electric plant operating at 33% overall thermal efficiency). (Source: Gehrs and others, 1981.)

CONCLUSIONS

The ecological implications of air pollution from synthetic fuels processing are relatively broad and somewhat poorly characterized at the present time due to limited ecological analysis in the vicinity of operating facilities. However, based on available evidence of injury, growth impacts, and crop reduction thresholds, potential fugitive emissions remain a concern because they are difficult to control. Additional concern is justified on the basis of pollutant mixtures, because of their high probability of occurrence and our lack of understanding of the consequences at the present time.

ACKNOWLEDGEMENT

Research supported in part by U.S. Department of Energy, Office of Health and Environmental Research, under contract W-7405-eng-26 with the Union Carbide Corporation and in part by the Electric Power Research Institute under contract number RP 1908-2. Publication No. 1787, Environmental Sciences Division, Oak Ridge National Laboratory.

LITERATURE CITED

Auerbach, S. I. (1979). Environmental Sciences Division Annual Progress Report for Period Ending September 30, 1979. ORNL-5620, Oak Ridge National Laboratory, Oak Ridge, Tennessee. pp. 62-63.

Bombaugh, K. J., G. C. Page, C. H. Williams, L. O. Edwards, W. D. Balfour, D. S. Lewis, and K. W. Lee (1980). Aerosol characterization of ambient air near a commercial Lurgi coal gasification plant, Kosovo Region, Yugoslavia. EPA-600/7-80-177, U.S. Environmental Protection Agency, Research Triangle Park, North Carolina.

Forney, A. J., W. P. Haynes, S. J. Gasior, G. E. Johnson, and J. P. Starkey, Jr. 1974). Analysis of tars, chars, gases, and water found in effluents from the Synthane process. In Symposium Proceedings: Environmental Aspects of Fuel Conversion Technology. EPA-650/2/74/118, Environmental Protection Agency, Research Triangle Park, North Carolina.

Gehrs, C. W., D. S. Shriner, S. E. Herbes, E. J. Salmon, and H. Perry (1981). Environmental, health and safety implications of increased coal utilization. In M. A. Elliott (Ed.), Chemistry of Coal Utilization, Second Supplementary Volume. John Wiley and Sons, Inc., New York. pp. 2159-2223.

Guderian, R. (1981). Terrestrial vegetation-air pollutant interactions: Non-gaseous air pollutants. In S. F. Krupa and A. Legge (Eds.), Air Pollutants and Their Effects on the Terrestrial Ecosystem. John Wiley and Sons, Inc., New York. (in press).

Heck, W. W., and C. S. Brandt (1976). Effects on vegetation: Native, crops, forests. In: A. C. Stern (Ed.), Air Pollution, Vol 2, 3rd ed. Academic Press, New York. pp. 157-229.

Heicklen, J. (1976). Atmospheric Chemistry. Academic Press, New York.

Jahnig, C. E. (1975). Evaluation of pollution control in fossil fuel conversion processes. Liquefaction: Section 2. SRC process. EPA-650/2-74/009/4. Environmental Protection Agency, Research Triangle Park, North Carolina.

Martin, A., and F. R. Barber (1971). Some measurements of loss of atmospheric sulfur dioxide near foliage. Atmos. Environ. 5, 345-352.

McLaughlin, S. B. (1981). SO_2, vegetation effects, and the air quality standard: Limits of interpretation and application. In W. W. Heck (Ed.), The Proposed SO_x and Particulate Matter Standard. Air Pollution Control Assoc., Pittsburgh, Pennsylvania. (in press).

Mudd., J. B., and T. T. Kozlowski (1975). Responses of Plants to Air Pollution. Academic Press, New York.

O'Hara, J. B., S. N. Rippee, B. I. Loran, and W. J. Mindheim (1974). Environmental factors in coal liquefaction plant design. In Symposium Proceedings: Environmental Aspects of Fuel Conversion Technology. EPA-650/2-74-118. Environmental Protection Agency, Research Triangle Park, North Carolina.

Reinert, R. A., D. S. Shriner, and J. W. Rawlings (1981). Responses of radish to NO_2, SO_2 and O_3 in all combinations of three concentrations. J. Environ. Qual. 10, in press.

Richmond, C. R., D. E. Reichle, and C. W. Gehrs (1976). Balanced Program Plan: Vol. IV. Coal Conversion. ORNL-5123, Oak Ridge National Laboratory, Oak Ridge, Tennessee.

Rubin, E. S., and F. C. McMichael (1974). Some implications of environmental regulatory activities on coal conversion processes. In Symposium Proceedings: Environmental Aspects of Fuel Conversion Technology. EPA-650/2-74-118, Environmental Protection Agency, Research Triangle Park, North Carolina.

Sather, N. F., W. M. Swift, J. R. Jones, J. L. Beckner, J. H. Addington, and R. L. Wilburn (1975). Potential Trace Element Emissions from the Gasification of Illinois Coal. NTIS-PB-241220. National Technical Information Service, Springfield, Virginia.

Saunders, P. J. W. (1971). Modifications of the leaf surface and its environment by pollution. In T. F. Preece and C. H. Dickinson, (Eds.), Ecology of Leaf Surface Micro-organisms. Academic Press, New York. pp. 81-90.

Shinn, J. H., and J. R. Kercher. (1980). Ecological effects of atmospheric releases from synthetic fuels processes. In K. E. Cowser and C. R. Richmond (Eds.), Synthetic Fossil Fuel Technology: Potential Health and Environmental Effects. Ann Arbor Science Publishers, Ann Arbor, Michigan. pp.199-203.

Shriner, D. S. (1976). Research needs related to air pollution from coal conversion processes. In Proc., 82nd National Meeting of the American Institute of Chemical Engineers. AIChE, New York.

Shriner, D. S., C. R. Richmond, and S. E. Lindberg (Eds.). (1980). Atmospheric Sulfur Deposition: Environmental Impact and Health Effects. Ann Arbor Science Publishers, Ann Arbor, Michigan.

Shriner, D. S. (1981). Terrestrial vegetation-air pollutant interactions: Nongaseous pollutants, wet deposition. In S. V. Krup and A. Legge (Eds.), Air Pollutants and Their Effects on the Terrestrial Ecosystem. John Wiley and Sons, Inc. New York (in press).

Shults, W. D. (1975). Preliminary Results: Chemical and Biological Examination of Coal-derived Materials. ORNL/NSF/EATC-18. Oak Ridge National Laboratory, Oak Ridge, Tennessee.

Spicer, C. W., and D. F. Miller (1976). Nitrogen balance in smog chamber studies. J. Air Pollut. Control Assoc. 26, 45-50.

Taylor, F. G. (1980). Chromated cooling tower drift and the terrestrial environment: A review. Nucl. Saf. 21, 495-508.

Treshow, M. (1968). The impact of air pollutants on plant populations. Phytopathology 58:1108-1113.

U.S. Energy Research and Development Administration (1975). Environmental Research Centers: A Recommended Complementary Program. ERDA 76-29, U.S. Department of Energy, . Washington, D. C.

CAN THE POOR COUNTRIES AFFORD BIOMASS ENERGIES?

V. Smil

Department of Geography, University of Manitoba, Winnipeg, Canada

ABSTRACT

Numerous environmental services provided by crop residues, above all the reduction of soil erosion, retention of moisture and recycling of nutrients, appear to be more valuable, especially for the poor countries, than the use of the residues for modern biomass energy conversions. Cultivation of energy crops in subtropics and tropics is mostly undesirable owing to the land and water demands of these plants, increased deforestation and monoculture vulnerability and, most likely, decreased availability of already scarce food.

KEYWORDS

Biomass energy; crop residues; rice straw; energy crops; sugar cane; cassava; environmental impacts; economic development.

INTRODUCTION

Proponents of renewable energetics believe that solar-based energy systems are especially appropriate for the poor countries as they "[are] ideally suited for rural villages and urban poor alike ... capitalize on poor countries' most abundant resources ... conform to modern concepts of agriculturally based eco-development from the bottom-up" and hence have the ability to "contribute promptly and dramatically to world equity and order" (Lovins, 1976). A critical look must show how simplistic and erroneous such views are (Smil, 1979a) and the case of biomass energy -- to which the renewable planners are assigning the essential roles of supplying liquid transportation fuels and generating electricity in small power stations -- provides an especially compelling illustration.

The question posed as the title of this paper can be interpreted in two ways. First, only a few less affluent populous nations, namely those for which the term 'developing' is a fair characterization rather than a euphemistic label (ie. countries such as South Korea or Taiwan) have acquired energy consuming structures akin to those of the rich world. Most of the poor world remains dependent on traditional biomass energies which provide over 90 percent of all fuel in many African countries, about half of energy in India and a third of China's total fuel use (Smil, 1979b; Smil and Knowland, 1980).

In all of these nations extensive use of forest fuels, crop residues and dung has led to serious environmental degradation -- deforestation, desertification, erosion, decline in soil fertility -- and so the question can be interpreted: can the poor countries afford the continuation of these perilous trends? A growing number of studies give an emphatically negative answer (see for example Eckholm, 1976; Myers, 1980; Smil, 1981a).

It is the second interpretation I want to answer here. Can the poor countries abandon the old ways and can they secure their energy supply -- partially or fully -- by modern biomass energy conversions? Can they establish large plantations of fast growing trees to be harvested on short rotations? Can they collect huge quantities of crop residues and burn them to raise steam for power generation, or pyrolyze them or ferment them anaerobically? Can they plant highly productive energy crops to use their sugars for alcohol fermentation?

Environmental consequences of such moves are, naturally, of fundamental importance as they can easily make the whole enterprise unsustainable and extremely costly in more than a monetary way. I will restrict my inquiry into such consequences to the two biomass energy resources which are, or could become, most widely available throughout the poor world: crop residues and energy crops.

CROP RESIDUES

Crop residues, especially cereal straws, are currently the most widely available source of biomass energy in the poor world. They are easily accessible, their quantities are huge, somewhere around one billion t are produced annually, and one of their main disadvantages, seasonal availability, is alleviated in many poor nations through multicropping or intercropping. Why not to use them as a fuel or feedstock? Because their numerous environmental services -- guarding against erosion, recycling nutrients, increasing organic matter and moisture holding capacity of soils and improving their structure and tilth -- are incomparably more valuable. All usual arguments against the removal of residues in countries with temperate climate are multiply more valid throughout the subtropics and tropics of the poor world.

Environmental Effects of Removing the Crop Residues

Rainfall erosion is, of course, a much more severe problem in the tropics because soil losses are closely related to the total kinetic energy of precipitation which is much higher in tropical than in temperate storms. Presence of continuous cover of plant residues is the best way to minimize soil losses by intercepting the drops, reducing the runoff and by increasing water retention.

Efficiency of residual mulches can be illustrated by recent Nigerian figures derived from experiments on a sloping alfisol field near Ibadan growing two crops of corn a year and receiving 1-1.6 m of rain during wet season (Osuji, Babalola and Aboaba, 1980). With local peasant practice (manual fieldwork, most residues remaining on the relatively undisturbed ground) soil losses in two years averaged 5.6t/ha while in conventional tillage (residues plowed in twice, harrowed once) they rose to 9.13t/ha and in bare fallow they reached 18.6t/ha.

Soil structure almost invariably deteriorates under cultivation in the moist tropics, mainly owing to intensive rainfalls accompanied by a rapid decrease in organic matter content (Sanchez, 1976). Application of organic residues is thus a must both for the maintenance of fertility and for the prevention of soil compaction. Significantly enough, in traditional tropical shifting cultivation the soils were never tilled and always protected, first with an ash mulch, later with overlapping canopies of intercropped species.

Sanchez (1976) reviews a rich literature on how residual mulching and incorporation perform many essential functions by controlling structural deterioration, increasing moisture retention, reducing leaching, runoff and erosion, lowering excessive soil temperatures and, in absence of synthetic fertilizers, supplying most of the needed nitrogen and sulfur. Increases in organic matter after residue incorporation may be a somewhat controversial issue in the temperate climate (Staniforth, 1979) but they have been repeatedly documented throughout the poor world (Wani and Shinde, 1980) where the absence or very low applications of chemical fertilizers and different environmental circumstances make the recycling of organic residues imperative.

To give just one dramatic example of organic recycling, careful experiments with applications of mushroom composts (that is partially digested rice straw) in Taiwan have shown that 20t of residues per hectare more than doubled the organic matter content of soil (from 0.61 to 1.41 percent), increased moisture content from 13.4 to 19.5 percent (by weight) and raised corn yields by as much as 33 percent (Joint Commission on Rural Reconstruction, 1977).

Soil's moisture holding capacity in dry tropics is greatly enhanced by residual mulching. A recent three year study of the effects of organic mulching on summer corn and the following winter wheat crop in India recorded yield increases up to 16 percent for corn and up to 18 percent for winter wheat (Prihar and others, 1979). Similar reports are available for groundnut, cow-pea and tobacco yields (Staniforth, 1979).

Recycling of nutrients is vastly more important in the developing countries whose economies cannot, and in foreseeable future will not be able, to secure adequate quantities of chemical fertilizers. Furthermore, residues can replenish nitrogen, usually the most needed element, in an especially effective way: carbohydrates in straw are used as a substrate for nitrogen fixing bacteria and extra 8-16 kg of nitrogen can be fixed per tonne of residues incorporated into soil (Staniforth, 1979).

Yet another interesting, and potentially very widespread, application of crop residues may be their use for reclamation of saline-alkaline soils whose areas have been increasing rapidly throughout the developing world. Indian experiments (Wani and Shinde, 1980) have shown that undecomposed straw incorporated into soils reduces pH slightly toward neutral and may thus become a welcome, cheap and accessible ingredient in a difficult task of controlling the soil degradation.

The inevitable conclusion: to maintain quality of soils compatible with permanent agriculture and to sustain reasonably high yields, crop residues in the poor countries should be treated as a critical ingredient of food production cycles, not as a waste material divertible to other uses, including energy conversions. To augment this general appraisal by a more detailed particular crop residue example I shall introduce some quantitative estimates for rice straw, by far the most abundantly available crop residue in the poor world.

Rice Straw

Straw-to-paddy ratios fluctuate with plant varieties, maturation periods, soil characteristics, fertilization, irrigation and weather conditions from as low as 0.4 to extremes over 2.0, but they are usually clustered between 0.8-1.2. Using 1:1 as the average, current production of rice straw in the developing world amounts to some 350 million t, with 90 percent of this total in Asia and nearly 40 percent in China alone.

Energy equivalent of this straw (assuming average of 3,500 kcal per kg of fresh
straw) is rather impressive: nearly 120 million t of oil equivalent or about ten
percent of the poor world's current consumption of all commercial fossil fuel energies. However, most of this straw is not, and I believe it should not become,
available for fuel conversions. Detailed utilization data are extremely rare but I
have been able to obtain two complete breakdowns typifying the situation in China:
one shows the nationwide utilization of straw from two crops of rice in Taiwan in
1978 (Table 1), the other details the disposition of the straw in a village in east

TABLE 1 Nationwide Utilization of Rice Straw in Taiwan in 1978*

Uses	First Crop (3,446 kg/ha)	Second Crop (2,832 kg/ha)
	percent	
Mushroom composting	16	9
Paper making	15	17
Plowed under	14	14
Fertilizer composting	13	15
Fuel	12	11
Animal feed	7	4
Fruit mulching	6	15
Unaccounted	17	15

* Joint Commission on Rural Reconstruction, Taipei, private communication.

Chinese province of Jiangsu, the heart of the developing world's most intensive
rice multicropping area (Table 2). Both examples share a very high degree of
recycling: adding composts (including those for mushrooms), plowing under, animal
bedding, feed (the last two to become eventually part of a compost), and mulching
results in 75 percent share in the Jiangsu case and in 56-57 percent share in the
Taiwanese mean.

TABLE 2 Utilization of Rice Straw in Production Team No. 1 of Baimao
Commune, Changsha County, Jiangsu Province in 1978*

	Tons	Percent
Harvested straw	82.9	100
Composted or plowed under	31.2	38
Pig fodder	7.9	9
Pig bedding	23.1	28
Fuel	19.2	23
Sold to State	1.5	2

*T.B. Wiens, private communication.

One-half to three-quarters of the rice straw in China are thus recycled, directly
or with a delay, to the soil -- and partial information available for more than a
dozen populous developing countries (including South Korea, Indonesia, Philippines,
Sudan, Colombia) indicates similar partitioning of use (FAO, 1979a). The remainder
is already used for household fuel, as well as for making a multitude of products
ranging from baskets, mats and ropes to paper, thatch and fences (Tanaka, 1973).
Consequently, there is precious little, if any, leftover straw to be used for

assorted biomass energy conversions.

Contribution of recycled rice straw to the maintenance of soil fertility in the developing world could be approximately calculated by assuming, rather conservatively, that just half of the rice straw produced is so used, and that the average N, P and K contents are, respectively, 0.6, 0.1 and 1.5 percent. Recycled rice straw would then provide theoretically about one million t of N, 180,000 t of P and 2.6 million t of K, or, respectively, about 7.1, 1.3 and 18.5 kg of N, P and K per hectare of rice fields.

In contrast, in the late 1970s the average applications of chemcial fertilizers per hectare of the poor world's agricultural land amounted to 6.3 kg of N, 1.2 kg of P and one kg of K (FAO, 1979b). Average figures for rice, if they would be available, would be undoubtedly higher than is the mean for all crops but the comparison is clear and staggering. Poor world's rice farming could not be sustained without extensive straw recycling.

Although rice straw has relatively low digestibility, low protein content (4.5 percent dry matter weight; consequently, animal's nitrogen balance cannot be sustained without supplements) and rather poor palatability, in poor countries it is often the only roughage available in large quantities and it contains enough cellulose (about one-third of dry weight) to make it potentially an excellent source of energy (Han and Anderson, 1974). Diversion of the straw for direct combustion or pyrolysis would thus take away often the only reliably available roughage for already not too well fed draft animals or pigs in many poor countries, affecting the availabilities of both the animate energies for fieldwork and animal protein for food.

There seems to be a way how to use the straw for animal feed, soil fertility maintenance *and* energy production: anaerobic fermentation. Straw can be fed to animals and the manure, digested with additional straw, yields both clean biogas and excellent organic fertilizer. However, this seemingly ideal theoretical solution is rather limited by practical considerations.

A key consideration in anaerobic fermentation is the necessity to maintain C/N ratio of the decomposing material at about 25:1-30:1 but cereal straws are both N and P deficient and have C/N ratios well over 80:1 (van Buren, 1979) and their use in digesters is dependent on the availability of manure to boost nitrogen content. Moreover, straw is very resistant to wetting, and can choke biogas digester (Subramanian, 1978), its cellulose is broken down only very slowly and lignin remains practically unaffected (Pfeffer, 1980) -- but these two compounds make up 40 percent of dry rice straw (Han and Anderson, 1974). So even when leaving the numerous technical and social difficulties encountered in biogas technology aside, the use of rice straw in anaerobic fermentation, perhaps the only ecologically acceptable use of the residue, will remain relatively small.

ENERGY CROPS

If the use of crop residues for energy conversions does not appear ecologically desirable, the case against energy crops in the poor world is even more persuasive. No matter what species would be cultivated they would need more land and more water than are currently used -- and they would inevitably cut down the availability of food. Yet land, water, and food are already in short supply throughout the poor world. I will deal briefly with the first two limitations and then concentrate on two major tropical energy crops, sugar cane and cassava.

Land

Per capita availability of arable land in the poor world averages just 0.21 ha and it is only 0.1 ha or less for many of the most populous nations; in contrast, there is over 0.5 ha of arable land per capita in the rich countries (FAO, 1980).

Moreover, this low availability is rapidly decreasing by widespread conversion of land to nonagricultural uses (Brown, 1978). To cite just one example: in China, the world's most populous poor nation, arable land decreased by about 12 percent between 1957 and 1980, and this, combined with population growth, cut the per capita farmland availability by half, to mere 0.1 ha (Smil, 1981b). Urbanization is advancing rapidly in virtually all populous developing nations and is a prime cause of cropland loss. For example, Brazil's share of urban population rose from only 15 percent in 1940 to 56 percent in 1970 (Wagner and Ward, 1980) and the shift is expected to continue.

If energy crops could make a significant contribution when cultivated on a small fraction of available farmland their promotion might be a rational strategy. However, even with the currently most productive energy crops grown in optimal environment the land requirements are huge. To substitute all liquid transportation fuel by sugar cane alcohol Brazil would need nearly 15 percent of its cultivated land -- and the share, in spite of future extensions of arable land, would grow to at least 20 percent by the year 2000 (Goldemberg, 1978).

Taking the existing land for energy crops would thus push the current food crops onto a less desirable land, further aggravating the already common problems of soil degradation, erosion and, in drier regions, desertification -- or it would bring about yet more deforestation. In either case negative ecological consequences would be swift and all too often irreversible.

Water

About half of the poor world's population -- from the Brazilian Northeast through the Sahel to India and large areas of China -- is living in regions with precarious water supply. To provide at least the rudimentary nutrition to their fast growing populations poor countries in these arid and semiarid regions, largely between $10°N-30°N$, expanded the area sown to grain by about half a million km^2 between 1950 and 1975 and increased irrigated farmland by some 70 percent between the early 1960s and the mid-1970s (Prentice and Coiner, 1980).

Over large areas demand for irrigation water during frequent and recurrent droughts is already surpassing the available surface flows and it is severely depleting the underground storages. Ironically, much of the precious irrigation water is wasted owing to the increasing salinity of soils throughout the world's arid and semiarid regions. With insufficient precipitation high evaporation concentrates salts in soils, chemical fertilizers compound the problem, modern field practices using heavy machinery compact the surface thus further impeding downward percolation, and the salt retained in the root zone starts cutting down the crop yields (Epstein and others, 1980).

No precise figure on global basis are available but conservative estimates show that 40-60 percent of all currently irrigated land -- or some 100-140 million ha -- have yields reduced owing to salt build-up (Polunin, 1979) and, consequently, increasing share of the so-called irrigation water must be used for vast removal rather than to supply the plants' needs. Obviously, the high yielding energy crops could only worsen this trend.

Sugar Cane

Sugar cane is the most efficient solar energy convertor of all the established crops and the long Brazilian experience with relatively large scale ethanol production -- which is now reaching new records under the ambitious PROALCOOL program -- has made the plant almost a model biomass energy source. However, to sustain high yields (at least 40 t/ha) sugar cane, besides requiring mild climate and large amounts of nutrients, must have good soils and plenty of water. Good land could be found most easily by displacing some established food crops and José A. Lutzenberger claims that this is already happening in Brazil (Daly, 1981).

Problem of water requirements is much more intractable. While the approximate range of seasonal evapotranspiration is only between 300-450 mm for small grains, 400-750 mm for corn and 500-950 mm for rice, for sugar cane it is between 1,000-1,500 mm (Doorenbos and Pruitt, 1977). Appealing 'solution' is to reclaim large natural swamps: this threat is already hanging over the Pantanal, the great swamp of Mato Grosso (Daly, 1981). High water requirements do not end with the harvest as the production of each m^3 of pure ethanol from sugar cane needs 13.6 m^3 of clarified water and 44.7 m^3 of cooling water (Yang and Trindade, 1978).

On the other hand, ethanol production from cane is the source of large amounts of stillage, usually 12-13 times the volume of distilled alcohol, which has a high water pollution potential when discarded. In Brazil in 1977, that is before the current boost of fuel alcohol output, pollution potential of stillage was already equivalent to wastes generated by 50 million people (Yang and Trindade, 1978). Stillage can be converted to fertilizer, feed additive or methane but the cost of these steps does not favor further processing and much, if not most, of it is dumped.

To obviate the land and water limitations of sugar cane other sturdier crops have been suggested for fermentation feedstock and cassava has emerged as the top candidate.

Cassava

Cassava *(Manihot esculenta, Euphorbiaceae)* has been promoted by many biomass energy enthusiasts as the ideal energy crop for the tropics: it does not need prime soils as sugar cane, it is relatively drought-resistant and high-yielding and the processing of its starch into alcohol may be extended over a longer period of time than with the sugar crops because mature cassava roots can be left unharvested in the ground for several months. Amory Lovins rushed to call cassava "eminently suited for making fuel alcohols" (Lovins, 1976).

Yet, large scale plantations of cassava for alcohol fermentation may diminish already precarious food availability in many tropical nations, cause additional destruction of forests, increase soil erosion, either severely deplete soil nutrients or necessitate huge synthetic fertilizer inputs, exacerbate already substantial pest and disease problems and even cause a demand for additional energy!

Cassava is today the most widely grown tropical root crop and most of the cultivation is done in small disjointed backyard or garden patches by individual families to supply a substantial portion of food carbohydrates in much of Latin America, equatorial Africa and South Asia. Five countries produce nearly three-quarters of the annual global crop of some 120 million tonnes: Brazil, India, Indonesia, Nigeria and Zaire (FAO, 1980).

Cassava starch is the major source of food energy for some 300 million people, mostly rural and poor: in Zaire nearly 60 percent of food calories come from cassava, in Nigeria about 35 percent (Goering, 1979). Any attempt to divert this crop into fuel production would drastically worsen the food supply in the countries which already rank among the worst fed in global comparison: average daily per capita intakes are just around 2,200 kcal in Zaire, and Nigeria (FAO, 1980). To avoid food losses new large plantations would thus have to be established and this would almost certainly necessitate further clearing of forest land -- and further destruction of tropical forests.

Establishment of cassava plantations entails complete clearing of forest ground, removal of all roots and obstructions from the soil and deep ploughing (Grace, 1977). Removal of all trees, destumping and burning of all wood in Africa is mandatory because cassava is highly susceptible to white-thread disease *(Fomes lignosus)* to which several trees are hosts (Onwueme, 1978). Weeds will start spreading soon (usually two weedings are necessary after planting) but until the canopy is closed the ground will be exposed -- for two to three months -- to potentially severe erosion.

During its growth, like all high-yielding carbohydrate plants, cassava has high nutritional requirements and it will exhaust soils very rapidly (Grace, 1977). Annual removal for a 40 tonnes per hectare crop is nearly 300 kg of N, 60 kg of P, 380 kg of K and 160 kg of Ca (Goering, 1979). Potassium requirements are especially high and essential because without high K levels nitrogen response drops sharply. Moreover, proper fertilization must be maintained over extended period -- yet the poor tropical nations do not have enough fertilizer even for their essential food crops. As the tubers have irregular shapes -- they may spread more than one meter and penetrate as deep as 60 centimeters -- mechanical harvesting with deep moldboard plows or with special harvesters is bound to cause much soil disturbance and, again, increase the erosion potential.

Cassava, even as cultivated today, is susceptible to a large number of pests and diseases. African cassava mosaic virus may wipe out as much as 90 percent of the harvest (Cook, 1978). Most severe South American disease is bacterial blight *Xanthomonas manihotis* which can cause even complete crop loss (Onwueme, 1978). Among the frequent pests are grasshoppers, mites, hornworms, scales and nematodes. In Brazilian state of Minas Gerais cassava is attacked by over 20 insect pests of which eight are potentially serious (Samways, 1979).

Root-knot nematode *Meloidogyne incognita* is a very serious pest throughout West Africa and a recent Nigerian study gave a strong support to a theoretical expectation that the infestation would get worse in extensive monocultures. When cassava is grown randomly mixed with okra, corn, tomato and pumpkin in a typical Nigerian garden plot -- which may be, moreover, fallowed for up to 4 years -- tiny nematodes with gigantic appetites are much less of a problem than in monocultural experimental plantings at Nigerian universities (Ogbuji, 1979).

Clearly, any massive monocultural extension of the crop in large-scale plantations for alcohol conversion would greatly exacerbate most of the pest and disease problems.

Because cassava tops, unlike sugar cane bagasse, are too moist (typically in excess of 72 percent) to generate the steam needed for the industrial processing of the crop, alcohol production from cassava is not fuel self-sufficient and it must draw on other energies. Obviously, the fuel most often elected to run the processing in a poor tropical country would be wood -- and cassava would thus further contribute to tropical deforestation.

Moreover, the tops cannot be easily recycled to maintain the soil fertility. Although a good 40 tonnes per ha crop will yield about 50 t of above ground residue, most of it is in the relatively woody stems which decay very slowly and sprout readily and unless the stems are all well shredded cassava becomes a weed in any subsequent plantings (Onwueme, 1978).

CONCLUSIONS

Even this brief survey makes clear that a large scale introduction of modern bio-mass energy conversions to rural areas of the world's poor nations is not only undesirable but that it could have harmful cummulative effects. The feedstocks would be crop residues or special energy crops and the removal of the former could lead to serious degradation of soils and food producing capacity, while the culti-vation of the latter could have similar undesirable environmental effects and it would also further reduce the already scarce supply of food.

REFERENCES

Brown, L.R. (1978). The Worldwide Loss of Cropland, Worldwatch Institute, Washington, D.C.

van Buren, A. (Ed.) (1979). A Chinese Biogas Manual, Interm. Technol. Publ., London.

Cook, A.A. (1978). Diseases of Tropical and Subtropical Vegetables and Other Plants, Hafner, New York, pp. 112-134.

Daly, H. (1981). Brazilian smears alcohol program, Soft Energy Notes, 4, 46-47.

Doorenbos, J. and W.O. Pruitt (1977). Guidelines for Predicting Crop Water Requirements, FAO, Rome, p. 36.

Eckholm, E.P. (1976). Losing Ground, W.W. Norton, New York.

Epstein, E., J.O. Norlyn, D.W. Rush, R.W. Kingsbury, D.B. Kelley, G.A. Cunningham, and A.F. Wrona (1980). Saline culture of crops: a genetic approach, Science, 210, 399-404.

FAO (1979a). Agricultural Residues: Quantitative Survey, FAO, Rome.

FAO (1979b). 1978 FAO Fertilizer Yearbook, FAO, Rome.

FAO (1980). 1979 Production Yearbook, FAO, Rome.

Goering, J.T. (1979). Tropical Root Crops and Rural Development, World Bank, Washington, D.C.

Goldemberg, J. (1978). Brazil: Energy options and current outlook, Science, 200, 158-164.

Grace, M.R. (1977). Cassava Processing, FAO, Rome, p. 11.

Han, Y.W., and A.W. Anderson (1974). The problem of rice straw waste. A possible feed through fermentation, Econ. Botany, 28, 338-344.

Joint Commission on Rural Reconstruction (1977). 33rd General Report JCRR, Taipei, Taiwan.

Lovins, A.B. (1976). Energy strategy: The road not taken? Foreign Affairs, 55, 65-96.

Myers, N. (1980). The present status and future prospects of tropical moist forests, Env. Conserv., 7, 101-114.

Ogbuji, R.O. (1979). Shifting cultivation discourages nematodes, World Crops, 31, 113-114.

Onwueme, I.C. (Ed.) (1978). The Tropical Tuber Crops, John Wiley & Sons, New York, pp. 109-163.

Osuji, G.E., O. Babalola, and F.O. Aboaba (1980). Rainfall erosivity and tillage practices affecting soil and water loss on a tropical soil in Nigeria, J. Env. Manag., 10, 207-217.

Pfeffer, J.T. (1980). Biological Conversion of Biomass to Methane, Dept. of Civil Engineering, Univ. of Illinois, Urbana, Ill.

Polunin, N. (1979). Conceivable ecodisasters and the Reykjavik imperative, Env. Conserv., 6, 105-109.

Prentice, K.C., and J.C. Coiner (1980). Agriculturally induced vegetation change, Hum. Ecol., 8, 105-116.

Prihar, S.S., R. Singh, N. Singh, and K.S. Sandhu (1979). Effects of mulching previous crops or fallow on dryland maize and wheat, Expl. Agr., 15, 129-134.

Sanchez, P.A. (1976). Properties and Management of Soils in the Tropics, John Wiley, New York.

Samways, M. (1979). Alcohol from cassava in Brazil, World Crops, 31, 181-186.

Smil, V. (1979a). Renewable energy: How much and how renewable, Bull. At. Sci., 35 (10), 12-19.

Smil, V. (1979b). Energy flows in the developing world, Am. Sci., 67, 522-531.

Smil, V. (1981a). In P. Auer (Ed.), Energy and the Developing Nations, Pergamon Press, New York. In press.

Smil, V. (1981b). Land use and land management in the People's Republic of China, Environmental Management. In press.

Smil, V., and W.E. Knowland (Ed.) (1980). Energy in the Developing World, Oxford University Press, Oxford.

Staniforth, A.R. (1979). Cereal Straw, Clarendon Press, Oxford.

Subramanian, S.K. (1978). In A. Barnett, L. Pyle, and S.K. Subramanian (Eds.), Biogas Technology in the Third World: A Multidisciplinary Review, IDRC, Ottawa, pp. 97-121.

Tanaka, A. (1973). Methods of handling the rice straw in various countries, Int. Rice Comm. Newsl., 22, 1-20.

Wagner, F.E., and J.O. Ward (1980). Urbanization and migration in Brazil, Am. J. Econ. Sociol., 39, 249-259.

Wani, S.P., and P.A. Shinde (1980). Studies on biological decomposition of straw, Plant and Soil, 55, 235-242.

Yang, V., and S.C. Trindade (1978). In Energy from Biomass and Wastes, Institute of Gas Technology, Chicago, pp.

ENVIRONMENTAL ASPECTS OF ALBERTA OIL SANDS DEVELOPMENT

K. R. Smith

Environmental Assessment Division, Alberta Environment, Edmonton, Alberta, Canada

ABSTRACT

The responsibilities of the Government of Alberta for the management of environmental impacts associated with oil sand development are examined. The review and approval of environmental aspects of surface mining and in-situ recovery technology are analyzed from the government perspective, highlighting the regulatory approval process. Special attention is focused upon the integration of environmental matters with energy conservation and economic appraisal of proposed developments.

KEYWORDS

Environmental impact assessment; oil sands; surface mining; in-situ recovery; regulatory approval process; energy conservation; economic appraisal.

INTRODUCTION

Alberta's vast oil sand resources are increasingly being viewed by many Canadians as a major source of hydro carbons to meet growing energy requirements. The extent of Alberta's non-conventional oil deposits, or oil sands, is immense. The four major Cretaceous Oil Sands Deposits are the Athabasca Wabiskaw - McMurray, Cold Lake, Peace River Bluesky - Bullhead, and Wabasca Deposits. Reserves of crude bitumen in-place in the four major Cretaceous deposits are estimated to be 967 billion barrels.

Surface mining and in-situ recovery techniques are the principle means of extraction. Of the four major Cretaceous deposits, it is expected that in-situ methods will be used to extract 78% of the reserves, while only 7% of the reserves will be surface mined, leaving 15% of the reserves uncertain as to extraction method. Of the 7% of the reserves expected to be mined, or 74 billion barrels, only 22.4 billion barrels have been classified as recoverable. Once extracted, bitumen upgrading is required to produce synthetic crude oil suitable for refining. Extraction, upgrading, and refining all result in impacts on both the biophysical and human environments.

The focus of this paper is on the responsibilities of the Government of Alberta for the management of environmental impacts associated with oil sand development. To place environmental management responsibilities in perspective, the institutional framework which establishes Alberta's oil sands resource development policies is briefly examined. The development approval process is outlined, indicating the responsibilities of The Department of the Environment and the Alberta Energy Resources Conservation Board, highlighting the Alberta Environmental Impact Assessment process as a key environmental management strategy linking planning activities to detailed environmental approval processes for individual projects. The integration of technical, environmental and economic aspects of oil sands proposals in a comprehensive regulatory procedure is discussed in the last section of the paper.

ALBERTA OIL SAND DEVELOPMENT FRAMEWORK

Canada's constitution vests the ownership and management of natural resources with the Provinces, and consequently oil sand resources are owned by the Province of Alberta. The Government of Alberta has encouraged and supported the development of non-conventional oil deposits, particularly in the face of declining conventional production. Recent uncertainties within Canada and on a global basis regarding the demand, supply, and price of oil have led the Province to adopt a basic policy approach to oil sand development that provides flexibility; development proposals are considered on a case-by-case basis.

The case-by-case project approval process takes place in the context of a number of regional planning and development initiatives. To improve knowledge of reserves, mineral lease holders are required to carry out exploration. To improve knowledge of environmental resources, a major Alberta Oil Sands Environmental Research Program was initiated. To encourage technology development, the Alberta Oil Sand Technology and Research Authority was established. To encourage decentralization of industrial development, The Energy Corridor policy was developed to guide the routing of products pipelines and the location of terminals and refineries outside the development area. To facilitate orderly community development, a regional plan was initiated through the office of a special area planning Commissioner. To accelerate improvements in reclamation technology, a major Oil Sands Development and Reclamation Study was completed. Increasing knowledge about oil sands development arising from these initiatives is applied in the project approval process. As technologies improve and resource demands increase, the future management of impacts resulting from oil sands development will be facilitated by these regional planning and development bodies.

THE DEVELOPMENT APPROVAL PROCESS

The Alberta Government has established a development approval process that is comprehensive in scope. The key agencies involved in the development approval process are: (1) The Department of Energy and Natural Resources with responsibilities for land surface and mineral rights, including royalties; (2) The Energy Resources Conservation Board with responsibilities for conservation in resource development; and the Department of the Environment with responsibilities for impacts on the biophysical and human environment. These agencies are responsible for setting the terms and conditions of development applied to individual development projects. The terms and conditions reflect the most recent changes in oil prices, reserve recovery economics, extraction and upgrading technology, and environmental knowledge.

In its simplest terms the development approval process involves the proponent making an application to the Energy Resources Conservation Board for permission to construct and operate an oil sands development. The application includes information on technical engineering and conservation features, environmental impacts, and economics. Following public hearings, the Energy Resources Conservation Board recommendations must be approved by the Government before approvals for the project can be issued.

THE DEPARTMENT OF THE ENVIRONMENT

The Department of the Environment Act came into effect in April, 1971, and gives the department and its minister responsibility for all matters pertaining to the environment. Statutes administered by the Department define this responsibility in specific contexts. Major statutes administered by the Department include The Clean Air Act, The Clean Water Act, The Water Resources Act, and The Land Surface Conservation and Reclamation Act. In order to ensure that oil sands proposals are examined in a comprehensive manner at the early stages of planning, proponents of oil sands projects are required to prepare and submit environmental impact assessment reports.

Environmental Impact Assessments (E.I.A.'s) are required for large scale oil sand development proposals and cover physical, biological and human components of the environment. The process for the preparation, review, and evaluation of E.I.A.'s is defined in policy guidelines. The Alberta approach to environmental assessment places priority on consultation between industry, community, and government. Emphasis is placed on ensuring that public participation is included in the assessment process and on ensuring that technical and locational alternatives are considered.

The Alberta Environmental Impact Assessment Process is a key environmental management tool linking regional development strategies to detailed approval processes for individual projects. Specific E.I.A.'s incorporate the results of improved environmental knowledge gained through the Alberta Oil Sands Environmental Research Program.

Regional data collection programs regarding bedrock and surficial geology, soils, climate, groundwater and surface hydrology, water and air quality, vegetation, terrestrial and aquatic fauna and other environmental parameters provide the framework for site specific data collection activities of proponents. Improved knowledge of baseline conditions and processes enhances impact prediction and the development of effective mitigative measures. Environmentally preferred alternatives for mining and materials handling; water supply, recycling, treatment and disposal; air emission controls; and reclamation and revegetation are examined through the E.I.A. process, including changes in best practical technology evolving through technology research and development programs. Alternative means of accommodating regional community impacts are examined, including dispersed and concentrated settlement patterns, phasing of construction activities, and associated infrastructure improvements.

Not only do E.I.A.'s incorporate the results of regional planning and development initiatives, they also provide a comprehensive framework for issuing detailed licences, permits and approvals for air emissions (governed by The Clean Air Act), water effluents (governed by The Clean Water Act), water diversions, and materials handling and reclamation activities by the Department of Environment. Maximum emission or effluent standards have been set which are used in combination with guidelines and technical criteria in reviewing plans, processes, and procedures. The E.I.A. provides the framework within which a licence will be considered but separate applications must be made for each licence under the Acts. This combined approach of E.I.A.'s and specific detailed licencing allows experience, operational flexibility, and changing technology to be incorporated throughout the regulatory mechanism.

THE ENERGY RESOURCES CONSERVATION BOARD

The Energy Resources Conservation Act establishes a quasi-judicial Energy Resources Conservation Board (E.R.C.B.) with broad powers regarding energy conservation. In terms of integrating environmental concerns with the technical and economic aspects of oil sands development, one of the powers of the E.R.C.B. is critical: The E.R.C.B. is empowered to control pollution and ensure environment conservation in the exploration for, processing, development and transportation of energy resources and energy.

The specific legislation administered by the E.R.C.B. which deals with oil sands projects is The Oil and Gas Conservation Act. The Act requires the approval of the E.R.C.B. before a scheme for the recovery of oil sands crude bitumen may proceed. Following a hearing and approval by the Lieutenant Governor in Council, the E.R.C.B. may approve the proposal. In practice, a comprehensive report on a development application is prepared by the E.R.C.B., dealing with the issues raised in the application and at hearings, recommending disposition, and if appropriate, recommending specific terms and conditions for an approval.

INTEGRATION OF TECHNICAL, ENVIRONMENTAL AND ECONOMIC MATTERS

In practice, The Department of the Environment and the E.R.C.B. have combined their respective requirements into a single but comprehensive review and approval procedure so that both agencies procedures are concurrent and avoid unnecessary delays while ensuring that environmental effects and values are identified in adequate detail so they can be compared to economic and technical analyses. This practice has resulted in the regulatory process becoming streamlined and more useful to decision-makers and the public in reviewing and approving oil sands proposals.

Alberta Environment and the E.R.C.B. have jointly prepared and issued guidelines respecting applications for commercial crude bitumen recovery and upgrading projects. These guidelines are generally known to bituminous sands lease holders in the Province. Applicants for approval of an oil sands project must prepare and submit a comprehensive application outlining the proposal, and the application is the subject of a public hearing.

Technical aspects of the application include: estimates of volumes of crude bitumen in place, a detailed description of the proposed mining or in-situ recovery techniques, a detailed comparison of extraction and upgrading processes, description of required utilities and energy sources and related off-site facilities, and material and energy balance information.

Environmental aspects of the application must include an environmental impact assessment ordered by the Minister of the Environment. Biophysical impacts requiring assessment include land use, surficial geology, soils and vegetation, fish and wildlife, water and air resources. Social impacts requiring assessment include population growth and distribution, direct and indirect employment, special population groups, public services, settlement patterns, economic base of the area, housing, and overall quality of life. Details of environmental monitoring programs for biophysical and social impacts are also required.

The economic aspects of the application must include: an evaluation of the commercial viability of the project; an economic impact analysis of the project at the local, provincial, and national levels; a benefit-cost analysis in terms of Alberta; and a broad description of the general desirability of the project.

Project proponents planning to proceed to the application stage normally contact the Board and Alberta Environment early in their project planning; in some cases five to seven years before construction is planned. Joint meetings with the proponent and the two agencies are held to discuss the required level of detail for a specific application, including technical, environmental, and economic aspects.

Public participation is an integral part of the application preparation and review process. Applicants are required to inform the public of the general nature of the project being proposed, the fact that an E.I.A. will be prepared, and the opportunities that will be available for public participation in the preparation of the assessment. As the assessment progresses, the public is kept informed of the specific areas of concern which are being addressed in the assessment. Proponents must ensure that the public has an adequate opportunity to review the E.I.A. in draft form, and to indicate to the proponent whether or not the E.I.A. is deficient. Proponents are required to obtain and incorporate the reaction of interested or affected persons, and provide appropriate documentation in the application.

Proponents are required to file their E.I.A. with Alberta Environment and with the E.R.C.B. as part of the application. Upon receipt, the Environmental Assessment Division initiates a review of the E.I.A. The purpose of the review is to determine whether or not the E.I.A. is suitable for a detailed examination carried out as part of the decision process. The initial review is not intended to evaluate the merits of the proposal, it is intended only to make a determination of whether or not there is sufficient information upon which to evaluate the merits. On oil sand projects, any deficiencies identified by Alberta Environment in the E.I.A. are coordinated with the E.R.C.B. review of the technical and economic aspects of the application. Deficiencies in the application are communicated to the applicant by the E.R.C.B. When the E.R.C.B. and Alberta Environment are satisfied that the application (including the E.I.A.) is suitable for an E.R.C.B. public hearing, the application is set down for hearing and appropriate notice given.

Upon receipt of the supplemental information responding to any deficiencies, the Environmental Assessment Division initiates a critical evaluation of the technical and scientific information and conclusions in the E.I.A. The critical evaluation of the E.I.A. takes place in conjunction with the E.R.C.B. public hearing process. The evaluation involves the identification of issues requiring cross-examination at a public hearing. The E.R.C.B. has a well defined public hearing process which provides the public with a full opportunity to express their views on the environmental implications of the proposed energy developments. In order to facilitate a single hearing process which examines the energy conservation, economic, and environmental implications of oil sand proposals, the Department assists the E.R.C.B. in the following manner: The Department advises the E.R.C.B. as to the suitability of the E.I.A. for public hearing processes and identified environmental deficiencies in the application; the Department assists the Board in cross-examination of the applicant and intervenors; the Department may sit on the hearing panel as an Acting Board Member; and the Department assists the E.R.C.B. in preparing its decision report and recommendations on the oil sands proposal. The application to the E.R.C.B. and approvals thereon must be referred to the Minister of the Environment for his specific approval of the application as it affects matters of the environment. Overall approval of all aspects of the application is given by the Executive Council or Cabinet.

The Minister has the responsibility regarding environmental matters to: give approval to proceed; give approval to proceed with the application subject to such terms and conditions as the Minister considers necessry, or refuse to give approval to proceed with the oil sands proposal.

The Minister's decision on the general acceptability of the project is indicated by the Ministerial Approval. The Department's position on the detailed environmental features is reflected in the terms and conditions of specific permits and licences issued later under Alberta Environment legislation. A project is not considered environmentally acceptable until the requisite permits are issued.

CONCLUSION

The Government of Alberta has a comprehensive and integrated regulatory procedure for the review and approval of oil sands projects, including the management of environmental impacts. Alberta's oil sands resource development policy stresses flexibility in order to accommodate changes in the demand, supply and price of oil as well as changes in technology. Development proposals are considered on a case-by-case basis within the context of a number of regional planning and development initiatives. The development approval process involves the proponent preparing an application to the E.R.C.B. to construct and operate an oil sands development. The application includes information on technical and energy conservation features, environmental impacts, and economics. Following public hearings, the E.R.C.B. recommendation must be apprcved by the Government before specific approvals for the project can be issued. In practice, the Department of the Environment and the E.R.C.B. have combined their respective requirements into a streamlined procedure that ensures environmental effects and values are compared to economic and technical analyses at the same time during the decision process.

CONFLICTS OF WATER USE AND ENERGY DEVELOPMENT IN THE OHIO RIVER BASIN

H. T. Spencer* and C. A. Leuthart**

*Department of Chemical and Environmetal Engineering, Speed Scientific School, University of Louisville, Louisville, Kentucky 40292, USA
**Liberal Studies Program, University College, University of Louisville, Louisville, Kentucky 40292, USA

ABSTRACT

The Ohio River is a major navigable waterway of the United States. This artery of commerce and development has now, with a system of 19 high-lift dams, opened up the entire American Midwest as a port of the Gulf of Mexico. The river links, through-out its 1579 kilometers, areas of intensive farming, abundant coal deposits, trans-portation corridors, and plentiful labor supply. Collectively these resources have resulted in sudden development of energy conversion facilities along the mainstem and its major tributaries. For the most part the development of energy conversion facilities in the states of the region has occurred independently, thus causing both political and institutional conflicts. Recent studies of these circumstances and possible means of their resolution are the topics of this paper.

KEYWORDS

Water-use management; Ohio River Basin; energy conversion facilities; navigable waterway management; river basin planning.

INTRODUCTION

The Ohio River Basin is situated between the U.S. eastern sea coast and the Great Plains States of the central U.S. It is a major component of the farming enterprise of the American Midwest and contains abundant energy resources (primarily coal), transportation facilities, labor, and water resources as well.[1] These attributes make the area a prime location for intensive energy development and a hope for a major contribution to U.S. energy independence. Various citizen and farm community action groups also located in the region have feared that such energy development would proceed without sufficient attention being given to the quality of life exper-ienced by its inhabitants.

---

[1]There is, within the ORBES region, coal reserves totalling 193×10^9 tons which is equivalent to 45% of the national total. This region contains 55% of the total amount recoverable nationally by underground mining techniques, and 23% of the reserves recoverable with strip mining techniques. (ORBES Main Report, 1981, p 63).

Lobbying efforts by these groups in 1976 led to the funding of a major government study of the situation. This investigation, known as the Ohio River Basin Energy Study (ORBES), was completed January 1, 1981 and is in part a topic of this paper. Issues raised during the ORBES project have subsequently led to additional studies, some funded by government and some by private capital. Possibly, the most important of the new studies will be that funded recently by the Hartford Foundation of New York, entitled "Siting of Major Energy Related Facilities Along the Ohio River". This project is being administered by the Ohio River Valley Water Sanitation Commission (ORSANCO) and The Council of State Governments.[2]

ORBES REGION AQUATIC RESOURCES

The region included for the ORBES report (Fig. 1) encompasses a quite varied topography. Mountainous areas are found in the eastern sector while level plains and wetlands are more typical of the western half. Because of these topographical differences, variations in aquatic habitats and water resources are encountered as well. These aquatic systems range from whitewater canoe and mountain trout streams to deep, clear recreational lakes, major navigable and free-flowing rivers as well as numerous wetlands and sloughs. More than 250 fishes occur in this area with 90 or more alone occurring in the Ohio River mainstem (ORBES Main Report, 1981, p 86).

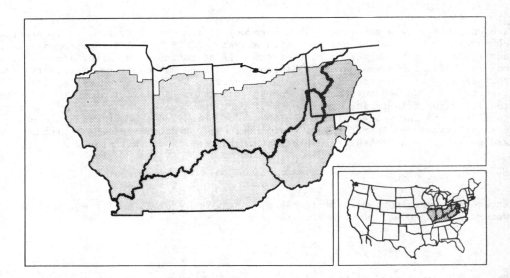

Figure 1. The ORBES study region.

[2]ORSANCO is a regional compact of eight states which became effective on June 30, 1948 with the addition of Pennsylvania and Virginia. The other compact states are New York, Kentucky, Ohio, West Virginia, Indiana, and Illinois. The initial charge given the compact Commission was to promote water pollution abatement within the signatory states. More recently the Commission has taken an interest in energy development and water-use conflicts.

The Council of State Governments is an organization created by the fifty states in 1933 to provide them with information and research on state programs and problems Its basic mission since its creation has been to strengthen state governments in the American Federal system. Collectively, ORSANCO and The Council of State Governments would seem ideally suited to manage a privately funded study of energy development in the Ohio River Valley.

The water supply that supports these varied aquatic systems is also heavily used by industries, municipalities, and energy conversion facilities. Under average conditions the ORBES region water supply amounts to 14,200 cubic meters per second but drops to 2,640 cubic meters per second at 7-day-10-year low flow. Indeed, at times the drop in flow under drought conditions has been sufficient to impede navigation on the lower Ohio River.[3] Water rights as promulgated in the American West are essentially unknown in the east and midwest. However, the time of conflict over water rights may now be at hand in the Ohio River Basin.

Water quality standards vary from state to state within the region but differences for the most part are minor. ORSANCO, through its Criteria Conflicts Committee, has been somewhat successful in resolving these differences over its thirty-three year existence but serious water quality problems still persist in the region. The upper reaches of smaller basin tributaries often suffer dessication at seasonal low flow while at the same time toxic agents become considerably concentrated in the remaining flow. The problems of eutrophication and dissolved oxygen depletion are also common to the region, although such problems are highly localized. To date the worst of these problems seem to occur in association with outdated and overloaded sewage treatment plants feeding into streams having 7-day-10-year low flows of zero. Unfortunately, such arrangements are common in the American Midwest.

The region's major rivers are also heavily impacted at 7-day-10-year low flow in terms of water quality. The Ohio River for example has a 7-day-10-year low flow of 1362 cubic meters per second at its confluence with the Mississippi but still contains along most of its length at low flow levels of total dissolved solids, cadmium, chromium, phosphorus, silver, copper, iron, lead, manganese, mercury, nickel, and zinc which violate standards (ORBES Main Report, 1981, p 93).

ENERGY DEVELOPMENT IN THE OHIO RIVER BASIN

The construction of large but widely dispersed power plants commenced in the Ohio River Valley at the end of World War II and, as might be expected for the times, their siting was influenced by considerations of national defense. All uranium enrichment took place at Oak Ridge, Tennessee, during World War II. This setting, at least to military strategists, appeared vulnerable and a decision at the top levels of government was made to build two additional facilities removed from each other and from Oak Ridge as well. The enrichment process current at the time was gaseous diffusion which is in itself an energy intensive operation requiring large (2000 MW) coal-fired power plants to support the size plants proposed. Subsequently the U.S. Atomic Energy Commission (AEC) contracted in the early 1950's with an investor-owned consortium of public utilities to build and operate what would for a time become the two largest coal-fired power plants in the world, both located on the upper Ohio River but at considerable distances from each other and from their dependent gaseous diffusion plant at Portsmouth, Ohio.

While these events were taking place in the upper Ohio River Valley, a coal-fired power plant of almost equal size was being built in the lower valley at Paducah, Kentucky, to supply the second gaseous diffusion plant. This second plant is located on property adjoining its associated coal-fired power plant and the two together still comprise one of the largest energy development facilities in the world.[4]

---

[3] The most recent occurrence of such an event was during August and September of 1980.

[4] This gaseous diffusion plant is currently being operated by the Union Carbide Corporation for the U.S. Department of Energy. The U-238 tailings now stored on this site after 25 years of U-235 stripping are estimated to be worth 35 trillion 1980 dollars as fast breeder fuel.

Thus, the development of energy industries in the Ohio River Valley started out on a large scale over a quarter of a centruy ago.

Growth in the electric utility industry occurred along with industrial development of the basin over the past 25 years. As an example of the impact of this expansion one need only inspect what has happened to regional air quality. In 1976 ORBES region coal-fired power plants produced 78 percent of the regional sulfur dioxide emissions from all sources. This constituted 52 percent of the total U.S. electric utility sulfur dioxide emissions and 32 percent of such emissions from all sources in the continental U.S. (ORBES Main Report, 1981, p 72). Long-range transport of this material to the northeastern U.S. and Canada, together with concommitant acid rains, have become a major national and international problem. Thus, current utility plans for continued expansion have quite expectedly encountered considerable opposition.

WATER POLICY ISSUES

Resolution of differences in water quality criteria among the states has been easy to achieve in comparison to resolving questions of siting major industrial and energy conversion facilities. However, differences in opinion among officials in adjacent states tend to center around air quality impacts more than water demand or, as is more often the case, around competition for industrial development sites. Also, the development of a nuclear power plant on one side of the Ohio River often draws criticism from state officials on the other side.[5]

Still, water policy issues are being raised with the initiation of intensive energy development. These issues include the amounts of water actually available for energy conversion facilities if other uses (municipal, industrial, agricultural, and transportation) are considered, and the decrease in water quality due to increases in solute concentrations resulting from all forms of consumptive use. Coal-fired and nuclear power plants of current design (evaporative cooling towers) do consume significant amounts of water from the basin but the synfuels industry now being developed in Western Kentucky may ultimately become the real contender for regional water resources. This synfuels industry, if developed as presently planned, would exceed in magnitude any collection of coal-based energy conversion facilities now in the basin.[6]

DISCUSSION OF ORBES RESULTS

The ORBES study concluded in January 1981 with the publication of a final main report along with some fifty-eight bound support documents, each covering a specific

---

[5]The Marble Hill nuclear plant being built by Public Service of Indiana at Madison, Indiana, is being opposed by state officials in Kentucky. Kentucky officials are also opposed to the Clarke Maritime River Port being built by the Indiana Port Authority on the Indiana shore of the Ohio River just above Louisville, Kentucky. Construction of an identical facility is now underway on the Kentucky side of the river just below Louisville, but Kentucky officials oppose the Indiana port on the basis of its conflicting with important recreational resources.

[6]To date the only draft environmental impact statement written for any of these western Kentucky synfuels plants is that for the Solvent Refined Coal-I Demonstration Project at Newman, Kentucky. Copies of this document may be obtained from the U.S.Department of Energy at Oak Ridge, Tennessee.

study area. Of special interest to this discussion are the reports by Leuthart and Spencer (1979) on aquatic resources and by McLaughlin (1981) on the legal and institutional aspects of interstate power development. The ORBES study and much of its support research was directed toward evaluating a broad range of year 2000 AD scenarios. It was not, as is often supposed, an attempt at predicting the future of regional energy demand. A major source of controversy in the basin has long centered around comparisons made between electric utility company projections of future demand which in recent years have tended to exceed the demand actually observed later. The building of a new power plant must, of course, be predicated upon its need and funding cycles for new construction often exceed a ten year period. Such estimates are understandably difficult to make over so long a period with any accuracy. Nonetheless, numerous citizen groups in the region have complained for years of inflated utility projections, and it is understandable that these groups would look to the ORBES study as a means of proving their case. It is interesting to note that even though the study personnel avoided ever voicing an opinion on this issue, the group did select as plausible growth rates which would have been considered by the utilities in the early 1970's as much below the expected for 2000 AD. Inflation, the Arab oil embargo of 1973, the rising cost of construction, conservation, the scarcity of capital and an awareness of increasing concern for the environment all contributed to the choice of these lower figures.

The key to study success was perceived as depending upon the scenarios being plausible in all respects. Otherwise evaluation of their impacts would have been a meaningless exercise. In all, a total of nine main scenarios were selected ranging from a conservation emphasis case with a growth rate in the electric power sector of 0.90 percent to a high electrical energy growth case with a growth rate of 3.90 percent. The base case, a scenario to which most others were compared, was assigned a growth rate of 3.13 percent as was an exact counterpart which differed only in terms of the degree of environmental controls applied.

Aquatic impacts turned out to be nearly scenario insensitive. At first this seemed an unreasonable finding, but closer inspection of the region's river systems revealed extensive base year (1976) violations of stream standards at 7-day-10-year low flow. These conditions, when carried through each scenario to the year 2000 AD, simply reappeared and thereby completely dominated the analysis. Thus, only the strict environmental controls scenario showed any real improvement between 1976 and 2000 in terms of water impacts. All others appeared much the same to impacts observed during the base period at the time of scenario initiation (ORBES Main Report, 1981, p 168).

Scenario differences for other parameters, particularly those related to air quality were considerably more dramatic. It should be noted, however that a paradox developed in terms of one problem's solution leading to the exacerbation of another. For example, the shifting of plant sites away from the main stem of the Ohio often resulted in lessened air impacts, a much desired result, but at the same time this caused considerable aquatic habitat and water quality impacts to occur in the smaller streams to which the plants were hypothetically moved.

To the authors, who were participants in the ORBES study from its beginning to its end, at least one conclusion seems clear. Something in the Ohio River Valley, be it the region's environment, its industrial development, or the building of more coal-based energy conversion facilities, has to give; it is also evident that cooperation among the states on these issues will be essential to any workable solution. The states are permitted under the U.S. Constitution to form compacts among themselves and indeed are sometimes encouraged by Congress to do so. It is probable that such a solution will eventually be found, although, given the long period (8 to 10 years) required for such a development to become formalized, one must anticipate that any new compact involving the ORBES states and energy development is far in the future.

REFERENCES

Leuthart, C.A. and H.T. Spencer (1980). <u>Fish Resources and Aquatic Habitat Impact Assessment Methodology for the Ohio River Basin Energy Study Region.</u> Grant No. EPA R804816. U.S. Environmental Protection Agency (Copies obtainable from U.S. Environmental Protection Agency. Cincinnati, Ohio 45268.

McLaughlin, J.A. (1981). <u>Legal and Institutional Aspects fo Interstate Power Plant Development in the Ohio River Basin Energy Study Region.</u> U.S. Environmental Protection Agency. (Copies obtainable from U.S. Environmental Protection Agency Cincinnati, Ohio 45268.

<u>Ohio River Basin Energy Study (ORBES) Main Report.</u> EPA-600/7-81-008 (1981). Office of Environmental Engineering and Technology, Office of Research and Development U.S. Environmental Protection Agency, Washington D.C. 20460. (Copies obtainable from U.S. Environmental Protection Agency. Cincinnati, Ohio 45268.

ESTIMATING THE FUTURE INPUT OF FOSSIL FUEL CO₂ INTO THE ATMOSPHERE BY SIMULATION GAMING

I. Stahl and J. Ausubel

International Institute for Applied Systems Analysis (IIASA)
A-2361 Laxenburg, Austria

ABSTRACT

Previous estimates of input of fossil fuel CO_2 into the atmosphere are reviewed, including those of NAS, IIASA, IEA, and Marchetti. Methods employed largely disregard that if CO_2-induced changes are indeed harmful then there may be efforts to prevent emissions. There is a need to include explicitly societal response to increasing CO_2 emissions in estimating future input as well as the strategic interaction among national energy policies. Traditional economic theory, game theory, and computer simulation (without humans) have disadvantages in this regard. Gaming, involving humans playing the roles of various nations, may be an illuminating approach to the problem. A simple game, focusing on coal, trade, and many nations is proposed as an initial effort.

KEYWORDS

Carbon dioxide; climate; coal; coal trade; energy forecasts; gaming; prisoners' dilemma; tragedy of the commons.

PREVIOUS ESTIMATES

A question of fundamental importance in evaluating the carbon dioxide issue is how much CO_2 is likely to be put into the atmosphere from burning of fossil fuels over the next 50 to 100 years. A wide range of estimates has been offered. While the numbers have varied a great deal, there has been a similarity of approaches.

Often the approach is simply to take the current level of fossil fuel combustion and multiply it by an assumed constant rate of change. Rotty (1977) estimated that historically CO_2 emissions from fossil fuel burning and cement manufacture have increased 4.3% per year except for periods of the two world wars and the global economic depression of the early 1930s. This figure of 4.3% has commonly been used to project future levels of atmospheric CO_2. For example, a JASON report (1979) opens with the statement, "If the current growth rate in the use of fossil fuels continues at 4.3% per year, then the CO_2 concentration in the atmosphere can be expected to double by about 2035..."

While the 4.3% figure seems to have been the one most often discussed in the literature (Munn and Machta, 1978; World Climate Programme, 1981), arguments are, of course, made for estimates on both sides of this number. At the low extreme, one finds the projections of Lovins (1980). With a slight decrease of use of fossil fuels, a CO_2 "problem" never comes about. With a zero growth rate, a doubling of atmospheric CO_2 is estimated to occur in 2119, almost 140 years from now. At the high extreme, one finds the 50 terawatt (TW) global energy scenario proposed as a limiting case by Niehaus (1979). This scenario analyses the consequences for atmospheric CO_2 if all of a high projected energy demand is covered by fossil fuels.

Most energy scenarios lie in between and project a continued growth of energy demand to between 20 TW and 35 TW over the next 50 years. (Estimated global primary energy supply in 1975 was about 8 TW (IIASA, 1981).) These include scenarios developed for studies by the US National Academy of Sciences (NAS), the International Institute for Applied Systems Analysis (IIASA), and the Institute for Energy Analysis (IEA). Whenever these scenarios do not project a large share of nonfossil energy, they lead to relatively serious concerns about climatic change in the next 50 to 100 years. Let us briefly examine the character of these projections which are the basis of much of the concern about CO_2.

NAS

Perry and Landsberg (NAS, 1977) project world energy consumption and emissions to the year 2025. The projections are for 11 geographic regions, which are sometimes large nations and sometimes aggregates of nations, based on estimates of the supply of energy resources of various kinds and energy demand. Demand is derived from projections of population, GNP, and the relationship of GNP per capita and energy consumption. Energy resources produced in a region are used to supply regional demand to the extent that production has been estimated to be able to meet demand.

Emissions are calculated for two situations. On the one hand, if regional demand exceeds regional production, an estimate is made assuming a new renewable, nonpolluting energy resource would be available to meet the deficiency of nonrewable resources. On the other hand, an estimate is made for the situation where regional deficiency would be met by coal, the fuel in greatest supply. Based on these assumptions, annual world CO_2 emissions in 2025 would be about 14 gigatons of carbon (Gt C) in the first case and about 27 Gt C in the second case, or about 3 to 6 times current levels. There is no feedback between environmental change and energy strategy, other than the possibility of being on one or another path at the outset.

IIASA

The IIASA Energy Program (Niehaus and Williams, 1979; IIASA, 1981) analyzed several hypothetical energy strategies for the period up to the year 2100 for their implications for atmospheric CO_2. Distribution of energy supply among coal, oil, gas, solar, and nuclear is derived from a global energy model developed by Voss (1977). This model is structured into six sectors: population, energy, resources, industrial production, capital, and the environment. There is no geographic disaggregation. Proportions of fossil fuels used are determined by the Voss model, with some additional consideration of available resources.

Among the scenarios explored (Niehaus and Williams, 1979) are four in which global demand levels out to either 30 TW or 50 TW in 2100. In both the lower and higher demand cases there is an analysis in which nuclear and solar energy play an important role and in which they do not. Table 1 shows the reserves of fossil fuels used in each strategy.

The scenarios with reliance on nuclear and solar energy lead to peak CO_2 emissions of less than 10 Gt C per year, while the scenarios with reliance on fossil fuels lead to emissions of about 22 Gt C and 30 Gt C in 2025, increasing somewhat thereafter. While consideration is given to available fossil resources at the global level, no more detailed study is undertaken. As in the NAS study, the only feedback between CO_2-induced environmental change and energy strategy is the one implicit in choice between paths at the outset.

IEA

For several years, Rotty and co-workers at IEA (Rotty, 1977, 1978, 1979a; Marland and Rotty, 1979) emphasized extrapolation of figures in the vicinity of the 4.3% estimate of historic annual increase in CO_2 emissions. Based on demand and fuel share projections made for six world regions, an annual fossil fuel release of CO_2 containing 23 Gt C for 2025 is calculated. Arbitrary global fossil resource usage rates are applied for very long term tests of sensitivity of atmospheric concentrations.

TABLE 1. Reserves of Fossil Fuels Used in Different Strategies

Strategy	Coal Gt C	Oil Gt C	Gas Gt C
30 TW with solar and nuclear	170	170	110
50 TW with solar and nuclear	230	210	130
30 TW fossil fuel	1980	190	120
50 TW fossil fuel	3020	230	140

SOURCE: After Niehaus and Williams (1979).

A more recent paper (Rotty and Marland, 1980) begins to evaluate constraints on fossil fuel use. Three kinds of constraints are discussed: resource, environmental, and fuel demand. With respect to the resource constraint, Rotty and Marland (1980) conclude that the fraction of total resources already used is so small that physical quantities cannot yet be perceived as presenting a real constraint. However, it is mentioned that unequal geographic distribution of the resources probably will continue to be a source of international stress. Climatic change as an environmental issue is also dismissed as a constraint to fossil fuel use. "Although global warming of about 1'C to 1.5'C over a 50-year time span is enough to cause concern among climatologists, it is probably not enough to cause a revision in policy of fossil fuel use -- especially at the beginning of the 50-year period, and when fossil fuels play a major role in the global economy." Arguments other than the uncertainty about possible costs of climatic change are not presented to back this conclusion.

In contrast, Rotty and Marland (1980) discuss at some length that slower growth in fuel demand dictated by social and economic factors will limit fossil fuel use. Reduced economic growth is projected as a result of problems with capital and escalating costs and shifts toward conservation and less energy intensive industries. It is unclear whether international stress and threat of environmental change are themselves elements in causing the reduction in demand. In any case, summing up estimates for about a dozen countries and half a dozen aggregate regions results in a CO_2 release in 2025 of about 14 Gt C, an annual growth rate of 2% per year over the current level.

Marchetti

One other projection, employing a quite different logic, must be mentioned. Marchetti (1980) has made a forecast of the amount of CO_2 which will be emitted to the year 2050 based on a logistic substitution model of energy systems (Marchetti and Nakicenovic, 1979). This model treats energies as technologies competing for a market and applies a form of market penetration analysis. A logistic function is used for describing the evolution of energy sources and is fitted to historical statistical data. The driving force for change in this model appears to be the geographical density of energy consumption, and the mechanisms leading to the switch from one source to another are the different economies of scale associated with each energy source. The approach downplays the causal importance of resource availability, political arguments, and prices.

With data on energy consumption back to 1860 and including both commercial and noncommercial (wood, farm waste, hay) energy sources, the slope of the fitted curve of energy demand implies an annual growth of 2.3% . (This contrasts with Rotty (1979b) who finds that commercial energy supply, excepting times of world conflicts and depression, has grown at a rate of about 5.3% since 1860.) Applying a future growth rate of 3%, Marchetti calculates energy consumption for the various sources for the period 1975-2050 based on the logistic equations. The model predicts a relatively rapid phaseout of coal, a quite important role for natural gas, and over the next 50 years a negligible role for new sources other than nuclear. The model predicts an increase in atmospheric carbon to the year 2050 of about 400 Gt, an amount below most other estimates, as well as a gradual reduction in atmospheric CO_2 thereafter.

COMPARING THE SCENARIOS

Comparison of the scenarios shows some features in common. Most prominent is an assumption that virtually all easily accessible oil and gas will be consumed. This assumption is difficult to argue with; the timing may be a point of contention, but these sources seem too attractive to remain underground.

When the scenarios differ, certain questions begin to arise. Is there some way to choose among them? Which is more likely? Is there a way to improve or bound the estimates generally? One way of choosing among the scenarios would clearly be to select on the basis of other assumptions, for example, about overall population or economic growth. Similarly, arguments can be made about lifestyle change or technical efficiency.

One perhaps more researchable means of both evaluating the scenarios and improving future forecasts may be to look for internal consistency. Do the various levels of emissions presuppose plausible patterns of international trade? Are they based upon distributions of natural resources which are in line with current estimates? Do they violate notions of national behavior?

A General Deficiency

The estimates above have a general deficiency when applied to analysis of the CO_2 issue. These are projections in which the estimates of ejected CO_2 are almost incidental; if the CO_2 issue is indeed trivial, then the estimates may be consistent. However, if the issue is more serious, than the method of estimation needs to take into account the changing level of CO_2 itself. That is, there is a need for feedback between environmental change and energy use.

If the emission of CO_2 is harmful in the long run for certain countries, possibly for many countries, then at some time interest in preventing CO_2 emissions will rise. Indeed, suppose that it becomes reasonably certain during the next one to three decades, either through results of more extensive research or by an actually experienced minor change in climate, that the effects of CO_2 emissions will be strongly negative for at least some countries. What would be the likelihood of reducing emissions? What forms of control might be feasible? What will be the effect of prevention efforts on emissions?

Focusing on this aspect of the issue indicates that a forecast of CO_2 emissions should not be a description of a mechanistic process but an analysis of the interaction between actual or perceived effects of CO_2 and the actions taken by various decision makers. Human decision makers largely determine how much CO_2 is added to natural sources and released to the atmosphere. In theory at least, governments and peoples could make the issue dissolve by deciding to reduce significantly the burning of carbon. It is their perception of the consequences of CO_2 emissions as well as their interaction with each other that will determine the decisions and thus the amount of CO_2 emitted.

The question of CO_2 emissions should not be regarded as only a technical one, but rather as one involving societal responses, where decision makers are a central focus. By focusing on the decision making processes and the effect of information on these processes, one can also hope to advance discussion of other important questions. For example, can we wait to take preventive action on CO_2 until we know with certainty about the effects of CO_2? Is the time lag among perceived effects, decisions, and adjustments so long that recommendations for action should be given quite soon?

WAYS TO STUDY SOCIETAL RESPONSES TO THE CO_2 ISSUE

Having established the reason for studying societal responses to the CO_2 issue, the question becomes what research strategy may shed light on the problem.

Do Not Study at All

The first possible answer is not to study the question at all. Many would stress the enormous complexity and long time perspective, implying that all findings would be uselessly hypothetical. Others would say, continue to leave the problem entirely as a concern of physical scientists. We do not, however, subscribe to these attitudes. That a problem is difficult, complex, and long-term does not mean that research will have no value. If the question is potentially important, serious attempts to explore it are worthwhile from several points of view. It is evident that we will not come up with correct predictions about what will happen half a century from now. However, we may make substantial progress over the scant picture currently available. And, by attempting better descriptions now, we may be able to develop useful means for organizing information so that as time evolves one can give gradually better and better answers. Moreover, having a more reliable forecast 5 or 10 years from today of impending climatic change will be of little value, if we have not made progress in analysing societies' concerns about and responses to climatic change.

Another group would say that at the present time it is simply unnecessary to worry about a long-term issue like CO_2. Before it becomes an acute problem, technological innovations like the "giga-mixer" (Marchetti, 1977) will take care of it. We are not certain that such a technologically optimistic view is justified. We by no means rule out the possibility of such solutions, but the probability of the absence of such solutions is large enough that thinking about a world with increasing CO_2 in the atmosphere is an important research task.

Hence we proceed with looking for concrete proposals on how to study societal responses to the prospect of high levels of burning of carbon.

Optimization

One approach is that of optimization, searching for the best possible outcome of exploitation of carbon and climatic resources. In simplest terms, one makes some best estimate of the impact of CO_2 and climatic change on economic activities and maximizes the present value of the benefits of burning carbon and the costs and benefits of the impacts caused by resulting levels of CO_2 in the atmosphere. Nordhaus (1979) has tried to determine an optimal path for the whole world to follow in burning of carbon with respect to the potential effects of CO_2 on the environment and economy.

There is a considerable problem with this approach. Global optimization implicitly assumes the existence of some kind of single benevolent world ruler, an entity capable of implementing the policies which the global objective function suggests. While we do not know the geopolitical configuration of the 21st century, this appears to be an unrealistic assumption, or at least an extreme one. It seems more likely that the world, even as integration and interdependence increase, will continue to consist of many sovereign states. Indeed, our major working hypothesis is that in the future there will be many independent nations, often having conflicting aims in regard to factors which form the CO_2 issue, particularly energy policy.

Game Theory

To advance our understanding of the likely evolution of the CO_2 question, it appears to be of fundamental importance to develop an analysis which can portray dynamically the conflict potentially inherent in the situation. Hence, we turn to the theory and playing of games, and the analysis of a situation involving conflicting interests in terms of gains and losses among opposing players. The conflict situation embodies at least two different kinds of games.

"The Tragedy of the Commons" game. One way of seeing the CO_2 issue is as a long term game involving a tragedy of the commons (Hardin, 1968), where the potential tragedy is use of the atmospheric common for waste disposal to the extent that a catastrophic change of climate takes place. For the sake of simplicity of exposition, this game can be represented as a kind of "Prisoners' Dilemma" (Rapoport 1974), with interaction between two players or nations. In the CO_2 case, each nation regards itself as small, with its actions having a minor effect on climate, and regards other nations as one big nation, "the outer world," with a large effect on global climate.

If we let nations A and B represent many nations in the eyes of the other party, the game can be studied in the form of a simple matrix. The numbers in the matrix imply a ranking of the outcomes for each player. A player prefers an outcome with as high a number as possible, that is, a payoff of 4 is best and a payoff of 1 is worst.

TABLE 2. Prisoners' Dilemma Game

We look more closely at the matrix, starting with A's decision. If A believes that B will burn, then A, seeing B as a big player, will think that the climate will be ruined anyway in the long run, regardless of whether he himself will burn. A faces a choice between an outcome of 2 for burning and 1 for not burning, and A will then burn and get the short term benefits of burning.

Since A regards himself to be small and hence only a marginal factor in the climate, then, if A believes that B will not burn, A will still burn. In this way A achieves his best outcome, 4, with the benefits of both burning and essential conservation of the climate. A prefers the short term benefits of burning to marginally affecting the climate in a beneficial way.

Since we assume that B has a similar view of the world, in this symmetric game the same decisions will be made by B. Each party will burn regardless of what he thinks that the other party will do. Hence, they will together reach the outcome 2,2 in spite of the fact that they would both prefer the result 3,3, that is, that no one burns.

This is the general problem of the tragedy of the commons. If the players were committed to pursuing a strategy of cooperation, in this case of not burning, everyone would be better off. If possibilities of forming binding agreements or establishing mutual trust are lacking, then each party will act contrary to the common interest.

An advantage of presenting the commons problem in the Prisoners' Dilemma format is that it focuses on the question of whether legally binding or morally committing agreements on not burning can be reached. Furthermore, it helps relate the analysis to the extensive experimental gaming work done on the Prisoners' Dilemma. (See, for example, Guyer and Perkel, 1972.) These advantages should outweigh the argument (see Dasgupta and Heal, 1979) that the Commons problem is not formally equivalent to the many player version of the Prisoners' Dilemma game in the sense that the strategies of the equilibrium solution are not generally formed by dominating strategies.

A Game of Opposing Interests. While the Tragedy of the Commons (Prisoners' Dilemma) game deals with the longer term problem of a catastrophic change of the climate, for example, in the form of a sea level rise induced by collapse of of the West Antarctic ice sheet, there is another game situation arising from impacts obtainable in a shorter time perspective, perhaps within a few decades. It seems quite likely that already at a global average warming of 1'C, considerable changes in agricultural conditions would be experienced. (See Flohn, 1980.) Such changes could well be beneficial for some regions and adverse for others.

As a hypothetical case, let us follow Gribbin (1981), who suggests the possibility that a global warming might lead in general to problems for areas producing wheat and corn, while improving prospects in rice growing areas. Such a scenario might hold if findings about warming being associated with drier conditions in middle and high latitudes (Manabe and Wetherald, 1980) turn out to be correct. If one looks at the situation from a global perspective and regards population as completely mobile, one might well find that the deteriorated conditions in some areas are offset by improved conditions in other areas. However, one must recognize that many attach a high negative value to large scale population movements. If one assumes limitations on or great costs of both internal and international migration, then the situation becomes one of opposing interests between players, that is, regions or nations.

Returning to our hypothetical case, a rice player may prefer burning of carbon, not only "domestically," but also elsewhere, since it may lead to a more favorable climate. A wheat player, in contrast, will not only dislike the burning carried out by others, but might even find his own burning detrimental, if he is a large enough player. This example leads to a matrix like the one in Table 3.

TABLE 3.   Game of Opposing Interests

Wheat

In order to emphasize the difference from the Prisoners' Dilemma matrix described earlier, we here assume that the player Wheat is relatively large. In this matrix we see that Rice prefers the situation when both burn carbon, while Wheat prefers the situation when nobody burns. The result will be that Rice burns, while Wheat does not burn.

Table 3 could very well have other forms. The important point is that in the medium term the CO_2 issue may have a different character than in the long term. In particular, in the shorter perspective some parties might actually prefer a higher level of CO_2 in the atmosphere. Both game situations described here and other plausible ones, involving, for example, distribution of benefits from direct CO_2 "fertilization" of plants or opening of arctic transport routes, stress that it is important to have an analysis of the issue that focuses on the fact that different countries have different perceptions regarding the character of a CO_2-induced warming and various societal responses may be warranted.

STUDYING THE GAME SITUATION

Having established that the CO_2 issue involves a game situation, the more specific question of how to study such a situation arises. In particular, focusing on the first kind of game problem, that of the Tragedy of the Commons, how does one begin to explore if there will be some agreement on international cooperation to limit carbon burning and if in the absence of such agreements a tragedy will occur.

Traditional Economic Theory

One possible approach would be to employ traditional economic theory, for example, of the general equilibrium type. We would then assume the existence of a great many, small, independently acting countries. This assumption in itself, however, implies the answer the analysis would give. With each nation acting independently, cooperation would not take place. The basic assumption precludes analysis of strategic interdependence existing between various nations.

It should be stressed that it is by no means without interest to carry out this type of economic analysis. It may provide an extreme value for what could happen if there is no cooperation at all. One might then come to understand better how serious a "tragedy" could result from the total absence of cooperation.

Game Theory

To incorporate considerations of interdependence among nations, we turn to game theory, which focuses on strategic relationships between rational actors. Broadly speaking, game theory can be divided in two categories, cooperative theory and noncooperative theory. (See, for example, Bacharach, 1976.)

Cooperative theory implies that the parties first find and decide on a jointly optimal strategy and then proceed (if side-payments are allowed) to divide the jointly optimal result. Much of cooperative theory consists of schemes for dividing the results. Cooperative theory could thus build on the global optimization model of Nordhaus and examine the question of how to share the results of a non-tragedy outcome. In contrast, noncooperative theory does not allow any commitment to agreements. As in the Prisoners' Dilemma situation presented above and traditional economic theory, an outcome with no cooperation, most likely leading to a tragedy, is obtained. A critical problem for exploring the CO_2 question is which body of game theory to apply, cooperative or noncooperative, when the reality might lie between the results produced by the two extremes of complete cooperation and complete lack of cooperation.

There is little meta-theory helping to identify which of the two main bodies of theory is appropriate. The little theory there is indicates that the total number of players is one factor of importance for determining whether stable cooperation is feasible or not. The fewer the players are, the more likely one is to get some cooperative agreement on a set of jointly optimal strategies. This has been shown mainly for games with players of equal size. The idea is that belonging to a cartel is advantageous, but that in certain cases it will be more advantageous to be a party outside of the cartel, when all other parties are in the cartel. In some very simple games, with players of equal size, simple demand and cost functions, and complete information, the critical number of players has been shown to be around 4 to 6. (Selten, 1973; see also Guyer and Cross, 1980, p. 131.) In games where there are players of different sizes and complex demand and cost functions and a dynamically evolving state of information for the players, the aid which game theory can provide in estimating the outcome is quite limited.

Simulation

Realizing the complexity of the CO_2 situation, we next consider using computer simulation, not involving humans as players. The advantage of computer simulation is that it allows complex cost and demand conditions, as well as stochastic characteristics regarding the state of nature. Furthermore, in contrast to game theory, one is not required to assume complete rationality and correct expectations by the actors. Finally, computer simulation models can be run a great number of times, testing how sensitive results are to changes in various parameters.

There is, however, a major obstacle to direct use of this type of simulation for studying the CO_2 issue. We do not know how to specify the equations representing the behavior of the nations to be represented. For example, how will a large country respond when some smaller country starts to defect from an agreement? Will it also defect to punish the other country? Or, will it continue to play cooperatively for awhile, hoping to bring the other country back to cooperation? Whether we get stable international cooperation or a tragedy of the commons can depend largely on how these behavioral equations are specified.

It has been shown in other game situations that it is difficult to formulate behavioral equations for players without studying first the actual behavior of human players in several runs of a game. Even for a game as simple as the Prisoners' Dilemma played many times in a row, it has been difficult to construct reasonable simulation models without using information from a great many actual gaming experiments. (See Stahl, 1975.)

Gaming

Gaming, the playing of games involving several humans (see for example Shubik 1975, 1980), appears to be a method well-placed for gaining insights into the CO_2 question. While maintaining several of the advantages of simulation and game theory, it overcomes the problem of performing a simulation without a behavioral basis. By involving human players in the game, we can observe human responses in specific situations and begin to learn about the critical variable, namely the propensity of humans in this situation to build trust and to respond to noncooperation. In the exploration of societal responses to the CO_2 issue, gaming thus would seem logically to precede simulation without humans. While gaming appears to have certain advantages as a research tool for the CO_2 question, it also has benefits from the points of view of education and collection of information (Ausubel and others, 1980; Stahl, 1980; Robinson and Ausubel, 1981).

Several objections can also be raised to the use of gaming. These center on the the question of whether a game played by a small number of players in an "experimental" setting in a short time can have any validity as regards a long-term problem of enormous complexity.

It must be stressed first that gaming can give only extremely tentative answers and that we see gaming as an appropriate method for our problem, not in an absolute sense, but in a relative sense as compared to other methods. With respect to the specific game proposed in this paper, we believe there are ways to respond to some of the most common criticisms of the methodology of gaming. These objections include the following.

[1] The relationship between the "level" of the actor in reality and the player in the game. The classic example of this problem in the gaming literature is the American college student playing the role of the Chinese foreign minister. The problem for the CO_2 game is probably less drastic. Given the international character of IIASA, it should be possible to have players from most of the nations covered in the game. Secondly, players will include both scientists in the energy field and people from government and industry, who, although not on the top decision level, have a good feeling for the real decision making process. In this regard experience with a IIASA regional water cost allocation game played with water and regional planners in Bulgaria, Italy, Poland, and Sweden is encouraging (Stahl and others, 1981). It should be mentioned that the positive experience in obtaining qualified participants is due to a great extent to the fact that the gaming experiments are designed to be completed in about three hours, or a single evening. The CO_2 game is being designed to be similarly attractive from the point of view of desirable players.

[2] The relationship between how a certain person would behave in reality and how he would behave in a game. How seriously does a person behave when playing in a game? In this respect we are again hopeful based upon experience with the water game. The fact that water planners in several quite different countries played in a similar manner (different from how students in these countries played) indicates that their playing was not random, but rather reflected careful professional thinking.

[3] The relationship between the real decision environment, for example, with respect to resources, technological development, and changing scientific information, and how this is covered by the institutional assumptions of the game. Clearly, there will be a tremendous discrepancy. We believe, however, that the game can still be used as an acid test for various hypotheses and theoretical models about the decision process. These hypotheses or models are generally built on at least as simplified a set of institutional set of assumptions as the game. If the players do not behave according to the model in the simple game with an institutional set up of the same simplicity as the model, they are not likely to behave according to the model in the more complicated reality either. The gaming should at the least be helpful in indicating what kind of hypotheses and models one should devote more research to.

MAIN CHARACTERISTICS OF THE PROPOSED GAME

While several games based on the CO_2 situation could be envisaged, the game under development focuses on the question of how much carbon may be burnt, as more information becomes available regarding the environmental and economic impacts of CO_2. The main structure of the game can be characterized by coal, trade, and many countries.

Why Coal?

The *sine qua non* of a severe CO_2-induced climate problem is the burning of coal.
We here take a doubling of atmospheric CO_2 as a level which warrants considerable concern. This choice is quite arbitrary. Both greater and lesser increases could have very costly -- or beneficial -- consequences. However, an increase of 50%, which might be associated with a degree of warming similar to the Medieval warm phase 1000 years ago, seems on the conservative side, while placing a threshhold as high as a tripling definitely seems imprudent, given possibilities for changes of sea level and so forth (Flohn, 1980).

As shown in Table 4, current estimates of total reserves and resources of oil, gas, coal, and other forms of carbon indicate that a doubling of the present level of atmospheric carbon dioxide within the next century can only be reached by substantial burning of coal. Even allowing for the considerable uncertainty of these estimates, it comes across strongly that non-coal carbon resources are not large or accessible enough to be exploited to a degree that is very threatening from a CO_2 perspective. Total oil and gas resources, even with the addition of high rates of deforestation, amount to only about half the 1500 gigatons of carbon (Gt C) roughly necessary for a doubling, given present models of the carbon cycle (NAS, 1977; Bolin and others,1979). It should be noted that some estimates of unconventional gas and oil resources, especially oil shale, are quite large (Rotty and Marland, 1980). However, exploitation of these resources on a scale which could be significant for CO_2 seems unlikely for until well past the year 2000 (Sundquist and Miller, 1980; IIASA, 1981). In contrast, coal could readily account for two-thirds or more of CO_2 emissions in a scenario of doubling in the early to middle decades of the next century (Marland and Rotty, 1980; Ausubel, 1980). Moreover, the extremely desirable characteristics of oil and gas make their exploitation appear less "optional" than exploitation of coal. Because coal plays this indispensable role in the CO_2-issue, it is logical to begin game development with the emphasis on coal.

Why Trade?

Why the game explicitly deals with trade in coal is evident from Table 5, which shows the approximate distribution of coal reserves and resources. More than four-fifths of the resources are held by three countries, the USSR, US, and China (henceforth regarded as "big" players). The total holdings of the remaining countries, including a few gigatons for

TABLE 4.   Potentially Available Carbon Resources

	Best current estimates (Gt C)	Upper limit speculation (Gt C)
Ultimately recoverable conventional petroleum resources[1]	230	380
Ultimately recoverable conventional natural gas[1]	140	230
Ultimately recoverable conventional coal[1]	3500	6300
Unconventional oil and gas[2]	hundreds of Gt	thousands of Gt
Biospheric (forests, etc.)[3]	—	~200

SOURCE: 1) Rounded from Rotty and Marland (1980); 2) See IIASA (1981), Rotty and Marland (1980); 3) Cumulative release in high deforestation scenario (Chan and others, 1980).

countries with holdings smaller than one gigaton, amount to around 800-900 gigatons, or somewhat more than half the carbon base required for a doubling. As it is unlikely that a major part of this will be used within the next few generations, it seems reasonably assured that a serious CO_2 problem will arise only with substantial use of the coal resources of the three big players. If these big players do not export large amounts of coal and also keep their own coal combustion low, a severe CO_2 problem should not arise. The CO_2 issue arises in scenarios, like the one proposed in the World Coal Study (WOCOL, 1980), where over the next 20 years roughly a ten-fold increase in coal trade is envisaged. Coal trade could become important for coal consumption in the way that oil trade is today for oil consumption. Without a substantial world coal trade, the physical quantities of carbon required for a CO_2 problem are unlikely to be used.

Trade in coal is also of importance for the estimation of future coal usage from another point of view. Coal usage will depend partly on the price of coal. Although total consumption of energy in the short run is fairly insensitive to price, consumption of a specific energy source in the long run will be sensitive to the price of the resource, since in the long run there are possibilities of substitution between various sources of energy. (See, for example, Nordhaus, 1977.) The extent to which there can be substitution between coal and other sources of energy is a major aspect of the CO_2 question, and our working hypothesis is that reasonably good possibilities of substitution exist. The price of coal traded on world markets thus becomes important for coal usage. If a country either has to import or can export coal, then the world market price of coal will influence the country's consumption of coal.

We foresee a world market price of coal, much in the same way as one talks today about a world market price of oil, not only in the market economies but also in the socialist countries. Three broad levels of coal prices can be envisaged.

TABLE 5. Approximate World Distribution of Coal Resources
              (in gigatons carbon)

Huge holdings		Large Holdings		Small Holdings	
USSR	3300	Australia	180	Czechoslovakia	12
U.S.	1700	FRG	170	Yugoslavia	7
China	1000	UK	110	Brazil	7
		Poland	80	GDR	7
		Canada	80	Japan	6
		Botswana	70	Colombia	6
		India	40	Zimbabwe	5
		South Africa	40	Mexico	4
				Swaziland	3
				Chile	3
				Indonesia	2
				Hungary	2
				Turkey	2
				Netherlands	2
				France	2
				Spain	2
				North Korea	1
				Romania	1
				Bangladesh	1
				Venezuela	1
				Peru	1

SOURCE: Based on data from World Energy Conference (1978).
           Very rough estimate of carbon wealth in Gt has been
           obtained by multiplying coal resources in 10^9 tons
           coal equivalent by carbon fraction of 2/3.

[1] A cost based price. Price would be close to long term average costs of marginal producers. This price is compatible with the idea that the coal market is characterized by
pure competition, that is, a market with a great many sellers. This resembles the oil
price situation prior to 1973.

[2] Monopoly price. The price would maximize the joint result of the producers. The sellers forming a cartel would be better coordinated than the buyers. This is a situation
similar to that of oil prices since 1973. For reasonable values of price elasticity, this
would imply a price several times higher than marginal costs. For example, if price
elasticity E = 1.5 (implying that a lowering of price by 10 per cent will increase long
term usage by 15 per cent), then with the the Amoroso-Robinson condition we would
have that p = E * MC / (E-1) = 1.5MC / 0.5 = 3MC, that is, price is three times marginal
cost.

[3] A price above monopoly price. Such a price could reflect concern for the environmental
effects of coal, in particular a CO_2 problem. On top of the ordinary price, there would be
a tax to lower the consumption of coal to that level which is optimal if the cost of coal
also includes the costs caused by the CO_2 effects on climate. This price is thus a kind of
shadow price. As pointed out by Nordhaus (1979) such a "carbon tax" could become
quite high, especially at the time when the CO_2 level in the atmosphere has gone up. (
Such a price can of course also be below the monopoly price, but always above the cost

based price, if the environmental effects of CO_2 are not so strong or immediately felt.)

Which kind of price prevails will be established largely by trade structure, that is, by the number and relative strengths of the exporters and importers.

Trade in coal is also of interest in connection with different schemes of international cooperation for reducing or preventing CO_2 emissions. The possibility for the larger countries to limit the supplies of coal either on the world market generally or to specific countries can give teeth to attempts at enforcing international agreements to limit the usage of coal.

Why Many Countries?

The basic reason for inclusion of more than a very few nations is that the CO_2 issue in reality concerns a world where many nations, able to act with some independence from one another, affect the problem. If we limit ourselves to only a handful of actors in all phases of development of the game, certain scenarios would be excluded where international cooperation is impeded by the actions of relatively small nations. This kind of possibility is also the reason that it is not appropriate to deal with aggregate energy regions, likely to contain several countries with quite different characteristics from the point of view of the CO_2 issue. Of course, it is also not necessary to represent every nation in order to capture the essence of the CO_2 issue.

Which nations or kinds of nations are critical? Obviously, the three big players are of great importance. At the same time, a major portion of energy consumption will be taking place outside these countries. Large future users of energy like Japan, Brazil, and Italy are among the nations with relatively small holdings of coal, and the behavior of such potentially large importers will be of interest. Players which can affect cartel efficiency are also of great significance. Even if the three big players account for around 80 per cent of total coal resources, the resources of some smaller holders are large from an absolute point of view. As Table 5 shows, another eight countries have substantial holdings, holdings currently estimated at more than 40 Gt. 40 Gt corresponds roughly to total global carbon emissions during the past decade. Such players can affect cartel efficiency.

Furthermore, there are some countries, although neither large coal consumers nor producers, that are important due to the fact that a severe CO_2 problem would be particularly imposing for them. For example, collapse of the West Antarctic ice sheet and ensuing rise of the world oceans (Schneider and Chen, 1980) could constitute a true catastrophe for low-lying countries like the Netherlands and Bangladesh.

While the "reality" of the situation is one reason to include more than just a few players, the other major reason is that answers to the question about likelihood of cooperation are dependent on the number of players involved in the game. For example, with only three players cooperation in the game is quite likely. As mentioned earlier, in some very simple games with all players of equal size 4-6 seems to be the boundary between few and many. In games with players of different sizes the boundary will probably lie higher.

Ultimately one would probably wish to include about twenty countries of different sizes and characteristics to catch fully the strategic problem. While the roles of only seven or eight countries would be played by human players, the remaining dozen nations would be played by computer programs, "robots." (See Ausubel and others, 1980.)

General structure of the game

The initial design of the game is oriented toward what may happen a generation from now, if there is confirmation of a CO_2-induced global warming. With the decades around 2010 - 2010 as critical, the game needs to cover a period of half a century and possibly longer. There will be up to about 10 rounds of decision, each representing 5 or 10 years. Since the game should be playable in roughly three hours, the number of decisions in each 15-20 minute round must be strictly limited.

The general structure of the game is shown by Fig. 1. Decisions are represented by dotted lines extending from the decision makers. Each player will have four principle decisions in a round.

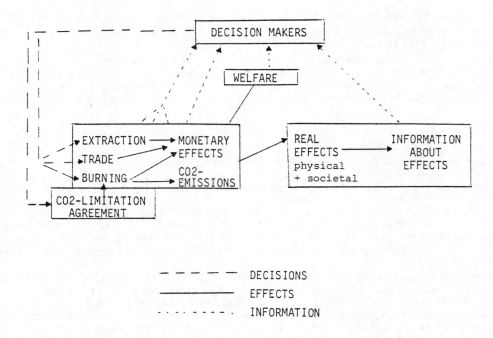

Fig. 1. Main Structure of the Game

[1] A mining decision: How much coal shall be extracted?

[2] A combustion decision: How much coal shall be burnt?

[3] A trade decision: How much coal shall be demanded or supplied on the world market at various prices?

[4] A prevention decision: How much shall one pay in order to get other nations to stop burning coal?

Extraction, trade, and burning decisions will have monetary consequences, while the burning decision also will have effects on climate. The welfare of the player is influenced by both these shorter term monetary effects and the longer term climatic effects of CO_2. Players will base their decisions not only on information about their own economic welfare, but also on information about mining, trade, and coal burning by other players and information which will develop gradually over time about actual and anticipated CO_2 effects. (For more information on game design see Ausubel and others, 1980; Stahl, 1980; Robinson and Ausubel, 1981.)

CONCLUSIONS

In the beginning of the paper various estimates for the input of fossil fuel CO_2 into the atmosphere were presented. These estimates may well be inadequate, since the methods underlying them do not take into account societal responses that CO_2-induced changes may bring about. Experimental gaming, focusing on development of international coal trade, may provide an approach which overcomes some of the deficiencies of other methods. On the basis of designing and playing a CO_2 and coal game with experts in energy and other fields, one should be in a better position to evaluate previous estimates of future CO_2 emissions and improve on these estimates.

Of course, one should not look to gaming for an authoritative forecast; the uncertainties inherent in the CO_2 issue will continue to mean that there is a foundation of uncertainty in all forecasts. Rather, the application of simulation gaming to this problem should be seen as part of the necessary contribution of many disciplines and methodologies to the building up of a satisfactory assessment of the CO_2 issue.

REFERENCES

Ausubel, J. (1980). Climatic change and the carbon wealth of nations. Working Paper 80-75. IIASA, Laxenburg, Austria.

Ausubel, J., J. Lathrop, I. Stahl, and J. Robinson (1980). Carbon and climate gaming. Working Paper 80-152. IIASA, Laxenburg, Austria.

Bacharach, M. (1976). *Economics and Theory of Games*. MacMillan, London.

Bolin, B., E.T. Degens, S. Kempe, and P. Ketner (1979). *The Global Carbon Cycle*. Wiley, Chichester.

Chan, Y.-H., J. Olson, and W. Emanuel (1980). Land use and energy scenarios affecting the global carbon cycle. *Environment International*, Vol. 4, pp. 189-206.

Cross, J. and J. Guyer (1980). *Social Traps*. U. of Mich. Press, Ann Arbor.

Dasgupta, P.S. and G.M. Heal (1979). *Economic Theory of Exhaustible Resources*. Nisbet and Cambridge University Press, Cambridge.

Flohn, H. (1980). Possible climatic consequences of a man-made global warming. Research Report 80-30. IIASA, Laxenburg, Austria.

Gribbin, J. (1981). The politics of carbon dioxide. *New Scientist*, April 9, pp. 82-84.

Guyer, M. and B. Perkel (1972). Experimental Games: A Bibliography 1945-1971. Communications Nr. 293, Mental Health Research Institute, U. of Mich., Ann Arbor.

Hardin, G. (1968). The Tragedy of the Commons. *Science*, 16, pp. 1243-1248.

International Institute for Applied Systems Analysis (IIASA) (1981). *Energy in a Finite World: A Global Systems Analysis*. Ballinger, Cambridge.

JASON (1979). The long term impact of atmospheric carbon dioxide on climate. Technical report JSR-78-07. SRI International, Arlington, Virginia.

Lovins, A.B. (1980). Economically efficient energy futures. In Bach, W., J. Pankrath, and J. Williams (eds.), *Interactions of Energy and Climate*. Reidel, Dordrecht. pp. 1-31.

Manabe, S. and R. Wetherald (1980). On the distribution of climate change resulting from an increase in CO_2 content of the atmosphere. *J. Atmos. Sci.*, 37, 99-118.

Marchetti, C. (1977). On geoengineering and the CO_2 problem. *Climatic Change 1*, p. 59ff.

Marchetti, C. (1980). On energy systems in historical perspective. IIASA, Laxenburg, Austria.

Marchetti, C. and N. Nakicenovic (1979). The dynamics of energy systems and the logistic substitution model. RR-79-13. IIASA, Laxenburg, Austria.

Marland, G. and R. Rotty (1979). Atmospheric carbon dioxide: implications for world coal use. In M. Grenon (ed.), *Future Coal Supply for the World Energy Balance*, Third IIASA Conference on Energy Resources, Nov. 28 - Dec. 2, 1977. Pergamon, Oxford. pp. 700-713.

Munn, R.E. and L. Machta (1979). Human activities that affect the climate. In WMO, *Proceedings of the World Climate Conference*, World Meteorological Organization, Geneva.

National Academy of Sciences (NAS) (1977). *Energy and Climate*. NAS, Washington DC.

Niehaus, F. (1979). Carbon dioxide as a constraint for global energy scenarios. In Bach, W., J. Pankrath, and W. Kellogg (eds.), *Man's Impact on Climate*, Elsevier, Amsterdam. pp. 285-297.

Niehaus, F. and J. Williams (1979). Studies of different energy strategies in terms of their effects on the atmospheric CO_2 concentration. *J. of Geophysical Res. 84*, pp. 3123-3129.

Nordhaus, W.D. (1977). *International Studies of the Demand for Energy*. North-Holland, Amsterdam.

Nordhaus, W.D. (1979). *The Efficient Use of Energy Resources*. Yale University Press, New Haven.

Rapoport, A. (1974). Prisoner's Dilemma - recollections and observations. In Rapoport, A. (ed.), *Game Theory as a Theory of Conflict Resolution*. Reidel, Dordrecht.

Robinson, J. and J. Ausubel (1981). A framework for scenario generation for CO_2 gaming. Working Paper 81-34. IIASA, Laxenburg, Austria.

Rotty, R. (1977). Present and future production of CO_2 from fossil fuels. ORAU/IEA(O)-77-15. Institute for Energy Analysis, Oak Ridge.

Rotty, R. (1978). The atmospheric CO_2 consequences of heavy dependence on coal. In J. Williams (ed.), *Carbon Dioxide, Climate and Society*, Pergamon, Oxford. pp. 263-273.

Rotty, R. (1979a), Energy demand and global climate change. In Bach, W., J. Pankrath, and W. Kellogg (eds.), *Man's Impact on Climate*, Elsevier, Amsterdam. pp. 269-283.

Rotty, R. (1979b). Growth in global energy demand and contribution of alternative supply systems. *Energy 4*, pp. 881-890.

Rotty, R. and G. Marland (1980). Constraints on fossil fuel use. In Bach, W., J. Pankrath, and J. Williams (eds.), *Interactions of Energy and Climate*. Reidel, Dordrecht. pp. xxx-xxx.

Schneider, S. and R. Chen (1980). Carbon dioxide warming and coastline flooding : a problem review and exploratory climatic impact assessment. *Annual Review of Energy, 5*, 107-140.

Selten, R. (1973). A simple model of imperfect competition, where 4 are few and 6 are many. *International Journal of Game Theory*, pp.141-201.

Shubik, M. (1975). *Games for Society, Business and War*. Elsevier, New York.

Shubik, M. (1980). State-of-the-art survey. In Stahl, I. (ed.), The Use of Operational Gaming as an Aid in Policy Formulation and Implementation. CP-80-6. IIASA, Laxenburg, Austria.

Stahl, I. (1975). The Prisoners' Dilemma paradox, bargaining theory and game theoretic rationality. Working Paper. EFI, Stockholm.

Stahl, I. (1980). An interactive model for determining coal costs for a CO_2-game. Working Paper 80-154. IIASA, Laxenburg, Austria.

Stahl, I., R. Wasniowski, and I.Assa (1981). Cost allocation in water resources - six gaming experiments in Poland and Bulgaria. Working Paper. IIASA, Laxenburg, Austria.

Sundquist, E. and G. Miller (1980). Oil shales and carbon dioxide. *Science*, Vol. 208, 16 May 1980, pp. 740-741.

Voss, A. (1977). *Ansätze for Gesamtanalyse des Systems Mensch- Energie- Umwelt*. Birkhauser, Basel.

World Climate Programme (1981). On the assessment of the role of CO_2 on climate variations and their impact. Report of a WMO/UNEP/ICSU meeting of experts in Villach, Austria, November 1980. World Meteorological Organization, Geneva.

World Coal Study (1980). *Coal - Bridge to the Future*. Ballinger, Cambridge, Mass.

World Energy Conference (1978). *World Energy Resources 1985- 2020, An Appraisal of World Coal Resources and Their Future Availability*. IPC Science and Technology Press, Guildford, UK.

IS THE LARGE-SCALE DESTRUCTION OF TROPICAL RAIN FORESTS NECESSARILY CRUCIAL FOR THE GLOBAL CARBON CYCLE?

N. Stein

Department of Biogeography, University of Saarland, Saarbrücken, Federal Republic of Germany

ABSTRACT

A re-examination of the biosphere, in particular of the tropical rain forests, is necessary since there are some contradictions in our present knowledge about carbon fluxes and carbon storage in various reservoirs. Traditional land-use systems as well as modern exploitation methods in the tropical rain forest areas need not be necessarily of such a magnitude that the global carbon cycle would undergo a major upset or even a climatic catastrophe. This conclusion is primarily based on the high photosynthesis rates of the humid tropics, on the very important role of tropical river systems for carbon transport between terrestrial and aquatic ecosystems and on the fact that there is, at least presently, probably a steady balance between all terrestrial ecosystems of the earth.

KEYWORDS

Sinks and sources for atmospheric CO_2; carbon fluxes between different reservoirs; land use practices in the humid tropics; role of rivers within the carbon cycle.

INTRODUCTION

In the last few years there has been a great number of research activities, conferences and publications on the global carbon issue. A great percentage of these works with corresponding public discussion has to be seen against the background of hypothetic negative climate change due to an anthropogenically increased CO_2 - content in the atmosphere. In conjunction with this, scenarios, based on reputed estimations of the future use of fossil fuels and on the calculated temperature increases, have been interpreted as an actual forecast for a climatic catastrophe for the next century.
In the meantime, however, these types of forecast are now being made more cautiously. The assumption that we are able to speculate, is certainly correct but we are not yet able to predict. There are several reasons for this. Most of the previous models are hydrostatic

heat balance models, which only take into account the thermal effect of various rates of an atmospheric CO_2-increase. Thereby the important role of further atmospheric factors, in particular water vapour content, clouds, precipitation, as well as various boundary conditions of the overall system earth atmosphere is being disregarded. The oceans and the biosphere which have a key position in the global carbon balance have not adequately been taken into consideration, at least not in those predictions based on a warmer climate due to the "green-house effect" by an increase of CO_2.

Since we are just not able to know what consequences are to be expected, every option is open to the speculator, ranging from a catastrophe over the status quo to climatic changes for the general benefit of mankind (Degens, 1981).

At this point, scientists of various disciplines are requested to minimize uncertainties through intensive research on CO_2 and climate. This aim can be achieved in two ways: 1) Precise inventory of all known facts should be considered, particularly the fact that carbon on earth is present in different forms and in different reservoirs and, thereby, in different quantities: as CO_2 in the atmosphere, as carbonates in the earth's crust, as carbonate ions in the sea, and in the many organic compounds of terrestrial ecosystems. 2) Metabolism and flow characteristics of these carbon bonds are not yet adequately understood in their global scale. As a consequence, intensive measurement studies must be initiated in order to have a closer understanding of this metabolism and flow characteristics within the global carbon cycle.

NECESSITY OF A NEW EXAMINATION ON THE IMPORTANCE OF THE TROPICAL RAIN FORESTS FOR THE GLOBAL CARBON CYCLE

Two problems, which are at present far from a satisfactory explanation, require a new and critical examination on the importance of the biosphere, in particular, of the tropical rain forests. These are: I) the discrepancy between the increase of atmospheric CO_2 during the last decades and the mean temperature curve during the same period of time (Fig. 1).

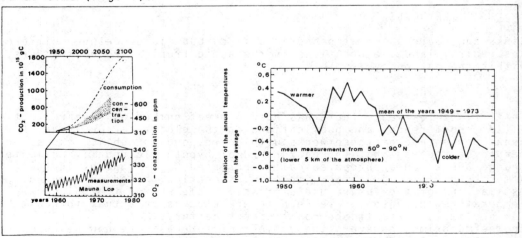

Fig. 1. Atmospheric CO_2-increase, predicted concentration of CO_2-increase and real temperature trend in the last decades (after Grassl, Kempe and Spitzy, 1981).

II) the assumption that yearly $7 \times 10^{15}g$ - $11 \times 10^{15}g$ C are being stored somewhere on our planet, commonly referred to as the missing carbon.

With regard to the former problem (I) the CO_2 - content of the atmosphere approximates now 340 ppm. Under the assumption that the models mentioned above would be correct, we should have expected a mean temperature increase of at least 0.5°C by now (Grassl and others, 1981). The temperatures in the northern hemisphere, however, have shown in the lower parts of the atmosphere a worldwide decrease since 1940.

There are several possible explanations for this fact which is contradictory to the models:

1. The models are inadequate or incomplete whereby, in particular, no room has been given for the metabolism of carbon as an extremely mobile element between reservoirs of considerably differing sizes.

2. Within the permanently existing natural climate fluctuations, we are now experiencing a cooling phase. Without a CO_2-increase there would have been an even greater temperate decrease.

3. Further man-induced changes in the atmospheric composition, e.g. the increase of atmospheric turbidity due to a higher aerosol load, counteract to the green-house effect due to IR-radiation losses.

4. The oceans reduce the temperature increase because of their great heat capacity.

5. The importance of the biosphere has not been properly judged.

6. Terrestrial ecosystems are also spatially open systems. In other words, material transport out of the system takes place. High loading with suspended substances, especially also particulate and dissolved C, signify transport from many forest ecosystems into seas and oceans. That means that the role of rivers in the global carbon cycle must not be ignored.

With regard to the latter problem (II) Woodwell (1978) and his work team came to the conclusion that apparently each year larger quantities of carbon enter the atmosphere than compared to those quantities which really are being monitored and measured in the global measurement network. In addition to the $5 \times 10^{15}g$ C, released by the burning of fossil fuels, $4 \times 10^{15}g$ C - $8 \times 10^{15}g$ C from deforestation is assumed to reach the atmosphere. Out of these $9 \times 10^{15}g$ C - $13 \times 10^{15}g$ C remain, according to the measurements, only $2.3 \times 10^{15}g$ in the atmosphere. The rest of $7 \times 10^{15}g$ - $11 \times 10^{15}g$ is probably stored somewhere on earth. At present a clear answer as to where this storage may occur has not yet been found. According to the calculations of the oceanographers the oceans can take only $1.5 - 3.0 \times 10^{15}g$ C per year. This would mean that $4 \times 10^{15}g$ - $7 \times 10^{15}g$ C remains unaccounted for. The oceanographers believe that this is be stored in the terrestrial biomass while the ecologists argue that it enters the oceans. Therefore oceanographers have had to re-examine their models.

In this connection it seems to be of utmost importance to review the role of the terrestrial ecosystems again. This applies, in particular, to the tropical rain forests since these forest systems have the highest percentage of carbon in the form of living organic substances.

On the other hand the tropical rain forest undergoes drastic landuse changes today, which may or may not upset the global carbon balance.

ANTHROPOGENIC INFLUENCES ON THE TROPICAL RAIN FOREST

There is little doubt that the tropical rain forests in their natural state are certainly a sink for atmospheric CO_2. The actual importance of the rain forests is found in the long-term storage of organically-fixed carbon in the phytomass. Furthermore, it seems to be very likely that the postulated balance between matter production and decomposition of organic substances, i.e. the restoring mechanism for CO_2 into the overall carbon cycle, is much more complex than is commonly believed to be.

There are different kinds of anthropogenic influence on the tropical rain forests, which may signify different degrees of impact on the carbon fluxes.

Shifting cultivation as a traditional and widespread form of cultivation in the humid tropics

Shifting cultivation (= swidden agriculture) is a common practice among many indigenous population groups in the humid tropics. The various forms of this land-use system are well-known and have often been described. In the course of this land-use system great quantities of dead wood remain on the ground. The decomposition rate of the dead tree trunks in the humid tropical climate is being estimated to be about $0.15-0.44$ $g/m^2/day$ (Odum and Pigeon, 1970). Probably 40% of the original phytomass would remain as organically fixed carbon in the form of wood and organic soil components (Lugo, 1980), provided that successional phases of vegetation development occur, as it is, indeed, the rule within the shifting cultivation system. This means that a temporary release of CO_2 into the atmosphere is possible but certainly no irreversible loss.

A characteristic phenomenon of the tropical shifting cultivation system lies in the fact that only certain areas are being used in the original forest. These areas are allowed to re-grow after a few years. The succession taking place after the period of agricultural use is characterized by rapid growth and, therefore, requires a high net photosynthesis, which ultimately means a net C-uptake from the atmosphere. Quantitative studies on this phenomenon are greatly lacking. A comparison with the temperate latitudes, however, shows that after deforestation, the biogeochemical re-establishment in the case of an undisturbed succession takes place very rapidly. As early as three years the net production of a hardwood forest in the White Mountains (U.S.A.) has reached the original production level of the 60-year-old stand before deforestation (Fig. 2).

Commercial timber extraction

In the case of commercial timber use, the practice of selective logging operations is frequently applied, i.e. only large, mature trees of a high commercial value are being extracted.
Table 1 shows an uneven distribution of the carbon storage in a Central Amazonian rain forest in accordance with the specific stand structure. The major part of the phytomass and of organic carbon is concentrated on the two upper stratums where usually the trees for commercial use are located. There is every reason to assume that the large pool of the trees from the lower stratums will start with a rapid growth and along with this require a high CO_2-net uptake from the atmosphere once the higher, shade-producing trees are removed.

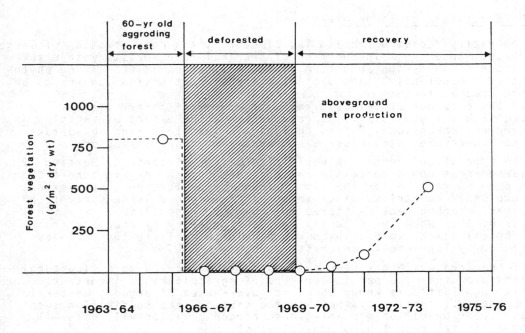

Fig. 2. Effects of deforestation aboveground net production in an experimentally
deforested northern hardwood forest ecosystem (after Likens and others, 1978).

TABLE 1   Vertical organization of aerial phytomass of dicotyledonous
trees and palms of a central Amazonian rain forest
(after Klinge and others, 1975)

Stratum	Individuals/ha Dicotyledonous Trees	Palms	Aerial phytomass (tons/ha) Leaves	Branches + Twigs	Stems	Total	Total %	Dry matter	C-content t/ha
A	50	0	2.3	48.7	139.2	190.2	27.6	85	38.3
B	315	0	7.1	123.1	269.3	399.5	58.0	190	85.5
C_1	760	15	3.9	26.1	47.3	77.3	11.2	35	15.8
C_2	2,765	155	2.0	3.6	10.0	15.6	2.3	7	3.2
D	5,265	805	2.2	0.7	1.8	4.7	0.7	2	0.9
E	83,650		0.6	0.2	0.6	1.4	0.2	0.6	0.3
Total	93,780		18.1	202.4	468.2	688.7	100	320	144
%			2.6	29.4	68	100			

Deforestation of large areas

The really decisive problem is, of course, the deforestation of large areas. Since the main function of tropical rain forests within the carbon cycle is the long-term storage of carbon in the living phytomass, the potential land use after forest destruction appears to be the crucial factor.
If there is not a total deforestation and if areas are reserved with the possibility of regeneration, the effect of rapidly growing successional stages could again lead to a high CO_2-uptake and, therefore, also cause a comparatively high C-fixation.

The type of deforestation would also be very important. Burning of large areas would certainly release more CO_2 to the atmosphere than a pure extraction for the timber industry or for the timber export into other countries. Since in the latter case a large part of the organic carbon remains stored in the wood, regardless of the final use of this wood, we urgently need an inventory of products from tropical timber woods, which would help to estimate the real loss caused by large-scale deforestations.

If the area which has been deforested is converted to agricultural land use, then the particular type of agriculture is important. Although little is known about turnover-rates of organic matter in managed cultivation systems, in the case of tree cultivation such as rubber or oil palm plantations, carbon fixation will naturally be again the result of photosynthesis.

THE ROLE OF RIVERS IN THE GLOBAL CARBON CYCLE

It has been mentioned above that all ecosystems are open systems and thus can never be expected to be perfectly balanced. The transport of particulate and dissolved carbon by rivers signifies a transfer from one reservoir (= terrestrial ecosystem) to another (= marine ecosystem) even though a changing percentage of the total organic carbon carried by a river may be stored in river sediments or on flood plains or may be en route oxidized (respired).

Meybeck (1980) estimated that the total organic carbon load from rivers to oceans is approximately 400×10^{12}g C per year. About half of it is supposed to be in the particulate form although in most of the rivers the particulate organic carbon is about half as abundant as the dissolved carbon.

Table 2 shows the input of total organic carbon from rivers to oceans on a worldwide basis. The tropical zone contributes 63% of the organic load, and the Amazon River alone reaches 14%.
The current estimates of organic export into the oceans suggest that rivers already in undisturbed terrestrial drainage basins may be significant in the global carbon cycle, particularly, with regard to increased atmospheric CO_2.

In the case of deforestations the rivers might even have a key position in the CO_2-issue. It can be expected that the total organic carbon will considerably increase, especially the particulate organic carbon (POC), including leaf litter, woody debris and soil organics. Since we know that just POC will reach the oceans to a large extent we must expect a comparatively high net transfer of carbon between forest ecosystems and the oceans, especially for the most unfavourable boundary conditions, i.e. a large-scale deforestation.

TABLE 2 Input of Total Organic Carbon from Rivers to Oceans
(after Meybeck, 1980)

A. Studied rivers

	Area $(10^6 km^2)$	TOC_{exp} $(t.km^2.year^{-1})$	Annual Load $(10^{12} gC)$
Siberian rivers	12.6	2.25	28.3
Nelson	1.15	0.62	0.7
MacKenzie	1.8	1.93	3.5
St. Lawrence	1.03	0.81	0.83
Mississippi	3.27	1.23	4.0
Amazon	6.3	8.7	54.8
Murray	1.1	0.28	0.3
Total known sample	27.25	3.4	92.4

B. Extrapolated sample

	Remaining Area $(10^5 km^2)$	Typical TOC_{exp} $(t.km^{-2}.year^{-1})$	Annual Load $(10^{12} gC)$
Tundra	4.4	0.6	8.6
Taiga	3.45	2.5	2.6
Temperate	17.7	5	88.5
Tropical	31.0	6	186
Semiarid	14.4	0.3	4.3
Desert	1.7	0.0	0
Total extrapolated sample	72.65	4.0	290

C. Total river discharge

	Total Area $(10^6 km^2)$	Total TOC Load $(10^{12} gC.year^{-1})$	% TOC Load
Tundra	7.55	4.5	1.2
Taiga	15.85	39.6	10.3
Temperate	22.0	93	24.3
Tropical	37.3	241	62.9
Semiarid	15.5	4.6	1.2
Desert	1.7	0	0
Total	99.9	383	

The question which arises now is what happens to this organic
material when it reaches the mouth of the rivers and the seas, and
how far can it be exported into the oceans, if at all?

Mulholland (1980) has shown that most of the organic carbon is
being deposited in the great delta areas of the world. Milliman,
Summerhayes and Barretto (1975) found that more than 95% of the
terrigenous sediment settles out before the salinity reaches 3‰,
and practically none of this escapes offshore (see also Wangersky,
1980). Similarly, Barretto and Summerhayes (1975) found that only
1% of the suspended load of the Sao Francisco River in Brazil escapes as far as 10 km from the river mouth. Based on these and

similar findings, Wangersky (1980) concluded that the evidence strongly favors the retention of almost all of the sedimented organic carbon within the estuary.

It cannot be excluded that this high net input into the coastal waters will, at least initially, stimulate matter production at the primary level in these waters. This effects could even be enforced by the anthropogenic release of nitrates and phosphates, which might lead to an unintentional fertilizing of the coastal waters. The resulting increase of algae growth might fixate an amount of carbon in the magnitude of some 10^{12}g C/a.

On the other hand, certain types of pollution may reduce the microbial decomposition of the organic material (Wangersky, 1980). Many of the pesticides and chlorinated hydrocarbons are preferentially adsorbed onto organic particulate matter. Because of this accumulation and the resulting slowing of the microbial decomposition of the organic material, there would then be less opportunity for diffusion of organics back into the water column.

We must admit, however, that we do not have accurate measurements and observations on this. We have arrived at a "research frontier" where detailed investigations should begin as soon as possible.

THE ROLE OF THE BIOSPHERE AS A WHOLE

In order to reach a more precise understanding of the role of the tropical rain forests in the global carbon cycle the importance of the other terrestrial ecosystems should also be considered. Loucks (1980) stated that biospheric contributions to CO_2 in the atmosphere represent an asymmetry of opposed effects from 1860 or earlier until the present.

"Asymmetry of opposed effects", according to Loucks (1980), refers to the fact that CO_2-source elements, such as temperate forest releases, depletion of soil organic matter, wetland releases, and others, increased or peaked prior to World War II and then decreased, whereas other sources (such as releases from tropical forest conversion) began to increase after these other sources had begun to decrease (Fig. 3).

A possible steady balance of all terrestrial ecosystems would, therefore, be the result of a particular historical constellation whereby a destruction of vegetation in the tropics would occur simultaneously to an increase of phytomass and also to a humus and peat accumulation in temperate and northern ecosystems. Miller (1980 and personal communication) stated that arctic and tundra regions would continue to be carbon sinks as CO_2 levels in the atmosphere rise.

By careful consideration of all facts which have been dealt with in this paper, one may come to the conclusion that the problem of the "missing carbon" is not of that magnitude as commonly believed. Hampicke (1980) does not exclude the fact that the tropics have a loss of 2 x 10^{15}g/a C. If this figure proves to be true, and the perspectives of this article tend in this direction, the overall steady balance of the terrestrial ecosystems would be, indeed, possible . The amount of 2 x 10^{15}g/a C could be compensated for by other terrestrial sinks. Even in the case that the release in the tropics would be higher than 2 x 10^{15}g/a C the rivers, the

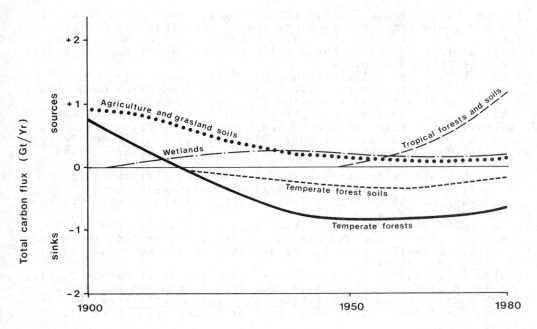

Fig. 3. Total carbon fluxes from 1900 (after Loucks, 1980).

flood plains and the delta areas could be another effective carbon sink.

From many points of view the destruction of tropical rain forests is without a doubt an undesirable event, but it need not necessarily lead to a major upset in the global carbon cycle or even to a climatic catastrophe.

REFERENCES

Ajtay, G.L., P. Ketner and P. Duvigneaud (1979). Terrestrial primary production and phytomass. In: Bolin, B., E.T. Degens, S. Kempe and P. Ketner (eds.), The Global Carbon Cycle, Chichester/New York/ Brisbane/Toronto, 129-181.

Armentano, Th.V. (ed.) (1980). The Role of Organic Soils in the World Carbon Cycle. Department of Energy, Conference 7903105.

Armentano, Th.V. and J. Hett (eds.) (1980). The role of temperate zone forests in the world carbon cycle - problem definition and research needs. U.S. Department of Energy CONF - 7903105, Washington, D.C.

Barretto, H.T. and C.P. Summerhayes (1975). Oceanography and suspended matter of northeastern Brazil. J. Sediment. Petrol. 45, 822-833.

Bolin, B. (1977). Changes of land biota and their importance for the carbon cycle. Science 196, 613-615.

Bolin, B., E.T. Degens and S. Kempe (1979). The global carbon cycle, Chichester/New York/Brisbane/Toronto.

Brinson, M.M. (1976). Organic matter losses from four watersheds
in the humid tropics. Limnology and Oceanography 21, 572-582.
Broecker, W.S., T. Takahashi, H.J. Simpson and T.H. Peng (1979).
Fate of fossil fuel carbon dioxide and the global carbon budget.
Science 206, 409-418.
Degens, E.T. (1981). Spekulation. Die Welt, January 21.
Gosz, J.R. (1980). Biomass distribution and production budget for
a nonaggrading forest ecosystem. Ecology 61, 507-514.
Grassl, H., S. Kempe and A. Spitzy (1981). Kohlenstoff: Klima-
wechsel, aber keine weltweite Katastrophe. Die Welt, March 21.
Hampicke, U. and W. Bach. (1980). Die Rolle terrestrischer Ökosysteme
im globalen Kohlenstoff-Kreislauf. Münstersche Geographische
Arbeiten 6, 37-104.
Klinge, H., W.A. Rodrigues, E. Brünig and E.J. Fittkau (1975).
Biomass and structure in a Central Amazonian rain forest. In:
Golley, F.B. and E. Medina (eds.), Tropical Ecological Systems,
New York/Berlin, 115-122.
Likens, G.E., F.H. Bormann, R.S. Pierce and W.A. Reiners (1978).
Recovery of a deforested ecosystem. Replacing biomass and nutrients
lost in harvesting northern hardwoods may take 60 to 80 years.
Science 199, 492-496.
Loucks, O.L. (1980). Recent results from studies of carbon cycling
in the biosphere. In: Schmitt, L.E. (ed.), Proceedings of the
Carbon Dioxide and Climate Research Program Conference, Washington
D.C., April 24-25, U.S. Department of Energy, CONF-8004 110 UC-11,
3-42.
Loucks, O.L. and Th.V. Armentano (1980). Problem Definition and
Feasibility Workshops. Terrestrial Biosphere Subcomponents on the
Global CO_2 Balance. DOE Report No DOE/ET/10040-4, The Institute
of Ecology, Indianapolis.
Lugo, A.E. (1980). Are tropical forest ecosystems sources or sinks
of carbon? In: Brown, S., A.E. Lugo and B. Liegel (eds.), The
role of tropical forests on the world carbon cycle. Gainesville,
Florida, 1-18.
Manabe, S. and R.T. Wetherald (1980). On the distribution of climate
change resulting from an increase in CO_2 content of the atmosphere.
J. Atmos.Sci. 37, 99-118.
Meybeck, M. (1980). River transport of organic carbon to the ocean.
In: Likens, G.E. and others (eds.), Carbon dioxide effects research
and assessment program: Flux of organic carbon by rivers to the
oceans. U.S. Dep. of Energy, Washington D.C. 219-269.
Miller, Ph. C. (ed.) (1981). Carbon balance in northern ecosystems
and the potential effect of carbon dioxide induced climatic change
(carbon dioxide effects research and assessment program), U.S.
Dep. of Energy.
Milliman, J.D., C.P. Summerhayes and H.T. Barretto (1975). Oceano-
graphy and suspended matter of the Amazon river, February-March
1973. J. Sediment. Petrol. 45, 189-206.
Mullholland, P.J. (1980). Deposition of riverborne organic carbon
in floodplains,wetlands and deltas. In: Likens, G.E. and others
(eds.), Carbon dioxide effects research and assessment program:
Flux of organic carbon by rivers to the oceans. U.S. Dep. of
Energy, Washington D.C., 142-172.
Odum, H.T. and R.F. Pigeon (eds.) (1970). A tropical rain forest.
U.S. Atomic Energy Commission. NTIS, Springfield, Va.
Richey, J.E., J.T. Brock, R.J. Naiman, R.C. Wissmar and R.F.
Stallard (1980). Organic carbon: oxidation and transport in the
Amazon River. Science 207, 1348-1351.

Siegenthaler, U. and H. Oeschger. (1978). Predicting future atmospheric carbon dioxide levels. Science 199, 388-395.

Stuiver, M. (1978). Atmospheric Carbon Dioxide and Carbon Reservoir Changes. Science 199, 253-258.

Wangersky, P.J. (1980). The fate of sedimented organic carbon in estuaries. In: Likens, G.E. and others (eds.), Carbon dioxide effects research and assessment program: Flux of organic carbon by rivers to the oceans. U.S. Dep. of Energy, Washington D.C., 294-313.

Wigley, T.M.L., P.D. Jones and P.M. Kelly (1980). Scenario for a warm, high CO_2 world. Nature 283, 17-21.

Woodwell, G.M. (1978). The carbon dioxide question. Scientific American 238, 34-43.

Woodwell, G.M., R.H. Whittaker, W.A. Reiners, G.E. Likens, C.C. Delwiche and D.B. Botkin (1978). The biota and the world carbon budget. Science 199, 141-146.

ENVIRONMENT AND HEALTH IMPLICATIONS
OF CANADIAN ENERGY POLICY

P. M. Stokes and J. B. Robinson

*Institute for Environmental Studies, University of Toronto, Toronto,
Ontario M5S 1A4, Canada*

ABSTRACT

Canada, in contrast to many industrialized countries, faces a rather wide choice in
her pursuit of energy policies. In the context of a large resource endowment,
several major energy resource projects have been proposed. Of these, the most
likely to proceed over the next decade include:
1. Frontier oil and gas exploration and development.
2. Tar sands and heavy oil extraction and upgrading.
3. Expansion of natural gas delivery systems to the East Coast.
4. Expansion of coal and uranium mining.
5. Expansion of coal fired and nuclear power plants.
6. Development or expansion of hydro electric schemes.
As well, various renewable resource projects have been proposed. The environmental
health impacts or risks for the conventional and nuclear projects have been widely
studied and while intagibles and indirect impacts continue to challenge researchers
and decision-makers, it can be claimed that the potential impacts of activities
1-6 can be reasonably evaluated at the project level. The impacts of the renewable
resource projects have enjoyed less attention in terms of objective evaluation but
are likely to be relatively low by comparison.

What is lacking is a methodology to assess the options at a general level, i.e.,
based on a comprehensive evaluation of the combined technologies. Compounding this
is the current dilemma over the future of Canadian energy policy. 1980 saw the
publication of a new National Energy Programme, significantly changing the
direction and focus of national policy but this has become embroiled in major
controversy over energy pricing and constitutional issues. This situation serves
to illustrate the complexity of the decision making process as it relates to energy
policy.

KEYWORDS

Canada; resource endowment; energy; environmental impacts; alternative strategies;
assessment methodology.

BEC 2 - V

INTRODUCTION

Canada's position on energy policy planning is paradoxical. Though endowed with many and diverse resources, including oil, natural gas, coal, uranium, water and wood, Canada is experiencing disputes between the federal and provincial governments over questions such as oil pricing, revenue sharing and taxation. As a result, the "energy problem" in Canada is not primarily one of resource supply but is one of policy.

The relative abundance of resources suggests that Canada has more choices than most industrialized countries concerning decisions on future energy development. In principle, these decisions can be made on the relative desirability of different mixes of supply sources or different policy paths. Because desirability should include evaluation of health and environmental impacts, there is a need for detailed assessments of the implications of energy strategies as packages or mixes rather than simply at the level of individual processes. As will become clear however, substantial difficulties exist in attempting to develop such assessments.

ENERGY RESOURCES

With respect to future energy supply and demand, Canada is in a secure position. There are substantial reserves of all fossil fuel resources, except conventional oil, and the prospect for future discoveries of reserves looks promising. For coal, measured reserves alone are estimated to be over one thousand times current annual production levels (Canada, Department of Energy, Mines and Resources, 1980a). Substantial reserves of conventional natural gas, approximately enough to last until the end of the century, exist in Western Canada (National Energy Board, 1979), and the results of deep drilling in unconventional formations in Alberta and of frontier exploration in the Arctic Islands and off the East coast have been encouraging. The tar sands and heavy oil deposits in Western Canada contain very large amounts of oil (Canada, Department of Energy, Mines and Resources, 1979) and a recent discovery south-east of Newfoundland is expected to provide much needed oil to the Atlantic provinces (Government of Newfoundland and Labrador, 1980). Existing hydro-electric capacity is large and can perhaps be doubled by the end of the century (Gander and Belaire, 1978), while substantial uranium reserves exist in Ontario and Saskatchewan (Canada, Department of Energy, Mines and Resources, 1980b). Finally, Canada has very large wood resources and studies have shown that a large proportion of Canadian energy demands could be met by renewable energy in the next century (Brooks, 1981).

However, with the possible exception of natural gas from Alberta, none of these future supplies is without its problems. Coal is dirty and expensive, and the supporting infrastructure necessary for its expanded use does not currently exist. Frontier oil and gas, and tar sands oil, are extremely expensive to develop and are hazardous to the environment. Nuclear power is subject to economic, environmental and political uncertainty while future large-scale hydro and biomass developments raise serious questions of environmental impact. Social considerations also restrict the rapid development of each major supply. Furthermore, the large scale of the projects required to develop these resources means that they can only be brought on stream as financial and labour resources permit.

All of this means that it is extremely important to plan future energy developments in order to minimize adverse health and environmental impacts. Planning should also ensure that the necessary resources are available to permit the development of such projects as they are required.

ENERGY PRODUCTION AND CONSUMPTION

In 1978, about 7.7 Exajoules[1] of primary energy were produced in Canada; about
2.5 EJ were exported. Canadians consumed about 7.4 EJ of primary energy, equivalent to 0.31 Terajoules[2] per person, one of the highest national per capita energy
consumptions in the world (Statistics Canada, 1980). The breakdown of how that
energy was produced and consumed in each province of Canada is shown in Figs. 1
and 2.

These Figs. illustrate the great regional diversity across Canada with respect to
both production and consumption of energy. They underline clearly the basis of
contemporary political problems over energy issues. For example, the province of
Alberta currently contributes about 71% of total Canadian production (over 85% of
oil and natural gas) but uses only 12% of total secondary consumption. Ontario
and Quebec, on the other hand, contribute only 7% of production but make up 60% of
Canadian energy consumption. The three Atlantic provinces of Nova Scotia, New
Brunswick and Prince Edward Island produce less than 1% of the Canadian total but
consume 6% of secondary consumption. This is mostly oil, all of which is imported.

From a resource aspect, the main short-term energy problem facing Canada is the
heavy reliance of Eastern Canada upon imported crude oil. Conventional supplies
of crude oil in the Western provinces are dwindling, and although very large tar
sands resources exist in the west, these can only be brought on stream fast enough
in the 1980's to meet demands in Western and Central Canada. To get Canadians
"off-oil" is thus a major concern of current federal policy (Canada, Department of
Energy, Mines and Resources, 1980c).

THE BASIS OF PLANNING

In Canada, the provinces own the energy resources within their borders and have
almost complete jurisdiction over supply and pricing of electricity. The strong
political decentralisation with respect to jurisdiction over those resources is
matched by the strongly regional character of energy supply and demand. National
energy policy planning can only succeed with inter-provincial co-operation as well
as federal-provincial agreements. At present, federal-provincial disputes over
energy pricing and revenue sharing are proving barriers to such cooperative
policy (Canada, Department of Energy, Mines and Resources, 1980c). While a large
number of major   energy resource projects have been proposed, it is still
unclear exactly what combination of them will be undertaken.

Table 1 contains a list of major proposed Canadian energy projects. Each of these
can be expected to have significant environmental and/or health impacts. Only
those projects which the authors consider likely to proceed within the next decade
are included but even with this restriction, it is clear that a large number of
major energy projects are expected to be undertaken over the next ten years.

Substantial problems exist, however, in trying to assess the environmental and
health impacts of such projects. It is neither possible nor appropriate in the
available space to itemise health and environmental effects of each type of energy
producing process; these have been discussed at length elsewhere (eg. Budnitz and
Holdren, 1976).

---

[1] Exajoule (EJ) - 10^{18} Joules

[2] Terajoule(TJ) = 10^{12} Joules

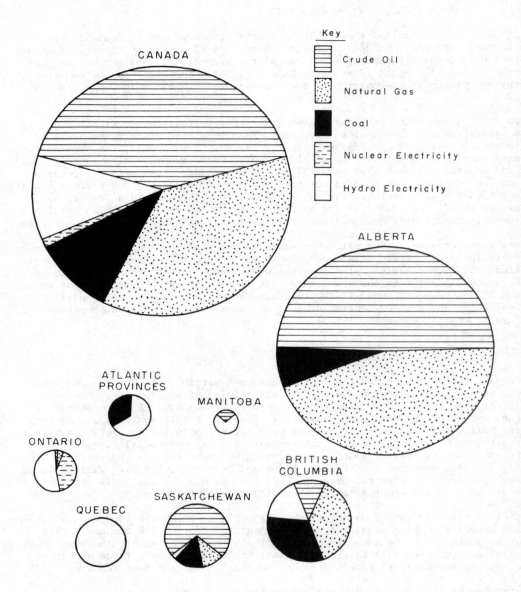

Fig. 1. Primary energy production: Canada and the provinces, 1978

Source: Statistics Canada Quarterly Report on Energy Supply-Demand in Canada,
          1978-IV, Ottawa, 1980.

Notes: Hydro and nuclear electricity valued at 3.6 MJ/KW.h
          Non-market energy (solar, wood, etc.) excluded
          Gas plant LPG's included in Natural Gas

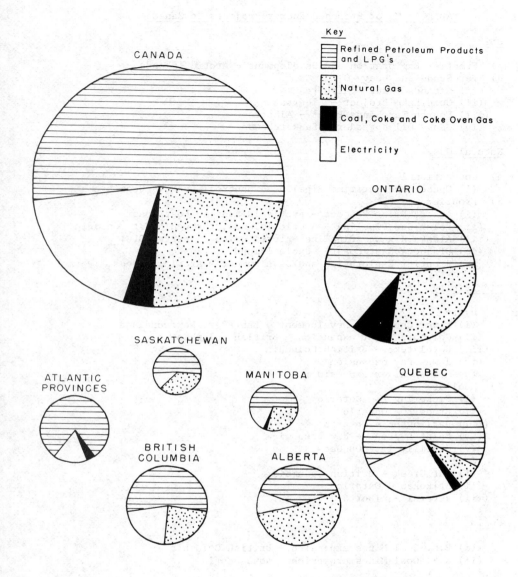

Fig. 2. Secondary energy consumption by source: Canada and the provinces, 1978

Source: Statistics Canada Quarterly Report on Energy Supply-Demand in Canada, 1978-IV, Ottawa, 1980.

Notes: Electricity valued at 3.6 MJ/KW.h
        Non-energy use included
        Non-market energy (wood, solar, etc.) excluded

TABLE 1 Major Proposed Energy Projects in Canada

1. Oil

    a) Frontier Exploration and Development – Arctic Islands
    b) Tar Sands and Heavy Oil
       (i) Alsands Plant – Alberta
      (ii) Cold Lake Project – Alberta
    (iii) Fourth Tar Sands Plant – Alberta
     (iv) Heavy Oil Upgrader – Saskatchewan

2. Natural Gas

    a) Conventional
       (i) Quebec and Maritime Pipeline – Montreal to Nova Scotia
    b) Frontier
       (i) Arctic Pilot Project – Arctic Islands to East Coast
      (ii) Polar Gas Pipeline – Arctic Islands, through NWT to Ontario
    (iii) Alaska Highway Pipeline – Territories, Alberta, B.C.
     (iv) Offshore East Coast – Sable Island
      (v) Frontier Exploration and Development – Arctic Islands

3. Electricity

    a) Hydro
       (i) Lower Churchill Development – Labrador, Newfoundland
      (ii) Columbia River expansion – British Columbia
    (iii) Revelstoke – British Columbia
     (iv) James Bay expansion
      (v) Nelson River expansion
    b) Nuclear
       (i) Pickering B – Ontario
      (ii) Bruce B – Ontario
    (iii) Darlington – Ontario
     (iv) Point Lepreau – New Brunswick
      (v) Gentilly II – Quebec
    c) Coal
       (i) Hat Creek – British Columbia
      (ii) Atikokan – Ontario
    (iii) several – Alberta

4. Coal

       (i) B.C. Coal Mines expansion – British Columbia
      (ii) N.S. Coal Mines expansion – Nova Scotia

5. Uranium

       (i) Rio Algom and Denison expansion – Ontario
      (ii) Cluff Lake and Key Lake mining – Saskatchewan

There are several situations which are of particular relevance to Canadian policy, either for reasons of climate or geography, or because of the nature of the resource per se. A small number of these have been selected as examples of the often complicated nature of the assessment process. The balance of the paper will discuss the process of assessing alternative energy strategies; in other words different mixes of supply sources.

ACIDIC PRECIPITATION RELATED TO COAL FIRED GENERATING PLANTS

In Canada and the United States, susceptible aquatic ecosystems are showing dramatic changes in chemistry and biology believed to be attributable to acid deposition (Likens and others, 1979). Soils are liable to be leached by acids, with
resulting changes in the receiving waters and the fertility of the soils themselves. The latter is likely to affect the long term growth of forest trees and
possibly crops as well. The sources of the acid deposition are multiple. The
phenomenon is related to oxides of sulphur and nitrogen emitted by thermal generators, non-ferrous smelters, industrial fuel combustion as well as transportation.
As shown in Table 2, thermal generators contribute 67% of U.S. sulphur oxide
emission and 15% of Canada's sulphur oxide emissions and 58% of the total S emissions for both countries are attributable to utilities.

TABLE 2 Current Emissions in the U.S. and Canada (in 10^6 Tons)

	USA (1980 Estimate)		CANADA 1979 (Inco 1980)		TOTAL	
	NO_x	SO_x	NO_x	SO_x	NO_x	SO_x
Utilities	6.2	19.5	0.3	0.8	6.5	20.3
Non-Ferrous Smelter	0.0	2.0	0.0	2.2	0.0	4.2
Transportation	9.0	0.9	1.1	0.1	10.1	1.0
All Others	7.1	7.3	0.8	2.2	7.9	9.5
Total	22.3	29.7	2.2	5.3	24.5	35.0

Source: Modified from US/Canada MOI
                Interim report February, 1980

Local effects of thermal generators are not too difficult to document and quantify.
Particulate and volatile metals, fluorides, oxides of sulphur and nitrogen and
radioactive materials are emitted. The local air quality is monitored for various
air pollutants and operations are required to conform to air quality criteria or
standards. If these criteria and standards are sound, and regulation is effective,
human health as well as plant and animal life are protected. A dose-effect
relationship can be determined on a local scale.

For air pollutants which are carried far from the source ie., these portions of
SO_x, NO_x and other gases particulates and aerosols which are not deposited locally,
the impacts are more difficult to assess and often impossible to attribute to the
original source. Furthermore, legislation put in place to protect local air
quality may be inadequate to control long range transport of pollutants. For
acidic deposition into the susceptible lakes on the Laurentian Shield in Canada,
contributions to acidity originate from both Canadian and U.S. industrial processes, but the relative contribution of each country to the acidity of the rain in
a specific region is the subject for continued debate. (Ontario Ministry of the
Environment, 1981)

Recognition of the truly international nature of this air pollution problem cul-
lminated with the signing of a US-Canada Memorandum of Intent in 1980. The
Memorandum has as its stated purpose:
    "To establish technical and scientific work groups to assist in preparation
       for and the conduct of negotiation on a bilateral transboundary air pollut-
       ion agreement."

The political and legal problems related to transboundary movement of air pollutants represent even more formidable barriers to resolution than do the scientific
questions which are still unresolved. Over the past decade or so, a great deal
has been learned about the long range transport, transformation and effects of air
pollutants in the context of acidic precipitation especially (eg. Drablĝs and
Tollan, 1980).

In the context of electricity generation, a cause-effect relationship for acidification of ecosystems is often impossible to establish for the following
reasons:
1. Absolute proof of acid related damage exists for relatively few instances and
these are in specific freshwater systems; while long term projections based on
current rates of change for lakes are regarded as sound and scientifically based,
there are still many uncertainties. This is partly because of the lack of
historical data, but also is a function of the individual characteristics of
specific study lakes and the fundamental nature of scientific investigation.
2. Negative impacts of acidic precipitation on terrestrial systems have not been
documented to the satisfaction of the scientific community; rates of change are
likely to be extremely slow in comparison to those occurring in aquatic systems,
and trends may be overwhelmed by other macro-scale variables such as climatic
factors and local events such as microbial or insect infestations.
3. The loss of a natural resource such as a lake, except for its fishery, cannot
easily be built into a cost benefit analysis, particularly when the resource is
remote from the source of air pollution and even separated from it by a national
boundary.
4. Human health effects resulting from acidic precipitation are still the subject
of speculation; there is the potential for direct effects from acid aerosols and
indirect effects from acidic drinking water which will solublise metals.
5. Thermal generation is not the only contributor to acidic precipitation;
relative contributions have to be estimated from models which use production
(effluent) numbers and estimate loading rates over long range. Furthermore,
additional estimates are required for different types of receiving surfaces.

This example shows how the evaluation of impacts from one specific process challenges the scientific community, even though considerable attention and funding
has been directed towards the problem.

For renewable or unconventional energy sources, we are even less well equipped to
evaluate environmental and health impacts.

IMPACTS OF SOME RENEWABLE ENERGY SOURCES

Renewable energy sources, such as sunlight, wind, falling water and biomass are
often perceived as benign or possessing "more attractive environmental characteristics than fossil fuel or nuclear fission alternatives" (Holdren and others,
1980). The literature on the actual or potential impacts of these sources is
modest in comparison with that on the conventional sources. Yet even a cursory
review of the processes and activities involved in the renewable areas indicates
that land use and production of large structures such as solar collectors, arrays
of windmills and dams can disrupt ecosystems, divert land from other significant
uses such as agriculture, and affect the health and safety of humans in the work
place.

Furthermore, most processes involving renewables, by their very nature, are
intervening in the natural hydrologic and energy fuel cycles of the biosphere
(Harte and Jossby, 1978). In a discussion of the environmental implications of
conventional, nuclear and renewable energy options for Ontario, Robinson (1977)

has emphasised the need for an increased level of effort (and therefore funds) to evaluate the impacts of less sophisticated small scale technologies as well as energy conservation. In comparison with the research and development effort devoted to nuclear technology, particularly in a relatively small market such as Canada, there has been scarce attention paid to alternative technologies. We are therefore not in a position to make decisions or choices based on the protection of the natural environment and human health.

"Concentration on high cost technological options of this nature tends to lock the future into a technological fix which may not be appropriate in the context of changing future realities such as the possibility of evolving social values. For example, there are indications of a shift to values which seek expression in a more decentralised society based on a smaller-scale, renewable energy resource technology, which could have substantial environmental benefits in the long term" (Robinson, 1977).

ALTERNATIVE ENERGY STRATEGIES

"The assessment of environmental impacts resulting from different stages of the fuel cycle is important in relation to policy-making and decisions about energy options or 'mixes' to be developed" (UNEP, 1979).

Any assessment tool which is designed to cover alternative strategies, and not just individual processes, has to deal with problems such as those already outlined for specific processes, but also with the combination and integration of these. The assessment method must be comprehensive. It should consider all significant aspects of the environment which could be affected, but should also strive to achieve balance - ie. the various components of the environment receiving impacts should be given a degree of attention commensurate with their importance. These will vary with the geographical region concerned. For Canada, Arctic development presents special problems of environmental sensitivity, technical aspects related to climate and temperature, and frequently social and health problems for native people. (Pimlott and others, 1972)

The assessment method should strive for a common currency in terms of quantitative evaluation of impacts of different components of the energy mix. For example, in Jain and Hutchings (1978) basic concepts are outlined for utilising energy and energy flow as a common denominator for analysing the impacts of different energy production technologies. Holdren (1980) suggests that comparisons should be based on environmental damage per unit of energy benefit.

CONCLUSIONS

A methodology to determine the environmental health and social impacts of various mixes or strategies of energy is the ideal goal for assessment of any energy policy. In Canada, while the choices are rather broad, decisions relating to energy policy are complicated by political problems. Furthermore insufficient emphasis has been placed on the impacts of the less conventional methods of energy generation. At present one has to conclude that while a methodology framework can be developed, a realistic assessment is only feasible on a fairly local and site specific scale.

REFERENCES

Brooks, David B. (1981). Zero Energy Growth for Canada. Toronto, McClelland and
    Stewart.
Budnitz, R., and J.P. Holdren (1976). Social and environmental costs of energy
    systems. Ann. Rev. Energy, 1, 533–580.
Canada, Department of Energy, Mines and Resources (1979). "Canadian Oil and Gas
    Supply/Demand Overview". Ottawa.
Canada, Department of Energy, Mines and Resources (1980a). "Discussion Paper on
    Coal", Report EP80-IE. Ottawa.
Canada, Department of Energy, Mines and Resources (1980b). Uranium in Canada - 1979
    Assessment of Supply and Requirements.
Canada, Department of Energy, Mines and Resources (1980c). The National Energy
    Program. Ottawa.
Drabløs, D. and A. Tollan (1980). Ecological Impact of Acid Precipitation. SNSF
    Project.
Gander, James and Fred Belaire (1978). Energy Futures for Canadians. Long-term
    Energy Assessment Program, Department of Energy, Mines and Resources, Report
    EP78-1.
Government of Newfoundland and Labrador (1980). "Submission to the National Energy
    Board", Hearing EHR-1-80, St. John's, Newfoundland.
Harte, J., and A. Jassby (1978). Energy technologies and natural environments:
    the search for compatibility. Ann. Rev. Energy, 3, 101–146.
Jain, R.K., and B.L. Hutchings (1978). Environmental Impact Analysis. University
    of Illinois Press, Urbana, Ill.
Likens, G.E., R.F. Wright, J.N. Galloway and T. Butler (1979). Acid Rain. Sci.
    Amer., 241, 43–51.
M.O.I. (1980). United States - Canada Memorandum of Intent on Transboundary Air
    Pollution. June 1980.
M.O.I. (1981). United States - Canada Memorandum of Intent Interim Report.
    February 1981.
National Energy Board (1979). Canadian Natural Gas-Supply and Requirements. Ottawa.
Ontario Ministry of the Environment (1981). The Province of Ontario: A submission
    to the United States Environmental Protection Agency opposing relaxation of
    SO_2 emission limits in state implementation plans and urging enforcement.
    130.
Pimlott, D.H., K.M. Vincent and C.E. McKnight (1972). Arctic Alternatives.
    Canadian Arctic Resources Committee, Ottawa, Ontario. 391.
Robinson, D. (1977). Comparing and assessing the environmental implications of
    conventional, nuclear and renewable energy options for electric power
    production in Ontario. Unpublished report, Environment Canada.
Statistics Canada (1980). Quarterly Report on Energy Supply-Demand in Canada,
    1978-IV. Ottawa, Supply and Services, Canada.
UNEP (1979). United Nations Environment Program. The environmental impacts of
    production and use of energy. UNEP-ERS-1-79, ERS-2-79, ERS-3-79, Nairobi.

ENVIRONMENTAL PROBLEMS BY ENERGY PRODUCTION IN DENMARK

P. B. Suhr

Energy Division, National Agency of Environmental Protection, Denmark

As a result of the energy crisis in 1973-74, most countries found it necessary to revise their existing energy policy.

One of the main objectives of the energy plan prepared by the Danish Ministry of Commerce in 1976 was to reduce the vulnerability of Denmark in the energy supply sector and, especially, to reduce our dependence on oil supplies, with due regard to the objectives set for the economic and social development and to the environmental quality requirements.

One of the instruments in this procedure was the creation of a diversified energy supply system. This implied i.a. a certain shift from oil to coal, and the introduction of new energy sources, i.e. nuclear power and natural gas. Moreover, the plan envisaged a limited use of renewable energy, for instance solar and wind power.

Many were surprised in Denmark to see that the shift from oil to coal in Danish power plants took place at such a rapid speed.

In the Ministry's 1976 plan, the pre-condition was set up that the 1985 consumption of coal would increase by approximately 60% as compared to 1975. As a result of the introduction of nuclear power envisaged in the plan, the consumption of coal would decrease after 1985, and in 1995 correspond approximately to the 1975 figure.

The actual increase in the consumption of coal in Danish power plants in the period 1975-79 was about 125% - from app. 2.7 million tons to app. 6.o million tons. To-day more than 8o% of the power production is based on coal, and the present total coal consumption in Denmark is well above 8 million tons/year.

Such a violent shift in such short time greatly affects the physical surroundings - the environment.

In connection with the follow-up of the 1976 energy plan, the Ministry of the Environment in April 1977 prepared a detailed program for the development and planning activities in the field of energy and environment.

The need was pointed out to initiate detailed studies of the impact of an increased consumption of coal in power plants, on the environment in general, and, in particular, on the atmospheric concentration of particles and sulphur dioxide. It was stated that by postponing the decision in principle regarding the introduction of nuclear power in Denmark, the consumption of coal for power production might increase further. The results of the studies were published at the turn of the year in 18 reports with the overall title "Coal Consequence Study".

A summary of the report was published by the Ministry of the Environment in a report entitled "Kulkonsekvensundersøgelse - luftforureningsmæssige konsekvenser af kulfyring på danske kraftværker" ("Coal Consequence Study - Consequences of Coal Firing in Danish Power Plants").

The main aim of the study was to investigate and assess the emission of pollutants from coal-fired power plants, in order to create a part of the basis of decision underlying the requirements and conditions laid down by future environmental approvals of coal-fired power plants.

In connection with the study, measurements were made at a Danish power plant (Studstrup), using both oil and coal. The measurements were made by firing of various types of coal. Further, emissions by coal firing and by oil firing were compared.

Moreover, analyses have been made of particles, in order to determine the grain size distribution and the contents of elements. Stressing mainly the emission values recorded, a number of calculations were made of the atmospheric concentration and deposition at selected power plants. The results have been used to estimate the impacts on health and the environment by a shift from oil-firing to coal-firing in Danish power plants.

In connection with the study, an investigation has also been made of the emission conditions to be expected by power production in the 1980's, based on the latest forecasts available.

As regards sulphur dioxide, it is a question whether, in the course of the 1980's, desulphurization equipment shall be installed in Danish power plants.

If only national considerations are made it is probable that, given the preconditions used, the coal consumption increase in the next few years may take place without a general obligation to install desulphurization equipment.

In the report the conditions in a number of other countries are described; for instance in the USA and Germany, where desulphurization equipment shall be installed in major power plants. However, it was found that the conditions in Denmark differ from these countries: the industrial structure, the meteorological conditions, lime in the soil and in lakes, and the air quality.

As regards air quality, the values recorded by measurements of sulphur dioxide in the Copenhagen area have followed a downward tendency. In 1969 the annual mean value was 60 $\mu g/m^3$, and in 1979

29 $\mu g/m^3$, or 5o% less. It is interesting to notice that a WHO working group has found that by annual mean values of sulphur dioxide
above loo $\mu g/m^3$, long term effects, with several symptoms, may occur. The working group also found that an annual mean value of
4o-6o $\mu g/m^3$ is sufficient to protect public health.

One of the elements to be considered in connection with sulphur dioxide is the acidification of the precipitation recorded in recent
years, attributed i.a. to the increased sulphur dioxide emission.

In spite of the reduction of the total sulphur emission in Denmark
in the 197o's, (which I shall explain in more detail later), the
average pH-value of the precipitation has fallen from 5.5 to 5.o
in the period 197o-77. Apart from unfavourable effects in a few
lakes of scenic value in Central and Western Jutland, the increased
acidification we have seen lately seems not to have caused major
problems in Denmark. The content of lime in most Danish lakes is
so high that they are able to neutralize the acid introduced; and
large quantities of lime are already added to agricultural soil in
order to neutralize the acidification resulting from fertilization.

In 1977 the power plants contrituted about 45% of the total sulphur
dioxide emission in Denmark. From calculations of the expected development of the sulphur dioxide emission from Danish power plants
appears that the emission will increase considerably in the course
of the 198o's, unless desulphurization equipment is installed. It
should be noticed, however, that if waste heat from power plants,
and natural gas, are used for heating instead of oil, as envisaged
in the Danish heating planning, the total annual emission of sulphur dioxide will be reduced in the 198o's, even if the power production follows the estimated increasing curve. The fall expected in
the 198o's is a continuation of the downward tendency in the 197o's.
One reason why the emissions were reduced in the 197o's is the limits, introduced long ago, to the content of sulphur in fuel oil
and heavy fuel oil. Another is the shift from oil to coal in large
power plants. And a third reason, the increased use of district
heating from power/heating plants.

These are some of the considerations on the basis of which the conclusion was drawn concerning sulphur dioxide that if account is
taken only of national conditions, and given the preconditions chosen, there is every probability that the growth of the coal consumption in the next few years may take place without general rules for
desulphurization equipment in Danish power plants.

Several considerations may, however, in the course of the 198o's
make it necessary to require the installation of desulphurization
equipment in major power plants. First, it remains to be seen whether sufficient quantities of low-sulphur coal may be purchased in
the future. Secondly, in special cases it might prove impossible
to observe the guidelines on the permissible sulphur dioxide pollution contributed by the individual power plants.

The national conditions should, however, be supplemented by considerations for our neighbouring countries, and the acidification of
soil and lakes, with the result that desulphurization equipment may
be made obligatory in Danish power plants.

To stress the importance we attach to this problem I want to mention 2 conventions with the aim of reducing transfrontier air pollution:

First, a convention prepared by the UN Economic Commission for Europe (EEC), signed by Denmark in 1979, and secondly, the Nordic Environmental Protection Convention, which entered into force in 1976.

Therefore, if in the light of both national and international considerations I am to answer the question: will desulphurization equipment be required in Danish power plants, the reply is: the NAEP is at present of the opinion that the development points to desulphurization requirements in certain plants in the course of the 198o's.

The question arose in practice for the first time when the NAEP considered the application for an extension of the Studstrupværk by "Blocks III and IV". In this connection the conditions regarding a maximum sulphur emission were limited in time, and may be amended in a period up to 5 years after Block IV starts operations; however, as regards Block III, not before it has been in operation for 3 years. Consequently, the plant must accept to install desulphurization equipment that may be required, and reserve the necessary space for it.

The question of desulphurization equipment or not, was accentuated in connection with the environmental approval of the planned Block III at the Amagerværk. The Swedish Environmental Protection Agency filed a complaint in pursuance of the Nordic Environmental Protection Convention, arguing strongly that because of the acidification of the Swedish soil and lakes, desulphurization equipment should be required at Amagerværket.

As regards nitrogen oxides, Denmark, just like several other countries, will only be able to reduce the emission of nitrogen oxides from power plants by improved firing technology.

The emission of nitrogen oxides from power plants represented about 3o% of the total emission in Denmark in 1977. However, the nitrogen pollution from power plants does not contribute significantly to the total nitrogen oxide pollution in Danish urban areas. What matters most in this connection is the pollution from motor vehicles.

From the Coal Consequence Study appears quite clearly that the guiding limits laid down by WHO concerning the total nitrogen oxide pollution are sometimes exceeded in certain Danish urban areas. This gives rise to great concern, and the NAEP plans to assess the possibility of establishing maximum limit values to the emission of nitrogen oxides from major new plants, with due regard to possibilities offered by the most advanced firing technology.

As regards suspended particles, I mentioned earlier that the study also covers measurements of the particle emission from Studstrupværket, and analyses of trace elements both by coal firing and oil firing. The particle emission situation in the 198o's is also calculated.

The levels of the emission of particles, including trace elements in the 1980's, do not warrant the introduction of guiding limit values that are more stringent than the values set by the NAEP.

To round off my account of environmental problems related to coal and power generation, let me mention the research and investigation activities the NAEP is involved in concerning various aspects of coal. A few examples:

We are at present carrying out a study of the use of coal in medium-sized coal-fired plants, e.g. power/heating plants, in district heating plants and in major industrial plants. Characteristically, such plants are located in or close to urban areas, and this results in special environmental problems. The study comprises part projects concerning environmental problems by coal storage, an assessment of the technical and economic aspects of flue gas cleaning to remove particles and sulphur dioxide, and an estimate of the emission and deposition of gases and particles from coal-fired plants.

Another example is a development project launched by the Nordic Council of Ministers concerning the deposition of fly ashes from coal-fired power plants. The main object is to provide knowledge on which to base the establishment of conditions for the deposition of fly ashes.

As already mentioned there has been a comprehensive shift in Danish power plants from oil to coal. Table 1 illustrates the development in energy consumption distributed on fuel types in the period 1972-80.

Table 1   Energy consumption distributed on fuel types in Denmark 1972-80 (in PJ).

	1972	1976	1980
Oil	717	627	532
Coal etc.	54	118	243
Garbage etc.	3	7	11
Net imports of electricity	8	3	–
Total	766	755	786

It appears from the table that more than 65% of the Danish energy consumption is still based on oil. Table 2 shows the use of energy as distributed on sectors.

Table 2   Energy use distributed on sectors in 1980 (in %).

	%
Non-replaceable electricity consumption	24
Room heating including electric heating	38
Process consumption	20
Transportation	18

It appears that more than 1/3 of Denmark's energy consumption is used for room heating, approx. 6o% of which is based on oil consumption.

It is in the light of these figures that we intend to introduce comprehensive heating planning in Denmark. The main object of heating planning is to carry out a drastic expansion with heat from heat/power stations and to introduce natural gas from the Danish section of the North Sea at an annual supply figure of 2.5 mia. m^3. With this part of the heating planning, which is expected to be carried through in 1988, 2/3 of the Danish energy consumption for heating purposes will be covered by collective heating systems.

The heating planning will have a number of favourable environmental effects. The shift from individual oil-fired boilers to power plant heating will bring about an increase in the efficiency of the total energy system, which again favourably affects the emission of air-polluting substances. In addition to this it will be easier to take emission-restricting measures, and local pollution may be prevented by using high stacks.

The introduction of natural gas will contribute to a reduction in pollution, first and foremost because natural gas has considerably lower emission factors than gas oil.

There still remains a number of unanswered questions in the Danish energy policy. The Danish Government had originally planned to present a new energy plan in the spring of 1981 (EP-81), but the preparations were so comprehensive that the presentation had to be postponed to the autumn of 1981. EP-81 is being prepared under the supervision of the Ministry of Energy.

EP-81 is intended to cover a period of 5o years and is divided as follows:

1. short view .... o - 5 years (1981 - 1985)
2. medium view .... 5 - 2o years (1986 - 2ooo)
3. long view .... 2o - 5o years (2oo1 - 2o3o)

On the long view (period 3) the main object of EP-81 will be to show whether sufficient energy resources are available which can be used in an environmentally acceptable manner and within reasonable economic limits.

In contrast to the former energy plan of 1976 EP-81 will present a spectrum of possible major decisions within the energy sector.

In EP-81 the main stress will be laid on an examination of the effects that economizing on energy will have on our energy resources and economy in general, meaning that energy-saving measures in the electricity and heating sectors should be combined with a more rational utilization of energy e.g. in industrial processes.

In addition to this, EP-81 will examine the social consequences of a variety of alternative possibilities for the future energy supply system in Denmark. The principal alternatives will be:

- Electricity production with and without nuclear power.
- Considerably increased natural gas supply from the end of the 1980's or the beginning of the 1990's.
- More priority to renewable energy.

The economic evaluations of supply and consumption alternatives also include an evaluation of the environmental consequences. During the summer the Ministry of the Environment shall prepare a report which is intended as a supplement to EP-81. For the number of examined energy supply systems it will include figures for the emission of sulphur dioxide, nitrogen oxides and particles.

In addition to this, the amounts of clinkers and fly ashes will be quantified. On the basis of these calculations the various alternatives for consumption and supply will be graded in accordance with environmental quality requirements.

Independent of the EP-81 report the NAEP has set to work on a report on safety of nuclear power plant operation. The report, which should be finished before the end of 1982, will inter alia contain the attitudes of the Danish authorities to the safety question seen in the light of experience reaped since the large plants were put into operation about 15 years ago.

In this connection I can mention that the safety report also will discuss which consequences the location may have for the safety of the population in case of accidents, especially in relation to the planning of emergency measures. In continuation of this investigation there will be a re-evaluation of the 15 sites now reserved for nuclear power plants.

I shall end my account be saying something about how the Danish authorities intend to follow the development in air quality.

According to the Danish Environmental Protection Act, County Councils, Local Authorities, and the NAEP shall obtain information about the air quality on a local and a national scale. Local Authorities and County Authorities shall inter alia apply this information in connection with approval of polluting commercial undertakings, also including power plants, in order to keep the local air quality on a satisfactory level.

In addition to this, the County Authorities are to map out sources of pollution and the condition of pollution with a view to regional environmental quality planning. Finally, the NAEP shall follow the developments on a national scale.

Local Authorities and County Authorities need effective tools for planning and administration in the field of air pollution. The NAEP has consequently co-operated with the local organizations on the preparation and implementation of a project concerning registration of air quality and atmospheric emissions in Danish urban areas - the so-called nation-wide air quality measuring programme.

The project is in its initial phase and will cover a period of 5 years with measuring and model calculations for 7 urban areas in selected county boroughs. The model calculations will also require the preparation of meteorological data set and emission reports. The project is expected to yield results which can be used in planning and administration for the entire country. In addition to this, the specific participating areas will obtain information about their immediate air pollution situation and its line of development.

THE LURGI RUHRGAS (LR) PROCESS AN ENVIRONMENTALLY AND ECONOMICALLY SOUND OIL SHALE RETORT

H. Weiss and F. Gonnert

Lurgi Kohle und Mineralöltechnik GmbH, Frankfurt am Main, Federal Republic of Germany

ABSTRACT

After a basic classification of the different types of oil shale retorts, a short summary is given of Lurgi's LR retort history.

The LR-Process is described and particular attention is given to its special feature which is the circulation of retorted residue as solid heat carrier with spent carbon combustion for generation of retorting heat.

Finally, some of the environmentally and economically relevant LR-Process reactions affecting emissions and product qualities are described.

KEYWORDS

LR-Process; oil shale; emissions; special reactions; oil shale products.

Basic Classification of Retort Types

Retorting oil shale entails the heating of a large mass of non-productive rock to pyrolyze the contained kerogen. The principal aim in retort design is to accomplish this heating at a low capital cost using the least amount of energy. Thus the method of transferring heat to the raw shale provides a convenient way to classify retorts. Three different methods of heat transfer may be distinguished.

CLASS 1 RETORT

Heat is transferred to the shale through a wall. A simple example of this type of retort is the Fischer assay apparatus used to measure shale quality. The Pumpherston retort mainly used before World War II is a typical example for a Class 1 oil shale retort. There are presently no major developments based on this type due to the expensive heat transfer and low thermal efficiency of this method.

CLASS 2 RETORT

Heat is transferred to the shale by passing hot gases through the shale bed. This class can be subdivided into:

- Class 2 a (internal combustion mode) whereby the hot gases are generated in the retort by combustion of residual carbon and a portion of the hydrocarbon product. Examples of this type are the Paraho retort, the Union A retort, the Superior retort, the Lurgi Spuelgas kiln, the Lurgi Schweitzer kiln and the Lurgi Hubofen process. This method of heat transfer is inexpensive and allows relatively low-cost retort construction. However, the yield of hydrocarbons is lower, and the retort gas is largely diluted by combustion products and nitrogen from the combustion air.

- Class 2 b (external combustion mode) whereby the gases are externally heated in a gas heater by combustion of a portion of the hydrocarbon product. Examples are the Petrosix retort, the Union B retort, the Lurgi Tunnel kiln and a variation of the Paraho retort. These retorts do require an external gas heater which may be expensive and cause operating problems. The retort has a higher hydrocarbon recovery from the shale, and the gas from the retort does not contain combustion products. On the other side the thermal efficiency is reduced as the residual carbon in the spent shale remains unused.

Common to most Class 2 retorts is their inability to process shale fines and their increasing problems with uniform shale size and gas distribution through the retort shaft area with increasing retort size, i.e. capacity.

CLASS 3 RETORT

Heat is transferred by mixing hot solid heat carriers with the fresh shale. This class can also be subdivided into:

- Class 3 a where the heat carriers are alumina or ceramic balls which are circulated and reheated by combustion of a part of the hydrocarbon product. Examples of this type are the Tosco II retort and the initial Lurgi Ruhrgas retort.

- Class 3 b where the heat carrier is the shale ash which is circulated and reheated by combustion of residual carbon. An example for this class is the current Lurgi Ruhrgas retort.

All Class 3 retorts have high oil yields, produce an undiluted gas and have no minimum shale size restrictions. An additional advantage of Class 3 b retorts is the high thermal efficiency due to the utilization of the residual carbon as retort fuel.

Lurgi's Oil Shale Retorting Experience

Some fifty years ago, Lurgi extended its coal carbonization activities into the retorting of oil shale and has taken an active part in the further development of the technology ever since. Several processes were developed over the years to a commercial stage and process contributions to all Class 2 and 3 retorts including in situ retorting were made. These developments, which finally led to the Lurgi Ruhrgas processes, are very briefly described below.

THE SPUELGAS KILN

The kiln is of the double-rectangular-shaft type and incorporates a drying, retorting and residual carbon gasification zone through which the shale travels from top to bottom. The shale is dried and retorted by recycling product gas

(Spuelgas) in counterflow to the shale. The Spuelgas is internally preheated by cooling the spent shale in the lower shaft section and finally superheated by mixing with hot combustion gases from an external product gas combustion chamber. This kiln yielded a highly diluted gas and reached an oil yield of up to 93%.

THE TUNNEL KILN

The oil shale is retorted in wagonettes moved forward periodically through a tunnel. The heat carrier in this case is retort gas which is passed through tubular heaters and then through the shale bed in the wagonettes. The kiln is particularly suitable for very high grade shales which temporarily become plastic during retorting. The oil yield from this Class 2 b retort reaches 95% of the Fischer assay.

THE LURGI-SCHWEITZER PROCESS

The principle of this process is the retorting of lumpy low grade oil shale in a descending gas stream. This retort consists of a cylindrical shaft with no refractory lining and is operated batchwise. After charging, the upper shale layer is ignited like a pipe and air is sucked through the shale bed from top to bottom whereby the combustion zone moves slowly downwards. The hot gases retort the shale below the combustion zone so that the heat required for the process is produced only by combustion of the residual carbon. On completion of the retorting process, the retort is tilted, discharged and subsequently recharged. This process gives a maximum oil yield of 95% of that indicated by the Fischer assay.

THE HUBOFEN

The original idea was to develop retorts with circulating grate but this operating method proved to be too complicated and too costly. This led to the development of the 'Hubofen' consisting of an inclined oscillating grate on which the shale is moved constantly and slowly downwards. Air is sucked through the shale bed creating an inclined combustion and retorting zone in such a way that retorting is just completed at the grate end.

This retort is remarkable because of its simple set-up, its substantially open construction and its favourable heat economy.

IN-SITU AND IN PILES RETORTING

During the Second World War, Lurgi participated also in the efforts towards the in-situ retorting of oil shale in Württemberg (West Germany). Chambers were formed in the shale seam by parallel adits. The lower layer of the seam was mined and the upper part of the seam was blasted by explosive charges. The retorting process was initiated by suction in one adit and, after ignition of the full chamber length, retorting proceeded to the parallel adit. Due to a lot of infiltrated air, oil recovery was low and the hazard of explosions great. Furthermore, it was not possible to enter the adits later on as the hazards of poisoning and explosion could not be sufficiently suppressed.

Due to the problems with the uneven grain size distribution in the in-situ retorts resulting in poor control and gas distribution, retorting in piles was tried out where more uniformly crushed lumpy oil shale is piled above long, perforated exhaust pipes resting on the ground, and the surface of the pile is ignited. Due to side wind, retorting was not sufficiently uniform and oil recovery was moderate.

The poor experience with in-situ and in piles retorting showed that oil shale retorting is economic and safe only in appropriately designed retorts above ground.

The Lurgi-Ruhrgas Process (LR-Process)

PROCESS DESCRIPTION

The Lurgi-Ruhrgas process (see Fig. 1) is a Class 3 b retort. Its heart is a solid heat carrier cirulating system named "loop", which includes:

- a liftpipe (1) to convey and reheat the circulating fine-grained heat carrier;

- a collecting bin (2) to separate the combustion gas from the hot heat carrier;

- a screw mixer (3) which mixes the hot heat carrier and the raw shale feed to induce retorting;

- a surge bin (4) which provides surge capacity and time to complete retorting.

On the product gas side of the loop are cyclone (5) and condensation systems (6). On the flue gas side of the loop are cyclone (7), shale flash preheating (8) and final electrostatic dedusting (9).

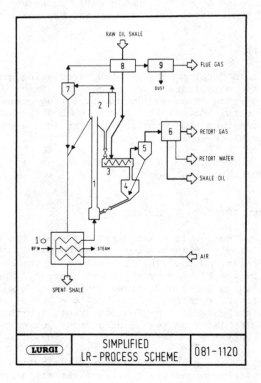

Fig. 1.   Simplified LR-Process scheme

Hot spent shale is withdrawn from collecting bin (2) and cyclone (7) and cooled in fluidized bed cooler (10) where process air is preheated and steam generated.

In the "loop" raw oil shale crushed to approximately 1/4 inch and preheated to approximately 150 to 200 °C is fed to the screw mixer where it is mixed with three to four times as much hot spent shale of 650 to 700 °C from the collecting bin. The raw shale is flash heated at about 500 to 530 °C and retorted within a few seconds.

The shale mixture leaving the mixer passes to the surge bin where retorting is completed and is then transferred to the lower section of the liftpipe. Combustion air preheated to about 450 °C is introduced at the bottom of the liftpipe. The air conveys simultaneously the shale to the top of the liftpipe while it burns the residual carbon from the spent shale.

The combustion gas and reheated spent shale are separated at about 650 to 700 °C in the collecting bin. The spent shale is returned to the screw mixer thereby closing the loop.

The overall process is outlined in some more detail in Fig. 2.

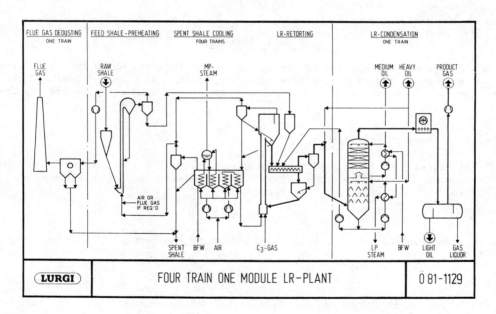

Fig. 2. Overall LR-Process scheme

The "condensation system" shown on the right side of the LR-loop consists mainly of a cat cracker fractionator type condensing tower, and a final air cooler-condenser. The condensing tower may serve several LR modules and consists of two stages. In the lower stage a heavy oil fraction is condensed at approx. 200 to 250 °C and residual dust is removed. Heavy oil dust concentrations can be easily kept low by recycling heavy oil to the mixer.

In the upper stage of the condensing tower the product vapours are cooled to 100 to 150 °C and a water and dust-free medium oil is condensed.

In the final air cooler-condenser light oil and gas liquor (retort water) is condensed leaving a high BTU retort gas at approx. 50 °C, which may be used as feed-stock for a downstream hydrogen production, as fuel.

The "shale preheater" consists of a vertical riser in which the feed shale is pneu-matically elevated and preheated by the hot LR flue gases. Coarser particles sep-arated from the gas upstream of the main feed shale collecting cyclone may be re-cycled to the riser.

Pressure at the bottom of the riser is controlled by a flue gas blower downstream of the feed shale collecting cyclone.

The flue gas is then finally dedusted in an "electrostatic precipitator" before being discharged to the atmosphere.

The 650 to 700 °C hot spent shale withdrawn from the LR system is cooled to approx. 200 °C in a multistage fluidized bed cooler in which the LR process air is pre-heated to about 450 °C and additional steam is produced. If the spent shale still contains some combustible carbon left unburnt from the LR system, then this carbon may be combusted in the first stage of the fluidized bed, so that no potential fuel is left unused in the process.

Special Reactions

In the LR system, several typical reactions can be observed which have a positive effect on the quality of the LR-flue gas, the retort gas, the shale oil and the spent shale. These reactions can be subdivided into the LR-liftpipe and the LR-mixer reactions.

LIFTPIPE REACTIONS

Most oil shales, as, for example, Colorado and Timahdit oil shale, contain calcium and magnesium carbonates which are partly decomposed at the liftpipe temperature of 650 to 700 °C according to the reaction

$$CaCO_3 \longrightarrow CaO + CO_2 .$$

The so formed oxides have a strong sulphur binding characteristic according to the reaction

$$CaO + SO_2 + \frac{1}{2} O_2 \longrightarrow CaSO_4$$

in which gypsium is formed and the flue gas is desulphurized.

Due to these reactions, low SO_2 flue gas concentrations are obtained if enough carbonates are available in the raw shale.

At normal operating conditions, a 30% carbonate decomposition rate was obtained for Colorado and Timahdit oil shale leading to very low SO_2 concentrations in the LR-flue gas.

Another important advantage is that fine granular spent shales with high CaO or MgO and low organic carbon contents will exhibit good cohesive strengths when moistened due to a reaction which is similar to cement formation.

Environmentally significant is also the formation of NO_x both from air and shale nitrogen in the LR-liftpipe.

The thermodynamic equilibria for the formation of NO in air show very low NO concentrations at normal liftpipe temperatures. Indeed, the NO concentration in the flue gas - NO_2 could not be detected - is normally below 100 ppm v/v.

MIXER REACTIONS

Two typical LR-mixer reactions can normally be observed. The first reaction, which reduces H_2S in the retort gas, depends on the content of carbonates in the raw shale and on the degree of carbonate decomposition in the liftpipe. This reaction can be described by:

$$CaO + H_2S \longrightarrow CaS + H_2O.$$

The unstable CaS is later on oxidized to $CaSO_4$ in the liftpipe.

The second reaction, which reduces the viscosity and pour point of the shale oil, is caused by a catalytic effect of hot spent shale in the mixer.

Most spent shales contain a considerable amount of silica SiO_2 which, contrary to clay Al_2O_3, has a pronounced visbreaking effect. In simple terms, SiO_2 catalytically tends to break C-C bonds while Al_2O_3 tends to break C-H bonds in the shale oil. Therefore, the LR-process has a distinct advantage in comparison to other oil shale processes because the oil has a lower viscosity and pour point.

Product Qualities from Colorado Oil Shale

SHALE OIL AND RETORT GAS

Typical qualities of shale oil and retort gas derived from Colorado oil shale are given in Tabel 1 and 2. It may be noted that the Colorado oil shale retorted had a total sulphur content of approx. 1.4% and a carbonate CO_2 content of approx. 18%.

Typical for LR-shale oil is the relatively low viscosity and pour point while the LR-retort gas has a high calorific value and is normally characterized by a low H_2S content.

SPENT SHALE, GAS LIQUOR, FLUE GAS

Typical qualities of spent shale, gas liquor and flue gas are given in Table 3.

While the size gradation of the Colorado spent shale is mainly in the silt range, organic carbon contents are as low as 0.2 wt %.

Due to the partial carbonate decomposition during retorting and combustion, the spent shale exhibits pronounced cementation properties when moistened. This leads to a high compressive strength and to very low permeabilities. They are key figures for economic spent shale disposal.

The quality of the gas liquor requires aftertreatment before disposal. Nevertheless, it may be possible to use steam stripped gas liquor for spent shale moisturization.

The flue gas is so low in SO_2, NO_x, CO and dust content, that it can be discharged into the atmosphere.

TABLE 1 Colorado Shale Oil Quality

		Condensed Oil Fractions	Uncondensed Naphtha Fraction
Percent of Total Oil	%	87	13
Density at 60°F	g/cm³	0.942	0.714
Grade API		18.7	66.7
Viscosity:	cst		
at 30°C		22.3	
at 50°C		10.3	
Conradson carbon	wt %	5.9	
Pour Point	°C	+18	
Bromine No.	g br/100 g oil	55	132
Ultimate Analysis:			
Carbon	wt %	85.16	86.13
Hydrogen	"	10.68	13.30
Oxygen	"	1.10	0.10
Nitrogen	"	2.16	0.16
Sulphur	"	0.90	0.31
LCV	kJ/kg	39 630	44 270
Boiling Analysis:			
IBP	°C	96	49
5 vol %	"	146	55
10 "	"	170	58
20 "	"	220	63
30 "	"	260	68
40 "	"	306	73
50 "	"	347	79
60 "	"	393	85
70 "	"	438	93
80 "	"	490	104
FBP (vol %)	"	508 (85)	150 (95)

TABLE 2 Colorado Retort Gas Quality

N_2	vol %	3.3
CO_2	"	29.9
H_2	"	24.9
CO	"	2.8
CH_4	"	17.0
$C_2 H_4$	"	6.4
$C_2 H_6$	"	6.5
$C_3 H_6$	"	6.7
$C_3 H_8$	"	2.5
		100.0
Density	kg/m_N^3	1.158
LCV	kJ/m_N^3	25 730
$H_2 S$	ppm	3 900
SO_2	"	45
COS	"	220
NH_3	"	45
AS	ppb	10

TABLE 3 Colorado Waste Stream Qualities

SPENT SHALE			
Grain Size	> 200 μ		12 %
	32-200 μ		13 %
	8 -32 μ		39 %
	< 8 μ		36 %
			100 %
Org. Carbon			0.2 %
GAS LIQUOR	pH-Value	9.2	
	Phenols	0.22	g/l
	Free Ammonia	6.0	"
	Fixed Ammonia	0.4	"
	Total Sulphur	0.22	"
	Fatty Acids	0.58	"
FLUE GAS	CO_2	16	vol %
	N_2	77	"
	O_2	7	"
		100	"
	SO_2	20	ppm
	NO_x	100	"
	CO	90	"
	Dust	0.075	grains/scf

INTERDEPENDENCE BETWEEN ELECTRICITY PRODUCTION AND WATER USE

D. Winje

Technical University of Berlin, Berlin (West), Federal Republic of Germany

ABSTRACT

The generation of electricity burdens the environment in many ways
with waste heat, and with water withdrawal and consumption.
The problems thus arising are found particularly in densely populated
industrial countries and in water-deficient countries.
Using the Federal Republic of Germany as an example, we will examine
the present strain of waste heat and water consumption on water
resources. Then we will see what environmental load can be expected
in the future.
We will examine the development of electrical energy use, types of
power plants, and various cooling systems, and then consider the
effect of all these on the environment.

KEYWORDS

Water withdrawal; water consumption; electricity generation; waste
heat; cooling systems; predictions.

INTRODUCTION

The generation of electricity burdens the environment in many ways,
e.g., with waste heat, and with water withdrawal and consumption.
The problems thus asising are found particularly in densely populated industrial countries and in water-deficient countries.
Using the Federal Republic of Germany as an example, we will examine
the present strain of waste heat and water requirements on water
resources. Then we will see what environmental load can be expected
in the future.
We will examine the development of electrical energy use, types of
power plants, and various cooling systems, and then consider the
effect of all these on the environment.

ELECTRICITY PRODUCTION

The generation of electricity in West Germany is dependent on a
number of uncertain determinants leading to widely diverging predictions. For the ensuing discussion the assumption is made that the
electrical power production of the public utilities of approximately
280 TWh in the year 1980 will increase to 525 TWh in the year 2000
and to 645 TWh in the year 2010.

COOLING SYSTEMS

Cooling systems are necessary in order to carry away the waste heat
of power plants. A representation of cooling systems is shown in
Fig. 1.

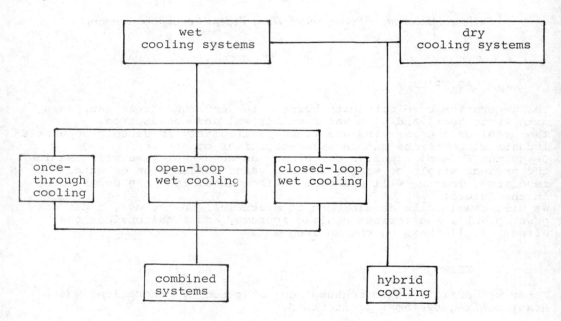

Fig. 1. Various types of cooling systems.

In wet cooling systems the entire volume of water which flows
through the condenser is drawn continually from a body of water and
mechanically freed of any impurities it may carry.
In the case of once-through cooling the heat captured in the condenser is given off entirely into a body of water.
In open-loop wet cooling a part of the heat is transferred to the
air through evaporation and convection, so that the load on the
body of water is reduced.

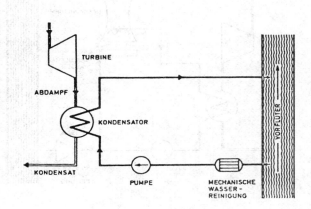

Fig. 2. Once-through cooling diagram.

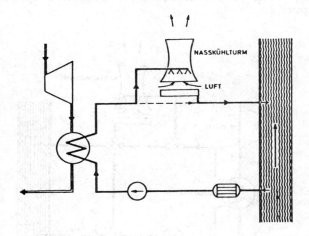

Fig. 3. Open-loop wet cooling diagram.

The use of closed-loop wet cooling enables the installation of electrical power units at limited water sources. This cooling technique allows the emission of practically the entire waste heat into the atmosphere through evaporation (ca. 75% at our latitude) and convection.
On the basis of good experience with closed-loop wet cooling this cooling method can be considered the state of the art even for the largest block capabilities at present of ca. 1300 MW with only one cooling tower.

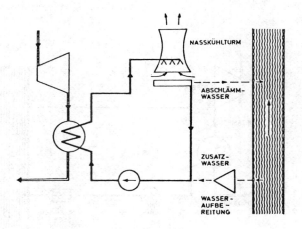

Fig. 4. Closed-loop wet cooling diagram.

Building combined cooling systems is a large-scale undertaking. Here the cooling system is designed for full performance in either once-through cooling, open-loop wet cooling, closed-loop wet cooling or a combination thereof.

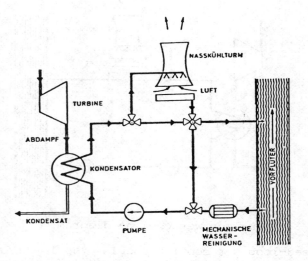

Fig. 5. Combined cooling diagram.

This method allows the choice of the optimal cooling technique under consideration of climatic conditions and of the immission capacity of the body of water.
As shown in Fig. 1, there are also other cooling systems. In addition to hybrid cooling (wet and dry cooling) the different methods of dry cooling should be mentioned. These methods are economically much less

suitable than the wet cooling techniques. For example, the transition to dry cooling would require the installation of about 7% more generating capability.
The costs of electricity generation are to a large extent dependent on the type of cooling system.
Costs increase by up to 20%, for example, when the transition is made from wet cooling to dry cooling.
The wet cooling methods will probably find the most wide-spread use in the near future.
Table 1 shows the specific parameters of water requirements and of waste heat discharge for fossil and nuclear power plants. Since data tend to vary, a tolerance range is indicated.
In addition to Table 1 it should be mentioned that the waste heat distribution between water and air is dependent on the cooling system. Here we make the assumption that the average waste heat distribution of open-loop wet cooling between rivers and atmosphere ist 25% to 75%.
It is assumed that the average rate of evaporation of river water is 0.2 m^3/s per GW discharged waste heat.
For the Federal Republic of Germany the further assumption is made that after the year 1990 only plants with closed-loop wet cooling will be licensed to operate.

WATER WITHDRAWAL AND WATER CONSUMPTION

Because of increasing electricity generation water withdrawal will rise from $18.8 \cdot 10^9$ m^3 in the year 1975 to $40.9 \cdot 10^9$ m^3 in the year 2010, as can be seen in Fig. 6.

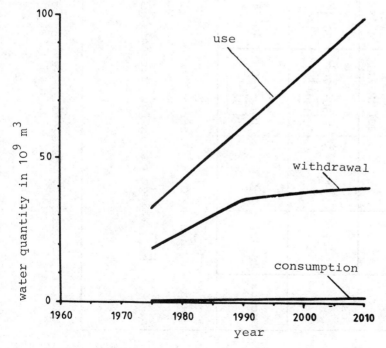

Fig. 6. Water requirements.

typ of power plant	effi-ciency %	energy loss %	waste heat power MW (MJ/s)	once-through and open loop cooling		closed-loop wet cooling		
				with-drawal m³/s	consumption m³/s	with-drawal m³/s	consumption m³/s	use m³/s
fossil	35-42	6-11	1130-1700 [1415]	30-40 [40]	0,3-0,4 [0,35]	0,35-0,7 [0,7]	0,35-0,6 [0,475]	[37]
nuclear	30-35	2-6	1675-2300 [1987]	40-55 [55]	0,4-0,55 [0,4]	0,5-4 [1]	0,5-0,8 [0,65]	[50]

Data for each 1000 MW electrical power

[ ] values chosen for calculations

Tab. 1. Efficiency, waste heat power and water requirements.

Water consumption will be $1.9 \cdot 10^9$ m^3 in 2010. These absolute figures on water withdrawal and consumption have real value only when seen in relation to the available ressources.

RESTRICTION - WATER CONSUMPTION

In West Germany the restriction on evaporation is estimated to be between 3% and 7% of the mean lowest annual runoff (1100 m^3/s) of streams.
Fig. 7. shows the trend of water consumption and the allowed restriction of 5% of the mean lowest annual runoff (55 m^3/s).

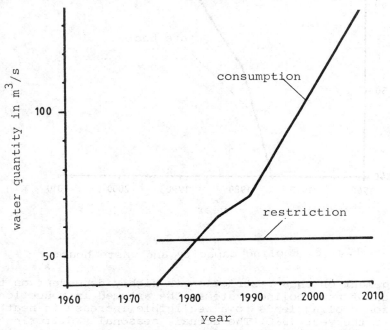

Fig. 7. Water consumption of public utilities.

This value can be regarded only as a reference point, which, because of its generality, cannot be strictly adhered to.
A seasonal and regional differentiation is necessary. Nevertheless, it is apparent from the foregoing that in West Germany the water losses through evaporation will be considerably higher than the restriction.

RESTRICTION - WASTE HEAT

The cooling capacity of a water body is determined by the possibility of heat transport over the border and/or into the sea as well as by heat discharge from the water surface.
The figures for the cooling capacity of the streams in West Germany vary from 29.5 GW to 79 GW. The Fig. 8 shows the trend of waste heat discharge into water bodies compared with a limiting cooling

capacity of 62 GW.

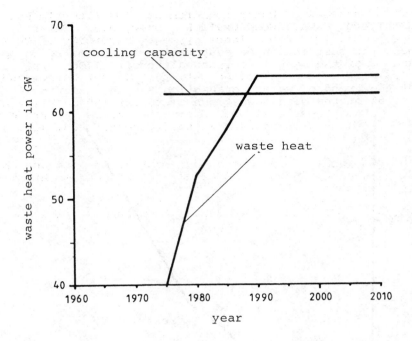

Fig. 8. Cooling capacity and waste heat.

The infringement on this restriction is highly dependent on the introduction of new cooling systems. The assumed introduction of recirculation cooling leads to a negligible increase in heat discharge after the year 1990. The actual, seasonally differing potential for heat descharge and the accompanying long-term effects must be determined separately for each body of water through ecological analysis and heat load calculations.

CONCLUSIONS

The results show that the generation of electricity not only substantially impairs air quality but also burdens water bodies through waste heat discharge and evaporation. For the purpose of an ecologically sensible utilization of water bodies it will be necessary to undertake their systematic management in regard to withdrawal, losses, and waste heat discharge.

761

REFERENCES

Gerking, E., Wärmekraftwerke und Umweltschutz, Frankfurt a.M. 1975.

Huston, R.J., An Overview of Water Requirements for Electric Power Generation, in Gloyna, E.F., Woodson, H.H., Drew, H.R., Water Management by the Electric Power Industry, The University of Texas at Austin 1975.

Kriesel, E., Wasserwirtschaftliche Aspekte thermischer Energiequellen, Darmstadt 1977.

Krolewski, H., Technische und wirtschaftliche Probleme zur Frage der Abwärme von thermischen Kraftwerken, Wasserwirtschaft, 63. Jg., 1973, H. 11/12.

Schikarski, W., Überprüfung der Abschätzung der zu erwartenden Abwärmemengen und des Kühlvermögens der Oberflächengewässer der BRD bis 1985, LAF-Notiz Nr.118 vom 13.12.1976.

Winje, D., Iglhaut, J., Der Wasserbedarf in der Bundesrepublik Deutschland bis zum Jahr 2010, Berlin 1980.

TOPIC D

New Energy Technologies

RECUPERATION DE CHALEUR PAR THERMOFRIGOPOMPE DANS L'INDUSTRIE LAITIERE

Y. Almin et G. Laroche

*Ingénieurs à la Direction des Etudes et Recherches d'Electricité de France,
Centre de Recherches des Renardières, France*

RESUME

L'utilisation d'une pompe à chaleur pour récupérer de la chaleur sur un fluide
(avec ou sans changement d'état) présente généralement un gros intérêt.
Cet intérêt est encore accru lorsque les besoins de chaud et de froid existent
simultanément, en pasteurisation par exemple. Il ne faut cependant pas oublier que
l'utilisation d'un échangeur de chaleur ne coûte rien en frais d'exploitation et
que l'association d'un échangeur et d'une pompe à chaleur est généralement la meil-
leure solution. Le bilan d'ensemble est encore amélioré lorsque les possibilités de
sous refroidissement du fluide de travail dans la pompe sont bien exploitées.

On présente deux exemples d'application de ces principes :

- Préparation d'eau chaude à la ferme par récupération sur les condenseurs
  des tanks à lait.

- Préparation simultanée d'eau glacée à 2°C et d'eau chaude à 95°C (pasteurisation)
  par couplage d'un groupe frigorifique et d'une pompe à chaleur haute température
  dans une laiterie de la région de NANCY.

I - PRODUCTION D'EAU CHAUDE A LA FERME PAR RECUPERATION DE CHALEUR SUR LE CIRCUIT FRIGORIFIQUE DES TANKS A LAIT

1.1 - Introduction

En France, environ 30 milliards de litres de lait sont produits annuellement et
les 2/3 de cette production sont collectés et dirigés vers des ateliers de transformation. Afin de limiter les coûts de transport (espacement des collectes) tout
en conservant au lait, matière première "fragile", ses qualités, des dispositifs
de stockage réfrigéré ont été mis en place sur les lieux de production.

Ces appareils dits tanks réfrigérés, sont équipés le plus souvent de machines
frigorifiques à compression de petite puissance à détente directe.

La chaleur de refroidissement du lait de 35 à 4°C (36 Wh/l), augmentée de l'énergie
nécessaire à cette réfrigération (20 Wh/l) est rejetée à l'atmosphère par le condenseur à air de la machine frigorifique et, de ce fait, reste inutilisée.

En supposant que seulement 80% du lait collecté soit réfrigéré, on peut estimer le rejet annuel de chaleur à l'atmosphère à 900 millions de kWh, soit environ 80.000 t d'équivalent pétrole.

Dans le but de réduire les consommations d'énergie électrique, les constructeurs ont, à juste titre, choisi des températures de condensation aussi basses que possible, et cet important rejet se situe donc à un niveau de 30 à 40°C qui le rend peu attrayant.

De nombreux dispositifs plus ou moins complexes peuvent être mis en oeuvre pour récupérer tout ou partie de cette énergie perdue et nous présentons ci-après certainement le plus simple d'entre eux ainsi que les résultats d'expérimentation tant en laboratoire qu'in situ.

1.2 – Montage d'un récupérateur de chaleur sur le circuit frigorifique d'un tank à lait réfrigéré

Le récupérateur consiste en un serpentin immergé dans un réservoir d'eau identique à celui d'un chauffe-eau à accumulation. Dans ce serpentin, inséré entre le compresseur et le condenseur, circule le fluide frigorifique comprimé donc surchauffé.

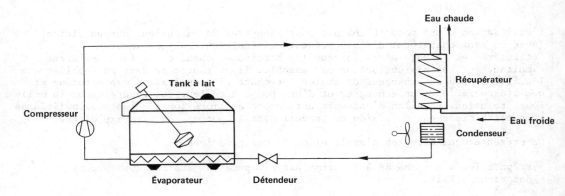

Figure 1 – Schéma de principe

Selon les conditions d'échange et de température d'eau, le fluide frigorigène se désurchauffe plus ou moins, voire même se condense dans cet échangeur. Ces conditions variant en permanence, il n'est pas possible de déterminer exactement la puissance instantanée de cet échangeur.

La fonction première du matériel étant la réfrigération du lait, le montage d'un récupérateur ne doit en aucun cas dégrader les performances de la machine frigorifique. En particulier, la durée de réfrigération de la capacité nominale du matériel ne doit pas être accrue.

Dans le but de déterminer l'impact d'un tel montage, nous avons conduit dans un premier temps en laboratoire (maîtrise plus aisée des paramètres) une expérimentation sur 2 tanks réfrigérés de constructeurs différents.

1.3 - Expérimentation en laboratoire

Les essais ont porté sur deux tanks à lait l'un de type ETSCHEID KW1200 2 Traites (capacité nominale 1200 litres soit 600 l/traite) équipé d'un récupérateur Frigocalor et l'autre de marque JAPY CF1680 2 ou 4 Traites (840 ou 420 l/traite) équipé d'un récupérateur RC210.

Globalement, les essais ont montré que le couplage d'un tel récupérateur avec une machine frigorifique équipée d'un détendeur ne modifiait pas les caractéristiques de la machine frigorifique qui continue à remplir sa fonction initiale : REFROIDIR et CONSERVER LE LAIT. Ces essais par contre, n'ont pas démontré l'intérêt réel du récupérateur, car les quantités d'énergies récupérées dépendent surtout des quantités d'eau soutirées, d'où l'expérimentation in situ que nous présentons maintenant.

1.4 - Expérimentation en exploitation agricole

1.4.1 - Caractéristiques du matériel installé

Tank de 6140 l de lait, 4 traites à détente directe, équipé de 2 groupes frigorifiques indépendants : compresseur hermétique CL5562 et récupérateur 200 l.

1.4.2 - Procédure d'essais

Ces essais avaient pour objet de vérifier, dans des conditions réelles d'exploitation, les résultats enregistrés en laboratoire. Les paramètres suivants ont été notés après chaque traite par le vacher : consommation des groupes frigorifiques, consommation du chauffe-eau à accumulation, débits d'eau des deux récupérateurs et et du chauffe-eau électrique et litrage du lait de chaque traite.

Un enregistreur de température permettait d'obtenir : les températures ambiantes, de l'eau froide et de l'eau puisée dans chaque récupérateur et dans le chauffe-eau à accumulation.

Deux montages ont été réalisés de Février à Décembre 1980.

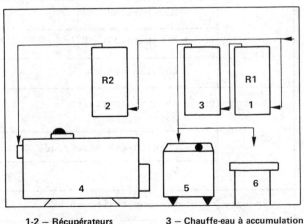

Montage 1 :

- Le récupérateur R1 est monté en série avec le chauffe-eau à accumulation, assurant ainsi le préchauffage de l'eau.

  Le chauffe-eau à accumulation assurant les besoins en eau chaude "tous usages" et l'alimentation du chauffe-eau à "eau bouillante".

- Le récupérateur R2 est utilisé uniquement tous les deux jours pour le lavage du refroidisseur.

1-2 — Récupérateurs   3 — Chauffe-eau à accumulation
4 — Refroidisseur   5 — Chauffe-eau «eau bouillante»
6 — Usages généraux

Figure 2 - Montage ①

Montage 2 :

Les deux récupérateurs et le chauffe-eau sont montés en série.

- Le puisage de l'eau s'effectue à la sortie du récupérateur R2 assurant ainsi les "usages courants ".

- L'eau du chauffe-eau à accumulation alimente le chauffe-eau "eau bouillante" et le système de lavage du tank à lait.

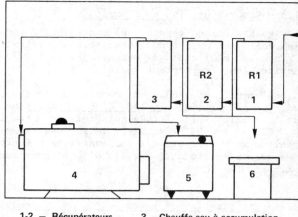

1-2 — Récupérateurs   3 — Chauffe-eau à accumulation
4 — Refroidisseur   5 — Chauffe-eau «eau bouillante»
6 — Usages généraux

Figure 3 - Montage 2

1.4.3 - Principaux résultats

Ils sont donnés dans le tableau suivant :

	Montage 1	Montage 2
Nombre de traites effectuées	364	56
Températures moyennes (°C)		
– Eau alimentation	15	12
– Récupérateur R_1	45	/
– Récupérateur R_2	80	52
– Chauffe-eau	78	78
Consommations en kWh par traite – Refroidisseur	17	9,5
– Chauffe-eau électrique	8	5
– Totale	25	14,5
Energie récupérée par traite en kWh	11,5	11,5
Taux de récupération %	24	39
Economie totale d'énergie au niveau du chauffe-eau par traite	13,2 kWh soit 60%	16,1 kWh soit 75%

Les montages ① et ② permettent de réduire les consommations d'électricité
nécessaires au chauffage de l'eau respectivement de 60% et 75% par rapport à
l'ancienne installation et ceci en récupérant respectivement 24 et 39% du potentiel
total récupérable.

Le montage ② qui permet d'utiliser au mieux la récupération est à préconiser ;
le montage ① ne se justifiant que lorsque les besoins d'eau aux températures
de 80 et 40°C sont respectivement faibles et assurés sans chauffe-eau d'appoint.

1.5 - Conclusion

Les différents essais effectués tant en laboratoire qu'en exploitation réelle
montrent que la récupération d'énergie sur les refroidisseurs de lait peut entraîner une économie d'énergie importante.

La rentabilité d'un tel investissement est difficile à évaluer, car fonction bien
évidemment de la quantité d'eau utilisée.

Malheureusement, les besoins sont souvent mal connus et avant d'envisager l'acquisition d'un tel récupérateur il est souhaitable de les estimer.

Par ailleurs, il faut souligner que le montage d'un récupérateur sur un refroidisseur doit être effectué par du personnel compétent afin que : les performances
du tank à lait ne soient pas modifiées et la fiabilité du matériel soit conservée.

II - RECUPERATION DE CHALEUR EN LAITERIE A UN NIVEAU THERMIQUE VOISIN DE 100°C

2.1 - Evolution technologique des PAC "Haute Température"

Les pompes à chaleur fonctionnant dans la gamme de températures du matériel frigorifique classique sont aujourd'hui largement répandues, et trouvent de nombreuses
applications : séchage du bois, logement, tertiaire... Certaines applications en
séchage ou en récupération de chaleur dans l'industrie, nécessitent des températures plus élevées (température d'évaporation supérieure à 20°C et de condensation
supérieure à 60°C).

L'adaptation technologique du matériel existant à ces nouvelles conditions a nécessité des études portant à la fois sur l'architecture des circuits, sur les compresseurs et leur lubrification, sur la stabilité des fluides thermodynamiques, sur
l'analyse des échanges de chaleur.

Les principaux résultats des travaux de recherche effectués par EDF dans ce
domaine sont les suivants :

- Le fluide frigorigène R114 est bien adapté au fonctionnement à haute température
(90°C ⩽ température de condensation ⩽ 130°C).

- Les problèmes de lubrification sont maîtrisés, et différents compresseurs donnent
satisfaction.

- Le coefficient de performance réel représente environ 40% du coefficient de performance théorique de Carnot calculé à partir des températures d'évaporation et
de condensation, en première approximation les principales irréversibilités sont
imputables : - au moteur (rendement = 90%)
                     - au détendeur (r = 75%)
                     - au compresseur (r = 70%)
                     - aux échangeurs (r = 90 à 70%).

Ce coefficient peut être nettement amélioré par l'utilisation du sous-refroidissement
du fluide frigorigène.

- Les coefficients d'échange mesurés au condenseur sont 3 à 4 fois supérieurs à
ceux calculés d'après la théorie de Nusselt.

- La densité de flux thermique, et le niveau du fluide frigorigène conditionnent
considérablement les coefficients d'échange dans un évaporateur de type noyé.

Ces résultats, d'une part, montrent que l'utilisation des pompes à chaleur jusqu'à
des températures de condensation de 130°C est possible industriellement dans la
mesure où l'on dispose d'une source froide à une température suffisante, d'autre
part, permettent un dimensionnement plus rigoureux des échangeurs, et ouvrent
la voie à la conception de matériel plus performant.

Les échangeurs représentent jusqu'à 40% du coût d'une pompe à chaleur, leur optimisation se traduit donc par une baisse sensible du prix des machines. Ce dernier
point est particulièrement crucial pour le développement de la technique des
pompes à chaleur.

Cette mise au point technologique des pompes à chaleur haute température, effectuée
par le CENTRE de RECHERCHES EDF des RENARDIERES, a conduit à repenser les applications industrielles des pompes à chaleur en utilisant un fluide frigorigène à
un niveau de température de 90°C à 130°C.

La condition principale autorisant le montage de ce type de machine est de disposer
d'une "source froide" à un niveau de température inférieur de 50 degrés au plus à
celui de la source chaude (respectivement 40 et 80°C à la sortie des échangeurs
pour des températures de 90 et 130°C au niveau de la source chaude).

Des sources froides à un tel niveau de température peuvent être constituées par des
rejets thermiques :   - air chaud,
                           - eau chaude,
                           - vapeur à basse tension.

Le montage d'une pompe à chaleur nécessite la continuité de la puissance délivrée
par la source froide, un stockage étant parfois nécessaire. Par ailleurs, la puissance moyenne disponible au niveau de la source doit correspondre (ou être supérieure) à la puissance récupérée par la pompe à chaleur.

Les pompes à chaleur "haute température" permettent la récupération de la chaleur
libérée par les groupes frigorifiques, le condenseur du groupe frigorifique cons-
tituant alors la source froide de la pompe à chaleur. Si les besoins de froid et
de chaleur se situent dans le même processus industriel, leur concomitance facilite
le couplage groupe frigorifique/pompe à chaleur.

L'ensemble groupe frigorifique/pompe à chaleur monté à la laiterie St HUBERT produit des frigories à 0°C au niveau de la boucle d'eau glacée et de la chaleur à
95°C au niveau de la boucle d'eau chaude industrielle. La concomitance de besoins
de froid à 0°C et de chaleur à 95°C environ est fréquente dans les industries agro-
alimentaires et permet de généraliser l'usage de thermofrigopompes à 2 étages.

2.2 - La thermofrigopompe installée à la laiterie St Hubert

2.2.1 - Problème posé

Nécessité d'une augmentation de capacité de 1 400 000 fg/jour (1 600 kWh/jour) d'eau à 0°C et de 2 500 000 kcal/jour (2 900 kWh/jour) d'eau à 95°C relative à la création d'un produit nouveau sur le site de l'usine de LUDRES (Meurthe et Moselle).

Les besoins de l'usine sont assurés par :

a - une boucle d'eau glacée alimentée par un bac à glace traditionnel (herse et groupe frigorifique).

b - une boucle chaude d'eau à 95°C produite par des chaudières fonctionnant au gaz naturel.

Le raccordement de l'installation liée à la nouvelle fabrication sur les boucles existantes s'avère impossible, les réseaux "chaud" et "froid" étant saturés.

2.2.2 - Solutions proposées

a - Installer une unité de production de chaleur et de froid propre à la nouvelle fabrication. Unité traditionnelle composée d'un groupe frigorifique et d'une chaudière.

Les inconvénients sont : décentralisation des vecteurs caloporteurs, multiplication des générateurs de froid et de chaleur et gaspillage d'énergie.

b - Installer une thermofrigopompe raccordée sur les réseaux existants, d'une capacité égale aux besoins de la nouvelle production.

Les avantages sont : aucune augmentation de capacité chaudière, alignement sur les vecteurs caloporteurs existants et surtout économie substantielle d'énergie.

C'est la deuxième solution qui a été retenue par la laiterie St Hubert pour l'équipement de son usine de Ludres.

2.2.3 - Principes de fonctionnement de la thermofrigopompe

La machine est composée de deux circuits frigorifiques dit "en cascade" (figure 4).

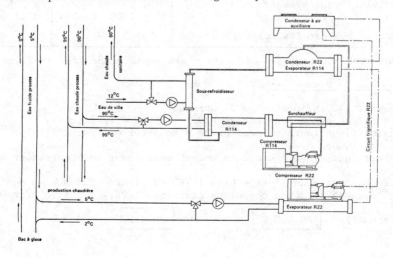

L'étage de production d'eau glacée est un ensemble motocompresseur TRANE, lié à un évaporateur classique, fonctionnant au R22.

La condensation du fluide frigorigène R22 s'effectue dans un échangeur intermédiaire, qui est à la fois le condenseur de l'étage de production de froid et l'évaporateur du second étage de production de chaud.

Cet étage équipé d'un ensemble moto-compresseur TRANE fonctionne avec du fluide frigorigène R114. Le condenseur classique de ce circuit produit les calories à 95°C.

On remarque immédiatement l'importance d'un ajustement précis de la puissance calorifique du condenseur R22 et de la puissance frigorifique de l'évaporateur R114.

Pour améliorer la souplesse de fonctionnement, nous avons toutefois prévu un condenseur à air auxiliaire sur le circuit R22.

L'étude du diagramme de Mollier montre que dans tous les cas, la récupération de l'énergie sensible de sous-refroidissement du fluide frigorigène améliore notablement les performances de l'installation. Pour une même puissance dépensée au compresseur, les gains de puissance frigorifique et calorifique peuvent atteindre chacun 20 à 30%.

Le montage d'un sous-refroidisseur sur l'étage R114 (sous-refroidisseur alimenté avec de l'eau de ville), augmente les performances de la machine et en outre délivre un troisième vecteur d'eau chaude à 55-60°C, qui est relié à la distribution sanitaire.

2.2.4 - Conditions de fonctionnement de la machine

Circuit de fluide frigorigène	Etage R22	Etage R114
Température d'évaporation	- 2°C	43°C
Température de condensation	48°C	100°C
Température à la sortie du sous-refroid.eur	-	65°C
Puissance frigorifique à l'évaporateur	73 000 fg/h	101 000 fg/h
Puissance calorifique au condenseur	101 000 kcal/h	104 000 kcal/h
Puissance calorifique globale (y compris sous-refroidisseur)	101 000 kcal/h	128 000 kcal/h
Puissance absorbée par les compresseurs	36 kW	38 kW

Les températures au niveau des circuits d'eau sont données sur la figure 4.

2.2.5 - Rentabilité économique

- Fonctionnement : 20 h/jour - 300 jours/an
- Energie récupérée sur l'évaporateur R22 : 1 460 000 fg/jour
- Energie calorifique fournie (condenseur + sous-refroidisseur R114) :
  2 560 000 kcal/jour
- Energie électrique consommée par le moteur entraînant le compresseur R114 :
  760 kWh/jour
- Consommation annuelle de l'étage R114 : 760 kWh/jour ⇒ 228 MWh/an ⇒ 65 500 F/an
- Consommation annuelle d'une chaudière classique : 96,4 T fuel n°2/an ⇒ 110 500F/an

- Economie annuelle   45 000 F   -   36 TEP

2.3 - Conclusion relative à la PAC installée à la laiterie St Hubert

L'utilisation de la pompe à chaleur R114 à la laiterie St Hubert doit entraîner d'importantes économies d'énergie et une réduction du coût d'exploitation.

L'économie d'exploitation espérée est de 45 000 F/an par rapport à une chaudière brûlant du fuel lourd n°2. Dans cette application, chaque kWh électrique consommé au niveau du groupe R114 se substituera à plus de 4 thermies consommées au niveau de la chaudière.

L'installation de St Hubert a démarré en mai 1981. Actuellement, nous ne disposons pas du résultat des essais conduits sur cette installation. Néanmoins le démarrage n'a soulevé AUCUN PROBLEME TECHNOLOGIQUE PARTICULIER, l'étude menée par le cabinet SERTH, le constructeur MONDIAL FRIGO, et EDF ayant conduit à retenir des solutions éprouvées dans les laboratoires EDF (départements Machines et Structures et Applications de l'électricité).

COGENERATION APPLICATIONS
IN COMMERCIAL FACILITIES

M. K. J. Anderson

ANCO Engineers, Inc., Santa Monica, California, USA

ABSTRACT

This paper evaluates the applicability of a number of cogeneration schemes for commercial facilities. The evaluation doesn't stop with the cogeneration equipment alone, but includes the central plant equipment and the air conditioning systems. In this way, the building mechanical system may be taken as a whole to evaluate its energy efficiency. The typical commercial requirements of supplying a large summer cooling load and a large winter heating load (with assistance from internal heat gain) are used.

The cogeneration equipment analyzed includes both topping cycles (noncondensing steam turbine) and Brayton cycles (gas turbine) with a waste heat boiler. The results of the analysis show that from both an economic and a supply energy standpoint, the Brayton cycle is a very desirable system when operated in a steam following mode. The topping cycle, on the other hand, is not always desirable from an economic or supply energy standpoint. In particualr, when coupled with an absorption chiller, which is an increasingly common configuration, the topping cycle results in no economic savings and an increase in the use of source energy compared to the conventional means of cooling, the centrifugal chiller.

KEYWORDS

Cogeneration; gas turbine; Brayton Cycle; steam turbine; topping cycle; chiller.

INTRODUCTION

The purpose of this paper is to study the applicability of various cogeneration approaches to commercial facilities. Typical loads are used for a large 100,000 m^2 (1,000,000 ft^2) commercial building such as a hospital, which has conditioning requirements around the clock. Various cogeneration systems are investigated to supply the average summer and winter loads. The annual energy use and utility bills for each system are calculated to determine which approach uses the least source energy and which costs the least to operate. The results of the study indicate which types of cogeneration schemes should be considered for a new or retrofit central plant in a commercial facility and which types of cogeneration should be avoided.

THE PROBLEM

The problem is to find the system which will most efficiently supply the average loads shown in Table 1.

TABLE 1 Average Load Requirements for Candidate Systems

- Summer:

  Steam - 5.3 GJ (5.0 MBtu/h) at 200 KPa (15 psig)

  Chilled Water - 6.3 GJ (6.0 MBtu/h or 500 tons) at 6°C (45°F)

- Winter:

  Steam - 15.8 GJ (15.0 MBtu/h) at 200 KPa (15 psig)

  Chilled Water - 1.3 GJ (1.2 MBtu/h or 100 tons) at 6°C (45°F)

These loads occur 24 hours a day, all year long. The loads are typical of a large complex facility which has a variety of demands for heating and cooling almost continuously.

The energy prices used in the analysis are the following:

- Electricity - $0.060/kWh
- Fuel   - $3.8/GJ ($4.0/MBtu)

These are average prices only and do not take into account hourly changes in Time of Use rates. It is anticipated that this will not create a significant error because the proposed systems do not significantly shift the hours of use.

This paper does not address the size of the equipment in the proposed systems. It simply assesses the overall efficiency of the generic types of equipment. Similarly, the paper does not address the economic desirability of each system as this is largely a function of local utility rate schedules. The capital investment will vary greatly from new construction to retrofit projects. Suffice it to say that for the given utility rates the gas turbine cogeneration options appear economically attractive even as a retrofit modification where a sufficient steam load exists (roughly 5.3 GJ or 5 MBtu/h).

The analysis considers the energy use of the heat and cold producing machines only. It does not account for any supporting pumps and fans or for the energy use of the rest of the facility. Any excess electricity production is likely to be used in these other areas within the facility.

THE ALTERNATIVE SYSTEMS

Case 1: Conventional System

The standard central plant consists of a low pressure boiler to meet the steam load and a centrifugal chiller to meet the cooling load. This is illustrated in Fig. 1. The thermal efficiency (First Law) of the boiler system is 75%. The coefficient of performance of the centrifugal is 4.4 kWh of cooling for every kWh of electrical energy input. (This represents an energy use of 0.8 kW per ton.)

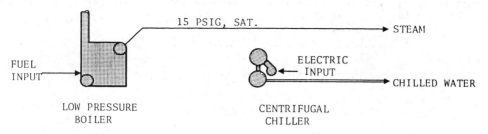

Fig. 1.   Conventional system

Case 2:   Steam Turbine with Centrifugal Chiller

In the second case the low pressure boiler is replaced with a high pressure boiler to generate steam at 1,380 kPa (185 psig) and 260°C (500°F), as shown in Fig. 2. The pressure of this steam is reduced through a back pressure steam turbine to its requirement of 200 kPa (15 psig). In this expansion energy is extracted at a rate of 186 kJ per kg (80 Btu/lb). It is assumed that in the conversion of this mechanical energy to electric energy, the generator efficiency is 90%. The efficiency of the boiler plant has been reduced to 70% because the steam temperature has increased by 121°C (250°F).

The chilling is provided by a centrifugal chiller, just as it was in Case 1.

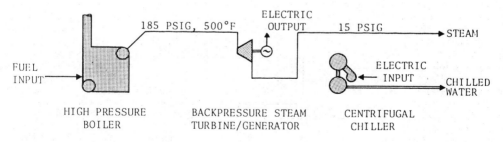

Fig. 2.   Steam turbine/generator, centrifugal chiller

Case 3:   Steam Turbine with Absorption Chiller

This case is similar to Case 2 except that the centrifugal chiller has been replaced with a single stage absorption chiller, as shown in Fig. 3. The capacity of the steam turbine has been expanded to handle the greatly increased requirement of low pressure steam.

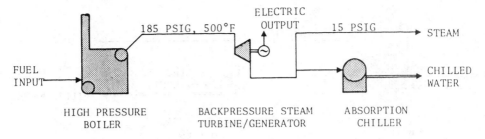

Fig. 3. Steam turbine/generator, absorption chiller

The boiler and steam turbine efficiencies remain the same as in Case 2. The efficiency of the absorption chiller is such that 1 kWh of cooling requires 1.5 kWh of
heat. (This is equivalent to 18 lb of steam per ton-hour of cooling.)

Case 4: Gas Turbine with Centrifugal Chiller

This is the first case to use a gas turbine, as shown in Fig. 4. The turbine directly drives the electric generator and the high temperature exhaust gases
(roughly 480°C or 900°F) pass through a waste heat boiler. In the production of
low pressure steam, 60% of the exhaust heat is transferred to the steam in the
waste heat boiler.

The chilled water is provided by a centrifugal chiller, as in Case 1.

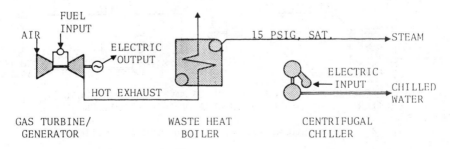

Fig. 4. Gas turbine/generator, centrifugal chiller

Case 5: Gas Turbine with Absorption Chiller

In this case a gas turbine generator and a waste heat boiler are utilized, as in
Case 4. However, instead of simply supplying low pressure steam to the normal
loads, steam is also used to operate an absorption chiller, as shown in Fig. 5.
This takes the place of the centrifugal chiller in the last case. The waste

heat boiler still produces low pressure steam with an efficiency of 60%.

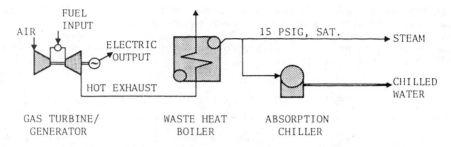

Fig. 5.   Gas turbine/generator, absorption chiller

Case 6:   Combined Cycle

The combined cycle utilizes both a gas turbine and a steam turbine to produce electricity as shown in Fig. 6.   The gas turbine drives a generator directly and sends
its hot exhaust gasses through a waste heat boiler.   In this case the waste heat
boiler generates medium pressure, superheated steam (1,380 kPa at 260°C, 185 psig
at 500°F).   The boiler recovers 50% of the exhaust heat which passes through it.

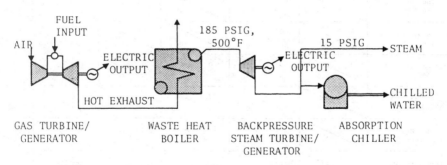

Fig. 6.   Combined cycle

After the waste heat boiler, the system is similar to Case 3.   The steam flows
through a back pressure turbine to generate electricity, dropping its pressure to
200 kPa (15 psig).   It then flows to the normal loads and to the absorption
chiller.

COMPARATIVE ENERGY USE

Using the assumptions stated in the description of each case, an analysis was performed on each system.   This analysis evaluated fuel use and electricity use or
production at two points—the average summer load and the average winter load.   In
all cases, the cogeneration systems are assumed to operate in a steam following
mode, producing as much electricity as the steam load permits.   It is assumed that

if this represents more electricity than the facility uses, the utility will buy the excess.

The detailed results of the analysis are shown in Table 2. This Table lists the critical energy flows for each system under summer and winter conditions. A summary of these results is shown in Fig. 7. It is based on the assumption of continuous operation (such as in a hospital) for half the year under summer conditions and half under winter conditions.

TABLE 2 Operation Variables of Alternative Cogeneration Systems

Summer / Winter	Conventional System	Steam Turbine, Centrifugal Chiller	Steam Turbine, Absorption Chiller	Gas Turbine, Centrifugal Chiller	Gas Turbine, Absorption Chiller	Combined Cycle
Steam Load (MBTU/H) ‡	5.0 / 15.0	5.0 / 15.0	5.0 / 15.0	5.0 / 15.0	5.0 / 15.0	5.0 / 15.0
Cooling Load (MBTU/H)	6.0 / 1.2	6.0 / 1.2	6.0 / 1.2	6.0 / 1.2	6.0 / 1.2	6.0 / 1.2
Boiler Fuel Use (MBTU/H)	6.67 / 20.0	7.71 / 23.1	21.6 / 25.9	0 / 0	0 / 0	0 / 0
Gas Turbine Fuel Use (MBTU/H)	0 / 0	0 / 0	0 / 0	10.4 / 31.3	29.1 / 35.0	37.8 / 45.4
Waste Heat Boiler Output (MBTU/H)	0 / 0	0 / 0	0 / 0	5.0 / 15.0	14.0 / 16.8	15.1 / 18.1
Steam Turbine Generation (KW) *	0 / 0	(105) / (316)	(295) / (354)	0 / 0	0 / 0	(295) / (354)
Gas Turbine Generation (KW) *	0 / 0	0 / 0	0 / 0	(549) / (1648)	(1535) / (1846)	(1994) / (2392)
Electric Chiller Draw (KW)	400 / 80	400 / 80	0 / 0	400 / 80	0 / 0	0 / 0
Absorption Chiller Use (MBTU/H)	0 / 0	0 / 0	9.0 / 1.8	0 / 0	9.0 / 1.8	9.0 / 1.8
Net Fuel Use (MBTU/H)	6.7 / 20.0	7.7 / 23.1	21.6 / 25.9	10.4 / 31.3	29.1 / 35.0	37.8 / 45.4
Net Electricity Use (KW) *	400 / 80	295 / (236)	(295) / (354)	(149) / (1568)	(1535) / (1846)	(2289) / (2746)
Net Energy Use (MBTU/H)	10.9 / 20.8	10.8 / 20.7	18.5 / 22.2	8.9 / 14.8	13.0 / 15.6	13.8 / 16.5
Net Cost ($/H)	50.7 / 84.8	48.6 / 78.4	68.7 / 82.4	32.7 / 30.9	24.4 / 29.2	13.9 / 16.6

‡ MBTU = 1.056 GJ

* Numbers in Parenthesis Represent Electricity Generation

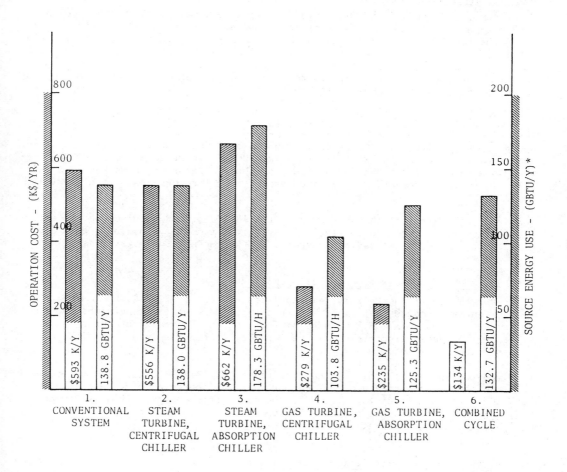

Fig. 7. Comparative source energy use and operating
cost of alternative cogenerative systems

*GBTU = 1055 TJ

In order to obtain the total energy use, electricity was converted to the equivalent fuel required for its generation at a rate of 11.1MJ (10.5 kBtu) of fuel per kilowatthour. The energy reported covers only the heat and cold producing machinery. None of the peripheral pumps and fans are included.

Case 1: Conventional System

Under summer conditions the conventional system requires an input of 7.1 GJ/h (6.7 MBtu/h) of fuel and an input of 400 kW of electricity. Under winter conditions the requirements are 21.1 GJ/h (20.0 MBtu/h) of fuel and 80 kW of electricity.

In a year of operation, this system requires 146 TJ (138.8 GBtu) of source energy. It costs $593,000 to purchase this energy.

Case 2: Steam Turbine With Centrifugal Chiller

Under summer conditions, this system requires an input of 8.1 GJ/h (7.7 MBtu/h) of fuel and an input of 295 kW of electricity. Under winter conditions the requirement is 24.4 GJ/h (23.1 MBtu/h) of fuel and the generator produces 236 kW of electricity. This represents an increased use of fuel for conversion to electricity on site.

In a year of operation, this system requires 146 TJ (138.0 GBtu) of source energy, a slight decrease from the conventional system. It costs $556,000 to purchase this energy. The annual savings compared to the conventional system is $37,000.

Case 3: Steam Turbine With Absorption Chiller

Under summer conditions this system requires an input of 22.8 GJ/h (21.6 MBtu/h) of fuel and it produces 295 kW of electricity. Under winter conditions the requirement is 27.3 GJ/h (25.9 MBtu/h) of fuel and the generator produces 354 kW. While this produces electricity all year round, it represents a significant increase in the fuel requirement.

In a year of operation, this system requires 188 TJ (178.3 GBtu) of source energy. This represents an increase of 28% compared to the source energy use of the conventional system. It costs $662,000 to purchase this energy. This system costs $69,000/yr more to operate than the conventional system. This is therefore not a desirable mode of cogeneration.

Case 4: Gas Turbine with Centrifugal Chiller

Under summer conditions this system requires an input of 11.0 GJ/h (10.4 MBtu/h) of fuel and it produces 149 kW of electricity. Under winter conditions the requirement is 33.0 GJ/h (31.3 MBtu/h) of fuel and the generator produces 1,568 kW of electricity.

In a year of operation this system requires 110 TJ (103.8 GBtu) of source energy. This represents a decrease of 25% compared to the source energy use of a conventional system. It costs $279,000 to purchase this energy. This represents an annual savings of $314,000 compared to the conventional system.

Case 5:   Gas Turbine with Absorption Chiller

Under summer conditions this system requires an input of 30.7 GJ/h (29.1 MBtu/h) of fuel and it produces 1,535 kW of electricity. Under winter conditions the requirement is 36.9 GJ/h (35.0 MBtu/h) of fuel and the generator produces 1,846 kW of electricity.

In a year of operation this system requires 132 TJ (125.3 GBtu) of source energy. This represents a decrease of 10% compared to the source energy use of a conventional system. It costs $235,000 to purchase this energy. This represents an annual savings of $358,000 compared to the conventional system.

Case 6:   Combined Cycle

Under summer conditions this system requires an input of 39.9 GJ/h (37.8 MBtu/h) of fuel and it produces 2,289 kW of electricity. Under winter conditions the requirement is 47.9 GJ/h (45.4 MBtu/h) of fuel and the generator produces 2,746 kW of electricity.

In a year of operation this system requires 140 TJ (132.7 GBtu) of source energy This represents a decrease of 4% compared to the source energy use of a conventional system. It costs $134,000 to purchase this energy. This represents an annual savings of $459,000 compared to the conventional system.

Summary of Analysis

The net results for all systems are shown in Fig. 7. The conventional system turns out to be more efficient and less expensive to operate than at least one cogeneration system.

The first alternative, adding a steam turbine to a conventional system produces some cost reduction but does not significantly reduce the overall source energy use. This may be a desirable addition to a conventional plant when backup generation capability can be provided by the steam turbine as well as base load power, eliminating the need for a diesel generator.

The second alternative, adding a steam turbine and absorption chiller is not a desirable option. The absorption chiller uses a large amount of steam but a steam turbine does not generate enough electricity with that steam to justify the use of the system. This is a significant finding because many such cogeneration systems are currently being planned and installed with the anticipation of energy and monetary savings.

Each of the three gas turbine options reduces the source energy use of the central plant equipment by from 4 to 25%. This represents a significant decrease in source energy use.

Note that cogeneration promoters speak in terms of increasing the efficiency of energy use from 33% (typical central plant generation efficiency) up to 80% (counting both generation and thermal energy use). This creates a false picture. Before cogeneration the efficiency of electricity production may be 33% but the efficiency of the thermal energy use at the facility is already in the range of 80%. (These are First Law efficiencies.) If you combine the two initial efficiencies in the proportion of the two energy uses, the actual initial efficiency will be in the range of 50% in a facility with a significant thermal load. Thus, the improvement from cogeneration will be raising the total efficiency from roughly 50% to roughly 80%. This is still a significant improvement.

While gas turbine cogeneration reduces source energy use somewhat, it reduces utility bills significantly. In the dases shown, the cost of operating the heating and cooling equipment dropped by from 53 to 77%.

This makes it very desirable for commercial customers. Cogeneration gives them an opportunity to buy large quantities of fuel (possibility at reduced rates and higher supply priorities) and generate electricity in place of the electric utility. Much of the viability of such a system may depend upon the rate of repurchase by the utility or the ability to wheel the power to other sites. This applies when the thermal load is significantly greater than the electrical load so that electricity is likely to be exported.

Note that with the introduction of absorption chilling and steam turbines to the gas turbine systems, the operating costs drop while the source energy use increases. This results because the greater thermal loads allow the gas turbines to generate more high priced electricity while using greater quantities of less expensive fuel. The most energy efficient system is not necessarily the least expensive to operate where the substitution of one fuel for another is involved.

CONCLUSIONS

The conclusions reached in this paper are the following:

● The use of a steam turbine in place of a pressure reducing valve (Case 2) generates small utility bill savings. It does not, however, significantly reduce the source energy required to operate the system.

● The use of a steam turbine and absorption chiller (Case 3) is not a desirable option. Compared to a conventional system, it increases the source energy use by 28% and in this example increases utility bills by $69,000/yr. This is a particularly significant finding because such systems are currently being installed under the assumption that they will save energy and money.

● The use of a gas turbine cogeneration system (Cases 4,5,and 6) reduces the source energy use by from 4 to 25%. In this example, annual operating costs are reduced by from $314,000 to $459,000.

● In this study, the more the gas turbine cogeneration system reduces the utility bills, the less it reduces the source energy use. The extreme example is the combined cycle (Case 6) which reduced the heating and cooling costs by 77% but only reduced source energy use by 4%.

● For any commercial facility with a significant annual heating and cooling load, gas turbine cogeneration appears to be an attractive approach to reducing energy use and utility bills. Steam turbine cogeneration does not offer the same magnitude of savings.

LARGE DEMONSTRATION SOLAR PROJECT IN ITALY

P. Baronti and G. Benevolo

ENI, Rome, Italy

INTRODUCTION

Solar energy has not "taken off" "in sunny Italy. Not too many projects, for the utilization of solar thermal energy at low temperature, have been carried out in Italy. Why? The general public is sensitive to the subject and single homeowners and public administrations are requesting bids for the application of solar systems for space heating and production of hot water. The government and government organizations, are pouring money into theoretical studies, demonstration projects and research. The government has proposed to the parliament further legislation to promote solar energy with financial incentives and subsidies. Still the solar energy industry is in its infancy and we do not see, in the market place, very many signs of a rapid growth of demand for solar systems, there are obviously reasons: the main reason is the lack of confidence in the economics of solar systems. Any economic analysis must be based on a number of assumptions. Two are the basic parameters: the rate of increase which are projected for the cost of the traditional fuels which are to be substituted by solar energy: the interest rate of the money to be borrowed for the construction of the solar systems. However, other important parameters are of a more technical and political nature; performance of the system. Durability or lifetime, maintenance on one side, and proper legislation, regulations and financial incentives, on the other side. Only large demonstration projects can help, at least in part, to provide the answers which are much needed. ENI is the leading energy Company in Italy, and one of the largest in the world. Its commitments to supply Italy with primary energy sources include oil, natural gas, coal, nuclear fuel and solar energy. In the field of solar energy, several companies of the ENI Groups are devoting their efforts to many applications, from solar thermal to photovoltaics, and are carrying out a number of projects. With regards to solar heating in the residential sector, a demonstration project, now almost completed, is outstanding for its size, scope, variety of applications. This is the projects that several companies of the ENI Group are financing and carrying out in the Region of Tuscany, in cooperation with local authorities and a large construction Company Consorzio Regione Etruria.

THE PROJECT

The project consists in the application of solar systems and in the application of new techniques of construction to 746 apartments located in 18 new apartment buildings, now under construction in several towns in the Region of Tuscany, in central Italy (Tab. 1). Tuscany has been chosen because of its solar characteristics. It is situated in the middle of Italy and its degrees/day vary between 1300 and 1800 (Fig. 1).

TABLE 1

LOCALITY	MODEL	APARTMENTS	MQ.	SYSTEM DESCRIPTION
1) Certaldo	1	18	1365,9	Passive direct gain
2) San Vincenzo	1	12	909,4	Passive Trombe wall 35 cm.
San Vincenzo	1	12	905,9	Passive Trombe wall 25 cm.
San Vincenzo	1	12	1031,9	Traditional of comparison
3) Forte dei Marmi	1	4	261,1	Passive saw-toothed Trombe wall
Forte dei Marmi	1	4	261,1	Traditional of comparison
4) Fucecchio	2	12	972,6	Air-water integrated heating system
Fucecchio	2	12	972,6	Integrated radiating panel heating system
5) Borgo a Mozzano	2	6	486,5	Air-water integrated heating system
Borgo a Mozzano	2	6	486,5	Traditional of comparison
6) Massa	2	12	971,1	Air-water integrated system
Massa	3	24	1901,5	Air-water integrated heating
7) Castelnuovo Garfagnana	2	12	985,9	Integrated heating system with heat pump
Castelnuovo Garfagnana	2	12	985,9	Air-water integrated heating system
Castelnuovo Garfagnana	2	12	985,9	Traditional of comparison
Viareggio	3	24	1801,5	Traditional of comparison
8) Torre del Lago	3	12	809,4	Integrated convector heating
9) Lucca	3	18	1428,9	Hot water integrated

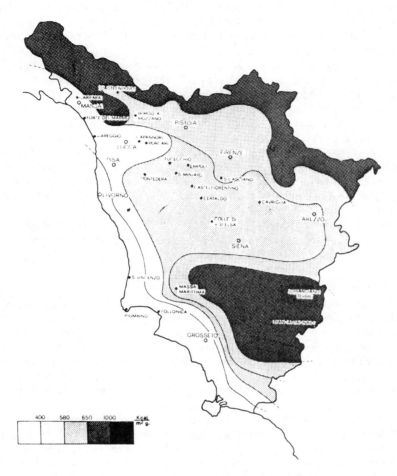

Fig. 1. Tuscan Region. Distribution of the differences
between thermal load and gaining solar energy for
a standard building in cold months.

The project considers 18 different types of applications which can be grouped,
for semplicity, into five major categories:
- integration of solar systems with the traditional ones (boilers and radiators)
  for space heating during the winter
- production of hot water for the whole year
- integration of solar systems with heat pumps (air-water heat pumps substituting
  the boiler)
- new forms of coibentation
- Trombe walls

There are various types of coupling among these five major categories, to obtain, as mentioned, a total of 18 applications. In addition the various solar systems, and the various designs and construction solutions are applied to apartments and to buildings of different typology and orientation and located in different climatic conditions. All together, about 1000 m2 or 10.000 FT2 of collectors are used. Most of them are flat plate collectors.

All the buildings are in a very advanced stage of construction and practically completed also, all of the crucial parts of the solar systems have been already incorporated in the building; insulations, special walls, I.E. Trombe walls, prefabricated panels, and, most important, piping. Flat palte collectors, storage tanks, and instrumentation are being installed in a number of apartments. It is expected that the installation will be completed by the spring of 1982.

Typical schemes of the several installation are shown in Figs. 2 and 3.

The systems will be monitored regularly for 2 years, by a company of the ENI Group. The same company will then monitor and provide not only for the solar heating but for the overall energy requirements, thus establishing the basis for the accounting and pricing of a global service for the supply of heat.

It is expected that the solar systems will provide, on the average, 80% of the needs for hot water, about 40% of space heating and about 40% energy saving in those apartments using Trombe walls.

CONCLUSIONS

The project is the largest of its kind in Italy and perhaps in Europe it is expected that it will provide important information, so much needed, to asses the performance of solar energy system, to provide design data and most important, to provide data for economic and feasibility analysis. The political, legal, promotional and social significance of the project are also of relevance. The project are also of relevance. The project is a joint cooperation effort among a large construction company, a Regional Government, the solar industry and the user (the homeowners).

In this cooperation, in fact, the construction Company has provided the buildings and has allowed for all the modifications of standard design and construction techniques for the application of active and passive solar systems. The solar industry has provided the solar systems and, at least in this case, has anticipated most of the investment capital which will be recuperated, if the project is successful. Within the foreseen scheme of princing of the entire energy supply, the Regional Government has helped with legislation and financial incentives. There are finally, the homeowners which have been already alerted and in some cases, trained in special conferences, for the operation of the new heating systems. The cooperation among all the different partners has been essential for bringing the demonstration project at the present stage of near completion. The cooperation will he also essential for the successful operation of the project, in the years to come. This project will constitute an interesting and important bench mark for future initiatives in Italy.

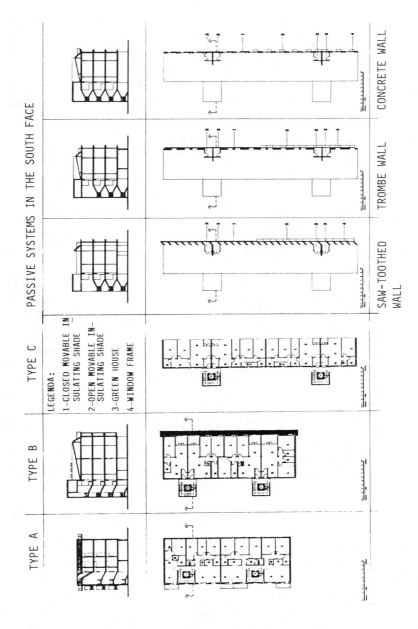

Fig. 2

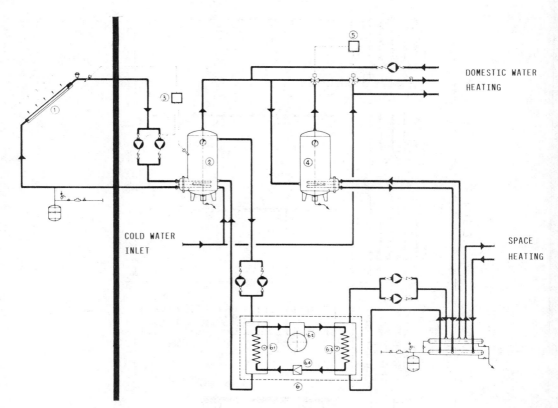

Fig. 3. Space heating and domestic water heating system with water to water heat
pump in series with the solar system.
Preliminary diagram.

ACKNOWLEDGMENTS

The project has been conceived and managed by AGIP S.p.A., of the ENI Group, together with the Regional Government of Tuscany and the construction Company "Consorzio Regione Etruria". Giusti Progetti of Florence, Arch. S. Los of Venise, and
INSO, another Company of ENI, have been the architect engineers for the design
and installations of active and passive solar systems.

CONTROL SYSTEMS FOR DISTRICT HEATING: SUBSCRIBER STATIONS AND HEATING INSTALLATIONS

J. Boel

Danfoss A/S, Denmark

ABSTRACT

The energy crisis and general subscriber needs are currently changing the demands on the control of district heating systems, especially in the areas of energy savings, comfort, and safety. Low-temperature district heating and suitable control systems for subscriber stations and heating installations are no doubt made to comply with these demands.

INTRODUCTION

Why Use Energy-saving Automatic Controls?

Since the energy crisis, the control of district heating systems has been brought into focus in order to reduce overall energy consumption.

The allocation of costs in connection with the production and the distribution of heat in a typical Danish district heating plant before the energy crisis in 1970 and in 1981 appear in the scheme below:

Allocation of costs	1970	1981
Fixed costs Operations Return of investments	70 to 75%	20 to 25%
Variable costs Oil transformed into heat	15 to 20%	55 to 60%
Heat loss in the distribution network	5 to 10%	20 to 25%

Fig. 1. Cost allocations: 1970 and 1981.

From a national-economic point of view, fixed costs in 1970 were three times as high as the variable costs of the heat production. In 1981, the picture changed completely. At this writing, variable costs are three times the fixed costs. Although actual heat loss in the distribution network has not increased, the rise in the price of oil has caused heat loss costs to increase fourfold. Under these conditions, heat plants must reduce the heat loss in the distribution network. During recent years, a number of analyses have been made concerning economic maximum flow temperature, especially in connection with alternative energy sources such as combined production of heat and power and waste heat production. In general, these analyses have concluded that the flow temperature is to be within the range 90 to 110°C, considering the plant installation costs for the network, circulating water flow, the automatic controls, and heat loss insulation. Higher temperatures:

- Increase the heat loss in the distribution network and thus demand additional insulation
- Require expensive installation because of increased safety regulation
- Make demands on the automatic controls which, consequently, make them more expensive
- Reduce the internal circulating water flow and thus the operating costs of the pump capacity

From the subscriber's point of view, the allocation of costs also show that the variable costs of the heat production and the loss in the distribution network are three times the fixed operating costs of the district heating plant. Consequently, the subscriber wants an optimum utilization of the heat in his installation (i.e., a maximum cooling of the district heating water in order to obtain a direct reduction of his own heat consumption). However, to derive additional benefits, the subscriber must also obtain an indirect decrease in return temperatures from the distribution network. This would reduce variable costs of the district heating plant.

Since the allocation of costs are 3 to 1 between variable and fixed costs, it is economical for both subscriber and national concerns to invest in energy-saving automatic controls.

What Requirements Does the Subscriber Make for His Installation and Consequently for the Automatic Controls?

- The installation must work automatically and with a minimum of maintenance
- The installation must work with an optimum of economy and energy utilization
- The installation must be capable of adapting itself to any heating demand and of ensuring constant comfort and room temperature

What Parameters Must the Subscriber Use to Control His Own Station?

- Flow temperature of the distribution network
- Pressure variations in the distribution network and the heating installation
- Method of payment

These parameters are discussed more fully below.

Flow temperature from the heating plant could be constant; as load changes, the circulating water flow is also changed. Flow temperature could also be controlled in proportion to the outdoor temperature. This would result in a constant flow of the circulating water.

Pressure variations in the heating installation might be the result of:

- Load changes in the distribution network caused by:
  - several dwellings being connected to the same distribution network
  - changes in the outdoor temperature which would affect heating demand
  - variations in the demand for hot domestic water during a period of 24 hours

- Load changes in the heating installation caused by:
  - control of the room temperature and thus of the circulating water flow by means of radiator thermostats
  - reduced pipe loss at decreasing load.

The method of payment might be an m^3 payment or an energy payment. In general, the subscriber needs an optimum cooling of the return water, whether the heat consumption is measured in m3 or in energy. An optimum cooling of the return water results partly in a reduction of the subscriber's variable costs and partly through a reduction of the heat loss in the distribution network. This also causes a reduction of the circulating water flow and assimilated pumping effect. The return temperature indicates the capability of the heating installation and the subscriber station to utilize the heat energy.

To summarize installation requirements, automatic controls in the subscriber station and in the heating installation must be able to control the following functions in order to obtain an optimum utilization of the available energy:

- differential pressure across the installation
- flow and return temperature
- time-limited heat supply (e.g., night set-back)

The subscriber station is the connecting link between the heating installation and the distribution network of the district heating plant. The subscriber station converts and adopts the temperature and the pressure conditions of the district heating network to the setting of the heating installation.

SUBSCRIBER STATION - HEATING INSTALLATION

There are two different constructions of subscriber stations:

- direct connection
- indirection connection

In direct-connection stations, water from the district heating plant is led directly into the heating installation where heat energy of the water is emitted; the water is then led back to the plant through the return pipe.

In indirect-connection stations, water from the distribution network of the district heating plant is led into the primary network of a heat exchanger. The heating installation is connected to the secondary network of the heat exchanger (i.e., the two networks are completely separate in pressure, temperature, and water flow).

The subscriber station also heats domestic water through either a storage water heater of a certain size that operates over a long period of time, or by an instantaneous flow water heater in which domestic water is heated immediately before it is consumed.

In the heating installation, radiators are equipped with radiator thermostats, which control the individual temperature in each room, prevent excess temperature, and utilize the solar radiation gains.

Automatic Controls

When developing automatic controls for district heating, both present and future market demands on self-acting and electronic automatic control systems must be considered.

Direct connection. A direct connection is the most economical installation in systems where the flow temperature of the district heating supply is controlled from the district heating plant and where the pressure level, which depends on the topographical conditions, makes it possible (Fig. 2).

The differential pressure control (AVD) ensures constant pressure conditions across the heating installation, irrespective of load variations. It is important that the network of radiators in the heating installation is well-sized to ensure optimum cooling. However, conditions might occur which make cooling insufficient. Such conditions will result in a return temperature, which cannot be accepted either by the subscriber or by the district heating plant. As a solution, a main return valve (FJV) is able to counteract such unacceptable high return temperatures.

In systems with low connection tightness and long branch pipes to the subscriber station and where the heat consumption is calculated according to water use, it might be necessary to compensate for the fluctuating flow temperatures by establishing a bypass at the most *unsuitably located* subscribers. However, a constant bypass might have an unfavorable influence on the return temperature in the system, which would be increased unnecessarily. As an alternative, thermostatic bypass valve (FJV) would improve this condition. The bypass would close when the consumption is high enough to ensure the flow temperature and would open only when the consumption is low.

As previously mentioned, the choice of automatic controls depends on several factors that affect each subscriber's heating payment:

- nominal installed radiator output
- nominal heated floor space
- maximum water flow

For the first two factors affecting payment, a main return valve (FJV) (mentioned in example 1) ensures minimal consumption. For the last factor, a flow limiter (AVD) ensures minimal consumption in the heating installation (Fig. 3).

In the two previous examples, the subscriber station is solely controlled by self-acting controls. Such controls are capable of controlling and limiting the differential pressure, the return temperature, and the water flow in the system.

A mixing system (Fig. 4) is recommended for district heating systems with a constant flow temperature from the district heating plant or for systems where an additional control of the flow temperature is required with a time-limited setback of the room temperature when the building is not in use.

The mixing system consists of a mixing bend, where the hot flow water from the district heating plant is mixed with the return water from the network of radiators. The flow temperature is controlled according to outdoor temperature by means of a weather compensator (ECT 601), which also limits the return temperature. That is, return water is not led out of the mixing network before it is cooled according to a certain max temperature set on the ECT 601. The built-in 24-hour or 7-day clock makes it possible to set back the room temperature at night or when the building is normally not in use.

Examples of Periods of Use in Various Building Categories

	Period of Use	Room temperature reduction in % of the total period
Dwellings	6:00 a.m. to 10:00 p.m.	33%
Rest homes	24 hours	0%
– rooms	8:00 a.m. to 10:00 a.m.	
– dining hall	12:00 a.m. to 2:00 p.m.	65%
	5:00 p.m. to 7:00 p.m.	
Schools	8:00 a.m. to 3:00 p.m.	65%
	7:00 p.m. to 10:00 p.m.	
	on weekdays not on Saturdays and Sundays	
Office buildings	8:00 a.m. to 4:00 p.m.	75%
	on weekdays not on Saturdays and Sundays	

It does not pay to control the flow temperature in centralized, controlled systems that are connected to a district heating plant. This is because the flow temperature to the buildings can never be controlled beyond 65 to 70°C for domestic water heating.

Indirect connection. Indirect connection is mainly used where the topographical siting, the pressure, and the temperature of a district heating plant require separate water networks, especially when connecting old dwellings. The heat exchanger of the indirect systems converts the high temperature/pressure of the primary network to an acceptable level in the heating installation. As the subscriber is anxious to secure the best possible economy and an optimum cooling, automatic controls, which adapt the flow temperature in the secondary network to the internal heating load in dependence of the outdoor temperature, are recommended.

Figure 5 shows an indirect system with a flow temperature control by means of an ECT 601. The flow temperature in the secondary network is controlled and the return temperature in the primary network is limited. Consumption and connection patterns change; consequently, it is recommended that the control valve (AMV 23) be connected with a differential pressure control (AVD) to compensate for pressure variations. In large buildings with a number of radiators in the same network, a constant flow control (AVDA) is also recommended. The AVDA maintains a maximum differential pressure according to radiator thermostats in the heating installation.

Indirect connection (Fig. 6) is used in both small and large systems, although the installation costs of an indirect system are higher for smaller systems than the installation costs of a direct system.

When a large number of independent dwellings are connected to the same system, it is advantageous to divide the system into small heat-exchanger units. The flow temperature to each individual dwelling is centrally controlled by the heat exchanger. To obtain an optimum control in the networks of radiators, each network is equipped with a differential pressure control (AVD), which compensates for pressure variations across the subscriber's stations.

OTHER CONSIDERATIONS FOR AUTOMATIC CONTROLS

Control of Hot Domestic Water

For domestic water, the following control parameters are necessary to obtain an optimum energy utilization:

- control of domestic water temperatures
- limit of the return temperature of the district heating water

The optimum control of the domestic water may vary depending on whether the domestic water is heated in a storage water heater or in an instantaneous-flow water heater and, further, whether the subscriber station is connected directly or indirectly.

Storage Water Heater

Direct connection. In direct connection (Fig. 7), the storage water heater is connected outside the automatic controls in the network of radiators. Optimum control is obtained by equipping the storage water heater with a self-acting temperature control (AVTB), which secures against excess temperatures in the storage water heater by means of the sensor.

To ensure an additionally optimum cooling of the district heating water, the return pipe from the storage water heater is equipped with a return temperature limiter (FVJ), which ensures a minimum heat consumption at large consumptions of domestic water.

Indirect connection. In indirect connections with a heat exchanger the storage water heater is connected to the primary network. Thus, it is possible to control the flow temperature beyond 65 to 70°C, which is necessary to obtain a domestic water temperature of 50 to 55°C. Consequently, a maximum energy saving is achieved when controlling flow temperature.

Instantaneous-flow Water Heater

The return water from an instantaneous-flow water heater (Fig. 8) is always cooled to a minimum; consequently, it is important to size and choose a control valve very carefully. For domestic water, self-acting temperature control with a sensor placed correctly on the outlet side ensures that the domestic water has only the necessary temperature.

As an instantaneous-flow water heater causes short-duration, high-volume loads on the installation and when the load on the network of radiators is also large, it is necessary to ensure an adequate flow capacity to the water heater. As an alternative to a network of radiators, differential pressure control ensures the pressure capacity to the water heater.

Preheated Domestic Water

In large systems or in systems with a large consumption of domestic water, it is advantageous to preheat the domestic water through the network of radiators or by a heat exchanger, either through a storage water heater with two heat coils or through two water heaters connected in series.

The example below (Fig. 9) shows a subscriber station with indirect connection and with a preheating function. The flow temperature in the secondary network of this system is controlled by a weather compensator (ECT 601), which also limits the return temperature in the primary network. The motor valve (AMV 40) controls the heating of the water in the storage water heater. The sensor belonging to this motor valve is placed directly in the network of the circulating domestic water.

With less load on the heating installation, the return water temperature from the heat exchanger is high. This surplus temperature is used for preheating the domestic water and at the same time the AMV 40 reduces the direct heating of the domestic water.

Heating Installation

Heat consumption in any facility depends on local climatic conditions, the construction of connected buildings, how the building is used, and the demand on room temperature/comfort level. Comfort and efficiency is very much influenced by the environment, and temperature is an important parameter. Consequently, temperature must be kept at a desired level throughout the heating season, irrespective of internal and external load conditions. However, an unintended high temperature level will result in a corresponding energy loss, which is not expedient from an economic point of view. To ensure a constant maintenance of correct, comfortable temperatures and an optimum utilization of available energy, each heating installation must be equipped with automatic controls.

Radiators equipped with radiator thermostats result in temperature control in each individual room, prevent excess temperature, and utilize *free heat gains*.

Demands on the Automatic Controls

To ensure optimum energy savings, automatic controls must comply with the following demands:

Pressure control must be able to compensate for possible pressure variations in the system and keep differential pressure in the system within close limits in order to obtain optimum operating conditions for the thermostatic control valves, irrespective of load changes.

Temperature controls must react quickly to reduce the energy consumption (especially when controlling small water flow) and for short-term consumption (e.g., in an instantaneous-flow water heater or in a heat exchanger).

Sizing and hydraulic balance are important considerations when sizing a heating installation; proper dimensions must be selected in order to ensure optimum control. Plans must also include a hydraulic balance for the system. This allows the heat emitting from each radiator to meet present load requirements. To ensure optimum energy savings, the radiator thermostats must also ensure that the radiators do not emit unnecessary heat at night set-back (i.e., when reducing the flow temperature and at venting). Both night set-back and venting result in lower room temperatures than the radiator thermostat setting; thus, the thermostatic element will try to maintain the desired room temperature by opening the valve in proportion to the falling temperature in the room. If the system is equipped with a valve of normal characteristic, the hydraulic balance in the system might be upset, because the water flow through the valve is considerably higher than calculated at sizing. Consequently, the characteristic of the thermostatic valve must rise gradually until it reaches the value and then becomes horizontal again. In practice, the water flow through a fully open valve will then not rise so much (i.e., the hydraulic balance in the system is not upset).

Automatic controls for both subscriber stations and heating installations must be adapted to each other in order to obtain optimum energy savings.

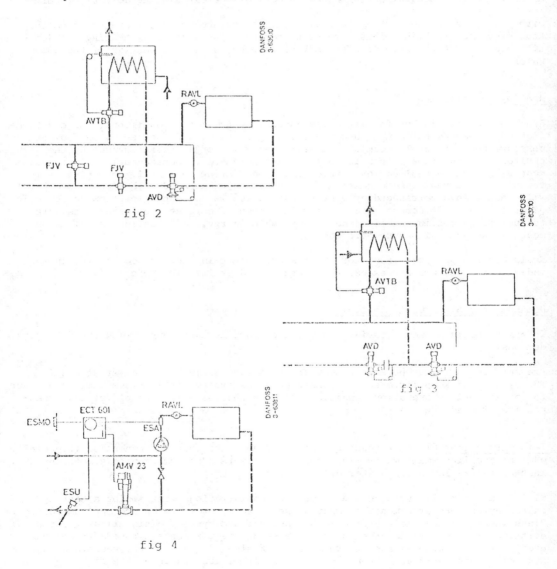

fig 2

fig 3

fig 4

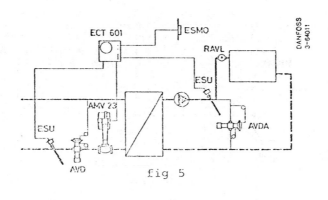

fig 5

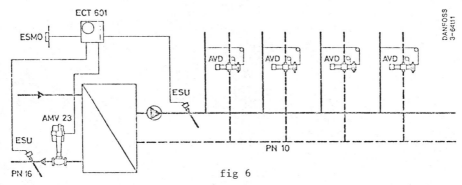

fig 6

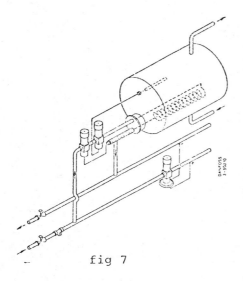

fig 7

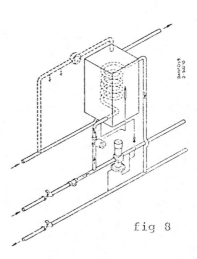

fig 8

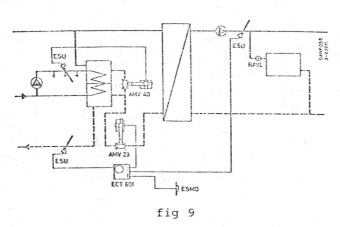

fig 9

THE ROLE OF THE PACKAGED ORC SYSTEM IN INDUSTRIAL WASTE HEAT RECOVERY

L. Y. Bronicki* and W. E. Rushton**

*Ormat Turbines Ltd., Yavne, Israel
**Swenson Division, Whiting Corporation, Harvey, Illinois, USA

ABSTRACT

Organic Rankine Cycle Heat recovery systems are available and can be installed economically on the recovery of waste industrial heat. Packaged skid mounted systems provide the necessary cost reductions to allow for the installation of these units at current power costs.

KEYWORDS

Organic Rankine Cycle; skid mounted equipment; waste heat recovery.

INTRODUCTION

Rankine cycle power generation equipment has been in existence for many years and is generally well known to the engineering community. Power generation using steam Rankine cycle equipment is probably the best known application. In an Organic Rankine Cycle (ORC) system an organic fluid is substituted for the water used in the steam cycle system. The steps that make up the cycle, as shown

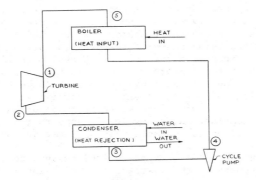

Fig. 1. Typical Rankine cycle flow sheet.

In Fig. 1, are expansion of the vapor across the turbine from point 1 to point 2. Condensation of the vapor and heat rejection occurs in the surface condenser between points 2 and 3.

The pressure of the condensed fluid is raised by means of a cycle pump from point 3 to 4. Heat input to the boiler/vaporizer results in vaporization of the liquid between points 4 and 5.

When the heat source is at a low temperature, the cycle obviously has a limited efficiency. It is therefore important to get as close as possible to the maximum theoretical Carnot efficiency.

To reach this objective, one has to introduce maximum amounts of heat into the cycle at the highest possible temperature. With steam, the boiling temperature has to be lowered so that the steam can be superheated to avoid damaging condensation during expansion in the turbine.

This superheating causes an important reduction of efficiency which, in steam plants, is corrected by multistage extraction at the expense of mechanical complexity. On the other hand, many organic fluids have a saturation line which allows superheating during an adiabatic expansion, thereby allowing better efficiency systems applied in a mechanically simpler system. The typical ORC systems applied in waste heat conversion will use the basic flow sheet as shown in Fig. 1.

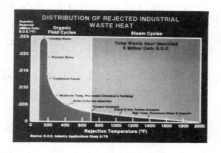

Fig. 2.   Distribution of rejected industrial waste heat.

Waste heat is rejected at various temperatures throughout industry. Figure 2 shows the distribution of rejected industrial waste heat in the U.S. according to a recent Department of Energy survey. As can be seen, waste heat is available at all temperature levels between ambient temperatures and 2000°F. Generally it can be said that the higher the temperature of a waste heat stream the greater its potential value. As the installed cost of an ORC system could be significantly reduced by the use of high temperature heat sources there is a great temptation to design an ORC system to operate in conjunction with these high temperature sources. However, in a well planned heat conservation program these higher temperature sources usually offer more cost effective recovery opportunities such as feedstock or combustion air preheating. As a result, it appears that the most effective market for ORC applications is in waste heat applications below approximately 400°F.

Another important consideration in applying ORC systems to industrial waste heat conservation is the source of the waste heat. Following is a table rating various waste heat sources.

Best Sources:

1. Low pressure steam.
2. Non-corrosive liquid cooling.
3. Partial condensation of hydrocarbons.
4. Clean flue gases.

Heat Sources to be Avoided:

1. Corrosive or fouling liquids.
2. Dirty flue gases.

Waste steam at pressures as low as atmospheric is an ideal heat source. Condensing steam provides high heat transfer coefficients along with a non-fouling service and as such probably provides for the lowest cost ORC installation. Cooling of liquids is also an excellent heat source for ORC systems. The only limitation with cooling of liquids is that generally the liquids cannot be economically cooled below 175°F and therefore if lower temperatures are required additional heat exchangers must be provided. Clean flue gases create problems in utilization due to the need for bringing the flue gases to a heat exchanger where the heat can be transferred to the organic fluid. This is most easily accomplished by installing a waste heat boiler in the stack and converting the flue gas heat to low pressure steam or a hot fluid stream. Corrosive liquids are generally avoided as the additional cost of corrosion resistant material in the boiler/vaporizer tends to make the system more difficult to justify economically.

From the above information we can sum up the following criteria as a guide for a cost effect ORC waste heat recovery system.

1. System must be capable of operating with heat sources of 400°F or lower.

2. System must be capable of operating with waste heat sources such as condensing vapors or cooling of hot liquids.

3. System to be designed to be compatible in non-corrosive environments.

4. Organic fluid must have suitable temperature-entropy diagram and be thermally stable to approximately 300°F.

5. Total installed cost of system must be cost effective.

With the above criteria in mind, the designer of an ORC system then has a number of decisions to make to determine how best to proceed. There are two factors that determine the economic viability of an ORC system -

1. Most likely future power savings.

2. Total installed cost of ORC system.

Predicting the likely power savings from a proposed ORC installation involves substantial uncertainty. However, it can be safely predicted that power costs will increase at a faster rate than the general rate of inflation. This means that the power costs will increase at a faster rate than the cost of ORC systems thus increasing their future attractiveness. We will not attempt to deal with predicting future power costs but it should be noted that using current power rates in evaluating a potential ORC investment is unrealistic.

The ORC system designer can do many things that have great impact on the installed cost of an ORC system. These systems can be assembled from various available components. However, designing systems which can be economically employed require a combination of multiple unit production and a rigorous optimization approach to design seldom required of other equipment systems. It, therefore, becomes necessary at an early date to decide whether or not it is best to proceed with a custom designed field erected unit for a specific application or to use a standardized and pre-assembled skid mounted unit. The standardized package has certain advantages over the custom designed unit. These are –

1. Engineering savings.
2. Low installation cost.
3. Multiple unit manufacturing cost savings.
4. High reliability.

The engineering cost for any unit accounts for up to 20% of the cost of the package. If a standardized pre-engineered package is used the engineering cost can be reduced to only a few percent of the package cost as the total cost of the engineering can be written off on the cost of ten or twenty units instead of only the specific unit being fabricated.

Fig. 3.   Skid mounted 300 KW ORC installation
in a refinery.

Installation costs of standardized package units can be dramatically reduced. The package unit can be completely pre-assembled on a skid before shipment. The only field work required for installation after the skid is placed on its base is connection of the unit to its heat source, cooling water source and the electrical grid.

Multiple unit manufacturing capabilities also help to reduce the cost of the system. Many of the components in the system can be purchased in lots of 10 or more for costs that are 10 to 20% below their single unit price. In addition, manufacturing or assembly problems that may occur on the first or prototype unit can be solved and the resultant savings passed on to all subsequent units.

Standardized units also offer a higher degree of reliability than custom designed units. The skid-mounted pre-assembled unit can be tested at the point of assembly so that it can be thoroughly proven before it is shipped. In this way all of the rotating equipment can be tested under full load and the control system can be tested and preset to reduce field startup time.

The standardized unit normally will have a slightly lower efficiency due to design compromises necessary for standardization. Our studies have indicated that standardized skid-mounted units are practical up to approximately a 2000 KW size. Above this size the equipment becomes too large to effectively place on skids and pre-assemble and, therefore, probably should be custom engineered for each application. With large equipment of this type it is probably best to engineer the system so that the components can be integrated into the users structure.

Another important consideration in the installed cost of the unit is the size of the installation under consideration. The U.S. Department of Energy has studied the impact of size on cost and has found that for a system designed for a 300°F inlet temperature liquid stream that a 1000 KW system would cost approximately 24% more per installed KW than a 3300 KW system. However, this same study indicated that a 300 KW system would cost approximately 74% more per installed KW than the 1000 KW system. From this analysis it can be seen that the use of ORC equipment for waste heat recovery will probably not be feasible on heat streams that produce less than 300 KW.

Fig. 4.   Skid mounted 300 KW ORC installation
in a fertilizer plant.

ORC systems tend to have very low ef ficiencies when the heat input to the system is compared to the actual power produced. The overall system efficiency is determined by the maximum theoretical or Carnot efficiency multiplied by a correcting factor which depends on the nature of the fluid used, particulars of the cycle (with or without superheat, reheat, etc.) and parasitic losses such as windage, friction, cycle pump losses, etc. The Carnot efficiency is a fundamental consideration. This is the highest theoretical efficiency obtainable by any heat engine operating between a given peak temperature and a given rejection temperature. Carnot efficiencies are higher with increasing peak temperatures or lower rejection temperatures. The graph in Fig. 5 shows the effect of increasing peak temperature with a fixed rejection temperature of 110°F on the Carnot efficiency. You will note that a Carnot efficiency of 10% is reached at peak temperatures of 173°F and a Carnot efficiency of 25% is reached at a peak temperature of 300°F. The actual cycle efficiency will be less than the theoretical Carnot cycle efficiency. The correcting factor indicated above by which the Carnot efficiency has to be multiplied to obtain the actual efficiency in well designed systems is 50 to 65%. The graph in Fig. 5 shows the effect of this reduced efficiency on the heat required to produce 1000 KW based on a 60% correcting factor. It can be seen at 173°F peak temperature that the system efficiency drops to 6% and at a peak temperature of 300°F the system efficiency drops to 15%. Practically these temperatures then illustrate the system efficiency range for the normal applications of an ORC system in waste heat recovery. These low efficiencies virtually dictate that the ORC system can only be installed in true waste heat applications where no value is placed on the heat input into the ORC system.

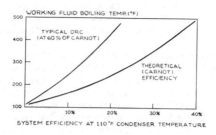

Fig. 5 Temperature-efficiency curve.

To this point we have outlined some of the basic reasons for believing that a packaged skid mounted ORC system is the correct approach in applying ORC systems to waste heat recovery. Our projections indicate that it should be possible to install a 1000 KW package unit for $1000/KW. Figure 6 shows the return on investment based on various power costs. Based on the $1000/KW installed cost a unit can be justified where power costs are 3¢/KWH or more. As we have shown size of the installation effects the economics of an installation and, therefore, the following approximations will help a potential user in evaluating a particular heat source.

Size Factors

Machine size has a significant effect below 1000 KW units and a reduced effect 1000 KW size adjustment factors –

Below 1000 KW   Cost Ratio – (Size Ratio)$^{0.4}$

Above 1000 KW   Cost Ratio – (Size Ratio)$^{0.7}$

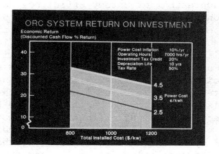

Fig. 6. ORC system return on investment.

Waste Heat Source

Assuming the same peak temperature and heat rejection temperature, different heat sources will impact on the total installed system cost. Following are approximate cost variations due to differences in the preheater/ vaporizer design based on different heat sources.

Hydrocarbon Liquid	No cost adjustment
Condensing Steam	20% reduction
Hot Water	10% reduction
Condensing Hydrocarbon	10% increase

Prepackaged skid mounted systems also reduce the cost of an installation. The following factors reflect the options previously described:

1. Prepackaged Standard Skid Mounted System   No adjustment

2. Single custom engineered-preassembled   20% over base case

3. Single Site Assemblied Custom Engineered System   30% over base case

We believe that cost effective ORC applications exist today. As power costs and industrial power pricing policies outpace the cost of producing ORC systems more applications will become cost effective. The systems are made up of well proven standard components and several vendors have initial units in commercial use. System standardization and multiple unit manufacture will further reduce costs and improve reliability. Standard units of 300 to 600 KW are in commercial use today from several suppliers. Larger pre-assembled units currently under design will provide further improved economics.

Figure 3 shows a skid mounted 30 feet long 300 KW Ormat ORC installed in a re-finery, cooling a heavy fuel oil stream from 370°F to 200°F. Figure 4 shows a 20 foot long 300 KW Ormat ORC installed on a 240°F waste wet steam stream in a fertilizer plant. Figure 7 shows an unattended fuel-fired ORC, 3000 of which have been manufactured by Ormat in the last 16 years.

Fig. 7. Unattended fuel fired ORC unit.

Properly selected applications in medium to high cost power regions can yield acceptable economic returns today using either very large scale custom systems or those packaged systems commercially available. Simple turbine cycles, low to moderate waste heat utilization rates, and economical heat exchanger surface area selections can yield attractive economics especially on condensing steam, hot water and liquid hydrocarbon applications which generally involve low cost in-stallation.

CHAUFFAGE DE SERRES A L'AIDE DES REJETS THERMIQUES DE CENTRALES EN CIRCUIT FERME

Y. Cormary et C. Nicolas

*Ingénieurs à la Direction des Etudes et Recherches d'Electricité de France,
Laboratoire des Renardières, France*

RESUME

Pour améliorer le bilan énergétique au niveau du chauffage des serres, il est possible d'envisager l'utilisation de la géothermie, du solaire de la pompe à chaleur, et enfin du rejet thermique des centrales.

Dans ce document, on examine le dernier cas pour une centrale refroidie par des aéroréfrigérants. Un modèle de calcul CLIMASER a été mis au point à cet effet, permettant de dimensionner thermiquement un projet de serres et d'en évaluer la rentabilité par rapport à un système classique de chauffage au fuel.

Le rejet, pour être rentable, ne doit pas exiger des températures minimales supérieures à 15°C, mais la nécessité de valoriser l'investissement impose une utilisation quasi-continue et en conséquence une fourchette de température très réduite.

Une application à la centrale de Dampierre en Burly disposant d'un débit de rejet de 2 m3/s est présentée pour un chauffage exigeant une température minimum de 14°C. L'installation envisagée permet l'implantation de 70 hectares de serre avec un temps de retour de 2,1 an pour le supplément d'investissement de ce système de chauffage par rapport au système classique.

I - INTRODUCTION

La présente étude concerne l'utilisation des rejets thermiques d'une centrale en circuit fermé pour chauffer des serres.

Sur le site de Dampierre en Burly, EDF met à disposition des agriculteurs 1 m3 d'eau disponible sur une paire d'aéroréfrigérant pour des usages agricoles. Compte tenu du niveau thermique obtenu sur les centrales en circuit fermé, l'utilisation est envisagée par passage direct de cette eau dans la serre qui pour des raisons de sécurité agronomique doit obligatoirement posséder un chauffage de secours.

Les hypothèses adoptées pour l'étude sont :

a) Température de l'eau de rejet

La température de l'eau résulte de la température extérieure humide et du régime de marche de la centrale. Les hypothèses admises dans cette étude sont données en annexe 1 (modulation systématique sur les trois régimes de marche de la centrale).

b) Serre et système de chauffage

L'étude concerne des serres en verre équipées d'un système de ventilation.
Le chauffage est envisagé :

- soit classiquement à l'aide d'une chaufferie au fuel lourd dissipant ses calories
à l'intérieur de tubes aériens lisses (0,3 m2/m2 de serre).

- soit à l'aide des eaux de rejets délivrées par la centrale et dissipées par un
système AERO-SOL (réf.1) composé d'un aérotherme à eau (0,1 m2/m2) et d'un système de chauffage du sol par tuyaux enterrés ainsi que d'un chauffage d'appoint-
secours constitué par un système air pulsé fuel (200 W/m2) fonctionnant lorsque
l'eau de rejet ne permet pas de maintenir la température de consigne fixée.

c) Calculs

Les déperditions de la serre sont calculées à l'aide du modèle CLIMASER (réf.2).
Les données météorologiques sont celles de la station de TRAPPES 1975 au pas de
3 heures.

d) Cultures

2 cultures sont essentiellement envisagées , celle demandant des températures minimum de l'ordre de 15°C (concombre, rose) et celle exigeant de l'ordre
de 8°C (tomate) et cultivée sur l'ensemble de l'année. Une rotation particulière -
laitue, tomate, concombre - est aussi envisagée.

II - RESULTATS THERMIQUES

Le modèle permet de connaître au pas de 3 heures, les consommations de chauffage
(classique ou appoint et rejet) et les niveaux de température atteints dans l'air
ou le sol de la serre, ainsi que la fréquence d'utilisation des rejets. Un exemple
détaillé des résultats obtenus est fourni en annexe 2 pour la rotation laitue-
tomate-concombre.

Tableau I : Consommation annuelle en kWh/m^2

Température minimum garantie		Fuel * seul	Rejet + appoint		
			Rejet	Appoint *	TOTAL Rejet + appoint
1	15°C	322	350	110	460
2	8°C	87	362	8	370
3	Rotation L - C - T 2° -8°C -15°C	174	228	38	266

*Consommation aux bornes de la serre hors rendement de chaudière et de distribution.

III - DISCUSSION DES RESULTATS

Le chauffage par rejet et appoint délivre globalement plus d'énergie que le chauffage au fuel car la régulation n'est pas la même.

Le fuel est utilisé seulement en-dessous de la consigne (8 ou 15°C) quel que soit le système (fuel seul ou rejet + appoint fuel).

Par contre, le rejet dont les frais proportionnels sont faibles est utilisé même au-delà de cette consigne avec un maximum cependant de 20°C dans le premier cas et de 16°C dans le second cas. Cette surchauffe induite par le rejet est généralement bénéfique aux cultures jusqu'à une certaine valeur dépendant du type de plante. Compte tenu de cette surchauffe, il est donc difficile de comparer l'effet thermique des rejets à celui d'un système fuel puisque les productions agricoles ne seront pas les mêmes avec un bénéfice plus ou moins grand pour le système rejet + appoint.

IV - PRINCIPALES PARTICULARITES

a) compte tenu de l'investissement important, il est nécessaire d'utiliser le rejet le plus souvent possible, aussi des cultures peu exigeantes en température comme la salade et ne valorisant pas les surchauffes, imposent d'arrêter la circulation du rejet pendant 50% du temps durant l'hiver (voir annexe II.B). Aussi ce type de culture est-il à rejeter près des centrales.

b) la surchauffe obtenue dans l'air de la serre par rapport à un chauffage classique au fuel est d'autant plus élevée que la culture peut supporter sans dégât un seuil minimum absolu de température plus faible (réglage du thermostat déclenchant le chauffage d'appoint). Au-delà de 14°C cette surchauffe est presque nulle et l'effet thermique du rejet est comparable à celui du fuel (figure 1).

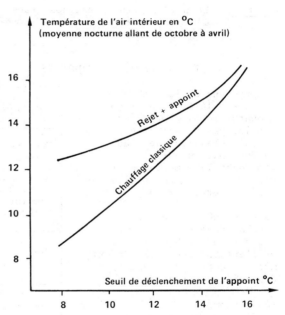

Figure 1 - Surchauffe obtenue grâce au rejet selon le seuil de déclenchement de l'appoint.

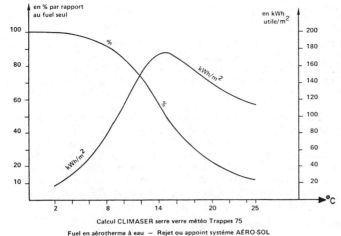

Calcul CLIMASER serre verre météo Trappes 75
Fuel en aérotherme à eau — Rejet ou appoint système AÉRO·SOL

<u>Figure 2</u> – Economie de fuel réalisée grâce à
l'emploi du rejet selon la tempé-
rature minimum à garantir.

c) si l'on élève le seuil
de température minimum à
garantir à la serre - déclenchement de l'appoint -, la
consommation de ce dernier
représente une part de plus
en plus importante dans l'ensemble des besoins. A 15°C,
le rejet ne représente plus
que 50% de la fourniture des
besoins thermiques globaux
(figure 2), mais la quantité
absolue de fuel économisée
est importante. Par contre
au-delà de cette valeur,
la quantité de fuel économisée
diminue et la valeur relative
chute considérablement.
En-dessous de 8°C le pourcentage d'énergie économisée est
très élevée mais la valeur
absolue est très faible.

V – RESULTATS ECONOMIQUES

Compte tenu de la faible surchauffe à 14°C, nous allons
comparer le système rejet +
appoint garantissant 14°C à un
système classique de chauffage
au fuel par tuyaux aériens
demandant la même température
et 467 kWh réellement dissipés
dans la serre par an pour 1 m2.

Dans les mêmes conditions,
le chauffage par eau de rejet
fournira 308 kWh utiles/m2
et l'appoint fuel 158 kWh utiles/
m2. Le détail des résultats
économiques est en annexe 3.

Tableau II – Bilan économique

F/m^2	Investissement chauffage	Charges annuelles chauffage
Fuel seul 14°C (fuel lourd)	74,80	76,32
Rejet 3 niveaux + appoint fuel lourd 14°C	136,50	46,72
(Rejet + fuel) –(classique)	+ 61,70	– 29,60

Substituer au système classique de chauffage au fuel l'emploi de l'eau de rejet
et d'un appoint entraîne un investissement de 1656 F pour économiser une TEP.
Le temps de retour de l'investissement est de 2,1 ans.

Ces valeurs sont très liées à la façon dont on utilise la serre. Les figures 3
et 4 montre l'évolution du gain des charges annuelles de chauffage par rapport
à la solution classique, selon la température de consigne fixée et le pourcentage
que représente ce gain de charges annuelles obtenues grâce au rejet par rapport
aux charges annuelles de chauffage dans le cas d'un système classique au fuel
lourd. En-deçà de 6°C l'emploi du rejet entraîne une augmentation des charges
annuelles, au-delà de 20°C, l'économie n'est que de 10% par rapport à la solution
classique ce qui est faible pour amener les serristes à se déplacer vers les centrales. Par contre entre 8 et 15°C l'économie sur les charges annuelles de chauffage atteint 30 à 35%. C'est donc vers des cultures valorisant un tel climat que
doivent s'orienter les serristes qui utiliserons les eaux de rejet.

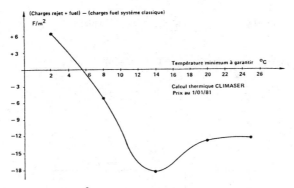

Figure 3 – Modification en F/m^2 des charges annuelles liées au chauffage (amortissement, fuel, électricité, entretien) lorsque l'on utilise l'eau de rejet à la place d'un système classique au fuel.

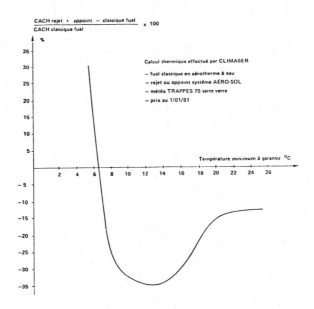

Figure 4 – Evolution du rapport $\dfrac{\text{variation des charges annuelles de chauffage}}{\text{charges annuelles systèmes classiques}}$

VI - CONCLUSIONS

L'emploi de l'eau de rejet présente donc un intérêt certain à la fois pour le serriste et pour la collectivité à condition que les cultures envisagées ne demandent pas de niveau de température minimum trop élevé (supérieur à 15°C) mais utilisent cependant largement les calories de l'eau tiède : les températures maximum à respecter doivent être supérieures à 15°C -20°C et les températures minimum supérieures à 8°C ; ceci pendant une partie importante de l'année.

814

ANNEXE 1

Température de l'eau de rejet (Centrale en circuit fermé)

Régime de marche admis pour la centrale

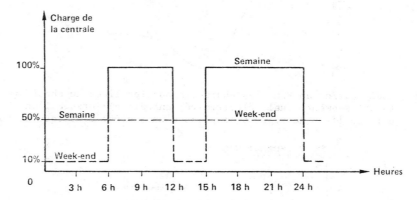

Évolution de θ rejet selon le régime de marche

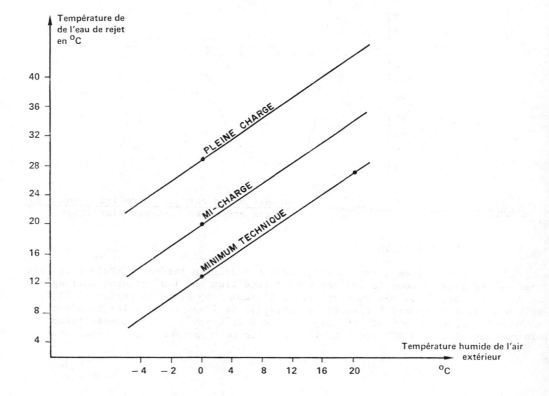

ANNEXE 2 — EXEMPLE DE RÉSULTATS DE PROGRAMME CLIMASER

Rotation Laitue — Concombre — Tomate

Chauffage par aérotherme et sol avec eau de rejet

Appoint air pulsé — météo : Trappes 75

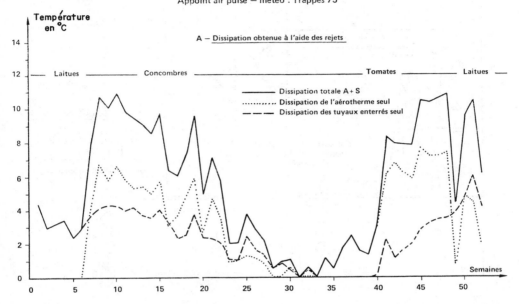

A — Dissipation obtenue à l'aide des rejets

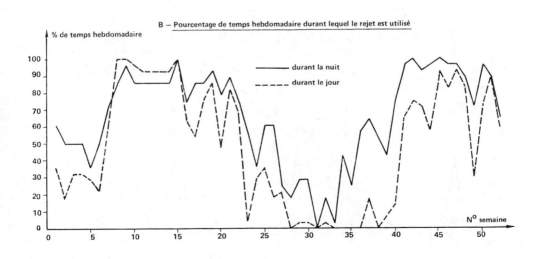

B — Pourcentage de temps hebdomadaire durant lequel le rejet est utilisé

ANNEXE 2 (suite)

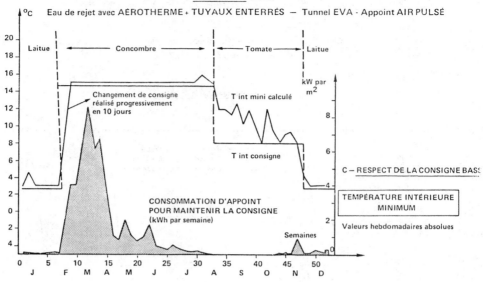

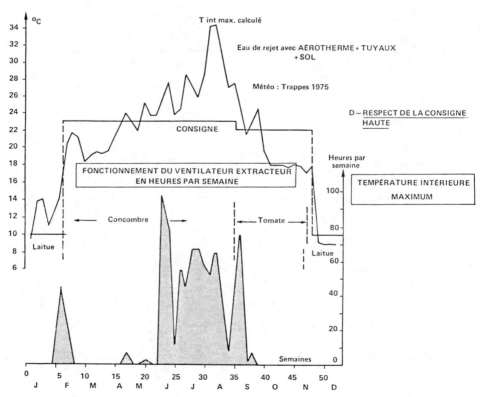

ANNEXE 3

BILAN ÉCONOMIQUE GLOBAL

Le chauffage assure une température de 14°C dans une serre en verre

F/m² de serre	Chauffage classique fuel lourd	Rejet + appoint fuel domestique	Rejet + appoint fuel lourd
Investissement	74,80	111,50	136,50
Variation par rapport au chauffage classique		+ 36,70	+ 61,70
Charges annuelles	76,32	60,58	46,72
Variation par rapport au chauffage classique		− 15,74	− 29,60
Temps de retour du surinvestissement		2, 3 ans	2, 1 ans
Francs par TEP économisée		1050	1656

Bibliographie

1 - Efficacité thermique et agricole de 2 types de chauffage de serre permettant d'économiser l'énergie C. NICOLAS. Congrès CIGR - Alméria - Fév. 81.

2 - How calculate greenhouse heat losses according to the plant comfort and the type of greenhouse heat exchanger Y. CORMARY - Congrès Dublin Septembre 1980.

3 - Chauffage de serre à l'aide des rejets thermiques de la centrale de Dampierre en Burly Y. CORMARY - C. NICOLAS note interne EDF W 1708.

HOFFMANN-LA ROCHE SLOW-SPEED DIESEL COGENERATION PROJECT

J. P. Davis and S. E. Nydick

Thermo Electron Corporation, 123 Second Avenue,
Waltham, MA 02254, USA

ABSTRACT

This paper describes the design and performance characteristics of a residual fuel fired cogeneration system to be installed at the Hoffmann-La Roche Inc., Belvidere, New Jersey plant. The basic system is described as follows:

- Residual fuel fired, slow-speed, two-stroke diesel engine rated at 23,300 kW.

- Supplementary fired recovery boiler on the diesel engine exhaust to provide 160,000 lb/hr 225 psig process steam to the plant.

- Cooling water heat exchangers to supply process hot water using the waste heat from the cylinders, pistons, and turbocharger.

- Basic power-only heat rate of 8690 Btu/kWh, improved to 3752 Btu/kWh with waste heat recovery.

- Overall fuel utilization efficiency in excess of 85 percent.

- Future operation on compatible coal-derived liquid fuels such as SRC-II.

The Hoffmann-La Roche facility is a prime example of a chemical plant where diesel cogeneration can totally eliminate the need for purchased electricity. The slow-speed, two-stroke diesel burns residual fuel oil, the liquid fuel most commonly used by U.S. industry and utilities. This type of diesel is used for electrical power production in many countries in Europe, Africa, and the Middle East. As a notable exception, none of these large units are found in the United States. The engine exhibits high power conversion efficiency, reliable economic performance, and the capability of burning coal-derived liquid fuels. The cogeneration system is scheduled for operation in 1982 with construction starting in 1981.

BACKGROUND

This diesel cogeneration program has the primary objective of providing near-term, high visible, and original hardware demonstrations to increase industry's interest in cogeneration systems and to expedite technology transfer. The chemical industry is currently a major user of cogenerated power. On the average, about one-third of the industry's power is produced by a combination of gas turbines and steam turbines. The gas turbines are primarily used for mechanically driven compressors and pumps, and the steam turbines for electrical generation.

Three state-of-the-art cogeneration approaches can be considered to increase the fraction of industry-generated power. The power-generating capability of existing boiler-steam turbine systems could be increased by raising turbine inlet pressures. With the average steam pressure in existing plants in the 450 to 800 psig range, an increase to 1200 to 1500 psig would result in power increases of about 25 to 30 percent. In general, however, the cost penalty of scrapping the existing boiler and installing a new high pressure boiler prohibits such modifications.

Additional gas turbines could be considered. With present power conversion efficiencies in the range of 25 to 30 percent and potential improvements to about 33 percent, gas turbines offer significant advantages over noncondensing steam turbines for meeting the power requirements of the industry. Fuel considerations, however, introduce serious questions regarding operation and maintenance costs. Natural gas or distillates are preferred fuels for gas turbines. Metallurgical limitations would demand extensive treatment of residual oil to remove impurities, such as sodium, vanadium, and ash, to acceptable levels. Coal-derived liquid fuels currently under development would probably also require further purification for gas turbine applications. High maintenance costs must be anticipated for the 24 hours a day, 7 days a week type of operation required for the industry.

The slow-speed, two-stroke diesel engine is the one state-of-the-art power conversion system that is uniquely qualified to meet the power requirements of the industry and that has the demonstrated capability to burn residual-type fuels. Diesel engines exhibit power conversion efficiencies in the order of 40 percent, thereby minimizing the quantity of fuel burned per unit of generated power. Engine performance remains essentially constant down to 50 percent load. This contrasts with gas turbine performance, which decreases significantly at part load. Effective engine reject heat recovery can be accomplished from the exhaust gas, the jacket, and the turbocharger air coolers. Extensive use of these engines in marine applications (one engine per ship) attests to the reliability and economics of slow-speed, two-stroke diesels under demanding operational conditions and with heavy petroleum fuels. The inherent characteristics of the engine and exploratory tests recently completed indicate that the engine is capable of burning coal-derived fuels when such fuels become available.

The cogeneration system demonstration for the chemical industry, therefore, is based on the use of a slow-speed, two-stroke diesel engine. Engines of this type are not currently in use in this country. The proposed effort will provide an industrial demonstration of such a system under typical U.S. operating conditions, and will clearly establish the technology, capabilities, and applicability of diesel cogeneration systems for achieving economic energy savings.

The diesel engine cogeneration system demonstration will be located at the Hoffmann-La Roche plant in Belvidere, New Jersey. The purchased plant electrical load is 20,000 -23,000 kW. Approximately 900 kW of mechanical drive power is currently cogenerated in the plant, using noncondensing steam turbines. Steam is produced in residual oil-fired boilers at 650 psig and 200 psig. Process steam is used at 200 psig and 85 psig. With the existing boiler and process steam pressures, the plant is operating close to its maximum steam turbine generating capacity and must purchase power from the local utility. The Belvidere plant is a prime example of a chemical plant where diesel cogeneration can totally eliminate the need for purchased electricity.

The cogeneration system includes the following:

- Slow-speed, two-stroke diesel engine rated at 23,300 kW.

- Supplementary fired recovery boiler on the diesel engine exhaust providing 225 psig process steam to the plant.

- Heat exchangers recovering engine reject heat from the jacket, pistons, oil and turbocharger and providing 170 F hot water to the plant.

The high fuel utilization in the supplementary fired recovery boiler significantly improves the overall cogeneration system energy utilization factor.

The design engineering is complete, the major components such as the diesel, generator, and boiler are already on order. The contracts for these components includes manufacture, delivery and erection and installation. The bid package for the general contracting, stack, and crane have been issued and contracts will be awarded shortly. Bid packages for the primary electrical, secondary electrical and instrumentation and controls will be issued shortly also. Construction is expected to begin as soon as the contract for general contracting is awarded. The plant is expected to start up in mid 1982.

SPECIFIC PROGRAM OBJECTIVES

The specific program objectives of the proposed cogeneration system demonstration program in the chemical industry are:

- Demonstration of Cogeneration Technology

- Fuel Conservation

- Economic Operation

- Demonstration of the ability to obtain all necessary environmental and fuel use permits

- Demonstration of the economic and regulatory benefits of the Public Utility Regulatory Policy Act

The overall success of the program will be measured in terms of the degree of success achieved for each objective.

Technology demonstration involves four major elements: The diesel-generator, the exhaust heat recovery boiler, the hot water heat exchanger, and operation in parallel with the local utility. The successful demonstration of each element of hardware is directly related to its performance and reliability under actual industrial operating conditions. Performance and reliability will be determined in terms of input-output measurements as a function of time, and the availability of the equipment to meet the demands of the plant.

Fuel conservation is determined as the difference between the fuel consumed by the cogeneration system and the total fuel (industry and utility) required to produce equal quantities of electricity and process heat in a noncogenerated system. Fuel consumption by the cogeneration system is directly measured. By measuring the electrical and thermal usage in the plant, calculated values of the noncogenerated industry and utility fuel consumption can be established. These calculations use the efficiency of the existing steam boilers and the effective heat rate of the utility currently serving the plant.

Economic operation of the cogeneration system is measured in terms of the annual costs of supplying the plant's total process energy needs. Comparison of calculated noncogenerated operational costs (considering measured energy consumption and current energy costs) with the costs of the cogeneration system established annual cost savings resulting from the use of the cogeneration system. The cost savings involve differential fuel costs, property taxes and insurance, and operation and maintenance costs. Combining the annual cost savings with the capital costs of the cogeneration system provides the first-year payback period. The payback period is transformed into after-tax return-on-investment (ROI), considering specific factors including federal and state tax rates, investment tax credit, and depreciation schedule.

A number of permits must be obtained prior to the start of construction. Of primary importance are the Fuels Use Exemption allowing the use of liquid fuel in the cogeneration system and the federal and state environmental permits. The Fuel Exemption is required as a result of congressional passage of the Power Plant and Industrial Fuel Use Act of 1978.

The applicable environmental permits are for Prevention of Significant Deterioration (PSD) and state boiler permits. PSD permitting is presently administered by the regional office of the U.S. Environmental Protection Agency, although the State of New Jersey is expected to assume its administration shortly.

The Public Utility Regulatory Policy Act (PURPA) eliminates a "qualified cogeneration" from much of the regulatory entanglement involved with the sale of power and entitles the cogenerator to sell all power generated at avoided costs and to buy the power required for plant operation at normal utility rate.

PLANT DESCRIPTION

The Belvidere plant is located along the Delaware River in Warren County, New Jersey. The 500-acre site was officially dedicated in 1972 and has a manufacturing capacity of almost 10,000 tons of Vitamin C a year.

A plot plan is illustrated in Fig. 1. The location of the proposed diesel cogeneration system is shown. The plant has adequate room for this system and convenient interfaces for power, fuel, and support services.

The total average steam output of the plant's boilers is 430,000 lbs/hr with about 65% generated at 650 psig/700 degrees F and 35% at 200 psig. The 650 psig/700 degrees F steam is utilized in existing back pressure turbines supplying mechanical power to the plant, reducing the steam to 200 psig, 85 psig, and 25 psig. Two hundred psig and 85 psig steam flows are utilized for process uses and 25 psig is utilized primarily for deaeration.

SYSTEM DESIGN

System Description
The diesel cogeneration system is designed to integrate with the existing plant and to be as energy efficient as possible. The system energy flow diagram is shown in Fig. 2.0.

Waste heat from the diesel engine is recovered from the exhaust gases, air cooler, and engine water cooling circuits. In order to maximize overall thermal efficiency, the temperature levels of the waste heat were matched to the plant thermal requirements. The exhaust gas, at a temperature of 550 degrees F, is the highest quality heat and is utilized to raise 225 psig saturated steam in a supplementary fired boiler. The boiler is supplementary fired because Hoffmann-La Roche has a requirement for 150-160,000 lbs/hr 200 psig steam, much greater than the amount that can be obtained without additional fuel input. Additional oxygen for combustion beyond that already contained in the flue gases is not necessary as a result of the large amount of excess air used in the diesel engine cycle.

The air temperatures entering and leaving the diesel air cooler are about 300 degrees F and 100 degrees F respectively. This waste heat source is utilized to preheat the feedwater for the supplementary fired boiler in an air to water heat exchanger from a nominal 70 degrees F to about 200 degrees F prior to entering the deaerator for oxygen scavenging. Of the sources of waste heat from the engine cooling circuits, the turbocharger is at the highest temperature, about 180 degrees F and the lube oil the lowest, about 120 degrees F. In order to maximize the amount of waste heat recovered for the engine cooling system and maintain a reasonable pinch point in the heat exchanger, only the turbocharger, jacket and cylinder water circuits were utilized. From these sources, about 260,000 lbs/hr of 70 degrees F water is heated to 170 degrees for use as feedwater in existing high pressure 650 psig boilers. Additional low temperature heat as well as the low temperature heat from the lube oil cooler is dissipated in a rooftop cooler.

As shown in Fig. 2, the continuous electrical output rating of the system is 23,300 kW (net). Residual fuel oil consumption is approximately 34.5 bbl/hr

(198.3 x 10^6 Btu/hr) in the diesel engine and 24 bbl/hr (138.4 x 10^6 Btu/hr) for supplementary firing. The energy content of the steam and hot water produced is

212 x 10^6 Btu/hr. If this process heat were produced in a separate boiler, an additional 43 bbl/hr of fuel oil would have to be consumed. A summary of the system performance data is presented in Table 1. The total energy utilization is 86% of the energy input. In terms of electrical generation, the effective cogeneration heat rate is 3752 Btu(LHV)/kWh.

Engine Cycle
The cogeneration system uses a slow-speed, two-stroke diesel engine burning residual fuel oil, the liquid fuel most commonly used by U.S. industry and utilities. These engines were originally developed by European manufacturers for marine applications and presently dominate the civilian marine propulsion field. The engines are also used extensively as stationary power sources in many parts of the world because of high conversion efficiencies and reliable economic performance.

A cross section of the large two-stroke diesel engine with an output power of approximately 23,300 kW is shown in Fig. 3. The engine has 10 in-line cylinders and operates at 120 rpm at rated power. The cylinder bore is 90 mm (35.4 in.), and the piston stroke is 1550 mm (61 in.). The brake mean effective pressure is 12.38 bar (180 psi). The engine has an overall height of 11,600 mm (38 ft.), a baseplate width of approximately 4,000 mm (13 ft.) and a length of about 21,510 mm (71 ft.). The net weight is 980 metric ton.

To achieve reliable operation with heavy residual fuels, the engine design is based on the two-stroke cycle to eliminate valves, and on the use of a crosshead piston to prevent contamination of the crankshaft lubricant with the residual oil combustion products. The crosshead constrains the power piston connecting rod to a linear movement. A seal surrounding the rod effectively isolates the upper fuel-burning section of the engine from the lower crankcase. To increase the power density of the engine, the inlet air is compressed by a turbocharger operated from the engine exhaust gases.

The engine is cooled by a closed, forced-circulation, water-cooling system. Individual circuits provide cooling for the lubricating oil, the pistons, the jacket (cylinder head and walls), and the charge air (after compression in the turbocharger). The engine is started using compressed air. Fuel oil is supplied by a fuel preparation system that removes water and solid impurities and steam heats the fuel to the required viscosity level.

Figure 4 illustrates the engine performance as a function of power output. Specific fuel consumption is relatively constant over a wide power range from 50 to 110 percent of rated load. At full load, fuel consumption is 149 g/hph, at three-quarters load 148 g/hph, and at half load 152 g/hph. Engine speed is automatically regulated by a governor. System control stations can be located either remotely or adjacent to the engine.

Sulzer's experience on the life of parts is summarized in Table 2 for heavy oil operation. Most parts have a useful life of 50,000 hr or longer. The parts that require replacement after 10,000 hr or less are the fuel nozzles and the sealing glands for the piston cooling circuit; parts with the shortest life are the least expensive. The recommended downtime for engine inspection and maintenance is 300 to 450 hr/yr. This downtime is equivalent to a plant availability of 95%.

Supplementary Fired Waste Heat Boiler, Exhaust System, and System Layout
The system layout is shown in Fig. 4. The exhaust gas from the diesel during normal operation flows through a pulsation dampener and then to the supplementary fired waste heat boiler. The purpose of the pulsation dampener is to eliminate any low frequency pressure pulsations which may eminate from the diesel engine.

The exhaust from the diesel engine is at a temperature of 500 degrees F and contains about 65-70,000 lbs/hr of O_2 (about 300% excess O_2) sufficient to more

than combust 138×10^6 Btu/hr required to raise 160,000 lbs/hr of 225 psig steam. In order to incinerate the small amounts of combustibles remaining in the diesel exhaust, introduction of the combustion media into the supplementary fired boiler is staged. About 55% of the diesel exhaust gases are combusted with 100% of the required fuel in a primary burner entering from the side of the furnace firebox. As a result of the large amount of excess O_2 in the diesel exhaust, combustion is

complete with a flame temperature of about 2400 F. Thus, all the unburned combustibles from the diesel engine are incinerated. The remaining 45% of the diesel exhaust gases enter the boiler as under fire air cooling the primary flame but providing sufficient retention time at high temperatures to incinerate the unburned hydrocarbons in the secondary flue gases.

A forced draft fan in the primary flue gas steam provides for the pressure drop across the primary burner. The remaining pressure drop in the system is supplied by the exhaust pressure from the engine. After exiting from the supplementary fired boiler, the exhaust gases flow through ductwork to a self drafting, 205 foot high, double wall steel stack. The boiler also acts as a silencer, dampening the high frequency of noise. The total available pressure head from the engine at 100% load is about 12 inches H_2O, sufficient to overcome

the pressure drop from the dampener, boiler, and ductwork. The boiler can be by-passed with the flue gases going directly through the pulsation dampener, silencer and self-drafting stack.

There are two steam atomizing oil guns in the burner, each capable of handling 100% of the steam load. Also, each can be removed while the other is operating. There is also a compressed air atomizing burner capable of utilizing heated No. 6 oil for a cold start with the capability of generating 25% of full-load steam flow. Ignition is accomplished with propane and an electrical igniter. Flame safety is provided by means of an ultraviolet flame scanner which quickly cuts off the fuel supply in case of loss of flame.

The boiler will be field erected, consisting of water-wall panels, a convective evaporator section and an economizer with refractory around the burner. Since flue gas temperatures will be a maximum of 1500-1600 degrees F, heat transfer by radiation in the water-wall sections will be a minimum, with the majority of the heat absorbed in the convective evaporator and economizer. The boiler has one large main steam drum and two small mud drums. A heating coil in the main steam drum heats the feedwater entering the economizer to 260 degrees F so that metal temperatures in the coldest section of the economizer are higher than the sulfur dew point (fuel 1% S) allowing operation of the boiler with an exhaust temperature of 300 degrees F to obtain maximum thermal efficiency and yet maintain reliable operation without acid corrosion.

Generator

The generator manufactured by Siemens, is a 60 pole, three phase, 13,800 volt, 60 hertz, synchronous machine directly driven by the 120 rpm diesel engine. It is rated at 29,750 KVA with a 0.8 lagging power factor for an output of 23,800 kW at 100% load. Under these conditions, the efficiency calculated by the method of separate losses according to IEEE and NEMA standards is 96.9%. Efficiency at 50% load is 96.0%. The generator has a single pedestal with the last bearing in the diesel (through the coupling) providing the other support. A single pedestal was selected rather than a two pedestal design to eliminate potentially harmful torsional critical frequencies in the 120 rpm range. The moment of inertia

(GD^2) is 2000 ton/m^2. The generator is a totally enclosed, self-ventilated, air-cooled design with the heat content of air transferred to a water cooling loop in an integral mounted heat exchanger.

Excitation of the generator is accomplished by a static regulatory exciter with collector rings. Voltage regulation is within $\pm$ 1% at 0.8 power factor when the generator assumes load from no load to 100% rated load.

The generator will be connected to the plant bus operating at 13,800 volts through a 15 KV switchgear/breaker. Existing transformers convert the 13,800 to 480 volts, 277 volts, etc. for plant use and to 34,500 volts for transmission of excess electricity into the utility electrical grid. Synchronizing equipment and the necessary protective relays are being added to the existing systems and to maintain plant and grid integrity and protection against faults, both within the plant and in the utility grid. A simplified schematic of the generator and the plant tie-in is shown in Fig. 6.

PERMIT APPLICATIONS

Prevention of Significant Deterioration Permit and Environmental Assessment

The most important environmental consideration for the Hoffmann-La Roche Diesel Cogeneration project is the impact on local air quality. The impacts on water quality are negligible since the quantity of water and the quality of the effluent from the Hoffmann-La Roche plant is not influenced by the presence of the cogeneration plant. Waste lube oil and sludge not combusted will be carted away to state-approved landfills.

The impact on local air quality is regulated by the various provisions of the Clean Air Act and its amendments. For areas which are in attainment for regulated pollutants, the PSD (Prevention of Significant Deterioration) regulations promulgated by the U.S. Environmental Protection Agency (EPA) apply. Presently in New Jersey, the regional office of EPA administers these regulations and issues

the required permits.[1] For areas which are non-attainment for the regulated pollutants, the emission offset policy applies and is administered by the state.

According to the definitions promulgated by the U.S. EPA in the August 7, 1980 Federal Register, the Hoffmann-La Roche lant is a major stationary source, as is most large industrial plants, and is thus covered by the regulations. However, the plant with the addition of the cogeneration facility has a "significant"

increase of only NO_x and SO_x emissions[2] with all other regulated pollutants below

the De Minimus value. (Data on the increase of plant emissions with the addition of the cogeneration facility are shown in Table 3). Thus the PSD applications permit contained an air quality analyses for only local NO_x and SO_x emissions.

The results of this analysis showed that for critical receptor points on local high terrain areas and at the plant boundaries there was actually a net decrease in ambient concentrations for NO_x and SO_x with the installation of the plant.

The U.S. EPA is presently evaluating the PSD permit application and approval is expected shortly.

The area is not in attainment for ozone, however, the increased emissions of its precursor, unburned hydrocarbons, is below the De Minimus level eliminating the need for consideration of emission offsets. The state boiler permits have also been submitted and their approval is expected shortly also.

Fuels Use Exemption
The Power Plant and Industrial Fuel Use Act of 1978 requires and exemption for a power plant on fuel burning installation with a heat input greater than 250 million Btu/yr. This act is administered, and exemptions granted, by the Economic Regulatory Administration (ERA) of the U.S. Department of Energy. Although according to current regulations, internal combustion engines generating electricity are not classified as major fuel burning installations (MFBI), the Hoffmann-La Roche cogeneration project requires a fuel use exemption as a result of the inclusion of a supplementary-fired boiler in the system. The cogeneration exemption, submitted to ERA in June, 1980, was applied for on the basis of a savings in natural gas and fuel oil and that demonstration of the technology is in the national interest.

[1] Gradually as state implementations plans are approved by the U.S. EPA, the states will administer and issue the PSD permits themselves.

[2] "Significant" increase for SO_x and NO_x is defined as greater than

40 tons/yr.

The Fuels Use Exemption Application included the Fuels Decision Report. The Fuels Decision Report contained a system description, technical discussions on why alternative fuels can not be utilized, a fuels balance, and an environmental report chapter. ERA is in the last phase of its analysis of the Exemption Application and the official granting of the exemption is expected shortly.

Qualifying Status
The Public Utility Regulatory Act (PURPA) is administered by the Federal Energy Regulatory Commission. The Act eliminates a qualified cogenerator from

regulatory entanglement, and allows the sale of all power at avoided costs[3], and allows purchase of power for plant operation at normal utility rates. Since avoided costs are by definition equal to or greater than average costs, the avoided cost principal improves the economic viability of cogeneration projects.

FERC is currently in the process of revising its definition of a qualified cogenerator to include, at a minimum, low-speed diesels. PURPA requires that state utility regulatory commissions administer the exchange of cogenerated power according to the provisions set forth in the act. The New Jersey Board of Public Utilities (N.J. BPU) has asked the state electric utilities to submit their plans for compliance to PURPA. Hearings on these plans are presently being conducted by the N.J. BPU and a final determination is expected in March of this year.

PERFORMANCE MONITORING

Engine-generator monitoring will include fuel consumption, power output, and operating temperatures and pressures. Heat rejection conditions will be measured in the exhaust-gas and water-cooling circuits to provide engine heat balance data and energy inputs to the heat recovery equipment. Specialized instrumentation will provide output conditions from the exhaust boiler and the hot water heat exchanger. The measurements will include flow rate, temperature, and pressure.

SYSTEMS ECONOMICS

Table 4 presents an economic analysis of the diesel cogeneration project. The

analysis is based on an "n^{th}" industrial installation free of all R&D related costs. Equipment costs are given in 1980 dollars (orders placed in 1980). Energy prices are representative of electrical and fuel costs in mid-1981. Pre-tax savings are estimated at about $4.4 million per year resulting in a pre-tax simple payback of 4.87 years based upon an investment cost of $2.15 million.

There are, of course, great uncertainties in the future costs for fuel and electricity. The above discussion is based upon a pre-PURPA scenario. With the adaption of the avoided cost principle, revenues from the sale of cogenerated electricity to the utility at above average costs should enhance the economic return. The full impact of Three Mile Island on the region's electrical costs have also yet to be determined and could impact system economics. This impact

[3]Defined as the cost of both the energy and capacity that a utility would avoid as a result of the generation of power by a cogenerator. Normally the utility could avoid producing its most expensive increment of power. Thus, avoided costs should be at minimum equal to, but most likely greater than, average costs.

can only improve the economic return on the cogeneration system. There is also great uncertainty in the cost of fuel. However, as a result of the system's favorable heat rates compared to utility heat rates and the utility's continued reliance on fuel oil, future increases in fuel costs should further improve the system payback.

TABLE 1

COGENERATION SYSTEM PERFORMANCE SUMMARY

Net Electrical Output	23,275 kW
Steam Output (225 psig)	160,000 lb/hr
Hot Water Output (180°F)	262,300 lb/hr
Residual Oil Consumption	
Diesel Engine	34.5 bbl/hr
Supplementary Fired Boiler	24 bbl/hr
Engine/Generator Electrical Efficiency	39.3%
Overall Energy Utilization	86.7%
Cogeneration Heat Rate (0.85 boiler efficiency - LHV)	3752 Btu/kWh
Engine Exhaust Conditions	
Flow Rate	478,000 lb/hr
Specific Heat	0.254 Btu/lb-°F
Temperature	550°F
Supplementary Fired Boiler Conditions	
Inlet Gas Temperature	550°F
Inlet Gas Flow	478,000 lb/hr
Outlet Gas Temperature	300°F
Steam Pressure (saturated)	225 psig
Steam Flow Rate	160,000 lb/hr
Feedwater Temperature	213°F
Energy Balance (LHV)	
Engine Fuel Input	198.1×10^6 Btu/hr
Supplementary Fired Boiler Fuel Input	138.4×10^6 Btu/hr
Total	336.5×10^6 Btu/hr
Steam Output	$186 \ \times 10^6$ Btu/hr
Hot Water Output	26.2×10^6 Btu/hr
Total	212.2×10^6 Btu/hr
Heat Exchange Losses	9.4×10^6 Btu/hr
Radiation and Convection Losses	3.0×10^6 Btu/hr
Auxiliary Electrical Load	1.6×10^6 Btu/hr
Stack Losses	30.0×10^6 Btu/hr
Total	44.9×10^6 Btu/hr
Electrical Power	79.4×10^6 Btu/hr
Total Useful Energy Output	219.6×10^6 Btu/hr

TABLE 2

SLOW-SPEED ENGINE COMPONENTS WORKING LIFE
HEAVY FUEL OIL OPERATION

Component	Useful Life (hours)
Crankshaft	No Limit
Crosshead Pin	No Limit
Main Bearing	50,000
Connecting Rod Bearing	50,000
Crosshead Bearing	35,000
Crosshead Slipper	60,000
Crosshead Oil Link Bearing	70,000
Canshaft Gear Drive*	70,000
Camshaft Bearing*	70,000
Fuel Cam, Plunger*	35,000
Fuel Pump Valve*	20,000
Fuel Nozzles*	10,000
Governor	35,000
Governor Drive	70,000
Turning Gear	90,000
Starting Air Control Valve	60,000
Starting Air Inlet Valve	30,000
Scavenge Air Valves	25,000
Piston Head*	35,000
Piston Skirt*	50,000
Piston Ring*	12,000
Piston Rod Stuffing Box, Rings*	15,000
Piston Cooling Telescopic Gland*	6,000
Piston Cooling Telescopic Pipes*	35,000
Outer Cylinder Cover*	50,000
Cylinder Head*	50,000
Cylinder Liners*	45,000
Cylinder Lubricators	50,000

*The above represent the lower limits when running under
particularly unfavorable conditions.

TABLE 3

SUMMARY OF SOURCE EMISSION CHANGES (tons/yr)

Pollutant	Without Diesel Cogeneration System	With Diesel Cogeneration System	Increase (Decrease)
Carbon Monoxide	83.5	89.5*	6
Nitrogen Oxides	638	2910	2272
Sulfur Dioxide	2600	3020	420
Particulate Matter	216	187*	(29)
Ozone (VOC)	16.7	26.5*	9.8
Lead	0.24	0.29	0.05
Asbestos	nil	nil	nil
Beryllium	nil	nil	nil
Mercury	<0.012	<0.014	<0.002
Vinyl Chloride	nil	nil	nil
Fluorides	nil	nil	nil
Sulfuric Acid Mist	40.8	47.2	6.4
Hydrogen Sulfide	nil	nil	nil
Total Reduced Sulfur	nil	nil	nil
Reduced Sulfur Compounds	nil	nil	nil

*90 percent of carbon monoxide and hydrocarbons and 37 percent of particulate matter formed in diesel engine are consumed in supplementary-fired heat recovery boiler.

TABLE 4
SYSTEM ECONOMIC SUMMARY

Installed System Costs

Installed Equipment	$19,017,000
Engineering and Construction Supervision	1,171,000
G & A	1,332,000
TOTAL	$21,520,000

Annual Operating Costs

Residual Oil at $3.33/10^6 Btu (HHV) ($0.89/gal)
$(0.95 \times 8760/hr \times 336.5 \times 10^6 \, Btu/hr \times 1.06 \times 5.90 \times 10^{-6} \, Btu)$ $17,513,408

Property Tax and Insurance (2% x $21,520,000)   439,400

Maintenance and Expendables (1.5% x $21,520,000)   322,800

Burdened Labor (8 x $25,000)   200,000

     TOTAL $18,466,608

Annual Value of Electricity Generated

Unit Value = $0.0509/kWh
$(0.95 \times 8760/hr \times 23,300 \, kW \times \$0.0509)$ $ 9,869,642

Annual Value of Process Heat Produced

Based on $3.33/10^6 Btu ($0.89/gal) Residual Oil and
   0.80 Boiler Efficiency (HHV)
$(0.95 \times 8760/hr \times 212 \times 10^6 \, Btu/hr \times \$5.90 \times 10^{-6}/0.8)$ $13,011,447

TOTAL VALUE OF ELECTRICITY AND PROCESS HEAT $22,881,089

Annual Pre-Tax Savings

   ($23,676,090 — $18,664,643) $ 4,414,481

Pre-Tax Simple Payback   4.87 Years

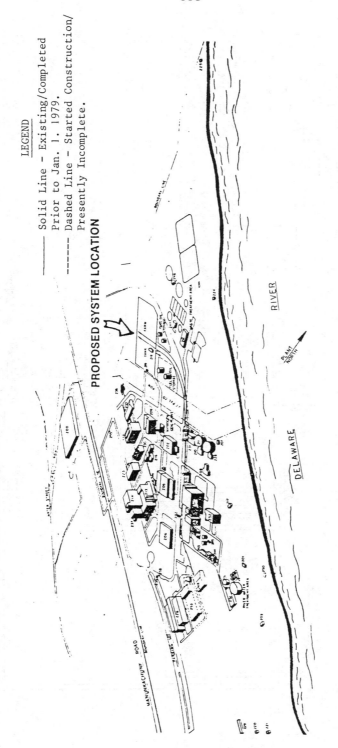

PROPOSED SYSTEM LOCATION

Figure 1. Plot Plan of Belvidere Plant

834

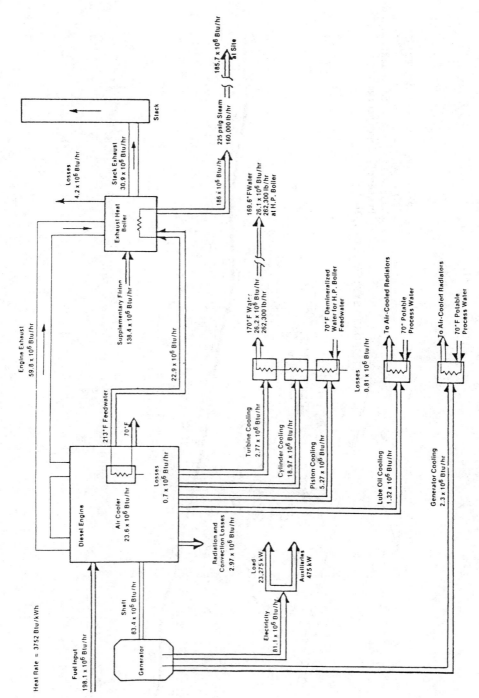

Figure 2. Cogeneration System Energy Balance (23.3 MW)
(Ambient Air at 68°F)

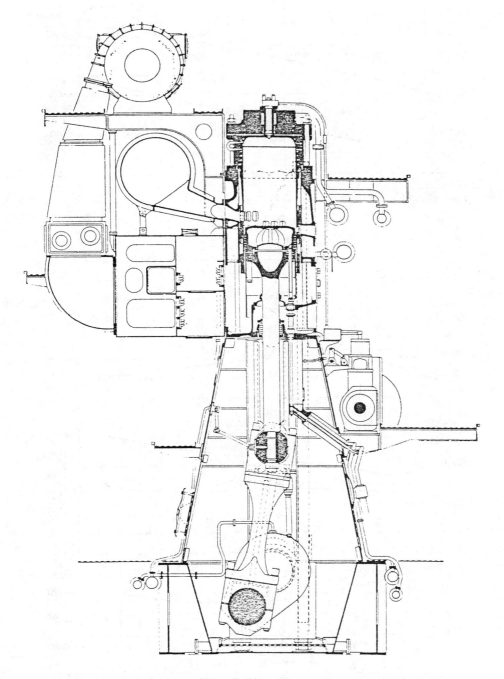

Figure 3. Cross Section of Sulzer Slow-Speed Diesel Engine

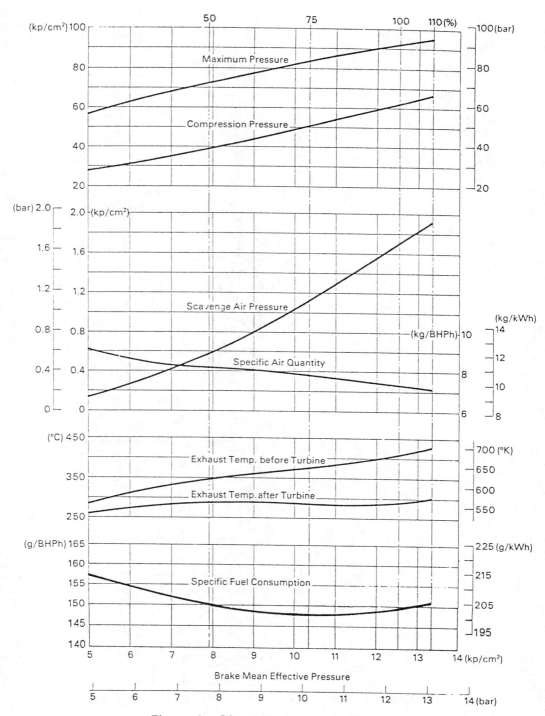

Figure 4.   Diesel Engine Performance

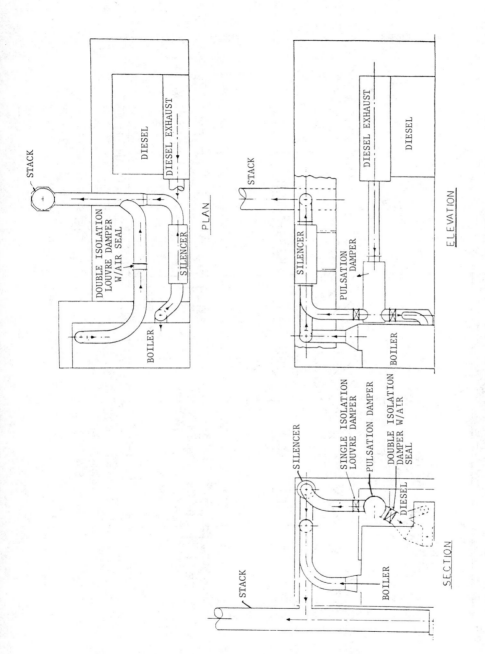

Figure 5. Hoffmann–La Roche Exhaust Gas System

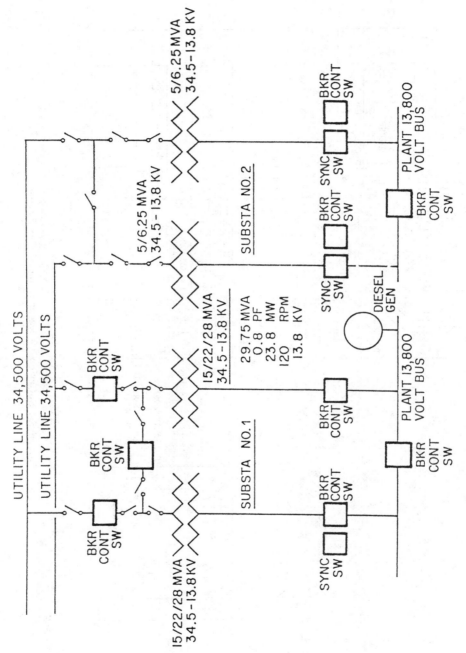

Figure 6. Schematic of Utility/Plant/Generator Tie-In

PRELIMINARY DESIGN STUDY OF UNDERGROUND PUMPED HYDRO AND COMPRESSED AIR STORAGE IN HARD ROCK

A. Ferreira* and P. E. Schaub**

*Electric Power Research Institute, Palo Alto, California, USA
**Potomac Electric Power Company, Washington, D.C., USA

ABSTRACT

This paper describes the studies performed as the bases for preliminary designs for Compressed Air and Underground Pumped Hydro storage (CAS and UPH) projects on an electric utility system. Acres American, Inc. did the design and cost studies for specific criteria of the Potomac Electric Power Company (PEPCo) in participation with and under the cooperative funding of the Electric Power Research Institute (EPRI) and the U. S. Department of Energy (DOE).

The study methodologies, cost estimates, and design details are applicable to other utilities where the geology includes competent rock suitable for access tunnels, shafts, waterway tunnels and for excavation of large caverns between 2500 and 5000 feet underground. CAS and UPH are today's technologies usable for economic and practical generation sources of replacement of oil-fired peaking capacity. Current state-of-the-art equipment, machinery, excavation techniques and controls can be utilized for confident design and construction of plants meeting utility criteria for load following, frequency regulation, and spinning reserve capabilities. Increased public acceptability of siting is demonstrated by relatively low environmental impacts and the licensing is straightforward with no major obstacles foreseen.

KEYWORDS

Pumped Hydro; Compressed Air Storage; gas turbines; underground reservoirs; high head pump/turbines; hard rock reservoirs; water compensated air storage.

INTRODUCTION

The studies were initiated in 1977, earlier efforts suggesting technical feasibility and commercial viability for energy storage concepts wherein underground storage of air was proposed for three geologies: salt, with caverns solution-mined out of a salt dome by a controlled water pumping process; aquifers, with air injected into the porous media to displace and be contained by the water; and hard rock, with caverns excavated by conventional techniques. Underground pumped hydro was limited by its nature to hard rock sites.

As a lead utility for the energy storage studies in hard rock, Potomac Electric Company (PEPCo), Washington, D.C., received a contract from EPRI and DOE. PEPCo commissioned Acres American, Inc. to serve as their principal engineers to perform the preliminary design analyses and evaluation. The work has been satisfactorily completed with EPRI now in the process of publishing the thirteen volumes and ten appendices which definitely document the preliminary designs of compressed air and underground pumped hydro in rock. (PEPCo and Acres American, Inc., 1980, Volumes I-XIII).

ENERGY STORAGE

One of the most practical methods in use today by the electric utility industry to manage energy from the supply side is the utilization of Energy Storage facilities (Kalhammer, 1979). In the practical sense, as used by utilities, Energy Storage is the process by which electricity is generated at the off-peak periods by base-load plants with relatively lower cost fuel and made available to the system during peak load periods, supplying both supplemental capacity and replacement energy. Electric utilities worldwide have accepted and are making excellent use of the Energy Storage concept known as hydroelectric pumped storage with well-known, literature-documented facilities in Asia, Australia, North and South America and in many Europeon countries. These facilities are operated as conventional hydroelectric plants during their generate modes (Vasilescu, 1976; Ferreira and Fisher, 1975). During off-peak periods lower cost, base-load electricity is generated to drive pumps that move water from a low level reservoir to a mountain-top (upper) reservoir.

Energy Storage saves oil and lowers overall production costs. During off-peak periods, the load level may be low enough so that coal-fired and even nuclear plants may have to be throttled to inefficient load levels. The power used in charging an Energy Storage system helps keep these base load plants near full load, and maximum efficiency, an important component in lowering production costs. Peak loads are typically met with combustion turbines and other fast-response, oil-fired units. Displacing peak period generation from these units, Energy Storage saves oil in addition to providing needed capacity. Production costs are lowered since the stored coal and nuclear generated storage energy is less costly than the oil-fired generation it replaces, even considering the efficiency cycle of the storage facilities.

Siting of conventional pumped storage facilities with both upper and lower reservoirs at surface level requires specific topographic terrain and greater alteration of land and water resources. Sites remote from load centers necessitate rather extended transmission line rights-of-way. Availability of suitable sites, growing environmental awareness and the need for long transmission lines may preclude this conventional pumped storage option for utilities. A number of alternative storage concepts are under development by EPRI and DOE but only two would appear ready for central station commercialization: Compressed Air Storage (CAS) and Underground Pumped Hydro Storage (UPH) (Kalhammer, 1979). Both concepts have attracted a growing interest over the past decade. The recently completed comprehensive study is the definitve evaluation, by means of preliminary design, of CAS and UPH to depths of 2500 feet and 5000 feet, respectively, in granitic gneiss quality rock.

Energy Storage plants will be built when, as for any other power plant, the capacity is needed, the cost to the consumer is the least among comparable alternatives, and the technology is practicable. The objective of the recent study was to specifically address these criteria for the utility industry with PEPCo as the illustrative developer. Under EPRI and DOE participation and direction, methodologies and results have been developed in a fashion to be readily usable by other utilities. The need and economics of both CAS and UPH are discussed elsewhere (PEPCo and Acres American, Inc., Vol. IV, 1980). The engineering practicabilities and summary cost estimates of each of the concepts are discussed in this paper.

STUDY APPROACH

The objective of the study was to perform a design of sufficient detail to provide firm estimates of cost, schedule and risks upon which a decision to construct a CAS or UPH facility could be based. The plan to address this objective included five tasks, generally sequential:

Task 1 - Design Criteria establishment of the overall guidelines for the engineering work and development of bounding assumptions for CAS and UPH facilities. The work consisted of a literature search and survey of precedent machinery and installations; then documentation, in a report, of basic assumptions, codes and governing design methods and formulas for systems and components, classified by engineering discipline. At this stage, a full expansion planning analysis was done to establish plant size, storage capacity, installation timing, relative economics and sensitivites to cost and performance characteristics (Driggs, 1980; PEPCo and Acres American, Vol. IV, 1980). The Design Criteria Reports (Herbst and Stys, 1978; Maass and Stys, 1980) were subsequently updated to reflect the changes in the state of knowledge following the preliminary design activity.

Task 2 - Site Selection and Investigation were included in the work to force site specificity into the planning and engineering functions. The value of the work is largely attributed to the real-world situations imposed, because these resulted in preliminary designs which address engineering obstacles. The first work step consisted of preparing a site selection methodology and evaluating a group of potential sites (Vasilescu, 1976). Since the key to UPH and CAS is sound geology, a program was next developed to explore the rock by a series of shallow boreholes and one deep borehole. A site suitable for drilling was chosen and both drilling programs carried out. The information obtained (PEPCo and Acres American, Inc., 1980, Volumes VI-VII) was used in preparing all designs.

Task 3 - Design Approaches comprised the bulk of the work. Using results from Tasks 1 & 2, various engineering alternatives for system and component design were evaluated. Specialist subcontractors and machinery manufacturers, in support of Acres' overall effort established coherent configurations for CAS and UPH and prepared draft construction schedules and cost estimates.

Task 4 - Environmental Studies was performed based on Tasks 1, 2, and 3 and allowed to proceed rather independently of the main design thrust. The result is an analysis of potential impacts and their relative magnitude (PEPCo and Acres American, Inc., 1980, Volume X). Further design efforts must address the concerns raised. Public safety issues were also addressed in this task.

Task 5 - Plant Design was primarily documentation of coherent designs for UPH and CAS based on Task 3 results. Final design decisions were reached and cost economics achieved. Consideration was given to construction procedures, site development, reliability and cost-risk. Overall control logic was addressed. Final cost estimates and construction schedules were prepared on the basis of final designs. A summary of costs, by major project component is shown in Table I.

COMPRESSED AIR STORAGE (CAS)

CAS operates on the open Brayton cycle. Unlike combustion turbines, however, compression and expansion are separate in time. The air is compressed by a motor-driven compressor, cooled to about 120° F, and stored in caverns excavated in rock underground at a pressure around 1000 psi. During the generation period the stored air is mixed with natural gas or distillate oil and expanded through a turbine-generator. The round-trip benefit is fuel savings. In the standard gas turbine cycle, compression work is about 60% of turbine work. With the turbine not driving

TABLE I. Summary of Capital Cost Estimates

924 MW Compressed Air Storage Plant

Item	Amount Mid-1979 $ x 10^6
Land, Site Access and Mobilization.........	8.7
Surface Facilities.........................	23.5
Storage System............................	55.9
Generator/Compressor System...............	172.2
Balance of Mechanical Plant...............	31.8
Switchyard................................	32.5
Electrical Plant..........................	22.4
TOTAL DIRECT COSTS........................	347.0 ($375.5/kW)
PEPCo Costs (15% of Direct Costs)..........	52.1
Engineering Costs (5% of Direct Costs).....	17.4
Construction Management Costs (10% of Direct Costs)..................	34.7
Contingencies (15% of Direct Costs)........	52.1
TOTAL	503.3 ($544.7/kW)

NOTE: Escalation and Interest During Construction are not included.

2000 MW Underground Pumped Hydro Plant

Item	Amount Mid-1979 $ x 10^6
Land and Site Access......................	6.2
Surface Structures........................	7.1
Upper Reservoir & Intake..................	31.2
Intermediate Reservoir....................	27.4
Lower Reservoir...........................	243.7
Shafts...................................	134.5
Miscellaneous Tunnels & Galleries.........	63.6
Powerhouse Civil Works....................	46.4
Pump-Turbines & Valves....................	56.8
Motor-Generators..........................	80.9
Transformers & Electrical Equipment........	62.3
Auxiliary Mechanical Equipment & Hoists....	42.3
Switchyard & Transmission.................	29.1
TOTAL DIRECT COSTS........................	831.5 ($415.7/kW)
PEPCo Costs (15% of Direct Costs)..........	124.7
Engineering Costs (5% of Direct Costs).....	41.6
Construction Management Costs (10% of Direct Costs)..................	83.1
Contingencies (15% of Direct Costs)	124.7
TOTAL.....................................	$1205.6 ($602.8/kW)

NOTE: Escalation and Interest During Construction are not included.

the compressor, it is free to deliver almost three times the electrical output for the same fuel input. The conceptual CAS layout is shown in Fig. 1.

Industry/research design studies are indicating that CAS is technically feasible with three kinds of underground storage. Air can be stored at pressure in caverns solution-mined from a salt formation (as at Huntorf), in caverns excavated in hard rock as in this PEPCo study, or as a "pressure bubble" within a porous aquifer. To minimize the amount of excavation in the PEPCo design for rock cavern storage, constant pressure storage is utilized. A water shaft from a small surface reservoir applies compensation pressure to the air storage in the cavern as the stored air is released during generation. The cost studies showed that the savings in excavation cost for the caverns is greater than the cost of the reservoir and water shaft.

Mechanical Design Issues

The initial work of the study was to examine, iteratively, the thermodynamics of CAS, the apparent cost sensitivities of performance parameters, and the available hardware. Emphasis was placed on identifying available rotating machinery so that a practical, near-term design could be developed.

Compression and expansion machinery available on a worldwide basis were reviewed. The world's first air-storage power plant constructed at Huntorf, Federal Republic of Germany is an example of Brown Boveri's approach (PEPCo and Acres American, Inc., 1980, Volumes IX, Appendix C, and Volume IX, Appendix D). The plant has been operating successfully since late 1978 at reported higher-than-guaranteed efficiencies. The 290 MW plant is owned and operated by the West German utility, Nordwestdeutsche Kraftwerke AG, based in Hamburg, and uses a modified gas turbine generating system. Its air storage caverns were solution-mined out of the area's underground salt deposits. Brown Boveri's turbine and Sulzer's compressor resources were applied to the PEPCo study, although the derived machinery train was different from that used at Huntorf.

Thermodynamics and project costs would dictate air storage at as high a pressure as possible since air volume and cavern cost decrease with pressure. Sulzer compressors, however, reach a threshold at about 1000 psi where a barrel type must be used in lieu of a split case design. The associated step increase in cost overcomes the pressure/cost advantage. Hence, 1000 psi storage was adopted.

Thermodynamic analysis also noted the advantage for exhaust gas recuperating during generation. Such a high pressure, high temperature, high flow recuperator has apparently never been built. However, the reduction in fuel consumption with the incorporation of such a component indicates good operating economics can be achieved. EPRI is continuing research on the practical development of this equipment.

Civil Design Issues

Geophysical testing of rock from the core drilling program was performed. This data and the results of in-situ rock strength testing (PEPCo and Acres American, Inc., 1980, Volumes VI-VII) were used in a finite element analysis of cavern stability (PEPCo and Acres American, Inc., 1980, Vol. VIII, E). In parallel a study of excavation techniques and costs was undertaken. The combined design optimization led to the 85 ft. high 66 ft. wide cavern cross-section. The resultant underground works are pictured in Fig. 2.

844

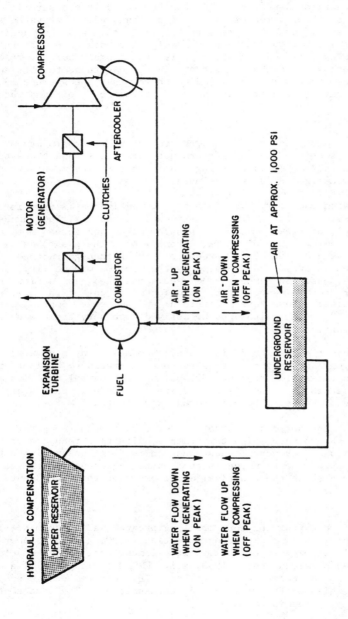

Fig. 1. Compressed Air Storage System

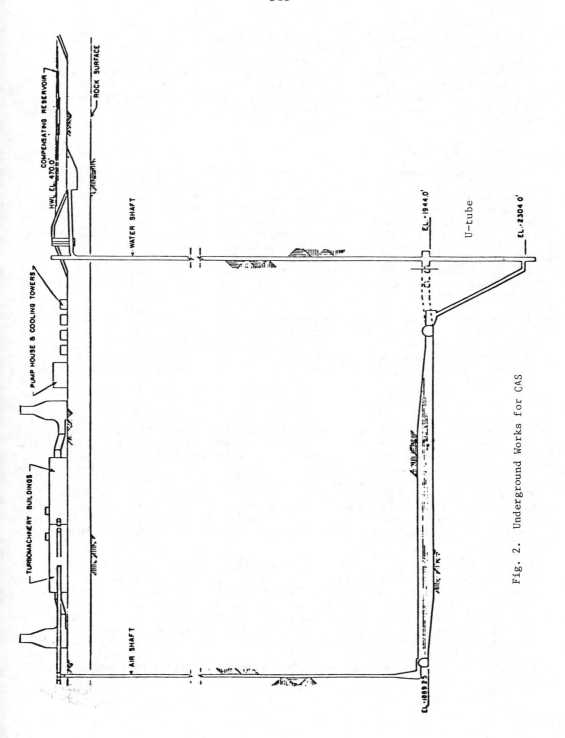

Fig. 2. Underground Works for CAS

Air is sealed within the air shaft by a steel liner (PEPCo and Acres American, Inc., 1980, Volume IX, Appendix A). Air is prevented from escaping up the water shaft by a U-tube in the bottom of the shaft. A question has been raised whether air dissolved in the water could escape up the water shaft (PEPCo and Acres American, Inc., 1980, Volume IX, Appendix B). The geometry of the shaft/cavern and shaft/reservoir intersections, the U-bend and restrictions on operation has been designed as an air-release preventative measure. EPRI is continuing further research on this project component also to provide better understanding and firmer design criteria for this phenomenon.

Control System Design Issues

The units must operate with fast-response, they must be self-starting with either emergency or station service power, and be able to operate over the entire load range and be available for synchronous condenser operation. During Task 3, it was determined that the basic operational capabilities provided by Brown Boveri and Sulzer, as part of their standard control packages, could be integrated into a coordinated control system for power generation, air compression and synchronous control system for power generation, air compression and synchronous condenser operation equal to the imposed criteria. Normal generation startup to full load could be accomplished in ten minutes, emergency startup in five and a half minutes, and changeover from full compression to full generation in nineteen minutes under emergency conditions (PEPCo and Acres American, Inc., 1980, Volume IX).

It is possible to start up a CAS unit in the compression mode by running up the entire machinery train with the turbine. Once the motor is synchronized, the turbine can be tripped and automatically declutched from the motor. This method is mechanically fatiguing to the turbine, however, and significantly reduces its life expectancy. The final plant design, therefore, calls for variable frequency starting.

UNDERGROUND PUMPED HYDRO (UPH)

A schematic of the Underground Pumped Hydro layout is shown in Fig. 3. During generation, water is released from the surface reservoir through a penstock shaft to the powerhouse elevation. The flow is split in horizontal tunnels on route to the hydro turbine-generators. The turbines discharge into a series of underground tunnel-caverns. An air vent shaft is connected to the cavern system to keep it at atmospheric pressure assuring maximum work from the water head. During the off-peak pumping cycle, the turbine-generators are reversed to operate as motor-pumps. The storage caverns are at a higher elevation than the powerhouse so as to provide positive suction head on the pumps thereby minimizing cavitation-erosion on the runners.

No fuel is consumed directly at site by UPH, this being one of the factors making it virtually environmentally benign. If pumping is accomplished with nuclear and coal-generated electricity, peaking with the UPH plant is accomplished without the use of oil. UPH is over 70% efficient; that is, only about 135% of the energy generated need be supplied as pumping power. As long as the peaking power alternatives to UPH are more costly to operate than coal or nuclear units by at least 35%, UPH maintains its high operating cost advantage. When this advantage is not off-set by any higher construction costs UPH is a generation option well worth pursuing from an economic as well as an environmental acceptability standpoint.

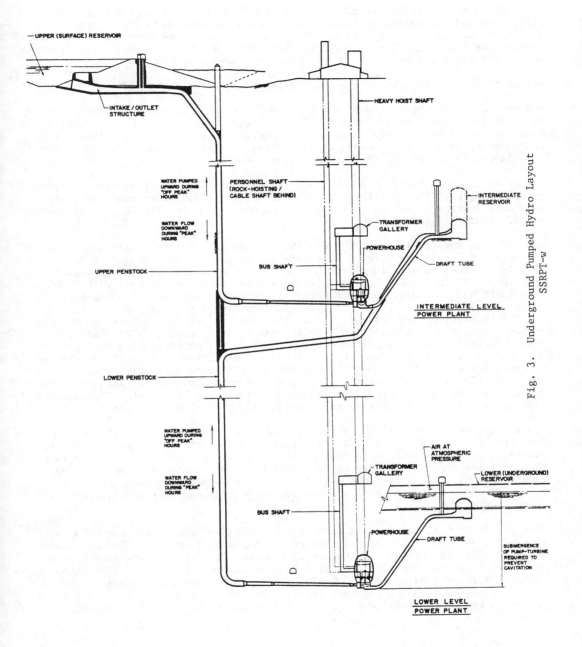

UPPER (SURFACE) RESERVOIR

INTAKE / OUTLET
STRUCTURE

HEAVY HOIST SHAFT

WATER PUMPED
UPWARD DURING
"OFF PEAK"
HOURS

PERSONNEL SHAFT
(ROCK-HOISTING /
CABLE SHAFT BEHIND)

INTERMEDIATE
RESERVOIR

WATER FLOW
DOWNWARD
DURING "PEAK"
HOURS

TRANSFORMER
GALLERY

POWERHOUSE

UPPER PENSTOCK

BUS SHAFT

DRAFT TUBE

INTERMEDIATE LEVEL
POWER PLANT

LOWER PENSTOCK

WATER PUMPED
UPWARD DURING
"OFF PEAK"
HOURS

AIR AT
ATMOSPHERIC
PRESSURE

WATER FLOW
DOWNWARD
DURING "PEAK"
HOURS

TRANSFORMER
GALLERY

LOWER (UNDERGROUND)
RESERVOIR

BUS SHAFT

POWERHOUSE

DRAFT TUBE

SUBMERGENCE
OF PUMP-TURBINE
REQUIRED TO
PREVENT
CAVITATION

LOWER LEVEL
POWER PLANT

Fig. 3. Underground Pumped Hydro Layout
SSRPT-w

Determination of Head & Machine Configuration

Since the power available from UPH is a direct function of the product of head times flow and excavation of the storage caverns involves over 30% of total plant direct costs, the volume of the caverns - and the water flow - should be minimized. This can be accomplished by increasing the station head for a given power require- ment. UPH heads between 2000 and 5000 feet are envisioned as practical.

Within this head range there are three types of pump/turbines.

o Single-Stage Reversible Pump/Turbines (SSRPT) are the conventional machines which evolved from Francis turbine and centrifugal pump experience. Currently heads are achievable in the range of 2000 feet. If two of these machines were set in tandem cascade, water storage at twice their individual head could be achieved. This layout is called the "2-step" and designated SSRPT-2.

The other major factors, of course, are the placement of the lower reservoir below ground along with the powerhouse and the connecting major water conduits. The remaining surface level reservoir is also reduced significantly in size by utili- zation of higher heads.

o Multi-Sate Reversible Pump/Turbines (MSRPT) are coming into common use in Europe and Asia. By combining several impellers in one casing, high heads can be achieved with less stress on each impeller than on that of a single-stage machine at the same head. It is, therefore, possible to achieve much higher heads with MSRPT than SSRPT while maintaining hydraulic and mechanical designs of the machines at comparable levels. Currently, a major disadvantage of these machines is the in- ability to provide flow control. Since they can neither load-follow nor operate at part-load, they are unable to meet the operating criteria of PEPCo and that most U.S. utilities are expected to opt for in such a comprehensive hydroelectric project. Manufacturers have proposed a two-stage machine (TSRPT) which could be flow-regulated, for heads in the range presently covered by MSRPT. No such units are currently in operation nor have they been comprehensively designed or tested and the concept must be considered as not available at this point. EPRI is, how- ever, actively pursuing this in an on-going new Research & Development effort.

o Uni-rotational Separate Pump Turbines (USPT) utilize a Pelton impulse turbine and centrifugal pump. These units are conventional state-of-the-art but the re- sulting plant configuration is more expensive than the other types since a long shaft between the pump and turbine is required. The turbine must operate above cavern water level and the pump must have a large submergence below that level. Options such as a booster pump or separate motor and generator add substantially to the cost.

Characteristic cost curves were developed for the head range achievable by each type. The selection of head was made on the basis of lowest cost for available machinery that met PEPCo operating requirements. The SSRPT-2 arrangement using equal drops of 2500 feet for a total 5000-foot dept was selected. Reference, "Volume VIII - UPH Design Approaches", contains this analysis including a thorough review of worldwide machinery experience. UPH pump/turbine equipment is also re- viewed in reference, Rodrique, P., June 1979. The characteristic cost curves are shown in Fig. 4.

Underground Civil Design

During Task 3, precedent underground work was studied. Excavation techniques for caverns were evaluated with the assistance of specialist contractors and consul- tants. From these evaluations, basic approaches and geometries were selected.

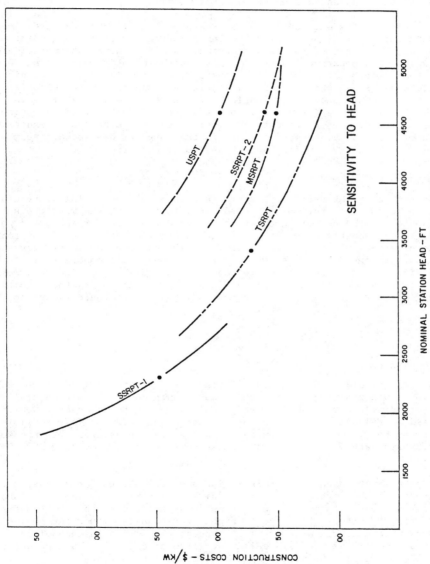

Fig. 4. Cost Comparisons of UPH Machinery Configuration

This information, combined with test results from the series of shallow holes (500 to 600 ft.) and the one deep hole (2500 ft.), was used to develop a finite element model of the caverns. Using this model, preliminary designs for the storage reservoir caverns and powerhouse chamber were prepared. An extensive effort, carrying into Task 5, was made to cost optimize the work while at the same time taking special concern to prepare a buildable arrangement (PEPCo and Acres American, Inc., 1980, Volume VIII, Appendix D and E, and Volume XI).

A total of five shafts were designed for the UPH plant (PEPCo and Acres American, Inc., 1980, Volume VIII, Appendix B). Consideration was given to sequential construction of the units, so the arrangement allows for concurrent operation of initial units with construction of the final units and storage caverns. During operation, as water flows through the penstock, cavern air is entering via or being released through the vent shaft. Normal access for men and material is through the personnel shaft. The cable shaft carries one 500 kV transmission circuit. The two others (providing, in total, one circuit per tandem pair of units) are carried, when construction is complete, in the personnel and heavy hoist shafts. The heavy equipment shaft is the largest, designed to permit lowering of the 18 kV/500 kV single phase transformers underground.

Heavy Hoist Equipment

The motor-generator rotors, transformers, and pump/turbine casings were discovered to weigh up to 300 tons for the selected UPH layout. Combined with the number of such loads to be handled during construction, and the need for speeds up to 250 ft./min. to allow reasonable turnaround during construction, this weight approaches the limit of precedence for 5000-foot lifts. The hoist arrangement, considered feasible, was designed by Acres, working with a specialist contractor. A perspective of the complex tunnel, shaft, and cavern configuration in the vicinity of the lower power facility is depicted in Fig. 5, underscoring the sensitivity of project costs to the selection and scheduling of underground excavation procedures.

STUDY RESULTS AND CONCLUSIONS

The UPH Plant

The engineering analyses and design efforts of the study resulted in the selection of the two-step arrangement, i.e., two power plants in series with a small intermediate balancing reservoir. Each step would have a 2500 ft. head (5000 ft. total head) based on reliable predictions of an appreciable advance in resersible pump/turbine technology. Six 333 MW pumped storage units operating at 720 rpm were selected, three each in the upper and lower power plants, providing 2000 MW total capacity.

Other major project components designed included the shafts for access, equipment handling and the high pressure water conduits. Electrical systems were studied and the arrangement with underground transformation step-up to 500 kV was adopted. For the motor-generators, high-speed, high-powered units with water-cooled rotors and stators were determined to be within acceptable limits of technology. Static converter starting equipment was selected. It was concluded that the derived UPH design would be within the proven practices of the industry.

The CAS Plant

In general, the CAS designs adopted for the study were based on precedent experience. Viewed against the only operational CAS plant (Huntorf), the PEPCo CAS design differs in that: 1. The use of hard rock caverns allows hydraulic compen-

851

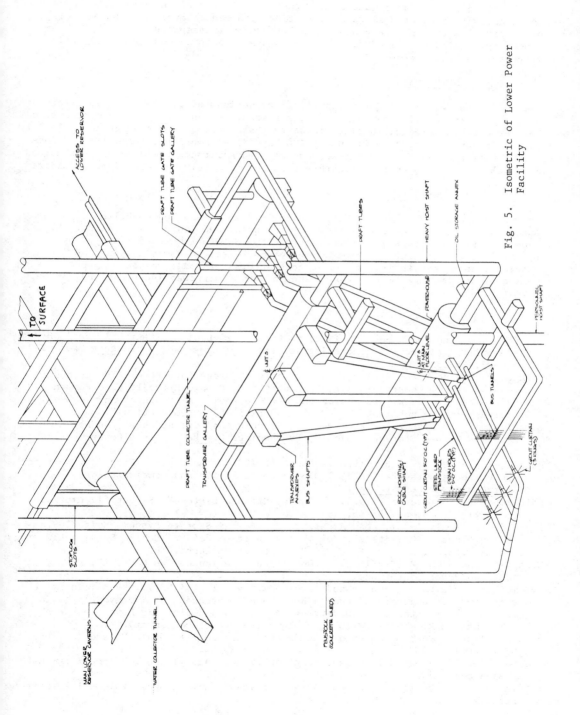

Fig. 5. Isometric of Lower Power Facility

sation to maintain near constant air pressure. This requires a surface compensating reservoir, a water shaft and facilities arranged to avoid accidental release of air bubbles through the hydraulic shaft system ("champagne effect"); 2. This study involved careful study of the benefits and costs of more advanced designs of heat recuperators.

CONCLUSIONS

The basic characteristics of the preliminary designs of the CAS and UPH plants are outlined in Table II. The engineering analyses, design efforts, and system studies reinforce to the electric utility industry that these energy storage technologies can be implemented with confidence (PEPCo and Acres American, Inc., 1980, Volume I). The technologies represent an economic and environmentally acceptable method by which the world's energy resources can be effectively and efficiently utilized.

ACKNOWLEDGEMENT

The authors express their gratitude to Edwin Shippey and Foster Pelton of Acres American, Inc., for furnishing information, data, and source material for this paper. Appreciation is accorded also to the U.S. Department of Energy, the Electric Power Research Institute, the Potomac Electric Power Company, Acres American, Inc., and their various subcontractors for their respective funding and well conceived work efforts in pursuance of this comprehensive study and design effort.

REFERENCES

Boivin, R. D. and Anderson, J. (1972). "Construction of the Churchill Falls Hydroelectric Development", Proceedings of the American Power Conference, Vol. 34, pp. 771-783.

Driggs, C. L. and Robb, P. W. (1980). "Energy Storage for PEPCO - The Economic Feasibility of Compressed Air and Underground Pumped Hydroelectric Pumped Hydroelectric Energy Storage on the PEPCO System", ASME Winter Annual Meeting, included in the publication G00195 - "Peaking Power Generation."

Ferreira, A. and Fisher, W. E. (1975). "Planning and Operation of the Northfield Mountain Pumped Storage Project", Proceedings of the American Power Conference, Vol. 37.

Herbst, H. and Syts, Z. S. (1978). "Huntorf 290 MW The World's First Air Storage System Energy Transfer (ASSET) Plant: Construction and Commissioning", Paper, American Power Conference.

Kalhammer, F. R. (December 1979). "Energy Storage Systems", Scientific American, Vol. 241, No. 6, pp. 56-65.

Loane, E. S. (July 1976). "An Assessment of Energy Storage Systems Suitable for Use by Electric Utilities", Vol. III, Report EM-264, by Public Service Electric and Gas Company for the Electric Power Research Institute.

Maass, P. and Stys, Z. S. (1980). "Operation Experience with Huntorf", Paper, American Power Conference.

Potomac Electric Power Company and Acres American, Inc. (1980). "Preliminary Design Study of Underground Pumped Hydro and Compressed Air Energy Storage in Hard Rock", for the U. S. Department of Energy and Electric Power Research Institute, EPRI Report EM-1589. Volume I - Executive Summary.

Potomac Electric Power Company and Acres American, Inc. (1980a). Volume II - UPH Design Criteria.

Potomac Electric Power Company and Acres American, Inc. (1980b). Volume III - CAES Design Criteria.

Potomac Electric Power Company and Acres American, Inc. (1980c). Volume IV - System Studies.

TABLE II. <u>Major Plant Characteristics and Design Parameters</u>

Parameter	CAS	UPH
Generating Capacity – Plant	1000 MW	2000 MW
Generating Capacity – Unit	225 MW	333/666*
Usable Storage Capacity	10 hr.	10 hr.
Approximate Cavern Depth*	2500 ft.	2500/5000 ft.
Electric Energy Ratio KWH input/KWH output	0.75	1.32
Fuel Heat Rate Btu/KWH Output	4250	0
Planned Outage Rate	2 wk./yr.	4 wk./yr.
Forced Outage Rate	10%	4%**
Black Start	Yes	Yes
Load Following	Yes	Yes
Fuel	ASTM No. 2 Fuel 0.1	---
Total Cavern Volume	811,000 yd^3	---
Intermediate Reservoir Volume	---	600,000 yd^3
Lower Reservoir Volume	---	7,800,000 yd^3

* Two-step system, two 333 MW units operating in series

** Combined outage rate for two units in series operation.

Potomac Electric Power Company and Acres American, Inc. (1980d). Volume V - Site Selection.

Potomac Electric Power Company and Acres American, Inc. (1980e). Volume VI - Site Investigation, Shallow Drilling.

Potomac Electric Power Company and Acres American, Inc. (1980f). Volume VII - Site Investigation, Deep Drilling.

Potomac Electric Power Company and Acres American, Inc. (1980g). Volume VIII - UPH Design Approaches.

Potomac Electric Power Company and Acres American, Inc. (1980h). Volume VIII, Appendix A - Upper Reservoir.

Potomac Electric Power Company and Acres American, Inc. (1980i). Volume VIII, Appendix B - Shafts.

Potomac Electric Power Company and Acres American, Inc. (1980j). Volume VIII, Appendix C - Heavy Hoist.

Potomac Electric Power Company and Acres American, Inc. (1980k). Volume VIII, Appendix D - Power Facilities.

Potomac Electric Power Company and Acres American, Inc. (19801). Volume VIII, Appendix E - Lower Reservoir.

Potomac Electric Power Company and Acres American, Inc. (1980m). Volume IX, CAES Design Approaches.

Potomac Electric Power Company and Acres American, Inc. (1980n). Volume IX, Appendix A - Air Storage System.

Potomac Electric Power Company and Acres American, Inc. (1980o). Volume IX, Appendix B - Champagne Effect.

Potomac Electric Power Company and Acres American, Inc. (1980p). Volume IX, Appendix C - Major Mechanical Equipment.

Potomac Electric Power Company and Acres American, Inc. (1980q). Volume IX, Appendix D - Mechanical Systems.

Potomac Electric Power Company and Acres American, Inc. (1980r). Volume IX, Appendix E - Electrical Systems.

Potomac Electric Power Company and Acres American, Inc. (1980s). Volume X - Environmental Studies.

Potomac Electric Power Company and Acres American, Inc. (1980t). Volume XI - UPH Plant Design.

Potomac Electric Power Company and Acres American, Inc. (1980u). Volume XII - CAES Plant Design.

Potomac Electric Power Company and Acres American, Inc. (1980v). Volume XIII - Preliminary Licensing Documentation.

Rodrique, P. (June 1979). "The Selection of High Head Pump/Turbine Equipment for Underground Pumped Hydro Energy Storage Application", Pump/Turbine Symposium, ASME-ASCE Conference.

Vasilescu, M. S., et al (1976). "Blenheim-Gilboa Pumped-Storage Project Design and Construction", Proceedings of the American Power Conference, Vol. 38, pp. 1006-1024.

FUEL CELL POWER PLANTS FOR ELECTRIC UTILITIES

A. P. Fickett, E. A. Gillis and F. R. Kalhammer

Electric Power Research Institute, Palo Alto, California, USA

ABSTRACT

This paper presents the major considerations underlying the U.S. electric utility R&D program in fuel cells, and it summarizes the significant progress achieved during the last five years in the development of long-lived electrode catalysts and supports, much-improved cell configurations, and plant designs of higher specific output. Together with the anticipated successful operation of a 4.5 MW fuel cell power plant in 1982/83, these developments are establishing the basis for the commercial introduction of fuel cell power plants into electric utility service by the mid-1980s.

KEY WORDS

Fuel cells; phosphoric acid cell technology; electrode catalysts; cell design; 4.5 MW fuel cell demonstrator; power plant characteristics

INTRODUCTION

The concept of the electrochemical fuel cell was first demonstrated almost 150 years ago, but serious attempts to develop the concept into a practical method of power generation began only in the early 20th century. At that time, the developing understanding of thermodynamic principles led to a full appreciation of the much higher energy conversion efficiencies that could be expected from fuel cells as compared to thermal engines based on combustion. But a focus on coal which proved to be an electrochemically untractable fuel, severe materials problems, and the inadequate understanding of electrochemical engineering principles combined to prevent practical success.

The renaissance of the fuel cell began in the 1950s when the concept was selected for development into the primary power source for the Gemini and Apollo space flight missions. Only a very limited amount of British and German work on key aspects of cell technology had been done to suggest possible practicality. Nevertheless, fuel cells were chosen over more fully developed energy conversion concepts because of their potential for excellent efficiency (minimum need for fuel) over a wide range of power output, high reliability (since no moving parts), freedom from noxious emissions (since not involving combustion), and the promise to achieve these desirable characteristics with power sources assembled from relatively small modules. The historic record is impressive: fuel cells met every one of these expectations and thus were a material factor in making extended space flight possible in the 1960s.

With the space application a practical success, the fuel cell began to attract interest as a terrestrial power generator. In the U.S., this interest led to major cooperative programs of fuel cell development, first with funding by the gas industry (described by V. Fiore in a companion paper in this Session), and since 1972 with support of individual electric utilities and then the Electric Power Research Institute. This paper presents the major considerations underlying the electric utility program in fuel cell development and summarizes the significant progress achieved in practically every aspect of fuel cell research, development and demonstration over the last five years.

FUEL CELLS FOR ELECTRIC POWER GENERATION

The achievement of very high efficiencies in converting the chemical energy of fuels into electricity was and remains a major incentive to developing fuel cells for power generation on the scale of electric utilities. While conventional utility power plants are limited to efficiencies of 30% to 40%, fuel cells operating on practical carbonaceous fuels are expected to reach 40% even in early commerical configurations, as much as 50% when fully developed and optimized as power plants. Equally important, fuel cells will be fully efficient over a wide range of power output, in contrast to thermal plants which can maintain peak efficiencies only at or near their design points. As a consequence, in peaking and cycling service fuel cells could be as much as 25%-50% more efficient than conventional power generating equipment. The technical basis for this expectation is discussed further below.

Space fuel cells typically use very pure hydrogen as fuel and oxygen as the oxidant. To be useful as utility power generators, fuel cells must use air instead of oxygen and, at least during the foreseeable future, carbonaceous fuels that are -- or, when considering coal-derived liquids and gases, are expected to become -- available to electric utilities at acceptable costs. The use of carbon dioxide-rejecting, acid fuel cell electrolytes has made it practical to use air and to employ fuel processing units in which the carbonaceous fuel is reacted with steam to produce a hydrogen-rich gas suitable for electrochemical conversion in the power section of a fuel cell power plant (see Figure 1).

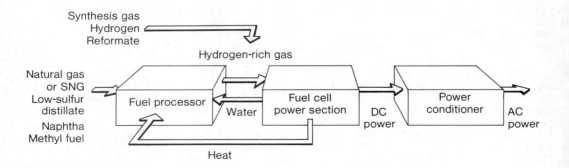

Fig. 1. Fuel cell power plant

The fuel processor is an integral part of a fuel cell power plant: not only does it produce electrochemically active hydrogen from inactive carbonaceous fuels but it utilizes waste heat and product water from the power section to do so. This feature is one important reason for the highly efficient and water-conserving operation of practical fuel cells. Another integral part of an electric-utility fuel cell power plant is a power conditioner in which the DC ouptut of the power section is converted to high-quality AC that can be fed into the electric grid.

Figure 1 lists a variety of fuels all of which are compatible with the fuel cell types (those based on aqueous acid and molten carbonate electrolytes, respectively) under development for electric utility applications. In Figure 2, fuel charac-teristics are presented together with the heat rates for utilization of these fuels in near-term phosphoric acid fuel cell power plants; heat rates ranging from 8360 to 8100 Btu per kWh correspond to efficiencies of 41% to 42%.

| | Light Distillate | SNG | Medium Btu Gas | | Methanol | Hydrogen |
			Lurgi	Texaco		
Composition	$C_7H_{14.3}$	CH_4	H_2 26.8% CO 61.1 CH_4 6.4 C_2H_6 0.2 CO_2 4.8	37.5% 50.0 0.3 — 10.7	CH_3OH	H_2
Higher heating value	20250 Btu/lb	1012 Btu/ft^3	351 Btu/ft^3	285 Btu/ft^3	9760 Btu/lb	325 Btu/ft^3
Lower heating value	18900 Btu/lb	911 Btu/ft^3	331 Btu/ft^3	266 Btu/ft^3	8585 Btu/lb	275 Btu/ft^3
Heat rate (HHV)	8300	8350	8360	8350	8210	8100*

*Lower when designed for H_2 (Source: Tennessee Valley Authority)

Fig. 2.   Phosphoric acid fuel cell power plant
(multi-fuel heat rates)

Utility fuel cell power plants will have very low levels of pollutant emissions, in part because electrochemical oxidation ("cold combustion") does not generate nitrogen oxides or particulates, and in part because sulfur compounds are removed from the fuel feed since they are detrimental to the fuel processor. The feasi-bility of removing sulfur and achieving very low emissions (at least ten times lower than the stringent standards in the U.S.) have already been demonstrated for experimental power plants.

Like other electrochemical devices (such as batteries or electrolysis cells), fuel cells are built up from individual cells. Because of this modular character-istic, the components of fuel cell power plants -- even of multi-megawatt power stations -- can be manufactured rapidly by mass-production techniques and assembled into truck-transportable modules at the factory. One benefit is that high con-struction financing costs and expensive field labor can largely be avoided. Another advantage is that utilities will be able to deploy and expand fuel cell installations rapidly and in relatively small increments. Modularity and absence of polluting emissions will permit fuel cell power plants to be located on urban and suburban sites close to electricity users. This, in turn, will reduce power transmission losses and can help utilities defer investments in new power lines. Finally, closeness of fuel cells to electricity users almost always means closeness to users of thermal energy. Thus, the fuel cell not only allows waste heat recovery without efficiency penalty (a characteristic of electrochemical energy converters) but offers prospects to keep the distances and costs of transporting heat as small as possible. This characteristic tends to make fuel cells attractive candidates for dual energy use system (DEUS) applications such as cogeneration and district heating which utilize the electric as well as thermal output of an energy converter.

Because of this combination of characteristics (which is unmatched by any of the conventional power generators) the fuel cell could serve utilities and energy users in several potentially important applications:

(1) Dispersed power plants of small (for example, 40 to 300 kW) capacity located in commercial and larger residential buildings. Such plants would use natural gas as fuel and meet the electric and thermal energy needs of the building with very high overall energy efficiency (well above 80% based on fuel energy input).

(2) Fuel cell power plants of perhaps 500 kW to 5 MW capacity on industrial sites, with the electric output going largely to the grid but the thermal energy to the industrial process. Natural gas would be an attractive fuel because of the convenience and high efficiency (above 80%) with which it would be used, but most of the fuels listed in Fig. 2 could be used as well. The power plant could be owned by the energy user, the utility, jointly, or perhaps even by a third party.

(3) Yet larger plants of perhaps 5 to 25 MW capacity could be located at the most suitable points on the electric grid to meet some or all of the utility's time-varying load. Such power plants might also be used to deliver their waste heat to large commercial or industrial customers. Depending on geo-graphic location and utility fuel supply, one or several of the fuels listed in Figure 2 would be used.

(4) Further in the future, fuel cell power plants could be integrated -- with respect to fuel and thermal energy -- with coal gasifiers into central-station power plants of perhaps 150 MW to 1000 MW capacity. Such plants would operate in the base load mode and have efficiencies of 45% to 50% based on the energy content of coal.

FUEL CELL DEVELOPMENT

The technological basis for the fuel cell applications sketched above has been pursued vigorously in several major R&D programs since the late 1960s. The residential/commercial and dispersed utility applications have received the most emphasis to date. Considerable technical synergism has occurred because both the gas and the electric utility application are focussing on the same cell technology: porous carbon structures catalyzed by small amounts of noble metals as electrodes, and phosphoric acid as electrolyte.

The major technical issues in the development of phosphoric acid fuel cell tech-nology have been to achieve sufficiently good durability and low cost. To some extent at least these goals are conflicting: increasing power density by means of higher temperature and pressure is a promising avenue to decrease costs but the same approach tends to reduce life. However, the successful development of more stable carbon supports and platinum alloy catalysts now indicates that electrodes of practical life are in hand. The data of Figure 3 extrapolate to catalyst and support durabilities of at least 40,000 hours at or above rated cell performance and voltage.

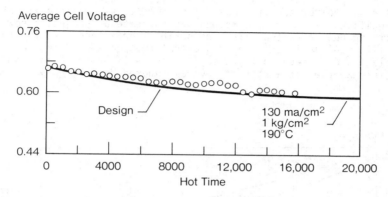

Fig. 3. Phosphoric acid cell stack endurance

Another key element of progress has been the development of improved cell configurations and cell stack designs. As shown in Figure 4, cell design has evolved from structures comprising thin porous electrode (cathode and anode) sheets and dense graphite separator plates, to configurations employing thin separator sheets and ribbed porous electrodes.

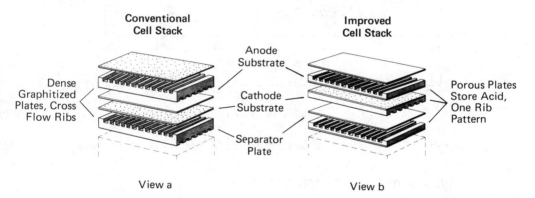

Fig. 4. Cell configuration comparison

In addition to substaining the electrochemical reactions, these electrodes also serve as gas distribution channels and as a reservoir for the phosphoric acid electrolyte. The new design permits storage of excess acid in the porous electrode/gas distribution structure, thus preventing premature dry-out of cells -- previously a major problem that limited cell and stack life to periods shorter than the improved endurance of electrode catalysts and supports. The new cell configuration also lends itself much more readily to continuous fabrication which promises to be less expensive than batch processing.

Major progress has also been made on the power plant level. The availability of more stable electrodes and catalysts is making it feasible to increase operating temperature to about 200°C and to pressurize cell stacks at around 6-8 atm., for a significant gain in cell performance and the rating of the major power plant components, with corresponding cost reductions as a result. Figure 5 summarizes the gains made in the past five years: power density has been increased by 75%, with an attendant reduction of specific costs for cell stacks, while cell voltage has increased significantly, with a corresponding increase of projected power plant efficiency. It can now be stated with considerable confidence that the technology base for utility fuel cells is in hand and that prospects are very good for fuel cell power plants to have competitive capital costs if produced in significant quantities (for example, at a rate of several hundred MW per year).

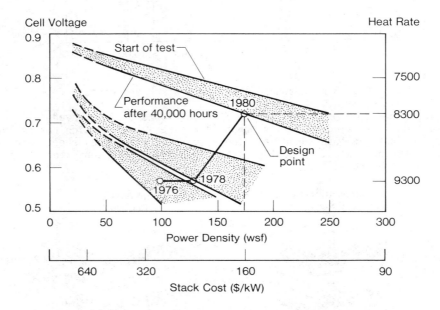

Fig. 5. Progress in phosphoric acid fuel cell technology

THE 4.5 MW FUEL CELL "DEMONSTRATOR"

In parallel with the development of improved fuel cell technology during the past five years, a major effort has been underway to validate the operating characteristics of a fully integrated fuel cell power plant connected to an electric utility grid. With support of the U.S. Department of Energy (DOE), the Electric Power Research Institute (EPRI), United Technologies Corporation (UTC, the developer) and a utility consortium led by the Consolidated Edison Company of New York, a 4.5 MW fuel cell power plant has been fabricated and installed on an urban site (Manhattan) within New York City. After a period of safety and acceptance testing this year, the demonstrator should begin operating and evaluation tests in early 1982.

An artist's conception of the fuel cell demonstrator site is shown in Figure 6. The nominal rating of the DC power section is 4.8 MW; output from the inverter (bottom right in Figure 6) will be 4.5 MW AC. The power section (Figure 7 and center of Figure 6) contains 20 fuel cell stacks of the type shown in Figure 8.

Fig. 6. 4.5 MW fuel cell demonstrator in New York

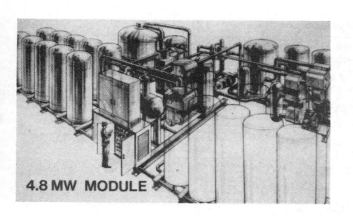

Fig. 7. 4.8 MW DC module

Fig. 8. 250 kW fuel cell stack

The 4.5 MW demonstrator is still based on the previous generation of cell technology (see Figure 4, left) and thus not intended to verify durability or cost characteristics. However, the demonstrator test is expected to validate the basic performance and efficiency characteristics of a fuel cell power plant and its operation by electric utility personnel on the power grid.

DEVELOPMENT OF COMMERCIAL CONFIGURATIONS

Phosphoric acid fuel cell development has reached the technological targets for commercial viability, and the 4.5 MW demonstrator should establish the feasibility of power plant packaging, siting and operation. The tasks remaining on the way to full commercial feasibility of fuel cell power plants are to:

- establish the specifications of primary importance to large and small utilities
- reduce costs through increases in performance, simplification of plant design and control, and identification of lower cost fabrication methods
- improve maintainability through design changes
- expand fuel capability to include virtually all clean carbonaceous liquids and gases.

These tasks are being addressed in a coordinated series of programs supported by DOE, EPRI and several utilties, with the goal being a commercial offering -- possibly a production run of about fifty 10-MW fuel cell power plants -- in the mid-1980s. In Figure 9, the characteristics of these commercial power plants are compared with those of the 4.5 MW demonstrator unit. Also included are tentative characteristics of a power plant concept in which a phosphoric acid fuel cell would be combined with a small, state-of-the-art coal gasification unit and the necessary gas cleanup equipment. This concept could become a practical route to the efficient and economic use of coal in smaller power plants.

	4.5 MW Demonstrator	Commercial Unit	
Fuel	Naphtha, natural gas, SNG	Clean coal liquids, gases, naphtha, natural gas	Coal
Module size (MW)	4.5	10–12	30–50
Heat rate (Btu/kWh)			
• Full load	9300	8300	10500
• Half load	9000	9000	11000
Projected life (yr)	20	20	20
Projected cost* ($/kW)	≈ 950	≈ 500	≈ 1500 (includes coal gasifier)
Efficiency (%) (cogeneration mode)	75–80	75–80	—

*1980 $ including IDC and installation; assumes production rate of 500 MW/yr

Fig. 9. Power plant characteristics (phosphoric acid)

In summary, recent progress in the development of fuel cell technology and power plant configurations has been very encouraging. Spurred by the success of programs conducted at UTC with DOE, EPRI and utility support, Westinghouse has initiated a development program in technical partnership with the Energy Research Corporation. This emerging competition should increase the prospects that a commercial power plant technology will become available to electric utilities. More than fifty U.S. electric utilities have translated their interest in this new technology into the formation of a Fuel Cell Users Group. With the coal of expediting the commercial introduction of fuel cells, this group has begun to establish desirable power plant specifications and the specific value of fuel cells to a variety of U.S. and foreign utilities. Prospects are indeed bright that the phosphoric acid fuel cell will become the first truly new power generating option for utilities since more than fifteen years, with good prospects for even broader applications in the future when advanced fuel cell technologies are expected to permit clean generation of power from coal with unprecendented efficiencies.

THE REALITY OF ONSITE FUEL CELL ENERGY SYSTEMS IN THE 1980'S

V. B. Fiore and R. T. Sperberg

Gas Research Institute, Chicago, Illinois, USA

ABSTRACT

A fuel cell is an energy device that converts fuel directly into electricity in a highly efficient and environmentally acceptable way using an electrochemical process. Although any hydrocarbon fuel can be used to operate a fuel cell system, the work that has been supported by the Gas Research Institute in Chicago, Illinois, U.S.A., has dealt with natural gas (pipeline quality) as the primary energy for onsite fuel cells.

The use of primary energy resources could be reduced if electricity were generated at or near the point of use, with the heat produced in the generation process being recovered and used onsite. Fuel cells are a means of satisfying this reduction if used onsite.

Onsite fuel cell power plants with heat recovery are highly efficient, extremely low in pollution levels, have a minimal noise level and are literally free of vibration. Their operation is automatic, requiring no attendant. Fuel cell power plants can be built in almost any size from a few kilowatts to several megawatts in capacity without a significant change in performance. In addition to an isolated operation, onsite fuel cell energy systems have characteristics ideally suited for cogeneration applications.

The gas industry has long been aware of the potential to develop an onsite fuel cell energy system for its high priority commercial and light industrial market sectors. In recognition of this potential, a major commitment was made by the gas utility companies and United Technologies Corporation (UTC) in the mid-1960's to develop a prototype onsite fuel cell energy system.

Today, a major field test of the onsite fuel cell concept and a supporting technology development effort in parallel with the field test is being supported by the gas industry, through the Gas Research Institute (GRI) and the utilities, in conjunction with the U.S. Department of Energy (DOE) and UTC. The field test will involve over thirty gas and combination utilities in the United States.

If the onsite fuel cell field test establishes the "market pull" and the technology development work advances the state of the art, fuel cell energy systems will be a reality in the 1980's.

KEYWORDS

Phosphoric acid fuel cell; cogeneration; onsite field test; TARGET; technology development; gas industry.

INTRODUCTION

The fuel cell concept which uses a simple, efficient and reliable electrochemical process to convert hydrocarbon fuels to power has been of considerable interest to the gas industry for a long time. As onsite generators of electricity and heat, which makes fuel cells cogenerators, fuel cells would provide the U. S. consumer with a new energy service option which could hold down fuel and utility bills, conserve energy resources, reduce the need for large capital investments in central power plants, and decrease pollution emissions to one-tenth the level produced by the cleanest gas-fired central electric generating plants now in operation.

The prospects for onsite fuel cells are exciting in light of recent legislation for cogeneration that will give a more realistic value to the electric power produced from cogeneration facilities and thus provide an economic incentive for their development and use. Although several potential cogeneration systems exist, such as gas turbines, diesels, and reciprocating engines, the gas industry has had a high interest in onsite fuel cells because of their unique characteristics. In addition, the gas industry sees onsite fuel cells providing an excellent business opportunity for them.

Onsite fuel cells generate electricity with an efficiency of about 40 percent. When combined with heat recovery in the form of either hot water or steam, the overall system efficiency is approximately 80 percent. Their virtual instantaneous response to transient demands for power gives them considerable flexibility. One of their most important features is their superior environmental characteristics which has created a great deal of interest in the United States for those cities where the air quality is of continuing concern.

The major onsite fuel cell effort currently being supported by GRI uses phosphoric acid as the electrolyte and will be the main topic of discussion for this paper. The operating temperatures of these fuel cells is approximately 350°F. There is also an advanced fuel cell technology, molten carbonate, that is being pursued in the United States and shows promise of having better performance characteristics and possibly lower installed costs per kilowatt. These cells operate at higher temperatures making them very suitable for industrial applications. The molten carbonate fuel cell, however, has a commercialization path that is longer than the phosphoric acid fuel cell and initially will probably not use pipeline gases as its primary fuel. TABLE 1 lists the characteristics of the two technologies.

TABLE 1 Fuel Cell Energy System Characteristics

Characteristics	Phosphoric[1] Acid	Molten[2] Carbonate
Overall Efficiency[3], %		
100% Capacity	80-85	85-90
Electrical Efficiency[3], %	40	40-60
Thermal Efficiency[3], %		
100% Capacity	40-45	45-30
Temperature Ranges of Thermal Energy, C (F)		
High Grade	135 (275)	371 (700)
Low Grade	82 (180)	121 (250)
Thermal Energy		
High Grade, %	20	80-100
Low Grade, %	80	20-0
Projected Power Plant Capacity Range, kW	20-1000	1000-10^6

1) Based on UTC 40kW power plant specification.
2) Based on project characteristics of the molten carbonate system
3) Based on lower heat value (LHV) of fuel.

Both fuel cell energy systems are similar in design with the major differences occurring in the power section construction and operating temperature. A block diagram of a general onsite fuel cell energy system is provided in Fig. 1. Each system consists of three integrated sections, each performing specific functions in the conversion of fuel to energy services. These sections are designed to process the incoming fuel, generate the power, and condition the power to uniform user specifications.

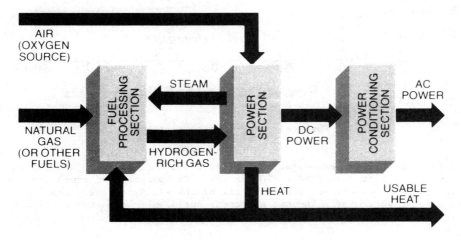

Fig. 1. Fuel cell power plant functional diagram.

BACKGROUND

The gas industry has an extensive history in the funding of fuel cell developments. Initial laboratory studies funded by the American Gas Association (A.G.A.) and Columbia Gas System Service Corporation led to the establishment in 1967 of a utility-supported group called TARGET, i.e., Team to Advance Research on Gas Energy Transformation. TARGET, the Federal government, UTC, A.G.A. and GRI have invested over $150 million in the technical development of onsite fuel cell energy systems, of which GRI and the gas industry have spent over $60 million, the Federal government $20 million, and UTC, as prime contractor, about $70 million. As part of this effort, an initial set of sixty-five units with a 12.5 kW capacity was field tested between 1971 and 1973 in some thirty-five installations.

The initial field tests, though successful, did not result in a commercial product. Test results did allow analysis of optimum product sizing and identified required technical improvements. TARGET continued its activities through the successful laboratory testing of a 40kW onsite fuel cell power plant. A GRI/DOE cofunded technical development program evolved from this work in 1977 with the objectives of further improving both performance and economics. This project resulted in the verification testing of an advanced prototype 40kW system at UTC facilities near Hartford, Connecticut; the system has operated for over 3,000 hours since its initial start-up in April of 1980. During that time, it has demonstrated high electrical efficiencies above the performance goal, as illustrated in Fig. 2, and the ability to operate under transient and steady state conditions. The longest continuous operation was over 1,200 hours; and, despite a few minor problems expected in any development system, the power plant has performed remarkably well. The overall fuel efficiency of the power plant as a function of the output power is shown in Fig. 3.

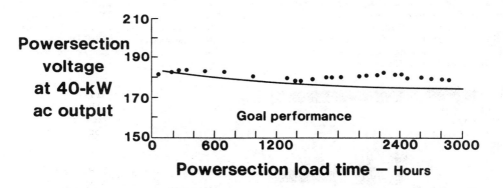

Fig. 2. Fuel cell powersection performance.

The 40kW fuel cell is a cogenerator that produces both electricity and hot water with an overall efficiency of 80 to 90 percent. For purposes of comparison, a 40kW power plant is capable of providing all of the electric power, heating, cooling, and hot water for a 16-unit apartment building in a northern climate using less gas than is now used for space heating only.

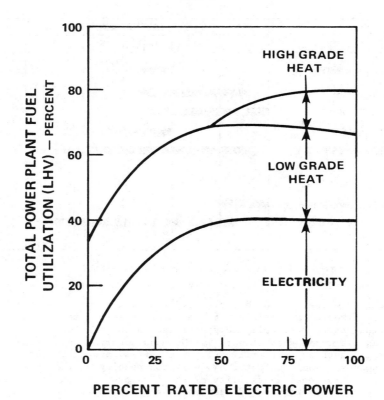

Fig. 3.   Electrical and thermal efficiency
characteristics for power plant.

CURRENT STATUS

With successful demonstration of the system's efficiency and durability, the
focus of the onsite fuel cell program at GRI has shifted. During the last year,
several utilities, DOE, and GRI have been constructing a program to address the
business uncertainties of onsite fuel cell energy service and to focus on the
needed system cost reductions.

The onsite fuel cell program is divided into major activities that will jointly
evaluate the fuel cell energy system as a marketable product. The first is a
major field test to evaluate the ability of the fuel cell energy system to
operate in a utility's environment under actual use conditions and to assess the
business issues related to utility onsite fuel cell energy service. The second
is a technology development effort with the manufacturer, UTC, to improve the
component/system cost and design. A schedule of these activities is illustrated
in Fig. 4.

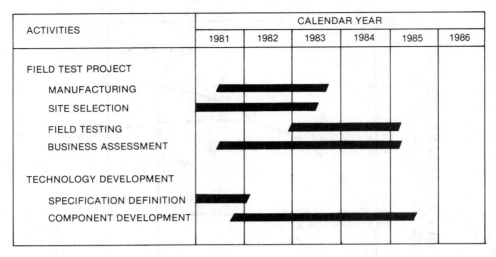

Fig. 4. Onsite fuel cell program schedule.

The field test is a market-oriented activity directed at generating the performance and market data sufficient to have the utilities make a commitment toward commercial onsite fuel cell energy service in the mid to late 1980's. This project has two major activities: 1) the manufacturing and testing of forty-five power plants at twenty-five to thirty sites, and 2) the utility business assessments. The manufacturing and field testing of the power plants is to verify the operational feasibility of the 40kW power plant in a variety of attractive early entry markets. The business assessment by the potential user utilities are to examine and define proposed business scenarios for commercial fuel cell service to be offered and to assess their meaning to each utility. To accomplish these primary objectives, the specific goals defined in TABLE 2 will need to be met.

TABLE 2 Major Goals of the Field Test Project

o   Operational Feasibility and Measured Reliability

o   Fuel Conservation and Environmental Characteristics

o   Institutional, Regulatory, Code and Legal Issues
    Evaluated

o   Variation of System Configuration, Applications and
    Conditions

o   Viability of Early Entry Markets

o   System Economic Performance

o   Public Exposure and Society Acceptance

o   Establish Commercial Service Scenarios

o   Identify Technical Power Plant Modifications

The manufacturing and field testing effort includes coordinated efforts of the manufacturer, utilities, and funding organizations (DOE and GRI). The manufacturer will be responsible for the fabrication and acceptance testing of the field test 40kW onsite fuel cell power plants. These power plants will have a capability of operating isolated or connected to the electric utility grid.

In addition to the business assessments being conducted by the utilities, they will also have prime responsibility in siting the field test units. Initially, approximately sixty to one hundred sites will be selected by each utility. These sites will be instrumented for the collection of generic building energy data which will be utilized in evaluating the compatibility of the onsite fuel cell energy system and to estimate the potential economics of onsite fuel cell energy service. After collection of approximately a year of data at each site, twenty-five to thirty sites will be selected by the utilities and approved by GRI and DOE for power plant installation.

The Technology Development Project of the onsite fuel cell program is a development oriented activity directed at providing the technology necessary for the manufacturer to make a commitment to supply onsite fuel cell energy systems at a price economical enough to penetrate the early entry markets established by the utilities. The technology development is conducted in parallel with the field test activities to coordinate the commitments of the utilities and the manufacturer to commercial onsite fuel cell energy service. This activity is divided into two major efforts. Initially, the specifications for the power plant will be examined, modified, and reevaluated so that the manufacturing of the power plant in small lot productions (fifty to two hundred fifty units per year) will likely result in an economical system which will penetrate the early entry market. It is projected that advanced technologies will be necessary to decrease the present cost of the hardware. The second phase of the technology development project will perform the development necessary to advance the technology of specific components and decrease the cost of power plant manufacturing. This effort could include component technology modifications for both performance and manufacturability reasons. This activity and the field test is projected to result in initial commercial onsite fuel cell energy service being offered in the mid to late 1980's.

COMMERCIAL PROSPECT

Despite the successful technical development to date, there remain significant questions concerning the commercial prospect for onsite fuel cells. These include questions of manufacturing and installation cost, improving technology, and market considerations. In the past year, GRI has established a Fuel Cell Users Group, consisting primarily of utilities participating in the field test, to help define the market and marketing strategies and to establish the specifications for the power plant to meet market operating demands. A key uncertainty in these studies is price, which is related to both cost and sales terms (warranty, advertising allowance, etc.) from the manufacturer. This in turn is related to a manufacturer's plans, capabilities and commitment to commercialize the fuel cell.

The field test, along with the parallel technology development effort discussed earlier, is a critical step in addressing the uncertainties raised. Both activities are funded with the leading manufacturer of phosphoric acid fuel cell systems, UTC, and directed at the utility industry as the initial commercial-izing organizations for onsite fuel cell energy service. The field test is directed at generating the performance, reliability, and market data sufficient to have utilities make a commitment toward commercial onsite fuel cell energy service. The Technology Development is a technology-oriented activity directed at advancing the component technology and manufacturing techniques to a sufficient state such that the manufacturer can make a commitment and have the technology necessary to supply onsite fuel cell energy systems at an economical price to early entry markets. These projects are coordinated to have the utility's and the manufacturer's commitments occur within the same time period increasing the probable success of the commercialization effort.

Currently, GRI is actively assessing the commercial potential of onsite fuel cell energy systems in the environment of future gas rate schedules, electric rate schedules, cogeneration policies, technology status, and required improvements and regulations. The result of this assessment will be used as a basis for finalizing the future level of effort by GRI in commercializing Onsite Fuel Cell Energy Systems.

CONCLUSION

The Phosphoric Acid Onsite Fuel Cell Program being supported by GRI is an integrated activity of two major projects, the Field Test Project and the Technology Development Project. If the field test is successful, it will establish a "market pull" for onsite fuel cells through the utilities. If the technology development effort reaches its goals, it will put the manufacturer in the best position to respond to that pull. Therefore, the reality of onsite fuel cells in the 1980's depends heavily on the overall success of these efforts. The probability of success will depend on the level of commitment given the program by all participants over the next four years.

EURELIOS, THE WORLD'S FIRST OPERATING SOLAR POWER TOWER PLANT (1 MWel)

J. Gretz

Commission of the European Communities, Joint Research Centre - Ispra Establishment, Ispra, Italy

ABSTRACT

The paper describes the helioelectric power plant, EURELIOS, sponsored by the Commission of the European Communities. This plant, of the mirror-field and central receiver type, was designed and built by a Consortium of European industries formed by ANSALDO and ENEL (Italy), CETHEL (France), MESSERSCHMITT-BOLKOW-BLOHM (F. R. Germany). The construction of the plant was completed by the end of 1980 and it was connected to the grid of the Italian National Electricity Generating Board, ENEL, at Adrano, Sicily, Italy, in April 1981.

The costs of the construction of EURELIOS are 10 million ECU, i.e. about 11 million US$.

KEYWORDS

Solar electricity generation; European Economic Community; power tower plant; EURELIOS.

INTRODUCTION

Helioelectricity can essentially be generated by thermomechanical and by photovoltaic conversion. Both technologies are today at about the same investment cost level of, say, 10 - 20 US$/W(el). Cost reductions are in view for both technologies, and only construction and operation of power plants will allow to evaluate the real energy costs of those technologies.

In the framework of its Solar Energy R&D Programme, the Commission of the European Communities decided in 1976 to build a helioelectric demonstration plant of rather large rating. In order to take advantage of the high exergetic potential of solar energy, it was decided to use high temperatures of the working fluid by means of high concentration in the power-tower concept. At that time, first evaluation studies indicated the cross-over point of investment cost and

energy costs between distributed systems and power towers at about 500 - 700 kW.
To be sure of the choice and in order to have a representative rating, the 1 MWel
seemed to be a sound choice.

A European Industrial Consortium has been set up for the layout and the construction of the plant, consisting of

- ANSALDO SpA and Ente Nazionale per l'Energia Elettrica (ENEL), Italy;
- CETHEL (combining Renault, Five-Cail-Babcock, Saint-Gobain Pont-à-
  Mousson and Heurtry S.A.), France;
- MESSERSCHMITT-BÖLKOW-BLOHM (MBB), Germany.

These firms completed the definition of the overall system and the engineering
design specifications for all subsystems of the plant including construction and
testing of prototype models by November 1978. Completion of construction of the
plant was scheduled for the end of 1980; this date was maintained.

The site of the power plant is ADRANO, a village 40 km West of Catania, Sicily,
Italy. It has an average elevation of 220 m, a North-South inclination of 5% and
lies near a small river. The plant will feed its electricity into the grid of the
Italian Electricity Generating Board, ENEL, who operate the plant and are co-
proprietor, together with the Commission of the European Communities.

SYSTEM CONCEPT

The plant design is based on the central receiver principle. 6200 m^2 of mirror
surface, mounted on 182 heliostats, reflect the direct solar radiation onto the

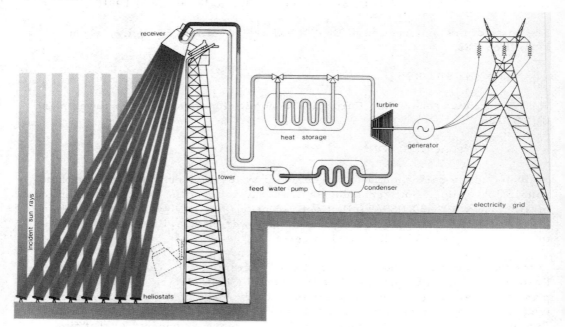

Fig. 1. System Concept

central receiver, located on a 55 m high tower. Water is passed through the receiver where it is converted into steam to drive a turbine coupled to a generator. The electrical energy generated is fed into the existing grid. A by-pass is used for start-up and shut-down procedures and a thermal buffer system is provided so that the plant can continue its operation without solar energy input for a period of 30 minutes.

TABLE I - Technical Parameters of EURELIOS

General Characteristics

Experimental plant of the central receiver-multiheliostat type
Location: 37.5 °N, 15.25 °E (Sicily, Italy)
Design point: equinox noon, assumed insolation of 1000 W/m^2.

Heliostat Field:	4800 kW(th) to receiver
Heliostats:	two types, two axes controlled, overall inaccuracy $\pm$ 4 mrad (1σ)
CETHEL type:	ca. 52 m^2, 8 focusing modules, 70 heliostats
MBB type:	ca. 23 m^2, 16 square elements, 112 heliostats.
Receiver/tower:	Cavity type receiver 4.5 m $\emptyset$ aperture in 55 m height, 110° inclination. Receiver outlet steam conditions: 512°C, 64 atm., 4860 kg/h (5346 kg/h possible).
Steam Cycle:	Turbine connected to receiver (no intermediate heat exchanger) Nominal power: 1200 kW(mech.) with steam of 510°C, 60 atm. Feed water temperature at receiver inlet: 36°C Cooling water temperature: 25°C maximum.
Thermal Storage:	Reduced electrical output for ca. 30 minutes.
Energy Storage:	vapour 300 kWh; Hitec: 60 kWh
Equipment:	pressurized (19 bar) water reservoir for 4300 kg; vapour produced from 19 to 7 bar. Two storage tanks, containing 1600 kg Hitec (overall capacity); Heat exchangers for 19 bar, 480°C and 410°C steam temperature.
Electrical System:	Power generation: alternator for 1100 kW min. for ca. 100 kW internal power and 1000 kW for external users; transformers, emergency power supply; Interface to grid: equipment to connect transformers to public grid; steam cycle control equipment; command, operation and monitoring is centralized in the control centre.

MIRROR FIELD

There are two types of heliostats placed within two subfields divided by a line going North-South. The MBB heliostats of 23 m^2 each are placed in the Eastern part, those of CETHEL of 52 m^2 each in the Western part. The aim of this is to test under field conditions the performance, behaviour and therewith the economics of heliostats of considerably different size.

THE MBB HELIOSTAT

The MBB heliostat consists of 16 mirror elements mounted on two supporting frames. These mirror elements have a fixed focal length of 190 m, their orientation can be adjusted using the fixation screws. The supporting frames are attached to the alt-azimuth drive mechanism by means of flanges. Each heliostat has a reflective surface of 23 m^2 and comprises 16 focused square mirrors, 1.20 x 1.20 m, made of 3 mm floatglass and sandwich structure.

Fig. 2. MBB Heliostat

THE CETHEL HELIOSTAT

The CETHEL heliostat consists of 8 modules, each one being made up of 6 mirrors of 1.8 x 0.6 m, so that the total reflecting area is 52 m^2. The 48 elementary mirror stripes are laterally flat and bent vertically and arranged in such a way as to envelop a sphere of the required bending radius, focusing at $100 < L < 200$ m. The glass is 6 mm thick. High stiffness is ensured by a triangulated iron structure which guarantees, together with the very precise tracking system, more than the required 4 mrad accuracy of the reflected beam, i.e. 2 mrad on the panel.

Fig. 3. CETHEL Heliostat

RECEIVER

The receiver, placed on the top of a 55 m high tower (see Fig. 4), is constructed by ANSALDO and is based on the results and experiences gained by G. FRANCIA and ANSALDO at the S. ILARIO Test Facility.

The once-through boiler type receiver consists of two parallel tubes, through which pressurized water flows, rolled up in a coil to form the walls of an opened conical body (a cavity type boiler) into which the solar radiation is focused. The tubing forms a preheating zone at the lower part of the cavity where there are highest solar fluxes, and an evaporating zone on the conical side walls of the receiver. The superheating zone is located in the central upper part of the boiler, more protected from the solar radiation.

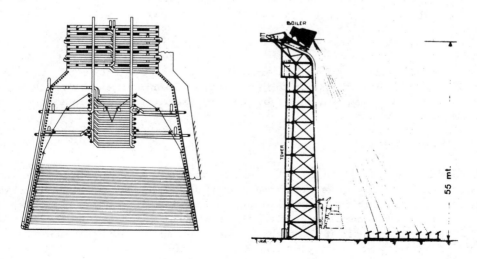

Fig. 4. Receiver and Tower

To maximize solar energy absorption, the tubes are finned, allowing the reflected energy to hit other zones of the boiler at least 2 or 3 times before being reflected outside. The tube surface is darkened by a heat treatment to approach as
much as possible a black body behaviour. In order to reduce the thermal energy
loss due to irradiation and convection, the receiver is equipped with antiradiating
structures, consisting of pyrex sheets, located from the inside over the tubes,
and transparent only to the incident radiation. At a first stage, the receiver will
be operated without the antiradiating structures in order to measure the effect of
it when operating with these structures later on.

THERMAL CYCLE

The feed water pump system, consisting of one centrifugal pump and a reciprocating pump, arranged in series, supplies water at 110 atm., while a plenum
chamber damps the peaks produced by the reciprocating pump. For reasons of
safety, the plant operates normally with one system, a second one being in
stand-by.

A turbine by-pass system raises the steam temperature during start-up and de-
pressurizes and desuperheats the fluid during shut-down. The thermal power
produced by the receiver varying with insolation, the control system keeps constant the temperature of the steam at the receiver outlet by controlling the feed
water pumps. This final steam temperature is maintained constant in transients
by means of two direct contact attemperators for every tube, the second one being
located near the receiver outlet in order to ensure a constant exit temperature.

A thermal buffer allows the plant to operate for 30 minutes at derated power without insolation in order to protect the turbine against thermal shocks when clouds
are passing over the heliostat field. The system consists of a tank of hot pressurized water at 19 bar and a molten salt system capable of superheating up to
$410^\circ C$ the saturated steam produced by the hot water tank. The tank can produce
saturated steam at a pressure decreasing from 19 bar during the discharge period.

COSTS

The design and construction costs of EURELIOS are 10 million ECU, i.e. about
11 US$/W(el). The Commission of the European Communities financed the project
for 50%, the rest is equally borne by the three participating member countries
France, Germany and Italy.

SUMMARY

EURELIOS, a project sponsored by the Commission of the European Communities,
is the world's first large-scale experimental solar power plant in the megawatt
range to produce electricity and feeding it into a national grid. It pioneers, by
that virtue, a new technology in the field of renewable energy sources. Its construction was completed in December 1980, the connection to the grid of the
Italian Electricity Generating Board, ENEL, in April 1981.

Although today´s investment costs of photovoltaics and thermomechanical converters are about equal, photovoltaics is likely to enjoy considerable cost reductions in the next decades. In order to become competitive, the costs of the heliostats have to be decreased considerably and the efficiency of the plant to be increased so that the mirror field size can be reduced.

The plant´s costs are 10 million ECU, of which 50% are borne by the Commission of the European Communities, the other 50%, equally, by the three participating member countries, France, Germany and Italy.

REFERENCES

Borgese, D., J. J. Faure, J. Gretz, A. Strub, H. Treiber, and L. Tuardich (1980). EURELIOS, the 1 MW(el) solar electric power plant: A European Community research project in the field of new energy sources. XI World Energy Conference, Munich, 1980.

BATTERIES FOR ELECTRIC VEHICLES

B. Hartmann

Brown, Boveri & Cie., Es, Heidelberg, Federal Republic of Germany

KEYWORDS

electric vehicles, near term batteries, advanced batteries

1. ABSTRACT

In order to guarantee the mobility of our society, storage of gasoline in automobiles has to be replaced by storage of other forms of energy. A big challenge is to use - at least partly - electric vehicles (EV's) instead of internal combustion engine driven vehicles.

In this case electrochemical batteries will be used as on bord storage systems and gasoline will be substituted by electricity generated from coal or unclear fuel. Since EV's do not produce noise and pollution, some environmental problems will also be solved.

A drawback of our today's EV's is that the lead acid batteries needed for propulsion are relatively heavy. Improvements are expected

- by utilizing "near term batteries" such as improved lead acid, nickel/iron or nickel/zinc batteries, which are in development and which will have approximately twice as much energy per unit of weight as today's lead acid batteries

- by turning to advanced batteries, which will possess once more twice as much energy per weight as the near term batteries. Zinc/chloride, lithium-aluminium/iron sulfide and sodium/sulfur batteries are considered as advanced systems.

Pros and cons and characteristic features of these battery systems will be discussed.

2. INTRODUCTION

While the battery propelled electric vehicle played an important role in the beginning of this century, the improvements which could be achieved with combustion engine driven vehicles were much faster, so that production of EV's was stopped after some time. An advant of EV's is expected, however, because of the increasing oil and gasoline prices and due to progress in the field of power electronics and battery technology. As a consequence of these changes it is expected that the properties of future EV's become technically and economically comparable or close to those of conventional vehicles. Near term and advanced batteries will allow to build many types of vehicles such as private cars for ranges up to several hundred km, commuter cars, vans, busses and locomotives for railways with attractive properties.

While the energy content of a battery for a small passenger car may be as low as 20 kWh it will be more than one order of magnitude higher for busses or locomotives. The requirements with respect to energy and power density are also depending very much on the kind of vehicle to be propelled. The requirements for passenger cars are specified in table 1.

TABLE 1   Goals for a BBC electric vehicle

maximum speed	130 km/h
range at 100 km/h	250 km
acceleration 0 - 50 km/h	7 s
life	> 100 000 km
mass of vehicle without battery	1200 kg

Due to a study of the Department of Energy 95% of all trips are within a range of less than 90 km, 98% within a range of about 250 km. In Europe the distances to be covered for one drive are even smaller. The outer appearance of an electrical car will be very similar to an ICE car. The optimal battery site should be the center of the vehicle causing only minute or no increases in height and weight.

3. CORRELATIONS BETWEEN VEHICLE AND BATTERY PROPERTIES

A study has shown, for example, that a 50 kWh battery with a power of 40 kW and a weight of 320 kg is able to propell a passenger car to a maximum speed of 130 km/h, to accelerate it to 50 km/h within 7 s and to achieve a range of 250 km at 100 km/h. The battery properties necessary to achieve these aims are shown in Table 2. They are compared with the requirements resulting from targets set by DOE (1) for a similar type of car. It can be concluded from the table, that the battery requirements depend very much on the properties aimed for the car.

The DOE targets may be achieved with near term batteries, where advanced batteries are necessary to meet the BBC targets.

The properties to be attained for batteries to propel other electric vehicles are much less severe than those for passenger cars.

The environmental conditions under which batteries have to operate are the same as for any other automobile component. The ambient temperature may vary between - 20 and + 40°C, the acceleration according to vibration amounts to 5 g and even 30 g in case of accidents.

TABLE 2   battery requirements according to

	DOE/ETV 1	BBC	(table 1)
energy		44,5	kWh
specific energy	56	140	Wh/kg
electric efficiency at C/3		80	%
specific power max.	104	135	W/kg
max. power		43	kW
mass of battery		320	kg
volume of battery		330	l
life	800	1000–1500	cycles
energy consumption		0,22	kWh/km

4. NEAR TERM BATTERIES

Although the lead acid batteries are known for more than 100 years large efforts are made for further improvements. The improved lead acid battery and other near term batteries such as nickel/iron and nickel/zinc are expected to approach the market in the middle of the eighties.

LEAD ACID BATTERY

Cell reaction: $Pb + PbO_2 + 2 H_2SO_4 \rightleftharpoons 2 PbSO_4 + H_2O$.

The battery operates at ambient temperature.

Problems are encountered at low and high temperatures and at high charging / discharging rates with respect to power, capacity decrease and cycle life.

As the industry for fabricating lead acid batteries is well established costs are relatively low, two thirds of it being material costs for lead.
Much progress has already been made with lead acid batteries.
Several steps in development are, however, necessary to further improve this type of battery. It is expected that these steps can be performed within the next several years. Main points of development are improved grid construction and utilization of the positive electrode. In any case the potential for energy density is low (50 Wh/kg). The abundancy of lead is low possibly causing additional cost increases.
Table 3 summarizes properties of near term batteries for comparison.

TABLE 3   goals for near term batteries

	commercial lead acid	lead acid goals 1986	nickel-iron 1986	nickel-zinc 1986
specific energy (Wh/kg)	30	50	60	70
specific power (W/kg) (20 s peak)	70	110	120	120
life (cycles)	240	800	1200	800
cost (% of lead acid)	100	100	170	> 170

NICKEL-IRON

Cell reaction: $Fe + 2 Ni(OH)_3 \rightleftharpoons 2 Ni(OH)_2 + Fe(OH)_2$

The system is very old as well, originally suggested by Edison. The theoretical energy density[*] is 265 Wh/kg. The system is thermodynamically unstable, but stabilizied by reaction overvoltage. The charging voltage causes 25-30% of the electrical energy converted to hydrogen as a byproduct. This results in a relatively low energy efficiency. Self-discharge occurs mainly due to chemical corrosion of the iron electrode in the strong alkaline electrolyte. The performance of Ni/Fe batteries at low temperature declines as the potassiumhydroxide content increases.

NICKEL-ZINC

Cell reaction: $Zn + 2 Ni(OH)_3 \rightleftharpoons 2 Ni(OH)_2 + Zn(OH)_2$

In this system the same expensive nickel electrode is used as in the nickel-iron battery. The zinc electrode gives additional problems namely growth of zinc dendrites and electrode shape change, resulting in a relatively short life time of a few hundred cycles.

Degradation of separator materials influences the life time and performance of all alkaline electrolyte batteries.

The goals for these near term batteries are compared in table 3.

---

[*] The theoretical energy density is defined as energy content of one kg of active materials.

5. ADVANCED BATTERIES

The advanced batteries which already reached a first engineering level are zinc-chloride, lithium-iron sulfide and sodium-sulfur batteries. The theoretical energy densities

TABLE 4   Theoretical energy densities

	lead acid	Ni/Fe	Zn/Cl	LiAl/FeS	LiAl/FeS$_2$	Na/S
theoretical energy density (Wh/kg)	180	265	510	460	650	760

The theoretical energy density of advanced batteries is by a factor two higher than the theoretical energy density of conventional batteries, but the realization is much more difficult. Due to new technologies needed the development of advanced batteries takes a longer period of time. The development began in the sixties.

ZINC-CHLORIDE BATTERY

Cell reaction: $Zn + Cl_2 \rightleftharpoons ZnCl_2$ (ag)

The zinc/chloride system consists of two subsystems, the cell stack where the electrochemical reactions take place and a separate container to store reactants. The current collectors are graphite sheets. In the completely discharged state a solution of zincchloride fills the cell. On charging zinc is deposited on one electrode and chlorine at the other. The chlorine has to be stored in form of chlorine hydrate which is only stable below 10°C. Auxiliary equipment such as the refrigeration system, pumps, heat exchangers and valves is necessary.

Problem areas for further development are growth of zinc dendrites, corrosion in heat exchangers, thermal management, life and auxiliary equipment with respect to weight and cost.

A pilot plant is under construction. The goals of this development are described in table 5.

TABLE 5   goals for advanced batteries

	Zn/Cl	LiAl/FeS	LiAl/FeS$_2$	Na/S
specific energy (Wh/kg)	80	100	150	150–180
specific power (W/kg)	90	180	180	135–170
life (cycles)	1500	1500	1500	1500

LITHIUM-IRONSULFIDE

Cell reaction I $2\ LiAl + FeS\ 2\ e^- \ Li_2S + Fe + 2\ Al$

Cell reaction II $4\ LiAl + FeS_2\ 4\ e^- \ 2\ Li_2S + Fe + 4\ Al$

The active materials in lithium-ironsulfide cells are lithium-aluminium alloy as the negative electrode and ironsulfide FeS or irondisulfide FeS$_2$, as the positive electrode. The electrolyte consists of a mixture of LiCl and KCl with a melting point above 350°C causing operating temperatures between 400 and 500°C. The positive mass is wrapped by yttriumoxide felt. Boron nitride felt usually serves as a separator and boron nitride pieces are used as electrical insulators. A multiplate cell design with e.g. five electrodes is commonly utilized.

The main problems to solve are cost of separator and feedthrough, electrode swelling and the sensivity to overcharging. The life time is limited by internal short circuits or decline in performance. Especially with FeS$_2$ electrodes corrosion problems have to be faced. Advanced thermal insulations are under development for all high temperature battery systems.

SODIUM-SULFUR

Cell reaction: $2 Na + xS \rightleftharpoons Na_2S_x$ $2,7 \leq x < \infty$

Central element of this system is the solid electrolyt, β-alumina, in the form of closed end tubes. The electrodes are liquid sodium and liquid sulfur soaked in a carbon felt.   to the melting point of the polysulfides the working temperature is 300 to 350°C. With additives or wetting gradients a sulfur utilization between 80 and 95% can be achieved. The power density is mainly dependent on the wall thickness of the βtube and its active surface area, the resistivity of the ceramic, sulfur electrode and container construction.

Further improvements are necessary in battery life, cell failure characteristics and thermal management.

All three types of advanced batteries incorporate new technologies and extraordinary materials. The high energy densities and new materials result in severe safety requirements which, however, are probably met technically by either of these systems. A long cycle life for single cells has been achieved with all systems. The demonstration of long battery life is still lacking, which is partly due to the early stage of development and the low reproducability in an almost laboratory type of fabrication.

6. CONCLUSIONS

Battery powered electrical vehicles seem to be able to solve an important part of future transportation problems. The advantages of electrical vehicles are the high overall efficiency and the possibility to utilize abundant primary energy resources to generate the electrical energy to charge the batteries. The quality of life in urban areas can be improved due to noiseless and pollution free traffic. In addition charging of vehicle batteries at night time can contribute to use base load power plant during off-peak period.

It is expected that near term batteries will allow to produce a first generation of electric vehicles with suitable properties for technical application. The introduction of advanced batteries will open new dimensions for electric vehicles. Principally it will be possible to use these cars for 98% of all trips.

Acknowledgement

Many helpful discussions with Dr. W. Fischer are gratefully acknowledged.

References

(1) Neng-Ping Yao, The Modern Engineering and Technology Seminar, Taipei, Taiwan, 1980

INDUSTRIAL PROCESS HEAT APPLICATIONS FOR SOLAR THERMAL TECHNOLOGIES

D. W. Kearney* and D. Feasby**

*Engineering Consultant, Del Mar, California 92014, USA
**Solar Energy Research Institute, Golden, Colordo 80401, USA

ABSTRACT

This paper presents an overview of selected solar industrial process heat (IPH) activities under development in the U.S. Included are a summary of the IPH field test program, status of solar thermal technologies, and results of specific technology/application matching and market studies. There is a large potential market for solar IPH in the United States. The near-term viability of solar technologies in the industrial sector is dependent upon both the economic and technical issues which vary depending on the application, plant site, and system selected.

KEY WORDS

Solar; industrial process heat; IPH field tests; market assessment; end-use matching; industrial applications.

INTRODUCTION

Current energy consumption in the United States is approximately 80.5×10^8 gigajoules (GJ) [76.3 Quads (Q)]: 28.8×10^8 GJ (27.3 Q) for residential and commercial buildings; 19.6×10^8 GJ (18.6 Q) for transportation; and 32.1×10^8 GJ (30.4 Q) for industry. Industry, the largest energy user, consumes 39% (Hooker and others, 1980) of the gross energy.

Industry is a diverse and complex market sector consisting of four subsectors-- manufacturing, mining, construction, and agriculture. The energy use within the sector is unevenly distributed (i.e., manufacturing (78.8%), mining (8.4%), construction (7.2%), and agriculture (5.6%). Energy use within the manufacturing subsector is highly concentrated: 60% of all energy use in the subsector is accounted for by the 10 largest manufacturing industries.

Approximately half of the total industry energy consumption is for industrial process heat (IPH)--thermal energy used in the preparation and treatment of manufactured goods (Brown and others, 1980). This represents a significant market for solar thermal technologies designed to provide thermal energy in the form (e.g., hot water, hot air, or steam) and duty-cycle required for a particular type of application, provided that the economic and performance criteria of industry can be met.

892

INDUSTRIAL USES

Large quantities of heated water at temperatures between 49°C and 100°C are used in industry for such processes as cooking, washing, bleaching, and anodizing. Hot water may be supplied by directly heating water in absorber tubes of flat-plate, evacuated tube, or concentrating collectors, and piping this water to the process. Alternatively, a separate fluid may be piped through the collector field and then used to heat potable water via a heat exchanger. Large solar ponds may also be used to provide large amounts of low-temperature hot water.

Many industrial processes require large quantities of relatively low-pressure saturated steam. Saturated steam at approximately 690 kPa, equivalent to a temperature of about 171°C, can be produced in a solar system in two ways: (1) pressurized water may be circulated in a collector field and then flashed into steam in a low-pressure chamber, or (2) a heat transfer fluid capable of higher temperature operation may be circulated in the collector field and then to a steam generator, where the fluid serves as a heat source to produce steam. Because of the higher temperature needed, industrial steam applications normally require tracking concentrating collectors, such as the parabolic trough, or certain types of nontracking high-performance collectors, such as the evacuated tube.

Low-temperature (less than 177°C) direct heat is used for crop drying and food processing as well as paint drying, curing, and mineral solution dehydration. Air may be heated directly in collector systems designed to handle air as the circulating fluid, or a liquid may be circulated through the collectors and pumped through a heat exchanger to heat ambient air.

High-temperature (greater than 288°C) direct process heat accounts for a large portion of all industrial heat needs. Industries such as petroleum refining, primary metals, Portland cement, and glass are the major users. Intermediate-temperature collectors (parabolic trough or evacuated tube) cannot be used for process heat at these temperatures. Other collector types, though, such as central receivers or parabolic dishes, have potential for direct solar heating in selected processes.

STATUS OF THE TECHNOLOGY

Figure 1 shows the types of solar technologies that can be used to supply process heat over a broad range of temperature levels. Solar ponds offer the potential of very low-cost process heat up to temperatures approaching 100°C. A solar pond is a large area of water, 1- to 3-m deep designed both to collect and store solar energy. Thermal loss mechanisms that occur in natural bodies of water are greatly reduced in solar ponds either by using a salt gradient to suppress natural convection or by using inexpensive plastic glazings. The largest pond (2000 m^2) in the U.S. provides heat for a city's recreational building and swimming pool. Israel, which is particularly active in this area, has four ponds varying from 1100 m^2 to 6400 m^2 in area to provide energy for industrial processes and to produce electricity. Continued development of solar ponds is needed to find better, cheaper, and more reliable methods of maintaining a stable, nonconvecting layer, of extracting the heat, and of solving the problems associated with material lifetimes of liners and glazings.

Energy at somewhat higher temperatures can be supplied by flat-plate collectors (up to 93°C) or nontracking concentrating collectors (up to 177°C), such as evacuated tubes and compound parabolic concentrators. Flat-plate and evacuated tube collectors are most widely used for residential and commercial heating, and are continually being improved as a result of the experience gained in the field.

The required features of high-temperature collectors are the ability to track the sun on one or two axes and to concentrate its energy many times by reflecting

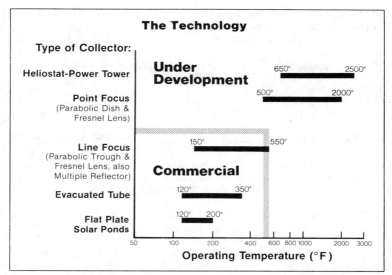

Fig. 1.   Typical Operating Temperature Range of Solar Collectors

incident solar rays onto a receiver.   The concentration ratio, defined as the
ratio of collector aperture area to receiver area, can vary from about 40 for parabolic troughs to over 2000 for heliostat/central receivers.

Fresnel lens collectors, which focus light through refraction onto an absorber
tube, and parabolic troughs or multiple reflector types, which concentrate by
reflection, are used to supply energy with temperatures of up to 288°C.   Parabolic
troughs can yield higher concentrations than the Fresnel lens types and have been
shown to operate at higher efficiencies than the multiple reflector types.   As a
result, they are the most common choice in the 154°C to 260°C intermediate temperature range.   All of these must track the sun, however, and will not collect any
significant amount of diffuse radiation.   Because of the strict requirements on
focusing accuracy, they are also relatively susceptible to contaminants such as
dust.   Most intermediate-temperature solar IPH applications today use parabolic
trough concentrators, but Fresnel lens collectors are available for this market.
The extensive use of parabolic troughs in DOE-funded field tests for steam applications, as well as a few commercial projects, has significantly improved the
design features and performance of these systems.

The point-focusing distributed receiver can achieve very high temperatures using a
parabolic dish concentrator to track the sun on two axes.   This technology is less
developed than line-focused collectors, though commercial products do exist and
can produce thermal energy at high temperatures (ceramic designs are projected to
operate as high as 1649°C).

High-temperature process heat requirements can also be met by a central receiver,
in which a field of individually guided mirrors (heliostats) reflect incoming
solar radiation to a single receiver mounted on a tower.   The working fluid--air,
helium, molten salt, liquid sodium, or water/steam--circulates through the
receiver and transports the collected energy to the process.

CURRENT APPLICATIONS

The U.S. Department of Energy has funded a series of field tests since 1977 to
gain operational experience in the application of solar energy to IPH requirements.   To date, 34 design studies or actual installations have been funded util-

izing technologies that vary from flat plates to line-focus concentrators to central receiver industrial systems. The types of solar systems include hot air, hot
water, and steam production applied to a broad spectrum of industrial processes.

Design studies for the first field tests were initiated early in 1977, with design
and construction of subsequent tests continuing to the present. As a guide to the
energy delivered by such a system, an array of 1860 m^2 located in Colorado, operating at 40% system efficiency, would deliver about 4641 million kJ/yr. Table 1
shows the construction and operational dates, the industrial application, field

TABLE 1. **Summary of Solar IPH Field Tests and Design Studies**

Location	Process	Collectors	Owner	Status
Hot Water (60°–100°C)				
Sacramento, CA	can washing	flat-plate & parabolic	Campbell Soup Co.	operational (June 1978)
Harrisburg, PA	concrete block curing	multiple reflector trough	York Building Products	operational (Sept. 1978)
La France, SC	textile dyeing	evacuated tube	Reigel Textile Corp.	operational (June 1978)
Hot Air				
Fresno, CA	fruit drying	flat-plate	Lamanuzzi & Pantaleo Foods	operational (Aug. 1978)
Canton, MS	kiln drying of lumber	flat-plate (liquid)	LaCour Kiln Services, Inc.	operational (Nov. 1977)
Decatur, AL	soybean drying	flat-plate	Gold Kist, Inc.	operational (May 1978)
Gilroy, CA	onion drying	evacuated tube (liquid)	Gilroy Foods, Inc.	operational (Sept. 1979)
Low–Temperature Steam (100°–177°C)				
Fairfax, AL	fabric drying	parabolic trough	West Point Pepperell	constructed (Sept. 1978)
Sherman, TX	gauze bleaching	parabolic trough	Johnson & Johnson	operational (Jan. 1980)
Pasadena, CA	laundry	parabolic trough	Home Cleaning & Laundry	construction
Bradenton, FL	orange juice pasteurization	evacuated tube	Tropicana Products, Inc.	construction
Intermediate–Temperature Steam (177°–260°C)				
Dalton, GA	latex production	parabolic trough	Dow Chemical	construction

TABLE 1. **Summary of Solar IPH Field Tests and Design Studies** (Concluded)

Location	Process	Collectors	Owner	Status
San Antonio, TX	brewery	parabolic trough	Lone Star Brewing Co.	construction
Ontario, OR	potato processing	parabolic trough	Ore-Ida Co.	construction
Hobbs, NM	oil refinery	parabolic trough	Southern Union Co.	construction
Large-Scale Hot Water				
Santa Isabel, PR	fruit juice pasteurization	evacuated tube	Nestle Enterprises	design
Santa Cruz, CA	leather tanning and finishing	flat-plate	Salz Leathers	design
Oxnard, CA	sodium alginate processing	flat-plate	Stauffer Chemicals	design
Des Moines, IA	meat processing	parabolic trough	Oscar Mayer	design
Shelbyville, TN	poultry processing	evacuated tube	Tyson Foods	design
Large-Scale Steam				
Haverhill, OH	chemical plant	parabolic trough	U.S. Steel	construction
Fort Worth, TX	corrugated board production	parabolic trough	Bates Container	construction
San Leandro, CA	pressurized hot water for washing	parabolic trough	Caterpillar Tractor	construction

test location, and generic system type. Operation of a large number of steam projects will begin in 1981, leading to a wealth of new operational data in 1982 and beyond.

Overall, the operational systems have shown good reliability but lower than predicted system thermal efficiency (Kutscher, 1980). An examination of the causes of lower efficiencies points to a need for better design to reduce thermal and parasitic losses, and for improvement of the solar/industrial process interface. Failure of routine nonsolar components (pumps, valves, controls) is a continuing problem. Results to date have also indicated that efficient and cost-effective reflector cleaning techniques must be developed for concentrating collectors to ensure good long-term performance.

TECHNOLOGY/APPLICATION MATCHING

Although it would appear that the best potential markets for solar energy in industry would be the highest energy-consuming industries, there are other considerations (e.g., temperature requirements, solar system development, plant location, and conventional energy costs) that preclude these from being viable near-term markets. Within the U.S. industrial market sector, there are hundreds of different processes spread over 350,000 manufacturing plants and 450 different four-digit Standard Industrial Code (SIC) industries. Additionally, industrial plants are widely distributed over the U.S. so that the performance of solar thermal systems varies considerably due to the geographic variations in insolation. Therefore, there is a need to match solar thermal technologies to specific types of plant and process requirements to help assure satisfactory system performance and economics.

Care must be taken to select a proper interface between the solar system and the industrial process. From a technical viewpoint, the interface should simplify the control system, minimize thermal storage requirements, and permit continuous solar system operation whenever the insolation level is sufficient.

To match solar thermal technologies to a specific industrial process, an analytical method and feasibility analysis model has been developed and implemented at the Solar Energy Research Institute(SERI). The model (PROSYS/ECONMAT) is a combined performance and economic model that is flexible and fast-calculating

TABLE 2. Factors Favoring the Application of Solar IPH Systems

Environmental Factors

- High insolation levels
- High ambient temperatures
- A pollution-free microclimate
- A polluted macroclimate or area with strict air pollution regulations

Process Factors

- Low-temperature process
- Continuous, steady operations (24 h/day, 7 days/week)
- Liquid heating application
- Built-in process storage
- Easy retrofit of the solar system
- Inefficient fuel usage, not easily rectified

Economic Factors

- High and rapidly escalating fuel costs
- Uncertainties regarding fuel supplies
- Available capital
- Long payback periods or low rates of return requirements
- High federal, state, or local solar tax incentives
- Energy-intensive industrial operation and high energy costs
- Energy conservation measures already incorporated
- Close, cheap land or a strong roof available
- Low labor costs
- New plant

Company Factors

- Desire to install a solar system
- A skilled maintenance and engineering work force
- Progressive management

(Stadjuhar, 1981). The performance element calculates the long-term annual energy output for several collector types including flat-plate, nontracking concentrators, one-axis tracking concentrators, and two-axis tracking concentrators. The economic element of the model provides solar equipment cost estimates, annual energy capacity cost, and optional net present value analysis.

The PROSYS/ECONMAT model has been used in a variety of in-depth analytical studies, specific plant case studies, and generic system studies. Case studies (Hooker and others, 1980) consisted of selecting and visiting a variety of industrial plants in different locations to collect detailed process data, to test the model on actual industrial plant applications, and to assess the viability of solar technologies for these applications. One of the results of the case studies was to develop a set of factors (Table 2) that favor the application of solar IPH systems. A solar IPH application advantageously meeting the majority of factors in Table 2 is potentially economical and may warrant a more detailed analysis.

Processes investigated for the case studies included crude oil production (dewatering), aluminum container manufacturing, corn wet milling, polymeric resin manufacture (paint production), fluid milk processing, and meat processing. Based on plant location, process energy requirements, and the solar thermal system selected, the economic viability varies considerably. Note that although a variety of industries were investigated, without substantial screening, there were two cost-effective applications out of thirteen case studies completed.

NEAR-TERM SOLAR IPH MARKETS

Several market assessments and studies (Battelle, 1977; Insights West, 1980; Intertechnology, 1977) have been completed for industrial processes to try to match the solar thermal technologies to specific IPH applications. As a result of these efforts, selected four-digit SIC industries have been ranked as near-term IPH markets (DeAngelis, 1980) and are shown in Table 3. These industries have little potential for use of extensive waste heat recovery and have temperature requirements that match available solar thermal technologies.

A computer mapping technique (DeAngelis and others, 1981), similar to the type used for land management studies, is currently under development at SERI to determine near-term geographic markets for solar IPH. Factors (e.g., conventional fuel costs and availability, collector performance, air pollution data, state tax credits, etc.) that favor the use of solar energy by industry are compiled state by state. These can be weighted as desired. Then composite maps of the United States are generated to identify the geographic markets.

The mapping technique, together with the market assessment data, will identify four-digit SIC industries in specific locations that are target near-term markets for solar IPH. Feasibility analysis tools (e.g., PROSYS/ECONMAT) can then be used on a plant-specific basis to evaluate the viability of solar energy.

SUMMARY

There is a large potential market for solar IPH. The near-term viability of solar technologies in the industrial sector is dependent upon both economic and technological issues that vary widely according to the application, plant site, and the system selected. The DOE-sponsored Solar IPH Field Test Program has been instrumental in getting solar systems tested in a variety of applications.

Certain criteria have been developed which, if met, favor solar IPH application. Also, analytical matching tools such as PROSYS/ECONMAT are available to perform preliminary feasibility analyses. Technology/application matching combined with

TABLE 3. Potential Near-Term IPH Market for Solar Thermal Technologies

Below 288°C	2895 – Alumina [*]
2048 – Prepared Feeds	2823 – Cellulosic Man-Made Fibers [*]
2051 – Bread/Bakery Products	2834 – Pharmaceutical Preparations
Below 177°C	2951 – Paving Mixtures/Blocks [*+]
2062 – Cane Sugar Refining [+]	3275 – Gypsum Products [*+]
2063 – Beet Sugar [*]	395 – Ground/Treated Minerals [*]
2075 – Soybean Oil Mills	**Below 100°C**
2077 – Animal/Marine Fats/Oils	2011 – Meat Packing
2085 – Distilled/Blended Liquor	2022 – Natural/Processed Cheese [+]
2421 – Sawmills/Planing Mills[*]	2261 – Finishing Plants – Cotton [*+]
2611 – Pulp Mills [*+]	2262 – Finishing Plants – Synthetic [*+]
2621 – Paper Mills [*+]	2435 – Hardwood Veneer/Plywood
2631 – Paper Boards Mills [*+]	2436 – Softwood Veneer/Plywood[*]
2653 – Corrugated/Solid Fiber Boxes [+]	2511 – Wooden Furniture [+]
2661 – Building Paper/Board Mills [*+]	2824 – Noncellulosic Fibers [*+]
2812 – Alkalies/Chlorine [*]	3271 – Concrete Block/Brick [*+]

[*]Industries with the largest energy costs as compared to their value of product shipments.

[+]Industries that are the largest users of petroleum.

market assessment data and geographic market data have established near-term solar IPH markets for specific four-digit SIC industries. However, even after the preliminary feasibility has been determined, there is a need to investigate the viability of solar IPH at the plant site because of factors such as process requirements, available insolation, and land constraints.

REFERENCES

Battelle Columbus Laboratories (1977). Survey of the Applications of Solar Thermal Energy Systems to Industrial Process Heat. National Technical Information Service, Springfield, VA.

Brown, K. C., D. W. Hooker, A. Robb, S. A. Stadjuhar, and R. E. West (1980). End-Use Matching for Solar Industrial Process Heat. SERI/TR-34-091. Solar Energy Research Institute, Golden, CO.

DeAngelis, M. (1980). Market surveys: potential solar IPH applications. Solar Industrial Process Heat Conference Proceedings and Presentations. Houston, TX.

DeAngelis, M., A. K. Turner, and J. Weber (1981). A geographic market suitability analysis for solar IPH systems. Solar Rising, The 1981 American Section of the International Solar Energy Society Proceedings. University of Delaware, Newark, DE.

Hooker, D. W., E. K. May, and R. E. West (1980). Industrial Process Heat Case Studies. SERI/TR-733-323. Solar Energy Research Institute, Golden, CO.

Insights West (1980). Solar-Augmented Applications in Industry. Gas Research Institute, Chicago, IL.

Intertechnology Corporation (1977). Analysis of the Economic Potential of Solar Thermal Energy to Provide Industrial Process Heat. National Technical Information Service, Springfield, VA.

Kutscher, C. F., and R. L. Davenport (1980). Preliminary Operational Results of the Low Temperature Solar Industrial Process Heat Field Tests. SERI/TR-632-385. Solar Energy Research Institute, Golden, CO.

Stadjuhar, S. A. (1981). PROSYS/ECONMAT User's Guide-Solar Industrial Process Heat Feasibility Evaluation. SERI/TR-733-724. Solar Energy Research Institute, Golden, CO.

TRENDS IN DISTRICT HEATING
(SOME FUNDAMENTALS OF DISTRICT HEATING ECONOMY AND PLANNING).

M. Larsen

Harry and Mogens Larsen I/S, Consulting Engineers, Denmark

Introduction

District heating is not only a matter of physics, ma-
thematics and local economics, it can also have considerable strategic importance because of its energy
conserving aspects and thereby diminish nations' dependence on imported energy resources.

In many countries guidelines have to be drawn up in order to make better use of energy resources as a part of
the security policies of those countries and as a means
of stabilizing finance and trade. District heating is
one measure which can facilitate energy conservation,
resource recovery and energy waste reduction, and is
therefore deserving of a place in these guidelines.

The planning criteria for district heating will, in
many cases for different reasons, be determined by political, financial and social factors rather than technical, but in this paper I keep mainly to the technical
side.

The economics of a DH. scheme are highly dependent upon
the initial investments and the capital costs' share of
the total operating costs. It is therefore a deciding
factor that low cost heat be moved, i.e. transported,
from the source to the consumer at the lowest possible
cost, and it should be borne in mind that a DH. mains
network, is first and foremost a transport system.

Lower cost heat is available in different ways, among others by the use of waste heat, and by using certain other alternative energy sources and raw materials in a more efficient and rational way than was the custom prior to the violent oil price increases of recent years.

Similarly, today it can pay to invest large capital sums to a far greater degree than ever before in plant and schemes capable of exploiting indigenous energy sources, including waste and surplus heat, whose price increases are not expected to have the same violent effect as the increases and instability of oil prices.

Conditions

Method and principle are of great significance when planning a DH. scheme, but it is also necessary to have a sound knowledge of a series of conditions having a decisive influence on the basis for design sizing, such as the thermal conditions and limitations of the available energy source, or the climatic conditions. In the case of the latter, the length of the winter and it's average temperature influences the degree of utilisation of the capacity of an installed DH. mains network in terms of the relationship between the network's maximum collective design loading per unit of time (peak load) and the total annual transfer of units of energy. This relationship can be expressed as the load factor.

Naturally, prior to undertaking any district heating project, it is also necessary to examine and determine conditions of an even more local nature, such as possible pipe routes, institutional barriers and financial limitations.

It can also pay to carry out sensitivity analyses in respect of the expected effects of energy price increases, general inflation, etc.

Heat source influence - and low temperature systems

In many countries district heating is seen primarily as
being associated with the generation of electricity,
but other forms of utilisation of surplus or waste
heat, such as the incineration of domestic waste,
straw, wood waste, etc., could be included as important
elements in planning the reduction of our dependence on
supplies of imported energy raw materials.

Geothermal energy will also be of interest in some
areas, as will wind energy, which is attracting a certain amount of interest in Denmark for example. Solar
energy may also be relevant in some countries.

Common to most is the fact that, the lower the temperature for district heating, the greater the degree of
utilisation of waste or surplus heat.

In a thermal power station, 40 percent of the heat
energy content of the fuel is at best utilized, whilst
in combined electricity generation and heat production
(co-generation), 80 percent utilisation is possible,
depending upon the load distribution relationship between the needed levels of electricity and heat production. The lower the extraction temperature at the turbine, the lower will be the quantity of additional fuel
required for heat production, i.e. when added to the
needs of electricity production.

In my home town of Odense, Denmark, it has been calculated that, if any district heating temperature higher
than the actual maximum of 203° F (95° C) had been selected as the extraction temperature from the combined
heat and power station, the annual fuel consumption -
in this case now coal - would increase by approximately
2,000 tonnes for each deg. C temperature increase.

Said in another way, if supply temperatures can be lowered from 302° F (150° C) to 203° F (95° C) the additional fuel requirement for combined heat and power production will be reduced by approximately 40 percent.

The peak demand in Odense in 1979 was recorded as 540 MW one day in February. Flow temperature at that time was 203° F and the return was 113° F (45° C), i.e., a temperature drop of 90 deg. F. The network of distribution mains of this utility is approximately 2 x 625 miles (1000 km).

Characteristic for district heating in Denmark are low temperature schemes with direct connections to the consumer systems; that is to say, the same water circulates in the distribution network and the house internal systems. The advantages of this combination are mainly to be found in the lower costs of installation and connection and in the efficiency of heat transfer.

Within the connected buildings the need for heat exchangers, circulation pumps and expansion vessels can be done away with, and the costs of retrofit are nearly always lower than for indirect connections, for example to higher temperature schemes. Further advantages can be gained with properties under construction. The available differential pressure of the street mains often means that pipes, valves, etc., of the internal system can be reduced in size.

Direct connection can also be used to provide an incentive to use the district heating supply sensibly, in cooling the water as far as possible, so restricting or reducing the scheme's pumping costs or achieving an increase in the reserve capacity of existing mains, or reducing the sizes of new mains.

Temperature differentials in the connected dwellings under peak load conditions can often reach 80 - 100° F (45 - 55° C). This means that the pipe sizes of the

mains system need only be slightly larger than for a
high temperature system, but insulation and installation capital costs, as well as mains' heat losses, are
usually far lower.

Heat densities and heat loads.

Energy costs are often related to the total population
per sq. mile or kilometre, and this points to lower
costs in the high her density city core areas, which in
turn often points to higher temperature DH. systems.

However, the Odense scheme is one of the most succesful
district heating schemes in the world, and has Europe's
lowest heat prices, yet it has a supply area of low
heat density and it is a low temperature system
throughout. Both characteristics are typical for nearly
all of the existing Danish schemes and provide what is
generally considered by many "experts" to be "the worst
possible basis for a viable district heating scheme",
however no one can doubt the existence of these
schemes, nor their viability. We are talking about fact
and not theory. We are also talking about low heat densities in the so-called "impossible" range of 140 - 280
x 10^3 Btu. per acre (0.1 - 0.2 MW per hectare). Fortunately, no-one told us it would be impossible.

Taking a basic look at the heating requirements of a
building, it varies hour by hour over the year, partly
due to the climatic influences and partly to the fact
that daily heat demands vary between buildings of different purposes, their degree of utilization throughout
the day and the particular properties of each building.

A heating system is designed for producing a certain
amount of heat in an hour, but there are very few hours
in a year when there is need for the system's full capacity. Over the remainder of the year the actual utilisation of the design capacity varies as a lesser per-

centage share of this capacity. The greater the difference between a year's high and low demands, the fewer
hours in a year will full utilization occur.

For example, in Denmark, a heat demand of more than 80
percent of a consumer's design peak demand only occurs
approximately 160 hours in a normal year and constitutes only 0.5 percent of the annual consumption. Heat
demands round about 75 percent of the design peak occur
over roughly 400 hours and constitute 2 percent of the
annual consumption, whilst 60 percent load corresponds
to about 14 percent of the total consumption.

As has been said, the degree of utilization of a supply
network's capacity can vary with different local climatic conditions. In Denmark the load factor is approximately 0.4, corresponding to a district heating network
sized to be able to meet the largest actual combined
heat load and used at only 40 percent of its capacity
on average throughout the year, compared with its potential of 100 percent load over the year's 8760 days.

The annual heat consumption is comprised partly of a
degree-days-dependent demand, and a non-degree-days-
dependent demand.
The "degree-days" demand corresponds to the heat demand
determined by the difference between the outside temperature and room temperature, corrected for sun and
other heat regains, while the non-degree-days-dependent
demand represents the heat loss in the net and the domestic hot water service consumption.

Figure 1 shows a typical duration curve for thermal
load in Denmark over the year's 8760 hours, with the
actual heat demand plotted on the vertical axis. 100
percent is the actual simultaneous peak demand and
should not be confused with the summated total heat requirements of each and every building, which is theoretical and only occurs if all buildings demand their
calculated maximum heat at one and the same time. The

ratio between these two conditions is termed the simultaneous demand factor, and should be given very serious
consideration during the design stage in view of its
importance in determining the sizes for the distribution mains network, and hence their costs, both in
terms of capital outlay and operation, i.e. mains' heat
losses.

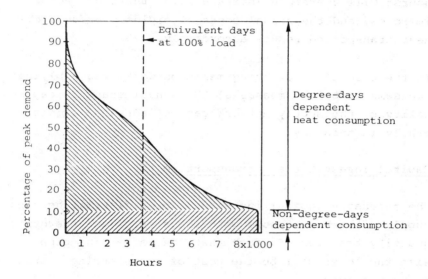

Fig. 1. Duration curve for thermal load.

The area below the curve in Fig. 1 is an expression of
the total heat consumption for a year, including pipe
losses and domestic hot water services, and must be
seen in relation to the outside design temperature,
which in Denmark is 10° F (-12° C), and the average
temperature throughout the locally relevant heating
season. The area is anological with the area to the
left of the dotted vertical line and corresponds to
about 3,650 hours of peak loading (100 percent). The
greater the number of these equivalent "max. hours" for
any project, the lower will be the annual cost of capital per transferred heat unit.

Load factor improvements

It follows therefore, that by replacing the consumption represented by the area at the top of the curve with an alternative and non-network transported heat supply (for example, "topping" or peak boilers at the consumer end), the new and lower peak will occur more often than the previous peak. The number of equivalent "max. hours" will therefore increase, i.e. the load factor improves, and the annual costs of capital per unit of heat transported reduce accordingly.

In the case of an existing mains network (and applying the same "topping" measures) the annual transfer capability of the supply network can be increased significantly in this way.

Capital investment as "transport costs"

The necessary capital investment for a district heating supply network decreases with the size of the energy quantity transferred, such that the larger the pipe size the lower will be the cost of transferring a unit of heat energy.

Fig. 2 shows an example of current "transport costs" in Denmark, in terms of capital outlay for the installation of twin mains of various sizes, related to the carrying capacity of each, and expressed as costs per metre twin pipe per m^3 water flow/hour.

In this example, normal flow velocities in the range 1.5 to 2.5 m/s. and flow and return temperatures of 194° F (90° C) and 104° F (40° C) respectively, were used as a basis for sizing.

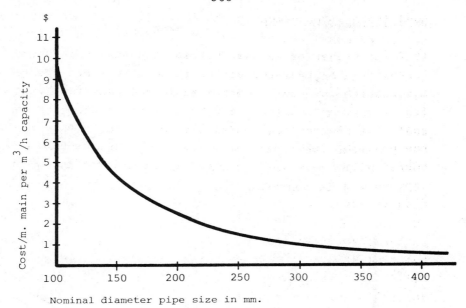

Fig. 2. Heat transport costs, expressed as costs of installed mains only (1981 costs).

The graph indicates that, when comparing the costs of
installed mains of about 6" – 8" diameter (150 – 200
mm) and upwards with their capacity increase for transported heat, the variation is impressively small. In
schemes having relatively large areas of supply, the
total mains-related capital costs (exclusive any service connections) will be dominated by the costs of
these larger mains and, in these instances, the ratio
of the total length of mains nework to the total quantity of transported heat will therefore be a valuable
guide to scheme economy.

The population density of the area of supply – or
rather, its heat load density – has a bearing on the
average distance over which a unit of heat is transported, and the heat demand per unit area can therefore be
another important factor in determining scheme economics for a given supply area on the basis of transport
distances. However, as Fig. 2 shows, the minor varia-

tions in the "heat transport" costs (in terms of capital investment) between the larger mains sizes, means
that the costs for larger distribution network schemes
can be held to be roughly proportional to the average
distance over which a unit of heat is to be transported.

Fig. 3 shows the approximate costs of installed twin
mains of the pre-fabricated and pre-insulated type, as
of mid-1981 in Denmark.

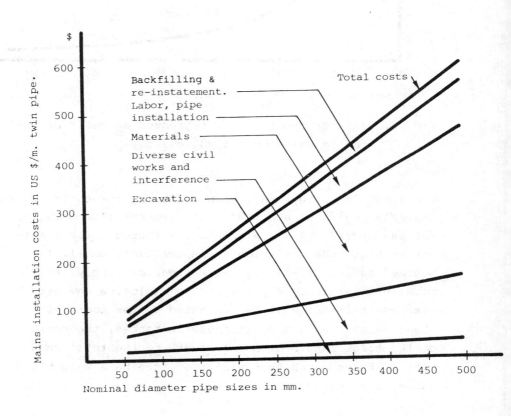

Fig. 3. Approximate costs of twin pre-fabricated mains installed
mid - 1981, Denmark.

District heating has a very important role to play today and in the future in most countries. It is a complex subject, but the application of common sense and
logic goes a long way to solving most of its problems.
It is a capital intensive system, but, as we know,
money can buy a lot of benefits, and, if those benefits
are gained at community and national levels as well as
by the individual, then it must be a sound investment.

THERMODYNAMIC CHARACTERISTICS OF ENDOGENOUS FLUIDS IN RELATION TO THEIR UTILIZATION ON A WIDER SCALE IN ELECTRICITY GENERATION

C. Latino and O. Sammarco

National Department of Mines, Grosseto, Italy

ABSTRACT

Electric energy is produced in Southern Tuscany, Italy, by forcing the hot fluids coming from the underground straight into a turbine. These fluids are mainly made up of steam.
In this note particular attention is paid to the thermodynamic characteristics of the endogenous fluids found in some steam, gas and mixed fields in Tuscany; these fluids have not always been exploited for electricity generation. Hypotheses   are forwarded for a rational utilization of the heat content of these fluids,   using energy production plants capable of recovering all the energy content of these fluids, whatever their status and nature, and with the highest possible termodynamic yield.
During exploitation of a field the enthalpy of the fluid is shown to vary greatly, not only as a result of pressure and temperature but also of the percentage variations of its components. Under these circumstances the plants must clearly be sufficiently flexible to permit eventual modifications to the different conditions of the field.

KEYWORDS

Geothermal field; heat content; total thermal energy; enthalpy; entropy; theoretical available heat drop; heat flow through the wells.

GEOTHERMAL RESEARCH IN ITALY

The endogenous fluids in Italy, consisting mainly of steam, have been utilized to produce electric energy since 1914. At the moment the annual geothermoelectric production represents 1.8% of the national electric energy production and about one quarter of the world's geothermoelectric production.
The steam exploited so far comes from two zones within southern Tuscany: the boraciferous region, lying west of Siena and south of Volterra, including Larderello, and the Amiata region, east of Grosseto along the southern margins of the Mt. Amiata volcanic apparatus. The geological situation is very similar in both zones. The predominant tensional tectonics of these areas is characterized mainly by Apenninic (NW-SE) fault system. The Amiata region, as opposed to the boraciferous region,

913

contains two volcanic edifices, that of Radicofani (Pliocene) and of Mt. Amiata (Quaternary). Their high permeability has effectively enlarged the recharge area but the intrusion of these apparata has also created radial faults that have contributed to splitting the reservoir.

The power-plants installed so far to exploit the fluid are nearly all direct steam intake, back-pressure (discharge-to-atmosphere)or condensing (mix condensors) types. The back-pressure plants can usually be converted to condensing cycles and are used in the initial stages of field exploitation or when the fluid contains a high percentage of uncondensable gases.

The geothermal research programme being implemented in Italy by ENEL alone or as an ENEL-AGIP joint-venture is aimed at extending research over wider areas and to greater depths. Until recently the boreholes rarely went below 1000 m.

THERMODYNAMIC CHARACTERISTICS OF THE ENDOGENOUS FLUIDS

In our opinion a geothermal field should preferably be classified according to the heat content of its fluid and the total thermal energy that can be extracted. For this type of classification one must first define the following parameters regarding fluid:
- composition
- pressure and temperature (where the fluid is gas, liquid or superheated steam)
- pressure, temperature and steam content (in the case of wet steam)
- flow-rate.

These parameters must be monitored from the start of production until the point at which it is possible to predict their future evolution.

A description will now be given of the characteristics, thermodynamic in particular, of the endogenous fluids of Sant'Albino, Bagnore and Piancastagnaio fields.

Sant'Albino

Ninety-seven percent of the fluid of this system consists of carbon dioxide; CH_4 and H_2S are present in much smaller concentrations of $1\ ^o/_{oo}$ and $1\ ^o/_{ooo}$ respectively.

The average pressure in the reservoir is about 10 ata. The fluid is not used to generate electricity but to produce liquid CO_2.

Exploitation of this field began in 1968 and since then neither the composition nor the pressure of the fluid have undergone variations (a mere 15,000 t/yr of CO_2 are extracted).

Some correlations have been noted between underground hydrodynamics and the gas exhalations.

The flow-rate of the CO_2 from the natural manifestation varies seasonally between 900 kg/h (in summer) and 1200 kg/h (in winter): the increased flow-rate is a result of a pressure rise in the gas-cap during a rise in the piezometric level. The flow of gas from one of the wells affects the flow of water in a nearby thermal spa building; an increase in the gas flow-rate brings about an increase in that of the water, which reaches a maximum value and then begins to decrease again. The water flow is affected by the formation of a water-gas emulsion, which is activated by the inflow of gas until the velocity of the gas contained in the water is so high as to the impede the formation of the emulsion itself.

A more thorough analysis of such natural fluidodynamic phenomena could contribute

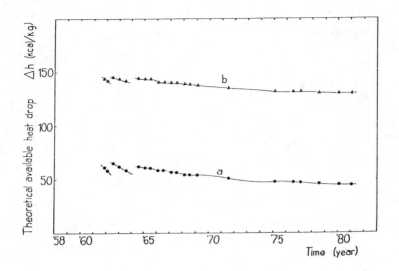

Fig. 1. Bagnore geothermal field. Theoretical available heat
drop versus time: a) by expansion to 1 ata and b) to
0.08 ata.

to the study of water and/or gas reinjection into the reservoir.

Bagnore

This geothermal field lies within the Amiata region.
The fluid, wich nowadays consists mainly of superheated steam, has been used to
generate electricity since 1959.
Initially the fluid from the first wells was almost entirely made up of uncondensable
gases, i.e., CO_2, CH_4, N_2, H_2, H_2S [1], but gradually became enriched in superheated
steam. The total content of uncondensable gases decreased abruptly at first and
then at a slower rate until, after 4 to 5 years, it reached values that approximate
the present regime.
The percentage by weight of the uncondensable gases is now constant (1980), with
slight irregular variations between 6 and 6.5%. The CO_2 represents about 90.4%
and the CH_4 3.4% of the uncondensable gases[2].

During the first 4-5 years the reservoir pressure, initially at 23.4 ata, decreased
at the same time and with the same trend as the decrease in the percentage of

---

[1] In geothermal research 'uncondensable gases' is a term commonly and erroneously
used in reference to the gaseous components accompanying the steam.

[2] These are again average weighted values of the percentages in the productive wells.

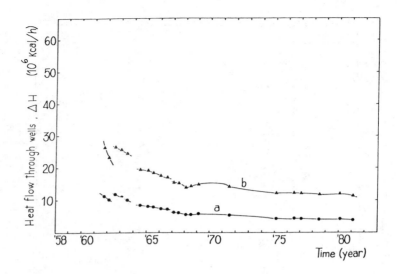

Fig. 2. Bagnore geothermal field. Heat flow through the wells
by isoentropic expansion of the fluid versus time:
a) to 1 ata an b) 0.08 ata.

uncondensable gases. However, the pressure, as opposed to the latter, continued decreasing thereafter, albeit at a more reduced rate, until it reached its current value of 4.2 ata. The drop in pressure unaccompanied by a reduction in the uncondensable gas content could be attributed to condensation after cooling of the fluid in the reservoir.

The thermodynamic evolution of the system has been expressed in graph form as the theoretical available heat drop of the fluid, Δh, and the total heat flow through the wells, ΔH, versus time and cumulative production.

The composition and parameters of state of the fluid discharged from the wells into the manifold upstream of the turbine being already known, the heat content was deduced, at regular time intervals, from tables and enthalpy-entropy diagrams. This heat content is the theoretical heat drop made available by the fluid in the case of an adiabatic-isoentropic expansion from initial state, respectively, to atmospheric pressure and to a pressure of 0.08 ata (Fig. 1).

Multiplying these values by the corresponding flow-rates in weight, q, we obtain the total flow of heat through the wells during production:

$$\Delta H = q \; \Delta h$$

In Figure 2 this entity is plotted versus time and in Figure 3 versus cumulative production.

These figures clearly show that all the fore-mentioned parameters decrease with both time and cumulative production; however, this decrease does tend to attenuate gradually with the increase in the abscissae values, although the situation varies

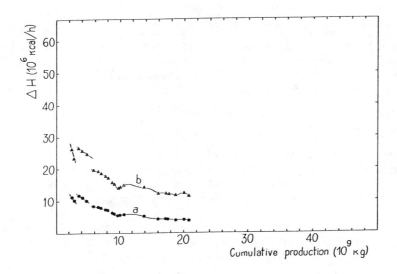

Fig. 3. Bagnore geothermal field. Heat flow through the wells
by isoentropic expansion of the fluid versus cumulative production: a) to 1 ata and b) 0.08 ata.

around the breaks in the lines that denote entry-into-production or shut-in of
wells.

The decrease in the ΔH values is more marked than in the Δh values, as the fluid
flow-rate diminishes with time.

The increases in ΔH_1^i and $\Delta H_{0.08}^i$, occurring at times with an increase in t and C,
are attributable to slight rises in the total flow through the wells. However,
these never reach the ΔH values obtained previously with the same flow-rates,
because of the decrease in ΔH versus time.

Note that the variations in $\Delta h_{0.08}^i$ are lower than the corresponding variations
in ΔH_1^i; this can be explained by the fact that steam entropy usually increases
with time and the isobars in Mollier's diagram diverge in the direction of the
increasing entropy.

The opposite occurs in the case of ΔH, as the bigger Δh is, the bigger will be
the value obtained when it is multiplied by q.

Piancastagnaio

This field, along with Bagnore, lies within the Amiata region. Since 1962 its
fluid, 80% of which comprises superheated steam, has been used to generate
electricity. Initially the fluid was more or less made up only of the uncondensable
gases also found at Bagnore, though in differing percentages.

The decline curve of the uncondensable gases has also roughly the same trend as at Bagnore: after 5-6 years the percentage of these gases tended to stabilize around values very similar to those of present-day.
95.4% by weight of these gases consist of CO_2 and 2.8% of CH_4.

The pressure drop in the Piancastagnaio field (from the initial 41 ata to the 17.6 ata of nowadays) was similar to, and accompanied that of the uncondensable gases in the early stages, but continued after the stabilization of the latter.
In order to determine the variations in the thermodynamic characteristics of the fluid caused by exploitation graphs were drawn of the Δh_1^i and $\Delta h_{0.08}^i$ values versus time, and of the ΔH_1^i and $\Delta H_{0.08}^i$ values versus time and cumulative production.

The same comments made for Bagnore are also valid in this case.
However, one might also add that the Δh values for Piancastagnaio, relative to expansion up to 1 ata and to 0.08 ata, show no tendency to decrease, despite their variations over the years. This factor should be referred to the reduction in flow-rate.
During exploitation there is a gradual drop in the heat absorbed by the fluid in the reservoir; in order to maintain the same heat content at well-head the fluid must therefore have a longer transit time in the area in which it absorbs this heat, and so its flow-rate must be reduced. In this way one can also prolong the life-span of a field, by reducing the thermal differences between the fluid and the rock, as well as the flow of heat.
Finally, note some obvious differences between these last two fields: the ΔH values of the Piancastagnaio system are always much higher than at Bagnore and the $\Delta H_{0.08}^i / \Delta H_1^i$ values much lower.

SUGGESTIONS FOR RECOVERING MORE FLUID FROM THE SYSTEMS AND MORE ENERGY FROM THE FLUIDS

A few examples only were sufficient to show how the composition and state of the endogenous fluids can vary greatly from one system to the next, even when these two systems are not completely separate. Each system obviously requires its own particular procedures and plants, in order to achieve an optimal fluid and energy recovery.
The tendency in future must be towards exploiting all of the endogenous fluids capable of providing energy, no matter their type and nature. Even the non-combustible and "uncondensable" gases can be exploited to produce electric energy, utilizing their expansion alone.
The Sant'Albino reservoir is not the only CO_2 system in Tuscany, as there is another similar to this at Pergine. ENEL recently discovered a predominantly CO_2 system [3] that is much larger and has higher pressures and temperatures than any of those described above. In the Bagnore and Piancastagnaio fields, and in

---

[3] ENEL intends using this system, replacing the gas with pressurized air the latter will be injected into the reservoir during periods of excess energy and utilized during periods of peak requirements of electricity to drive compressed gas turbines.

others in which the fluid is mainly superheated steam, there are some wells producing uncondensable gases only; only rarely are these gases passed into the turbine.

For the moment there are no reservoirs containing uncondensable gases only that are exploited to generate electricity. Yet the uncondensable gases from wells near those producing steam or from gas fields near vapour-dominated fields could be used for two important objectives:
- to generate even small quantities of electricity;
- to create cold springs for condensation, as these gases are usually gaseous mixtures of mainly CO_2.

The second objective could become of primary interest in situations in which, due to the distance from the sea or from rivers, it is difficult to obtain the evaporation calories from the exhaust.

At all events the solution to the problem of exploiting gas wells, even in steam fields, must always account for any eventual effects of gas production on the flow of steam. One positive aspect is undoubtedly that of the rapid decrease in the percentage of these uncondensable gases (caused by gas drainage), as this could lead to a faster increase in the heat recovery by condensation[4].

Reliable forecasts of the evolution of ΔH are felt to be quite important when choosing the type of power-plant, both as regards expansion to 1 and to 0.08 ata (presumed pressure in the condensor).

Condensation plants are not recommended in the case of:
- very rapid decreases in these parameters;
- small differences between the $\Delta H^i_{0.08}$ and the simultaneous ΔH^i_1 values (curves very close together: low entropy steam).

The temporal evolution of the gas/steam ratio is also important when selecting the power-plant. Where this ratio is expected to be always high the condensation plants are to be avoided. Should the gas percentages decrease in time, however, the power-plants should be capable of operating by condensation whenever these percentages are felt to have reached the maximum limit.

Variations in fluid flow-rate and its parameters of state must also be borne in mind when designing turbines capable of operating at high efficiency in any degree of exploitation of a reservoir.

CONCLUSIONS

An exhaustife study of the thermodynamic behaviour of an endogenous fluid provides valuable information for understanding phenomena that would otherwise be inexplicable; it also offers important guidelines for the choice of utilization scheme most fitted to the regime prevailing in each system.

As in the case of heat engines the thermodynamic cycle of the fluid, passing from natural to artificial thermodynamic conversions, must be modified to resemble

---

[4] During the early stages of exploitation of a vapour-dominated reservoir it is sometimes more convenient to discharge the exhaust directly into the atmosphere rather than transfer it through the low-pressure turbine and condensor, as the excessive amount of gas would require more power to be discharged from the condensor than into the atmosphere.

Carnot's as closely as possible.

The first (natural) cycle conversions in geothermics are highly variable from one field to the next and even within the same system, as exploitation gradually proceeds. In order to maintain high efficiency figures the second cycle conversions must be adapted to the first, which are represented by the conditions of state of the fluid in the reservoir. We must therefore have a knowledge of the initial thermodynamic conditions of the reservoir and of their evolution.

The information dealt with in this paper belongs to a wider-reaching study aimed at defining the total evolution of a geothermal field by analysis of the initial trend of some specific thermodynamic parameters (TAHD and HFW). Two reservoirs have been examined so far: crucial data have been obtained but are still insufficient for our purpose. Many other systems must be studied, possibly from the beginning of their exploitation to the end. The more varied and complete our file of statistics, the more reliable will be any forecasts of new systems, interpolating between reservoirs whose characteristics most resemble those of the new system.

Specific energy differences in fluids that have just recahed the surface have been taken into consideration when designing the utilization plants. One useful approach when planning future drillings and utilization schemes would be to define, by means of existing wells, the energy level of the fluid at various points of the reservoir and its modifications, so as to optimize continually the exploitation methodology.

ACKNOWLEDGEMENT

The autors wish to thank the Unità Nazionale Geotermica (ENEL) of Pisa and the Gruppo Minerario (ENEL) od Larderello for their kind and helpful assistance.

REFERENCES

Atkinson, P. and others (1978). Analysis of reservoir pressure and decline curves in Serrazzano zone, Larderello geothermal field. Geothermics, 7, 133-144.

Atkinson, P. and others (1978). Thermodynamic behaviour of the Bagnore geothermal field. Geothermics, 7, 185-208.

Berry, P., C. Cataldi and E.M. Dantini (1978). Geothermal stimulation with chemical explosives. Energy Tecnology Conference & Exhibition, Houston, Texas.

Bottinga, Y. and V. Courtillot (1976). Le transfert d'ènergie termique à travers l'ècorce terrestre. l'Industrie Minérale, 58, 373-376.

Brighenti,G. (1978). Sul meccanismo di produzione di vapore ed acque calde dal sottosuolo. l'Industria Mineraria, 29, 11-17.

D'Amelio, L. (1960). Corso di macchine. Treves, Napoli .

Facca, G. and F. Tonani (1964). Theory and technology of a geothermal field. Bulletain Volcanologique, 27, 143-189.

Rinehart, J. (1970). Heat flow from natural geysers. Tectonophysics, 10, 11-17.

Sammarco, O. (1976). Ripercussioni di un allagamento in miniera sulle venute di gas. Proceeding 1st Italian-Sovietic Mining Meeting in Sardinia, Cagliari.

Sammarco, O. (1979). Il riscaldamento delle correnti di ventilazione nei pozzi e nelle vie di entrata d'aria: indagini preliminari su bilanci termici in miniere con flussi di calore variabili. l'Industria Mineraria, 30, 267-289.

Trevisan, L. (1951). Una nuova ipotesi sull'origine della termalità di alcune sorgenti della Toscana. l'Industria Mineraria, 3, 41-42.

Willoughby, W.W. (1978). Steam rate: key to turbine selection. <u>Chem. Eng.</u>, <u>11</u>, 147-154.

AIR STORAGE GAS TURBINE POWER STATIONS
AN ALTERNATIVE FOR ENERGY STORAGE

J. Lehmann

Brown Boveri (BBC), Federal Republic of Germany

Summary

This paper gives an overview about air storage gas turbine power
plants. After describing briefly the operation experience with the
world's first plant in Huntorf an outlook is given for future development. The development of advanced systems up to integrated
use of PFBC boiler systems is explained.

1.   Introduction

The power demand in an electricity distribution grid follows the need
of the connected users. Depending on daytime, weekly sequence and
seasonal development, the power demand varies significantly. In some
certain electricity grid the peak power demand goes up to two times
the minimum demand during night time. (Fig. 1,2)

Different methods have been applied for peak power production. Beside
the application of special peak power plants like i.g. gas turbine
power plants, storage schemes have been developed. The most common
storage scheme is the hydro pumped + storage scheme, mostly installed
in mountain or hilly areas.

Over more than 30 years back engineers tried to adopt this idea to
another working medium. Instead of using water as energy storage storage medium, compressed air should take this task so stored compressed air forms the base for intermediate energy storage.

2.   Basic principle of air storage GT power plants

The normal gas turbine process is characterized by the compression of air and the expansion of hot combustion gas at the same time. Out of the power released from the expansion of the combustion gas, almost 2/3 is consumed for air compression.

The basic idea of the air storage gas-turbine power plant is to separate the two sub-processes into a timely sequence. By doing that it is possible to compress the air during low-load demand in the grid.

The compressed air is then used during peak-load time as combustion air for the gas turbine combustion process. At that time no energy is needed for air compression and so the total power released from the combustion gas expansion can be supplied to the electrical grid as peak power. (Fig. 3)

3.   The first air storage gas turbine power plant in operation

The first air storage power plant in the world is operating since December 1978 on a commercial basis in Huntorf near Bremen, West Germany. This area is supplied with electricity by the Nordwest-deutsche Kraftwerke AG (NWK), one of the largest utilities in West Germany. The overall installed capacity of this utility is some 5 000 MW. The load distribution over time shows significant peaks, especially in winter time. So NWK started to investigate the possibility to install an energy storage scheme in the late sixties. Due to the lack of an hilly area, other systems then the hydro-power schemes had to be applied.

The supply area of NWK is located in the very flat north-west region of West-Germany. For several years this area was used for underground storage of fuel oil and industrial gases. The storage had been undertaken in underground salt formations which form a very tight and secure surrounding. From that experience it was an evident idea to store compressed air in underground salt cavernes, too.

The main data of the Huntorf power plant are shown in figure 4. The turbo-set has been deviated from proven elements out of gas turbine and steam turbine technology to minimize first-of-a-kind problems.

The charging ratio of 1:4 was equivalent to the acutal situation in the NWK grid in the early seventies. Due to actual experience with the operating plant and the changed grid situation, a charging ratio of 1:2 would be preferable today.

4.   Operating results

The Huntorf power plant operated very successfully after overcoming the first start-up problems. The diagram (fig. 5) given from the actual operation on a winter day in 1979 and the cumulative operational data given in figure 6 and 7 show the excellent response of this plant to the net requirements.

In 1980 the plant has been operated less due to increasing prices for natural gas. Therefore the additional installation of a recuperative heat exchanger is under discussion. In this context instalment of an additional compressor set for changing the charging ratio is under consideration.

The cumulative data as per 31.03.81 is given in Fig. 8. These values show that the plant is used preferably for load curce leveling. Large amounts of peak power were rarely required, but utilization of off-peak power helped to stabilize the daily load curves.

The gas turboset operates with air flow control. So the part load efficiency is quite better than in a normal gas turbine powerplant. Taking this as an advantage, the average output of the Huntorf Stations is approx. 100 MW compared to the installed capacity of 290 MW.

All first-of-a-kind difficulties have been resolved; there is only one problem left. The connecting pipe lines between the underground storage caveries and the surface liner to the power station are

suffering significant corrosion. So rust is transported to the air
inlet filters of the plant. Sometimes blocking the air flow. NWK is
trying to overcome this problem by replacing the ferritic steel
piper therough by fiberglass polyester piper. This measure will be
complete soon.

5.   Different charging ratio

As mentioned in connection with the Huntorf power plant other charging ratios may be interesting according to the local grid situation.
Therefore Brown Boveri has studied in the past different charging ratios and evaluated the technical possibilities. This have been done
for ratios of 1:4, 1:2, and 1:1. The studies have shown that all
these systems are feasible. In these studies the gas turbine or so-
called thermal block has been unchanged. The adoption of the charging
ratio has been fulfilled by varying the compressor set (Fig. 9). The
investment cost increase is estimated to be appr. 5 % of the total
plant investment costs.

The charging ratio 1:1 offers the possibility to operate the plant
also in a direct gas turbine cycle. This operation does not result
in any economical advantages in terms of fuel consumption. But it
shows advantages in terms of system reliability in case of storage
facility failures. According to the investigations underway a trend
to charging ratios of 1:1 can be observed.

6.   Improvement of the thermodynamic cycle

The basic intention for improvement of the thermodynamic cycle is the
saving of additional fuel. In time of increasingly limited fuel availability on the world market special attention is given to this aspect. As mentioned before for the Huntorf power plant discussion is
underway to install an additional recuperative heat exchanger between
turbine off-gas and inlet-air. This step offers a 30 % saving for the
fuel needed.

Furthermore systems up to completely adiabatic systems are studied
all over the world. Zaugg et al. have compared different systems
following the investigations in the United States.

The adiabatic system (Fig. 10) offers maximum possiblities to save
primary energy because no additional fuel is necessary. The process
temperature level is limited by the acceptable compressor outlet
temperature and the heat storage effectiveness.
Feasible large heat storage facilities are actually not available.
The application of this system needs further research and development.

The diabatic system (Fig. 11) is a feasible approach to advanced
systems compared to the Huntorf plant. It may operate with or without HP combuster. Both alternatives are equivalent thermodynamically.
The decision for one alternative has to be made concerning specific
criteria for the project.

7.   Charging cycle

The Huntorf power station has been designed for a daily cycle. As has
been shown in th United States in might be more advantageous to operate the plant on a weekly cycle (Fig. 12) thus using the low-load
demand time over a week-end for recharging the air storage.

The design for a weekly cycle gives a determinant input on the size
of the air storage facility. The overall volume differs for constant
and sliding pressure systems (respectively rock and salt formations).
As shown in figure 13 the size of the facility - given here for a
sliding pressure salt-storage facility - depends mostly on the actual
time of power operation, charging ratio and adaption to the weekly
cycle. The weekly cycle requires more than two times the valve needed
for a daily cycle. Specific calculation have to be made for every
project.

8   Air storage power plants with PFBC

Finally one development shall be mentioned here which might get more importance in the future. Based on the saving of gases and fluid fuel, investigations have been started in the United States and Sweden to use a pebble-bed fluidized combustion system with an air storage scheme. This process gives the possibility to use locally available fuel like hard-coal or lignite to be burnt in the power plant.

The air stored in the underground facility may then be used as comustion air in a PFBC boiler to generate coal gas, which drives a gas turbine. The respective draft sketch is given in Fig. 14. Alternatively the air may be heated up by the PFPC-boiler and drive a hot-air turbine (Fig. 15). In both cases some critical areas have to be investigated thoroughly.

The off-peak time in the grid is again used for recharging the air storage facility. Further investigations will be needed to study all effects on the PFPC boiler system.

9   Summary

Conventional battery systems will not contribute in a larger extent for intermediate electricity storage. The development of these battery systems will concentrate an application in the medium and low-voltage systems (Fig. 16). On the high-voltage systems the hydro-pump storage scheme will keep its place as one dominant system. But due to the fact that the local possiblities to construct hydro-pump storage schemes may be limited in the future, the air storage systems will be a competitive alternative.

Advanced systems using gas turbine with or without additional fuel or using PBFC-boiler system will give an opportunity to the utilities to realize an economic peak-load coverage. Therefore the air storage power plants will constitute in the near future a major contribution for intermediate energy storage.

References:

- Mattick, Haddenhorst,Weber, Stys, "Huntorf- the world's first 290 MW gas turbine air storage peaking plant", proceedings american power conference, 37, 1975.

- Herbst, Stys, "Huntorf 290 MW - the world's first air storage system energy transfer (ASSET) plant: construction and comissioning", american power conference 1978

- Hoffeins, Romeyke, "Huntorf - initial operation experience on the world's first air storage gas turbine", gas turbines, April 1979

- Schüller "Pneumatisches Speicherkraftwerk ohne Brennstoffzufuhr", Elektrizitätswirtschaft 76, Jahrgang 21, November 1977, Heft 24

- Zaugg, Hoffeins, "Brown Boveri air storage gas turbines", Brown Boveri review, volume 64, number 1, Januar 1977

- Pellin, "European Utilities assess compressed-air-storage schemes", Electrical review international, volume 207, number 2, July 1980

- Crotogino, Quast, "Compressed-Air Storage Caverns at Huntorf", Rockstore Conference Stockholm, June 1980

- Zaugg, Stys, Hoffeins, "Air storage power plants with special consideration of USA-conditions", 11th world energy conference Munic, September 1980

- Hoffeins, Romeyke, Hebel, Sütterlin, "Die Inbetriebnahme der ersten Luftspeicher-Gasturbinengruppe", Brown Boveri Mitt. 8-80

- Maaß, Stys, "Operation Experience with Huntorf, 290 MW World's First Air Storage System Energy Transfer ("ASSET) Plant", American Power Conference, Chicago, April 21-23, 1980

- Herbst, Große Luftspeicher-Gasturbinen-Anlagen in Verbundnetzen", Wärme, Band 86, Heft, 1980

- Lehmann, "Luftspeicher-Gasturbinen-Kraftwerke/Aufbau, Wirtschaftlichkeit, Einsatzmöglichkeiten", BBC-Informationstagung Energieerzeugung, Mannheim, Juni 1980

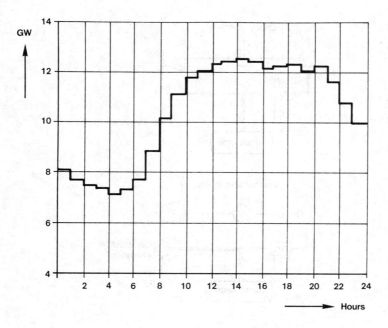

Fig.1 Daily Load Curve

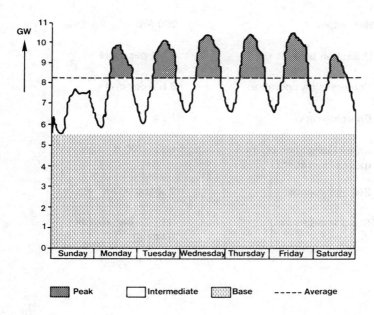

Fig.2 Weekly Load Curve

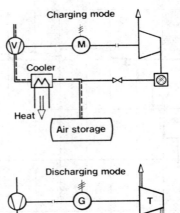

Charging mode

Cooler

Heat

Air storage

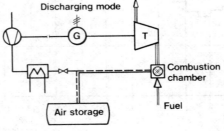

Discharging mode

Combustion
chamber

Air storage

Fuel

Fig.3 Air storage GT power plant
Principle

Net output	290 MW
Utilization time	2 h / per cycle
Compression operation	8 h / per cycle
Charging ratio	1 : 4
Annual utilization hours (power operation)	600 - 700 h
Storage volume	2 x 150.000m^3 approx.
Storage pressure	55 - 75 bar approx. (sliding)

Fig.4 Air Storage GT-Power Plant
Main Data of the Plant

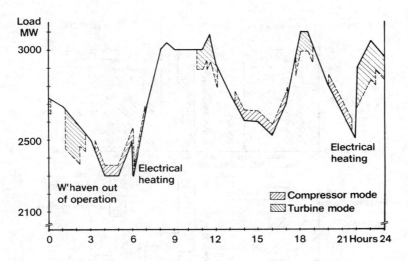

**Fig.5 Air storage GT power plant Huntorf
Operation on Monday 1/22/79**

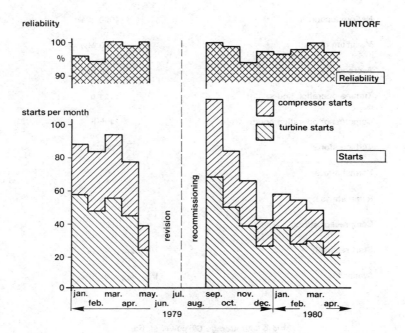

**Fig.6 Air storage GT power station
Start Number and Reliability**

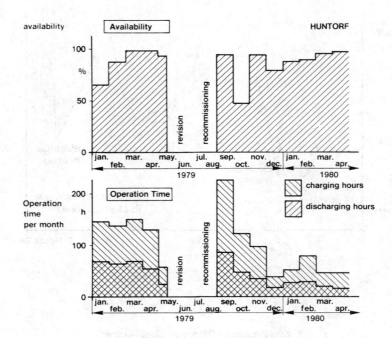

Fig.7   Air storage GT power station
Availability and Operation Time

Net production	114000 MWh
Maximum output	290 MW
Average output	~ 100 MW
Turbine operation hours	1121 h
Compressor operation hours	2496 h
Turbine starts	1040
Normal starts	1005
Rapid starts	35
Compressor starts	697
Start reliability	I/81: appr. 99 %
Availability	I/81: appr. 97 %

Fig 8   Air storage GT power station
Operation Data up to 31.3.1981

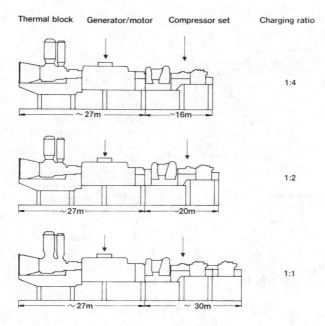

Fig.9 Air Storage GT Power Plant
Variation of charging ratio

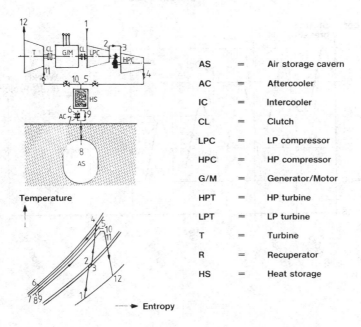

AS	=	Air storage cavern
AC	=	Aftercooler
IC	=	Intercooler
CL	=	Clutch
LPC	=	LP compressor
HPC	=	HP compressor
G/M	=	Generator/Motor
HPT	=	HP turbine
LPT	=	LP turbine
T	=	Turbine
R	=	Recuperator
HS	=	Heat storage

Fig.10 Air Storage GT Power Plant
Adiabatic System

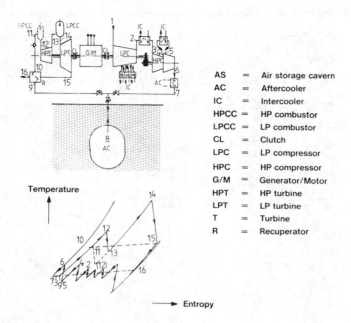

AS	=	Air storage cavern
AC	=	Aftercooler
IC	=	Intercooler
HPCC	=	HP combustor
LPCC	=	LP combustor
CL	=	Clutch
LPC	=	LP compressor
HPC	=	HP compressor
G/M	=	Generator/Motor
HPT	=	HP turbine
LPT	=	LP turbine
T	=	Turbine
R	=	Recuperator

Fig.11 Air Storage GT Power Plant
Diabatic System

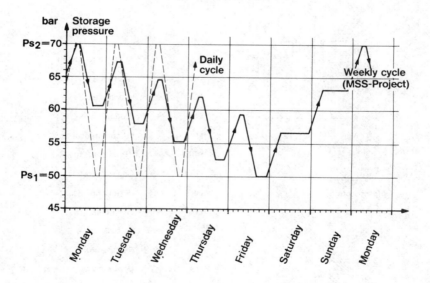

Fig.12 Air storage GT power plant
Operation cycles

	Constant volume/ sliding pressure	Operation time	Variable volume/ constant pressure
Daily	~10 × 10⁵ m³	11,4 h Daily Turbine operation	~ 3,5 × 10⁵ m³
	~ 12 × 10⁵ m³	13 h Daily Turbine operation	~4 × 10⁵ m³
Weekly	~ 27 × 10⁵ m³	13,3 h Daily Turbine operation (5 days per week) 9,5h Daily Compressor operation (7 days per week)	~ 8 × 10⁵ m³

Fig.13 Air Storage GT Power Plant
Storage volume for charging ratio 1:1/290 MW

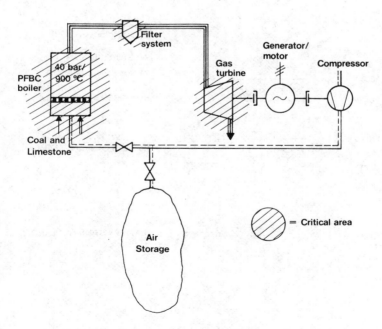

Fig.14 Air Storage GT Power Plant
Application of PFBC (1)

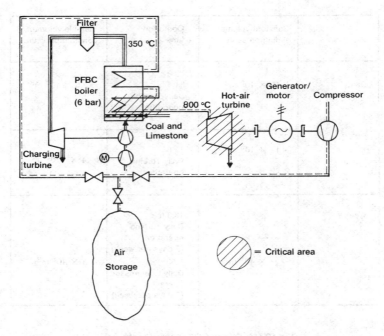

**Fig.15 Air Storage GT Power Plant
Application of PFBC (2)**

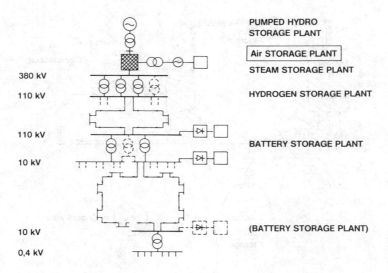

Fig.16 Different Storage Plants

THE U.S. DEPARTMENT OF ENERGY
HIGH TEMPERATURE TURBINE
TECHNOLOGY PROGRAM

G. Manning and J. Neal

U.S. Department of Energy, USA

ABSTRACT

Of the various advanced energy conversion technologies for use of coal by the electric utilities, the gasification-gas turbine combined cycle (GCC) shows the most promise. Within the United States Department of Energy, the major advanced technology development program for the use of coal-derived fuels in the GCC is the High Temperature Turbine Technology Program. This paper compares estimates of the environmental, thermodynamic, and economic performance of the GCC system with other advanced coal burning systems. The High Temperature Turbine Technology program is described and the current status is summarized.

KEYWORDS

Gasification combined cycle; advanced energy conversion; coal-derived fuel technology; integrated combined cycle; coal fueled gas turbine.

INTRODUCTION

One of the principal concerns with the increased use of coal is the environmental impact of its burning. New technologies must be developed in order to assure that utilities (which account for 75 percent of annual U.S. coal consumption) can continue to rely primarily on coal and can retain the flexibility to site coal-fired plants in areas that now, because of environmental problems, are candidates only for oil- and gas-fired generation plants. It is in the nation's best interest that the utility sector be provided with the technological options that will allow coal to be burned more cleanly than current systems can achieve economically. In addition, environmentally acceptable use of coal in the utility sector, by technology transfer, will allow more coal to be used in the industrial sector where substantial oil and gas fuels are also consumed.

Of the various advanced technologies for the use of coal, the gasification-gas turbine combined cycle (GCC) promises to most significantly increase thermodynamic efficiency and reduce levels of pollution.

Figure 1 presents a process flow diagram for a typical GCC plant. Crushed coal, water and compressed air are fed into the gasifier. Hot gases from the gasifier are passed through a cleanup system for removal of particulates, sulfur compounds, and other chemical contaminants, and are burned in a combustor prior to driving the gas turbine electric generator. From the turbine, the expanded hot gases are routed through a steam generator relinquish some of their remaining heat, and are exhausted to the atmosphere. Steam thus generated is expanded through a steam turbine driving an electric power generator. Exhausted steam is condensed, and a pump arrangement feeds the fluid back to the steam generator.

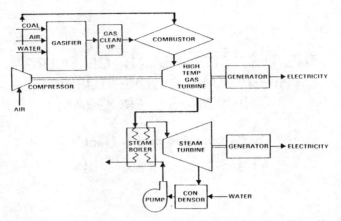

Fig. 1. Gasification-gas turbine combined cycle process flow diagram.

Estimates of the environmental, thermodynamic, and economic performance of this system are compared with other advanced coal systems in Table 1. It should be noted that the higher thermodynamic efficiency of advanced GCC results in this technology requiring 23 percent less coal than conventional technology. This efficiency improvement not only has implications for lower cost electricity (especially if delivered coal prices rise more rapidly than expected) but also results in reducing the environmental and potential health impacts associated with coal mining, transportation, disposal, and air emissions. Commercial availability of this technology would also improve the marketability of the higher sulfur Eastern coals, thus broadening the U.S. base of usable coal reserves.

TABLE 1 Estimated Environmental and Cost Peformance of Utility Technologies

	SO$_x$ Removal (Weight %)	NO$_x$ Remaining (Lb/MMBtu)	Particulates Remaining (Lb/MMBtu)	Efficiency (Percent)	Electricity Generated Costs (mills/kwh)
Gasification Combined Cycle	95-99	0.03-0.19	.01	38-44	46
Conventional Flue Gas Desulfurization and Particulate Control	90	0.5-0.6	.03	34	52
Advanced Scrubbers and Particulate Control	95	0.3-0.4	.03	33-35	49-53
Atmospheric Fluidized Bed	85-92	0.3-0.5	.03	35	46
Pressurized Fluidized Bed	90-95	0.08-0.12	.03	37-41	46
Molten Carbonate Fuel Cell	95-99	0.03	.01	44-48	42

Considering the advantages of the advanced GCC system, the Department of Energy initiated the High Temperature Turbine Technology (HTTT) Program in 1976. This technology effort seeks to increase the overall firing temperature of utility size gas turbines from today's 2000°F (1093°C) to 2600-3000°F (1427°C-1649°C) and allow use of coal-derived fuels containing a broader range of contaminants than current gas turbine specifications permit. Such increases in turbine temperature levels, if successfully developed, will result in GCC systems with overall power generation economics 20 to 30 percent better than current technology pulverized-coal steam systems with flue gas desulfurization (see Fig. 2). Development of advanced turbine technologies is particularly important because the private sector has made a commitment to demonstrate gasification combined cycle at full scale using current technology turbines. While this will prove the superior emissions characteristics (i.e. better than oil-fired plants) of GCC, the private sector activities do not include any support of high temperature turbines and associated components necessary for advanced GCC to have the significant economic advantages shown in Fig. 2. Thus, the Federal research and development effort to develop high temperature turbine technology is critical in order for the turbine manufacturers to have the high-risk technology base necessary to commercialize second generation turbines that will allow GCC systems to attain superior economic performance as well as superior environmental performance.

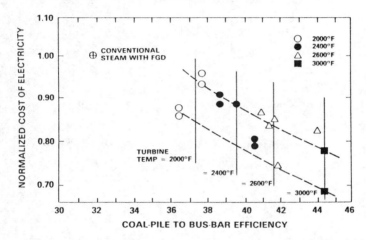

Fig. 2. Comparison of gasification-gas turbine combined cycle economics with other advanced technology.

The High Temperature Turbine Technology program is structured in the following phases: Phase I - System and Program Definition; Phase II - Technology Testing and Test Support Studies; Phase III - Technology Readiness Verification Testing.

Technology readiness is defined as "that point in the development cycle after which minimal risks would be involved in the development of the turbine subsystem for use in a full-scale open cycle gas turbine system." In other words, it is the point at which all technical problems appear to have a reasonable solution and no barriers appear to exist which would prevent taking the next developmental step into commercializing the subsystem. Technology readiness does not mean that the turbine subsystem is actually ready for the commercial marketplace.

HTTT PHASE I

The Phase I system and program definition was begun in 1976 and was completed in 1977. Four contractor teams participated in this phase. The major characteristics of the integrated gasifier combined cycle plants they proposed to develop are shown in Table 2.

The results of Phase I increased the U.S. Department of Energy's confidence in the GCC system concept. Particularly encouraging was the fact that the contractors identified several approaches to the development of a high temperature turbine suitable for electric utility use.

TABLE 2 Major Characteristics of Phase I GCC Plants

Contractor	Curtiss-Wright	General Electric	United Technologies	Westinghouse Electric
Capacity (Mw)	749	513	955	925
No. of Gasifiers	12	12	4 (molten sodium carbonate)	4 (fluidized bed)
Gas Cleanup System	Hot	Cold	Cold	Cold
No. of Gas Turbines	4	2	4	4
Turbine Inlet Temperature (°F)	3000	2683	2600	2600
Turbine Cooling	Air	Water	Air & Water	Air
Efficiency (%)	47.3	41.4	43.3	40.9

HTTT PHASE II

The principal engineering issues to be addressed in Phase II of the HTTT program center around the high temperatures necessary to obtain the desired efficiency.   At metal temperatures above 1200°F (649°C), the phenomenon known as hot corrosion is experienced. Additionally, particulate carryover from the gasification process will contaminate the hot gas stream powering the turbine. The combined effects of accelerated corrosion and erosion must be overcome to make a high temperature turbine reliable and therefore acceptable for use by electric utilities. The objective of Phase II of the HTTT program has been to design and test two of the most promising methods for mitigating the effects of the harsh operating environment of the GCC system.

Two of the Phase I contractors, Curtiss-Wright and General Electric, were selected to participate in the Phase II development and testing of advanced turbine technology. Curtiss-Wright was selected to investigate the potential for transpiration air cooling of the turbine; General Electric was selected to investigate a novel water cooling system. This phase of the program was started in 1977 and is expected to be completed by the end of 1981.

Curtiss-Wright Phase II Testing

Several years before initiation of the HTTT program at the Department of Energy, the U.S. Navy had sponsored a program at Curtiss-Wright to develop aircraft jet engine turbine cooling using transpiration air. This same concept (see Fig. 3) was proposed by Curtiss-Wright for the DOE high temperature turbine. Using engines on loan from the U.S. Navy, the Curtiss-Wright program involves complete engine tests at temperatures above 2600°F (1427°C).

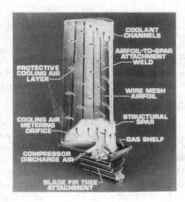

Fig. 3. Transpiration air-cooled turbine blade concept.

The first series of tests made use of a low pressure ratio (3.5 to 1) engine. The object of this testing was to measure the changes in cooling air flow caused by various levels of contaminants in the engine fuel and not necessarily to prove the efficiency of the cycle. This engine was run on both distillate oil and simulated low-Btu gas.

After the first 50 hours, the fuels were contaminated with ground fly ash of various sizes. This engine has been operated for over 800 hours in a 2600-3000°F (1427-1649°C) gas stream environment with the following results:

- Cooling air flow reduction is related to particle size. Contaminants of 2.7 microns and less caused no more flow loss than unconventional fuel.

- No mechanical or thermal distress has been noted on any of the turbine components.

- No evidence of hot corrosion has been noted.

- Blades and vanes have sustained foreign object damage without mechanical failure.

The second series of tests use an engine called the Turbine Spool Technology Rig. This engine has a higher pressure ratio (7 to 1) than the first, slightly different hot section vanes and rotors, and has been assembled with much tighter clearances to provide better aerodynamic efficiency. The tighter clearances of this engine allow much more precise performance measurements to be taken. The engine will be calibrated on clean fuel and then run on contaminated fuel to determine the resulting performance degradation. Completion of this test series will prove both the performance efficiency and durability of the transpiration air cooling concept.

In parallel with both turbine tests, Curtiss-Wright has been conducting tests of various combustor configurations and materials compatability tests. Over 50,000 hours of corrosion and erosion testing have been conducted.

General Electric Phase II Testing

The Phase II testing performed by the General Electric Company focuses on the engineering feasibility of components as opposed to complete engine tests. Their confidence in proceeding with such an approach was established in earlier programs, sponsored by General Electric and the Electric Power Research Institute, which involved tests of small water-cooled turbines.

General Electric's approach to eliminating hot corrosion is based on reducing the turbine nozzle and bucket temperatures below 1000°F (538°C) with a novel cooling water system. For the stationary nozzles, a conventional circulating water path is used. For the rotating turbine buckets, however, General Electric has designed a "once-through" cooling water system. This design overcomes the turbine imbalance problems caused by uneven heat flux across the turbine wheels that have plagued earlier attempts to develop water-cooled turbines.

In the HTTT Phase II effort, separate hot gas tests have been conducted on a sectoral combustor assembly, a composite, water-cooled, first stage turbine nozzle, and a monolithic, water-cooled turbine nozzle to determine material durability and compatability. The first stage of the turbine is subjected to the highest heat flux and must therefore have a higher heat transfer capability from the skin to the flowing cooling water. General Electric has designed a first-stage nozzle constructed of a copper alloy with Nitronic 50 cooling water tubes and structural spars cast in place. The nozzle is clad with Inconel 617 for corrosion/erosion resistance. The second and third stage nozzles and turbine buckets are subjected to a lower heat flux and can thus be constructed of Inconel 718 clad with Inconel 617. Cooling water passages are drilled in these monolithic nozzles and buckets. Figure 4 shows the first stage turbine nozzle and bucket.

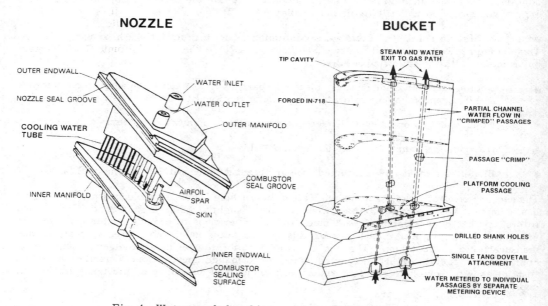

Fig. 4. Water-cooled turbine nozzle and bucket concept.

Heat transfer experiments have been conducted on the nozzles and buckets using aerodynamic, air turbine, and shock tunnel tests. These tests have allowed refinement of the design of the cooling system employed.

*General Electric trade name. Similar to a Lurgi except that the coal fines are fed into the gasifier by means of a patented screw feeder device.

In the first U.S. test of a fully simulated gasification combined cycle, General Electric connected a small Gegas* gasifier to a cleanup system, ignited the gas in a combustor, and routed the hot gas through a cascade turbine simulator consisting of a set of first-stage, water-cooled, composite nozzles. The primary objective of this test was to evaluate the durability of the nozzles in an actual low-Btu coal-gas environment. A secondary objective was to prove the capability of the cleanup system employed. Particulate contamination was reduced in two water wash stages and a Benfield sulfur removal stage eliminated 93 percent of the sulfur from the gas stream.

The final test in this series will use a mock-up of the cooling water delivery system to demonstrate the ability to get water to the turbine rotor, distribute it to all turbine blades, achieve even cooling, and discharge the water from the rotor.

SUMMARY AND DISCUSSION

The harsh operating environment of a gasification-gas turbine combined cycle plant dictates the use of durable materials and novel cooling methods. Two distinctly different approaches to the design of high temperature turbines have been taken by the contractors selected for the High Temperature Turbine Technology program at the U.S. Department of Energy. At this stage in the program it would be premature to predict whether either of the concepts under investigation will become a commercial product. However, results to date from both contractor efforts are extremely encouraging and it is expected that the objective of Phase II of the HTTT program, proof of technical feasibility, will be achieved. All of the technical results of the program to date have been published and are in the public domain. Thus, manufacturers not involved in the first phases of the program may also be able to conceptualize an advanced design turbine, increasing the probability of commercializing this important new technology.

Upon successful completion of Phase II, the U.S. Department of Energy expects to proceed with Phase III of the HTTT program. This phase will seek to demonstrate the overall durability of advanced cooling methods in actual turbines of sufficient size and configuration to establish the suitability of the technologies for the electric utility industry. Phase III will be a staged effort lasting approximately six years. Through open competition, the Department of Energy expects to sponsor multiple systems engineering and component development efforts leading to the design, fabrication, and testing of one or more Technology Readiness Vehicles (TRV). It is anticipated that successful TRV testing will encourage manufacturers to take the high temperature turbine concepts through the final development stages and into commercial production.

U.S. GASIFICATION TECHNOLOGY

C. L. Miller

Division of Coal Gasification, United States Department of Energy, USA

ABSTRACT

The large variations in coal rank, physical characteristics, and chemical composition in the extensive coal reserves of the United States require that advanced gasification concepts be developed before this entire resource base can be considered as available for conversion through gasification into alternate clean fuel gases, liquids, and chemical feedstocks. The development effort in progress to achieve that goal, and some of the advanced gasification concepts, are reviewed.

KEY WORDS

Coal Gasification; Gasification Processes; Gasifiers; High-Btu Gas; Medium-Btu Gas; Synthesis Gas; Entrained-Bed Gasifiers; Fluidized-Bed Gasifiers; Alternate Fuels.

INTRODUCTION

Beginning about 1961, a great deal of effort and money has been spent in the United States to develop a number of unique coal gasifiers, to further advance gasification systems, and, in general, to extend the state-of-the-art of coal gasification. Since coal gasification has been practiced at varying levels of activity on a commercial scale since the early 1920's, the purpose and need for these efforts has often been the subject of much discussion.

Of course, the justification for this continuing effort lies in the potential that coal gasification offers for successfully converting the extensive indigenous coal reserves of the United States into gaseous and liquid fuels and chemical feedstocks. Coal gasification technology presents viable options for consideration in establishing energy policy and priorities. Once this technology is developed, the capability will exist to reverse trends in the supply and demand of energy within the United States.

However, the full potential of coal gasification can be realized only if gasifiers and gasification systems are developed that will:

- be able to process the wide range in types (ranks) of coal available as potential feedstocks,

- permit realization of the full flexibility of coal gasification as a conversion technology,

947

● operate at the high capacities required.

A number of gasification concepts that offer that potential to achieve these objectives are under development in the United States.

ENERGY RESOURCE AND CONSUMPTION PATTERNS AS BASIS FOR DEVELOPMENT OF TECHNOLOGY

The development of coal gasification technology in the United States has been characterized by periods of intense activity followed by longer periods of apparent disinterest. The most recent period of developement activity was initiated when it was believed that the nation was soon to experience shortages of some types of fossil fuels (e.g., natural gas), and it was recognized that the future availability of energy resources in the United States depended on reversing the trends being experienced in energy consumption.

Analyses of energy supply and demand trends have established that the percentage of the total energy being supplied by oil and gas has increased while the percentage supplied by coal, hydro, and wood has decreased. Indeed, oil and gas now supply over 75% (gas 26%, oil 49%), of the energy consumed.

These shifting patterns in preference for the various forms of energy supply have little meaning until they are evaluated within the context of available energy resources and national objectives. While the oil available to the U.S. through the management of its own resources represents less than 25% of its indigenous fossil fuel reserves, oil supplies 49% of the energy demand. Approximately half of that oil is imported. By contrast, coal represents 56% of the Nation's fossil energy resources but supplies only 18% of the energy demand. Such comparisons make it obvious that if the United States is to move toward energy independence, it must develop the technology to make full use of its coal resources.

In deciding how best to use these coal resources, the need to consider the magnitude of the resources, its distribution and characteristics, and the potential market to be satisfied is self-evident. In evaluating the resource base, we find that coal is distributed in several major coal fields scattered across the country. The coals in these fields have distinct differences in physical properties. In general, the coals in the West are more reactive, non-agglomerating, and low in sulfur, while the Eastern coals are less reactive, agglomerating, and comparatively high in sulfur content. Thus, to use this resource fully, gasification technologies had to either exist or be developed that could process a feedstock with a high degree of variability in its physical properties, including coal reactivity; tendency to cake or agglomerate when heated; ash content and its properties; heating value per unit weight and also per unit volume; amount and characteristics of volatile materials; and finally sulfur content.

An early comparison of these characteristics of the coal resource base with the operating capabilities of existing gasifiers, the projected size of the desired conversion industry, and the spectrum of products required demonstrated that existing gasifiers each possessed severe but differing limitations when considered for use with some coals. It was determined that the coal conversion industry required by the United States needed more advanced, flexible gasifiers if all U.S. coal reserves were to become potential sources of alternate fuels.

It was decided to develop a number of gasifiers nearly simultaneously. Because of a projected shortfall in the available supplies of natural gas, the initial effort was concentrated on developing gasifiers particularly suited for producing a substitute natural gas from coal. Subsequent efforts have been expanded to develop the more versatile medium-Btu gasifiers. The full flexibility of these gasifiers and the role that these systems can have in creating viable options for most of the energy consuming sectors of the economy are shown in Fig. 1. It can be seen that the coal gasification development program has included the study of many concepts of gasification and gasification processes.

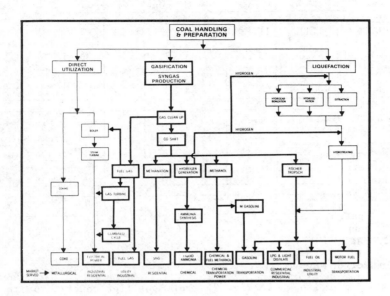

Figure 1 The Flexibility of Coal Gasification.

GASIFIERS AND GASIFICATION TECHNOLOGY BEING DEVELOPED

It is recognized that existing first generation gasifiers (e.g., Lurgi), can be used to gasify lignite and subbituminous coals (i.e., Western coals). Indeed, it now appears that all the regulatory and institutional barriers delaying the construction of the first commercial high-Btu gasification plant using Lurgi technology have been removed. Great Plains Gasification Associates project, designed to produce 125 million cubic feet per day of pipeline quality gas from North Dakota lignite, is now in the final design stage with pre-construction activities underway. However, only through the development of advanced second generation gasifiers and gasification processes can the total coal reserves available in the United States be considered as suitable feedstocks for conversion into alternate fuels. To reach this goal, a number of promising concepts that, in addition to being able to process caking as well as non-caking coals, have the potential to lower the product production cost, increase the efficiency of operation, and result in greater operational reliability are under development.

In what follows, a number of these concepts that are either at or have been developed beyond the Process Development Unit (PDU) stage are reviewed; some of the distinguishing characteristics of the technology are presented; the objectives of the development effort are listed; and the current status of the project is given. The gasification concepts that will be discussed are presented in Table 1.

Although this table groups gasification concepts by an oversimplified designation of its major fuel gas product (i.e., high, medium, or low Btu), it should be remembered that operating conditions can be selected that will permit the majority of the gasifiers to produce a variety of fuel and/or synthesis gases.

TABLE 1
CHARACTERISTICS OF COAL GASIFIERS UNDER DEVELOPMENT

PROJECT	DEVELOPER	PLANT LOCATION	BED TYPE	TYPES OF COAL FEED	CAPACITY (TPH)	PRESSURE (PSI)	PRODUCT GAS TYPE	STATUS
HIGH - BTU GASIFIERS								
HYGAS PILOT PLANT	INST OF GAS TECH	CHICAGO	FLUIDIZED	ALL	3 0	1000	HIGH	PILOT PLANT OPERATION COMPLETED
BIGAS PILOT PLANT	STEARNS ROGER	HOMER CITY PA	ENTRAINED	ALL	5 0	1200	HIGH	DEVELOPMENTAL OPERATION
CATALYTIC COAL GASIFICATION PDU	EXXON	BAY TOWN TEXAS	FLUIDIZED BED	ALL	0.05	500	HIGH	DEVELOPMENTAL OPERATION
HYDROGASIFICATION PDU	ROCKWELL INTERNATIONAL	CANOGA PARK, CA	ENTRAINED	ALL	0 75	1000 1500	HIGH	DEVELOPMENTAL OPERATION
MEDIUM - BTU GASIFIERS								
PRESSURIZED, AGGLOMERATING FLUIDIZED BED PDU	WESTING HOUSE	MADISON PA	FLUIDIZED	ALL	0 6	100 225		DEVELOPMENTAL OPERATION
U-GAS AGGLOMERATING FLUIDIZED BED PDU	INST OF GAS TECH	CHICAGO ILL	FLUIDIZED	ALL	1 0	15 60	MEDIUM	DEVELOPMENTAL OPERATION
HIGH RATE PRESSURIZED ENTRAINED - PDU	MOUNTAIN FUEL RESOURCES	SALT LAKE CITY, UTAH	ENTRAINED	ALL	1 25	200	MEDIUM	CONSTRUCTION
MOLTEN SALT PDU	ROCKWELL INTERNATIONAL	CANOGA PARK, CA	MOLTEN BATH	ALL	1 0	15 300	LOW/ MEDIUM	DEVELOPMENTAL OPERATION
LOW - BTU GASIFIERS								
ATM ENTRAINED - BED SLAGGING PILOT PLANT	COMBUSTION ENGINEERING	WINDSOR CT	ENTRAINED	ALL	5 0	15	LOW	DEVELOPMENTAL OPERATION

HIGH-BTU GASIFIERS

BIGAS Coal Gasification Process

The BIGAS gasifier is shown in Fig. 2. This gasifier consists of three elements in a sigle vessel: slag quench zone in the bottom; Stage 1 where the char burners are located, and Stage 2 where the coal feed is injected. Operating as a high pressure (1000 - 1500 psig) entrained-bed slagging gasifier with temperatures of 2700 - 3000°F in Stage 1 and 1500 - 1700°F in Stage 2, this system is being developed principally as a producer of a high-Btu gas. The high operating pressures favor the kinetics of methane formation directly in the reactor, where approximately half of the methane in the substitute natural gas product is made. As an entrained system high, throughput capacities per unit of reactor volume can be achieved and the conversion of the ash to a slag insures high carbon conversions.

Construction of the facility was completed in 1976 and shakedown activities were begun shortly thereafter. Since that time, the development program has been underway. Early in this period, major problems were encountered with the design of the char burners, and failures were experienced that at least on one occasion, resulted in significant damage to the surrounding facilities and instrumentation. Since that time, a significant redesign effort has been successfully completed, resulting in a char burner that now operates in a routine manner with very little operational difficulty.

As would be expected of a facility of this size (5 tons per hour), a considerable amount of time was required to complete these shakedown activities and to begin the test program. However, at this time, considerable success is being achieved in the development program. Notable achievements made to date include:

- Between January 21 and April 2 of this year, the facility was operated for a total of 352 hours. These operations were conducted as three successive runs of 58, 122, and 172 hours respectively (hours of operation are based upon continuous coal feed and char recycle to the gasifier).

- All tests were voluntarily terminated under controlled conditions.

- No major mechanical problems or process upsets were encountered.

- An instrumentation interlock system has been installed and successfully operated that will permit optimization of operating variables.

- The char burner design continues to prove its reliability and operational efficiency.

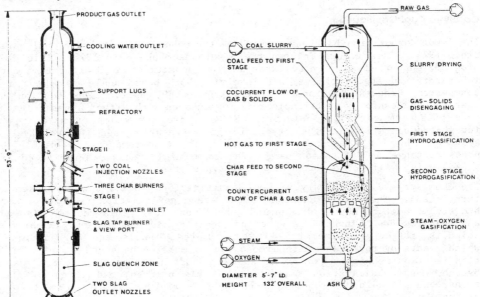

Figure 2 The BI-GAS Gasifier.

Figure 3 The HYGAS Gasifier-
With Steam Oxygen Gasification.

It is currently planned to operate the BIGAS facility through 1982, when the status of the development program will be evaluated. This evaluation will also include consideration of the total facility and its potential value to the gasification program. Other test programs that can be performed at the site will be considered.

HYGAS Coal Gasification Process

Research on the HYGAS process was initiated in 1964, based on earlier efforts completed during the 1945 through 1964 period. The development work progressed from laboratory scale through bench scale and culminated in the construction of a 75 ton per day pilot plant, which was dedicated in October 1970. Coal was first introduced into the reactor in late 1971. Fig. 3 shows the hydrogasifier vessel developed as part of this effort. This vessel has four gas-solids contacting stages internally connected with transfer lines. The four sections from top to bottom are slurry drying, first stage hydrogasification, second stage hydrogasification, and steam oxygen gasification. The reactor has been operated at pressures from 500 to 1000 psig and at temperatures that range from 600°F at the discharge of the slurry drying system to 1800°F in the steam oxygen gasification section.

The HYGAS process offers an integrated system for high pressure gasification to produce a high-Btu gas from coal. Up to 65% of the methane in the SNG product is formed in the gasifiers. The operability of this concept on all U.S. coals has been demonstrated in over 11,000 hours of operation.

The development activity on this concept through the pilot plant stage was completed in late 1980, and a conceptual commercial design has been prepared. The final report has been submitted. A comprehensive review of the data available from the

HYGAS Pilot Plant program by representatives of both the Department of Energy and the Architect and Engineering firm of Pullman Kellogg confirms that adequate information has been produced, with some qualifications, for supporting the design of a HYGAS demonstration plant. Any continued further development of this concept of coal gasification is considered to be the responsibility of the industrial sector.

Catalytic Coal Gasification

The Exxon Corporation has been working on the concept of catalytic coal gasification since 1971. This effort entered the development phase in 1978. This development program, supported jointly by Exxon and the Department of Energy, is a coordinated laboratory and engineering effort involving both bench scale equipment and a one-ton-per-day Process Development Unit. The PDU gasification section has been operated successfully on Illinois coal impregnated with either a potassium hydroxide or a potassium carbonate solution. Efforts are now underway to demonstrate the integrated operation of the cryogenic gas separation and the catalyst recovery systems. The fluidized-bed reactor operates at a temperature of 1275°F with a pressure range of 265 to 485 psig. The commercial process, anticipated to operate at 485 psig, will be able to handle all types of coal.

The chemistry of the Catalytic Coal Gasification process can be broken down into three steps. The first reaction is the steam gasification reaction, where carbon and steam react to form hydrogen and carbon monoxide. This reaction is endothermic and requires heat input. The second reaction is the water gas/shift reaction which is slightly exothermic. The third reaction is the methanation reaction which reacts hydrogen and carbon monoxide to make methane and steam. The methanation reaction is highly exothermic and gives off about as much heat as was consumed by the steam gasification reaction. In this process, all three of these reactions take place in one reactor, reducing to a minimum the requirement for additional heat.

Recently, Exxon announced its intention to proceed to a 100 TPD pilot plant to be built in Amsterdam, the Netherlands. The PDU will continue to be operated in support of this pilot plant effort.

Hydrogasification — Short Residence Time PDU

The Hydrogasifier reactor concept has been under development since early 1977. This process employs a rapid non-catalytic coal hydrogenation technique termed flash hydropyrolysis in an entrained flow reactor to produce the coal/hydrogen reaction. As a direct hydrogenation scheme, the process is theoretically capable of producing approximately 20% more high-Btu gas per ton of coal than a process using an intermediate synthesis gas step. Since the process operates in a regime of rapid kinetics, it allows the use of an entrained flow reactor system. An entrained flow hydrogenation reactor uses pulverized coal with a large active surface area, and it can operate at high enough temperatures (1600 - 1900°F) so that the reactions occur in milliseconds. These rapid reactions make it possible to process a high caking coal without pretreatment.

As shown in Fig. 4, this reactor is a single-stage, short residence time entrained flow system for the production of methane by the direct hydrogenation of bituminous and subbituminous coals. The reactor train consists of two sections, the first of which is the injector-reactor tube section. The flash hydropyrolysis of coal occurs immediately downstream of the injector, and the reactor tube provides the residence time for hydrogenation of the liquids and char. The second section, the recuperator, is composed of two stages. In the first stage, the heat in the products of hydrogasification is transferred by counterflow heat exchange to the hydrogen used as input to the injector. The resulting hot hydrogen is fed to the injector. This reactor operates at 985 psig with a gas residence time of 1.8 seconds and a reactor exit gas temperature of 1800°F.

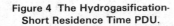
**Figure 4 The Hydrogasification-
Short Residence Time PDU.**

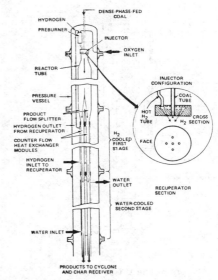

The engineering-scale tests conducted to date were performed at coal feed rates of 1/4 and 3/4 tons-per-hour in 15 foot long reactor tubes ranging in diameter from 2 to 6 inches. By the end of July 1980, a series of short duration tests had been completed in an engineering-scale test unit at a coal throughput of 3/4 ton-per-hour. The construction of a 3/4 ton-per-hour Integrated Process Development Unit (IPDU) is expected to be completed in FY 1982.

The work completed to date has established the advantages of this gasifier, which include the ability to process virtually any type of coal, the ability to adjust the process parameters over a wide range of gas/liquid ratios, rapid efficient mixing of the reactants, and a reduction in size of the reactor as a result of the fast reactions. Continued successful development of this concept of coal gasification will depend upon demonstration in the IPDU that all the associated subsystems can be operated efficiently. These subsystems include the separation of the methane and liquid byproducts from the gas stream and efficient utilization of the char to provide the hydrogen required for the hydrogasification reactions.

MEDIUM-BTU GASIFICATION

Agglomerating Fluidized-Bed Gasifiers

As noted in Table 1, and as shown in Fig. 5, the U-Gas gasifier, and Fig. 6, the Westinghouse gasifier, two such gasifiers are now under development. The gasifiers differ in the way that the reactants (coal, steam, and oxygen) are introduced into the reactor and in the operating environment used to initiate and use the technique of ash agglomeration as a means of removing the ash from the system. Ash agglomeration is the key processing principle for both systems. In general, this process occurs in the following manner. Fresh, unpretreated coal is fed into the gasifier, where it is combusted in a stream of oxygen. Steam fed with the oxygen reacts with the coal and char to form hydrogen and carbon monoxide. As the bed of char circulates through the jet, the carbon in the char is consumed by combustion and gasification, leaving particles that are rich in ash. The ash-rich particles extrude through the pores to the surface of the char, where they stick to other liquid droplets on adjacent particles. In this way, ash agglomerates are formed that are larger and denser than the particles of char in the bed. The agglomerates defluidize, migrate to the annulus around the feed tube, and are continuously removed.

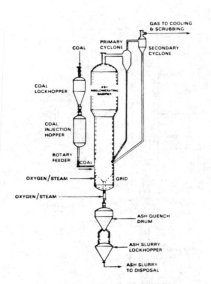

Figure 5 The U-Gas Gasifier.

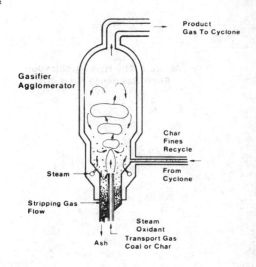

Figure 6 The Westinghouse Gasifier.

The raw product gas, containing methane, hydrogen, carbon monoxide, and gaseous impurities, exits the reactor at about 1800°F. Cyclones are used to remove char particles from the raw gas before it is quench cooled in a scrubber that also removes most of the remaining particulate matter. These fines are reinjected into the gasifier along with the fresh coal. All of the fines collected and recycled are consumed by the combustion, gasification, and agglomeration processes within the reactor.

The operating conditions for these systems include gasification temperatures that range from 1750 - 1900°F, raw gas temperature before quench of 1700 - 1850°F, and gasification pressures of 50–340 psig. Pressure can be as high as 600 psig if the raw gas is to be subsequently converted to a substitute natural gas.

Both processes are nearing the end of their development period. The U-Gas system is being incorporated into the design of a industrial fuel gas demonstration plant, and the Westinghouse gasifier is being proposed for demonstration status as part of several projects submitted to the Synthetic Fuels Corporation. Either process can be designed and operated for use in a combined cycle power generation facility or can be used to manufacture medium-Btu fuel gas or a synthesis gas.

Pressurized Entrained Down-Flow Gasifier

Development of a pressurized high-rate entrained downflow gasification process has proceeded from extensive laboratory testing to the detailed design of a 30 tons per day Process Development Unit. In this gasifier, the pulverized coal is fed to the top of the gasifier and is entrained in a stream of product gas. It is reacted at a temperature of approximately 2850°F, or well above the melting point of the ash, with preheated oxygen and superheated steam in a refractory lined downflow reaction chamber. The reaction products are partially cooled by radiant heat transfer to the walls of a primary heat exchanger located immediately below the reaction chamber. An abrupt increase in flow area at the entrance to this heat recovery unit is provided to cause the slag flowing down the walls of the reactor to separate from the refractory surface and fall as droplets through the radiant heat recovery section into a slag accumulation chamber. The slag and larger ash particles are quenched with a water spray, accumulated, and discharged. The gases, entrained soot, and fly ash pass through a convective exchanger and a scrubber.

The final product is a cooled clean gas of intermediate heating value that is suitable, after sulfur removal, for use as a fuel in firing industrial boilers and furnaces, or for use as a synthesis gas.

This reactor can handle all types of coal, and it operates at approximately 185 psig. At present, the construction of the 30 ton per day unit has been started. It is currently anticipated that this effort will be completed in late FY 1982 with shakedown and startup activities to begin immediately thereafter. The unit will be operated to confirm results obtained in earlier work that indicated the reaction of pulverized bituminous coal could be performed in this system at through-put rates of between 500-1000 lbs. coal per hour per cubic foot of reaction-zone volume. In addition, data required for scale-up to larger gasifiers (600 tons per day) will be generated.

Molten Salt Gasifier

In Molten Salt Gasification, the gasification reactions occur within a pool of molten sodium carbonate. This melt acts as a dispersing medium for both the coal being gasified and the oxygen used for the gasification; it acts as a heat sink, with high heat transfer rates, for absorbing and distributing the heat of oxidation; it acts as a heat source for the pyrolysis and distillation of the volatile matter of the coal; it reacts chemically with, and absorbs the sulfur from, the coal; it provides an environment in which the sulfur compounds formed act as catalysts for the partial oxidation of coal; and it retains physically the ash present in the coal.

At present, the developmental work on this concept is being performed in a 24 ton per day PDU. The reactor vessel, as shown in Fig. 7, is a 4 ft. O.D. (3 ft. I.D.), 16 ft. high stainless steel vessel lined with fused cast alpha-alumina refractory. In operation, coal and carbonate at a ratio of 38% coal to carbonate (at reference design conditions) are injected into the gasifier through four nozzles fed by pressurized lockhoppers.

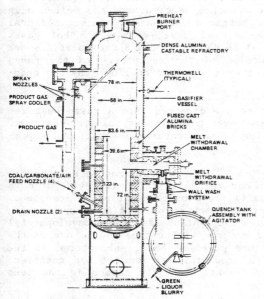

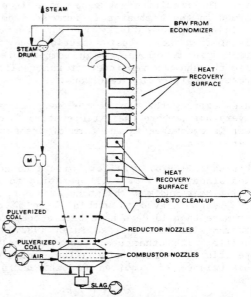

Figure 7 The Molten Salt Gasifier.   Figure 8 The Atmospheric Entrained Flow Gasifier.

These reactants enter the 1800°F pool of molten salt, where the gasification of the coal occurs. The product gas exits the gasifier vessel from the top and is cooled to 350°F in a spray cooler and reduced in pressure. A small stream of melt is continuously withdrawn through a side overflow port for ash removal, sulfur removal and sodium carbonate regeneration. The hot melt falls into the quench tank, where soluble components are dissolved and insoluble ash is suspended in the resulting aqueous slurry. This slurry is then processed to separate ash and sulfur from the carbonate. As noted, the carbonate is regenerated and recycled.

This process is capable of gasifying all coals. Since integrated operation of the PDU was started in mid-1979, a number of runs have been completed. These runs have been characterized by operating periods of varying length, from 14 hours to over a week. Although substantial progress has been made in solving problems and in the development of this concept, several significant problems still exist (e.g., unpredicted plugging of melt withdrawl outlet, improved measurement and control of the melt level in the gasifier, etc.). Efforts to resolve these problems and routinely achieve integrated operation of this PDU will be continued in FY 1982.

LOW-BTU GASIFICATION

Atmospheric Entrained Flow Gasifier

This gasifier is being developed to produce a low-Btu gas for combustion in an electric power generating station. The gasifier, as shown in Fig. 8, is designed for operation at atmospheric pressure. A combustor section having tangentially oriented fuel nozzles is at the bottom of the structure. Directly above it is the reductor section where additional coal is fed into the gasifier. Hot gases in these two sections entrain and gasify the feed coal as it passes vertically through the unit.

In operation, pulverized coal and recycle char are fed through the combustor fuel nozzles and oxidized in the combustor zone with a near stoichiometric quantity of air. The resulting hot gases rise into the reductor section, and the molten slag formed in the combustor is removed from the bottom and quenched. Pulverized coal is injected tangentially through the reductor fuel nozzles into the hot gases rising from the combustor section. Feed coal is devolatized and the volatiles are cracked in the lower, high temperature section of the reductor. As the gases rise through the remainder of the gasifier, they are cooled to 1700°F by the endothermic gasification reactions.

Gases exiting the top of the reductor section are directed into a waste heat recovery unit, where their temperature is reduced over tubular heat transfer surfaces. Heat is recovered in the form of preheated feed water, saturated steam, and superheated steam.

This experimental gasifier, with a capacity of 5 tons per hour, has been in operation since mid-1978. All operation to date has been air blown with Pittsburgh #8 coal. Heat and mass balances around the gasifier and other major plant components have been achieved to within 1% to 5% closure. Periods of continuous gas making for more than 13 days have been accomplished. The last phases of the developmental program will be completed in FY 1982. This program will establish the use of the gasifier with other coals and will solve some problems being encountered in the operational balance between the combustor and reductor sections of the gasifier. This balance is required to achieve full utilization of the carbon in the feed coal.

SUMMARY

This discussion has concentrated on the specific characteristics of new gasification concepts being developed in the United States as part of a number of research and development programs. These new concepts incorporate a number of advanced processing techniques and attempt to eliminate the deficiencies believed to exist in state-of-the-art gasifiers. It should be noted, however, that these development programs are but a small part of the effort now in progress worldwide to advance the state of the art. As part of this work, significant activities are underway to relate the U.S. program to other national and international efforts. The goals of this coordination of effort and information exchange are to prevent duplication and to achieve implementation of a gasification industry that has a maximum ability to use all coals as feedstocks at a minimum cost and a maximum flexibility in output product.

INSTALLATION DEMANDS AND TECHNIQUES FOR DISTRICT HEATING PIPES

H. Chr. Mortensen and J. Christiansen

Danish District Heating Association, Denmark

Demands on the Installation and Techniques of District Heating Pipes

Centralized generation of heat is advantagous only if an economic system of distribution to the consumer is available. Because of this requirement the technical design of a distribution network is a very decisive factor in the economics of the district heating operations.

In Denmark - where district heating, the collective heating form, has been known and utilized for the last 50 years - hundreds of kilometres of distribution network in various concrete duct designs, established on site, were installed since the number of district heating schemes increased rapidly 15-20 years ago.

Although the soil conditions in Denmark appear to be ideal for installation of district heating pipes, many of the distribution networks installed in those days, especially in small town areas, suffer from the consequences of unforeseen deteriorating factors, poor workmanship or incompetent supervision.

As regards larger heating mains, experience resulted in advanced concrete duct constructions with effective in situ insulation of the heat carrier pipes. The obtained know-how was also utilized by the manufacturing plants in order to design new generations of modern distribution pipe systems with a high degree of prefabrication, reducing the manual work on job site.

Rising labour costs too contributed to this trend and gave birth to several modern pipe systems highly insulated with rigid polyurethane foam, having a coefficient of thermal conductivity half of that of mineral wool and one fourth of that of cellular concrete insulation.

The Composite Pipe Design

A rational production of complete district heating pipe
systems in the factory became possible when highly insu-
lative polyurethane foam types were developed in the early 60ies, which were able to fill narrow cavities several
metres long.

Rigid polyurethane foam is generated by mixing the two
components, polyole and isocyanate, utilizing freon as a
blowing agent. Provided a proper formulation and processing at the factory a temperature resistance of $120\text{-}130^{\circ}C$
can be maintained throughout the lifetime of a pipe sys-
tem.

The old concrete duct was thus modernized and replaced by
a simplified composite construction consisting of a steel
service pipe, a highly efficient polyurethane insulation
material, and a watertight, robust outer casing of poly-
ethylene.

Today the majority of all preinsulated pipes produced and
installed in Europe belong to the so-called bonded system
where the three components are securely bonded together
during manufacture.

Expansion forces,arising when the steel pipe is heated,are
transmitted through the polyurethane foam to the outer casing, and the entire composition will move in the surrounding soil.

Such a pipe design is suitable for simplified pre-stressed
installation techniques where compensators are omitted and
the number of fixed points substantially reduced

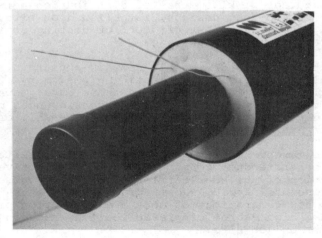

Preinsulated
pipe of the bon-
ded design with
copper wires for
electronic sur-
veillance against
leakages.

National Demands

By now preinsulated district heating pipes have been used
in the distribution networks in Denmark for several years,
in some cases for transmission lines. Especially in the
small dimensions the use of these pipe systems has increased to such a degree that almost exclusively preinsulated pipes are used for branch pipes and dimensions up to
3-400 mm, with an upward tendency.

In 1979 the Danish Parliament passed a heat supply law. A
consequence of this law is that in future requirements will
be made on the installtion, the operation, and the maintenance of pipe systems for district heating, among them the
requirements for supervision of the installations.

Resulting from the requirement of the law for a nation-
wide heat planning, investments in the extension of district heating, combined power/heat transmission and reno-
vations of the network are expected to take place in the
next few years. The investments are expected to amount to
more than 100,000 million d.kr. (corresponding to 300 $
for each inhabitant).

Previously there have been no general technical provisions
or regulations concerning pipe systems for district heating, but by the end of this year the norm "Distribution
Networks for District Heating" of the Danish Engineers'
Association and the standard "Distribution Networks - Preinsulated Pipes" of the Danish Standards Association are
expected to be terminated. Norm and standard can thus take
effect soon, and this will probably increase the quality
of the already approved preinsulated district heating
pipes.

Optimum Insulation.

The increasing prices of energy
raw material have caused a corresponding demand for energy
saving pipe constructions,
e.g. pipes with a better insulation. For several years
the insulation level has
remained unchanged, but
some Danish manufacturers
of preinsulated pipes
have supplemented their
range of products with
extra insulated pipes,
and they have ascer-
tained that the de-
mand for this kind
of pipe has been
greatly increasing.

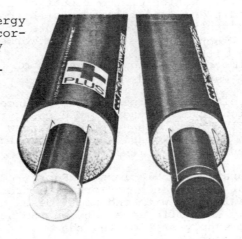

In order to acquire an extensive knowledge of the para-
metres determining the optimum thicknesses for district
heating pipes, the Danish District Heating Association,
supported by the Energy Ministry, have 1980 initiated an
investigation project comprising following phases of work:

1. Collection of background material, e.g. information
   about pipe types, theoretic methods and basis for the
   calculation of heat losses, determination of heat losses for all standard dimensions (steel pipes), and for
   the total network as well as heat losses in the existing networks, among them a reference network.

2. Conditions for energy-economical calculations, e.g. the
   technological possibilities to increase the insulation
   thickness, collection of prices and elaboration of cost
   models as a function of changed insulation thicknesses,
   determination of energy raw materials, calculation interest, time of depreciation etc.

3. Calculation of optimum insulation thicknesses, e.g.
   setting up an economic method of valuation and a sensitivity valuation on the basis of the various components' share of the total price.

4. Proposals for requirements of highest allowable heat
   loss (k-value), e.g. minimum requirement of different
   types and dimensions and the possible dependance on the
   prices of energy raw material. Consistency valuations
   for the economy of the public, the private, and the energy sector as a result of a possible implementation of
   the requirements and a valuation of the possibilities
   of the implementation of the requirements.

5. Supplementary investigations, e.g. measuring and calculation of the heat loss in a well-defined reference
   network, valuation of the result of improved operational conditions, and relation to the consumption measuring. Furthermore any requirements of insulation level
   at the renovation of existing networks.

For the time being following results of the above programme can be mentioned:

- As to preinsulated pipes the dependance of the optimum
  insulation level on the assumed energy price is not un-
  expectedly high. When using oil as raw material the insulation level is considerably above the usual standard
  at present.

- The theoretic calculations (EDP-programmes) of the heat
  losses of a reference network with preinsulated pipes
  is quite in line with the heat losses determined indirectly as the difference between the heat produced at
  the heat supply station and the amount of heat measured
  at the consumers.

- A great need for technical assistance and guidance has been ascertained, especially investigation methods, in connection with the renovation of already existing net-works.

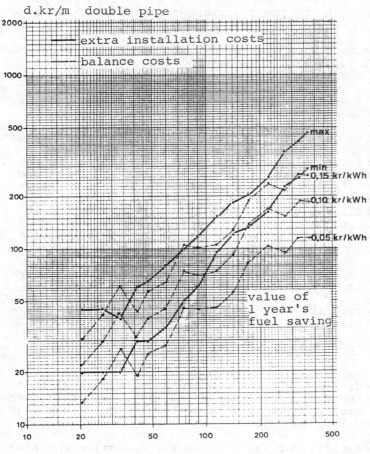

d.kr/m double pipe

steel pipe diameter

Extra installation costs and balance costs for outer ca-sing dimensions 1 step larger than standard. Calculation interest 9%, lifetime 25 years, fuel saving 4% annually.

Approximate value of spared heat loss, 1 year:

CHP-operation	0.05 d.kr./kWh
coal fired boiler	0.10 d.kr./kWh
oil fired boiler	o.15 d.kr./kWh

Damage File

For a number of years the Swedish District Heating Association have issued an annual report about defects to district heating pipe systems.

A summary of 10 years' reports gives an interesting picture of the state of the district heating network in a period with a heavy development of Swedish district heating.

Usually the causes of damage to district heating pipelines are untight joinings of the outer casing, earth settlements, traffic loads, and outer influence on job site.

12% of the defects on prefabricated pipe systems were discovered through electronic systems.

The work with the Danish draft standard gave rise to problems with finding corresponding supporting material for the documentation e.g. of the kind of damage done to the established system and the size of the damage. Therefore a project was started with the aim of forming a countrywide damage file to the future benefit of the district heating trade. The file is made through registration of information in an EDP file. The information is collected from all members of the Danish District Heating Association, which provides approx. 90% of all Danish district heat.

This EDP file enables us to get a quick view of various interesting "key values". Thus systematic faults at joints, installation faults etc. will become obvious after a certain time.

Joints

Preinsulated pipes have now reached a quality level that makes them extremely resistant to all impacts during transport and storage, an it is well suited for quick and easy installation in trenches.

The great stability of the pipe during operation has also been proved, and thus its lifetime is only influenced by extreme soil settlements and mechanical damage at later excavations.

The correct joining of pipes on site is therefore of crucial importance as a complete tightness of the medium pipe and of the outer casing must be ensured for the whole lifetime of the pipe system.

A high quality steel pipe with improved welding properties is still preferred as carrier pipe for the hot circulated water in prefabricated systems, not least because of the obvious advantages to the designer of the network, but also because the joining of steel pipes by welding is a

globally accepted technique ensuring a high degree of
tightness and mechanical strength of the joint through
well-known control methods.

Although preinsulated pipes of the bonded type show a low
rate of faults the statistics of faults clearly show that
far the greatest risk of water ingress into and deterioration of the polyurethane foam will occur when joining
of the outer casing has been made badly.

Therefore it is of great importance that the various designs of muff couplings are able to withstand any impacts
in the soil.

The instruction must be easily comprehensible, and the
making of a correct joint should not ask for too much ex-
pertise.

The development of electronic surveillance systems,which -
via copper wires moulded into the insulation material -
indicate the presence of any moist, has led to an improvement of the quality of the joint.

This kind of surveillance system has so to say enabled us
to "look into" a district heating system in operation and
tell if any constructive details may be improved.

Experiences made on site have thus resulted in important
improvements of the quality of the joints, and a large
number of surveillance systems in operation have proved
that tight and reliable joints can be maintained throughout the lifetime of the distribution system.

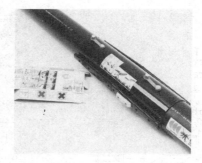

Preinsulated pipes during   A robust sealing of the
installation in excavated   protective casing with a
channels.   taper lock muff.

Testing of Joints

Requirements of the mechanical stability and tightness of joints form an essential part of the future Danish standard for preinsulated pipe systems.

Extensive tests with different joint types used on the Danish district heating market have been carried through, and fruitful exchanges of experiences have taken place between development and testing engineers, a dialogue that will contribute to the further development of even better joints.

With the purpose of developing relevant test methods for joints in preinsulated district heating networks extensive tests with different joint types have also been made within the framework of a West German research programme.

This programme is expected to be terminated in 1981, and valuable knowledge will be published for discussion among experts.

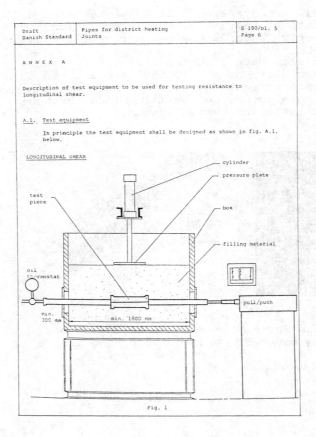

Draft Danish Standard	Pipes for district heating Joints	S 190/bl. 5 Page 6

A N N E X A

Description of test equipment to be used for testing resistance to longitudinal shear.

A.1. Test equipment

In principle the test equipment shall be designed as shown in fig. A.1. below.

LONGITUDINAL SHEAR

Fig. 1

Draft Danish Standard	Pipes for district heating Joints	S 190/bl. 5 Page 8

ANNEX B

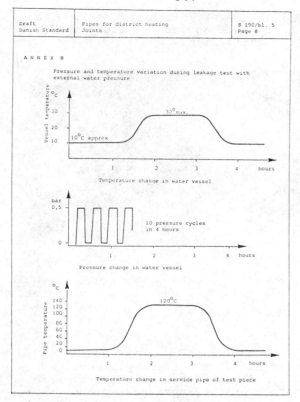

Pressure and temperature variation during leakage test with external water pressure

Temperature change in water vessel

Pressure change in water vessel

Temperature change in service pipe of test piece

Field Insulation of Joints.

The correct way of installation on site depends greatly on the local weather conditions, the organization of the site and the correct adaptation of all components to the working operations.

Insulation of joints with polyurethane liquids is a process that especially claims observance of the correct temperature conditions and mixture proportions.

Thus a new, industrialized part of the system, the foam pack, has met with great interest on the Danish district heating market.

The foam is supplied in robust packings consisting of separate chambers containing exact dosings of liquids for each size of joint.

When mixing the foam liquids a partition is removed, and after mixing the contents of the packing intensively with the built-in mixing stick, the foam is pressed into the joint.

The method developed is extremely clean and meets with
all environmental requirements as the procedure takes
place in a completely tight system.

Foam pack with the exact   The mixed liquids are
dosing of foam liquids for   pressed into the joint.
each size of joint

Training

The use of preinsulated district heating pipe systems in
Europe has increased rapidly during the last few years.
Therefore - in order to meet the increased demand - the
manufacturers rearrange their project oriented production
of pipe parts depending on the customers' specific requirements towards a more industrial production of a range
of high-grade stocked system parts.

In this way the customer obtains considerable advantages
such as a high supply capacity and short times of delivery.

The manufacturer thus strives to produce and develope his
product which by means of detailed instructions is installed by contractors who have increasingly specialized in
the installation of preinsulated district heating systems.

The practical training of installers on site is one of the
supplier's important jobs before work can be started. In
spite of the low rate of faults of preinsulated district
heating pipes of the bonded type that have been in operation for several years, the continuous training of installers is an essential factor for the further reduction of
the fault rate for this type of pipes.

In Denmark the first permanent training school has now
been established. For several days the installers receive
instruction about the technique of the system, the correct
way of joining the pipes and installing the various components of the system.

After the instruction course each installer receives a certificate documenting his participation in the course.

A recently elaborated Danish norm for installation of district heating distribution systems states the importance of correct installation and prescribes that the making of joints must be done by installers who have received a training in the pipe system technique concerned.

Class room training.   Training school certificate.

Quality Requirements

The increasing use of prefabricated district heating pipe systems in Europe has increased the interest in their construction and quality in many countries, and several national committees at present prepare quality requirements for such pipes and their accessories.

In spite of dissimilarities in requirements for a district heating network, distributing the heat from CHP-stations with elevated temperatures and pressure and from oil-fired district heating plants with low temperatures and pressure the creation of a European collaboration among district heating technicians, material specialists and pipe manufacturers succeeded the object of which was to co-ordinate different national points of view.

The great advantage of a broad European co-operation is that the entire knowledge from future test work and the development of test equipment will become available to all partcipating sountries.

Thus, advanced studies of the thermal ageing of different PUR-foam types and of the influence of leaking water on the pipe insulation layer are at present taking place at testing institutes in Denmark, West Germany and Holland.

After five sessions of this European committee the creation of joint European minimum requirements for preinsulated pipe systems is almost succeeded. The aim is to secure a quality product to the customer, but not to act restrictive to further development.

PIPE-IN-PIPE SPECIFICATIONS – NATIONAL BODIES	
Swedish District Heating Association Terms of Delivery, October, 1980 Revisions during 1980 due to general comments and European discussions	Valid
West-German AGI Minimum Requirements May, 1980 with general specifications of pipes and accessories, terms and general conditions for installation of pipe systems	Valid
Dutch Specifications, October, 1980 The present 2nd draft proposal has been composed by representatives from GEB Rotterdam GE Den Haag Pegus Utrecht N.V. KEMA VEG-Gasinstituut N.V. and it is now open to comments from manufacturers, raw material suppliers and other users	Draft
British Standard BS 4508, Part 3 & 4 This standard was issued in 1977. A general revision of the entire BS 4508 has been initiated.	Valid
Finnish District Heating Association Terms of Delivery 1. draft proposal was issued January 6, 1981.	Draft
Danish Standards Association National Standards for Preinsulated Pipe-in-pipe Systems In October, 1980 5 draft proposals were circulated for comments until January 1, 1981: 1. General requirements – 2. Steel pipes – 3. PUR- foam – 4. Casing – 5. Joints	Draft

PRESSURIZED FLUIDIZED BED COMBUSTION

S. Moskowitz

Curtiss-Wright Corporation, Wood-Ridge, New Jersey, USA

ABSTRACT

In order to reduce our long term dependence on petroleum fuels, the utilities and the electricity intensive industrial sector must embrace a strategy which includes utilizing coal. Since synthetic liquids and gaseous fuels derived from coal are expected to be reserved mainly for transportation and residential/commercial sectors, the option of direct combustion of solid coal must be used for the production of electric power. Pressurized fluidized bed (PFB) combustion is a technology that can provide power in an environmentally clean and efficient manner, even when using high sulfur coal. This paper presents the PFB concept for utility baseload operation and describes some alternative applications. The 13 MWe PFB Pilot Electric Plant sponsored by the U.S. Department of Energy is discussed and an overview of the technology support test results that has provided an optimistic outlook for early commercialization of this concept is presented.

KEYWORDS

Coal; fluidized bed; combustion; power plant; gas turbine.

INTRODUCTION

Many industrial nations have been plagued with problems which affect utility options for the production of electric power. These problems include (1) increased levels of air pollutants including acid rain, (2) high cost of electricity to the rate payers from escalation in oil costs, (3) brown-outs from the outage of large non-modular plants, (4) limited water resource, (5) uncertain projections for load growth caused by energy conservation and economic conditions coupled with (6) high cost of borrowing for capital investment.

While electric utility demand will increase over the next several decades, the options for base and intermediate load plants have become limited. Coal, an abundant and low cost fuel clearly is a route that can be taken by the utilities for new or retrofit plants. However, there are several requirements that must be addressed. Firstly, the combustion of coal must be in an environmentally acceptable manner. Secondly, high sulfur coal indigeneous to the region must be primarily used in a highly efficient system to maximize the advantage of low fuel cost. Finally, the system must have a moderate capital cost and offer operating flexibility.

Fluidized bed combustion (FBC) is a technology that can satisfy these requirements. It is an economic and environmentally attractive alternative to flue gas desulfurization (FGD) for base load or intermediate duty electric utility applications as well as for steam production. Recognizing this, the U.S. Department of Energy initiated a program in 1976 with Curtiss-Wright to design, construct and operate a PFB pilot electric plant using high sulfur coal.

PLANT PROCESS

Fluidization is the suspension of solids within an upward moving gas stream. The process involves the combustion of crushed coal in a fluidized bed of finely sized limestone or dolomite. The bed material in the combustion chamber through which air is blown behaves very much like a liquid.

A homogenous turbulent motion occurs in which there is a large area of surface contact and interaction between the air and the suspended bed particles. This results in uniform bed temperatures, even distribution of particles and almost instantaneous heat transfer between solids and gases. A heat exchanger is immersed within the bed for heat extraction purposes so that a constant combustion temperature can be maintained.

The sulfur in the coal is captured by the chemical reaction with the constituents in the sorbent. Calcined limestone or dolomite reacts with sulfur dioxide and oxygen in the combustion gas to form $CaSO_4$, a dry grannular environmentally benign material. Consequently, flue gas desulfurization is unnecessary.

Numerous advantages are offered by PFB over the conventional pulverized coal fired steam boiler plant such as: fuel flexibility, higher plant efficiency, lower SO_2 and NOx emissions, easier to dispose ash, reduced water requirements, lower capital cost, lower cost of electricity, and a modular system which offers shorter construction time, increased availability, and operating flexibility.

Figure 1 presents a simplified diagram of how a PFB combustor is integrated into a combined cycle system. A portion of the air from the gas turbine's compressor is used to fluidize the bed and to support combustion at 1650°F. The remaining portion of the compressor air is indirectly heated essentially to bed temperature by flowing through the tubular heat exchanger which is fully immersed in the bed. The hot gas resulting from the combustion process flows through particulate cleanup stages and subsequently, joins with the clean heated air from the bed's heat exchanger. The mixed flow passes through the gas turbine to drive the compressor-turbine. Next, the gas flow enters the power turbine which is coupled to the generator to produce electric power. The hot gas leaving the power turbine enters a waste heat recovery boiler which produces steam. This steam flows to a steam turbine-generator, producing additional electric power. In this system 60% of the total plant power is produced by the gas turbine and 40% is produced by the steam system. The reduced steam power results in a 50 percent reduction in water consumption. The coal-pile-to-busbar efficiency is in the range of 40%, about 5 points higher than conventional pulverized coal fired boiler systems.

Operating flexibility is an important advantage of the air heater PFB cycle. To accommodate off-design or part power operation, some of the heat exchanger inlet air can be diverted to the turbine inlet flow stream. This results in a lower mixed gas temperature at the turbine. With reduced heat extraction air through the in-bed heat exchanger, the coal flow to the bed may be reduced to maintain constant bed temperature at a level which promotes high combustion efficiency and excellent sulfur capture. Load following can be accomplished rapidly since a change in bed temperature or height is not required.

SYSTEMS DESCRIPTION

A baseload 500 MW$_e$ combined cycle plant consists of three modules each generating 100 MW$_e$ by a double ended generator driven by a pair of gas turbines. A single condensing steam turbine generator produces 200 MW$_e$.

Each gas turbine has a nominal rating of 50 MW$_e$ with an airflow capacity of 625 lb/sec, and a turbine inlet temperature of 1600°F. Each gas turbine has a single PFB unit containing a 28 ft ID combustion section. The in-bed heat exchanger consists of air tubes with a vertical height of 16 ft, equal to the bed depth.

As a result of the modular construction of the plant and the complete independence of the gas turbine units, plant power can be reduced by initially decreasing turbine inlet temperature as described above. Secondly, one or more gas turbine units can be shut down completely with the remainder of the units generating at the design point. This permits a very large turndown capability with little sacrifice to overall plant heat rate.

The modular construction also provides a high availability of plant power during such events as forced outage or scheduled maintenance periods. For instance, if there is a loss of either one gas turbine or a PFBC or a waste heat recovery boiler, over 83% of plant rated power is still available. The loss of one gas turbine alternator still permits two-thirds of rated power to be generated at nearly design point heat rate. With the loss of the steam turbine or its generator, 60% of the plant rated power is still available but at a simple cycle heat rate.

With high cost of borrowing capital, lengthy periods from planning to commissioning plants, and uncertainties in economic growth, the utility's dilemma in planning for future needs is eased by the modular system which permits the building block addition of power to meet the increments in energy demand.

The economic feasibility of an electric generating unit is established from the total cost of generating power. This involves costs for capital investment, fuel, operation and maintenance, and costs associated with system reserve requirements and replacement energy. The concept of load-carrying capability is used to assess the difference in reserve requirements for the system. The replacement energy costs are associated with the forced outage rates of major components. The levelized costs over 30 years for alternative plants for a given capacity factor, usually 70 percent, are established.

Studies comparing the PFB combined cycle with a pulverized coal fired steam plant with scrubber indicates that the PFB is 15% lower in capital cost, 10% lower in fuel cost and 65% lower in variable operating cost. The cost of electricity for the PFB cycle is over 20% lower than the conventional plant.

The use of the baseload plant configuration described above is also feasible for intermediate duty since PFB plant shutdown and start-up are more rapid than conventional steam plants. The shutdown of the air heater PFB plant does not require bed cool-down to protect the in-bed air heat exchanger tubes since the tubes are designed to operate at bed temperatures. Bed material removal is also not required to facilitate start-up where vertically oriented heat exchanger tubes are used. Experience has shown that loss of bed temperature during shutdown is about 150°F over a 24 hour period. Thus, bed light-off on coal after a weekend shutdown is accomplished without preheat procedures and full power may be developed over a short time.

The basic gas turbine, PFB combustor and waste heat boiler arrangement shown for the baseload plant may also be used for cogeneration applications with the widely varying steam requirements for different industries.

The coal burning fluidized bed combustor can also be used for heating air in compressed air energy storage systems (CAES) for peaking type power generation. This type of system involves the storing of compressed air energy during off-peak power load periods for use in generating electrical power during peak load periods.

Figure 2 illustrates the application of the coal-fired PFB to the CAES system. For the compression cycle (off-peak power time), the motor/ generator acting as a motor, drives the compressors to pump air into an underground cavern for storage at 125°F.

During the blowdown or power producing cycle, 600 lb/sec of air from storage passes through a regenerator before entering the high pressure turbine. This is a conventional steam turbine but uses air as the working fluid. Before entering the PFB unit, the air splits into two paths, one for combustion and the other for the cooling tubes. The PFB combustor design is essentially the same as described for the baseload plant. After passing through the PFB, the combustion gas and heat exchanger air flow to the low pressure turbine. This is a component of a conventional combustion turbine.

The power generation work by the combined low and high pressure turbines is 196 MW$_e$ for an air storage pressure of 800 psia. The work done by the turbines is over 80% of the fuel energy. If the electrical power used by the compressor drive motor is charged to the CAES system, the overall net power output efficiency is reduced to about 30%. This efficiency can be increased to 40-50% by recovering heat from the compressor intercoolers and aftercooler to produce process heat, heating boiler feedwater or for district heating.

TECHNOLOGY STATUS

The outlook for early commercialization of the PFB combined cycle is most promising since relatively few major components in the overall plant require technological development. Separately, the fluidized combustion system and the combined cycle power generation system have been in commercial use for many years.

The marriage of the two technologies introduced three basic areas requiring component evaluation. These include the performance or durability of (1) PFB combustor and in-bed heat exchanger tubes, (2) hot gas cleanup system, and (3) turbine blades when exposed to the gas stream contaminants. The evaluation of these component technology areas was performed in various rigs for 8000 hours and in a PFB Technology Plant for over 3000 hours. Following this, a Pilot Plant will be used to evaluate the effect of scaling-up components and the dynamic control and response of the PFB and the gas turbine when integrated into a system subjected to load following.

The PFB Technology Plant operating conditions are compared in Table 1 with the Pilot Plant and one gas turbine module of a Commercial Plant.

The Technology Plant (Fig. 3) includes a 3 ft diameter version of the Pilot Plant PFB combustor with nine heat exchanger tubes, a recycle cyclone and ash removal system, a hot gas cleanup system, a small gas turbine/electric generator (SGT), coal and dolomite receiving, storage, transport and injection systems, computer control and data logging system, on-line gas analysis monitoring, etc. Testing in this plant demonstrated that many technological areas of uncertainty were successfully resolved.

TABLE 1   PFBC Design Conditions Comparison

Plant Cond.	Technology 1 MW	Pilot 13 MW	Commercial 80 MW
Outlet Temp, °F	1,650	1,650	1,650
Inlet Temp, °F	506	506	516
Pressure, psi	94.4	94.4	100
Fluid. Vel. fps	2.7	2.7	2.7
Bed Airflow, pps	2.3	38.5	213
Cooling Air/Tube, pps	0.43	0.43	0.43
Comb. Excess Air, %	30	30	30
Coal Flow, pph	585	10,242	55,100
Bed Height, ft	16	16	16

The emissions generated in a PFB system by the solids storage and handling and the steam cycle are typical of any coal fired boiler plant. The emissions from the solid residue of the PFB process consists of ash removed from the bed to maintain bed height constant during operation and fly ash removed by the hot gas particulate separators in front of the turbine section. The ash is not considered a hazardous waste and can be disposed of as sanitary land fill with appropriate consideration for leachate properties. A chemical analysis of a typical sample of ash from a fluidized bed using dolomite and Pittsburgh #8 seam bitumous coal provided the following: 7.9% CaO, 52.8% $CaSO_4$, 1.6% $CaSO_3$, 19% MgO, .03% C, and 18.7% ash.

Tests conducted in the Technology Plant showed the NOx emission to be consistently below 0.2 lb/10^6 Btu compared with U.S. environmental limit of 0.6 lb/10^6 Btu. This was achieved without any special NOx control. The air heater PFB design operates at conditions that are well suited for low NOx production. The bed temperature is 1650°F and has a variance of only $\pm$ 20°F throughout the bed. This is a far departure from conventional furnaces where stoichiometric temperatures reaching above 3000°F occur.

Tests also showed that with 3% sulfur content coal (HHV = 13,000 Btu/lb), the SO_2 emission was below 0.3 lb/10^6 Btu with the dolomite feed rate based on a calcium-to-sulfur mol ratio of 1.5. A capture efficiency of 95% was achieved at this ratio. The selection of a (a) low fluidizing velocity for high combustion efficiency and lower elutriation of particles, and (b) a deep bed to incorporate sufficient immersed heat exchanger surface provided increased gas and sorbent contact time which results in the improved SO_2 capture.

It was originally thought that the particulate carryover from the bed would require cleaning to a level below the standards for air quality in order to prevent erosion of turbine blades. However, a 1000 hour test on a small gas turbine coupled to the coal fired PFB combustor was conducted in the Technology Plant which showed excellent turbine results with particulate loadings higher than the air quality requirement. The hot gas cleanup train consisted of three stages of cyclone separation located between the combustor and the turbine. The first stage is designed to remove the large particulate comprising 75% of the grain loading with a 6 inches H_2O pressure drop. The second stage is a high efficiency unit designed to clean up the gas to an acceptable level for the turbine at a $\Delta P = 75$ inches H_2O. The third stage will act as a failsafe device and polishes the gas at a $\Delta P = 27$ inches H_2O.

Initial testing on a shakedown turbine for 150 hours revealed unacceptable erosion at (a) the vane trailing edges as shown on 3 vanes at the 7 o'clock location in Fig. 4 and (b) the blade trailing edge tips as shown in Fig. 5. The hot gas cleanup system was subsequently modified and the following results were observed. The mean particle size was reduced from 3.4 to 1.3 micron, the maximum size was reduced from over 22 microns to below 10 microns, and the loading entering the turbine was reduced by about 40% to 85 ppm.

A rainbow turbine was installed and the evaluation was conducted for the 1000 hour test without blade changes (Fig. 6). The stator vanes were cast materials using both the cobalt (FSX-414) base and the nickel (IN738) base alloys. Three types of aluminide diffusion coatings and three types of vapor deposited M-Cr-Al-Y claddings (where the M constituent is either nickel, cobalt or iron) were applied to the vanes. The rotor used the nickel (U-720) base wrought alloy with two types of aluminide diffusion coatings. The results of the metallurgical analysis after testing demonstrated that the Fe-Cr-Al-Y coating on either cobalt or nickel alloys offered excellent protection from alkali metal corrosion effects. These materials also offered a high degree of resistance to erosion when the limit on the gas stream particulate size was maintained below 10 microns. The absence of trailing edge erosion on the vanes and only minor tip erosion on the blades was a significant departure from the 150 hour turbine results. Useful life of 25-50,000 hours is predicted from these test results before refurbishment of the .020 inch cladding may be required on industrial size blades.

The heat exchanger tubes fabricated from iron or cobalt-base alloys also showed excellent corrosion and erosion resistance. There was no evidence of erosion on the tubes after 2000 hours of operation on the same set of tubes and test coupons. There was less than 1 mil of sulfide penetration on the iron and cobalt base alloy during this test. The results indicate that life expectancy of the heat exchanger tubes will exceed boiler tubes.

With these encouraging results the DOE program has now proceeded to the next phase – construction of a PFB combined cycle pilot electric plant. The Pilot Plant design configuration consists of an industrialized gas turbine with an airflow capacity of 120 lb/sec. The power turbine is gas coupled to the gas generator and mechanically coupled to a gearbox and generator, which produces 7 MW of electric power. The gas leaving the power turbine through a bifurcated duct enters the waste heat recovery boiler which produces a steam flow rate of 58,000 lb/hr. This steam is used for process and heating purposes. This plant will be a modification to an existing total energy system at Wood-Ridge, New Jersey. The equivalent electric power output for the Pilot Plant is about 13 MW. An artist's concept of the Pilot Plant under construction is shown in Fig. 7.

Coal stored in an existing 1700 ton storage bunker is milled to 1/8 inch and dried to a surface moisture below 0.5%. The coal is pneumatically conveyed to the bed from a lock hopper. Dolomite is stored in a 600 ton silo and conveyed in the same manner as the coal.

The bed ash removal system consists of a water cooled fluidized ash cooler column with a lock-hopper at the exit for depressurization. A pneumatic transport system conveys the dry ash to a 100 ton ash hopper.

The construction of this plant is expected to be complete in early 1983.

CONCLUSIONS

The PFB combustion technology can provide clean and cost competitive electric power generation using high sulfur eastern coals. The air heater PFB concept provides modular system benefits and can address to a range of duty requirements. Demonstration of the technology in the 1 MW Technology Plant and the Pilot Plant now under construction, enhances the prospects for early commercialization of PFB Combined Cycle Plants.

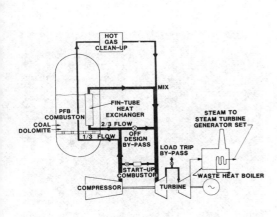

Fig. 1.   Flow diagram air heater PFB cycle.

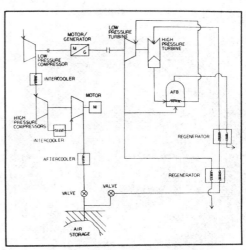

Fig. 2.   Coal fired CAES.

Fig. 3.   PFB technology plant.

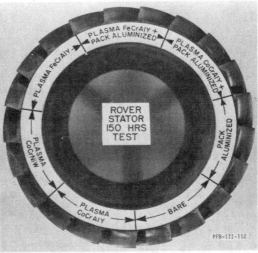

Fig. 4.   Turbine stator after 150 hour test.

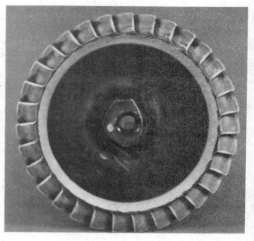

Fig. 5.   Turbine rotor after 150 hour test.

Fig. 6.   Turbine after 1000 hours operation with coal fired pressurized fluidized bed.

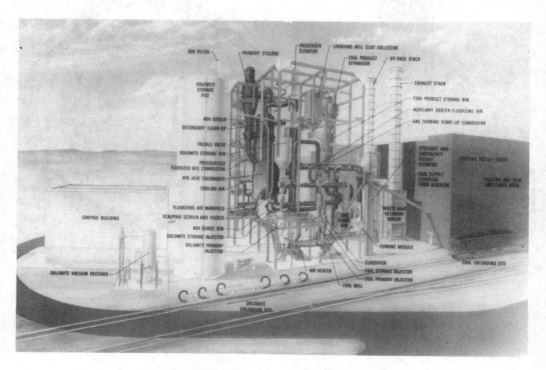

Fig. 7.   PFB pilot plant, Wood–Ridge site.

DISTRICT HEATING METERING SYSTEMS IN DENMARK

N. Nedergaard

Herning Municipal Works, Denmark

1. Introduction.

In a district heating (DH) system the heat meter is the basis of the settling of accounts between the consumer and the supplier. Due to the rising energy prices the heating bill plays a fairly important part in the individual economy. The whole complex of tariffing and metering has thus claimed increasing interest during recent years.

As the meter forms the basis of the heat bill it must be fairly exact and reliable. But furthermore the whole settling system should:

- ensure a fair distribution among the consumers

- promote a pattern of consumption, that gives the best possible economy of the DH system as a whole

- encourage energy saving.

2. Individual or collective settling.

A number of investigations of late years have revealed that the heat consumption with collective settling without meter is somewhat higher than with individual metering and settling, typically 15-20%. Increasing energy prices and problems in the energy supply make this unacceptable. Therefore the common opinion in Denmark is against future establishing of systems with collective metering and settling.

Accounts between the heat station and for example a block are often settled on the basis of a main meter. The building society then distributes the account between the individual appartments on the basis of some kind of metering.

Normally, metering is no big problem as regards bulk consumers: blocks, institutions, business premises etc., because there is basis for spending the nescessary money on the metering equipment.

Most meters, however, are small individual meters. The Herning

Municipal Works, thus, have 9,000 individual out of 10,000 meters, and as mentioned above there is a trend towards even more individual meters.

Small individual meters should

- be cheap in first cost
- be reliable in the operation
- be easy to install
- claim only little maintenance
- have a relatively wide measuring range (20-700 l/h)
- have a small pressure drop
- have good measuring exactitude (normally, the claim is $\pm$ 2%, at small loads however $\pm$ 5%).

The claims partly clash and are thus difficult to meet at one and the same time. The following considerations primarily aim at small meters because notoriously the individual metering of a small consumption gives the biggest problems not least due to the large number of meters.

3. Optimization of a DH system.

On installation of a DH system the various desires often clash:

desires	claims to the system
low network installation costs	large temperature differences, high pressure
low consumer installation costs	direct supply, high flow temp., small cooling
low heat losses	low temperatures

The local conditions will determine what is optimum. In Denmark it has turned out that a reasonable distribution in most cases will:

- have flow temp. about $80^{\circ}C$
- have direct supply (max. 6 bar operating pressure)
- have a settling system ensuring good cooling ($\Delta t = 40-45^{\circ}C$).

To encourage a good cooling at the onsumers, water quantity meters have been used up till now. Due to variations of the flow temperature in the network the water quantity meter will cause some discrimination between the consumers.

If the flow temperature from the heating stations is $80^{\circ}C$ and if it is ensured - through circulation - that the flow temperature in the street pipes does not get below $75^{\circ}C$, it is possible to give an example of the discrimination

	A	B
Heat consumption Gcal/year	25	25
Flow temperature $^\circ$C	80	75
Return temperature $^\circ$C	30	30
Number of m^3/year (billing)	500	555

Consequently B pays about 10% more than A for the same amount of heat.

The rising heat prices increase the demand on equity, and in Denmark there is a growing interest in replacing the measuring of the water amount by a way of measuring the amount of heat.

4. Branch system.

A reasonable settling system will usually contain three elements:

1. a fixed rate for each meter.

   Covers purchase and maintenance of meter, reading and issuing of heat bill.

   Is typical 200-400 d.kr./meter and gives 2-3% of the income of the district heating utility.

2. a fixed "load dependant" rate.

   In Denmark this tariff unit is usually connected to heat area in kcal/h or to the size of dwelling (m^3 cubic content or m^2 floor space). Maybe the electronic meters will make it reasonable in future directly to use the individual consumer maximum load.

   This "load depending" rate must cover capital costs, reserves, maintenance etc. and today it gives typically 15-25% of the income of the district heating utility.

3. a variable rate depending on the consumption.

   Here the tariff unit is either m^3 water or Gcal.

   The variable rate must cover the consumption dependant costs of the district heating station, that is purchase of fuel and heat, electricity, water etc.

   Gives typically 70-80% of the income of the district heating utility.

5. Measuring of the heat consumption.

In Denmark the consumption of heat has so far mainly been charged according to the water amount consumed. However, there are some district heating stations charging in accordance with the amount of heat consumed.

m^3 meter and the calorie meter both have their advantages and drawbacks:

	advantages	drawbacks
m^3 meter	- low first cost - simple and robust - makes the consumer interested in a good cooling - the individual consumer pays himself the heat loss in his own service pipe.	- differential treatment of the consumer because of varying flow temperatures - it becomes necessary with several cirkulations in order to ensure a fairly uniform flow temperature.
calorie meter	- compensates for varying flow temperature - greater variations in the flow temperature may be omitted. Possibly the flow temperature in general may be lower.	- more expensive in first cost - needs more maintenance - the consumers have no interest in keeping a low return temperature. - the heat loss in the service pipes is payed jointly.

With the increasing energy prices it becomes very important to reduce the heat loss in the pipe network as much as possible. This can be obtained by means of a better insulation but also through lower water temperatures, both flow and return temperatures. Out of consideration for the possibility to apply more low-valued heat it is also desirable to keep temperature as low as possible (in practice hardly below 70°C flow temperature).

A meter type that better than m^3 and calorie meter makes allowance for the wish for low temperatures is the heat equivalent meter. Quite simple it is a calorie meter with a fix built in return temperature.

The effects of the three types of meters can be illustrated by some simplified examples:

	A (m^3)	B (calorie)	C (calorie)	D (heat equivalent)
Heat consumption Gcal/year	25	25	25	25
Flow temperature $^{\circ}$C	85	85	70	70
Return temperature $^{\circ}$C	30	45	45	30
Water consumption m^3/year	455	625	1000	625
Return temperature to the station	$\sim$ 50	$\sim$ 50	$\sim$ 50	$\sim$ 35
"Circulated amount of water" m^3/year	$\sim$ 725	$\sim$725	$\sim$1250	$\sim$ 725
Settling: m^3 meter m^3/year	455	625	1000	625
Calorie meter Gcal/year	25	25	25	25
Heat equivalent meter Gcal/year (return = 30° fixed)	25	34,4	40	25

It is obvious that the use of the heat equivalent meter compensates for varying flow temperatures but it also makes the consumer interested in keeping a low return temperature.

In the examples it is assumed that the consumers' installation are dimensioned well enough to hold the same return temperature even when the flow temperature is lowered.

The example under A may be typical for a consumer with m^3-measuring. But the return temperature to the station will be substantially higher owing to the needed circulation, e.g. 50°C. This means that the circulated amount of water for each consumer is not 455 m^3 but 714 m^3.

The calorie measuring (B and C) will have a tendency towards high return temperature and thus either a high flow temperature or a high amount of circulated water. Also when measuring the calories it will be necessary to have some circulations in order to ensure a fair flow temperature at the extreme ends when there is low load.

The heat equivalent measuring compensates for varying flow temperatures and thus makes it reasonable to hold a rather low flow temperature. As the return temperature is fixed the consumers are interested in a good cooling. This means that the principle offer a possibility of low temperatures in the network (small heat losses) and at the same time a reasonable amount of circulated water.

Advantages and drawbacks of the heat equivalent in comparison with the above survey for m^3 and calorie meters:

	advantages	drawbacks
heat equivalent meter	- makes the consumers interested in a good cooling	- rather expensive in first cost
	- compensates for vary- ing flow temperatures	- requires some maintenance
	- allows greater variation in flow temperature and therefore some circulati- on can be ommitted. Pos- sibly the flow temperatu- re in general may be lo- wer.	- the heat loss in the service pipes is payed jointly

With the new electronic meters that are under development questions like purchase price and maintenance will probably be of less impor- tance in the future.

6. Meters

The m^3-meters applied in Denmark are mechanical meters with dif- ferent constructions. Today the mostly applied meter is a multi- ple-ray vane wheel meter with magnetic transfer of power between vane wheel and roller componenter, fig. 1. The meters are inex- pensive and robust, but problems may arise when measuring very small loads, especially gradually as the meters become dirty.

Until a few years ago the calorie meters were completely mecha- nical. As the water meter thus also has to draw the calorie comp- tometer, the meter became still more inert at low loads.

During recent years electronic calorie meter heads have been intro- duced into the market. They are built onto a usual vane wheel me- ter and do not strain it. Current supply may be delivered from a built-in battery, fig. 2.

Many firms also experiment with the development of purely elec- tronic meters. The principles (induction, ultra sound, fluidi- stor) have been apllied for many years, but up till now they have been too expensive in the small sizes.

Several firms have presented electronic meters for the reading of DH consumption:

the fluidistor meter (fig. 3)
the induction meter (fig. 4).

The meters are still relatively expensive and are hardly quite developed. But many practical tests are carried out (fig. 5) and it is likely that the electronic meters will replace the mecha- nical meters by and by.

A special problem of the purely electronic meters are their con- nection to the electricity supply network. This means that the meter will stop if there is an interruption of the supply. In Denmark , electricity shall be supplied via a fuse. If the cu- stomer just losens this fuse the heat consumption will not be registered.

To impede deception some manufacturers have built in an hour meter to ascertain whether the meter has been connected the whole
year. Proposals have also been put forward of building in a magnetic valve, turning off the heat when there is an interruption
of supply. A prerequisite of this solution would be a reliable
electricity supply in the area.

Below are stated some typical data of heat meters. Furthermore,
there refers to the manufacturers' catalogues.

	water quantity meter	calorie meter (water quantity meter with an electronic calorie head)	calorie meter fully electronic.
price	450	1,500	3,000
installation approx. Dkr.	40	80	300
max.load m3/h	1.0	1.0	
cont. load m3/h	0.4	0.4	2.7
+ 5% from 1/h	40	40	
exactitude			
+ 2% from 1/h	60	60	30
power connection	none	battery (5 years)	220 V.A.C.

7. Conclusion

As mentioned the trend is towards fully electronic meters. But
they are still relatively expensive and have not got over all
teething troubles yet. Thus it is wisest to postpone the replacement of existing meters by electronic calorie meters.

To reduce heat loss and circulated water quantity is should be
considered to use a calorie meter with a fixed, built-in return
temperature, a heat equivalent meter.

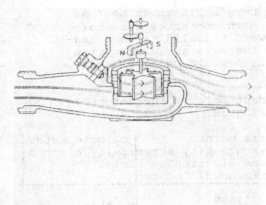
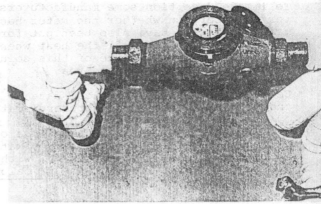

Fig. 1. m^3-meter

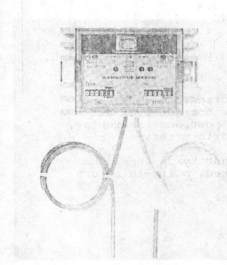
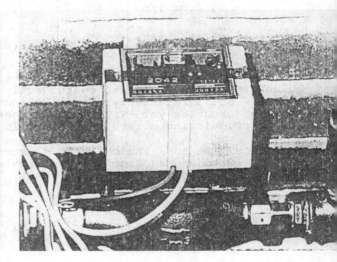

Fig.2.   m^3-meter with electronic calorie-head

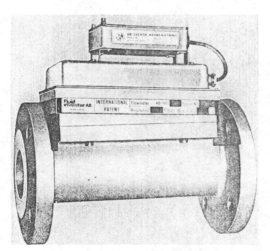
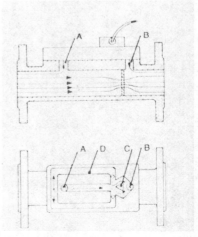

Fig.3.   Fluidistor meter

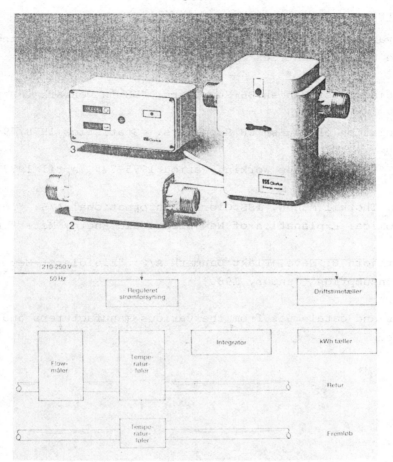

Fig.4. Induction meter

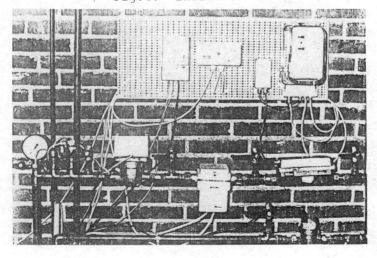

Fig.5. Test of different meters in a one family house

988

Literature

The Nordvarme Committee on Delivery Questions: "DH Tariffing in the Nordvarme Member Countries" 1980-02-06.

The Association of Danish DH Suppliers: "DH in Denmark". May 1979.

The Association of Danish DH Suppliers: "Statistics 1978/79"

UNICHAL: "Report of the Working Period 1975-77", April 1977.

Grad.eng. Thorkil Munch, ISS Clorius International A/S: "Technical and Economical Explanation of New Electronic Energy Meter" Nov. 79.

Eng. Thorbjørn B. Neve, Fläkt Danmark A/S: "Fluidistor Metering of Energy Consumption" January 1980.

Brochures and catalogues from the various manufacturers and suppliers of meters.

FLUIDIZED BED COMBUSTION AND ITS POTENTIAL USE FOR A VARIETY OF SOLID FUELS

J. N. Nikolchev

ANCO Engineers, Inc., Santa Monica, California, USA

ABSTRACT

The decreasing supply of the world's energy resources has increased the interest in more efficient utilization of these resources. The development of new technologies continues to expand the types of fuels that can be efficiently used for production of useful energy. Fluidized bed combustion (FBC) is a relatively new technology which provides a viable method of burning a wide variety of fuels in an environmentally acceptable way. It has the potential of being very useful to both developed and developing countries, by allowing them to efficiently burn a combination of fuels that are most readily available and that usually would not be used as fuels. Possible candidate fuels include: solid and liquid municipal waste; various industrial waste; wood waste; a variety of agricultural waste and residues such as rice hulls, almond shells, walnut shells, orchard waste, bagasse, potato peels, and pits from numerous fruits and vegetables; peat, low-grade coals, oil shale, and many other fuels. This paper provides a brief description of several promising FBC systems with an overview of developing programs in the United States and some European countries. Use of various fuels is examined with a discussion on further research that is required to develop these as viable energy alternatives.

KEYWORDS

Fluidization; Atmospheric Fluidized Bed Combustion (AFBC); Pressurized Fluidized Bed Combustion (PFBC): Circulating Fluidized Bed Combustion (CFBC); coal; biomass; municipal solid waste.

INTRODUCTION AND BACKGROUND

In recent times, fluidized bed combustor developments have shown significant progress and this combustion method is becoming more accepted as a viable way of burning a wide variety of fuels in an environmentally acceptable manner. There are several very important reasons for this strong interest in fluidized bed combustion systems. In particular, they have the capability of burning a very wide range of fuels, from high-quality coal to very low-grade combustibles, with the capability of reducing the emissions of such pollutants as SO_2 and NO_x to acceptable standards.

A fluidized bed reactor consists of the reactor vessel, which contains the solid fluidized material through which passes the fluidizing fluid (either liquid or gas). Figure 1 shows the different forms of fluidization possible. The principle of a fluidized bed operation is based on the fact that a fluid forced upward through a bed of particles will reach a point at which the drag forces levitate the particles. At this time, the bed expands, permitting an increase in the flow rate of the fluid. The particles can now move freely, the bed acts like a heavy liquid, and the total drag force equals the weight of the particles. Further increases in the fluid flow rate increases the spacing of the bed particles and then the formation of "bubbles" begins. These bubbles of fluid move vertically through the fluidized solid bed. They enhance the mixing of the bed and give it the appearance of boiling liquid. At very high fluidizing gas velocities, the bed motion becomes quite violent and starts slugging and splashing.

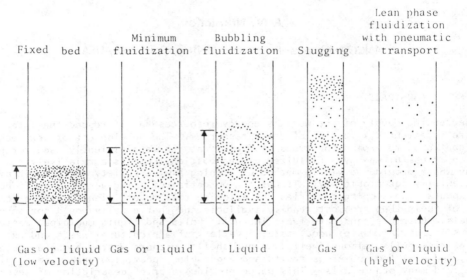

Fig. 1. Various kinds of contacting of a batch of solids by fluid.

Although, for the past 30 years, fluidized bed reactors have primarily been used by the chemical engineering industry as highly effective devices for heat transfer between solids and gases, such as the fluidized catalyst cracking units, the first large-scale, commercially significant use of fluidized beds was made by Fritz Winkler for the gasification of powdered coal. The patent for this process was awarded in 1921 (U.S. Patent No. 1687118), and the first gas production unit started successful operation in 1926. Around 1940, chemical engineers started to use fluidized beds in catalytic cracking processes. Since then, fluidized beds have been used for transportation of solids, mixing of powders, drying, heat exchange, and many other applications. During the 1950s, several major U.S. and European manufacturing corporations became active in the examination of fluidized bed combustion (FBC). The British National Coal Board also became interested in FBC at that time. Since then, many other countries have become involved in developing FBC. The development effort expanded to particularly large proportions in the United States after it became apparent that federal air pollution control regulations could not be met easily unless very expensive pollution control devices were used. In addition, the recent jump in oil prices made the use of alternative fuels

—coal in particular—much more desirable (Anson, 1976) . The increased cost of
waste disposal, combined with the decreasing amount of available disposal sites,
have also increased awareness in the United States of the potential benefits possible from efficiently burning waste using well-designed FBC systems.

GENERAL DESCRIPTION OF FBC SYSTEMS

The recent emphasis on increasing the amount of energy generated by fuels other
than oil has made coal the most popular fossil fuel for the near future. Fluidized
bed combustion has thus emerged as the most acceptable way to burn the increasing
amount of coal that will be required. Because of this, more work has been done
on coal-burning FBC systems than on any others. Examples of such systems will be
used to briefly describe the operation of FBC plants.

Figure 2 shows a schematic of a typical atmospheric fluidized bed combustor (AFBC)
boiler power plant. The AFBC unit consists of the combustor structure which contains a bed made of inert material and coal. The inert material is usually limestone or dolomite which has the capability of capturing the SO_2 emitted during
coal combustion. Thus, there is no need to have an expensive and efficiency-re-
ducing flue gas desulphurization device. The amount of coal in the bed is rarely
more than 1-3% of total material weight. The excellent heat capacity of the limestone provides very uniform temperature distribution and, through the submerged
heat transfer surfaces, the temperature can easily be controlled(usually in the
neighborhood of 900°C). This allows the temperature to be kept low and, consequently, to reduce NO_x formation. Submerging the heat transfer surfaces in the
bed provides high heat transfer coefficients, thus reducing the size of the heat
exchangers and their cost. In addition, these submerged tubes help break up the
bubbles formed during fluidization. These bubbles reduce the amount of oxygen contact with the fuel, so it is very desirable to eliminate them or at least reduce
their size. The design of the FBC allows many grades of coal to be burned without
significant problems. The long residence time makes it possible for slow-burning
coals, such as anthracite, to be used. The large heat capacity of the bed allows
for a wide divergence in the fuel heat content, ash content, moisture content,
volatile content, etc.

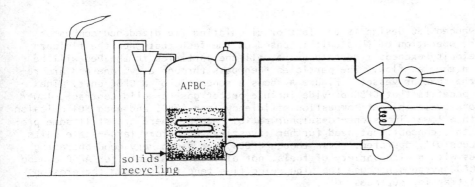

Fig. 2. Typical AFBC-boiler power plant.

In addition to the AFBC units, there are several advanced FBC unit designs being considered. Of these, pressurized, fluidized bed combustors (PFBC) have shown, by far, the greatest potential for sucessful develoment in the near future. Figure 3 shows a typical PFBC plant. In this plant, the combustion is achieved in a pressurized vessel; thus, the combustion gases leave the unit at high pressures (more than 10 atm) and temperatures (normally about 900°C). The gases go through a particulate cleaning device such as an electrostatic precipitator, cyclone, or a filter, and then through a turbine. In addition, steam is produced by the submerged tubes in the bed and passes through a turbine to produce additional electricity. The efficiency of such a plant could approach 55%. However, there are still some serious obstacles that have to be overcome before PFBC becomes completely accepted on a commercial scale, the most pressing of which is the ability to sufficiently clean the combustion gases before they reach the turbine and cause damage to the trubine blades.

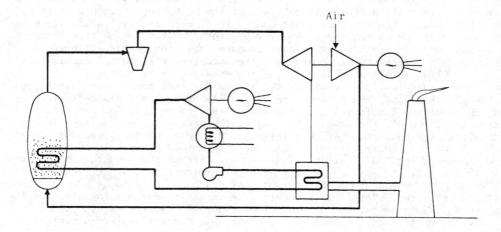

Fig. 3. PFBC combined cycle plant.

Another advanced FBC design is the fast or circulating fluidized bed combustor (CFBC). The operation of this unit is based on the fact that when the fluidazation gas velocity exceeds a certain value, all the particles in the bed will be entrained in the gas flow. The particles then pass through a cyclone and are returned to the reactor vessel. Figure 4 shows a schematic of a CFBC unit. The additional benefits that CFBC provide include better gas-to-solid contract, lower NO_x emission levels, higher combustion efficiency than AFBC, and good SO_2 emission control with a lower limestone-to-sulphur ratio. Again, there are still some problems with this concept that need further investigation before large-scale units can be successfully operated. Even though CFBCs have shown very good operating capabilities with a wide variety of fuels, not all fuels suitable for AFBC can be used with CFBC systems because the high velocities tend to increase the erosion of the heat transfer surfaces.

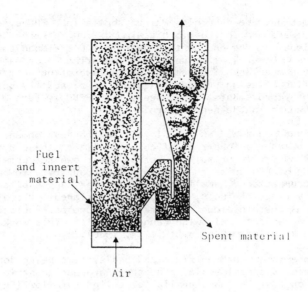

Fig. 4. Fast or Circulating Fluidized Bed

OVERVIEW OF CURRENT FBC PROGRAMS

Many countries in the world are currently involved in programs for the development of fluidized bed combustion. The United States and Europe are currently in the lead in the amount of money spent for these programs. Also, it appears that the People's Republic of China has had a very large program in fluidized bed reactor development since 1960; presently, there are over 2,000 FBC boilers in China. These boilers have capacities ranging from 4 to 50 tons of steam per hour. FBC boilers with 130 tons per hour capacities are currently being tested (Zhang, 1980). It will be very beneficial to find out more about their programs, since many of the FBC systems are using very low-grade fuels and have been in operation for a significant amount of time.

The United States is currently following a development program designed to provide fuel flexibility, multi-fuel firing capability, high combustion efficiency, and overall plant efficiency and reliability within environmental complinace (Byam, 1980). The first commercial prototype unit built in the United States is the Georgetown University FBC, which had been operating since August 1979. It produces 50 tons per hour of saturated steam at 1,723 kPa with the capability to go to 4,306 kPa. In addition, there is the 30 MWe Riverville experimental unit, which has been used to run long-term tests of different FBC system components. In regard to FBC applications in commercial work, the Tennessee Valley Authority has taken the lead by planning to install a 20 MWe AFBC pilot plant by the end of 1981. Future plans include increasing this capacity to 200 MWe, with the long-term goal of eventually producing 800 MWe FBC systems. In addition to these large units which burn coal, there are numerous smaller experimental units for other biomass fuels. Babcock and Wilcox operate a 2 MWe AFBC for the Eelctric Power Research Institute which was recently used to provide useful data for possible improvements in FBC performance.

Johnston Boiler Company has installed several small packaged commercial units, which can operate on a wide range of biomass fuels and coal. In California, Sun

Diamond Growers have been burning walnut shells in a FBC system since October 1980. Tri-Valley Growers and the California Almond Growers Exchange will also have FBC units operational by the end of this year that use almond shells and olive, cherry and peach pits. Other smaller units are operating in or being planned by many food industry companies. The success of the operating units have shown the economical benefits of using FBC systems and the reliability possible after implementing recent improvements. It appears that PFBC systems will enjoy the same commercial success in the near future.

The United Kingdom has had a long history in the development of FBC systems and continues to be in the forefront of commercializing these systems. Currently, there are more firms in Britain investing their own resources in the development and production of FBC units than in any other country. In particular, there are quite a few pre-packaged, small FBC boilers that can be purchased now. These units burn a variety of fuels, but most are designed for coal. In addition, there is still a considerable amount of work being done on PFBC and the development of data on burning an extremely wide range of fuels. This work is being done in the British Babcock boiler at Renfrew (Willis, 1980).

There is also a lot of work on fluidized bed systems being done in Germany, Denmark, Sweden, Finland, Czechoslovakia, and other European countries. Finland's efforts have focused on the CFBC unit available, called Pyroflow (Engstrom, 1980). This system claims to have efficiencies that are significantly better than those achieved by AFBC units. Denmark has also done significant work in this area. Primarily, their efforts have been toward the use of the FBC systems for combinations of locally available fuels (coal, wood waste, etc.) for steam or hot water production (Kellerup, 1980). Germany is involved heavily in the design and building of several large demonstration plants (to come on-line in 1982). These include AFBC, PFBC, and CFBC (Holighaus and Batch, 1980).

There are many systems also being installed in Japan, India, Brazil, and many other countries, too numerous to describe here. It is clear that, while current FBC developments are producing many significant changes and improvements, fluidized bed combustion has become widely accepted as a viable method to utilize many different fuels, fufilling a wide variety of needs.

FBC APPLICATION TO VARIETY OF ADVANCED FUELS

Although fluidized bed combustion developments have been primarily directed towards their potential use in direct coal combustion, it has achieved significant progress recently in its acceptance as a viable method of burning many other fuels. The types of fuels that have been experimented with in FBC units include: low-grade coals, residues from coal mines and coal working operations, municipal waste, wood waste, wood residues from paper mills, and a huge variety of agricultural wastes. Most of these fuels are condisered to be part of the general category of biomass fuels. The increased use of biomass as an energy resource will alleviate the transition from gas, oil, and other fossil fuels to renewable and less expensive energy sources.

Biomass Combustion

Biomass represents a diverse energy resource. Currently, biomass fuels are being used in only a few large-scale operations, primarily in the forest produce and sugar cane industries. These industries use paper mill by-products and bagasse in direct combustion boilers. It is in these applications that fluidized bed combustion is receiving its widest acceptance and use. Although the developed

countries are taking the lead in establishing biomass use programs, a real potential for biomass may be in the developing countries where the cost of fossil fuels
is prohibitive but biomass is plentiful. Fluidized bed combustion may be the best
direct combustion method available to them.

Biomass can be divided into the following categories: wastes are materials representing a disposal cost (includes municipal waste); residues have either some positive value or represent zero disposal cost; and fuel crop is biomass specifically
cultivated on energy farms for its energy content (Crane and Williams, 1979). All
of these biomass fuels can be successfully burned in fluidized bed systems. Presently, there are quite a few small-scale AFBC plants and pilot plants operating
around the world. Almost all have had exceptional success in demonstrating AFBC
possibilities. The basic principle of operation is not different from the coal-
using systems, except that the inert bed material is different, since there is
usually no need to absorb any SO_2 emissions.

Biomass fuels are particularly attractive because they are cheap and usually do
not cause as much pollution as the more common fossil fuels. The restricted use
of these fuels on a large scale, up until now, has been mainly due to the inherent variations in the fuel quality. Many residues have a high moisture content
and are very hard to burn by traditional means; other residues, such as walnut
shells, almond shells, peach pits, etc. require a long time to burn because of
their low moisture and volatile content. FBC is uniquely suited to burn both of
these fuel types efficiently becaue the long residence time for the fuel particles
in the bed and the associated turbelence provide excellent conditions for complete
combustion of many different types of fuel. Low heat content fuels could also be
effectively utilized, if they are mixed with some coal in order to ensure stable
combustion. Several such systems have been tested in the United States (Copeland
1980) and the United Kingdom. There are many small plants operating now which
have sufficiently shown that FBC systems using a variety of biomass fuels can be
economically and technically successful. There is a need to build new larger
plants that are using more of the recent developments and fully utilize the benefits that FBC offers, such as high heat transfer coefficients for submersed heat
transfer surfaces.

Municipal waste has been used to generate energy in Europe for many years, but the
use of standard incinerators has been very costly due to pollution, corrosion, and
erosion problems. The use of FBC systems provide a viable method of reducing
these problems and increasing the use of municipal waste of fuel around the world.
The few experimental systems tested in the United States and other countries in
Europe have had very promising results (Preuit and Wilson, 1980).

ADDITIONAL RESEARCH AND DEVELOPMENT WORK REQUIRED

In order to assure the complete acceptance of FBC in the near future, and demonstrate its significant benefits, there are several major areas that need further
development and testing. A critical issue is the reliability of the methods for
injection and removal of material from the bed. There is no general agreement on
whether it is better to inject the fresh material above the bed or below it and
achieve good fuel distribution. Both methods have been used in the past, but not
enough research has been done which looks into the distribution of the fuel. The
present approach to solving distribution problems has been by trial and error.
Different baffles and other immersibles are tried until something works.

One of the largest sources of combustion inefficiency in AFBC boilers has been the
loss of unburnt carbon fines by entrainment in the exhaust gases. There is a need
to reduce this loss as much as possible. The current most popular reduction methods are: the recycling of carbon fines; and secondary carbon burn-up beds. Both

methods increase systems efficiency significantly, but their use increases the overall complexity, costs, and maintenance requirements to unacceptable levels. In particular, the coal handling metering valves and controls are prone to frequent failures. There is a significant amount of research being done in this area together with development of advanced elutriation control technologies. ANCO Engineers and several other companies are currently invovled in developing such technologies. However, there is no one method that has been tested on a sufficiently large scale for a long enough time, in order to determine its perfomance and reliability.

The major obstacle for building large scale PFBC systems has been the need to clean the high-pressure, high-temperature combustion gases from particulates before passing through a turbine. The methods for doing this that are being investigated include: high-efficiency cyclones; electrostatic precipitators; and a variety of high-temperature filters. These studies are far from showing any conclusive results. Part of the reason is due to the lack of information that will determine the degree to which these gases will have to be cleaned in order not to cause any damage to the turbine blades.

Due to the unique design of FBC heat transfer surfaces, there is a need for more information on how the variation in bed material effects corrosion and erosion of these surfaces. This is especially critical to CFBC systems, because of the high velocities involved. Additional work is also required on plant load response characteristics, and the possible disposal problems of the spent bed material. There are several recycling methods under investigation using limestone for SO_2 absorption and its possible use for fertilizer is also being studied. In general, there is a need for more long-term and large-scale tests in order that the system reliability performance may be well investigated and confirmed.

CONCLUSIONS

Fluidized bed combustion is currently the best available method for direct combustion of a very wide variety of fuels in an environmentally acceptable way. Although there are still areas that need further development, the current state of the technology of AFBC is sufficient to warrant serious consideration for building commercial plants to provide electricity, industrial steam, or hot water. There are already several commercially available, small, packaged units, although there are no operational units that produce over 100 MWe of power. (Several such units are expected to be available by 1990.)

PFBC systems are not yet at the same stage as AFBC systems. They show potential for higher overall efficiency, but there are several serious obstacles in their development that have to be overcome before commercial acceptance will be reached. In particular, an operating, high-temperature, high-pressure, gas-cleaning device will have to be demonstrated.

While CFBC is closer to commercial acceptance, there are several areas that could cause serious problems for long-term operation of such units. For example, there is a need to evaluate possible erosion problems as a function of operation time.

Despite the apparent obstacles that will have to be overcome before large scale FBC systems become commonplace, with the imminent shortage of non-renewable energy sources and the need to increase the use of the renewable energy resources, more and more such systems are being planned and built. Use of such innovative technology is the only possible way to deal with the growing demand for energy in the world. In particular, fluidized bed combustion will provide a great opportunity for developing countries to become more self-sufficient by depending on diverse, locally available fuels for their energy requirements.

REFERENCES

Anson, D. (1976). Fluidized Bed Combustion of Coal for Power Generation. Prog. Energy Combust. Sci., Vol. 2, pp. 61-82.

Byam, J. (1980). Fluidized Bed Combustion; a Status Check. Proceedings of the Sixth International Conference on Fluidized Bed Combustion, Atlanta, Georgia, pp. 18-22.

Copeland, G. G. (1980). Operating Experience with a Variety of Low Grade Fuels in Fluidized Bed Combustion. Proceedings of the Sixth International Conference on Fluidized Bed Combustion, Atlanta, Georgia, pp. 579-583.

Crane, T.H., and Williams, R. O. (1979). Energy from Biomass - the Realities. Proceedings of the Second International Conference on Energy Use Management, Los Angeles, California, pp. 1748-1755.

Engstrom, F. (1980). Development and Commercial Operation of a Circulating Fluidized Bed Combustion System. Proceedings of the Sixth International Conference on Fluidized Bed Combustion, Atlanta, Georgia, pp. 616-620.

Holighaus, R., and Batsch, J. (1980). Overview of the Fluidized Bed Combustion Programme of the Federal Republic of Germany. Proceedings of the Sixth International Conference on Fluidized Bed Combustion, Atlanta, Georgia, pp. 30-35.

Kollerup, V. (1980). Atmospheric Fluidized Bed for Coal and Wood Waste Combustion Especially for District Heating. Proceedings of the Sixth International Conference on Fluidized Bed Combustion, Atlanta, Georgia, pp. 334-342.

Preuit, L. C., and Wilson, K B. (1980). Atmospheric Fluidized Bed Combustion of Municipal Solid Waste: Test Program Results. Proceedings of the Sixth International Conference on Fluidized Bed Combustion, Atlanta, Georgia, pp. 872-882.

Willis, D. M. (1980). An Overview of the Progress in Fluid Bed Combustion in the United Kingdom. Proceedings of the Sixth International Conference on Fluidized Bed Combustion, Atlanta, Georgia, pp. 23-29.

Zhang, Xu-Yi (1980). The Progress of Fluidized-Bed Boilers in People's Republic of China. Proceedings of the Sixth International Conference on Fluidized Bed Combustion, Atlanta, Georgia, pp. 36-40.

1000MW COAL GASIFICATION-COMBINED CYCLE COOL WATER PROJECT

D. R. Plumley

*Coal Gasification Combined Cycle Projects, General Electric Company,
Schenectady, New York, USA*

ABSTRACT

This paper describes the evolution and present status of the Cool Water Coal Gasification Program. The Program will build and operate a 100 MW coal gasification combined cycle power plant to be located in southern California and planned for startup in late 1983.

Plant design features such as configuration, performance, and environmental characteristics are discussed. It is anticipated that this plant will demonstrate the viability of the integrated coal gasification combined cycle technology and will be the first step toward commercialization. A projected schedule and the performance of a possible first commercial plant is presented.

KEYWORDS

Coal gasification; combined cycle; power generation.

INTRODUCTION

The United States is committed to dramatic increases in the use of coal for power generation. The basic reasons for this are clear when it is understood that the energy content of U.S. indigenous recoverable coal resources are known to be twenty to thirty times that of U.S known recoverable resources of natural gas or petroleum, respectively (Fig. 1). The simple fact of the resource base doesn't direct coal usage to utilities, but two other associated factors do. First, coal cannot substitute directly for the energy requirements of the transportation industry, and second, coal is environmentally dirty and cannot be cleaned up economically in the small usage segments of the energy infrastructure. These three simple facts lead to the conclusion that utilities in the U.S. will increase their use of coal for power generation in spite of the occasional predictions that more indigenous natural gas and petroleum will be found than is now believed to exist. For, even if it is found, the price will preclude utility usage in base-load generation. It is, therefore, hard to believe that there is any realistic scenario that could change the present trend toward coal for power generation in the foreseeable future.

The present method of converting coal to electricity is, of course, conventional coal-fired boilers coupled with steam turbine generators. However, environmental requirements which have increased the cost and reduced the efficiency of these conventional plants have provided a "niche" for a new technology which promises to be less expensive to build, more efficient to operate, and, given those two factors, will be able to generate electricity at a lower cost.

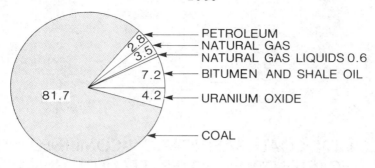

SOURCE: COAL DATA BOOK, THE PRESIDENT'S
COMMISSION ON COAL,U.S. GOVERNMENT PRINTING OFFICE,
WASHINGTON,D.C. FEBRUARY 1980.

Fig. 1. Known recoverable U.S. energy reserves (Btu content-%).

This new technology is the integrated coal gasification combined cycle plant (IGCC). It results from the coupling of second-generation gasification systems with relatively conventional combined cycle plants. The pressurized gasification process allows the removal of sulfur from the coal before combustion at a volume that is one three-hundredth of that which is experienced in the stack gas cleanup systems used in conventional coal plants. So, in addition to all the advantages shown in Fig. 2, there is every reason to believe that cleanup system costs which are volume-sensitive will experience lower incremental costs when environmental standards become more restrictive.

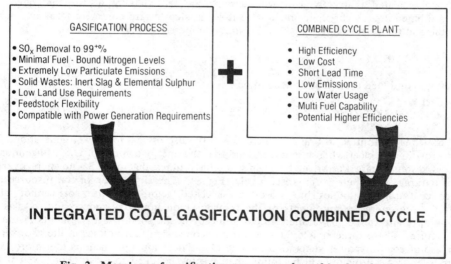

Fig. 2. Marriage of gasification process and combined cycle plant.

The combined cycle plant, which combines a gas turbine, a heat recovery steam generator, and steam turbine, brings its high efficiency, low cost, short lead time, low emission levels, and lower water usage to the marriage. The combination provides a plant with attractive environmental characteristics (Fig. 3) and, General Electric believes, economic advantages (Fig. 4) over conventional plants which are available with today's gas turbine firing temperatures, and which will show even greater advantages with increased firing temperatures that General Electric is planning for the future.

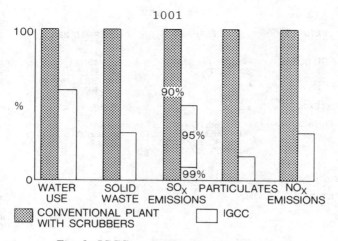

Fig. 3. IGCC environmental advantages.

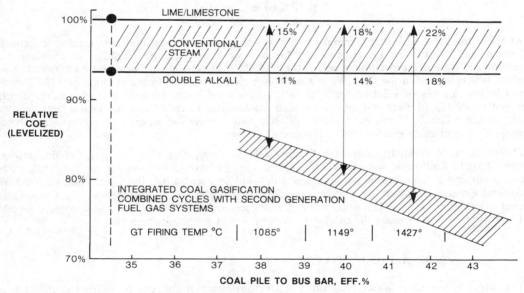

Fig. 4. Cost of electricity vs. efficiency comparison.

BACKGROUND

During the period from 1976 to 1979 we conducted a number of studies not only to develop the economics of the IGCC but also to select the most promising gasifier concept. There are three basic types of gasifiers in various stages of development—the entrained flow, the fixed bed, and the fluidized bed gasifier. Some important characteristics of each type are shown in Fig. 5. Although the advantages of the integrated coal gasification plant are available with any of these gasifier systems, the studies led General Electric to favor the entrained flow because of its inherent simplicity, its ability to handle a wide range of coals, its fast response times, the fact that it has effluents that are easily handled, and, most importantly, its impressive experience base.

DESIRED FEATURES	ENTRAINED		FIXED BED		FLUIDIZED BED
DESIGN & OPERATIONAL SIMPLICITY	INHERENT SIMPLICITY		RELATIVELY COMPLEX - GRATES, STIRRERS, CLEANUP		SENSITIVE TO OPERATING PARAMETERS
COAL TYPE FLEXIBILTY	YES		LIMITED		LIMITED
BROAD EXPERIENCE BASE	MONTEBELLO RUHRCHEMIE HARBURG WINDSOR LOCKS	15 T/D 150 T/D 150 T/D 120 T/D	LURGI BRITISH GAS GEGAS	480 T/D 400 T/D	WALTZ MILL 15 T/D
ENVIRONMENTAL CHARACTERISTICS					
PHENOLS & TAR SOLID WASTE	NO		YES		TRACE
— ASH	OK		OK		OK
— SPENT DOLOMITE	NONE		NONE		OBJECTIONABLE
LIQUID WASTE	TREATABLE		TREATABLE		TREATABLE

Fig. 5. Gasifiers for IGCC.

At about the same time that General Electric was conducting studies, Southern California Edison, Texaco, and the Electric Power Research Institute were studying the feasibility of conducting a commercial-scale demonstration program for a 100 MW integrated coal gasification combined cycle plant, which was called the Cool Water Coal Gasification Program. In the summer of 1979 Texaco and Southern California Edison formally initiated the Cool Water Program and were subsequently joined by EPRI, the Bechtel Corporation, and the General Electric Company in the joint venture partnership. This $275 million project will build and operate the nation's first large-scale power generating plant using gasified coal to produce electricity.

The plant, to be located at Southern California Edison's existing Cool Water generation station near Daggett, California, will gasify 907.2 metric tons per day (1000 tons per day) of bituminous coal producing a medium-Btu synthesis gas which will be used as fuel for a conventional 100 MW General Electric combined cycle power plant. Fig. 6 shows an artist's concept of the Cool Water Coal Gasification Plant. The success of the project, structured as a practical commercial-scale demonstration, will expedite the commercialization of coal gasification technology.

PROGRAM OBJECTIVES

The intent of the Cool Water Program is to demonstrate that coal can be utilized as a fuel (via gasification) for combined cycle plants in an environmentally superior manner while satisfying the electric utility system requirements for reliability, economy, and compatibility.

PROCESS DESCRIPTION

Coal will be delivered to the Cool Water plant by rail in unit trains and will be unloaded from each hopper car and conveyed to silos for storage. The coal will then be crushed and pulverized to the required size, slurried with water, and fed to a Texaco gasifier. A flow diagram of the Cool Water Integrated Gasification Combined Cycle Process is presented in Fig. 7.

In the gasifier, the coal and water will be reacted with oxygen at between 1260 C and 1538 C (2300 F and 2800 F) and 41.4×10^5 Pa-gauge (600 psig), producing a synthesis gas consisting primarily of hydrogen (H_2) and carbon monoxide (CO) with a specific heat of combustion of approximately 8.059×10^7 J per m^3 of gas at 15 C and 1.01×10^5 Pa (270 Btu/SCF). Within the gasifier

Fig. 6. Cool Water Coal Gasification Plant.

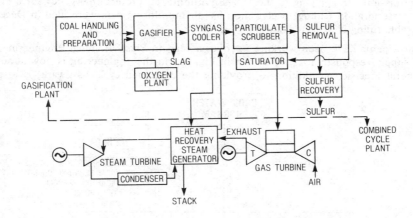

Fig. 7. Flow diagram of Cool Water Integrated Gasification Combined Cycle Process.

the coal ash will be melted into slag, quenched with water, and removed through a pressurized lockhopper system as glassy, gravel-like pellets for disposal. The synthesis gas produced will be cooled in the gas cooler to produce saturated steam, which will be superheated in the heat recovery steam generator, thus recovering and utilizing much of the process waste heat.

After cooling, the synthesis gas will pass through a wet scrubbing system to remove any remaining particulates. Gaseous sulfur compounds, consisting primarily of hydrogen sulfide (H_2S), will be removed in a Selexol® system. A Claus-type* sulfur recovery system will receive a concentrated H_2S stream from the Selexol® unit for conversion to elemental sulfur for disposal or sale.

® Allied Chemical Corp.

* As developed by Ford, Bacon & Davis Company

The synthesis gas from the sulfur removal system will then be directed to the combined cycle plant. The combined cycle will include a 65 MW gas turbine generator which will be designed to operate primarily on the synthesis gas although provisions are made for startup on distillate oil. The 55 MW steam turbine generator will receive superheated steam produced in the heat recovery steam generator, thereby increasing the overall output and efficiency of the plant.

The Cool Water Coal Gasification Program is a five-phase program, with initial production of synthetic gas scheduled for October 1983, as shown in the Program Schedule (Fig. 8). The program is well advanced, with Phase I, Preliminary Engineering and Permits, completed and Phase II, Final Engineering, under way. Phase III, Procurement and Construction, began in April 1981 and site work started in July.

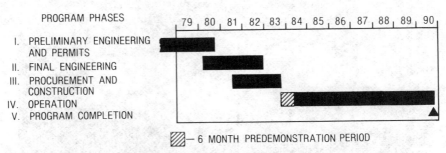

Fig. 8. Cool Water Program schedule.

The program received approval from the California Energy Commission and the California Public Utility Commission. In applying to the U.S. Environmental Protection Agency for a Prevention of Significant Deterioration Ruling, twelve months' air monitoring data were filed in December 1980, and a favorable ruling is expected shortly.

Current participants have been working for over a year on Phase II, Final Engineering, in keeping with their supply responsibilities, as shown in Fig. 9, and the engineering is now about 40% complete. General Electric, in addition to providing the combined cycle hardware, is responsible for

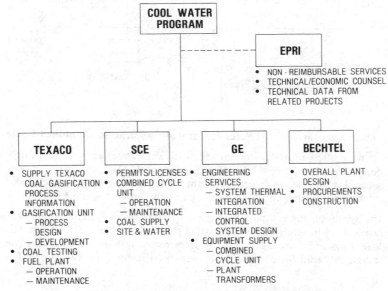

Fig. 9. Participants' program responsibilities.

the design of the overall plant thermal configuration and the integrating control system required to marry the gasification plant with the combined cycle.

PLANT PERFORMANCE

The criteria being applied in the design of the integrated system recognize the need to incorporate the novel features of the process necessary to demonstrate commercial viability. At the same time, features which have already been applied commercially and whose effect on plant performance can be extrapolated with reasonable confidence, such as steam turbine reheat, are not included (Fig. 10). Thus, the initial expected performance levels of 96 MW (net) with an approximate 11.49 KJ/WH (10,900 Btu/KWH) heat rate should not be considered an optimized configuration. As the plant matures, gasifier operating parameters will be modified, which will achieve higher gasification efficiency and improved gas quality. These improvements will result in a greater fuel plant output with a corresponding increase in plant electrical output to 106 MW and a net coal pile-to-busbar heat rate of approximately 10.86 KJ/WH (10,300 Btu/KWH).

	COOL WATER	COMMERCIAL PLANT
Characteristics		
Output (MW)	96-106	500
Heat Rate (Joule/KWH)*	9.76 - 10.33	8.20 - 8.39
Efficiency(coal pile to busbar,%)	31 - 33	38 -40
Reheat Steam Cycle	no	yes
Gas Turbine Firing Temp. (°C)**	1085	1149
Slurry Ratio(coal to water by weight)	0.60	0.67
Coal Sulfur Content(%)	0.46 - 3.5	3.5
Components		
Gasifier Trains	1	4
Cleanup Systems	1	2
Gas Turbines	1	4
Steam Turbines	1	1
Heat Recovery Steam Generators	1	4

* 1 Joule/KWH = 1055 Btu/KWH
**°C = (°F -32)/1.8

Fig. 10. Comparison of Cool Water with commercial IGCC plant.

Commercial plants 500 MW and larger will be more energy-efficient with heat rates in the neighborhood of 9.12 to 9.33 KJ/WH (8650 to 8850 Btu/KWH). The design features necessary to achieve this efficiency level will result primarily from the use of larger, more efficient reheat steam turbines and improvements in the coal slurry ratio in the gasifier. As stated earlier, these commercial IGCC units can be expected to be superior to the conventional coal-fired boiler with flue gas desulfurization, where heat rates approximate 10.43 KJ/WH (9900 Btu/KWH). In addition, the performance level estimated for future commercial plants with higher gas turbine operating temperatures and advanced gasification technology will be in the range of 8.64 to 8.85 KJ/WH (8200 to 8400 Btu/KWH).

From an environmental performance standpoint, the Cool Water plant is being designed to satisfy California's stringent emission standards, using a 3.5% sulfur coal, though the program coal is low-sulfur western bituminous. This is to allow flexibility to participants in selecting coals that they may wish to have tested. The cost to the program of providing this flexibility is modest, highlighting a key aspect of IGCC systems — the ability to handle more stringent environmental requirements with relatively little incremental investment.

PROGRAM FUNDING

The Cool Water Coal Gasification Program has been organized as a joint venture of "participants" and "sponsors" who will own an interest in the project equivalent to their capital cost contribution.

Participants contribute a minimum of $25 million and sponsors contribute a minimum of $5 million to the Program.

The present cost estimate for design, construction, and startup of the facility is $275 million. Each of the current participants, Texaco, Southern California Edison, Bechtel, and General Electric, has committed $25 million to the venture. The Electric Power Research Institute, through a separate participation agreement, will contribute $50 million; thus overall funding of $150 million has been contractually committed to date. Two other strong candidates are presently pursuing participation with their commitments expected shortly. Discussions are taking place with a number of potential sponsors, and it is anticipated that additional funding will be identified in the very near future so that the Program can proceed on schedule.

COMMERCIAL AVAILABILITY

As previously indicated, the Cool Water schedule calls for initial energy production in October 1983, with long-term "commercial" operation occurring after a six-month shakedown period. Reasonably successful performance coupled with the commercialization capabilities and interests of the Cool Water participants could lead to IGCC plant orders (Fig. 11) in the 1984-86 time frame with operating dates approximately four years later. Studies by EPRI have indicated that once a commercial-size unit such as Cool Water has demonstrated satisfactory performance levels, a significant portion of utility generation expansion would be satisfied by IGCC plants.

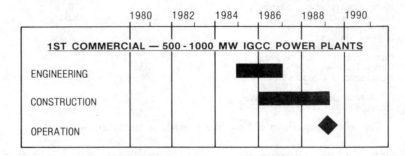

Fig. 11. IGCC commercialization schedule.

In recognition of this potential, a number of site-specific feasibility studies for commercial size IGCC plants have been initiated. The forerunner of these is the Central Maine Power Company study. In this project, funded under DOE's Alternate Fuel Production Program, a 500 MW IGCC plant is being compared with a conventional coal plant to be located at Sears Island, Maine. The 15-month feasibility study will be completed late this year with a decision on the next step to be made in 1982.

SUMMARY

General Electric Company experts see coal gasification as holding the greatest promise of any of the synthetic fuels processes for power generation. The Cool Water IGCC plant scheduled to be completed in 1983 is meant as a prototype for plants 500 MW or larger that could compete in cost with plants that burn coal directly but do not have their pollution problems.

Successful demonstration of this technology will provide the utility industry with a more economical, more environmentally acceptable option for generating electricity from coal in the late 1980's and beyond.

EPRI RESIDENTIAL SOLAR DEMONSTRATION

G. G. Purcell

Electric Power Research Institute, P.O. Box 10412,
Palo Alto, CA 94303, USA

ABSTRACT

In this project, solar heating and cooling systems that have electrical auxiliary
power were designed and tested for the purpose of demonstrating systems that
minimize the total cost of supplying the heating and cooling energy needs of
residences. These systems were designed on the basis of a realistic examination
of costs on both sides of the electrical meter. In this way, configurations could
be identified which are compatible with the needs of both the consumer and utility.
Background of the project and early test results are presented.

KEY WORDS

Solar heating and cooling; Residential Experiments; Preferred solar systems, Solar
Simulation; Solar economics

INTRODUCTION

The United States has devoted considerable effort to the development and demonstration of solar heating and cooling systems for both residential and commercial
applications. For the most part, these efforts have focused on the needs of the
consumer alone. However, in the search for the most cost-effective -- and preferred (1) -- solar heating/cooling systems the cost of providing backup power
must be considered along with customer needs for energy and the desire to conserve
fuel. Clearly, an integrated, minimum-cost approach will have the best prospects
for market viability.

This paper presents both background and selected results of an ongoing solar
heating and cooling research project sponsored by the Electric Power Research
Institute. The purpose of the project is to experimentally verify a previously
developed analytical methodology that can evaluate options for solar heating and
cooling systems. In the form of a commercially available computer program (EMPSS),
this methodology can be used to determine economically preferred systems for any
given utility service area.

THE "PREFERRED SYSTEM"

A preferred system is defined as one that minimized the total cost of meeting the energy needs for a specific application. The cost includes all energy-related costs at the point of use such as installation, energy conservation measures, and operating costs, as well as the costs of supplying auxiliary energy, including investment in generating capacity at the power plant, fuel costs, and the costs of transmission and distribution if electricity is used as the auxiliary energy source.

Designing for preferred systems involves an optimization of cost versus benefit of systems installed at the point of use and of the energy supplied by the utility. The major steps include consideration of (1) energy conservation at the point of use (2) efficient utilization of power plant generating capacity and (3) the reduction in auxiliary energy usage. These factors are related in that attempts to reduce overall energy demand at the site by energy conservation measures, (such as adding insulation to the building envelope), may also reduce peak demand at the site and thus reduce the required capacity of HVAC equipment in a new building. Thus an economic benefit that accrues due to a reduction in demand at the site may also include a benefit via a more efficient utilization of power plant system capacity. On the other hand, installation of a solar system for space conditioning or domestic hot water may reduce energy inputs but not necessarily the demand component of energy supply. Solar augmentation without efficient utilization of energy supply via load-managed storage cannot lead to dependable savings in generating capacity requirements. In fact, utility load factors could be worsened by the requirement that solar houses have reliable but little used back up capacity coincident with utility system demand peaks.

Early in this work it became clear that a computer methodology that incorporated all of the aforementioned characteristics would be required for doing a preferred systems analysis, in cooperation with a number of utilities. This analysis would include two sites, Long Island, New York and Albuquerque, New Mexico where experimental systems would be built and tested. The program, known as EMPSS (EPRI Methodology for Preferred Solar Systems) is available through EPRI's code distribution center. (2)

PREFERRED SYSTEM METHODOLOGY

In developing a methodology for determining preferred solar systems, the key need was to establish the true cost of supplying the backup electricity. Existing rate structures normally lag the incremental cost of energy supply, either when used as the primary source or as backup for the solar system. For the analyses undertaken by EPRI, it was therefore decided to include in the methodology the capability of determining the true cost of supply.

Cost of supply for an electric utility includes:

· Energy-related costs, primarily for fuel and operation of generating equipment.

· Demand-related costs associated with the capital requirements for generation capacity, transmission, distribution and other equipment to meet temporal power demand; and

· Fixed costs including all other costs not directly dependent on demand patterns

Figure 1 illustrates the computation approach graphically and the type of input data that can be specified by electric utilities. The procedure for computing the cost of supply incorporates hourly energy demand at the residence and at the utility. The calculated energy demand at the residence depends on the thermal characteristics of the house, the HVAC system and climatic data obtained from a weather tape for the particular region.

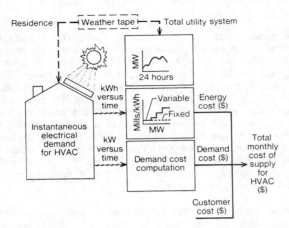

Fig. 1. Computation of Cost of Supply

The weather tape affects both the building load and the utility system wide load. In this way, appropriate economic accounting is made for energy resources and capacity used at times when the utility systems and the building load are simultaneously peaking because of weather conditions.

SYSTEM SIMULATIONS

With the cost-of-supply model and specific utility-related inputs provided by various utility participants, computer simulations were carried out for reference HVAC systems in single-family residences. The reference systems were developed to be representation of:

· Generic System Categories. Conventional; conventional with load management; solar augmentation and solar with load management. Note that "conventional" reference systems also distinguished between heat pump and resistance space-heating auxiliary.

· HVAC Functions. Domestic hot water; domestic hot water and space heating; and domestic hot water, space heating and cooling.

The four generic system categories for single family residences were defined as:

· Conventional System. Draws electric energy on demand to satisfy hot water and space conditioning needs. This system would be commonly used without load management storage or solar supplement and includes heat pump and/or electric resistance heating.

· Load Management System. Through use of storage, electrical demand is shifted to off-peak periods in a manner favorable to utility operation.

· Solar System. System displaces energy resources but electrical energy is drawn when solar energy needs to be supplemented by the backup energy supply.

Solar and Load Management System. System displaces energy resources, and shifts usage of electrical energy to off-peak period by utilizing the thermal energy storage.

Figure 2 indicates in a single matrix how utility characteristics influence the choice of preferred systems. The two characteristics that influence the choice of preferred systems are (1) cost of electrical supply and (2) cost variation with time. The cost of electrical supply is indicative of the energy cost component for a specific utility while the cost variation with time is caused by the time variation in equipment use and cost for generating electricity. If a utility's cost of energy for generating electricity is low and its cost variation with time is also small, it will be difficult to justify any departures from a conventional system. On the other hand, a high cost of supply and large variation in cost with time would justify consideration of systems with both solar and load management features. An intermediate case would be one quite common for a number of utilities, low cost of electrical supply and a large variation in cost with time, thus favoring load management systems.

Although the figure conceptually illustrates the main factors governing system choices, the most cost-effective ("preferred") system can only be identified through a quantitative analysis, for example, using the EMPSS code, with the appropriate input data supplied by the local utility.

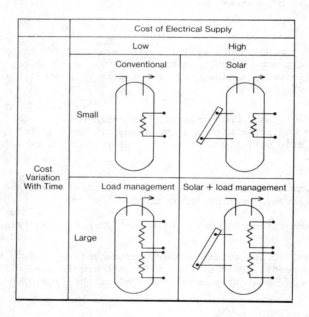

Fig. 2 Preferred System Choices as a Function of Utility Characteristics

SIMULATION RESULTS

Figure 3 illustrates EMPSS simulation results for use of solar hot water systems.
The cost analysis results are expressed in terms of two economic indices, (1)
annualized life-cycle cost and (2) pay-back period. These results were computed
using cost-of-supply data furnished by the Public Service Company of New Mexico
and the Long Island Lighting Company (for 1975). The studies were made as a part
of an effort to define various experimental solar/load management system combinations for installation in five single-family residences that were being constructed
with EPRI support in Long Island, New York, and five in Albuquerque, New Mexico.
(3) From comprehensive instrumentation systems incorporated in the homes, data
was obtained during 1979, 1980 and 1981 for the purpose of monitoring the operation
of the systems and to verify the computer methodology described above.

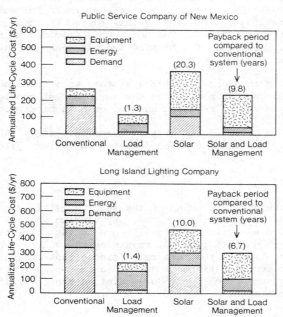

Fig. 3   Life-Cycle Total Costs for Baseline Domestic Hot Wa

Similar simulation analyses were also produced in this project for other types
of systems such as hot water and space heating systems, with resistance heat or
heat pumps; and combined heating and cooling systems. Using this methodology,
we projected costs for future periods considering changes in generation mix,
inflation rate, fuel cost evaluation, etc. and determined how the cost relationships among the options change and how the preferred system may change.

These computer simulations were repeated with input data from twenty-four electric
utilities that provided a diverse mix of geography, climate, generation mix and
total generation capacity. The results were reviewed with utility participants.
It was concluded that the methodology contained the major elements required for a
cost analysis, allowed peak costs to be allocated according to the specific needs
and preferences of the utilities, and was flexible enough to meet the needs of all
the utilities. The computer program is available through EPRI.

VALIDATION OF METHODOLOGY

The objective of the final phase of this work was to provide experimental data for
the purpose of validating the computer program and to obtain operating experience
on solar systems in a real world environment. In this manner, a high degree of
confidence could be gained in support of the type of analyses shown in Figure 3.

In order to maximize the test information and minimize the total number of buildings
instrumented, solar/HVAC systems were designed so that as many as 80 different
system configurations could be tested during the project test period. This
allowed 10 fully instrumented houses (in two locations) to provide test information on two types of solar equipment; liquid and air collectors; for two functions;
space heating and domestic hot water. These 10 houses also provided test data
on thermal storage for load management of space heating and cooling, and domestic
hot water systems, as well as separate or integrated tanks for solar and load
management. Back up heating and cooling systems tested included, electric resistance and several heat pump designs. The heat pumps provided multiple functions
of air-to-air heating, solar-assisted heating and off-peak cooling of thermal
storage.

Test plans for the major candidate preferred systems were drawn up prior to each
heating and cooling season. In most cases the various test configurations could
be switched remotely via telephone lines and the on-site microprocessor controller.

Analysis of test results continues through 1981 and into early 1982. An example
of the test results to date for the solar domestic hot water systems are presented
below.

SOLAR HOT-WATER RESULTS

Figure 4 illustrates the general design of the solar domestic hot water systems installed in three of the ten EPRI solar homes. By closing valve V1, opening valve
V2 and disconnecting the electric heating elements in the solar tank the separate
two tank arrangement could be tested. By reversing the valving and reconnecting
the solar tank heating elements the integrated single-tank arrangement could be
simulated. These test configurations along with a test plan that specified two
lengths of off-peak versus on peak periods, maximized the diversity of experimental
equipment and test conditions.

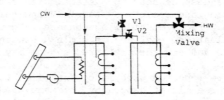

Fig. 4 General Design of the Load-Managed Solar Hot Waters

Experimental system results were compared with predictions of how a conventional, demand water heater would have performed under the same conditions of water delivery rate and temperature. (The EMPSS computer model was used to generate the conventional water heater predictions.) A typical 52 gallon water heater was selected as the conventional system in contrast to a 200 gallon experimental load-managed solar water heating system.

Two types of test results were of particular interest. First, the shifting of peak period loads and the attendant effect on solar system performance, and second the energy consumption and savings.

SHIFTING OF PEAK PERIOD LOADS

Data in Figure 5 below, clearly shows the effectiveness of a two-tank design in shifting peak-period loads. Less obvious was the reduced benefit of collected solar energy caused by the greater measured standby losses (5.46 KWH total) of the two-tank design when compared with those indicated in the analysis of the single-tank conventional water heater (1.3 KWH) having less surface area.

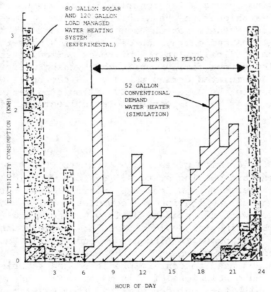

Fig. 5 Electricity Consumption on an Average Weekday-
Long Island House No. 3 - March 1981

When comparing the results of winter and summer test data for this same system it was noted that due to a lower average daily water usage rate and increased solar collection in the summer, the load managed solar water heating system completely displaced the peak period load that's characteristic of a conventional water heater.

The single-tank test results showed that with roughly similar water use rates and greater solar radiation the single-tank design was not as effective in shifting peak period loads as the two-tank design. This advantage stemmed from the added storage capacity of the second tank even though the two-tank design had greater standby losses.

One other important test result (but not unexpected) was noted when a new family moved into one of the test houses. Because of the new family's unusually high consumption of hot water (140 gallons per day), the added capacity of the second 120 gallon load-managed tank was insufficient to avoid substantial peak period use of electricity. This shows that individual customer behavior must be considered if a truly optimum system is to be specified.

ENERGY CONSUMPTION AND SAVINGS

The primary purpose of water storage as used in the context of this project is the shifting of elecrrical energy usage to the off-peak period. However, if load-managed systems are to be economically viable care must be taken that standby heat losses are not allowed to negate the benefit of shifting the peak period load. The net energy savings from load-managed solar water heaters, will have to justify the added cost of the added solar equipment. Clearly, the test data for the solar load-managed systems in EPRI's proejct do indicate that there was enough savings from the solar contribution to more than offset the increased standby losses even with two-tank systems.

In measuring energy saving by the solar contribution it was noted that in general the solar contribution was higher in the single tank case compared to the two-tank systems. The explanation is that, for the conditions tested, there was a better coincidence of thermal loads and energy sources for the single-tank water heater: The hot water produced electrically (in the off-peak period) just matched early morning needs of the Albuquerque family, while the evening needs were met well by the solar-heated water accumulated during the day. Since there was less hot water in reserve during the day in the case of the 80 gallon single-tank system, there was less standby heat loss as well as better solar collection efficiency due to a cooler storage tank.

SUMMARY

A method was developed to identify preferred -- that is, truly lowest-cost -- residential solar heating and cooling systems. Systems were identified that were compatible with utility operations in that they minimized the total cost of meeting the energy needs for a given application. A computer model was developed to test the method in cooperation with two primary utility partners and fourteen additional utility participants. Experimental designs of solar and load management systems were developed for five test homes in Albuquerque and five in Long Island. The homes were built and test plans drawn up for the experimental phase. Test data was taken for two years and used to confirm predicted characteristics. Final publication of results and conclusions of this major EPRI-funded project is expected in early 1982. With the confidence gained from experimental validation, the EPRI-developed computer code should help electric utilities in assessing the impacts of solar heating and cooling systems in their own service territories, advising their customers on cost-effective approaches, and communicating with equipment designers and suppliers.

ACKNOWLEDGEMENT

The author would like to express his gratitude to Dr. John Swanson of the Arthur D. Little Company for his contribution to this paper. Much of the background for this paper is based on early work on this project under the direction of Mr. Dan Nathenson of Arthur D. Little Co. and Dr. John Cummings of EPRI.

REFERENCES

(1) D. Nathenson and J.E. Cummings, "Solar Heating & Cooling: An Electric
     Utility Perspective", ASHRAE Journal, December, 1978.

(2) Richard L. Merriam and D. Nathenson, EPRI Report ER-771, Electric Power
     Research Institute, Palo Alto, CA   (1978).

(3) D. Nathenson, J. Burke, R. Merriam, J. Swanson, "System Definition Study:
     Solar Heating and Cooling Residential Project", EPRI Report
     ER-467-SY, Electric Power Research Institute, Palo Alto, CA   (1977)

MULTICOMPONENT RENEWABLE ENERGY SUPPLY FOR LOW TEMPERATURE DISTRICT HEATING SYSTEMS

H. C. Rasmussen and J. Jensen

*Energy Research Laboratory, Odense University, Campusvej 55,
DK-5230 Odense, M, Denmark*

Summary

Performance data based on dynamic modelling of a village co-
operative energy system leads to a technico-economic evaluation
of a local district heating system. Multicomponent renewable
energy supply is shown to provide both a better energy utilisation and a better overall economy than systems with single
component supply. The project is partially EEC-funded.

1. Introduction

In Danish local communities a marked interest for the use of
renewable energy sources has emerged. A collaboration between
the Energy Research Laboratory and Båring-Asperup "folk high
school" resulted in an enquiry among the 267 households in Båring-
Asperup concerning their present energy consumption for heating,
electicity and transport. Further, attitudes towards energy
conservation were investigated in the enquiry. An 80 % reply
was obtained and the results indicated a very positive attitude
towards energy conservation.

A static model for the supply of the heating demand was established
leading to the following conclusions :

- the individual usage of renewable energy in single households
  could lead to a reduction in the present oil consumption by
  60 %, at the expense of an increase in the use of electricity
  by 56 %, with the use of heat pumps. In Denmark electricity
  is at present produced almost totally by coal-fired power stations;

- that a multicomponent renewable energy supply for a local district
  heating system would lead to a reduction in oil consumption of
  up to 78 %, with unchanged use of electricity.

From an economic viewpoint the latter also seemed more attractive
mostly due to the employment of larger units.

The analysis presented in this paper thus concerns a performance
analysis based on dynamic models of the latter system, where
renewable energy sources wind, solar, straw and biogas, as supply
for a district heating system, are included in combination with
heat pumps and heat storage. Fig. 1 shows a sketch of one of the
modelled systems with the energy flows indicated.

---

x) will present the paper

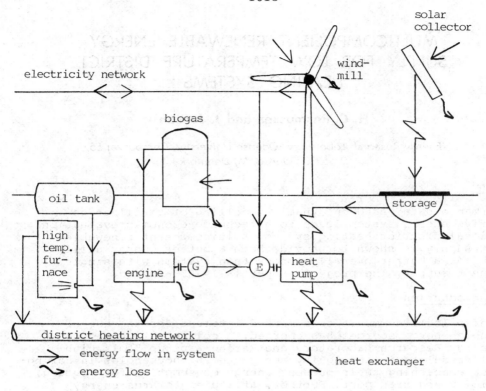

Figure 1. Energy flow in supply system for Båring-Asperup
          (System Al)

Activities within the local community led to the establishment of
Båring-Asperup Energy Cooperative with the purpose of :

planning, establishing and running energy supply with contribution
from renewable energy sources. The energy cooperative should use
renewable energy to the extent, that in the long term is regarded
economical feasible, and should be responsible for spreading
information to interested groups.

The energy cooperative has obtained partial financial support from
the EEC to establish a local low temperature district heating
system based on the system analysis presented here.

2. System component models

Heat demand

The total heat demand from the supply system is estimated from
the results of the enquiry; the demand for house heating is
reduced by 25 % due to better future insulation. The distribution
over the year is based on data available for Denmark with regard
to external climate (1). The hot water supply and the loss in the
distribution system is taken as constant throughout the year.
The total demand over the year is shown in Figure 2.

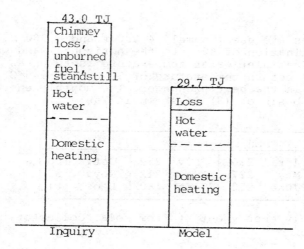

Figure 2.   Model of heat demand from the supply system

Heat pump

The model describes an electric heat pump with a 10°C temperature
difference in the heat exchangers and with an efficiency (ε), of

$$\varepsilon = 0.5 \; \varepsilon_{carnot} = 0.5 \; \frac{t_d + 283°C}{t_d - t_s + 20°C} \qquad [1]$$

where

t_d (°C)   is the district heating water temperature after the heat
         pump

t_s (°C)   is the temperature of the fluid from the heat source before
         the heat pump.

The total efficiency of the heat pump is 84 %, so the energy
delivered by the heat pump (Q), with an electricity expenditure
of U, is

$$Q = \varepsilon \cdot 0.84 \cdot U \qquad [2]$$

Seasonal heat store

The model describes a hemispherical water store with a top
insulation of 200 mm mineral wool. The yearly temperature variation
is to a good approximation (2) sinusoidal with an amplitude of
30°C and a yearly average of 30°C (minimum temperature in January).
During the simulation the monthly losses to the earth and through
the top insulation are calculated. Storage tanks of 50,000 m³ and
75,000 m³ are modelled.

Solar collectors

Two collector sizes are used, namely 4,000 m^2 and 6,000 m^2, facing
south with an inclination of 60° to the horizontal and with two
layers of glass. The inlet water temperature is taken as 35^oC
established by the return temperature of 30^oC with a 5^oC temperature difference in the heat exchanger. The monthly energy output
calculated on the basis of (3) is shown in Table 1.

	MJ/m^2 · month		
Jan. 86	April 220	July 252	Oct. 108
Febr. 112	May 227	Aug. 212	Nov. 50
March 202	June 252	Sept. 180	Dec. 61

Table 1. Monthly energy output from solar collectors

Bio energy (biogas and straw)

The yearly potential of biogas from the agricultural sector in the
areas immediately surrounding the village is 300 · 10^3 m^3/yr with
an energy content of 6.9 PJ/yr (4). The biogas is converted for
use either by direct combustion with an overall efficiency of
68 % (including 6 % provided as process electricity) or in a
combined heat and power unit with a net electricity production
of 24 % and a net heat delivery of 23 %.

The estimated straw surplus in the municipality amounts to
1,400 ton/yr (4). With a heat content of 13 MJ/kg and a combustion
efficiency of 80 % the energy potential is 14.6 TJ/yr. In the model
it is supposed that half of this can be collected.
Both the biogas and the straw potential is taken as evenly
distributed over all months of the year.

Wind

The windmills, with a sweep area of up to 3,000 m^2, are connected
to the public electricity network thus establishing a free
exchange of energy. The average power is fixed to 55 W/m^2 according
to production statistics for a number of smaller windmills in
Denmark (5).

Back-up system

A peak demand oil fired boiler with 90 % efficiency is used.

District heating distribution system

The system is modelled for simulation according to the values in
Table 2 with a fixed return temperature of 30^oC and delivery
temperature and flow calculated according to the demand.

month	Variation of delivery temperature			
	50 - 80°C		50 - 70°C	
	temp. °C	flow m³/h	temp. °C	flow m³/h
January	80	34	70	43
February	77	34	68	42
March	75	33	67	41
April	66	30	61	35
May	53	21	52	22
June	50	18	50	18
July	50	18	50	18
August	50	18	50	18
September	52	20	51	21
October	64	29	59	34
November	72	32	64	39
December	78	33	68	42

Table 2.   Delivery temperatures and flow in district heating system

3.   Energy supply system models

Dynamic models of energy supply systems with and without heat storage are simulated on a monthly basis. The monthly energy production for different combinations of renewable energy sources covering the heat demand is calculated. Only systems that from a technical viewpoint will be practicable in the near future are investigated.

In the systems modelled the energy distribution to the single households consists of a low temperature district heating system. In Denmark the distribution network normally supplies hot water with a temperature of 80-95°C. A reduced delivery temperature implies a better overall system performance caused by higher energy efficiencies of solar collectors and heat pumps and reduced thermal losses, especially from the distribution network.

Model dynamics

The inlet water for the solar collectors is, via a heat exchanger, taken from the return water from the district heating prior to other energy sources thereby ensuring maximum efficiency of solar energy utilisation.

The following points should be taken into consideration when estimating the capacity of the heat pump :

- the heating demand that is not covered by solar energy
- the waterflow in the district heating system.

The latter is important since the temperature difference between the condensor with a temperature of 60°C and the district heating water after passage through the heat pump must at least be 10°C.

System simulation

An overview of the systems modelled is shown in Table 3.

	District heating system delivery temp. range	Straw	Biogas	Heat Pump kW	Wind-swept area m²	Solar collec-tor area m²	Seasonal heat store m³	Back-up supply % of total energy
A1	50-80	–	CHP	58	70	4,000	50,000	65
A2	50-80	–	H	58	1,320	4,000	50,000	58
A3	50-80	–	H	112	2,000	6,000	75,000	44
B1	50-80	–	–	100	240	4,000	50,000	55
B2	50-80	–	–	250	470	4,000	– " –	38
B3	50-80	–	–	500	510	4,000	– " –	36
B4	50-70	–	–	100	240	4,000	– " –	55
B5	50-70	–	–	250	520	4,000	– " –	35
B6	50-70	–	–	500	870	4,000	– " –	25
C1	50-80	–	H	250	3,000	4,000	– " –	23
C2	50-80	H	–	250	2,750	4,000	– " –	12

Table 3.   The supply systems modelled. CHP = Combined heat and power
unit. H = heat unit

Three classes of system are simulated :

A - Solar collectors are connected to the district heating pipeline
via a seasonal heat store and a heat pump. The necessary electricity production for the heat pump and electric equipment
is delivered from a number of windmills and from the combined
heat and power biogas unit if used. A sketch of system A1,
with the energy flow indicated, is shown in Fig. 1. The
monthly coverage of the heating demand as calculated by the
dynamical simulation of the system A2 is shown in Fig. 3.

B - Solar collectors are connected to the district heating pipeline
via a short term heat store and a heat exchanger. The temperature
is further raised by heat pumps with the ground as heat source.
Different delivery temperatures in the distribution network
are studied. Electricity is supplied from windmills. The monthly
coverage of the heating demand of systems B4-B6 is shown in
Fig. 4.

C - Components as in the B class supplemented with biofuels (biogas
or straw). The monthly coverage for system C2 is shown in
Fig. 5.

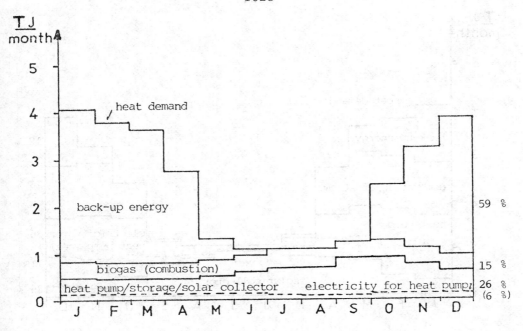

Figure 3. Monthly coverage of the heating demand in the A2 model

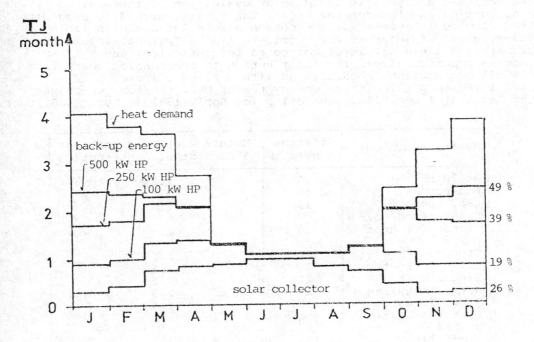

Figure 4. Monthly coverage of the heating demand in the B4, B5 and B6 models

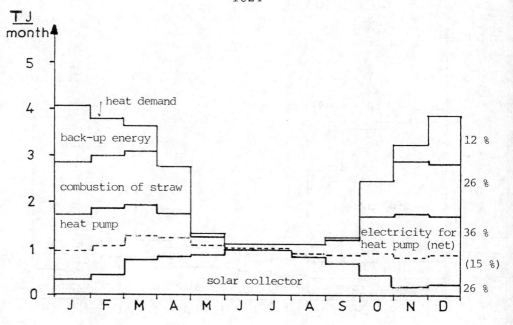

Figure 5.   Monthly coverage of the heating demand in the C2 model

4.   Economics

The economic analysis is based on an evaluation of the cost of
construction and the running costs. The annuity method is used with
the expected lifetimes of the components as indicated in Table 4.
Three levels of interest are considere : 18 % (present level in
Denmark), 8 % present level corrected for inflation) and 5 %
(socio-economic value). The contribution to the running costs from
the oil consumption is calculated with a fixed oilprice of
Dkr. 1,760/ton, including an energy tax of Dkr. 410/ton, which was
the level at 1 Dec. 1980 (the oil price April 1981 is Dkr. 2,800/ton).

System component	Lifetime years	Maintenance cost per year % of cost of construction
Building	20	1.5
Boiler	15	2.5
Heat pump	15	2.5
Underground collector for heat pumps	20	1.5
Storage	20	1.5
Solar collector	20	2.5
Biogas unit	20	2.5
Manure handling	10	2.5
Straw furnace	15	2.5
District heating system	20	2.5
Windmills	20	2.5

Table 4.   Expected lifetime of system components and maintenance
           costs as a percent of cost of construction

The relative cost of construction is shown in Table 5. In 1980-Dkr. the value 100 amounts to Dkr. 27,500,000. The relative running costs are shown in Table 6 where the value 100 amounts to Dkr. 4,000,000 (1980).

Model	Fig.	Building	Boiler	Heat pump	Heat pipes	Heat storage	Solar collector	Biogas unit	Manure handling	District heating system	Windmills	Straw furnace	Total
A1	1	7	4	1	–	12	16	9	2	24	1	–	76
A2	3	7	4	1	–	12	16	7	2	24	6	–	79
A3	–	7	4	2	–	18	25	9	2	24	9	–	100
B1-B3	–	7	4	5	3	–	16	–	–	24	8	–	67
C1	–	7	4	5	3	–	16	8	2	24	13	–	82
C2	5	7	4	5	3	–	16	–	–	24	13	2	74

Table 5. Comparison of the relative cost of construction for the different energy supply systems

Model	Fig.	Staff	Maintenance	Back-up energy	Straw	Interest and depreciation			Total running cost		
						I	II	III	I	II	III
A1	1	6	10	20	–	84	46	37	120	82	73
A2	3	6	10	24	–	88	48	38	128	88	78
A3	–	6	13	19	–	111	62	48	149	100	86
B1-B3	–	4	9	18	–	74	41	33	105	72	64
C1	–	6	12	7	–	91	50	40	116	75	65
C2	5	6	10	4	5	82	45	36	107	70	61

Table 6. Comparison of the relative running cost for the different energy supply systems
Note : I   - Interest 18 %
       II   - Interest 18 %, 10 % inflation
       III - Interest 5 %

5. Discussion

The calculated yearly coverage of the heating demand from the different energy sources from dynamic modelling can be interpreted as
the expected long term average performance of the actual multicomponent system. The results obtained can be used for comparison
of the individual systems but not to describe how the actual
heating demand is covered hour by hour.

Dynamic modelling shows that the amount of oil presently used for
domestic heating can be reduced to 10 % of its present level by
introduction of energy saving measures along with the establishment
of multicomponent district heating systems based on renewable energy.

For systems with a solar collector area of up to 4,000 m^2 and a short
term heat store it will be possible, even in the summer months, to
use all the collected solar heat in the district heating system.
Although a greater solar collector area leads to a reduced oil
dependende it is not economical attractive with the present oil
prices. Reducing the collector area below 4,000 m^2 will make room
for other energy sources i.e. biogas which the analysis indicates
to be the most economically attractive when used in CHP mode rather
than the combination of windmills and a direct combustion of biogas.

Compared with systems including solar collectors, those with heat
pumps in combination with either biogas or straw are attractive
both from an energetic and an economic viewpoint. An increase in
the heat pump size exceeding 250 kW seems only attractive if the
delivery temperature in the district heating system is reduced
below 70°C.

The economic optimal system is shown to be highly dependent on the
rate of interest. It is shown that the high interest rates in
Denmark are unfavourable to the introduction of renewable energy
sources. It would therefore be desirable to aim for a reduction
in interest rates in this area if the present policies of the
EEC and the Danish Department of Energy concerning independence
from oil are to be achieved.

Acknowledgements

This work was supported in part by the Danish National Council of
Technology and the Danish Board of District Heating. The authors
wish to thank N. Land for valuable cooperation on the preliminary
analysis.

References (all in Danish)

1. "Oversigt over graddage i København for sæsonerne fra 1936/37
    til 1978/79", Teknologisk Institut, Tåstrup, 1980

2. H. Lawaetz og P.N. Hansen : "Solvarmesystem med sæsonlagring-
    et projektforslag", R78-39, Thermal Insulation Laboratory,
    Technical University of Denmark, 1978

3. K. Krægpøth : "Explicit udbyttefunktion for solfangere", Med.
    nr. 52, Thermal Insulation Laboratory, Technical University of
    Denmark, 1977

4. J. Jensen, N. Land og H.C. Rasmussen : "Analyse af kombinerede energisystemer for fjerde zone baseret på vedvarende energi. Del I : Landsbyen Båring-Asperup med nærområde", Energy Research Laboratory, Odense University, 1980

5. P. Rasmussen : "Status over energiproduktionen fra mindre vindkraftanlæg", Risø I-26, 1980

ENERGY AND MICROCOMPUTERS:
AN OVERVIEW

R. B. Spencer

ANCO Engineers, Inc., Santa Monica, California, USA

ABSTRACT

The advancements of digital computer technology leading to microcomputers has made many potential applications economically feasible, particularly in the area of energy use management. Knowledge of the status of computer technology will provide potential users with a basis for applying these analysis and control "machines" to their specific needs. Examples of successful computer and micro-computer applications and indications of potential applications may stimulate energy-using businesses to incorporate microcomputers into their activities and thus improve the efficiency of their operations.

This paper presents a brief summary of the status of computer technology and provides both examples and concepts of applications of computers and microcom-puters directed at improved energy management other than a few purely analytical applications, the emphasis is on the use of the newly emerging microcomputers. The future trend of advancing computer development and reduced costs will expand the use of microcomputers; this will result in an accelerated improvement in energy use efficiencies and potentially reduced energy costs.

INTRODUCTION

Computer technology has experienced rapid advances in the past ten years. As a result of recent advances leading to the full commercial development of micro-computers, applications to many energy-using situations are now considerably more attractive (Spencer, 1979a). Due to significantly improved reliability, increased capacity and lower costs, digital computer technology, in the form of microcom-puters, today is within the means of almost any energy-using business regardless of business type or size. This paper summarizes the status of computer techno-logy, emphasizing microcomputers, and highlights some of the many existing and potential applications of computers and microcomputers for improving the effic-iency of energy-using activities.

STATUS OF COMPUTER TECHNOLOGY

There are three major categories of digital computers: mainframes, minicomputers, and microcomputers. At different times, each type has been applied to improving the efficiency of energy-using activities. The earliest applications of main-frame computers in the process industries appeared in the United States during the late 1950's (U.S. Department of Energy, 1979). Minicomputers developed in the mid-1960's with a computational power comparable to that of the early mainframe computers but with costs ranging from $10,000 to $50,000. The first microcomputers became available in the early 1970's with a computational performance comparable to that of a small minicomputer. The early 1980's has seen the advent of a "computer on a chip." Recently, very large-scale integration (VLSI) has produced a single-chip 32-bit microcomputer containing the equivalent of up to 450,000 transistors (Hewlett-Packard, 1981). Ultra-large-scale integration (ULSI) promises to continue the trend: historically, component density per single chip has doubled each year. Figures 1, 2 and 3 summarize the trends in: 1) computer chip development, 2) price of computers (all three categories), and 3) specific costs of computer memory (Buchanan and Sheldon, 1980).

Associated with the smaller size and cost of recently developed computers is the added benefit of improved reliability. Today's microcomputers are extremely reliable and perform efficiently under a wide range of vibration, temperature, humidity, and pressure conditions, particularly when compared to the closely controlled environments required for computers just ten years ago.

Computers have been advancing at a rapid pace, resulting in increased performance, higher reliability, smaller size, greater availability, and lower costs. Currently, computer software has not kept pace with the rapid hardware developments. However, as computer use becomes more widespread in the 1980's, it is inevitable that the present lag of computer programs, application codes, and simpler, more accessible languages for unsophisticated users will expand accordingly.

The appearance of mini- and microcomputers has led to revised thinking in the use of computers for energy management. In particular, these smaller and less-expensive computers have fostered the concept of distributed control as opposed to central control of plants and processes. It is now economically and technically feasible to dedicate a mini- or microcomputer to an individual process for local control and, if desired, to control a distributed network of local computers from a central computer under the direction of a single operator. The distributed control schemes and associated hardware are much simpler and usually less expensive to develop and implement. The distribution of control functions among a number of computers results in the added benefit of an increase in overall system reliability. Figure 1 shows an example of a distributed computer network used in the control of manufacturing, fabrication, and testing of computer chips.

Applied computer technology has a proven potential for reducing energy use and associated energy costs. The opportunities for computers applied to energy management are the result of four contributing factors (Spencer, 1979b): 1) direct energy savings -- minimizing energy use in unit operations through improved management; 2) raw material yields -- energy savings derived from the fuel value of previously wasted raw materials or energy required to produce the raw materials; 3) improvements in production capacity -- energy savings derived from better quality control and an increase in production, and 4) indirect savings -- energy savings derived from improved design and process analysis.

APPLICATIONS

Computer application can be divided into two major categories: analysis and control. Both can provide direct or indirect improvement in the efficient use of energy. The following applications grouped by major category illustrate the broad range of possibilities for effectively utilizing computers (particularly microcomputers) for energy management.

ANALYSIS APPLICATIONS

Engineering Analysis

One obvious application is the direct use of computers for engineering analysis of energy-using systems and components (Smith, 1981). Digital computers provide rapid calculations with greatly improved accuracy over hand-analysis techniques. In addition, iterative solutions to converge on final results are handled in a straightforward manner by use of computers. Engineers can include a greater number of variables and second-order effects into their analysis that would not be practical otherwise. Numerous computer programs are available for analyzing a wide range of energy systems, such as boilers, chillers, turbine-generators, and solar systems. With the ease of using higher level computer languages, custom codes for simple calculations can be written relatively easily by engineers with only limited training in computer programming.

Simulation Analysis

Computers, particularly mainframes and minicomputers, are well-suited for simulating large, complex, energy-using systems such as buildings, power plants, and industrial processes. A thorough engineering and design analysis of a complex energy-using system can be performed on a digital computer to establish critical design parameters, select the best alternatives, and optimize the system for minimum energy use. As an example, Table 1 shows several commercially available simulation codes that analyze buildings for a wide range of environmental conditions and types of heating, ventilating, and air-conditioning equipment. As another example, a computer simulating fuel consumption for the various stages of a manufacturing plant can estimate total energy use. This model could then be exercised for various mixes of fuels and energy forms (depending on price and availability) to select the lowest cost way to operate the plant for a given output.

Computer-Aided Design and Manufacturing (CAD/CAM)

The use of computers for aiding design and manufacturing of products will greatly increase in the future resulting in reduced energy use. Systems are, or soon will be, available which will: 1) model a part using interactive graphic input; 2) perform a detailed stress analysis on the part; 3) optimize the design to minimize weight or minimize machining time; 4) produce a tape suitable for use on a numerical controlled tool machine, and 5) automatically machine the part to the proper specifications. A subset of these computer-aided design activities is in the design of facilities, where drawings are contained in computer storage media and modified drawings can be quickly redrawn using large bed plotters. The advantage of this is that a single design change may require changing a number of drawings and all these changes can be made automatically by the computer using the proper drawing indices. Speedup of drawing, illustration, and plotted output will be enhanced by the use of high-speed laser printing and plotting. Copy machine companies are already connecting duplicating machines directly to computers to print "a page at a time" at rates of current high speed copy machines. These activities save energy and human resources by producing products more efficiently.

CONTROL APPLICATIONS

Process Monitoring

Process monitoring is usually the first step in developing a plant-wide computer

network and often the most expensive. It is also the most important. Once the
system is installed and is operational, a wide variety of additional applications
are quickly identified.

Process monitoring is possible with a mini- or a microcomputer. The system can
potentially have the following capabilities: 1) data acquisition, filtering, and
storange in retrievable, usable form; 2) high- and low-limit variable alarm; 3)
advanced warning of possible problems; and, 4) informative multicolor graphic
displays of real-time process data, as well as historical trends. Energy maintaining of a building's environmental systems (HVAC) is also a common application,
particularly in commercial buildings.

On-line Process Analysis

Once the mini- or microcomputer system has been installed, data collected through
monitoring can be used in several ways. For example, a series of engineering and
real-time calculations can be made to determine process performance, efficiency,
and energy consumption by using energy and material balance programs. Process
analysis is possible on-line, in the same mini- microcomputer used for monitoring,
or off-line in another computer. This affords a more rigorous approach to process
analysis without interfering with the monitoring process.

Process Control

Two strategies generally associated with computer control are supervisory and
direct digital control. Of the two, supervisory control was the first to become
accepted by the process industry and is now used in HVAC system control.

Supervisory control, also known as set-point control, calculates and transmits
certain process set points or targets to dedicated controllers or microprocessors
by monitoring various process and/or desired conditions, comparing the current
values of controlled variables with their respective targets, calculating the
error, and sending a corrective set point to a specific controller or dedicated
microprocessor. The supervisory computer then resets the particular local
controller at the new set point, and the controller acts on it as if it had been
manually set by the operator. In the past, industrial automatic control was
predominantly analog; therefore supervisory control by computer became the
popular choice over direct digital control.

In direct digital control, the microcomputer performs the control function, as well
as monitoring pertinent process conditions, calculating new set points, and
correcting errors. This means that, if the microcomputer has to be taken off-line
for reprogramming, maintenance, or other reasons, the process has to be operated
under manual control. With the cost of microcomputers declining, a viable
solution is to have a backup computer to take over when the primary one is temporarily out of service.

Some typical applications of microcomputer systems successfully implemented for
monitoring and control by energy-using businesses are: 1) cycling and control
for HVAC systems; 2) electrical load control in small factories; 3) peak demand
limiting and control for vacilities; 4) boiler efficiency optimization, and 5)
irrigation monitoring and control.

Hierarchical Computer Systems

One application of computers to energy management which has received considerable
use recently is a multicomputer total energy management and control system (EMCS)
for large multibuilding facilities (e.g. school campuses, large hospital complexes,

and city administration building clusters. Figure 5 illustrates the general concept of a multicomputer hierarchical EMCS. The equipment used for EMCS ranges in complexity from a small microprocessor (for a single-building local control system) to a minicomputer which can handle multiple control points in larger buildings, to a *tree* -distributed processing system employing a central mini-computer coupled to a number of remotely located microprocessors. This is a rapidly expanding field, with systems commercially available in a cost range of a few thousand dollars to a few hundred thousand dollars.

CONCLUSIONS

In the future, computerized process and energy monitoring and control systems will become more distributed and less centralized. Digital technology will continue to overtake analog technology in dedicated control function applications. In addition, control software will continue to become more specialized and easier to use.

The ever increasing cost of energy, the rising cost of raw materials, tighter environmental regulations, and a shortage of competent technical personnel are causing a re-evaluation of some of the existing methods to control energy use. The newest versions of computers represented by the microcomputer revolution offer a realistic means of achieving significant reductions in energy use without sacrificing industrial production and human comfort.

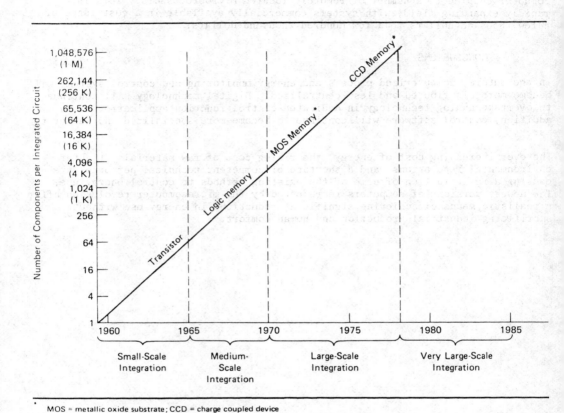

Fig. 1 Complexity of Electronic Circuitry

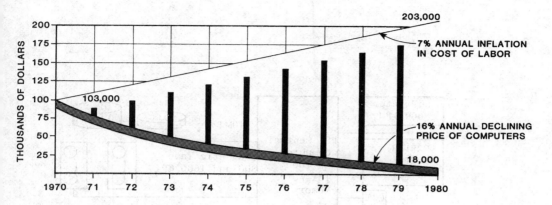

Fig. 2   The Rising Cost of Labor vs.
The Declining Cost of Computer Hardware

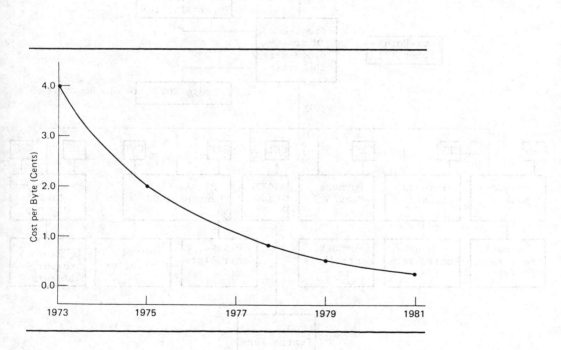

Fig. 3   Trend in Cost of Computer Memory

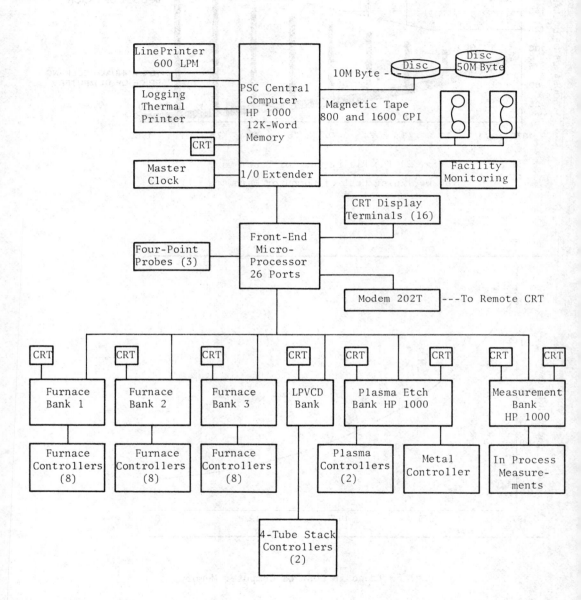

Fig. 4 Distributed Computer System Example

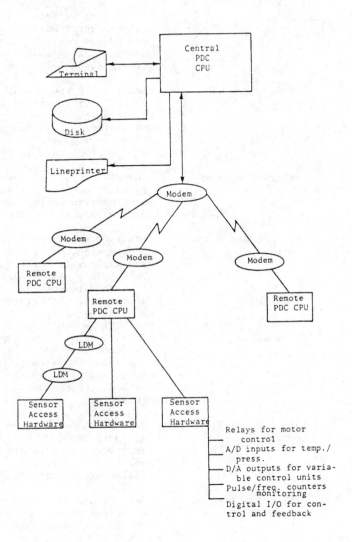

Fig. 5   EMCS Conceptual Sketch

TABLE 1: TYPICAL COMPUTER PROGRAMS FOR ENERGY SYSTEM SIMULATION

Code Name	Originator	Applications
AXCESS	Edison Electric Institute	Comparison of a building's energy requirements using alternate mechanical/ electrical systems. Allows designer to compare total building energy usage and demand for each subsystem using any combination of available energy sources (i.e., electricity, gas, oil, coal).
ECUBE 75	The American Gas Association	Analysis of new buildings and various retrofit or replacement projects for existing buildings by predicting impact of thermostat setback and setup, ventilation air reduction schedules, and equipment shutoff in economic terms.
TRACE and Version 200	The Trane Company	Energy analysis of new and retrofit building designs for alternative HVAC equipment to achieve energy use reduction and life cycle cost optimization.
ESOP	NASA-Johnson Space Center	Analysis of energy requirements, water consumption, solid waste consumption, and waste water effluent from Integrated Utility Systems by calculating facility load requirements and evaluating yearly operational characteristics.
MACE	McDonnel Douglas Corporation	Determines monthly and annual energy requirements for a proposed building. Performs cost analysis of various heating and cooling systems, building construction, and prime energy sources using methods recommended by ASHRAE whenever possible.
NECAP	NASA-Langley Research Center	Determines building energy consumption and performs cost analysis. The set of six programs calculates a variety of energy use aspects for buildings following standard ASHRAE procedures.
NBSLD	National Bureau of Standards	Heat load analysis of new or existing buildings which can simulate a variety of indoor temperature modes: fixed temperature, night setback, night setback with equipment limitations or floating temperature.
DOE-2 (Formerly CAL-ERDA)	U.S. Department of Energy and University of California, Berkeley	A public domain program with streamlined input procedures and shorter running time than NBSLD. It is a calibration for other programs to be used for building heat load analysis in California.

REFERENCES

Buchanan, J.R. and Sheldon, G.L. (1980). Using Small Computers as Management
    Tools. American Management Associations Extension Institute.
Hewlett-Packard Journal (1981). Vol.32, No.6.
Smith, C.B. (1981). Energy Management Principles. Pergamon Press, New York.
Spencer, R.B. (1979a). The Present and Future Role of Computers in Energy Use
    Management, An Overview. Proceedings of the Second International Conference
    on Energy Use Management, Los Angeles, California.
Spencer, R.B. (1979b). Transportation, Communication, and Computers. In
    Efficient Electricity Use, C.B. Smith (Ed.), 2nd Ed. Pergamon Press, New York.
U.S. Department of Energy (1979). Computer Technology: In Potential for Indus-
    trial Energy Conservation. DOE/CS/2123-T2, the Technical Information Center,
    U.S. Department of Energy.

THE GERMAN DEVELOPMENT PROGRAM IN COAL GASIFICATION AND LIQUEFACTION

H.-J. Stöcker, R. Holighaus and J. Batsch

Kernforschungsanlage Jülich GmbH, Projektleitung Energieforschung, D-5170 Jülich, Federal Republic of Germany

ABSTRACT

On overview of the German development program in coal gasification and liquefaction is given. The technologies are briefly described. The major projects, their objectives and their status are presented. Economic aspects are discussed.

KEYWORDS

Coal gasification; coal liquefaction; coal conversion; synthetic fuels; coal hydrogenation; Fischer-Tropsch synthesis; Mobil process

INTRODUCTION

World energy consumption habits do not mirror world energy reserves: although coal is the world's most significant fossil fuel, accounting for approximately 70 per cent of total economic energy reserves (1), it covered only 33 per cent of demand in 1977. This disparity between energy reserves and utilization is due to the relatively low cost of extracting and utilizing oil and natural gas as opposed to coal. In the long term, however, the world-wide balance between reserves and consumption will have to be restored, unless other energy sources such as nuclear energy, solar energy, and fusion energy are able to cover a considerable proportion of demand. Coal is particularly suitable for substitution for oil and natural gas. It can be transformed into other energy sources in a variety of ways: gasification and liquefaction processes can be used to convert coal into gas substitutes, fuels, hydrogen, and electricity. Coal is, therefore, in a position to satisfy all major types of energy demand provided new technologies are developed.

COAL GASIFICATION

In coal gasification oxygen and steam react with carbon. At the same time, the hydrocarbons present in the coal are released, decomposed by heat, and, depending on the reaction conditions, also converted with oxygen and steam. The reaction between the steam and the oxygen, which is essential for the process, consumes a great deal of heat.

There is a great variety of possible applications for gasification products: town gas, synthesis gas, low BTU-gas and synthetic natural gas (SNG).

In low BTU-gas production the oxygen required for heat generation is introduced in the form of air. The gas produced thus contains nitrogen from the atmosphere, and its heating value is therefore only about one-sixth that of natural gas, which is why it is called 'low BTU'. It can be used for heating purposes and in power stations. For the other three types of gas, pure oxygen has to be used in the generation process.

In principle the types of gas mentioned here can be produced by all the known gasification methods if appropriate process stages are included before or after gasification. However, for the production of gases with a high heating value, compared with natural gas, it is advantageous to use a process in which hydrocarbons - especially methane - are formed in the gasifier. On the other hand, for synthesis purposes it is advantageous if the raw gas leaving the gasifier is free of hydrocarbons.

Three basic processes are used for coal gasification (figure 1):

(a) Fixed-bed gasification. There is an exothermic reaction between the oxygen in the gasification medium and the residual carbon, leading to a sharp rise in temperature. The gas gives off much of this heat as it passes through the close packed coal, and this supplies the heat required for the gasification reaction, and also the heat for drying and heating the coal in the reactor. Since the gas leaves the reactor at a relatively low temperature, and the coal is completely converted in the high-temperature zone at the highest overall yields. The Lurgi process is based on this principle.

(b) Fluidized-bed gasification. The intensive mixing in the fluidized bed means that the temperature and solids distribution is virtually the same throughout the bed, and so inevitably when the ash is removed some of the coal goes with it. Over and above this systematic loss of yield, some of the coal is carried out directly with the gas. This is the fine particle fraction which has not been completely converted because the residence time in the gas stream was too short. This system is used in the Winkler gasifier.

(c) Entrained gasification. The advantage of this method is that the ash is removed to suitable parts of the gasifier either as molten slag or in solid form after cooling. Because of the extremely short residence time in the reactor gasification has to be carried out at high temperatures (1500° C) and the coal has to be finely ground in order to avoid unacceptably high coal losses. A considerable proportion of the coal, converted with expensively-produced oxygen, is used to heat up the gas and coal and although most of this expensive energy can be recovered and used in the process - as electrical energy, for example - it does reduce the overall yield of the gasification process. This method is used in the Texaco, Saarberg-Otto, and Shell-Koppers gasifiers.

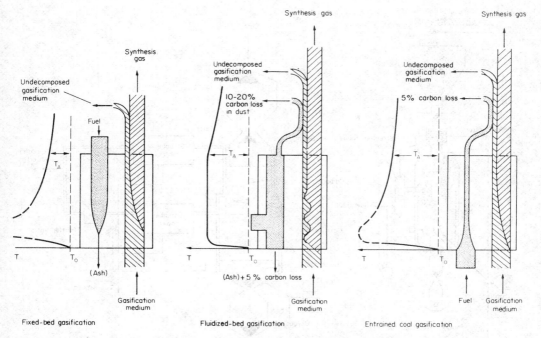

Fig. 1. Comparison of the three fundamental processes for coal gasification

The German development program in the area of coal gasification is concentrated on the further development and improvement of existing processes of all three mentioned principles. Of all the projects undertaken, the further development of Lurgi pressure gasification (figure 2) is most closely dependent on commercially available technology. As the Lurgi process is dependent on increased pressure (30 bar) in any case, the problems of high-pressure operation still facing the other processes have already been solved. By further increasing the process pressure to 100 bar, the cost, performance and efficiency of the process can be improved. In particular, the amount of methane in the product gas can be considerably increased. The experimental plant for testing this improved process (Ruhr 100) is designed for a maximum coal throughput of 7 tonnes/hour. It has been in operation since mid 1979. Up to now, a pressure of 85 bar has been reached during operation. At these conditions the amount of methane in the raw gas is about 13 - 14 per cent.

In the past, Winkler gasifiers - which can use most kinds of coal except highly caking coals, but normally operate with lignite - have operated only at atmospheric pressure. but it is thought that operation under pressure would bring considerable improvements. An experimental plant has been built for the development of this technique, processing 1 tonne/hour of lignite. This plant has been in operation since mid 1978. In the framework of the Coal Refining Program the realization of a plant of commercial size is being discussed (25 tonnes/hour).

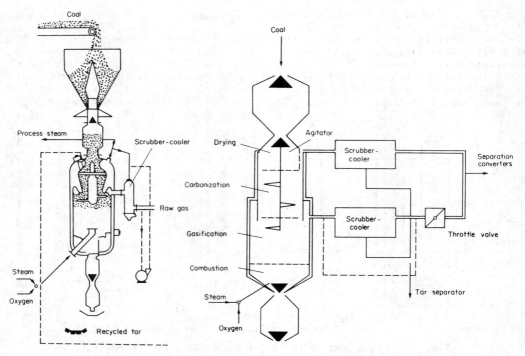

Fig. 2. Lurgi pressure gasification , conventional and Ruhr 100

Particular priority has been given to the development of entrained coal gasification because the success of gasification is to a large extent independent of the proper- ties of the coal processed. As a result, three different processes working on this basic principle have been used in experimental plant (by Texaco, Shell-Koppers, and Saarberg-Otto). The pilot plant for the Saarberg-Otto process has a capacity of 11 tonnes/hour of coal. In the operation of this plant, considerable problems have been encountered, in particular with regard to uniform coal feeding and gas cleaning. The time required to solve these problems has been considerably longer than expected.

The Texaco gasification process relies heavily on the technology of oil gasification. The finely-ground coal is suspended in water and injected into the gasifier as though it were a liquid. The advantages of this simple system were demonstrated by the im- pressive results obtained in April 1978 with a pilot plant processing 6 tonnes/hour of coal. Up to now, this plant has been in operation for some 8,000 hours.

In the Federal Republic of Germany, completely new gasification processes are being developed to use process heat from nuclear reactors. The advantage of such processes is that all the coal processed can be converted into gas, as the necessary heat is obtained from an external source. Two lines of development are being pursued: in the first, the heat is introduced by heat exchangers which are situated in the reactor; in the second, the energy is obtained by splitting methane and using the hydrogen thus formed as the gasification medium. Both processes have been developed far enough to be tested in semi-technical plants with carbon processing rates of approximately 0.2 tonne/hour. The gasification processes will then have to be tested on a pilot scale before a prototype linking a high-temperature reactor with a gasification plant can be built.

COAL LIQUEFACTION TECHNIQUES

Coal can be liquefied in two different ways. In Fischer-Tropsch synthesis the coal is first broken down into basic components (CO and H_2) by gasification, and then re-synthesised into the desired range of products. Coal hydrogenation, on the other hand, is based on the principle that coal can be liquefied by splitting the coal molecule into the desired size and at the same time adding hydrogen. This process is not aimed at complete conversion of the moleculare structure, and has thermo-dynamic advantages.

COAL HYDROGENATION

Coal hydrogenation in its new   version of the IG Farben process is illustrated in figure 3. The finely-ground coal and added catalyst are mixed with recycled oils and fed through several heating units by a paste pump before entering the reactors. Hydrogen is mixed with the paste before it reaches the heating units. In the original process, the reaction took place at the extremely high pressure of 700 bar and at a temperature of 450° C.

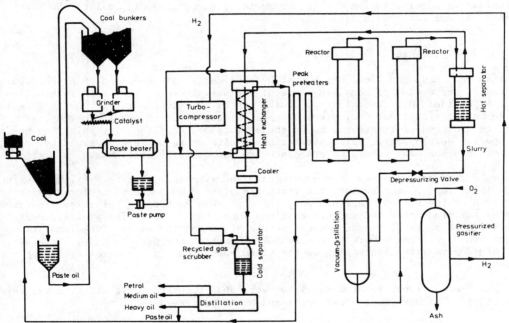

Fig. 3. IG-Neu coal hydrogenation process

Hydrogenation can now be carried out at only 300 bar, thanks to alterations in the process, and the conventional IG hydrogenation process has been modified to take account of these and other advantages. The basic technology for thes 'New IG' (IG-neu) process was developed in continuous laboratory plants. Before it can be applied on an industrial scale, the process will be further developed in two pilot stages. A plant with a capacity of 6 tonnes/day in Völklingen was completed at the end of 1980 and is now in the commissioning phase. A second plant with a capacity of 200 tonnes/day is being constructed in Bottrop.

The main improvements incorporated in the New IG process, as compared with the older one, are as follows:

1. Higher coal processing rate;
2. Lower hydrogenation pressure (300 instead of 700 bar);
3. The slurry formed is thickened by distillation to form a residue which can be handled by pumps, whereas in the earlier process, the residue from the centrifuges was transformed into coke in a low-temperature carbonizing furnace which would not be environmentally acceptable today;
4. The ash is produced in a form which can be dumped without further treatment, whereas it was difficult to dispose of the low-temperature coke from the earlier process without some damage to the environment;
5. The slurry is utilized directly to produce hydrogen, which is obtained at a pressure of between 80 and 100 bar, thus saving expenditure of energy for compression.

In the earlier process, the hydrogen was produced by a pressureless process from expensive coke. All the sulphur compounds in the slurry are converted into hydrogen sulphide, which is easy to handle, and gasification of the slurry therefore provides an ideal solution for environmentally acceptable disposal of the inevitable residues of coal hydrogenation.

INDIRECT LIQUEFACTION

Unlike coal hydrogenation, which must be considered and developed as a whole because of the system of recycling some of the oil produced, indirect liquefaction can be divided into several stages which can be treated separately, as it entails a series of processes: Coal preparation, Coal gasification, gas scrubbing, synthesis and product treatment. Among these processes only the synthesis step needs a special development. Coal gasification is already being intensively developed in other areas. The remaining processes are commercially available.

To make the whole process more economical, the synthesis phase offers interesting possibilities of improvement. The synthesis plant does not account for a major portion or the overall capital cost, for more than 70 per cent of costs are taken up by gas production and gas preparation alone. But the efficiency of synthesis does affect profits, as they depend on the yield and the value of the end product. The development work on the synthesis part of the process must, therefore, be aimed primarily at stepping up the yield and producing high-quality hydrocarbons.

In the Federal Republic of Germany two versions of indirect liquefaction are being developed: Fischer-Tropsch synthesis and Mobil-process.

The work on Fischer-Tropsch synthesis is aimed at producing high-quality chemical feedstocks with high yields and selectivity. A wide-ranging research programme has been set up to develop new catalyst systems and identify the optimum conditions for reactions. Considerable progress has been made since research was resumed, but even more progress will be necessary before the process can be applied commercially. Fischer-Tropsch synthesis alone will not produce the whole range of raw materials now derived from mineral oil. It can synthesise olefins (straight or branched chains) but aromatic (cyclic) compounds are also needed for the chemical industry; these can be produced by coal hydrogenation or by the Mobil process.

The Mobil process produces methanol from synthesis gas in an intermediate step. Using zeolith catalysts this methanol is converted to low boiling aromatic hydro-carbons at almost stoichiometric yields. Beside several projects which deal with the further development of the catalysts, a project which envisages the planning, construction and operation of a demonstration plant with a capacity of 100 bbl/day is being carried out. This project is jointly funded by the U.S. and German Federal Governments.

ECONOMIC ASPECTS

Extensive studies on the economics of coal refining processes have been performed. Fig. 4 summarizes results of these studies. The numbers show that the production of SNG and liquid products from coal are not economic at the present time expect on the basis of very low coal prices, which is in fact an unrealisitc assumption. The hard coal prices in Germany are about 6 DM/GJ for imported coal and 9 DM/GJ for domestic coal.

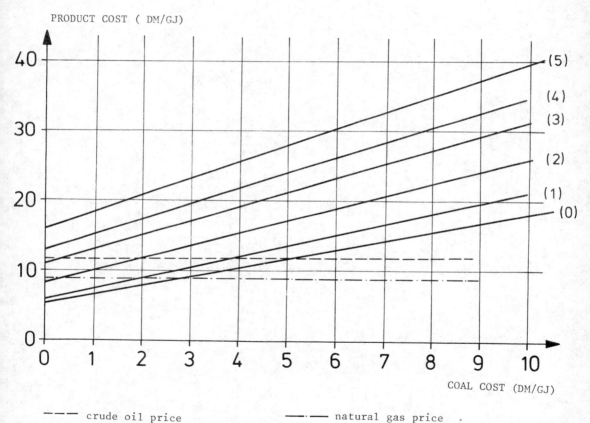

Fig. 4. Product costs vs. coal costs

The production of synthesis gas from German brown coal, however, is expected to become competitive in the near future. This seems to be possible because of the combined effect of the low price of the raw material (3 - 4 DM/GJ) and relatively low investment costs (as compared with complete SNG plants).

Under the assumption that the average increase of the coal price will be significantly lower than the increase of price for crude oil and natural gas, some authors expect that the production of SNG may become competitive within 10 to 20 years. The production of liquid fuels from coal is expected to be competitive somewhat earlier. It is essential to have a mature technology available at that time. For this reason, the German Federal Government supports the development of coal refining technologies.

Most of the projects mentioned in this article are funded in part by the Bundesminister für Forschung und Technologie (BMFT).

A MICROCOMPUTER BASED SYSTEM FOR POWER DEMAND MONITORING AND CONTROL

A. Traca de Almeida and A. Gomes Martins

*Departamento de Engenharia Electrotécnica, Universidade de Coimbra,
Portugal*

ABSTRACT

Digital computers are increasingly being used for power systems control. One of
the functions that can be carried out by a digital processor is to monitor electrical loads and to perform load management, making a positive contribution for a
higher efficiency of operation and an increased performance of existing power system
networks.

This paper presents a microprocessor-based system for power demand monitoring and
control. The microcomputer chip is an all-in-one processor having built-in program
and data memory, timer counter, clock, input/output ports for interfacing, and a
central processing unit, enabling a simple and low-cost configuration to be
achieved. The control system is located at the transformer station, where the data
necessary for demand evaluation are obtained through a group of meters simultaneously monitored by the computer. The loads are grouped on a programmable priority
basis, having a separate meter for each group of loads. For each demand period,
the power demand is evaluated and compared with the user-programmable target demand
level. The necessary load-shedding actions begin whenever this target level tends
to be overpassed. Another possible operation mode of this device is to cause loads
to be shed under emergency conditions. If the device detects a voltage drop above
a certain level through magnitude comparison with the normal value or if the frequency drops, nonessential loads are shed to facilitate network recovery from the
disturbance. The frequency is measured counting the pulses of the processor interval timer between the positive-zero crossing points of the voltage waveform. The
device can also receive an emergency ripple control as an interrupt to perform load
shedding.

During the day, the power consumption for each group of loads is sent to separate
zones of an additional data memory, allowing a single recorder to register the
different load patterns at the end of the day. These records can be used for consumption analysis and to check potential saving opportunities.

INTRODUCTION

There are strong incentives to level the load curves. Due to the daily and seasonal
variations, to meet the peak load demand and to have a reserve margin capable of
satisfying the reliability requirements, it is necessary to install extra generating
and distributing capacity. These additional investments are going to make electricity more expensive and thus create more inflationary pressure. In distribution

networks, especially in radial feeders, the peak loads produce unwanted reductions
on the supply voltage, causing undesirable performance of equipment connected to
the grid. For the same energy which is delivered to the load, the line losses tend
to increase when the load factor decreases. During the peak periods, it is necessary to use the less-efficient generating stations or the ones which use more
expensive fuels. Besides capital savings, energy conservation is thus possible
with proper load management. In Portugal, these problems are particularly serious
since 20% of the peak load has a duration under 400 hours. Also, the ratio between the annual peak and the average power is 1.75, whereas the average value for
other European countries is 1.45.

In power systems, when there is an excessive overload on which the spinning reserve response and load frequency control are slow and inadequate, load-shedding
may be necessary. The normal approach uses underfrequency relays to cut the
supply to the loads progressively. This procedure can be disruptive, as both
essential and nonessential loads are disconnected.

The authors propose that a microcomputer-based load management system could monitor
power demand of several groups of loads that are ordered on a priority basis according to their importance to the consumer. The overall demand is compared to a target
level preset by the consumer, and if it tends to be exceeded loads are shed, starting with the less important ones. In this way, the load profile is leveled with
the corresponding benefits to both consumer and electric utility. The voltage and
frequency are also monitored to detect abnormal drops which indicate an overload in
the system. If this is the case, nonessential loads are shed to ease the recovery
of the power system while supply to essential loads is maintained. The supply is
restored to the shed loads when the emergency period is over; that is, when the
frequency and voltage return to normal values.

LOAD MANAGEMENT

Electricity is a premium type of energy, and its generation and distribution are
expensive. Due to the increasing problems relating to the current availability of
primary energy sources, price fluctuations, environmental constraints with their
associated costs, the huge investments required, it becomes increasingly more
important to achieve efficient electricity use (Smith, 1975). All electricity use
must be modified through consumer education and automatic regulating equipment,
such as thermostats and light-activated switches. For a given application, electricity must be compared with suitable alternatives to make sure that it is the
wisest choice.

In addition to considering methods that save electrical energy, it is important to
reduce peak power demand (Isaksen, 1980). Peak demand reduction is not only beneficial to the consumer as his electricity bill is decreased, but also benefits the
utility in its efforts to relieve some of the stress associated with capacity
expansion. Furthermore, a utility that achieves such a reduction has a higher overall efficiency. The principle of load management is illustrated in Fig. 1.
Cutting wasteful use, especially in peak periods, and transferring part of the
loads can substantially reduce power peaks. Although careful planning of heavier
load operation can improve the load factor, it is important to monitor on-line
power consumption of the different loads and take corrective steps if necessary to
avoid overshooting peak demand for a preset target level. The consumer must understand that any economic benefit may be offset by the transfer of loads, possible
cost increases, or additional losses of production associated with the rescheduling
of some operations. In some cases, these negative effects can be meaningless. The
type of loads which are best suited for load control normally operate in response
to long-time constants, do not operate continuously, or perform as nonessential

auxiliary equipment; thus, their transfer is not disruptive. Typical plants and equipment which can be shed include space heaters, soil heaters, immersion heaters, induction heaters, arc furnaces, infra-red dryers, air-conditioning plants, cold room refrigerating plants, fans, compressors, pumping equipment, battery chargers, plating baths, electrolytic separators, nonessential lighting, and rolling mills. Where some heating loads (such as drying and baking activities) make a significant contribution to the peak demand and cannot be easily rescheduled, it may prove more economical to use a different energy source.

SYSTEM CONFIGURATION

Figure 2 represents the load monitoring and management system configuration. The backbone of the system is the 8748 (Intel Corporation, 1977) single-chip micro-computer which features an eight-bit CPU with 2.5 µs cycle time, 1 K bytes of RAM, three eight-bit input/output ports, a clock, and a timer counter. The 8748 has an efficient instruction set and has an eight-level subroutine nesting capability. Due to its features, this microcomputer allows a simple and inexpensive configura-tion to be achieved. Another advantage derives from the fact that the 8748 is available in the Complementary Metal Oxyde Semiconductor (CMOS) technology which presents very low power requirements, good noise immunity, wide supply tolerance, and extended temperature range. The low-power circuitry used throughout the system needs only modest supply and battery backup requirements.

The microcomputer is interfaced through an input/output expander with the load meters and breakers. The loads are divided into eight groups, each group having a separate meter. For each unit of energy consumed, a digital counter associated with the meter is incremented by a pulse produced either through electromechanical means in the integrating mechanism or through an optoelectronics device which moni-tors the rotation of the disc. The digital counters are scanned by the microcompu-ter to process the overall level of demand. Each group of loads is assigned a priority level which can be set by the order of connecting to the back panel plugs. Essential loads have the highest priority and nonessential loads are graded in the other seven levels.

To generate the subperiods and the metering period which is 15 minutes (in Portugal), a clock IC whose input is the 50-Hz voltage waveform is used. This also provides the drive for a time display. A backup oscillator feeds the clock when the power fails, enabling the correct time to be maintained.

Both the instantaneous maximum peak power demand and the power demand level in each metering period can be set through programmable BCD thumbwheel switches in the front panel. The actual power demand is shown in a display that provides real-time infor-mation of the plant performance. A separate buffer memory of 2 K bytes is used to store the power consumption of the different loads for each 7.5-minute intervals during the day. This memory is divided into eight blocks, one for each group of loads. At the end of the day, the recorded data can be registered in a byte-oriented plotter which scans the memory in an auto-increment address mode. The records of energy consumption can be used to analyze the energy flow and potential conservation opportunities.

A voltage magnitude comparator is used to detect abnormal voltage drops which indi-cate an emergency or an overload situation. When a voltage drop greater than 26% is detected, an interrupt is generated to cause all nonessential loads to be shed. This type of control can be especially useful in rural areas, where in periods of peak load, serious undervoltage could prevent single-phase motors from starting and cause television pictures to fade and lights to fluctuate severely (Burton, 1980). In rural areas, poor load factors make the installation of extra capacity an unattractive investment. However, if large local consumers are offered incentives, they may agree to curtail part of their load when the voltage drops to a designated level. As another alternative, remote teleswitching can be accomodated through a ripple filter, whose signal could cause an interrupt. There is a scope of improvement in the use of main borne carriers for load teleswitching (Zahavi, 1980; Morgan, 1979; Murray, 1980; Schaefer, 1977). This allows the electric utility to control the time and volume of demand switching in both normal and emergency situations following any sudden loss of plant. However, the consumer must be given attractive tariffs in order to accept this scheme.

SOFTWARE ORGANIZATION

The programs which are carried out by the 8748 microcomputer are deposited in the 1 K bytes of EPROM memory available. The program of the main power demand routine and the frequency check routine (Fig. 3). The frequency is checked every 1s to detect abnormal drops which are going to influence the outcome of the load management control.

Power Demand Routine

Figure 4 is a flowchart of the power demand routine. The program starts with the reading of the power demand limits, instantaneous peaks, and peaks that occur within the metering period. These are manually set in the front panel. The counters, which receive impulses from the meters, are scanned and their contents stored in the 8748 user's registers. After reading, the counters are reset.

The value of the power demand is calculated using the data gathered, and sent to the power display. If the calculated value tends to overshoot the power peak instantaneous preset limit, load-shedding is initiated starting with low-priority loads. If not, the available energy for consumption in the next sampling interval is calculated, taking into account the power limit for the demand period, the power already used in the metering period started, and the time available until the end of the metering period. Then, the instantaneous power demand is compared with the energy available for the actual sampling interval and, if it is greater, priority load-shedding is initiated. If it is smaller, load reconnection may start providing the frequency is in its normal state.

Every 7.5 minutes of the program's run, power consumption for each load group is deposited in a separate block of the buffer memory. At the end of the day, the plotter is initiated to receive the load profiles of the eight groups in sequence.

Frequency Check Routine

Frequency is a good indicator of the power balance within a power system. When there is an emergency due to a sudden loss of plant, voltage reduction, underfrequency tripping, or even manual disconnection can be considered (Baldwin, 1976; Maliszewsky, 1971). Load-shedding must be carried out in several steps to avoid overshedding, frequency transients, and overshoots. Load restoration occurs even more gradually than a load-shedding sequence and with a sufficient number of time delays to avoid repetition of load-shedding (Mendarozqueta and Santiago, 1980; Ghai and Verma, 1980).

Frequency measurement. Load management control must integrate existing frequency conditions to be fully effective. For this reason, our equipment measures the frequency every second to detect any abnormal situation. The frequency is measured using the 8748 internal timer/counter which provides an accurate 80 μs timing cycle derived from the microcomputer's 6-MHz crystal. Five cycles of the voltage waveform are used to monitor frequency in order to absorb any noise or interference. When the frequency is 50 Hz, the five cycles correspond to 100 ms. The frequency is measured in the following steps:

- The positive-zero-crossing of the voltage waveform is detected.
- The internal timer is started to generate a 90-ms delay which corresponds to 1125 timing cycles.
- When the time delay finishes, the internal counter is started. Each counting corresponds to 80 μs.
- The next positive-zero-crossing of the voltage waveform is detected which marks the end of the five cycles and makes the counter stop. The frequency is then given by $f=(1000-N \times 0.8)/(90/5)$, where N is the result of the counting. For example, if the frequency is 50 Hz, the counting will be 125; whereas for 49 Hz, it will be 150.

The frequency check routine flowchart is represented in Fig. 3. If the measured frequency drops below a critical value (as in the chart, 49 Hz), the nonessential loads are shed and the frequency status flag is set to 1, which means an abnormal condition. If the frequency is less than 49.7 Hz but greater than 49 Hz, no loads are shed but the frequency status flag is also set to 1. If the frequency is greater than 49.7 Hz, the frequency status flag is set to 0, indicating a normal condition. In the load management program, the frequency abnormal status condition prevents the connection of nonessential loads. Additionally, after an abnormal condition has occurred, nonessential loads are only reconnected in gradual steps and start with those of highest priority.

CONCLUSIONS

A low-cost and simple structure load monitoring and management system based on a single-chip microcomputer has been developed. This system should be attractive to small and medium consumers leading to savings on their maximum demand tariffs, thus enlarging substantially the field of application of power demand control. In addition, this device can be operated either in normal or in emergency conditions to make load-shedding of nonessential loads. In emergency situations, the selective load-shedding can be much less disruptive than large voltage reductions or complete blackouts. It is hoped that the large-scale use of this device can reduce the capital and fuel efforts of the utilities, and at the same time provide a less expensive electricity service to the consumer.

Although voluntary load control for peak demand control is well-established for large consumers through the use of economic incentives and deterrents offered through the electric rate structure, the same does not happen with the direct load control. In this last case, it is up to electric utilities to control specified customer loads in peak or emergency periods. Besides ripple control, customer-installed equipment can also be used to detect abnormal drops in voltage and frequency and contribute to releasing some of the burden on the power system. Some research effort is necessary to develop standard equipment and procedures, taking into account the social and economic factors involved.

REFERENCES

Baldwin, M. S., and H. S. Schenkel (1976). Determination of frequency decay rates during periods of generation deficiency. IEEE Trans. on Power Apparatus and

Systems, PAS-95, 26-36.

Burton, D. P., and D. Hefferman (1980). Optimiser: A microcomputer based current controller to improve voltage levels in rural locations. In IEEE Proceedings, 127, pt B, no. 1.

Ghai, A. K., and H. K. Verma (1980). Microprocessor controlled automatic load shedding and restoration. IEEE Power Summer Meeting, Minneapolis, USA, pp. 13-18.

Intel Corporation (1977). MCS-48 Microcomputer User's Manual.

Isaken, L. (1980). Bibliography on load management. IEEE Summer Power Meeting, Minneapolis, USA.

Maliszewsky, R. M., and R. D. Dunlop (1971). Frequency actuated load shedding and restoration philosophy. IEEE Trans. on Power Apparatus and Systems, PAS-90, 1452-1459.

Mendarozqueta, J. A., and N. Santiago (1980). Automatic voltage restore for high and very high voltage substations. IEEE Conference on Development on Power Systems Protection, London.

Morgan, M. G., and S. N. Talukdar (1979). Electric load management: some technical, economic, regulatory and social issues. IEEE Proceedings, 241-312.

Murray, B. E. (1980). Utility requirements for load monitoring and control. IEEE Conference on Power System Monitoring and Control, London.

Schaefer, J. C., and L. C. Markel (1977). Estimating operating cost benefits of load management. IEEE Power Engineering Society Winter Meeting, Paper A-77, 031-8.

Smith, C. B. (1975). An energy-constrained society. In C. B. Smith (Ed.), Efficient Electricity Use. Pergamon Press, Oxford.

Zahavi, J., and D. Feiler (1980). The economics of load management by ripple control. Energy Economics.

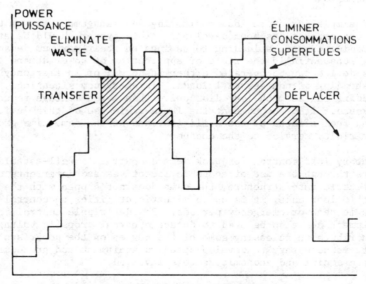

Fig. 1. Reduction of peak demand.

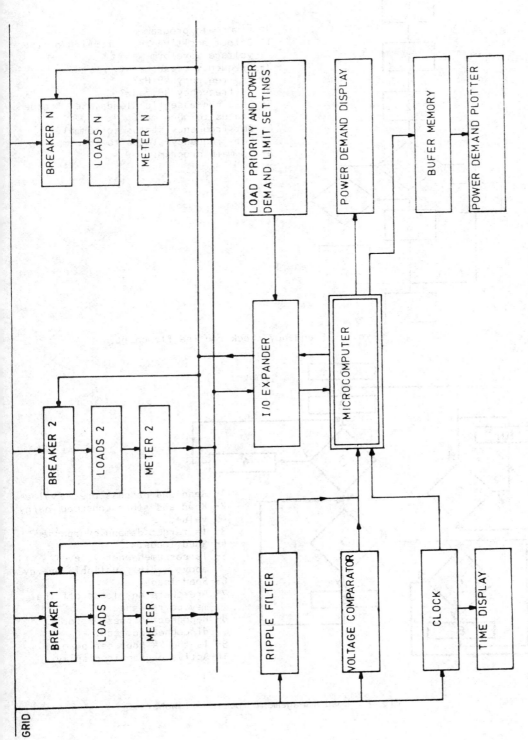

Fig. 2. Load monitoring and management system configuration.

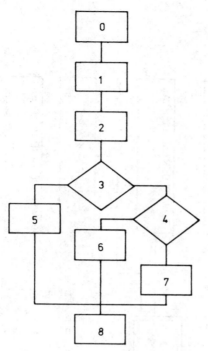

0- From main program
1- Detect positive zero-crossing of the voltage waveform.
2- Frequency measurement.
3- Is frequency 49 Hz?
4- Is frequency 49.7 Hz?
5- Shed non-essential loads. Set frequency status to abnormal.
6- Set frequency status to normal.
7- Set frequency status to abnormal.
8- To main program.

Fig. 3 Frequency check routine flowchart.

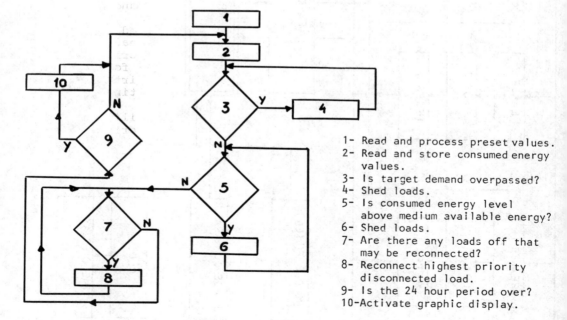

1- Read and process preset values.
2- Read and store consumed energy values.
3- Is target demand overpassed?
4- Shed loads.
5- Is consumed energy level above medium available energy?
6- Shed loads.
7- Are there any loads off that may be reconnected?
8- Reconnect highest priority disconnected load.
9- Is the 24 hour period over?
10- Activate graphic display.

Fig. 4. Load management routine flowchart.

PROJECT—PROCESS SURPLUS HEAT— HIRTSHALS CO-OPERATIVE DISTRICT HEATING SOCIETY

M. Uhrskov

Hirtshals Co-operative District Heating Society, Denmark

On Thursday, 21st August, 1980, Hirtshals Co-operative District Heating Society and Hirtshals Co-operative Fish Industries tested the first stage of the large process surplus heating plant which represents so much new thinking and so many economies within the field of local supply.
The client was Hirtshals Co-operative District Heating Society and Mr. Mogens Uhrskov, works manager, Hirtshals Co-operative District Heating Society was in charge of both detail planning and construction.
Prior to this "future project" the 1970's have witnessed several debates in Hirtshals as to whether it was economically feasible to utilize the process surplus from the town's fishmeal factories.
Mr. Knud Størup, mayor, Hirtshals, took the initiative for setting up a team of technicians from the Borough og Hirtshals, Hirtshals Fishmeal Factories and Hirtshals Co-operative District Heating Society.

Their assignment was to prepare a report on the possibilities of utilizing the process surplus heat from the Fishmeal Factories located at the harbour of Hirtshals.
The result of the group's work and negotiations with various parties and authorities was that Hirtshals Co-operative Fish Industries and Hirtshals District Heating Station agreed on carrying through the project on 1st March, 1980.
Immediately after the detail planning was commenced and on 7th April the first Løgstør pipes were laid.
Already at that time Løgstør Rør had been chosen as the pipe supplier.

To be able to utilize the process surplus heat (which is present during the evaporation process in drying drums) for heating purposes, air and steam heat must be converted into water in a counter-current condenser and pe pumped through the primary side of a plate heat exchanger.
On the secondary side of the heat exchangers the water from the district heating station is heated from approx. 50°C to 80°C when the return water from the consumer is pumped from the district heating station to the fishmeal factories via a heat exchanger installed there, then back to the district heating station and from there to the consumer.

Such a heat recycling plant with a capacity of 5 Gcal/h is installed at every fishmeal factory.

Between the factories at the harbour and Hirtshals District Heating Society a 1.500 m (273mm) double pipeline and 500 m (219mm) of service pipes have been laid.

The area under Hjørring Private Railway's 5 loading tracks and 2 main tracks was forced by pressing 630mm steel casing pipes under the tracks. The distance exceeds 80 m.

To take the largest possible quantity of heat from the factories a coupling line between Hirtshals and Emmersbæk heating stations was constructed.

This pipeline, approx. 1.100 m was ready to be put in operation before the winter of 1980/81.

In connection with the process surplus heat plant an advanced computer system is installed the purpose of which is to supervise and control the heat recycling processes at the fishmeal factories and the district heating station.

A central data centre has been built at the District Heating Station. This system comprises a computer which collects and processes various data, as e.g. temperature, pressure, waterquantities, calories, operating hours, etc.

The data processed is written in hard copy so that the service department can quickly obtain information about all technical installations.

All heat-technical units both at the fishmeal factories and at the district heating station can be controlled from the central data centre.

Subcentres will be installed at the fishmeal factories which independently record all measurements and operating conditions.

In addition, the subcentres and the central computers are provided with selfsupervision units.

This high level of safety means that information about possible defects in the system is immediately available and therefore the defects can be corrected without delay.

The second stage, which comprises the coupling line between Hirtshals and Emmersbæk heating stations, increasing the number of process surplus heat suppliers by including Skagerak Fishmeal Factory Ltd. and Hirtshals Fishmeal Factory Ltd. (Superfoss, Northern Jutland), regulation and installation of a computer unit, was finished by February 1981.

With the factories operating, a quantity of heat, which is far byond the capacity of Hirtshals Co-operative District Heating Society to take, can be produced.

It will certainly be an interesting assignment to explore the possibilities of storing surplus heat in a heat accumulation tank and then consume it when the factories are out of operation.

Conclusion:

The process surplus heat plant will be able to supply at least 15.000 - 20.000 Gcal annualy providing that the supply of trash fish remains unchanged. This means savings on oil of approx. 2.400 t. per year. On a longer view investigations will be conducted to find out whether the process surplus heat and the lower price of heat render it reasonable to expand the distribution system to the neighbouring towns under the auspices of Hirtshals District Heating Society.

TOPIC E

Industrial Productivity
and Development

ENERGY ANALYSIS IN SCRAP - BASED MINI-MILLS. AN EVALUATION OF DIFFERENT METHODS OF ELECTRIC ARC FURNACE OPERATION

A. Borroni*, C. M. Joppolo**, B. Mazza*, G. Nano* and D. Sinigaglia*

**Istituto di Chimica-fisica, Elettrochimica e Metallurgia,
Politecnico di Milano, Italy*
**CESNEF, Istituto di Ingegneria Nucleare, Politecnico di Milano, Italy

ABSTRACT

This paper deals with an analysis of direct and indirect energy inputs in the electric arc furnace steel-making process. The model which has been developed provides a coherent frame of reference for comparing various technological options. The following significant cases have been analyzed utilizing operational data from industrial plants:

(i)   conventionally operated electric furnaces;

(ii)   electric furnaces in which oxy-fuel burners are used as an auxiliary heat source during the melting-down phase;

(iii)   electric arc furnaces in which scrap is continuously charged and smelted after having been pre-heated in a separate furnace (the BBC-Brusa process).

On the basis of our analysis, we proposed a multiple criteria evaluation of the energy impact of various methods of electric furnace operation.

KEYWORDS

Industrial energy conservation; electric arc furnace steelmaking; process energy analysis; total energy requirement; electricity substitution; electrical load leveling; evaluation of technology options; multiple criteria preference space.

INTRODUCTION

In 1979, Italy used 8,700 GWh of electric energy in the production of crude steel in electric furnaces. This figure represents 44.7 per cent of the electrical energy used in the iron and steel industry (19,395 GWh), which in turn represents 20.2 per cent of the electrical energy consumption in all industrial sectors (96,125 GWh) and 12.1 per cent of the total consumption of electrical energy (160,012 GWh).

The production structure of the Italian iron and steel industry is actually quite unique in comparison to other industrialized countries, and is characterized by the widespread development of so-called "mini-mills" where steel is produced from scrap iron in electric furnaces. As a matter of fact, in 1979, out of a total of 24.25 million tons of steel, 12.9 million tons (53.3 per cent) were produced by electric furnaces, and 10.2 million tons (42.0 per cent) were produced in basic oxygen furnaces (the BOF process) through refining of the pig iron produced in blast furnaces (the remaining 4.7 per cent was steel produced in open-hearth furnaces by the Martin-Siemens process).

In comparison, the production figures for 1979 for electric steel in other industrialized countries were the following : 34.2 per cent in Great Britain, 24.6 per cent in the U.S.A., 23.6 per cent in Japan, 15.3 per cent in France and 14.0 per cent in Germany (F.R.).

METHODOLOGY

In this study we will preliminarily examine an energy analysis model which applies engineering process analysis techniques.[1] This model should provide a coherent frame of reference for evaluating energy conservation measures and deciding upon various alternatives.

It should be stated that this study will only analyze energy consumption as measured in physical units, and that only economic analysis techniques would permit us to fully evaluate other production factors such as raw materials, work and capital.

Our engineering process analysis should provide us with useful information on total energy requirements for the production of a given quantity of electric steel in different plants. In general, the total energy requirement is the sum of all the energy inputs directly and indirectly needed in the production of one ton of steel (i.e., all energy needed in the general production system, and not just in the steel plant itself). We must therefore consider not only the final stage of the process (in our case, the steel plant), but also the preceding stages, in order to identify the energy needed for the production of material, energy, and equipment utilized in the steel plant. The "backtracking" process to be followed is represented in Fig. 1, where we have arbitrarily defined four levels of regression in establishing the boundaries of the system (Long, 1978).

We decided not to consider energy inputs beyond the second level of regression. We will separately show the results obtained from an analysis considering only the first level, or the first and second levels together.

Figure 2 shows the scheme that we have followed in our analysis and all the inputs that we have attempted to quantify. It should be noted that we did not consider energy requirements for transport, since they are quite difficult to evaluate and are not fundamental to a comparison of various technological alternatives in already existing plants. Furthermore, we did not consider specific preliminary treatments of scrap, such as shredding, cryogenic shredding, scrap pressing, etc.

HYPOTHESES, GUIDELINES, AND PARAMETERS UTILIZED IN
THIS STUDY

1.   In the quantification of direct energy inputs, i.e., fuels, electrical energy, exothermic reactions, we followed various guidelines according to the energy source used. Thus, for fossil fuels (primary energy

---

[1] A more detailed description of the energy analysis model and of its practical application to electric steel plants is given in a previous paper (Borroni and others, 1981).

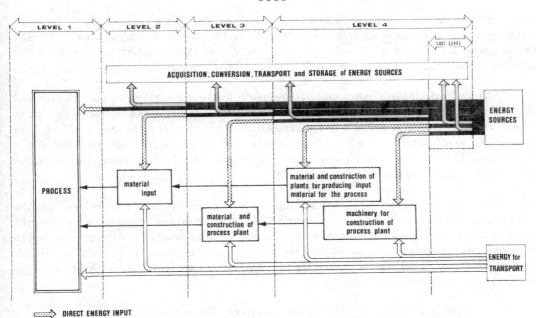

Fig. 1. Energy analysis scheme for a general process: specification of different inputs and definition of possible system boundaries.

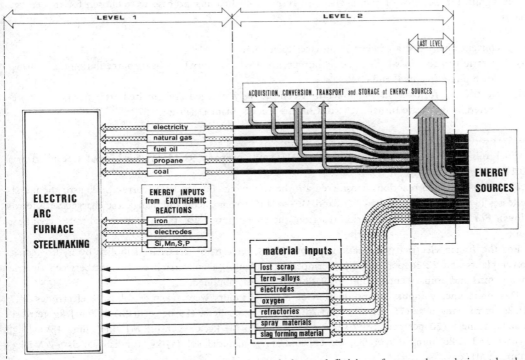

Fig. 2. Energy analysis scheme applied to electric steel plants: definition of system boundaries at levels 1 and 2 and specification of direct and indirect energy inputs.

sources) the quantity consumed was converted into energy units, using the net heat values.[2] Electrical energy input was converted in MJ according to the ratio of 3.6 MJ/kWh. Finally, for exothermic reactions occurring in the arc furnace process, we applied the value of ΔH^O at 1,600° C to the quantity of substance reacted upon.[3]

2. To quantify the indirect energy inputs related to the supply of direct energy inputs, we evaluated all of the losses occurring in each step of the conversion process from the primary energy sources. In the case of inputs utilized as electrical energy, it was necessary to distinguish the amount deriving from hydroelectric power plants and that from thermoelectric plants. This is the only correct method for evaluating all of the energy requirements needed in the conversion from the primary energy sources, respectively dam water and fossil fuels. [4]

3. To quantify the indirect energy inputs related to the consumption of materials, we considered the direct and indirect energy inputs in their production processes. [5] Furthermore, we considered only the energy required for the production of the amount of material used. Special attention must be paid to the energy input related to scrap. In this case, the consumption is only the quantity lost in the process. In other words, we considered only the energy embodied in the amount of scrap needed to cover the metallic losses and the steel plant recycling.

ENERGY ANALYSIS OF THE OPERATIONAL DATA OF VARIOUS STEEL PLANTS

Three significant varieties of the electric arc furnace steelmaking process were chosen for this energy analysis :

(i)   conventionally operated electric furnaces (plant A);
(ii) electric furnaces in which oxy-fuel burners are used as an auxiliary heat source during the melting-down phase (plants B and C);[6]
(iii) electric arc furnaces in which scrap is continuously charged and smelted after having been preheated in a separate furnace (the BBC-Brusa process, plant D)(Brusa, 1977).

---

[2] Fuel oil 41.06 MJ/kg, natural gas 34.57 MJ/Nm^3, propane 91.13 MJ/Nm^3, coal (recarburizers) 30.17 MJ/kg.

[3] The main oxidation reactions considered are the following: Fe $\longrightarrow$ FeO (corresponding to the metallic losses in the furnaces), Si $\longrightarrow$ SiO_2 and Mn $\longrightarrow$ MnO (from silicon and manganese charged with ferroalloys). For the electrode combustion reaction, the energy input was taken to be equal to the coal heat value.

[4] For the Italian electricity generation system (27.74 per cent of electricity supplied by hydroelectric power plants and 72.26 per cent by thermoelectric plants), a conversion coefficient between electric energy units and primary energy units of 9.22 MJ/kWh was obtained.

[5] The direct energy inputs for production of material inputs were taken as follows : electrodes 34.7 MJ/kg, ferro-manganese (75 per cent Mn) 7.2 MJ/kg, ferro-silicon (75 per cent Si) 34.9 MJ/kg, ferro-silicon-manganese (20 per cent Si, 70 per cent Mn) 14.6 MJ/kg, oxygen 6.5 MJ/Nm^3, lime 4.9 MJ/kg, furnace and ladle lining 9.9 MJ/kg, cement and spray material 4.1 MJ/kg, tundish panels 5.2 MJ/kg, scrap 3.2 MJ/kg.

[6] Case (ii) represents two plants using burners fed by fuel oil (B) and natural gas (C) respectively.

Case (ii) was chosen in order to evaluate the suitability of integrating electric energy with other sources in already existing plants. Case (iii) was chosen because it represents a new plant in which an energy conscious design substitutes a larger amount of electric energy, carrying on the pre-heating phase and the smelting phase in separate furnaces.[7]

Table 1 shows the main features of steel plants A, B, C and D. It should be noted that all four plants represent comparable alternatives from an energetic point of view, despite their differences in scale, operation and design. In fact, not only the raw material used (100 per cent scrap iron) and the steel produced (carbon steel for bars, rods, beams, etc.) are practically identical in the different cases, but also the electric furnaces are characterized by significant technological parameters (i.e. specific electric power and productivity) which rank all of them among the HP furnaces.[8]

Table 2 shows the relative figures for energy and material consumptions in producing one ton of billet, as ascertained from the four plants of our study. For plants A, B and C, the data presented here refer to a normal production period of one year (1979); for plant D (which applies the BBC-Brusa process), we refer to a start-up period (in 1979) of two months.

Table 3 shows the results of the energy analysis in terms of energy requirements per ton of billet at levels 1 and 2.

ENERGETICS MULTIPLE CRITERIA FOR IDENTIFYING PRE-FERRED ALTERNATIVES

Now that we have presented the results obtained from an energy analysis of the four plants, we shall deal with two probleems, i.e., (i) how to evaluate different solutions regarding technology and plant design, and (ii) how to choose a preferred alternative.

With an energy outlook limited only to considering the final stage of the process (in our case, the steel plant), a decision maker would be led to establish a scale based on the amounts of direct energy inputs (level 1). In this case, the scale would be B, C, A, D, as can be seen in Table 3. However, the following observations should be made :

1. the direct energy input required by the "least efficient" plant (D) is 44.5 per cent higher than that of the "most efficient" plant (B): plant B would seem to be by far the best;

2. the direct energy inputs required by plants A and C are practically identical (plant A requires 0.16 per cent more energy than plant C): therefore, the choice between one or the other would be indifferent.

On the other hand, if we continue to maintain an outlook limited exclusively to energy, but take into consideration indirect energy inputs too (as would be the case, for example, for a decision maker who has the responsibility of administering the energy use on a larger scale, i.e. regional, national, etc.), the following order of preference would be established: B, D, C, A. However, it should be pointed out

---

[7] Since this case represents great differences in design and construction of process plant, it would be more appropriate to extend our energy analysis to level 3 of Fig.1, to include the direct energy inputs for material and construction of equipment. Nevertheless, by limiting the energy analysis to only the first two levels, the conclusions resulting from our study are not substantially altered.

[8] Generally speaking, the following distinction is made between HP and UHP arc furances:
— HP(High Power) furnaces: specific electrical power $0.25 \div 0.35$ MVA/t, productivity $15 \div 20$ t/h (increasing to a maximum of almost 30 t/h when using auxiliary burners);
—UHP (Ultra High Power) furnaces: specific electrical power $0.45 \div 0.55$ MVA/t, productivity : approximately 40 t/h (increasing to more than 50 t/h when using auxiliary burners).

TABLE 1   Description of the Analyzed Steel Plants

Description	HP furnace	HP furnace with oxy-fuel auxiliary burners		HP furnace with scrap pre-heating furnace (BBC-Brusa process)
		Fuel oil	Natural gas	
	(A)	(B)	(C)	(D)
Pre-heating furnace				
Features	-	-	-	See footnote*
Burners number and power (MW)	-	-	-	20 x 2
Fuel	-	-	-	Natural gas
Pre-heating temperature (^{o}C)	-	-	-	1,200
Productivity (t/h)	-	-	-	80
Electric arc furnace				
Features	See footnote **			See footnote***
Number and capacity (t)	2 x 50	3 x 33	1 x 50	2 x 100
Rated electric power (MVA)****	15	12	15	30
Specific electric power (MVA/t)	0.30	0.36	0.30	0.30
Auxiliary burners number and power(MW)	-	2 x 2	2 x 2	-
Working time of auxiliary burners (min)	-	30 ÷ 45	45 ÷ 60	-
Tapping temperature (^{o}C)	1,665	1,665	1,665	1,715
Tap-to-tap time (h and min)	3 : 10	2 : 20	3 : 10	3 : 30
Productivity (t/h)	15.8	14.2	15.8	27.4
Productivity (all furnaces) (t/h)	31.6	42.6	15.8	54.8
Continuous casting				
Machines and lines number	2 x 3	2 x 3	1 x 4	2 x 3
Semis (square billets) size (mm)	115	115	90 ÷ 130	160
Products	Round bar for reinforcing concrete	Round bar for reinforcing concrete,wire rod	Round bar for reinforcing concrete	L and U iron, HE and IPE beams

*   Movable hearth furnace; continuous operation; heat recovery from exhaust gases with combustion air pre-heating.

**   Conventional type furnace; suction of exhaust gases through a fourth hole in the roof; no heat recovery.

*** Furnace rotating on its own vertical axis, with a fixed roof; continuous charging of the pre-heated scrap and suction of exhaust gases both of which occur through a fourth hole in the roof; heat recovery with pre-heating of the combustion air fed to the scrap pre-heating furnace.

**** Allowed overload for rated power : 20 per cent.

that :

1.   the energy input required by the "least efficient" plant (A, in this new order) is only 7.2 per cent greater than that of the "most efficient" plant (B): the very slight difference in their energy requirements indicates that an advantage regarding direct inputs is, to a certain degree, cancelled out by the disadvantages stemming from the indirect inputs;

2.   plant D, which was at a great disadvantage at the first level, has now an energy input only 3.5 per cent greater than that of plant B;

3.   viewing the situation from this wider energy outlook, plant A and C no longer appear equal (the

energy required by A is 1.6 per cent greater than that for plant C).

In the present production situation, especially in Italy, the total energy requirement in and of itself does not seem to offer a sufficient energy criterion for evaluating technological solutions. Rather, in the decision-making process regarding energy, in addition to the total requirement, one must also take into consideration that fraction of energy provided as electricity. In fact, the concept of replaceable electrical usage is often found in literature: by this some authors mean that since electricity is a particularly valuable energy form, it should not be utilized for furnishing heat, particularly at low temperatures.

TABLE 2   Consumptions per Ton of Billet

Consumptions	HP furnace	HP furnace with oxy-fuel auxiliary burners		HP furnace with scrap pre-heating furnace(BBC-Brusa process)
		Fuel oil	Natural gas	
	(A)	(B)	(C)	(D)
Electricity (kWh)	682	575	637	530
Electricity for electric furnaces (kWh)	627	528	587	450*
Electricity for accessories (kWh)	55	47	50	80
Fuels				
Fuel oil (kg)	0.1	3.4	-	-
Natural gas (Nm^3)	-	-	6.6	44.8
Propane (kg)	0.2	0.2	-	-
Recarburizers (coal) (kg)	1.1	0.7	0.4	5**
Electrodes (kg)	6.5	4.6	5.3	5.5
Ferro-alloys				
Fe-Mn 75 per cent (kg)	12	14	12	-
Fe-Si 75 per cent (kg)	6	7	6	3
Fe-Si 20 per cent - Mn 70 per cent (kg)	-	-	-	12
Oxygen (Nm^3)	0.5	11.5	10	6**
Slag forming material				
Lime (kg)	35	40	37.5	20
Limestone (kg)	8	3	5.5	40
Fluospar (kg)	2.5	2.5	2.5	-
Refractories				
Furnace lining	8	10	10	12***
Dolomite	13	15	15	3.8
Spray material	2.5	2.5	2.5	-
Cement and others	4.5	4.5	4.5	4
Ladle lining	6	7	6	4.5
Tundish panels	4	4	3	2.4
Lost scrap **** (kg)	116	116	116	63
Scrap ***** /billet ratio	1,163/1,000	1,163/1,000	1,163/1,000	1,111/1,000

*   It should be noted that this figure includes a portion of about 35 kWh/t due to the over-heating between 1,665 and 1,715 °C (see Table 1).
** The products of this steel plant (see Table 1) require a refining phase in which the steel is first thoroughly decarburized with an high consumption of oxygen, and then recarburized.
*** Pre-heating furnace included.
****Metallic loss in furnaces and steel plant recycling.
*****Charged, including foreign matter.

TABLE 3   Energy Requirements per Ton of Billet at Levels 1 and 2

Energy requirements (MJ)	HP furnace	HP furnace with oxy-fuel auxiliary burners		HP furnace with scrap pre-heating furnace (BBC-Brusa process)
		Fuel oil	Natural gas	
	(A)	(B)	(C)	(D)
DIRECT ENERGY INPUTS				
Electricity and fuels	2,501	2,240	2,534	3,608
Electricity	2,453	2,069	2,294	1,908
Fuel oil	4	140	-	-
Natural gas	-	-	227	1,549
Propane	10	10	-	-
Coal	34	21	13	151
Exothermic reactions	680	656	642	578
Iron	273	273	273	202
Electrodes	197	138	159	166
Others	210	245	210	210
TOTAL LEVEL 1	3,181	2,896	3,176	4,186
INDIRECT ENERGY INPUTS				
Acquisition, conversion and transport of energy sources used in the final process stage	3,831	3,242	3,585	3,017
Electricity	3,827	3,228	3,579	2,976
Petroleum distillates	1	12	-	-
Natural gas	-	-	5	28
Coal	3	2	1	13
Direct energy inputs for production of material inputs	1,260	1,366	1,302	1,005
Lost scrap	370	372	370	201
Electrodes	227	159	183	191
Ferro-alloys	296	345	296	280
Oxygen	3	75	65	39
Slag forming material	172	196	184	102
Refractories	192	219	204	192
Acquisition, conversion and transport of energy sources for production of material inputs	1,271	1,398	1,326	1,009
Lost scrap	550	553	550	299
Electrodes	227	159	183	191
Ferro-alloys	431	503	431	408
Oxygen	5	117	101	61
Slag forming material	19	22	20	11
Refractories	39	44	41	39
TOTAL LEVEL 2	6,362	6,006	6,213	5,031
TOTAL 1 + 2	9,543	8,902	9,389	9,217

In electric arc furnace steelmaking process, heat is supplied within a very wide temperature range, between room temperature and temperatures exceeding 1,700 $^{\circ}$C: therefore, from a theoretical point of view, it is particularly important that the most suitable combination of primary and secondary energy sources be identified for use in various temperature ranges. On the other hand, due to the restraints on availability, practical considerations have also contributed to the great interest in limiting electrical energy consumption. This is true both for electrical energy consumed (such restraints are presently very tight in Italy, particularly in areas where the great majority of steelworks are located), and for rated electrical power (this last constraint has led to the adoptation of auxiliary burners in the smelting phase as a means of leveling out the electrical load diagram).

Therefore, let us assume that our hypothetical decision maker adheres to multiple energy objectives or applies multiple energy criteria (i.e., total energy requirement plus electricity consumption). Conflict thus arises and a compromise must be made, i.e., a preferred alternative must be chosen. It should be noted that there is no fundamental conflict between these multiple objectives, but rather between them as a whole and the technological, economic, social and other limits that do not allow for their full and simultaneous implementation.

Zeleny's article (1977) has been our primary source with regard to defining the methodology in order to combine multiple criteria.

Let us look at the preference space in Fig. 3. Axis x shows the ratio between the minimum consumption of electrical energy evidenced in the four plants and the consumption of electrical energy in any one of the plants (E.E. min/E.E.). Axis y shows both the ratios between the minimum total energy requirement and the total energy requirement of any one of the plants, calculated by considering separately the first level by itself, and the first level plus the second one [(T.E. min/T.E.)$_1$ and (T.E. min/T.E.)$_2$].

In defining the axes, we have explored the limits achieved along each particular attribute of importance in the available set of alternatives. The highest achieved scores with respect to the two attributes assessed in this way form an ideal alternative (I). [9]

On the basis of this definition, the decision maker will consequently prefer an alternative which is as close as possible to the ideal (I). In other words, he will employ the Euclidean measure of distance (i.e., $d = [(x_I - x)^2 + (y_I - y)^2]^{1/2}$) to provide a ranking.

Keeping in mind the criterion of selection used and applying it to the total direct energy inputs, we can see (in Fig. 3) that the order of preference is : B, C, A, D.

The conclusions would be quite different if we considered both the direct and indirect energy inputs. In that case, the plant D would become the preferred one, moving from fourth to first place, while the relative positions of the other three plants would remain substantially the same. This is due to the fact that with steelworks D we have plant design features which aim at substituting electrical energy with other energy. In the other steelworks, however, we are concerned with adjustments in plant installation features which aim at integrating electrical energy with other energy.

ACKNOWLEDGEMENT

The authors would like to express their gratitude to G. Fantinelli and U. Brusa who provided the operational data referring to plants A,B,C and to plant D respectively.

----

[9] Note that point I can be displaced depending on changes in the available set of plants. It would there - fore become a mobile target.

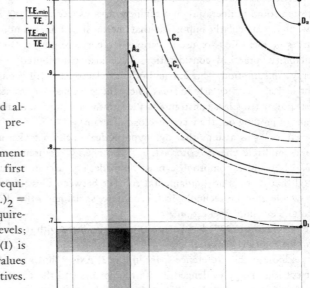

Fig. 3. Identification of the preferred alternative via multiple criteria preference space.

E.E. = electric energy requirement (always considered at the first level); $(T.E.)_1$ = total energy requirement at the first level; $(T.E.)_2$ = = combined total energy requirement for the first + second levels; min = minimum. The ideal (I) is chosen as the best x and y values from the competing alternatives.

Alternatives	Case 1			Case 2		
	Attribute values		Euclidean distance from the ideal	Attribute values		Euclidean distance from the ideal
	$x = \dfrac{E.E.min}{E.E.}$	$y = \left[\dfrac{T.E.min}{T.E.}\right]_1$		$x = \dfrac{E.E.min}{E.E.}$	$y = \left[\dfrac{T.E.min}{T.E.}\right]_2$	
A	0.778	0.910	0.240	0.778	0.933	0.232
B	0.922	1	0.078	0.922	1	0.078
C	0.832	0.912	0.190	0.832	0.948	0.176
D	1	0.692	0.308	1	0.966	0.034
I	1	1	0	1	1	0

REFERENCES

Borroni, A., C.M. Joppolo, B. Mazza, G. Nano, and D. Sinigaglia (1981). Metallurgia Italiana, 73. Accepted for publication.

Brusa, U. (1977). Metallurgia Italiana, 69, 1-3.

Long, T.V. II (Ed.) (1978). Energy analysis and economics. IFIAS Workshop Report. Resources and E - nergy, 1, 151-204.

Zeleny, M. (1977). Adaptive displacement of preferences in decision making. In M.K. Starr, and M. Zeleny (Eds.), Multiple Criteria Decision Making. TIMS Studies in the Management Sciences, Vol. 6. North- Holland Publishing Company, Amsterdam. pp. 147-157.

CONVERSION OF REFUSE TO STORABLE FUEL

G. P. Bracker and H. Sonnenschein

ABSTRACT

Mannesmann Veba Umwelttechnik GmbH (MVU) had developed a technology of conversion of refuse to a storable briquetted fuel the so called ECO-BRIKETT. The characteristic data related to German conditions are
- yield of ECO-BRIKETT from refuse   4o %
- heating value of ECO-BRIKETT   16 GJ/t.

This technology has been proved with satisfying results in a pilot plant with a throughput of 5 t/h in about 2ooo testing hours. More than 7oo t ECO-BRIKETTS are produced for combustion trials.

The ECO-BRIKETTS can be used in furnaces run by communities such as school centers, heating power stations as well as by the industry. Here in Europe is a high demand of alternative energy especially in the cement works. So there is an application of the ECO-BRIKETTS granted. Besides this fuel trading companies are very interested in taking over the ECO-BRIKETTS.

The first plant in an industrial scale is under construction in the so called Rohstoffrückgewinnungs-Zentrum Ruhr (RZR). The throughput will be 3oo,ooo t/y MSW in 2 lines with 25 t/h each. The extraction of fuel will be nearly 12o,ooo t/y by which about 5o,ooo t/y fuel oil can be substituted. This plant will go into operation in 1982.

KEYWORDS

MSW; refuse; energy; storable fuel; briquetts; waste economy; resource recovery; Resource Recovery Center Ruhr (RZR).

INTRODUCTION

On April 27th, 1979 the pilot plant of MVU conversion of refuse to storable fuel was started by minister Dr. Volker Hauff in Herne, West Germany. The erection of this pilot plant which received governmental support by the Ministry of Research and Technology meant a technological challange of

- ever-growing refuse dumps
- a simultaneous shortage of raw materials.

This is another step forward from the traditional waste disposal towards waste economy.

The pilot plant has the object of testing the ECO-BRIKETT-technology developed by MVU. After the testing program was carried out the results confirmed that the organical fraction of the refuse (MSW) which was separated by shredding, screening and classifying can be briquetted without using any binding agents. It is, however, essential to stick to certain grain sizes and degrees of moisture and not to exceed their limits.

The quality of these ECO-BRIKETTS, i.e. limitation of the contained inert material and noxious components is guaranteed by the technology tested during more than 2,ooo operating hours in the pilot plant.

TECHNOLOGY

Figure 1 shows the scheme of an "ECO-BRIKETT" production plant.

First of all the refuse is screened in a trommel revolving screen, where the material which already has the desired particle size and the glass are screened out and passing the mill. By dividing the stream of material the shredding facility will be relieved.

In order to improve the efficiency of the screen and to extend the duration of the material in the trommel special fittings were installed. By means of these fittings a selective size reduction is possible which increases the plastic components in the trommel overflow and big pieces of glass are nearly completely crushed so that they can be screened out.

Steel scrap is separated from both fractions by a magnetic separator and is gained as a by-product.

The screen overflow is then fed into a hammer mill with a grating. This mill is a suitable shredding device since the crushed material is mainly decomposed and disintegration which is advantageous for future processing. The disintegration degree is largely influenced by the right sized meshes of the grating.

Drying of the middle fraction with an initial moisture of about 38 % down to an exhaust moisture of lo - 15 % takes place by a stream dryer. Simultaneously the light fraction is separated from the heavy fraction. The criteria of separation serves as a substitute to meet the specific requirements for separating combustible material from non-combustible.

A cyclone filters the flyable material from the classifying air. The exhaust air is cleaned from dust in a filter unit and partially reconveyed to the air heater. From the material which is separated by the cyclone the finest fraction with a high content of inerts is undergoing a second screening. The screen overflow is milled and pressed into briquettes.

The fuel yielded from German MSW comes up to approximately 4o % ECO-BRIKETTS (see schedule 1).

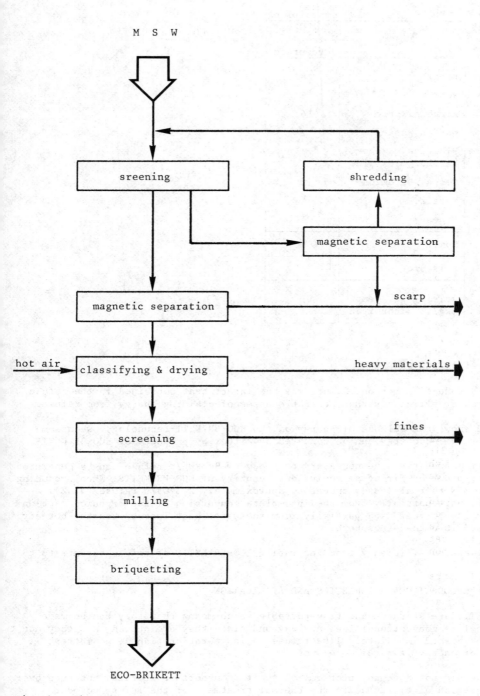

Fig. 1   Flow-sheet

SCHEDULE   1   MSW composition

	Weight Percentage
Paper	28
Wood	5
Kitchen Waste	16
Garden Waste	2
Plastics, Rubber, Leather, Textiles	7
Glass, Keramics	16
Metals	5
Fines	21
Summa	1oo
Moisture	25 - 4o
Arising per year (kg/y · inhabitant)	35o

BALANCE OF ENERGY

This fuel production method can utilize the latent heat contained in the refuse with a greater efficiency than other processes of claiming energy from refuse.

Figure 2 shows the balance of energy of the ECO-BRIKETT-technology. We assume that the required energy can be convered by either by-products or ECO-BRIKETTS.

We further assume that the necessary currency of 9o kWh/t refuse can be generated in a power plant by firing an equivalent quantity of ECO-BRIKETTS. When producing fuel from MSW thermal losses amount to approximately 2,3 GJ/t refuse (31 %). These originate partially from the incomplete conversion of the organic fraction of the refuse to fuel, and partially from the waste heat losses when the hot air is released into the atmosphere.

The total energy efficiency from generating ECO-BRIKETTS of refuse amounts to 63 %.

PRODUCT CHARACTERISTICS AND APPLICATION

ECO-BRIKETTS are storable and transportable which means that they can be made available like conventional fuel any day and place they are needed, and they don't have to be burned immediately like refuse in incineration plants, no matter, if the heat or energy is required or not.

So far, we did not have any problems of their transport or storage, not even over a longer period of time. Due to the thermal treatment of the refuse nc odor annoyances have occured.

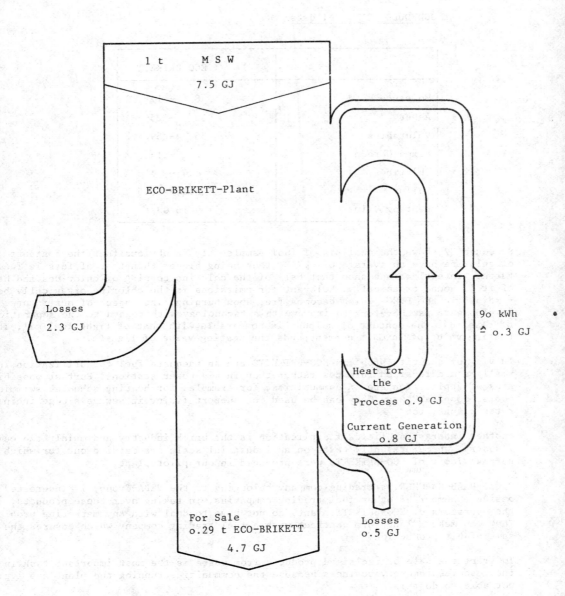

Fig. 2: Sankey-Diagram of the energy balance of the ECO-BRIKETT-process
refering to German MSW (average)

SCHEDULE 2 Fuel data

	ECO-BRIKETT
Moisture	8 – 12 %
Ashes	1o – 25 %
Volutables	55 – 59 %
Fixed Carbon	12 – 14 %
Chlorine	o.4 – 1.o %
Sulfur	o.2 – o.5 %
Heating Value	16 GJ/t

Schedule 2 shows the analysis of fuel samples. It is obvious that the contents of chlorine have an average of o.7 %, thus being higher than that of fossile fuels. Further investigations show that half of the chlorine consist of chlorides and half of it is bound organically. Relevant for emissions is the chlorine organically bound – mainly in the PVC – which becomes free upon burning. The object of our future developments is therefore to improve this technology with regard to the separation of plastics. The content of sulphur is comparable with that of light fuel oil. The heating value of 16 GJ/t corrensponds the heating value of lignite.

In view of quantity problems ECO-BRIKETTS are an adequate fuel for utilization in small or middle-sized furnaces rather than in big power stations. Further possibilities of application are in communities, for example, for heating schools, swimming-pools, hospitals etc. They can be used for support firing in sewage sludge incinerator plants, too.

Another interesting field of application is the brick industry and mainly the cement industry, where presently tests on an industrial scale are carried out for which purpose 7oo t of ECO-BRIKETTS were produced in our pilot plant.

Since RAAB-KARCHER, a trading company belonging to the VEBA-Group, is interested beside a number of other fuel trading companies, in taking over these products, the operators of ECO-BRIKETT plants do not have to deal with any marketing problems. They can make a long-term contract with a fuel trading company which ensures the sale of his production.

To grant the sale of reclaimed products from waste is the most important task in the resource recovery business because the communities running the plant are i.g. not able to do this.

RESOURCE RECOVERY CENTER RUHR (RZR)

The Kommunalverband Ruhrgebiet (KVR) is an association of at present eleven towns and five districts. Its tasks are to concern the region as a whole, predominantly in the field of environmental protection, landscape preservation and leisure facilities. KVR has placed the order for the planning and basic design and supervision of building site for the RZR with MVU.

In the first stage - besides other waste - 3oo,ooo t/y of MSW are to be processed in two lines each with a capacity of 25 t/h ECO-BRIKETTS. The first two ECO-BRIKETT lines will start operation in 1982.

Two more lines with a capacity each of 25 t/h will be established as soon as we have made some experiences.

During the first stage nearly 12o,ooo t/y of ECO-BRIKETTS will be produced according to the described technology, by which 5o,ooo t/y light fuel oil can be substituted.

DEVELOPMENT OF A COMMERCIAL FLUID BED PACKAGE BOILER

L. Brealey and J. H. Wilson

*Engineering Development Dept., NEI Mechanical Engineering Limited,
Sinfin Lane, Derby DE2 9GJ, UK*

ABSTRACT

Development work in NEI during the past five years was directed towards adding fluidised bed boilers to the existing range of package units. It was the aim to develop simple and efficient equipment which would operate automatically from light-up to shut-down, incorporating standard components suitable for the burning of a wide variety of solid fuels. This necessitated the development of equipment for coal feeding, recycling material lost from the bed during operation, and for removing large sized ash from the bed.

Extensive use was made of aerodynamic models to optimise gas flow patterns and of rigs to investigate combustion behaviour. Two horizontal shell boilers were also used in the development. The first, a boiler designed and previously used with a chain grate, was converted to fluid bed firing and used for fundamental work on design of bed, bed materials and bed rating. Evaporation rates up to 2000 kg/hr were consistently obtained against the stoker rating of 1360 kg/hr steam.

The second boiler was specifically designed for fluid bed combustion and used to supply works steam during the winter of 1979/1980. It is designed for a maximum output of 7710 kg/hr steam and with a half size bed installed achieved up to 4660 kg/hr. The major problem of bed material retention and recycling was solved.

Six commercial boilers are currently in the process of construction or installation.

KEYWORDS

Fluidised bed; shell boiler; packaged boiler; coal combustion; steam generation.

INTRODUCTION

Very large numbers of horizontal shell boilers in the range 0.5 MW to 30 MW are used for space heating and the production of low pressure process steam. Their popularity stems from the fact that they are easily operated, can be supplied as a finished package, do not require complex water treatment and, above all, are

relatively inexpensive. NEI Cochran Ltd., is the largest shell boiler
manufacturer in Europe and its total output of about 1000 units a year is greater
than the combined production of all other boilermakers in the United Kingdom.
Boilers are manufactured to operate on oil, gas, oil-gas combinations and solid
fuels, and a large range of combustion appliances are included in the available
designs. The Company pursues a policy of continual development of all equipment
and within this context it was decided, five years ago, to investigate the
possibility of adding fluidised bed combustion to the other methods available for
firing the boilers.

During the development one main objective was to retain the simplicity
associated with shell boilers so that the system could be understood and operated
by those who had become accustomed to other combustion methods.

Preliminary work was started in 1974 using a water cooled pot. The results
obtained produced information on bed materials, the sizes and varieties of coal
and other fuels which could be burned, methods of air distribution and light-up
procedures, and encouraged the Company to proceed to the construction of a boiler.

SHELL BOILERS

Most people are familiar with the modern packaged version of the horizontal shell
boiler as fired by oil, coal or gas. The majority sold over the last twenty
years are of the "wet-back" type, having a fully water cooled combustion chamber
at the end of the furnace tube. They are generally used for producing process
steam at pressures up to about 20 bar, and with outputs up to around 30 000 kg/hr
steam. A high pressure hot water version is also produced, but the mechanical
details are almost identical to those of the steam boiler so that it is
unnecessary to consider it separately.

The construction of the shell boiler is quite straightforward, Fig. 1.

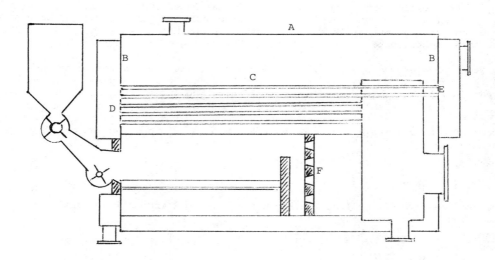

FIG. 1. HORIZONTAL FLUIDISED SHELL BOILER

It consists of an outer horizontal cylindrical pressure shell A, with flat end plates B, restrained by tubes C, which form the gas convection passes. Usually there are three gas passes, a first pass through the furnace tube into the combustion chamber, a second pass through convection tubes to the boiler front into a gas transfer box D, and a third pass of convection tubes taking the gases to the rear end of the boiler into the rear smoke box E. Boiling is initiated in all the immersed heat transfer surfaces and takes place at different rates, depending on the temperature and flow rate of the gas and whether the internal heat transfer is by radiation or convection. Steam is collected within the boiler shell in a space above the free water level and extracted through a conventional stop valve.

When fired by coal many suitable stokers are available depending on the coal type and ash properties, and the method selected for removing ash from the boiler. The main difference however between oil and gas fired shell boilers and those fired by coal is the size of the boiler itself. A shell boiler fired by coal is significantly larger than the oil fired equivalent which leads to a correspondingly higher cost of the pressure parts. It is the ancillary equipment however, such as that for combustion, ash collection and removal, fuel storage and fuel handling, which accounts for a large proportion of the higher capital cost of coal firing over oil firing in shell boilers.

INSTALLATION OF FLUID BEDS IN SHELL BOILERS

The geometry of the furnace tube of a horizontal shell boiler is not the easiest in which to install a fluid bed. Of necessity the bed must be "shallow" and it is difficult and expensive to immerse water cooled tube surfaces to maintain an acceptable bed temperature. It is possible to utilise the lower half of the furnace tube to cool the bed, requiring the fluidisation of a semi-circular area in section, using horizontal sparge pipes to carry the fluidising air. Another method is to use a flat distributor plate, the space underneath acting as a wind box, but with either method the free-board above the bed is limited to half to two-thirds the furnace tube diameter. A bed retaining wall normally constructed from refractory brick defines the end of the bed and because all the combustion gases must pass over this wall at high velocity the elutriation of ash and bed material is considerable and unavoidable. In order to maintain the bed depth within acceptable limits a supply of new or recycled material must be continuously or frequently added. This is a problem which assumes greater significance in a horizontal shell boiler than in any other type and must be solved before satisfactory operation can be claimed.

RIG WORK

The earliest work in N.E.I. was carried out in a cylindrical water cooled pot of 500 mm diameter and a maximum heat rating of 0.5 MW.

From tests with various bed materials alumina, in the particle size range 0.7 mm to 1.0 mm, was chosen from both a practical and theoretical standpoint, and all subsequent work has been done using this material. The main reason for the choice was that it is very dense, almost twice that of sand, allowing higher air velocities through the bed and therefore higher burning rates for particles of a given size. This property is well illustrated in Fig. 2, which shows the terminal velocities in air at 900°C for particles of different size and density. The obvious disadvantage of alumina lies in the high cost relative to sand, emphasising that the bed material must be conserved in the boiler system and not allowed to pass out with the coal ash.

Various methods of coal feeding were studied and top feeding was shown to be satisfactory.

Experiments with various solid fuels ranging from anthracite to high volatiles coals and including pelletized refuse derived fuel, showed that all these materials could be burned successfully. Combustible plastic waste materials could also be burned provided they were fed to the bed in a pelletized form.

Bed temperature control was studied and it was found that the simplest procedure was to set the air flow through the bed and then control the fuel supply rate to achieve the desired bed temperature. These tests also showed that in spite of the water cooled bed periphery from 50% to 100% of excess air was required to attemperate the bed.

The investigation of methods of bringing air into the bed resulted in a preference for a flat bed plate fitted with vertical capped standpipes, each having a number of radial holes drilled near the top. This method provides a space below the plate which acts as a windbox so that air distribution to the standpipes, and therefore fluidisation, is very uniform, thus avoiding local high temperatures during operation, and the problems which can arise from this cause.

Several methods of light-up were considered but the selected one which provides minimum start-up time is to admit gas into the standpipes where mixing with air takes place and the premixed gas is ignited over the bed. By suitable design, incorporating necessary interlocks, a completely safe procedure has been developed by which means the bed reaches operating temperature in a matter of 8 to 12 minutes.

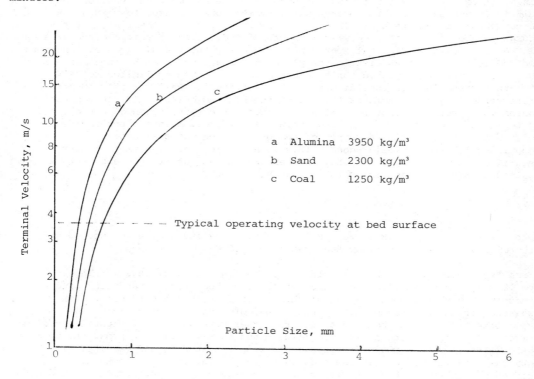

a	Alumina	3950 kg/m³
b	Sand	2300 kg/m³
c	Coal	1250 kg/m³

— Typical operating velocity at bed surface

FIG. 2.   TERMINAL VELOCITY OF PARTICLE

EXPERIMENTAL WORK WITH THE DEMIPAC BOILER

In 1976 it was decided that the rig had served its basic function of supplying design information and that future work would be carried out on a boiler. Accordingly, a boiler which had previously operated with a conventional chain grate stoker, giving an evaporation rate of 1360 kg/hr of steam, was modified. The boiler, known commercially as the Demipac, is a conventional 3-pass wetback horizontal shell boiler.

The fluid bed had an area of about 0.75 m² and was constructed using the standpipe principle. Coal feed arrangements consisted of a simple star feeder, below a coal hopper, allowing the coal to fall down an angular chute through the boiler front onto the end of the bed. No bed material recycling equipment was fitted, and it was assumed that ash would be totally elutriated from the bed and pneumatically transported through the convection passes to a cyclone collector fitted immediately prior to the I.D. fan. Light-up arrangements using propane gas were as previously described.

One problem which became immediately apparent was the poor quality of fluidisation at the side edges of the bed. This was attributed to the slope of the cylindrical furnace wall at the point of intersection with the distributor plate, which was some 230 mm below the centre-line. Following model tests a special type of nozzle was developed for the side edges of the bed and this provided a satisfactory solution to the problem. Good fluidisation is obviously very important at the furnace wall to promote heat transfer and avoid ash sintering.

Difficulties were also experienced with the coal feed arrangements. The problem was mainly experienced at high coal feed rates when sintering took place immediately below the coal feed chute. Clearly, the horizontal transport of coal on the bed was not at a sufficiently high rate to avoid elevated temperatures at this point. The problem was solved by incorporating a rotary spreader unit which intercepted the free fall of the coal and distributed it over the bed surface.

Although an acceptable coal feed system had now been developed, the feeder was only satisfactory for singles coals, and it was susceptible to foreign bodies in the coal.

Performance of the boiler in terms of output proved quite acceptable. The fluid bed operated satisfactorily with specific coal burning rates around 200 kg/hr m² (1.6 MW/m²) and a boiler output of 1500 kg/hr steam. This meant that the rated output of the boiler when operated with a chain grate had already been exceeded. Although bed material was being elutriated continuously, and was collecting in the furnace tube and combustion chamber, it could easily be returned to the bed by hand after the boiler had cooled down following a test run. In the meantime a bed recycle system was being developed. The maximum output eventually achieved on this boiler was 2000 kg/hr and the results encouraged the Company to design and manufacture a large boiler.

THE COALMASTER FLUID BED BOILER

The name Coalmaster is applied commercially to the range of coal fired shell boilers manufactured by NEI Cochran Ltd., normally fired by chain grate stokers of their own manufacture. The fluid bed prototype was designed to include a maximum bed area of 4 m². Based upon a specific heat release of 1.6 MW/m² which had previously been achieved, and assuming a boiler efficiency of 78%, a rating of 7710 kg/hr evaporation was expected. Heating surfaces were designed accordingly and, together with the bed itself, accommodated in a shell size which, if fired by a chain grate, would give an output of 4540 kg/hr steam.

To ensure that satisfactory air distribution over the bed surface could be achieved a full scale isothermal model of the furnace tube and combustion chamber was constructed in wood and perspex, and the boiler distributor plates were fitted and tested using the fans which would later be fitted to the boiler. With this facility it was possible to check the uniformity of fluidisation and the performance of the coal feed system.

During work with the Demipac boiler some of the later experiments suggested that higher specific burning rates could be achieved, and in order to provide a means of investigating the limiting value without exceeding the designed evaporation rate, it was decided that the boiler would be fitted first with a 2 m² bed.

Coal Feeding

In order to take advantage of the capability of a fluid bed to burn a wide range of fuels the feeder must be able to deliver any coal at a controlled rate. In view of the limited performance of rotary valves or screw feeders, particularly when faced with wet small coals, a vibratory feeder was developed and fitted. Its performance with singles was good, throughput being acceptably linear with signal. When using smalls however the feed rate was erratic, being sensitive to moisture content of the coal. A novel feeder was therefore developed and has proved capable of handling all types of coal; it provides a continuous delivery at a controlled rate and, in common with other types used, drops the coal onto a rotating spreader which distributes the fuel fairly uniformly over the whole area of the fluid bed.

The Fluid Bed

The bed comprises a flat plate with vertical and capped standpipes at 50 mm square pitch; each standpipe has six holes near the top of the vertical side. Thermo-couples are fitted to measure temperatures in the fluid bed at five points and of the bed plate. Provision is made to measure air pressure drops through the bed plate and the bed itself. The bed material used is alumina, bed depth is pre-selected to between 60 and 100 mm and is controlled to ±3 mm of the required depth during operation. The sides of the bed are in contact with the wall of the furnace tube and the adjacent standpipes are designed to ensure good fluidisation of the bed which is in contact with the heat transfer surface. The rear end of the bed is retained by a refractory brick wall, the height of which has been optimised to minimise the amount of bed material carried over the wall.

Bed Material Recycling

The velocity of gases over the retaining wall far exceeds the terminal velocity of the particles of bed material so it is inevitable that material, together with small sized coal ash, is carried over from a bubbling bed during operation: this material must be collected and returned to the bed. It was at first thought that it would be necessary to separate the coal ash from the bed material before returning the latter to the bed, but this has proved to be unnecessary. The major problem is to ensure that all elutriated bed material collects in the combustion chamber and does not enter the second pass tubes. This is achieved by the installation of a gas diffuser wall constructed in a refractory material behind the bed retaining wall. The diffuser wall (F in Fig. 1) has the effect of producing a more uniform gas velocity profile in the furnace tube and combustion chamber, thereby eliminating the high velocity gas stream over the bed wall and allowing bed material to drop out.

From the combustion chamber it is returned to the bed by a dense phase conveying system. This system operates automatically during operation, the signal to return the bed material being derived from the pressure drop through the bed. In this way the bed depth is controlled.

Coal Ash Removal

To collect the fine ash a multicyclone is fitted at the rear of the boiler. Means are also provided for removing large sized ash from the bed because if such ash is allowed to accumulate to exceed 25-30% of the bed volume fluidisation is impaired, hot spots develop and sintering results. This ash is therefore not allowed to exceed 20%, bed material and ash are removed, the ash sieved out and the bed material returned. To date this procedure has been carried out off load and, where an operating regime allows it, is preferred because the coal has been burned out and carbon loss is minimised. Commercial boilers are now incorporating the means of bed clean-up during operation.

Operating Experience

After commissioning and short term operation during the autumn of 1979 the boiler was used to supply works steam throughout the winter. Most problems encountered were associated with ancillary equipment such as coal feeders and spreader and the bed recycling system rather than the boiler and fluid bed itself. These problems were however resolved and reliable operation was achieved.

The maximum output obtained was 4660 kg/hr steam representing a heat release from the bed of 1.9 MW/m^2. The bed temperature was controlled at 950°C and excess air was between 70% and 80%. Carbon loss was always very low, at about 2%, when burning singles, and boiler efficiency was calculated at around 80% with exit gas temperatures of about 190°C. Because of works steam demand achievable turn-down was not explored fully, although it was confirmed that at least 2:1 was obtained. Light up time was eight to ten minutes to bring the bed from ambient to operating temperature when the requirements of the boiler itself imposed no restriction on rate of light-up.

The control system, based on modulating the air supply from a steam pressure signal, and regulating the coal feed rate from bed temperature signals through a three-term controller, worked satisfactorily.

Later work on a commercial hot water boiler showed that turndown ratios well in excess of 5:1 can readily be achieved using an on/off method. When load reduction is required both air and fuel are shut off for a period and then brought in again, this cycle being repeated for as long as is necessary. On the boiler which was used the air and fuel were cut off for periods up to 15 minutes and brought on again for periods of 3 minutes. Utilising this cycle the bed temperature did not fall below 780°C during the 15 minutes "off" period, and recovered to 900°C during the 3 minutes "on" period. The procedure is made possible by the relatively large energy store of a shell boiler, the actual length of the "on" and "off" periods will be determined for a given hot water or steam system by the range of output conditions which are acceptable to the user.

CONCLUSIONS

The development of a fluidised bed boiler having the same general form as existing horizontal wet-back shell boilers presents problems not encountered in configurations where a larger freeboard space is available above the bed.

(a) Bed material projected upwards, particularly at the rear end, meets a high velocity horizontal gas stream which carries the particles away from the bed. The effect can be reduced by using high density bed material, but this is relatively expensive and must be recovered for recycling.

(b) The bed itself is of necessity shallow and the depth must be controlled within small limits at all times.

(c) Fuel must be spread over a wide area of the bed in order to avoid sintering because horizontal mixing becomes increasingly limited as bed depths decrease.

These problems have been investigated fully and solutions found. In addition, a new coal feeder and spreader, suitable for all types of coal, has been developed, together with material handling equipment suitable for the recycling of bed material.

There is no doubt that fluidised bed combustion techniques, as applied to horizontal shell boilers are a viable alternative to conventional stokers and will have several advantages to offer, not only the ability to burn a wider variety of fuels, but also a convenience and simplicity of operation.

ENERGY CONSERVATION THROUGH THE USE OF GAS TECHNOLOGY

I. M. Coult

Shell International Gas Ltd., Shell Centre, London, UK

ABSTRACT

Gaseous fuels can make an important contribution to the conservation of our fossil fuel resources. Significant developments in heating technology have become possible through the more effective use of the special characteristics of gaseous fuels. Examples of this technology for metal reheating show how its adoption can lead both to large energy savings and to improved process productivity and product quality. In this way gas is shown to have a value considerably above the cost of alternative primary energies. Further examples show how gas technology makes large savings in energy economically attractive in widely used processes such as water heating.

KEYWORDS

Gas technology; rapid heating; jet impingement heating; latent heat recovery.

INTRODUCTION

The finite nature of the world's fossil fuel resources has been well publicised in recent years and industries in most developed countries are only too well aware of the rising cost of fuels. Nevertheless the adoption of energy saving methods by industry in general has been slower than expected. For example, in the UK during the decade prior to the 1973 oil embargo, the net energy consumption per unit of output in industry (energy intensity) declined steadily with an average annual decrease of 1.6%. In the two years immediately following the oil embargo the annual decrease in industrial energy intensity averaged 3.3%, but in the remaining years to 1979 the average annual decrease reverted to 1.6%, thus continuing the trend established before energy conservation entered the spotlight.

Many factors can be argued to contribute to the slow response of industry to the national need for energy conservation, such as limited availability of capital and doubts about future energy prices and inflation. However, it seems clear that in those industries in which energy consumption does not contribute a large proportion of final product cost, but which collectively consume nearly half of all industrial energy, investment in energy saving has not been popular. Nevertheless, significant savings of energy can be achieved by adopting new technology in which energy conservation is combined with other process improvements. In these circumstances it is the overall improvements such as increased output or improved product quality which yield a competitive advantage in the market place and thus provide the main incentive for investment.

For industrial process heating, adopting new technology involves not only the selection of improved equipment but also the selection of the most advantageous fuel. Whilst energy savings can be made by improving the thermal efficiency of an existing process, for example by installing feed back controls, other, and often much more significant savings can be found through the selection of improved heating technology based on an optimised fuel selection. In this respect gaseous fuels such as natural gas offer many advantages over alternative fuels. Gas when associated with the right heating equipment, can lead to significant improvements in both process performance and energy utilisation. The important characteristics of gaseous fuels are:

- cleanliness of combustion gases, thus avoiding contamination of processed material and having a low impact on the external environment;

- controllability of heating conditions leading to fast response, precise location and close repeatability of heat application;

- flexibility to operate with low air-fuel ratios to provide clean, non-oxidising furnace atmospheres;

- suitability for automation and low maintenance requirements.

Apart from the ability to provide controlled atmospheres in a furnace, electrical heat also possesses these important characteristics but has the major disadvantage of being a secondary form of energy and generally leads to a much lower efficiency of primary energy utilisation than does the direct use of gas. It is sometimes argued that the secondary nature of electrical heat is becoming much less important with an increasing contribution from nuclear sources. Whilst this may prove true in the long term, for the next decade or two at least hydrocarbon fuels will remain the predominant source of electricity in most countries and nuclear electricity will continue to be viable economically only at very high load factors.

The difference in characteristics of gaseous fuels from other sources of heat energy, leading as it does to different designs of heating equipment, also results in the very important but often overlooked conclusion that the best fuel cannot be selected on the basis of a cost per unit of energy alone. Instead the choice must be made taking into account all aspects of the process which are affected, such as differential capital investment, changes in productivity and reductions in energy consumed. For a process where gas is advantageous it can be shown that a significantly higher price may be justified per unit energy supplied than for alternative fuels. Thus a premium value is established for gas which is specific to the type of application and alternative fuel considered and the ratio of this gas value to the price of the alternative fuel is defined as the value factor. Provided the actual gas price to customers is less than the product of the value factor and the alternative fuel price, the selection of (or conversion to) gas will result in better process economics.

The use of 'value factor' in this way has the merit of including the effect of fuel choice on several factors (e.g. fuel consumption, process productivity, product quality, material wastage) in so far as these are reflected in the process costs. It therefore presents a much broader view of the consequences of the choice than one of these factors alone could provide. Although energy consumption is only one of the factors included in the calculation of the premium value of gas, it is observed in most cases that where the use of gas results in a high value factor, a saving of process energy is simultaneously achieved, although there may be exceptions.

One factor which may involve considerable energy savings is material waste. If an improvement in processing can result in a reduction in the reject rate of the processed material, e.g. steel billets or ceramic plates, then the energy saved is the total energy consumed in bringing the metal or ceramic part to its rejected state from its initial state as an ore. This is referred to as the acquired energy of the wasted material.

Examples are given below of a number of process heating operations in which the use of gas results both in energy saving and in attractive economics for the process overall.

REHEATING METALS FOR HOTWORKING

Convective Heating

Conventional radiant heating still dominates most metal heating processes related to forging, rolling and extrusion. However, substantial advantages are offered by convective heating based on very high speed gas burners (Soulages, 1969; Winter, 1971) and in recent years many such systems have been installed.

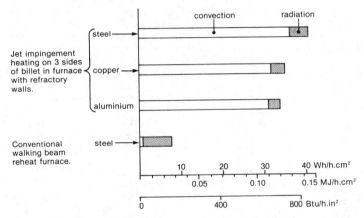

Heat flux; average values over temperature rise from cold to final temperature.

Fig. 1. Comparative heating rates for jet impingement and conventional furnaces.

Figure 1 demonstrates how, in a typical metal reheating operation, the use of gas-fired jet impingement heating can achieve heating rates throughout the cycle which are several times higher than are obtained in a conventional radiant furnace.

The high heating rates enable a given production rate to be obtained from a smaller furnace and greater space can be given to recuperative preheating of the steel with consequently large savings in energy. For example in a typical jet impingement furnace using the Shell 'URG type T' burners described below and for a given rate of production and heating efficiency, the length of the furnace is only about a quarter of that of a conventional radiant furnace. If more length is accepted, considerably increased efficiency can be achieved but still with a shorter length overall than the conventional furnace. Furthermore, the use of a predominantly convective mode of heat transfer enables the furnace to be designed with a low thermal inertia. This results in large energy savings during start up, shut down or changes in heating rate, as well as providing a heating system which has a rapid response to changes in production rate or component temperature.

Component heating times are also reduced significantly by the jet impingement technique and this decreases the amount of decarburization and scaling that occurs, with a consequent improvement in product quality.

A quantification of the savings in energy obtainable with such rapid heating systems is provided in two examples. The first is for a steel billet furnace with an annual throughput of 2000 tonnes in which the billets are raised to 1200°C at the rate of 1 tonne per hour. In this case the saving in energy to produce each tonne of hot metal is over 50%. This however does not include the energy saved as a result of reducing the reject rate of the furnace from 2% to 0.5%. The acquired energy contained in the rejected metal is equivalent to 0.6 GJ/tonne or 12% added to the energy consumption of the conventional process.

The second example is of an aluminium billet reheating furnace with an annual throughput of 6000 tonnes in which billets are raised to 500°C at a rate of 3 tonnes per hour. In this case the alternative method of heating is the electric induction furnace and whilst the consumption of energy as electricity is less than as gas, the assumption of an overall efficiency of 30% from primary energy to delivered electricity leads to a saving in primary energy consumption for the process of over 65% when using the gas firing technique.

In these two examples, apart from the very considerable energy savings achievable, there is a strong economic incentive to adopt the gas fired process. In both cases the capital employed for gas fired equipment is less than for the alternative fuel, and for the steel reheating furnace a gas value factor around 4 is found relative to fuel oil. For the reheating of aluminium the gas value factor is just over 1 against electricity which with the generally much higher price of electricity makes gas very attractive for this application.

Recent developments in the technology of high speed convective heating have concentrated on improving further the burner performance to enhance the possibilities for designing furnaces with lower thermal inertia and high overall efficiency. For example the Shell 'URG type T' range of burners are now capable of providing jets of fully burnt gases at jet speeds between 120 and 180 m/s. The burners which have rated throughputs between 8 and 116 kW and turn down ratios up to 20, are normally arranged in a multiple burner panel which becomes a section of the furnace wall. This arrangement results in a very uniform heat distribution as well as a very high rate of heating of metals.

These high speed convective heating techniques can also be applied effectively in larger scale furnaces (5-100 t/h) than those given in the examples above as was demonstrated in a development between Shell and the British Steel Corporation (Winter, 1976). This development proved the effectiveness of jet burners with generally similar performance to the URG type T for the rapid heating of steel billets for rolling. The furnace used incorporated roof mounted burners and a walking beam transport system. It was demonstrated that even without heat recovery, a thermal efficiency (on lower heating value) of 43% could be achieved, which is comparable to the efficiency of conventional furnaces with recuperation. At the same time the furnace length was reduced to less than 25% of the conventional furnace and the low thermal inertia and rapid heating characteristics led to a marked improvement in furnace operating efficiency.

Radiant Heating

Significant energy savings can be achieved in reheating furnaces where either
jet impingement techniques are considered unsuitable or a high degree of radiant
heating is desired. In this case the improvements in energy utilisation efficiency and also in overall process efficiency come from reducing the thermal
inertia of the furnace. Even with the best currently available insulating
refractories for use in furnaces to heat the metal to 1000°C or above, there is
a large mass of hot refractory in the walls.

It is possible to reduce this high thermal inertia substantially if the hot
radiating surface can be confined to specially designed burners which provide
the source of radiation without the need to raise a large mass of refractory to
a high temperature. This permits the radiant heat to be rapidly varied, and to
be rapidly switched on or off, according to demand.

The Shell URG type RCS burner has been developed to meet these requirements and
it has been shown that in a typical application (e.g. steel heating), about 30%
of the heat transferred to the charge is by radiation from the burner surface,
and the rest is by convection from the gases in the furnace. Relatively little
heat is received by the surrounding furnace refractories.

Applications have been achieved in both the metals and the ceramics industries.
An example is illustrated in Fig. 2, which shows a furnace for batch heating
steel plates prior to pressing at 1050°C. The plates each weigh 180 kg and have
a thickness of 10 mm. For a batch process, with some dead time occurring between
successive charges, this furnace shows remarkable efficiency levels. The average
thermal efficiency, based on lower calorific value, over each working shift,
including stoppages and varying working rates, has been measured at around 41%.
Compared with a conventional oil-fired furnace which this system replaced, and
which had an average efficiency around 5%, this represents an energy saving of 88%.

The average time to heat each plate in this furnace, using 42 burners, is about
15 minutes. If the dead time between charges is minimised, the efficiency can
be raised from 41% to about 52%.

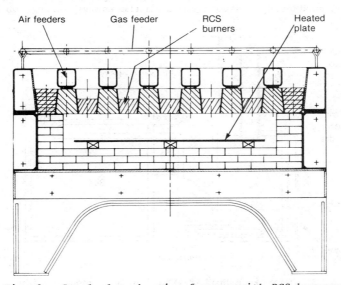

Fig. 2.   Steel plate heating furnace with RCS burners.

WATER HEATING

Another important contribution to energy saving made possible through gas technology is in process water heating. Hot water is used in very many industrial
operations, such as in the textile, paper and food processing industries, as
well as for space heating throughout the industrial and commercial sectors. In
water heating, the benefits that stem from using gaseous fuels arise from the
possibility of recovering the latent heat of condensation.

The savings in energy amount to considerably more than the latent heat alone
(which is about 10% of the combustion heat), because additional sensible heat is
recovered by the reduction of the flue gas temperature to the condensing level
from the normal level employed when condensation in the flue system is deliberat-
ely avoided. In many practical cases more than 20% additional combustion heat
is reclaimed when flue gas condensation is applied.

Although the required heat recovery could be achieved by the use of a highly effective heat exchanger with extended surfaces (e.g. finned tube), this approach is
very expensive and economically difficult to justify in industrial applications.
It is generally more economic to recover the latent heat by direct contact between
the flue gases and a stream of water, usually in the form of a spray of small
droplets.

An example of this principle is shown in Fig. 3, in which the recuperator operates
in conjunction with a conventional gas-fired boiler. Several such installations
have demonstrated the favourable economics of latent heat recovery. For a
typical situation based on costs in Germany and considering an installed capacity
of around 1 MW, the payout time for the additional investment needed for the
unit with latent heat recovery, compared with a conventional gas boiler, has been
around 9 months. When the choice is between an oil fired system and this gas
fired system with latent heat recovery, the improved economics of the gas fired
system result in a value for gas around 20% higher than for distillate oil.

The use of gas rather than oil for water heating can lead to reductions in operating and maintenance costs for the associated equipment, but the main advantage
of the gas fired condensing system stems from the energy savings obtained.

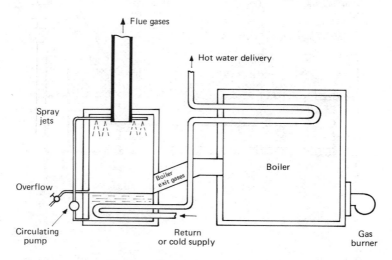

Fig. 3.   Gas-fired boiler with flue gas condensation.
(Courtesy of Gebr. Fröhling)

OTHER PROCESSES

There are numerous examples, including those described above, in which the introduction of recently developed technology based on the use of natural gas produces
significant energy savings. Figure 4 indicates the savings achievable in a range
of such industrial processes. For most of the processes, particularly those
involving the reheating or melting of metals and the firing of clay products the
use of gas gives significant process improvements in addition to energy saving.
In particular Fig. 4, shows how a reduction in the amount of reject metal can
add substantial savings in acquired energy to that saved directly in the heating
process.

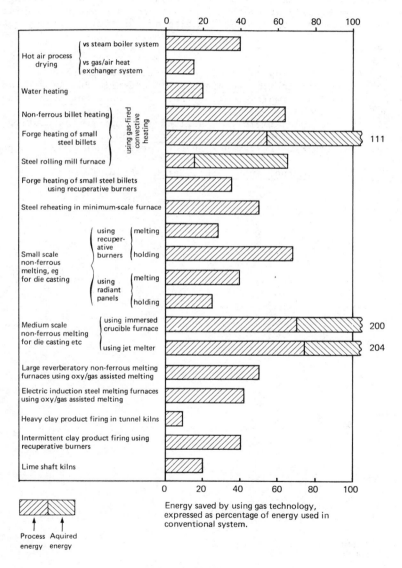

Fig. 4.   Energy savings by the use of gas technology.

CONCLUSIONS

Examples have been given to show how, in a wide range of industrial processes, the introduction of new technology based on gaseous fuels can achieve the twin objectives of improving the process performance and conserving energy. There is less incentive for energy conservation to be widely adopted in the sectors of the process heating industry where energy intensity is low than in, for example, the primary metal and chemical industries. However, the incentive is increased where improvements are obtained in other aspects of production, such as improved productivity or product quality, and reduced losses of acquired energy in the processed materials. In many cases, the introduction of energy-saving gas technology is economically attractive overall, and is capable of making a significant contribution to the improved utilisation of our limited energy resources.

REFERENCES

Soulages, J. (1969). Développement des utilisations du propane en métallurgie. Conference of the Association Technique de l'Industrie du Gaz en France.
Winter, E.F. (1971). Jet impingement. First conference on natural gas research and technology, Chicago.
Winter, E.F. and K.F. Coles (1976). Gas fired jet impingement heating of steel rolling billets. Shell International Gas report no. SIG 76/1.
Dept. of Industry, (1978). Energy use in the pottery industry, UK Department of Industry.
Winter, E.F. (1978). Journal Inst. Fuel, March, 46-58.

AN ENERGY ANALYSIS OF THE
RECONSTRUCTION OF A CAR STARTER MOTOR

T. R. Cox and M. E. Henstock

*Department of Metallurgy and Materials Science, University of
Nottingham, University Park, Nottingham NG7 2RD, UK*

ABSTRACT

The refurbishing of items of vehicle electrical equipment has traditionally been
carried out to reduce the cost of their replacement.

An analysis of the direct energy cost of dismantling a starter motor and of reconditioning the major components under industrial conditions shows that such components may be put back into service at an energy investment of only about 2% of
that required for the manufacture of new, replacement, items.

KEYWORDS

Energy; automotive; rebuilding; materials; component; analysis.

1. INTRODUCTION

For many years the rebuilding of unserviceable automotive electrical equipment has
provided the consumer with a practical and economical alternative to the purchase
of items manufactured from new components. At current prices, a factory-rebuilt
starter motor for a popular type of family saloon might, typically, cost £55, a
total saving of some £25 over the cost of the brand-new unit.

The practicality of such rebuilding depends, at its simplest, on the relative costs
of new components and of the labour and consumables associated with rebuilding.
Traditionally, labour costs have been kept relatively low through the use of power
tools in disassembly. Since 1973, however, changes in the cost of energy have been
such as to prompt an energy audit of the rebuilding process, to establish whether
the re-manufacture of units is energy-efficient, in addition to being cost-
effective.

Various types of vehicle electrical unit are remanufactured on an industrial scale
in Britain; they include alternators, dynamos, starter motors and windscreen wiper
motors of both domestic and foreign design. It was decided to examine one specific
type of starter motor, the Lucas M35J, which is fitted to a large sector of the
domestic car population.

The object of the study was to analyse all the processes used to rebuild unservice-able motors of this type and, after taking account of the yield of re-usable com-ponents, to compare the energy of rebuilding with that of the initial construction of the motor from new parts.

2. ECONOMIC ASPECTS

There is a substantial difference, usually about two orders of magnitude, between the retail cost of a finished product and its eventual value as scrap materials (Lund, 1975). It has long been recognised that benefit accrues to the consumer through an extension of the product life or, alternatively, through the identif-ication of a secondary use.

Electrical units used in vehicles are assemblies of parts, most of which may be replaced when worn out or broken. The service life of such a unit can, in principle, be extended indefinitely. In some instances rebuilding to extend product life may be necessary because the equipment is no longer in production or because currency restrictions preclude the import of new parts or replacement vehicles. In other, probably the majority, of cases the stimulus for rebuilding is the immediate saving in cost and the degree to which rebuilding is done depends upon the cost and availability of energy, labour and materials.

It has now become clear that monetary economics, although still the dominant factor in manufacturing industry, is increasingly being linked with the energy specificity of a process. In the field of energy analysis there are at least two, conflicting, schools of thought.

1. Energy analyses are independent of economic analyses; choices may be made on the basis of either or both, and this implies that any material resource can be exploited given sufficient energy. Energy should thus be treated as a unique and essential resource.

2. Energy is only one of the primary natural resources that control technology.

The second of these is, initially, the more plausible since not only is it impossible to assign an energy concept of value but it is clear that other primary inputs, such as capital, labour, land and minerals are also indispensable (Veach Long, II, 1975).

The concept of efficient allocation of resources can be defined in numerous ways. It is convenient, here, to give those that relate to physical efficiency and economic efficiency:

1. Maximum physical efficiency is attained when, through the efficient applic-ation of energy to a system and the careful husbandry of materials throughout, neither is wasted.

2. Economic efficiency is attained when given resources of capital, labour and all forms of natural resources are so combined that a higher output of any desired product is possible only at the cost of a reduced output of some other desired product.

3. METHODOLOGY

The starter motor examined in this study is rebuilt in the factory according to a schedule of operations and inspections. The experimental section of the present investigation comprised:

3.1 Timing of each operation in the dismantling cycle.

(a) Removal of terminal nuts, by impact wrench.
(b) Removal of commutator end bracket fixing screws. The bracket is examined for damage to it, to the brushbox, brush springs and brushes. Good assemblies are sent for cleaning whilst unserviceable ones are dismantled further.
(c) The remaining assembly is clamped and the pole-shoe retaining screws removed by impact wrench. The drive end bracket fixing bolts are removed and the yoke assembly removed for examination. Serviceable yoke assemblies are degreased by hand, machine, chemical cleaning and by automatic shot blasting. The four pole shoes are separated, and cleaned. Unserviceable assemblies are separated, by removal of the earth rivet, into yoke and field coils.
(d) The remaining assembly is placed in a jig to compress the main spring, facilitating removal of the circlip that retains the drive parts on the armature shaft. The drive assembly is sent for sub-dismantling. The drive-end bracket is removed from the shaft and the bearing brush withdrawn for machining or cleaning, depending on its condition.
(e) The armature assembly is then tested. Serviceable units are re-skimmed before re-use; rewinding of unserviceable units is preceded by cutting off the ends of the copper hairpins, burning of the insulating material, and pressing the old windings and commutator out of the laminations.

Observations included the time of operation and the power rating of the tools, elapsed time for each operation, labour input, and power ratings of all cleaning, degreasing, abrasion and grinding processes.

3.2 Determination of the yield of reusable components obtained by dismantling.

3.3 Determination of the weight of each component and identification of the material and process route used in its manufacture.

3.4 Determination of the electrical, mechanical, thermal and human energy expended at each phase of the dismantling process.

It was assumed that the assembly process for the rebuilt motor was the same as that used in its original manufacture, such that the energy values for these may be omitted from the calculations. Observations were limited to establishing that this was in fact the case. In any event, any error will be in favour of reconstruction, since not all assemblies are completely dismantled.

4. RESULTS

4.1 Table 1 shows the amount of material used in the major components of the motor. The weights were determined in the present study, whilst the materials were from the official specification.

4.2 Table 2 gives data for the dismantling operations.

4.3 Table 3 gives the yield of the principal components.

5. STANDARD DATA

Power consumed by the equipment employed in rebuilding is given in Table 4.

An average figure for the energy expended through human effort in light industry is 14.5 kJ/min (.870 MJ/h) per 65 kg man; other values apply at certain stages and have been used for the appropriate calculations.

Data for the production of materials used in the motor are given in Table 5.

6. CALCULATIONS

Calculations are divided into three groups:

(a) Dismantling
(b) Cleaning
(c) Manufacture of components

The work to date has not provided detailed data on all cleaning processes or on yields for all components. Thus, calculations will be made for only those major components for which adequate data are to hand. Further, since it was not possible to determine the precise process route for all components, standard data for the most likely route have been used.

6.1 Dismantling

6.1.1 Power tool operations. The compressor generates 28.316 m^3/minute and is powered by a 140 kW motor. Therefore 1 m^3/min requires 3.96 kJ/s.

Energy required to remove terminal nuts

= Power tool rating (m^3/min) x energy requirement for 1 m^3/min (kJ/s) x time of operation(s)

= 0.706 m^3/min x 3.96 kJ/s x 2.72 s   = 0.0076 MJ/unit

Commutator end bracket screws

= 0.406 m^3/min x 3.96 kJ/s x 4.85 s   = 0.0078 MJ/unit

Drive end bracket bolts

= 0.706 m^3/min x 3.96 kJ/s x 3.66 s   = 0.0102 MJ/unit

Earth rivet

= 0.424 m^3/min x 3.96 kJ/s x 3.65 s   = 0.0061 MJ/unit

Pole shoe screws

= 0.565 m^3/min x 3.96 kJ/s x 6.43 s   = 0.00002 MJ/unit

Total for power tool operations   = 0.046 MJ/unit

6.1.2 Energy expended by dismantlers

General figure for light industry: 12.5-16.7 kJ/min/65 kg man

Total time for dismantling operation = 163.7 s/unit

High value = $\frac{163.7}{60}$ x 16.7 =   45.56 kJ/65 kg man

Low value $= \dfrac{163.7}{60} \times 12.5 =$ 34.10 kJ/65 kg man

Average value   39.83 kJ/65 kg man

Total labour   = 0.080 MJ/unit

Total energy of dismantling   = 0.126 MJ/unit

6.2 Cleaning of yoke assembly

Length of conveyor belt inside degreaser	:	42 m
Velocity of belt	:	3.96 m/min
Average residence time	:	10.6 min
Length of shift	:	8 h
No. of units processed/8 h shift	:	4 000

6.2.1 Handcleaning prior to degreasing

Standard value of energy expended	=	13.8 kJ/min/55 kg woman
Average cleaning time	=	0.95 min/unit
Energy expended	=	13.8 x 0.95 = 0.013 MJ/unit

6.2.2 Degreasing

Heat energy	= 790 MJ/h x 8 h		$= 6.32 \times 10^3$ MJ/shift
Extraction	= 3 hp x 750 J/s x 3 600 x 8 h	= 64.8 MJ/shift	
Conveyor	= 1.5 hp x 750 J/s x 3 600 x 8 h	= 32.4 MJ/shift	
Cooling water pump	= 7.5 hp x 750 J/s x 3 600 x 8	$= 0.162 \times 10^3$ MJ/shift	
Cooling tower	= 10 hp x 750 J/s x 3 600 x 8 x 50%	$= 0.108 \times 10^3$ MJ/shift	

Total energy used in degreaser $= 6.590 \times 10^3$ MJ + 97.2

$= 6.687 \times 10^3$ MJ/shift

Capacity of degreasing belt   = 4 000 units/shift

Energy per unit (at maximum capacity) $= \dfrac{6.687 \times 10^3}{4 \times 10^3}$ = 1.67 MJ/unit

6.2.3 Chemical plant

Extraction by 2 x 7.5 hp motors

Energy = 2 x 7.5 J/s x 3 600 x 8 = 324 MJ/shift

Extraction by 1 x 10 hp motor

= 10 x 750 x 3 600 x 8   = 216 MJ/shift

Heating of vats: 7 vats x 3 heaters x 20×10^3 J/s x 3 600 s/h x 8 h

= 12 096 MJ/shift

```
Total               =  12 636

Energy/unit         =  12 636
(at 4 000 units/       ───────  =  3.159 MJ/unit
     shift)             4 000
```

6.2.4 Automatic shot blasting machine

This is driven by one three horsepower motor, for approximately 5 min, with an average load of 95 units.

$$\text{Energy} = \frac{3 \times 750 \times 60 \times 5}{95} = \underline{0.007 \text{ MJ/unit}}$$

6.2.5 Total energy expended in cleaning and degreasing of yoke assemblies

$$= 0.013 + 0.007 + 3.159 + 1.67 = 4.849 \text{ MJ/unit}$$

However, only 53% of yokes from the dismantling step undergo the above cleaning process. Thus, total energy

$$= 4.849 \times 0.53 = \underline{2.57 \text{ MJ/unit}}$$

6.3 Machining of yokes

Machining, i.e. the drilling and tapping of holes to update to the latest specification, requires an average of 4.35 min on a 7 hp machine.

$$\text{Energy expended} = 7 \times 750 \times 60 \times 4.35 = 1.37 \text{ MJ/unit}$$

Since only 39% of yokes undergo machining

$$\text{Total energy} = 1.37 \times 0.39 = \underline{0.534 \text{ MJ/unit}}$$

6.4 Rewinding of armatures

6.4.1 Burning off of insulation

The burn-off plant processes 1 200 armatures/week for a total energy cost of 12 therms (1.2×10^6 Btu, 1 265 MJ).

$$\text{Energy expended} = \frac{1\ 265}{1\ 200} = \underline{1.054 \text{ MJ/unit}}$$

6.4.2 Rewinding of armatures

All armatures are rewound by hand, and a standard energy value of 9.2 kJ/min/65 kg man was assumed.

```
Average rewinding time  =  23.4 s/armature
                            23.4
Energy expended         =  ──── x 9.2  =  0.0036 MJ/unit
                             60
```

6.4.3 Soldering of armatures

Armatures are soldered at $170\,^{\circ}\text{C}$ for 2h using a 2 kW heater and with up to 10 armatures in the solder at any one time.

Energy expended $\qquad = \dfrac{2 \times 10^3 \text{ J/s} \times 3\ 600 \times 2}{10}$

$\qquad\qquad\qquad\qquad = \underline{1.44 \text{ MJ/unit}}$

6.4.4 Cleaning of laminations

The laminations are cleaned using a 5 hp scratch brusher for an average of 44 s.

Energy expended $\qquad = 5 \times 750 \times 44 = \underline{0.165 \text{ MJ/unit}}$

Total energy for rewinding of armatures $= 2.663 \text{ MJ/unit}$

However, only 61% of armatures undergo rewinding. Hence,

Total energy expended $\quad = 2.663 \times 0.61 = \underline{1.624 \text{ MJ/unit}}$

6.4.5 Reskimming of armatures

The commutator faces are reskimmed on a 1.5 hp lathe for an average of 63 s.

Energy expended $\qquad = 1.5 \times 750 \times 63 = \underline{0.071 \text{ MJ/unit}}$

The reskimmed armatures are also cleaned as above.

Energy expended $\qquad = \underline{0.165 \text{ MJ/unit}}$

Only 17% of armatures from the dismantling process go through the above reskimming operation

Total energy expended in reskimming armatures $= (0.071 + 0.165) \times 17\%$

$\qquad\qquad\qquad\qquad\qquad\qquad = \underline{0.04 \text{ MJ/unit}}$

7. YIELD OF REUSABLE PARTS

The lowest figure for the yield of reusable major components will be the limiting factor in the numbers of motors that may be rebuilt. This limiting value is the 64% of reusable commutator end brackets. Hence, 157 motors will, on average, be dismantled to yield sufficient parts to rebuild 100 units.

8. ENERGY COMPARISON BETWEEN ORIGINAL AND REMANUFACTURED COMPONENTS

8.1 Total energy of remanufacture

Dismantling	0.126 MJ/unit
Cleaning	2.570
Machining	0.534
Rewinding	1.624
Reskimming	0.040
	4.894 MJ/unit

8.2 Total energy of original manufacture

For the components for which yield and dismantling data are available

Yoke (including welding)	74.9 MJ/unit
Pole shoes	23.0
Field coil	83.0
Armature coil	11.6
Commutator	9.4
Armature shaft and laminations	69.1
	271.0 MJ/unit

9. DISCUSSION

This investigation has been limited to an analysis of the direct energy associated with the reconditioning of certain major components of the starter motor. Of the components not so far considered the drive end bracket represents the largest initial energy investment. No attempt has been made to assess indirect energy consumption, such as factory heating and lighting, since these would presumably also be involved in any assembly process with which reconditioning is being compared.

The errors in the study are associated with the process routes in the manufacture of the component parts of the motor. However, even in the case of the most extreme assumption, the energy requirement will still be within 15% of the values determined. Such errors are insignificant in relation to the energy difference between reconditioning and initial construction:

$$\frac{\text{Energy of reconstruction}}{\text{Energy of initial construction}} = \frac{4.894}{271} \times 100\% = \underline{1.8\%}$$

This difference, of almost two orders of magnitude, does not reflect the fact that components that are, during reconditioning, discarded as unserviceable will often be reclaimed as materials rather than as parts. Although such reclamation and recycling will clearly result in energy economies relative to metal extraction from ore, these savings relate to the general advantages of recycling and not to the specific activity of reconditioning. (Chapman, 1974)

10. CONCLUSIONS

The reconditioning of the starter motor has been shown to confer considerable energy savings relative to the provision of new component parts. Savings are of a magnitude such that the economic advantages of the process are unlikely to be affected by forseeable rises in energy price.

More than 50% of the energy requirements (including manpower) of reconditioning arises from cleaning and degreasing of yoke assemblies; the most energy-intensive operation is the heating of the cleaning vats, which consumes about 34% of total direct energy.

TABLE 1   Materials used in the starter motor

Component	Material	Process	Weight(kg)
Yoke	Low carbon mild steel	Hot rolled non-rimming strip; welded	1.508
Pole shoes	Low carbon mild steel	Drawn	0.121 (for each of the set of 4)
Field coils	Aluminium	Drawn and annealed	0.156
Commutator disc	Electrolytic copper	Sheet or strip rolled and annealed	0.071
Armature coil	Electrolytic copper	Drawn wire or section strip	0.105
Armature laminations	Low carbon steel	Strip or sheet cold rolled or reduced	0.842
Shaft	0.3% carbon steel	Bar, as drawn case hardened	0.842
Drive end bracket	10% silicon aluminium alloy	Chill cast	0.188
Commutator end bracket	Low carbon steel, extra deep drawing	Cold rolled and annealed	0.158
Pinion	Carbon-Mn steel, fine grained, high tensile	Bar, bright drawn	
Barrel	Low carbon, deep drawing	Coiled strip cold reduced	0.176
Screwed sleeve	Low carbon free cutting steel	Bar, bright drawn	0.033
Main spring	Low carbon steel	Bar, cold drawn, case hardened	0.088
Brushes	76% copper plus lead, carbon and manganese	Powder metallurgical process	-

TABLE 2   Times for power tool operations in dismantling

Action	Time per unit(sec)	Power rating (m³/min)
Remove terminal nuts	2.72	0.706
Remove C.E. bracket fixing screws	4.85	0.406
Remove D.E. bracket fixing bolts	3.66	0.706
Remove head of earth rivet	3.65	0.424
Remove pole shoe screws	6.43	0.565
Remove circlip	*	0.005

*
1 operation of 15 cm cylinder with a 20 cm stroke per unit

TABLE 3   Yields of principal re-usable components

Component	Yield (%)
Armatures suitable for reskimming only	17
Armatures requiring rewinding	61
Yoke, serviceable assemblies	53
Yoke only	39
Commutator end bracket	64
Drive end bracket	69

TABLE 4   Power supplies to reconditioning plant

Degreaser:	Heat input	790 MJ/h	
	Extraction	8.1 MJ/h	(3 hp motor)
	Conveyor belt	4.05 MJ/h	(1.5 hp motor)
	Cooling water pump	20.25 MJ/h	(7.5 hp motor)
	Cooling tower pump	27.0 MJ/h	(10 hp motor, working 50% of the time)
Chemical plant:	Extraction	40.5 MJ/h	(2 x 7.5 hp motors)
	7 heated vats	0.42 MJ/h	(3 x 20 kW heaters each)
Automatic shot blasting:		8.1 MJ/h	(3 hp motor)
Compressor for power tools:		504 MJ/h	to produce 28.3 m³/min (1 699 m³/h or 60 000 ft³/h)

TABLE 5 Energy data for the production of components
used in the starter motor (Boustead, Hancock, 1979)

Component	Energy requirement (MJ/kg)	Weight (kg)	Energy expended (MJ)
Yoke	49.6	1.508	74.8
Yoke welding	-	-	9.4×10^{-2}
Pole shoes	47.5	0.121	23.0
Field coils	532.1	0.156	83.0
Armature coil	110.2	0.105	11.6
Commutator disc	132.7	0.071	9.4
Armature Laminations	51.3	1.683	69.1
Armature shaft	30.8		
D.E. bracket	276.1	0.188	51.9
C.E. bracket	55.0	0.158	8.7
Pinion	53.1	0.176	19.0
Barrel	54.7		
Screwed sleeve	51.5	0.033	1.7
Main spring	51.5	0.088	4.5

(The yoke is welded using submerged arc welding with double deoxidized copper coated mild steel at 280 amps and 28 volts at a speed of 100 cm/min over a length of 20 cm.)

REFERENCES

1. Boustead, I., and Hancock, G.F., Handbook of Industrial Energy Analysis, Chichester, Ellis Horwood (1979)

2. Chapman, P.F., The energy costs of producing copper and aluminium from primary sources, Metals and Materials, 8, (2), pp. 107-111, (February, 1974).

3. Chapman, P.F., Energy conservation and recycling of copper and aluminium, ibid, 8, (6), pp. 311-319, (June, 1974).

4. Long, Thomas Veach, II, Workshop report: International Federation of Institutes for Advanced Study, Workshop on energy analysis and economics, Lidingö, Sweden, (June 22-27, 1975).

5. Lund, Robert T., Incentives for longer product life: a case study, Working paper, Center for Policy Alternatives, Massachusetts Institute of Technology, CPA/WP-75-15, (October, 1975).

ACKNOWLEDGEMENT

The authors wish to acknowledge the cooperation and assistance of Lucas Electrical Ltd., which was accorded in the preparation of this paper.

THE COMPOSITION OF THERMOECONOMIC FLOW DIAGRAMS

R. A. Gaggioli* and W. J. Wepfer**

*Marquette University, Milwaukee, WI, 53233, USA
**Georgia Institute of Technology, Atlanta, GA, 30332, USA

ABSTRACT

The commodity called exergy provides a rational measure for assigning unit costs (e.g., ¢/KJ) to "energy" utilities, whereas energy does not. The methods of cost accounting on the basis of exergy is presented in an easily comprehensible manner. The resulting Thermoeconomic flow diagrams for actual case studies of industrial power and chemical process complexes are presented, along with the procedures for composing the diagrams.

KEYWORDS

Second Law, Thermoeconomics, Costing, Exergy, Available Energy.

INTRODUCTION

It is the exergy content of the commonly labeled "energy" comodities, not their energy, which really represents their potential to cause changes such as to work, heat, convert chemicals, etc. Consequently, it is the exergy of these commodities that is used up as a process proceeds, and that is worth paying for. It thus makes sense to do cost accounting on the basis of exergy rather than energy (Gaggioli & Wepfer, 1980a).

The exergy concept goes back to Maxwell (1871) and Gibbs (1875) and has had many names such as available energy, essergy, availability. It was Keenan (1932) who first emphasized that cost accounting should be based on availability not energy. Obert and followers have made straightforward applications of exergy costing in some uncomplicated cases of practical importance (see Gaggioli & Wepfer, 1980a, for example), while Evans, Tribus and followers have made important theoretical contributions (e.g., Evans and El-Sayed, 1970). The aim, here, is to provide a good understanding of the principles and the realm of exergy accounting.

EXERGY ACCOUNTING

Exergy accounting in "energy" systems is important in regard to a number of types of engineering-economic decision making. The types to be considered here are (i) cost accounting, and (ii) design improvement.

Money Balances. The first step in costing "energy" flow streams is the application of money balances to the system of interest. The cost (C) of the product of an energy-converter equals the total expenditure made to obtain it-- the fuel expense, plus the capital and other expenses:

$$C_{products} = C_{expended} = C_{fuel} + C_{capital}, \text{ etc.}$$

The average unit cost, c_p, of a single product is the total cost, $c_{product}$ divided by the output power: $c_p \equiv C_{prod}/P_{prod}$. Similarly, the fuel cost can be expressed as $C_{fuel} = c_f P_{fuel}$ where c_f is the unit cost of fuel and P_{fuel} is the power input to the converter. The money balance then yields

$$c_p = c_f P_{fuel} / P_{prod} + C_{capital} / P_{prod}$$

The first term on the right reflects fuel costs, and the second represents the rate of amortization of capital (and other) expenses.

Cost Allocation Methods. Consider an energy-converter that takes in one exergy supply and delivers a single product, for example a boiler that takes in fuel and produces steam. For single-output devices it makes no difference which measure of product output is chosen, the same result is obtained for the product cost.

As in the case of a back-pressure turbine, for example, some energy-converters yield multiple outputs. The money balance on such a turbine is

$$c_{LP} P_{LP} + c_{shaft} P_{shaft} = c_{HP} P_{HP} + C_{turbine}$$

where HP and LP represent high-pressure (supply) steam and low-pressure (output) steam. Generally the right-hand side and all the power flows (P's) are known. This results in one equation with two unknowns: the average unit costs c_{LP} and c_{shaft}. Another equation is needed. There are various methods for obtaining a second equation. Which method is appropriate is outside the province of thermodynamics, and depends on cost-accounting considerations of the specific facility and of the economic decision-making purpose of the costing. Two typical cost allocation methods for obtaining additional cost equations will now be presented. For concreteness, the context will be co-generation turbines.

The extraction method assumes that the sole purpose of the turbine is to produce shaft power. Therefore, the shaft work is charged for the complete capital cost of the turbine and for the steam exergy used by the turbine to produce the work. With this rationale, the additional equation is obtained by equating the unit costs of high- and low-pressure steam exergy $c_{LP} = c_{HP}$. The money balance yields

$$c_{shaft} = c_{HP}[P_{HP} - P_{LP}]/P_{shaft} + C_{turbine}/P_{shaft}$$

The result is that the shaft work bears the entire burden of the costs associated with the turbine process and capital expense.

Insofar as the purpose of a turbine is to produce shaft work, the extraction method is appropriate. However, if there were no turbine, the efficiency of producing any required low-pressure steam would be lower and, hence, its unit cost higher. So when both steam and shaft work are required it is not inappropriate to charge this steam proportionately for the costs attributable to the turbine. The equality method charges the shaft work exergy and that of the low-pressure steam exergy equally, for the cost of high pressure steam and turbine capital: $c_{LP} = c_{shaft}$. Then, the money balance on the turbine gives

$$c_{shaft} = [c_{HP} P_{HP} + C_{turbine}]/[P_{LP} + P_{shaft}]$$

The choice of a costing method depends upon the object of economic study and/or upon the circumstances under which it is being made.

Up to this point the methods discussed involve only <u>average</u> unit costs of flow streams. A different problem is one in which, for example, an economic study has the purpose of ascertaining the desirability of increasing (or decreasing) the manufacture of some product, which results in additional power requirements. <u>The incremental method</u>--which charges the additional units of power required for the cost of producing them--is then inappropriate. The use of <u>average</u> cost of producing all of the units would be inappropriate (except in the special, though not infrequent, cases when incremental costs are close to average costs).

Another instance when incremental costs are appropriate is the case of <u>design optimization</u>. For example, suppose that a co-generating industrial utility were being designed to meet specified low-pressure steam needs, whereas the electric tric power output was not specified. What electric power output should the system be designed to deliver? Say that whatever electric power is produced could be sold at a market price of π. Power production capacity can be changed by varying the temperature T_{HP} and/or pressure p_{HP} of the high-pressure steam for examples. The optimal T_{HP} and p_{HP} would be those for which the profit $\Pi = \pi P_{elec} - C_{total}$ would be maximized; that is (if T_{HP} and p_{HP} are unconstrained), when

$$0 = \pi \frac{\partial P_{elec}}{\partial T_{HP}} - \frac{\partial C_{total}}{\partial T_{HP}} \quad \text{and} \quad 0 = \pi \frac{\partial P_{elec}}{\partial p_{HP}} - \frac{\partial C_{total}}{\partial p_{HP}}$$

In other words, the profit would be optimal when the incremental costs $\partial C_{total} / \partial P_{elec}$ of the power equal the price; e.g.

$$c_{elec,T} \equiv [\partial C_{total} / \partial T_{HP}] / [\partial P_{elec} / \partial T_{HP}] = \pi$$

Evans, Tribus et al (e.g., 1970) have shown how, using incremental costs per unit of exergy, <u>each</u> of the units in a complete plant can be optimized in isolation (virtually) from the rest, with assurance that the overall plant is optimal-- assurance that is achieved, practically, by iteration. One advantage over the usual overall optimization methods is that the designer is not kept remote from the individual units, but can study each one.

<u>Flow Diagrams</u>. Whatever the intended purpose of a Thermoeconomic analysis--for cost accounting, system operation, or design--flow diagrams are useful. A pre- requisite is an exergy flow diagram, showing all the flow rates of exergy to and from the plant and between the various units, as well as the exergy consumption within each unit. (Such diagrams are valuable in themselves, showing the <u>real</u> inefficiencies--opportunities for improvement.) Using the exergy flow diagram, money balances may be applied unit by unit, along with costing methods like those described above, to get (i) the rate of money flow with the various exergy flows, and (ii) the unit costs thereof. Then money flow diagrams and unit-cost diagrams can be constructed, showing the money flow rates and unit costs associated with each of the exergy flow streams. This information can be superimposed on the exergy flow diagram. For example, Figs. 1 and 2, to be discussed in detail presently, are composite exergy-flow and exergy unit cost diagrams.

COMPOSITION OF FLOW DIAGRAMS

Exergy Flow Diagrams are composed in the same manner as conventional energy flow diagrams. Depending upon the commodity with which exergy is being carried, the appropriate, usual thermodynamic property evaluations are made; for example, see Gaggioli and Petit (1977 and 1980), Rodriguez (1980). Exergy flows are shown on simplified diagrams for two co-generating power plants (Wepfer, 1980) on Figs. 1a and 1b, and a coal gasification plant (Gaggioli and Wepfer, 1980b) on Fig. 2.

Money-Flow and Unit-Cost Diagrams. Figures 1 and 2 also show unit costs of exergy in parentheses. The determination of these unit costs will now be described.

Consider first Figure 1a. For convenience, presume that capital expenses are irrelevant because either (i) the plant has been fully depreciated, or (ii) the current decision to be made calls for the viewpoint that the capital has been sunk. The costs of fuel are 0.52¢/kwh of exergy in natural gas and 0.55¢/kwh in coal. Typically, the costing analysis would be started by applying money balances to each of the two boilers, in order to calculate the costs of the 10.5 Mw of exergy in 600 psi steam from the HP boiler and the 1.5 Mw in 175 psi steam from the LP boiler. However, the unit costs of electricity and of the condensate -- "feedback" streams -- are unknown; these unit costs are among the results to be obtained from the costing analysis. Therefore an iterative procedure is called for, starting with assumed unit costs for the "feedback" streams. Since, in this case, the exergy content of these streams is very small, the accuracy of the initial guesses is more or less inconsequential. (Typical means for improving the initial estimates, when more important, will be illustrated with Figure 1b.)

Once the first-iteration unit cost of the 10.5 Mw boiler output is obtained, a money balance may be applied to the turbine. Presuming that the role of the turbine is, equivalently to produce steam efficiently at 50 and 175 psig as well as shaft work, the unit costs of exergy in three outputs would be equated: $c_{shaft} = c_{50} = c_{175}$. Then these could be obtained from the turbine money balance. Next, for example, a money balance could be applied to the generator to get the first-iteration unit cost of electricity.

The 0.2 Mw of exergy in the condensate returning from the plant can be credited to the plant at the same unit cost as the exergy supplied to it in 50 psig steam; that would be to charge the plant for the exergy which it extracted from that H_2O. In turn, the unit cost of the condensate supplied to the feedwater heater would be determined with a money balance on the condensate pump, employing the first-iteration unit cost of electrical exergy for the power to the pump motor. Then a money balance on the heater (including the pressure reducing valve) can be used to get the unit cost of the condensate supplied to the boiler feedpumps, and a money balance on each feedpump yields the first-iteration unit cost of the respective boiler feedwater supplies.

With the first-iteration unit costs for electricity and feedwater inputs to the 600 psi boiler, the second-iteration electricity and feedwater costs could then be determined. The iterations could continue until these costs converged satis- factorily. Then a money balance could be applied to the 175 psi boiler to get the final unit cost of its 1.5 Mw output. And, finally, a balance on the mixing of this 1.5 Mw with the 1.0 Mw in 600 psi steam and the 0.5 Mw in 175 psi turbine back-pressure steam will yield the final unit cost of the 2.7 Mw delivered to the plant in 175 psi steam, completing the costing.

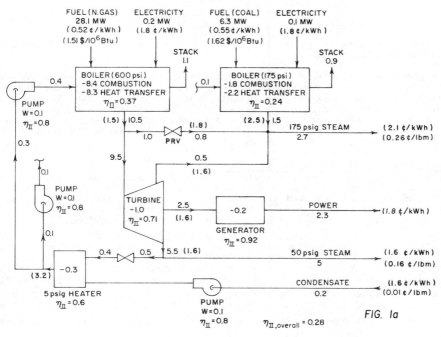

FIG. Ia

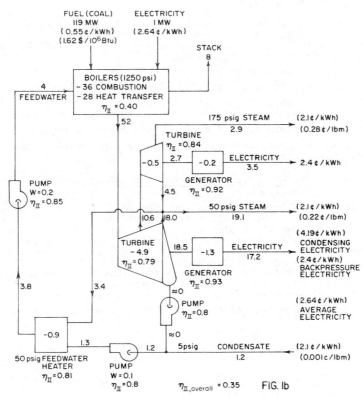

FIG. Ib

The procedure described is among several alternative paths leading to the final results shown on Figure 1a.

Figure 1b shows the results of a thermoeconomic analysis of another co-generating facility. In this case the facility was not already in existence, but was among alternatives being contemplated for installation. On the one hand it was desired to establish the unit cost of steam and electricity, and on the other hand comparison of the cost of electricity produced by the high-pressure and intermediate-pressure turbine stages with that produced by the low-pressure ("condensing") stage. The costing used was the following:

Preliminary Costing of System Exergy Outputs

1. Preliminary costing of output steam and HP and IP shaft work: (a) Assume that all equipment -- except the generators, the low pressure (50 psig to 1" Hg) turbine stage, the condenser and the condensate pump -- is devoted to producing 175 psig and 50 psig steam and shaft work on an equivalent basis; i.e. with $c_{175} = c_{50} = c_{SWHI}$. (b) Apply a money balance to all the equipment, including $182.31/hr for capital ($17,000,000 amortized over 20 years at 8.5% after taxes, and accounting for income and property taxes) plus $33.94/hr estimated operating expenses excluding "fuel". Assuming a "fuel" cost of 2.35¢/kwhr for input electricity -- the cost at which power could be purchased from the local utility -- solve for $c_{175} = c_{50} = c_{SWHI}$. (c) Apply a money balance to the generating equipment driven by the HP and IP turbine stages, with $22.65/hr plus $0.39/hr for capital and operating expenses and c_{SWHI} for the unit cost of shaft work, to get the unit cost of preliminary costing for electric power c_{EHI}. This completes the first iteration of the c_{EHI}, c_{175} and c_{50}.

2. Preliminary costing of LP shaft work and power: Apply a money balance to all of the equipment attributable to the production of "condensing electricity" (i.e., the equipment excluded in Part 1), amortized at the rate of $14.88/hr for capital assuming operating costs of $0.07/hr, and using a "fuel" cost at the rate c_{50} from Part 1; solve for c_{EL}. This completes the first iteration for preliminary c_{EL}. Finally, the average first-iteration cost of electric power from the whole plant can be evaluated: $c_E = [c_{EL}P_{EL} + c_{EHI}P_{EHI}]/[P_{EL} + P_{EHI}]$. With this cost for electric "fuel", the second iteration for preliminary values of $(c_{175}, c_{50}, c_{EHI})$ and of (c_{EL}, c_E) can be determined. The unit costs shown in Figure 1b are these preliminary costs, after three iterations (differing from those of the second iteration by less than one percent).

Final Costing of Intra-System Exergy Flows and System Outputs.
The foregoing preliminary costs give, for example, only the average cost c_{50} of the 19.1 Mw of 50 psig steam, without differentiating between the costs of the 4.5 MW from the IP turbine and the 18.0 Mw from the IP stage of the condensing turbine (which together yield this 19.1 Mw output plus the 3.4 Mw supplied to the feedwater heater). If it is desired to obtain the unit costs of such intrasystem exergy transfers, they can be determined by applying money balances to the individual components -- boiler, HP stage of the condensing turbine, IP stage, IP turbine, etc. -- in the manner described for the system of Figure 1a. The unit costs from the preliminary costing can serve as a point of departure for this more detailed costing, using the preliminary c_E for the unit cost of electric "fuel" to the boiler and pumps, and $c_{175} = c_{50}$ for the unit cost of condensate exergy. (Note

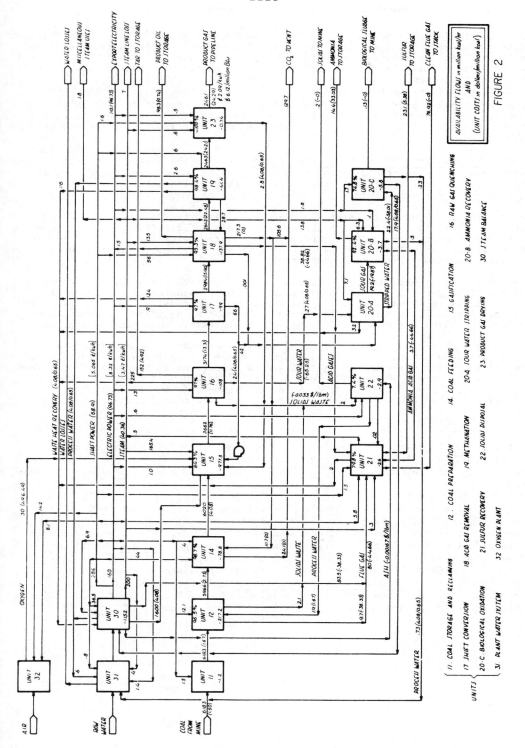

FIGURE 2

that such preliminary costing could have been done to get starting values for unit costs of condensate and electric "fuel" for Figure 1a; the preliminary costing would have been more straightforward for Figure 1a than 1b.) After the iterative determination of the intrasystem costs, the results can be of value, for example, as guides to design modifications (e.g., Gaggioli and Wepfer, 1980b).

Also, it should be mentioned that values of c_E, c_{EL}, c_{EHI}, c_{175} and c_{50} obtained from the detailed costing could differ slightly from those of the preliminary costing; e.g., the unit capital costs attributable to the exergy outputs of the HP and IP stages of the condensing turbine could differ from those for the IP turbine.

Summary of Results, and Practical Applications. With the simple systems portrayed by Figures 1a and 1b, the typical steps have been shown for the composition of thermoeconomic flow diagrams. Figure 2 shows the results obtained, using analogous procedures, for a more complex process – synthane coal gasification.

The company which owned the system of Figure 1a was being allowed to sell any excess electricity (over and above its plant needs) at 1.8¢/kwhr -- lower than the cost of producing it -- and hence necessarily overcharging for the steam going to its own processes. The foregoing results were used to argue, with the Public Service Commission, for a price increase.

When this company was short of electric power, it had to pay 2.35¢/kwhr. During the course of planning for expanded steam and power capacity at another mill, the system of Figure 1b was one of the alternatives being considered. One conclusion was that the large turbine should not have the condensing stage, because power derived therefrom was much more expensive, at 4.2¢/kwhr, than purchased power.

The conceptual design shown in Figure 2 led to inordinately high unit costs for the utilities -- steam, shaft work, electric power. Straightforward revision of the design of the utilities subsystems led to reductions in these unit costs from 3.47 to 2.15, from 5.06 to 2.70, from 8.32 to 2.35¢/kwhr. In turn, and most importantly, the unit cost of product gas was reduced from 6.12 to 4.97 dollars/10^6 Btu. See Gaggioli and Wepfer (1980b) for more details.

REFERENCES

Evans, R. and El-Sayed, Y. (1970). "Thermoeconomics and the Design of Heat Systems," Trans. A.S.M.E., J. Eng. Power, 92, 27-35.

Gaggioli, R. and Petit, P. (1977). "Use the Second Law First," Chemtech, 7, 496-506.

Gaggioli, R. and Wepfer, W. (1980a). "Exergy Economics," Energy, 5, 823-838.

Gaggioli, R. and Wepfer, W. (1980b). "Second Law Costing Applied to Coal Gasification," AIChE Tech. Manual, VI, 140-145.

Gibbs, J. W. (1875). See Collected Works, vol. 1, p. 77, Yale U. Press, (1948).

Keenan, J. (1932). "A Steam Chart for 2nd Law Analysis," Trans ASME, 54, 195.

Maxwell, J. (1971). Theory of Heat, Longmans Green, London, (later editions).

Petit, P. and Gaggioli, R. (1980). "Second Law Procedures for Evaluating Processes," Thermodynamics: 2nd Law Analyses, ACS Symposium Vol. 122, 15-38.

Rodriguez, L. (1980). "Calculation of Available Energy Quantities," ibid., 39-60.

Wepfer, W. (1980). "Applications of Available Energy Accounting," ibid., 161-186.

INTEGRATED MULTI-TASK ENERGY SYSTEMS

K. Illum

Institute of Development and Planning, Aalborg University Center,
P.O. Box 159, 9000 Aalborg, Denmark

ABSTRACT

In this paper it is demonstrated that the prevalent concept of national or
regional energy systems is closely tied up with the concept of energy as a
commodity that is consumed by segregated sectors of the economy in quantities
that may be immediately measured in some standard unit (joule, calories etc.).
When the traditional systems are reorganized in order to enhance their thermodynamic efficiency and to utilize with better advantage the potentials sustained
by solar radiation it appears, however, that the concept of energy as a commodity flowing through the economic system towards the points of end use is
no longer relevant.   Therefore, an alternative paradigm is suggested for the
conception and description of integrated systems that are able to sustain
required states of thermodynamical disequilibrium with respect to the atmosphere.

It is shown that in urban areas efficient performance may be obtained more
readily with integrated multi-task systems than with separated systems, each
performing a single task.

KEYWORDS

Energy system structure; cogeneration; computer simulation.

THE POINT OF DEPARTURE

The prevailing classical techniques applied in our affluent societies require
abundant amounts of fuel because they were established during the period when
fuel resources seemed to be abundant.   It seems a reasonable assumption at our
point of departure, that engineering and organizational ingenuity, concerning the
exploitation of available potentials at maximum efficiency with respect to the
accomplishment of the entire set of tasks in human societies, has not been
nurtured during the period of cheap fossil fuels, and that it may be possible in
future to establish another kind of human affluence within a framework of
energetic structures very different from the presently existing systems.

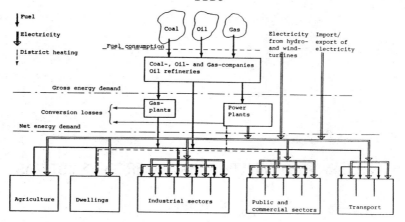

Fig. 1. The traditional concept of energy demand systems.

From the point of view of economics and social organization the traditional energy
systems may be characterized as sectorial consumer systems as shown by Fig. 1.
Each sector consumes a certain amount of fuel and a certain amount of electricity
per year.   The total consumption of those commodities may by simple bookkeeping
be added up to what is called the net energy use, all quantities being measured
by their calorimetric heating values.

As long as the cogeneration of electricity and thermal effects for district
heating in the public utility power plants only plays a minor role, the
calorific analysis of energy flows suffices for the evaluation of the effects of
potential energy conservation measures in each sector upon the gross energy
(i.e. fuel) consumption in a system as shown by Fig. 1.   For example, the
effects on gross fuel consumption of a certain improvement of the heat insulation
of buildings, the application of solar panels for hot water supply or the
replacement of an old drying plant by a less inefficient one may immediately
be determined.   Also the change in fuel consumption caused by a change in the
consumption of electricity may readily be computed.

The simplicity of this analysis, which allows for the use of simple calorimetric
concepts as "energy consumption in industry", "energy demand" etc., is however
a characteristic property of the extremely simple structure of the traditional
system configuration and not a general property of energy conversion systems.

Setting out to explore the prospects of technological innovations that may be
implemented in order to attain human needs through more efficient utilization
of available potentials, we must extend the analysis to cover alternative
systems which are structurally more complex than the traditional setup and
consequently recognize the need to establish a more generally valid conceptual
basis and methodology for the analysis.

Papers by prominent authors in this field of applied science - notably Keenan
(1973), Reistad (1975), Gaggioli (1977), Gyftopoulos (1977) among several
others - and study group reports such as API's "Efficient Use of Energy"
(1975) have been instrumental in bringing about a rational understanding of
the problem of efficiency in the construction of energy systems.   Reports
like "Cogeneration of Steam and Electric Power" (1978), "Potential Fuel
Effectiveness in Industry" (1974) and the article by Williams (1978) have
demonstrated the practical feasibility and economic advantage of improving
process efficiency.

A PARADIGM FOR THE DESCRIPTION OF INTEGRATED MULTI-TASK SYSTEMS

When the constraints embedded in the present economical and organizational
structures are relaxed in order to examine technical prospects as to the fulfil-
ment of human demands through efficient utilization of available potentials, a
system description paradigm in which subsystems are identified with regard to
their thermodynamic characteristics appears to be appropriate.

Within an environment described by the varying climatic and meteorologic
conditions in the atmosphere and the pattern of human activities, the technical
system may in general be conceived as two interacting subsystems: a thermodynamic
process system and a general demand system, see Fig. 2.

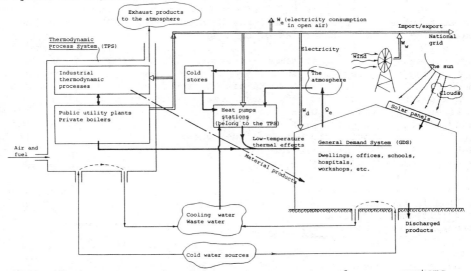

Fig. 2. A general paradigm for the description of energy systems.

Processes involving chemical reactions, phase transitions, heat interactions at
high temperatures and controlled expansion or compression of gases are grouped
with the thermodynamic process system, for short denoted TPS in the following.
The TPS thus comprises heat engines, heat pumps and refrigeration plants, boilers
and industrial production involving thermodynamic processes.

The general demand system, denoted GDS in the following, represents the immediate
physical environment around human indoor activities, i.e. the buildings
enclosing workshops, offices, shops, private homes etc. together with electrically driven mechanical machines for manufacturing, lamps for illumination,
office machines and communication devices, electrical appliances in households
etc.   Solar panels may be considered to be parts of the building structures
which serve to increase the absorption of solar radiation.   The physical composition of the GDS together with the human pattern of behaviour and the climatic
variations determine the variations in the demand for electrical effects and
thermal effects from the TPS.

In addition to electricity and heat the TPS may supply materials produced through
thermodynamic processes to the GDS.

Note that this conceptual grouping should not be interpreted as a geographical
partitioning of the TPS from the GDS.   Thermodynamic process machinery may be
located in buildings belonging to the GDS and may thus contribute to the heating
of rooms by direct heat interaction.   Note also that traditional energy system

structures fit into this paradigm as special cases of the comprehensive class of structures covered by it.

The loss of available work attached to the supply of low temperature thermal effects from the TPS may be kept at an acceptable low level when those effects are drawn from condensers at temperatures only slightly above the temperature required in the central heating systems.   Either the condenser of a heat engine or the condenser of a heat pump may be used.   An efficient TPS that will meet the varying demands for simultaneous power and heat effects from the GDS may thus be constructed as shown in principle by Fig. 3.

The temperatures of the high-temperature reservoir, which is sustained by the continuous combustion of fuel, may be higher than 2000°C.   For metallurgical reasons the actual temperatures of smoke gases circulated through boilers or turbines must be kept somewhat lower.   Modern gas turbines with ceramic blades may, for example, be run at temperatures up to about 1500°C.   When heat effects in the range from say 100°C to ca. 600°C are required for some industrial process, the direct application of hot smoke gas to provide those effects implies a loss of the potential work that could have been produced by some heat engine that utilizes the thermal potential between the hot smoke gas and the substances circulating in the production machinery.   This loss may be avoided by inte-grating the production process into the general utility system as shown in principle by Fig. 4.

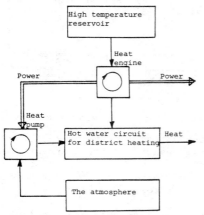

Fig. 3. The principle of efficient generation of low-temperature thermal effects in combustion systems.

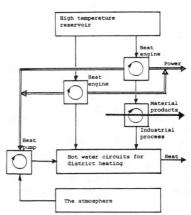

Fig. 4. The principle of efficient generation of thermal effects for industrial processes.

A substantial improvement of the efficiency of the thermodynamic process system with respect to the set of tasks determined by the GDS, including material production requirements, may thus be obtained by an integrated process system instead of several machines each serving a particular task.

ON THE FORMAL DESCRIPTION AND COMPUTATIONAL ANALYSIS OF DISCRETE CONTROL MODELS OF INTEGRATED SYSTEMS

A system of the type given by Fig. 2 may be modelled as a discrete control system which performs the task of sustaining certain quasistatic states of disequilibrium with the atmosphere in some of its subsystems.

A general demand system (GDS) may be described physically as a composition of simple systems (in the thermodynamical sense) with finite or infinite heat capacities, which are pairwise coupled by energy transducers. The atmosphere may be represented by a simple system with an infinite heat capacity and with a temperature which is an exogeneously determined function of time. The interior of rooms, the components of building structures, water tanks etc. may be represented by simple systems with finite heat capacities and with temperatures that vary at rates determined by the immediate thermal effects across the transducers. The transducers may represent heat conductors, heat convection circuits, heat radiation, electric conductors, absorption of solar radiation and converters from kinetic wind energy to electricity. The wind may be considered to be a work reservoir from which work is delivered to the electric grid at a varying, exogenously controlled rate.

A thermodynamic process system (TPS) may be described as a network of coupled process controllers: heat exchangers, condensers, boilers, turbines, compressors, pumps etc., each of which represents a function that maps one or two input states onto one or two output states, possibly with work or heat as additional inputs/outputs. Examples of possible process systems structures are given by Figs. 5 - 8. A formal language for the description of such thermodynamic systems in a linear notation has been developed together with a computer program for the compilation of computer models and the computation of stationary modes of operation on the basis of system descriptions in this language.

In a discrete control model the state of the system is evaluated and the appropriate control actions executed at discrete points of time $t_0, t_1, \ldots, t_i, t_{i+1}, \ldots, t_{i+1} = t_i + \Delta t$. The control actions at time t_i may consist in changing the mode of operation of the TPS and the positions of valves that control heat convection circuits. When a circuit is opened(closed) its power-consuming pumps will be started(stopped).

Assume that the TPS has a finite number of possible modes of operation, let the members of the set $M = \{m_1, \ldots, m_n\}$ denote the different modes and let $M(t_i) \in M$ denote the actual mode of operation during the time interval from t_i to t_{i+1}. Then

$$M(t_i) = C(M(t_{i-1}), \bar{T}(t_i), T_a(t_i), \bar{V}(t_i), \bar{P}(t_i)) \tag{i}$$

where C is the discrete control function that determines the next mode of operation $M(t_i)$ from the previous mode $M(t_{i-1})$ and the values of

$\bar{T}(t_i) = (T_1(t_i), \ldots, T_r(t_i), \ldots)$: the vector of endogenously controlled temperatures of simple subsystems S_r, $r = 1, 2, \ldots$, with finite heat capacities.

$T_a(t_i)$: the temperature of the atmosphere.

$\bar{V}(t_i) = (V_1(t_i), V_2(t_i), \ldots)$: the vector of wind velocities at the locations and altitudes of installed windmills.

$\bar{P}(t_i) = (P_1(t_i), \ldots, P_j(t_i), \ldots)$: an exogenously controlled boolean vector containing information as to which industrial processes are required to be in operation at the moment, i.e. $P_j(t_i) = 1$ if process no. j is operating at time t_i and $P(t_i) = 0$ otherwise.

$M(t_i)$ may depend upon $M(t_{i-1})$ because it may be desirable that the frequency of mode shifts be kept at a minimum.

The temperature values $T_r(t_i)$ result from integration over the time interval from t_o to t_i of the set of simultaneous differential equations

$$\frac{dT_r}{dt} = \frac{1}{c_r} \left[\sum_s k_{rs}(T_s(t) - T_r(t)) + \sum_q B_q(t_i) \cdot \kappa_{qr} \cdot \bar{T}_q(t_i) + k_{ra}(T_a(t) - T_r(t)) \right.$$

where
$$\left. + \bar{x}_r(t_i) - \rho_r(T_r(t)) \right] \tag{ii}$$

c_r is the heat capacity of the simple system S_r.

k_{rs} is the coefficient of thermal conduction between the simple subsystems S_r and S_s.

q is an index to a heat convection circuit that interacts with S_r and a number of other subsystems.

B_q is a binary function with value 1 if the circuit q is open and 0 otherwise. $B_q(t_i)$ is determined by a discrete control function in the same manner as $M(t_i)$:
$$\bar{B}(t_i) = (B_1(t_i), \ldots, B_m(t_i)) = B(M(t_i), \bar{T}(t_i), T_a(t_i)).$$

$\kappa_{qr}(\bar{T}_q(t_i))$ is the thermal effect from the circuit q to the subsystem S_r, computed from the temperatures $\bar{T}_q(t_i)$ of the subsystems that q passes through and the efficiencies of the respective heat exchangers included in the circuit.

k_{ra} is the coefficient of thermal conduction between S_r and the atmosphere.

$\bar{x}_r(t_i)$ is the vector of exogenously controlled effects to S_r: solar radiation effects, depending on date and hour of the day, degree of cloudiness and geographical location; electrical effects controlled by human action; heat convection effects caused by the drawing of hot water for ablution and other purposes; heat from human bodies; etc.

$\rho(T_r(t))$ denotes heat radiation to the atmosphere.

For a given set of functions $(T_a, \bar{V}, \bar{P}, \bar{x}_1, \bar{x}_2, \ldots, \bar{x}_r \ldots)$, representing the exogenously controlled variables, and a given set of control functions $\{C, B\}$, the sequence $M(t_i)$, $i = 1, 2, \ldots$, is determined by (i) and (ii). As the rate of fuel consumption $f(t_i)$ during the time interval from t_i to $t_{i+1} = t_i + \Delta t$ is determined by $M(t_i)$: $f(t_i) = \phi(M(t_i))$, the annual fuel consumption F_{ann} is given by

$$F_{ann} = \sum_{i=0}^{N-1} f(t_i) \cdot \Delta t = \sum_{i=0}^{N-1} \phi(M(t_i)) \cdot \Delta t$$

where N is the number of time steps over the year.

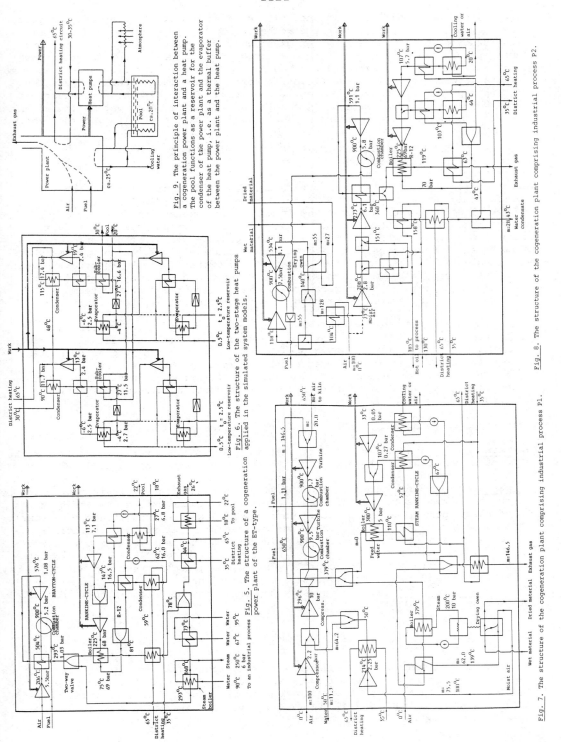

Fig. 9. The principle of interaction between a cogeneration power plant and a heat pump. The pool functions as a reservoir for the condenser of the power plant and the evaporator of the heat pump, i.e. as a thermal buffer between the power plant and the heat pump.

Fig. 6. The structure of the two-stage heat pumps applied in the simulated system models.

Fig. 5. The structure of a cogeneration power plant of the ET-type.

Fig. 8. The structure of the cogeneration plant comprising industrial process P2.

Fig. 7. The structure of the cogeneration plant comprising industrial process P1.

To facilitate the detailed formal definition of such systems in an easy-to-grasp linear notation that may be interpreted by a computer program, a special formal system description language has been developed. The corresponding computer program comprises algorithms for the compilation of a computer model and the subsequent simulation of any system described in the language. The system description contains information on the specifications of the physical properties of subsystems and transducers, possible modes of operation of the TPS, control functions and time series for the exogenously controlled variables. The system simulation consists in the stepwise solution of (ii) and computation of (i) over the year.

ON THE A PRIORI ANALYSIS OF THE TRANSITION TO MORE EFFICIENT SYSTEMS

In integrated multi-task systems, where the functioning of one component in the thermodynamic process system TPS (Fig. 2) depends on the simultaneous functioning of several other components involved in multi-task operations, physics provides no answers to questions about how much fuel is consumed to perform a particular task. Computations of the amount of energy used to provide certain goods and services, as recorded in energy analysis reports, are exclusively related to systems where each single task is performed by one particular machine. When integrated multi-task systems are included in the set of considered technical alternatives, the economic comparisons - in monetary as well as in resource economic terms - must refer to the entire system comprehended by the integrated-system alternative.

During the period of transition from traditional systems to more efficient structures, a series of investments in technical changes should ideally be planned in such a manner that the greatest possible advantage is achieved at each step. At any time during the process when a choice has to be made as to which of a number of possible alternative changes should be carried out next, each alternative should be evaluated with regard to the reduction in fuel consumption obtained as a result of its implementation.

Three types of technical changes should be considered:

A: structural changes, i.e. changes that involve the structure of the TPS or the control structure. Referring to the formal description above, B, C or M may be altered.

B: parametric modifications in the GDS, i.e. changes that result in new values of the coefficients of thermal conduction, heat capacities, window areas, rates of electricity consumption etc. Such changes may be brought about by improving the thermal insulation of buildings, replacing existing electrical apparatus by more efficient ones, etc. Referring to the formal description above, the values of c_r, k_{rs}, k_{ra}, $\bar{x}_r$ or κ_{gr} may be altered. An increase in the electricity production capacity from windmills or in solar panel areas may also be considered to be parametric modifications.

C: an expansion of the set of tasks by the inclusion of an additional industrial process in the TPS.

A case study: Danish urban areas

To evaluate fuel consumption reductions from improvements in the technical systems, models of average Danish urban areas with about 25,000 inhabitants have been constructed. In each case the variations in the annual fuel consumption caused by certain technical changes of the types A, B and C have been computed by the simulation of a number of system model versions by means of the above-mentioned computer programs.

In those cases the GDS represents buildings used for dwellings, offices, schools, hospitals, shops, industrial workshops etc., with the addition of electricity consuming machines and devices located in open air and therefore not contributing to the heating of rooms, i.e. electrically driven vehicles, street lamps, pumps etc.

Thermal effects from the TPS to buildings are assumed to be supplied via district heating systems.   In some system versions additional thermal effects to hot water tanks in dwellings are obtained by means of solar panels.   Moderate supplies of electricity from windmills are taken into account in some system versions but work reservoirs for storage of electricity have not been included.

The control structures which govern the functioning of the TPS have been constructed in accordance with the following criteria: (1) there should be as few shifts in the mode of operation (including stops and starts) as possible;   (2) the import/export of electricity from/to the national grid should be reduced to a minimum;   and (3) the currently available potentials from solar radiation, wind and possibly industrial processes should be exploited as far as possible.

Three alternative types of TPSs - denoted TT, ET and EE in the following - have been considered as supply systems for the GDS:

TT : a traditional thermal system with separate plants for the production of electricity and heat respectively, apart from 15% of the district heating effects taken from cogeneration plants.   Electric power is assumed to be produced at large modern steam power plants operating with an efficiency of 42%.   Thermal effects are assumed to be supplied from boiler stations with a thermal efficiency of 92% at a maximum temperature of 90°C in the district heating circuits.

ET : a (nominally) efficient thermal system consisting of four power plants and a number of heat pump units with structural compositions as shown by Figs. 5 and 6.   The interaction between the power plants and the heat pumps is shown in principle by Fig. 9.   This process system configuration is taken to be representative of the most efficient systems that may be constructed using components available on the market today.

EE: a (nominally) efficient electrochemical system (a fuel cell) working in cooperation with heat pumps in the manner shown by Fig. 9.   Sixty per cent of the fuel input to the fuel cell is assumed to be converted to electricity while 22% is assumed to be available as thermal effects for district heating.   This system is taken to be representative of future plants for efficient conversion from chemical to electrical potentials.

The fuel consumption required to deliver certain power effects and thermal effects to an industrial process is an undefined quantity in cases where the industrial process is integrated in a multi-task TPS. It has, therefore, been chosen to evaluate the increase in the total system fuel consumption caused by the addition of a certain industrial process to a reference system. In each case the reference system has been taken to be a version of the presently considered GDSs together with a TPS of one of the above-mentioned types without industrial processes.

Indicative results for reference systems without industrial processes.   In Fig. 10 the columns marked "exist." display the computed relations between the annual electricity and heat consumption in the GDS (left columns) and the annual fuel consumption (right columns) in two urban area models called X-city and Y-city.

The fuel consumption reductions obtained by the structural improvements brought

about by replacing a TT-process system by an ET- and an EE-system respectively
are shown by the right column partitions marked ET and EE.   It appears that the
fuel consumption is reduced by ca. 35% when the TT-system is exchanged for an
ET-system while the GDS remains unchanged.

The columns marked "new" in Fig. 10 show the relative fuel consumption values
for TT-, ET- and EE-systems respectively after the electricity and heat demands
from the GDS have been reduced by technical improvements, i.e. parametric modifications of various kinds as mentioned above.   It appears, for instance, that
the fuel consumption may be reduced by ca. 60% by implementing the structural
change from TT to ET and the parametric change from "exist." to "new".

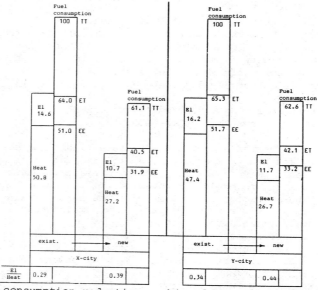

Fig. 10. Fuel consumption reductions achieved by structural and/or parametric
changes in the technical system.

From the analysis of results obtained by computer simulation of a larger number
of system versions, linear differential relations were found to hold between the
annual fuel consumption F_{ann} and the following annual quantities (see Fig. 2):

W_d : the electricity consumption in the GDS;

Q_e : the sum of integrals of uncontrolled heat effects between buildings in
the GDS and their environment, i.e. a quantity that may be reduced by
improved heat insulation etc.;

W_e : the electricity consumption outside buildings;

W_w : the electricity production from windmills.

The following differential relations were found to hold within the considered
domain of variations from the existing systems:

For TT-systems: $dF_{ann} = 1.70dW_d + 1.18dQ_e + 2.70dW_e - 2.57dW_w$

For ET-systems: $dF_{ann} = 1.49dW_d + 0.65dQ_e + 2.11dW_e - 2.00dW_w$ (iv)

For EE-systems: $dF_{ann} = 1.31dW_d + 0.52dQ_e + 1.75dW_e - 1.66dW_w$

Observe that the change in the annual fuel consumption F_{ann} caused by certain changes of W_d, Q_e, W_e and W_w - i.e. by parametric modifications of the GDS - depends significantly on the efficiency of the TPS with respect to cogeneration of electricity and heat. In particular the pay-back in terms of fuel consumption reductions from investments that reduce Q_e depends strongly on the process system.

<u>Addition of industrial thermodynamic processes to the reference systems.</u> Table 1 below gives examples of computed increases in fuel consumption when industrial processes are added to the reference systems. Three examples of industrial process requirements are considered:

Process P1:
- 10 tons/hour of steam at 250°C/6 bar. The condensate is assumed to be re- turned to the TPS at 90°C.
- 120 tons/hour of hot water at 95°C. The water is assumed to be returned to the TPS at 65°C.

This process is assumed to be in operation from 0630 to 1500 hours every work- day (monday to friday).

Process P2:
- 15000 Nm³/hour of hot air at 650°C to a kiln.
- 140 tons/hour of steam at 200°C/10 bar to heat the interior of a drying oven. Moist air at 100°C is discharged from the oven at a rate of 56600 m³/hour.

This process is assumed to be in continuous operation from monday at 0000 hours till friday at 2300 hours every week.

Process P3:
- 50000 Nm³/hour of hot air at ca. 525°C to be circulated through a drying plant. Moist air at 140°C is discharged from the plant.
- A thermal effect of 5.3 MW delivered to a process from a hot oil circuit with a temperature of 185°C at the inlet to process plant and 100°C at the outlet.

This process is assumed to be in continuous operation from monday at 0000 hours till friday at 2300 hours every week.

The processes may be driven from separate plants: furnaces for the supply of hot air; ordinary fuel fired boilers for the supply of steam, hot water and hot oil. In those cases the increase in fuel consumption when the process is added to the reference system has been directly computed, assuming a thermal efficiency of 92% in the boilers.

Alternatively the industrial processes may be integrated with the reference systems TPS in accordance with the principle illustrated by Fig. 3. In the given examples integration with ET-systems is considered. In the case P1, steam and hot water is drawn from an ET-system as shown by Fig. 5. In the cases P2 and P3 one of the four ET-systems used in the reference system is replaced by the systems shown by Fig. 7 and Fig. 8 respectively.

The results given in Table 1 show that substantial fuel savings may be obtained by integrating industrial processes with the TPS instead of using separate plants: 30% of the fuel required to run a separate plant may be saved in the case P1, 42-43% in the case P2 and 40-42% in the case P3.

THERMODYNAMIC PROCESS SYSTEM TPS			GENERAL DEMAND SYSTEM GDS			
SYSTEM TYPE	INDUSTRIAL PROCESS		X-City "exist."		Y-City "new"	
	Separate	Integrated (cogeneration)	Abs. GWh	Rel. %	Abs. GWh	Rel. %
			F_{ann}			
TT	none		539	100	374	100
ET	(reference system)		345	64	252	67
			ΔF_{ann}			
TT or ET	P1		26.9	100		
ET		P1	18.7	70		
TT or ET	P2		100.2	100	100.2	100
ET		P2	57.3	57	58.5	58
TT or ET	P3		111.2	100	111.2	100
ET		P3	64.8	58	66.7	60

Table 1. The increases (ΔF_{ann}) in the annual fuel consumption (F_{ann}) when industrial process requirements P1, P2 and P3 respectively are added to the reference system demands in the urban area models X-City and Y-City (see Fig. 10).

CONCLUSION

The "energy demands" in a society may of course be defined in terms of the electricity and heat flows to the various points of "end use". There are, however, no generally valid relations between those energy flow quantities and the rate of fuel consumption in the society.

The examples given in this discussion demonstrate that in concrete cases relations between changes in the "end use" energy demands and changes in the rate of fuel consumption may be found by computer simulation of a discrete control model of the entire energy system in its climatic and social environment. In particular it appears that the relations between thermal effect demands to industrial processes and the rate of fuel consumption depend strongly on the structures of the thermodynamic systems in which those effects are produced.

REFERENCES

Gaggioli, Richard A.(1977). Proper Evaluation and Pricing of "Energy". Proceedings of the International Conference on Energy Use Management. Tucson, Arizona. October, 1977.

Gyftopoulos, Elias (1974). Potential Fuel Effectiveness in Industry. Ballinger Publishing Company. Cambridge Mass., 1974.

Gyftopoulos, Elias (1977). Effective Energy End-Use - Oppotunities and Barriers. Proceedings of the International Conference on Energy Use Management. Tucson, Arizona. October, 1977.

Keenan, J.H. (1973). The Fuel Shortage and Thermodynamics. Proceedings of the MIT Energy Conference. MIT Press, Cambridge Mass.,1973.

Reistad, G.M. (1975). Available Energy Conversion and Utilization in the United States. Journal of Engineering for Power, july 1975, pag.429.

Williams, Robert H. (1978). Industrial Cogeneration. Annual Review of Energy, 1978, vol. 3, pag. 313-356.

Ford, K.W. (ed.). Efficient Use of Energy. American Institute of Physics. New York, 1975.

Noyes, R. (ed.). Cogeneration of Steam and Electric Power. Noyes Data Corporation, Park Ridge, New Jersey, 1978.

ENERGY SAVING IN CONTINUOUS CASTING PROCESS BY NEW CONCEPTS IN THE DESIGN OF EQUIPMENT

B. Indyk* and R. Wilson**

*Dept. Metallurgy, University of Strathclyde, Glasgow, UK
**Timex, Dundee, UK

ABSTRACT

Significant improvements in quality of the finished product, versatility and productivity of equipment and large savings of energy achieved by redesign of equipment based on analysis where energy was given primary significance.

KEYWORDS

Process analysis, separation, integration, reusage, versatility, casting rate,
savings.

INTRODUCTION

Since the dawn of civilisation mankind looked for metals. The stone tools were
not providing good enough service so they had to be replaced by something tougher
and more flexible. However, metal was not so readily available as the stones;
it had to be paid for dearly by effort, time and most important energy - even for
its most primitive use. As the progress went from early bronze through gold and
silver to the present vast variety of metals and alloys, this requirement of effort and energy associated with every stage of their production, has increased.
Energy and metals have become two cardinal pivots which decide the pattern of development of every society.

Energy is required to produce metals; starting from the mines where metal ore is
extracted, through the various stages of preparation, to smelting, purification,
alloying, casting, working and heat treatment, to the finished product. Equally
metals are required to produce and transport energy for our consumption; from
mines or hydro-electric turbine generators to electrical cables, tankers and internal combustion engines. The more technically advanced the society the more
energy and metals it consumes. Invariably savings can be achieved almost at every
stage of production, provided the proper analysis of the process is conducted, and
in particular during the melting, casting and hotworking stage.

Until quite recent times, the conventional production of steel involved:
    1. refining in molten state at high temperature,
    2. transporting and casting into ingots,
    3. "soaking" of ingots
    4. rolling into billets or sheet,

5. remelting the unavoidable rejects.

Introduction of continuous casting changed the situation quite considerably. The overall time spent in processing the steel from refining stage to the finished product has been drastically reduced by eliminating the longest stage of all, namely the "soaking" of ingots. Energy associated with this part of the process is therefore saved. Savings are also achieved by drastically reducing the amount of rolling and rejects, since it is possible to cast much closer to the shape of finished product. Direct and indirect savings are achieved at the ingot casting stage namely : heating and cooling as well as production of the container material.

With all these benefits, one is tempted to ask why continuous casting was not introduced earlier into production of metals - it is after all a process which was known for one and a half centuries. However, as often happened, the path between an obviously bright idea and its practical application was long, tortuous and thorny.

Having finally arrived at the practical solution of the very difficult technical problem, we cannot however rest contented with laurel wreaths on our heads. Once the justified euphoria, caused by a significant industrial achievement fades, a detailed analysis of the process stage by stage again shows that it is wasteful and further economies and improvements can be achieved. It is the purpose of this paper to show that such analysis is both possible and urgent. It is not necessary to wait another century and a half before these theoretical postulates are taken seriously into practical consideration, because by then few conventional energy sources will be available. A practical example researched and tried out by the authors shows that the ideas put forward here are a workable proposition.

ENERGY IN CONTEMPORARY CONTINUOUS CASTERS

There are many types of continuous casters in current use, depending on what metal is produced and in what circumstances. Nevertheless, from the point of view of energy analysis it is convenient to group them into two major classes:
(1) vertical caster, (2) horizontal casters

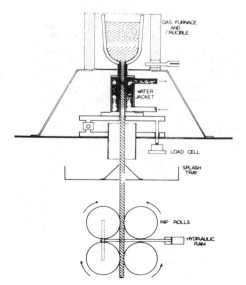

Fig. 1. Closed head vertical unit.

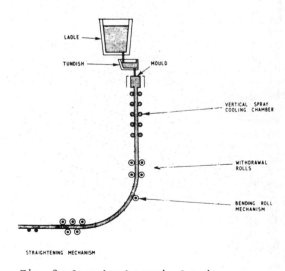

Fig. 2. Open head vertical unit.

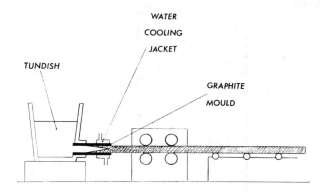

Fig. 3. Horizontal caster.

The vertical casters can be further grouped into the (3) closed-head system, (4) open head system.

This last grouping is somewhat misleading therefore it requires clarification; a closed head system is the one where there is a positive column of metal above the mould in which solidification takes place, and an open head is the one where the metal is poured directly into the mould. Since it would be difficult to inject metal into the horizontal mould without a preceding tundish therefore the horizontal casters, of necessity, have to be closed head system.

The merits of open versus closed head systems depend on the actual circumstances, type of metal and safety considerations. It is easy to see that where reciprocating mould is essential, closed head presents a problem which so far defies solution; equally, that the benefit of the reciprocating mould has to be paid for by greater complexity in construction and control, higher energy consumption and problems with oxidation or similar chemical drawbacks, which in some cases (e.g. steel) are embarrassing.

The absolute minimum energy required for the continuous casting process depends to a large extent on the starting point of the metal. Where solid scrap or otherwise prepared metal is used, a large amount of energy is consumed in melting, superheating and transferring to a tundish or mould prior to solidification. Where the metal is refined separately and brought to the casting plant in the already molten state it is necessary to make sure that the amount of superheat is sufficient to allow for the losses during the transportation and subsequent pouring into the tundish or mould. To this direct amount of energy in the form of heat, there must also be added the indirect energy in the form of power required to move the molten metal, its massive containers and subsidiary equipment. The larger the distance between the refining and casting locations the larger the energy required for transportation, and to maintain the superheat.

It is generally agreed that the level of molten metal above the solidification front is one of the important parameters in continuous casting. However, the precise correlation of the dependence of this parameter on others is still a question of dispute and is likely to remain so for some time. One of the main difficulties in the interpretation of experimental results is the fact that the hydrostatic pressure exerted by the liquid metal at the solidification front is not a single valued quantity but is a function of the depth of the liquid sump in the metal, which in turn depends on the properties of materials concerned, the design of equipment as well as operational variables. The depth of liquid sump can vary

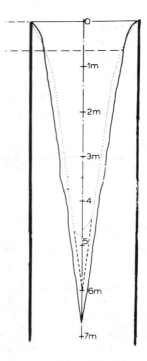

Fig. 4. Deep liquid sump.

from a fraction of a centimeter, for example in casting of copper tube, up to about ten meters when casting large steel billet. Yet it is in the case of steel that a tundish with a large horizontal area is introduced between the ladle and mould in order to maintain steady flow of liquid metal. Maintenance of this tundish at the required temperature to prevent the metal from freezing is a major drain on energy. From the point of view of quality of the final product, it introduces a source of oxygen, after a lot of trouble has been taken and energy spent to remove it. Though in closed head systems this particular energy drain is avoided, still the liquid metal is poured openly into exposed tundish giving rise to a somewhat smaller heat loss.

Having thus passed by any of the devious routes, all of which are wasteful, the liquid metal finds itself in the mould, the part of the equipment which is most essential to the process. Here the solidification begins and has to carry on at least to the point where the solidified skin is strong enough to withstand the pull of the withdrawal mechanism and the hydrostatic pressure of the liquid. Large amount of heat must be extracted from the metal rather quickly therefore, enforced cooling is essential. Thus the mould is usually cooled by water flowing in the surrounding jacket. Whether the water is supplied directly from the mains or from a recirculating system, it enters the cooling system at a fairly low temperature and leaves it a few degrees hotter. The heat carried out by water is either lost in the drainage or discipated in an entirely careless manner.

The internal structure of the metal as well as the possibility of overheating of the withdrawal mechanism frequently necessitate secondary cooling. This is achieved by a spray of water directly onto the solidified billet as it emerges from the mould. With vertical casting, and in particular in case of aluminium and steel, this is the stage where extraordinary precautions for safety have to be taken, for the consequences of a sudden breakout of molten metal may be tragic. As previously, the heat removed by the cooling water from the metal is lost.

The finished product leaves the casting plant having very little excess energy. Disregarding the energy consumed in refining the metal, the total energy extracted from the metal during primary and secondary cooling is the absolute minimum required for an efficient process. This minimum quantity, Q min, can be this used as the comparison for different types of processes or different casting units in order to assess how far they depart from the ideal.

SUGGESTED REMEDIES

In trying to put forward remedies for the reduction of waste and possible improvements in the production practice and the quality of the finished product, one is faced with two different problems:
1. No major alterations are likely in an existing plant which is already in production, unless it is misbehaving seriously.
2. Caution is the rule in installation of new plants; production record is given preference to new ideas.

Though no sudden revolution is envisaged in the near future, it is necessary to keep the discussion alive, hoping that gradual progress shall be maintained.

In existing plants it is fruitful to analyse every stage of production concentrating on energy losses. Frequently the unnecessary energy losses are associated with other undesirable problems. As an example let us consider that liquid metal exposed to an atmosphere loses a lot of heat by radiation and convection. This is particularly true when it is poured, say from the ladle to the tundish, in a highly turbulent and dispersed stream. At this stage oxygen is again picked up to the detriment of the finished product. Elimination or reduction of heat loss is likely also to eliminate or reduce the oxygen pick-up. Shielding and insulation should therefore be used wherever circumstances permit. Closed container has the advantage over an open one that the atmosphere in it can be controlled. Distances traversed by the liquid metal should be made as small as possible and time as short as possible.

In installation of new plants or introduction of alterations to old ones it might be useful to adopt the following guidelines:-

a) Integration of the process as a whole. Wherever possible, the transportation of liquid metal should be eliminated altogether. The last statement, when applied to metal like steel, seems completely unrealistic, yet theoretically there is no reason why the steel should not be cast directly from a suitably constructed converter in which stirring was achieved not by physical movement of the converter as a whole, since it is the least effective anyway, but hydraulically or electromagnetically. It is of course realised that such a system is too revolutionary at the moment to be taken seriously by the designers and producers and only academics can have such "unlikely" dreams. Nevertheless, if taken seriously, the system would have enormous advantages. No ladle, no tundish, flow or metal controlled by both heating and cooling of the flow path, and in emergency by a physical stopper, no oxidation, no equipment carrying about liquid metal.

b) Separation of the primary cooling and solidification from the bulk liquid and from the secondary cooling. As the succeeding example shows this enables control not possible otherwise as well as saves energy.

c) Integration into the rest of the plant or works. What happens to the metal before it reaches the casting unit as well as when it leaves should be envisaged in the design. If, for instance, it is required to hot work the metal, it should not be cooled to the extent that reheating is necessary etc.

d) Reusage of wasted heat. The practical limit on the rise in temperature of the cooling water in primary cooling, is dictated by the onset of the so called "bumping" usually caused by the air dissolved in water. A smoothly operating cooling system therefore requires rather high volume of water. It seems thus that the heat carried away by the water has to be wasted due to relatively low temperature. However, every plant or factory nowadays is provided with both cold and hot water for personnel use. Equally the offices dining halls and other accommodation of similar type have to be heated during the cold periods. If the energy contained in the cooling water is transferred to central heating or hot water system, even 10°C above the mains water temperature makes a substantial saving in the fuel bill. Yet this can be achieved with a very small capital expenditure.

PRACTICAL EXAMPLE: TIMEX RECYCLING OF BRASS SCRAP

In production of watch bezels Timex was faced with a problem that about three quarters of brass used had to be scrapped as cuttings and swarf. Large savings were to be made in this scrap could be remelted and cast into bars of the original size. A decade ago an integrated continuous caster designed by United Wire Company was installed in the plant. The caster delivered what it was meant to deliver: almost all the scrap was recovered, so that Timex purchase of new brass dropped to one quarter of what it was originally.

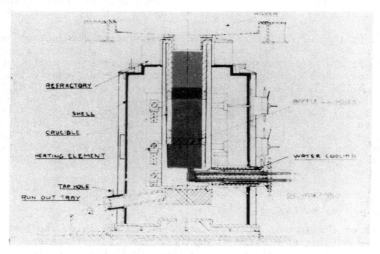

Fig. 5. Production Unit.

However, the success story was not complete. The following problems became apparent after several years of experience: (a) lack of versatility and control, (b) limitation on production rate, (c) zinc-rich phase on the bar. The production caster was fitted with a mould with four outlets so that four bars could be cast simultaneously. Practice has shown that it was not possible to increase the casting speed above 17-18cm/min. The surface of the cast bar was covered with regions of zinc-rich phase which, being much harder than the bulk of material had to be removed before the bar could be passed on. A thorough analysis of the process was therefore undertaken and a decision was made that a programme of experiments would be carried out in conjunction with the University of Strathclyde on a specially built smaller experimental unit.

As a result of the analysis, the lack of versatility and control, at least in part could be ascribed to the asymmetry of solidification front, which was evident from the surface isotherm marks on the bar, and which is a characteristic feature of conventional horizontally cast bar or billet. The limitation on the speed of casting could not be explained in a simple manner, though a method of cooling was suspected to be the main factor. Several mechanisms for the occurrence of the zinc-

Fig. 6. Zinc-rich phase on the surface

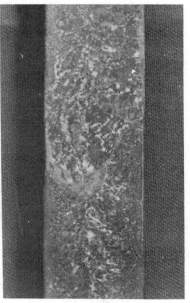

Fig. 8 Asymmetrical
solidification.

Fig. 7 Experimental Unit.

rich phase were possible: 1) diffusion through the graphite mould from hotter to colder region, 2) diffusion through the solid skin of brass from the liquid sump in the bar, 3) inverse segregation of zinc, 4) evaporation in hotter regions of the "air gap" and condensation in the colder region. These could only be sorted out by experiment.

It is noted that the essential difference in design of the experimental unit is that the die feeder, the die and the die cooler have been placed in an extended section outside the main casting furnace housing, thus minimising heat transfer from the crucible and furnace assembly to the die cooler. A separate heater ensured adequate and controllable flow of liquid metal through the feeder into the mould. An entirely novel system of "internal" cooling of the mould was devised. Ref. It consists of six pairs of concentric copper tubes which could be pushed into six holes in the mould symmetrically spaced around the bar outlet. The position of each copper tube could be adjusted separately so that heat transfer from the metal to water in the tube could be controlled to a much larger extent than is possible by the water flow rate alone.

Subsequent experiments have demonstrated that the asymmetry of the solidification front inherent in the horizontal continuous casting can be overcome by adjustment of the position of cooling probes.

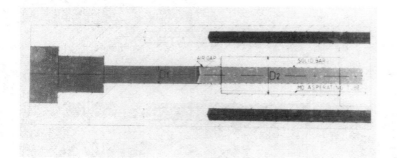

Fig. 9.

Internally cooled die.

Fig. 10.

Symmetrical solidification.

It was also possible to control the position of the solidification front within the mould, so that it could be moved at will just by pushing in or withdrawing the probes a certain amount. This possibility extends the useful life of the mould considerably, because wear is not confined to one portion of the mould or die.

Though the achievement of symmetry in heat flow pattern produced a much better surface finish of the bar, it did not eliminate the zinc rich phase. Nevertheless, it has been noticed that the amount of this varied inversly to the speed of casting; i.e. the faster the casting rate, the smaller amount of zinc rich phase. The casting rate could be increased up to the physical limit of the withdrawal mechanism and the rate at which the metal could be melted. Speeds in excess of 80 cm/min were achieved, which meant that from the single outlet in the small experimental caster more metal could be cast than from four outlets in the big production unit.

In order to eliminate the zinc rich phase a further series of experiments was performed. Scanning electronymicroscope analysis of the metal bar eliminated the possibility of inverse segregation of zinc and diffusion from the liquid sump. Examination of mould also eliminated diffusion of zinc through graphite. The remaining possibility of evaporation and condensation within the "air gap" was confirmed by enlarging the gap and asperating the vapours contained in it by means of a sampling tube. Samples obtained contained zinc only. It was possible to eliminate the zinc rich phase by blowing nitrogen into the air gap and so removing the zinc vapour, before it condensed in the cooler parts of the mould.

In consequence of this investigation an additional observation was made, namely that a bar of several sizes could be cast in a single mould, by moving the solidification front to a predetermined section of the mould. As an extension of this, a possibility of casting a tube and a bar from the same mould was investigated and proved practical. Indeed, owing to the difficulty usually

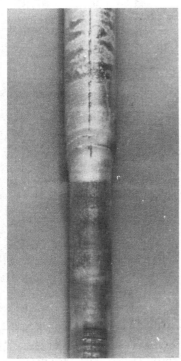

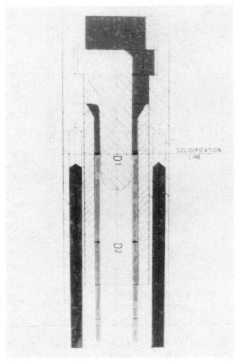

Fig 11. Two sizes of bar in a single cast

Fig 12. Bar-tube casting arrangement

encountered in starting a cast of tube due to the frequency with which the mandril breaks, it is advisable to start with a bar and then to switch over to tube, every time the casting process has to be interrupted for some reason. This way the mandril is surrounded by liquid metal during start up and there is no danger of it breaking.

This extraordinary flexibility, which a cautious designer or operator would not even dream of, was achieved by careful analysis of the process and an experimental investigation which followed.

TABLE 1

Assymmetrical heat flow through internally cooled die.

Segment No.	TOP				BOTTOM			
	θ_1	θ_2	$\Delta\theta$	$\frac{dQ}{dt}$	θ_1	θ_2	$\Delta\theta$	$\frac{dQ}{dt}$
1	1026	1025	1	3	1026	1020	6	18
2	1014	1011	3	9	1014	998	6	18
3	1002	998	4	12	1002	984	18	54
4	991	985	6	18	991	961	30	90
5	979	973	6	18	979	943	36	108
6	967	958	9	27	967	922	45	135
7	955	945	10	30	955	901	54	162
8	943	933	10	30	943	883	60	70
9	931	919	12	36	931	883	48	34
10	919	905	14	42	919	882	37	47
11	908	892	16	48	908	882	26	57
12	896	882	14	129	895	881	14	139
13	885	880	5	120	885	880	5	120
14	826	823	3	90	826	823	3	90
15	776	773	3	90	976	773	3	90
16	729	726	3	90	729	727	2	60
17	685	683	2	60	685	683	2	60
18	643	641	2	60	643	641	2	60
19	604	602	2	60	604	602	2	60
20	567	566	1	30	567	566	1	30
21	532	531	1	30	532	531	1	30
22	500	499	1	36	500	499	1	30
23	470	469	1	30	470	469	1	30
24	441	440	1	30	441	440	1	30
				1132				1622

Total $\frac{dQ}{dt}$ = 2754 cal sec^{-1}

Regarding the energy balance it has been found that symmetrical solidification front improved the situation markedly as far as the mould itself was concerned. On comparison to the production caster it was found that the energy saving was considerable. It seemed that two thirds of the power supplied to the production caster was expended in heating water unnecessarily.

TABLE 3

Comparison of heat balance of production
unit to those of internally cooled die.

Type of Cooler/Die Assembly	Production Unit	Modified Cooler Die Assembly	Modified Cooler Die Assembly
Bar Diameter cm.	1 x 2.064 + 3 x 3.175	1 x 2.064	1 x 2.540
Casting Speed V.	17.0 cm min^{-1} 0.283 cm sec^{-1}	44 cm min^{-1} 0.733 cm sec^{-1}	14.51 cm min^{-1} 0.242 cm sec^{-1}
Water Flow (1.sec^{-1})	2.750	0.301	0.104
Water Inlet Temp (^{o}C)	21	21	22
Water Outlet Temp (^{o}C)	32	29	30
$\Delta\theta_w$ (^{o}C)	11	8	8
$\frac{dQ_w}{dt}$ Cal. sec^{-1}	30250	2408	832
$\Sigma \frac{dQ}{dt}$ Cal. sec^{-1} (Summation of Segments)		2754	967
Q_{min} Cal. sec^{-1}	7968	1940	778
$\frac{dQ_w}{dt}$ / Q_{min}	3.80	1.24	1.07
$\Sigma \frac{dQ}{dt}$ / Q_{min} (segments)		1.42	1.24
Excess heat waste to water	280%	24%	7%
Heat waste to atmosphere		42%	24%

TABLE 2

Symmetrical heat flow through internally cooled die.

Segment No.	θ_1 Centre Temp $^{\circ}$C	θ_2 Surface Temp $^{\circ}$C	$\Delta\theta\,^{\circ}$C	$\dfrac{dQ}{dt}$ cal.sec^{-1}
1	980	973	7	6
2	972	963	9	7
3	963	955	8	6
4	954	945	9	7
5	944	935	10	7
6	935	925	13	8
7	925	912	14	10
8	914	900	20	11
9	905	885	23	105
10	890	867	23	181
11	868	848	20	158
12	841	828	13	103
13	812	805	7	55
14	787	782	5	39
15	762	758	4	32
16	738	735	3	24
17	715	712	3	24
18	692	688	4	32
19	670	667	3	24
20	649	647	2	16
21	628	625	3	24
22	610	606	4	32
23	592	588	4	32
24	573	570	3	24

$$\text{Total } \frac{dQ}{dt} = 967 \text{ cal sec}^{-1}$$

It was realised that majority of the technical problems described before, would have been solved automatically if a proper consideration was given to energy balance and heat transfer in the first place.

CONCLUSION

To conclude this discussion, it is necessary to add a few sobering thoughts. We did not entirely manage to do what we preached. We have not achieved anything in the way of reuse of the wasted heat. Conservatism here is much stronger than even the enthusiasm of an innovator.

Reference (1).   Specific aspects of die and cooler design are covered by Worldwide Patent Applications by Timex Corporation, Middlebury, U.S.A. through whom non exclusive patent and know-how licences are available.

ACKNOWLEDGEMENT
The Authors wish to acknowledge Timex Corporation, Middlebury, U.S.A. and the University of Strathclyde for permission to publish this paper.

ANSWERS: A COMPREHENSIVE SOLUTION TO THE ECONOMIC, TECHNICAL AND INSTITUTIONAL PROBLEMS OF SOLID WASTE REUSE

P. F. Mahoney* and G. L. Sutin**

*Smith and Mahoney, P.C. 79 N. Pearl Street, Albany, New York, USA
**Gordon L. Sutin & Associates Ltd., 124 James Street South, Hamilton,
Ontario, Canada

ABSTRACT

In the United States, the processing and recovery of solid waste is a relatively
new and struggling industry. Recognizing the irreversible ills of landfilling and
the potential energy resource contained in garbage, the City of Albany, New York
requested our firms to investigate alternate methods of solid waste disposal over
ten years ago. ANSWERS, the Albany, New York Solid Waste Energy Recovery System,
is the first profit-making resource recovery program in the U.S. A $26 million
dollar project, it includes material recovery and the production of a solid waste
fuel as well as a combustion system utilizing steam boilers fired by the processed
fuel. Under a 20 year agreement, the City is to process 750 tons/day of municipal
solid waste from the area producing 600 tons of Refuse Derived Fuel (RDF) for purchase by the State of New York. RDF is produced by shredding the waste and magnetically removing ferrous metals. The fuel will be burned in two waterwall boilers
fitted with spreader stokers, due for start-up in mid-1981. The ash will be processed to recover non-ferrous metals and the remaining inert material will be used
as a gravel substitute. The processing facility has been operating at maximum capacity successfully since its start-up in February, 1981.

KEYWORDS

ANSWERS; Albany, New York; solid waste; resource recovery; shredding; refuse derived fuel; ferrous metals; steam; conservation; ash reclamation.

INTRODUCTION

For many years there has been a clearly identifiable need in the United States
for a more acceptable method of getting rid of solid waste than simply creating
one sanitary landfill after another. There have been various attempts at a solution, and such activities have intensified in recent years with our awakening to
the increasing dollar value of the non-renewable resources our society is throw-
ing away. Until recently, however, the most large-scale attempts at solving the
solid waste problem have proven uneconomical, technically flawed, or environmentally unacceptable.

THE PROBLEM - THE CHALLENGE

As the general problem of waste disposal became more pressing, the number of acceptable options decreased. Many people became concerned with the misuse of land for waste disposal, and were, for a time, at least partially committed to the destruction of large amounts of waste by incineration. Incineration, however, soon raised serious questions of atmospheric pollution and this alternative became unpopular. Potential solutions to the solid waste disposal problem became subject to ever tightening constraints requiring attention to the preservation of the environment.

While waste disposal questions were being pondered, serious related issues began to emerge. It became apparent that we were discarding huge amounts of non-renewable materials in the form of society's cast-off products. Each aluminum can, copper radiator core, or steel washing machine shell bulldozed into the earth represented another wasted portion of a finite supply of those materials.

The basic search for acceptable methods of solid waste disposal began to turn in part toward reuse of society's discards. However, for many years salvage operations were limited by their generally uneconomical nature. Quite simply, it remained cheaper - in some cases, much cheaper - to use new material than to separate the usable fraction from the thousands of tons of waste picked up at our nation's curbsides every day.

At the same time escalating prices and dwindling supplies were making material reclamation more attractive, the so-called energy crisis of 1973, with its fuel cost increases and threatened supply shortages, intensified the search for alternative fuels. Since anything which will burn is convertible to energy, the objectives of the quest for an acceptable method of solid waste disposal became: develop a method for economically producing energy from solid waste, recovering non-renewable resources as feasible, and eliminating or greatly reducing the waste disposal problem.

With the foregoing objectives in mind, Smith and Mahoney in conjunction with Gordon L. Sutin developed a solid waste processing system which includes material recovery and the production of a solid waste fuel as well as a combustion system utilizing steam boilers fired by the processed fuel. The system is ANSWERS, the Albany, New York Solid Waste Energy Recovery System, and it may hold the answers to many of today's questions about solid waste disposal.

BACKGROUND

Development of ANSWERS actually began in 1970 when we were assigned to investigate alternative methods of solid waste disposal for the Albany, New York metropolitan area. The primary focus, at the time, was the possible elimination of sanitary landfills and their growing problems. Even as early as 1970, it appeared that energy would be a reasonable expectation of any such process, since economics suggested that the bulk of municipal solid waste was suitable only for burning. However, after investigating all alternatives, we concluded that none were economically attractive. Indeed, in 1970 when #6 fuel oil was selling for seven cents per gallon, the best alternative for producing fuel from solid waste would have been a break-even situation only if fuel oil sold for twenty-two cents per gallon.

Fuel oil prices rose sharply during the past decade attributable to the oil embargo of 1973 and 1974. The January, 1981 price of #6 fuel oil in the Albany area was seventy-nine cents per gallon, and our area is not atypical of the nation. Albany proved to be ideally suited to the development of a system such as ANSWERS.

An alternative to landfilling was being sought by the City of Albany, this effort growing out of a strong municipal commitment to the preservation of the environment and the conservation of energy. Also, there was a major fuel customer willing to enter into an agreement to use a processed refuse fuel for the sake of energy conservation. It remained only for ANSWERS to offer an economically attractive process.

The institutional problems associated with such a venture in the U.S. are potentially overwhelming. Drastic change is involved, and coping with this change requires interested, patient, and competent public officials who will get involved and take the time necessary to understand the economics and technology of the process. Albany's long time mayor, Erastus Corning, 2nd, took the time required to fully understand the process in all its aspects and provided the support and impetus necessary to focus all interests on a solution truly in the best interests of the community at large.

DESIGN OBJECTIVES

Adopting the concept of recovering energy and materials from solid waste as an alternative to landfilling, in 1973 the City of Albany established the following design objectives:

o Provide an economical and environmentally acceptable alternative to landfilling.

o Economically produce a competitively marketable fuel or energy product.

o Economically recover recyclable materials for which there is a market.

o Design a system free from environmentally undesirable waste or by-products.

o Use only proven or existing technology in the system itself.

TOWARD A SOLUTION

The economic problems encountered in pursuing the design objectives of ANSWERS have been relatively minor. As mentioned earlier, the project went forward only after it was established that there was a reasonable chance of the undertaking paying its own way.

One major shortcoming of other solid waste recovery projects has been the necessity to charge significant front-end fees for refuse dumping to permit operation on a self-sustaining basis. Many times these fees constituted increases in waste disposal costs which communities could ill afford. ANSWERS, however, will require only a nominal dumping fee and thus generate disposal cost savings for several communities.

The ANSWERS approach to resource recovery is simply: nothing is recovered unless it can be sold for more than the cost of recovery. Recovery concentrates on the items of greatest demand, those which produce the most revenue. It becomes a matter of examining the by-products at various stages in the process and asking, what can be used? For what portion of this material is there likely to be a market?

Economic considerations, however, often run a far second to the institutional problems arising from such a venture. ANSWERS was moved effectively toward realization by the interest of Albany's Mayor Corning. Certainly the mayor was a key factor in establishing the necessary relationship between City and State as supplier and consumer of refuse derived fuel, and as a result ANSWERS has a long-term contract with New York State for the use of the fuel.

It was necessary to make arrangements with a number of surrounding communities for securing their solid waste. This meant that several communities had to be stimulated to consider alternative means of refuse disposal. Again, Albany's Mayor was most effective in bringing all parties together on a common solution – including some communities which may not have felt they yet had a "problem" of any appreciable dimension.

Problems of considerable weight are common when it comes to the "where" of such an undertaking. As few people are likely to want a new landfill next door, so are few likely to want a fuel producing plant adjoining their property. On this point ANSWERS represents a significant compromise. The ANSWERS plant itself is located on a portion of a former landfill, while the fuel will be burned at a new steam generating plant some nine miles away in Albany. Strict engineering judgement would suggest that production and consumption of the fuel take place on the same site. This, however, would have meant working for construction of the fuel plant in a solidly urban area and looking forward to a parade of refuse trucks to and from the plant. Rather, use of the site selected for the plant represents an upgraded use of landfill property. Also, the processed fuel is considerably more compact than the unprocessed waste and can be transported in fewer, more generally acceptable vehicles.

The technical problems of ANSWERS have largely been solved through the use of existing technology. These are more appropriately dealt with in a later section.

ANSWERS: THE SYSTEM

The Albany, New York Solid Waste Energy Recovery System is a regional resource recovery program designed to initially process 750 tons of municipal solid waste per day, producing a fuel suitable for steam generation and recovering all economically recyclable materials. A municipal commitment to preserve the quality of the environment and conserve energy, together with a fuel customer willing to use processed refuse fuel have been the key ingredients in the project. The amount of fuel required by the State is approximately equal to the amount of fuel the ANSWERS plant can produce on a day-to-day basis. Further, ANSWERS has been designed as a profit-making venture and as such is economically attractive to all concerned.

The project is a joint venture of the City of Albany and the State of New York. In October 1976 these two arms of government signed a unique twenty-year agreement under which the city produces fuel to be purchased by the State's Office of General Services (OGS). The OGS has constructed two refuse fired boilers to generate steam to heat and cool State buildings in downtown Albany. The fuel product will be sold to the State at a 20% savings over the market price of the fuel currently in use (#6 fuel oil).

Presently, the OGS is burning a daily average of 25,000 to 40,000 gallons of fuel oil to heat and air condition State buildings. Since the amount of fuel which can be created from the Albany area's solid waste is approximately equal to the current OGS fuel demand, the agreement of City and State to develop and use this resource is a natural basis for a total energy and resource recovery system.

The ANSWERS plant will reduce some 750 tons per day of solid waste to 600 tons of refuse derived fuel for shipment to the boiler plant. The fuel product, having a heating value of 5,000 to 5,600 Btu per pound, will take the place of some nine to thirteen million gallons of fuel oil per year while consuming only one thirty-second of that amount of energy in its production. Converting considerably more than 200,000 tons of refuse per year into usable fuel, ANSWERS will be the first refuse to energy project in the U.S. to produce income for its sponsoring community.

ANSWERS accepts refuse from fourteen Albany area communities committed to hauling their wastes to the processing plant. The plant operates on a one shift schedule seven days a week.

The raw refuse is weighed on arrival and a dumping fee of $2.50 per ton charged. On a 1,800 ton capacity tipping floor, presorting and segregation of materials take place. At this point certain materials are removed - demolition and construction debris, noxious and explosive materials, segregated recoverables (bundled cardboard or cartons of cans or the like), white goods, and bulky items which might damage the plant's shredders. Segregated recoverables are set aside in a designated area within the receiving building. White goods are removed for immediate sale to scrap dealers. Large objects, such as tires, tree stumps, furniture, and carpets are held for end-of-day processing and disposal. A small portion of the current landfill is maintained for limited disposal of those few non-recoverable items.

At this point, the day's 750 tons of incoming refuse is reduced to some 620 tons. This remainder is then pushed onto a vibrating conveyor which discharges onto an apron conveyor feeding the shredders. Objects potentially hazardous to the shredders are removed at a manual picking station.

Two horizontal-shaft, 50 ton/hour shredders reduce the raw material to a nominal three inch particle size. Ferrous materials are magnetically removed and routed to open containers. The remaining material, some 600 tons of refuse derived fuel, is compacted into closed transfer trailers for shipment to the New York State OGS generating facility in downtown Albany. (A simple cut-away view of the processing plant appears in Fig. 2.)

At the boiler plant, the fuel is stored in trailers and in a one day surge storage pit. It is burned in two new spreader, stoker type boilers, designed specifically for this fuel product and capable of generating steam at a rate of 200,000 lbs./hr. In addition to the steam, the combustion process generates a good quality sterile ash consisting of less than two percent unburned carbon and one half of one percent putrescibles. Combustion gasses pass through three stage electrostatic precipitators to produce emissions well within the U.S. Environmental Protection Agency standards. An estimated ninety tons per day of ash will be processed to recover aluminum, coins, glass cullet, and other materials for which markets exist. The residue will be used as a gravel substitute under new City specifications for sidewalk and highway construction.

ANSWERS AND THE ENVIRONMENT

ANSWERS offers several environmental advantages for the Albany area:

o Ten sanitary landfills in the Albany region will be closed permanently and the land reclaimed for other use. The plant itself occupies less than one-fourth of the area of one present landfill.

o Eighty acres of the present Albany landfill, the portion not required by the processing plant, will be reclaimed. This area will be restored as an integral part of the

Albany Environmental Park, which also includes two
hundred and eighty acres in Albany's ecologically-
significant Pine Bush once earmarked for sanitary
landfill use.

o Each day more than seventy-five tons of non-renewable
natural resources will be conserved through recovery, and
enough fuel oil to heat 8,000 to 12,000 single family homes
will be saved each year.

o Emissions quality at the steam plant will be improved over
current operating conditions. Since the refuse fuel is of
low sulphur content and electrostatic precipitators are
employed to remove particulate matter, the result will be
cleaner air for downtown Albany.

o For every unit of energy expended by ANSWERS to produce
and deliver the fuel, thirty-two units of energy will be
conserved through the use of the fuel product (see Table 2).

DOLLARS AND CENTS

The ANSWERS operation cost $26 million, about half, $11.5 million, was spent by
the City for the processing plant and $14.5 was spent by the State for the boiler
plant. Approximately 45% of Albany's share was funded under the New York State
Environmental Quality Bond Act of 1972.

In guaranteeing the State a 20% saving on the cost of fuel oil, the City is actually helping the State pay the cost of constructing the RDF boilers. The minimum
net saving for the State on 11 million gallons of fuel oil (using an early 1980
price of fifty-eight cents per gallon) will amount to two million dollars per year.

In addition to providing a less expensive fuel for the State, ANSWERS is making
possible lower disposal costs - reduced dumping or "tipping" fees - for more than
a dozen municipalities. Solid waste disposal costs for communities participating
in the undertaking has been reduced from a range of $4.00 to $12.00 per ton to a
consistent $2.50 per ton. The net savings to these communities will amount to
some $220,000.00 per year.

After operating expenses and amortization, the City of Albany will realize net income from the sale of fuel and recoverable material of approximately $500,000.00
per year. Since these and the other figures just presented are estimates predicated on the first year's operation, savings will increase as the costs of fossil
fuels and energy in general continue to increase.

The major economic significance of ANSWERS, of course, arises from the overall
petroleum situation. The combustible portion of solid waste has always had a certain energy value, but it has taken the rising cost of fuel oil to make this energy value worth the recovery effort. In 1973 the combustible fraction of a ton of
solid waste was worth an average $4.00, when reckoned in terms of the cost of ob-
taining the same total heating value from #6 fuel oil. In 1981 the value of this
same combustible fraction of a ton of solid waste reached $35.00. The values of
certain other recoverables have also risen (see Table 1), though perhaps not so
dramatically as the value of energy, making solid waste a considerable "resource"
in its own right.

SPECIAL FEATURES OF ANSWERS

One of the outstanding features of the ANSWERS operation is its simplicity. Although production of fuel from solid waste is not a widespread activity, the ANSWERS processing plant involves no new technology whatsoever. All of the plant's components may regularly be seen in operation elsewhere; all are standard and each is readily available from more than a single manufacturer.

In moving refuse on the plant's several conveyors, for instance, the motors employed are all of the same relatively small size and type. Likewise, conveyor belt sizes have beeen standardized. This design simplicity limits the stock of essential spare parts that must be maintained in the plant and also serves to minimize maintenance costs.

Since simplicity and economy were prominent among the system's design objectives, ANSWERS employs manual labor when it appears to be the most economical way to perform a task. Much of the presorting at the start of the reclamation process is manual. No doubt some of this could be automated, but only upon the certainty of higher operating costs.

One particular problem receiving special attention was the design of the combustion system, since refuse derived fuel cannot be burned efficiently in just any existing equipment. The boilers were designed especially for the fuel. The RDF will be fed into the boilers some eight to twelve feet above the grates where some three-fourths of the combustible fraction - the light material, including paper, leaves, plastics, and so on - will be burned in suspension at temperatures up to $2500^{\circ}F$. The remainder of the combustibles drop to a moving grate where they will burn at a maximum temperature of $750^{\circ}F$ (see Fig. 3).

More Special Features

A common problem with refuse fuels is storage of the fuel product between production and burning. This has been solved in ANSWERS by the use of compactor trailers which serve the dual purpose of storage and transportation containers. These hold a day's fuel in readiness and reduce the volume of storage required for this difficult-to-store material.

The shredders are driven by diesel engines, a considerable energy saving over electric motors, and in keeping with the fundamental purposes of ANSWERS the heat from the diesel drive engines provides 80% of the space heat for the processing plant.

Part of ANSWERS' economic attractiveness results from the engineering approach to the problems of fuel production and consumption. The receiving building does not require a crane, so the structure is lower and expensive receiving pits and high maintenance lifting equipment are avoided. Extensive front-end preparation of solid waste is not required. Single stage shredding and magnetic separation are among the features holding the operation and maintenance costs for this refuse preparation and combustion system to perhaps half those of a mass burning system.

The boilers are designed especially for the fuel product, yet they can burn oil or gas as well. They require some 75% less floor area than those employed by mass burning processes and are one-fourth less in height than mass burning incinerators, so overall boiler housing is considerably reduced. The boilers have no refractory lining, so a potentially costly maintenance item is avoided. The boiler grate, subject to much lower temperatures than mass burning equipment, is simple and does not require expensive special steels. Finally, the amount of air required for burning refuse derived fuel is far less than is needed for mass burning, so fans, ducts, stacks, and precipitators are smaller and require less electrical energy.

Except for pre-process sorting and magnetic ferrous metal removal, saleable materials are recovered at the ash end of the process. As the light fraction of the fuel is burned, items such as glass, coins, and aluminum, copper, brass, and other metals drop to the moving grate where the 750°F temperature is insufficient to cause melting or slagging. Thus it is possible to recover these resources from sixty-six cubic yards of ash per day at the end of the process rather than from 4,000 cubic yards of refuse per day at the beginning.

ANSWERS FROM ANSWERS

The ANSWERS processing plant was ready for start-up in March of 1980. It has been operating at design capacity continually since February, 1981, shortly after the shakedown period. The State's boiler plant is scheduled to be operational in the latter half of 1981, due to delays. For efficiency of operation of the processing system, the City has leased the plant to a private firm for the twenty-year term of the arrangement with the State.

ANSWERS was made possible by the vision and environmental concern of Albany's Mayor Corning and the cooperation of many municipal and State agencies and officials. ANSWERS is more than a simple acronym for Albany, New York Solid Waste Energy Recovery System. It is also a positive response to many of today's questions concerning waste disposal, environmental protection, energy shortages, resource recovery and the economics of energy conservation.

TABLE 1   Values of Solid Waste Recoverables

Composition	% By Weight	Value of Component Per Ton of Solid Waste 1973	1981
Combustibles	76%	*4.00	**35.90
Ferrous metals	7%	1.50	1.50
Non-ferrous metals	1%	2.00	5.00
Inerts (glass, dirt, rocks, etc.)	16%	.50	.65
		8.00	43.05

*Fuel oil (#6) price 9¢/gallon
**Fuel oil (#6) price 79¢/gallon

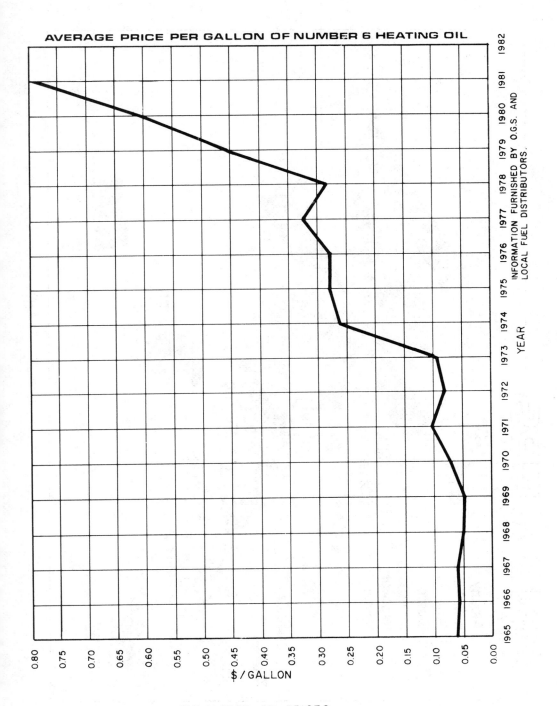

FIG. I - FUEL OIL PRICES

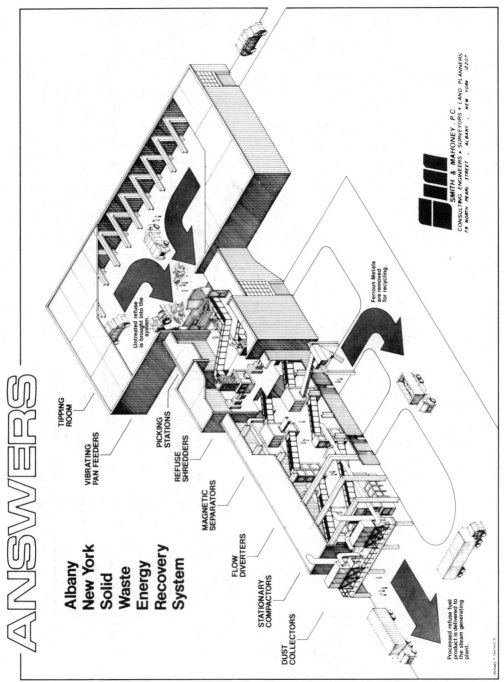

ANSWERS

Albany New York Solid Waste Energy Recovery System

TIPPING ROOM

VIBRATING PAN FEEDERS

PICKING STATIONS

REFUSE SHREDDERS

MAGNETIC SEPARATORS

FLOW DIVERTERS

STATIONARY COMPACTORS

DUST COLLECTORS

Untreated refuse is brought into the system.

Ferrous Metals are removed for recycling.

Processed refuse fuel product is delivered to the steam generating plant.

SMITH & MAHONEY, P.C.
CONSULTING ENGINEERS • SURVEYORS • LAND PLANNERS
79 NORTH PEARL STREET , ALBANY , NEW YORK 12207

FIG. 2 – ANSWERS CUT-AWAY

Table 2
Energy Balance—Fuel Product

Energy Input Description	Total Weekly Input (Million B.T.U.)
Two front end loaders (100 h.p. each), fuel total 6 gallons per hour	35.63
Vibrating pit conveyor	7.74
Inclined feed conveyor	11.87
Refuse shredder	470.40
Shredder discharge conveyor	4.79
Belt conveyor to magnetic separator	4.79
Belt conveyor to flow diverter	4.79
Belt conveyor to stationary compactors	4.79
Stationary compactors	14.82
Round trip to steam plant (18 miles)	155.65
Miscellaneous plant and building inputs	112.56
Ash removal (18 miles round trip)	22.05
Total weekly B.T.U.	849.88×10^6

FUEL PRODUCT HEAT VALUE (weekly)

= 4,200 tons X 10 million B.T.U. per ton X 0.65 boiler efficiency

= $27,730 \times 10^6$ B.T.U.

NET ENERGY GAIN = $26,880 \times 10^6$ B.T.U. per week

RATIO: $\dfrac{\text{Energy recovered}}{\text{Energy input}} = \dfrac{32}{1}$

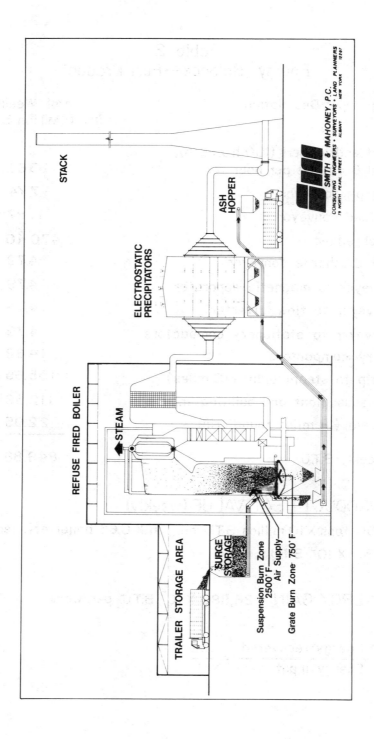

FIG. 3 - SCHEMATIC ELEVATION OF STEAM GENERATING PLANT

ENERGY-CONSERVING INDUSTRIAL PLANT DESIGN—AN ENERGY EFFICIENT BREWERY

J. M. Newcomb

ANCO Engineers, Inc., Santa Monica, California, USA

ABSTRACT

Industrial processes account for roughly 40% (29 x 10^9 GJ) of the annual energy consumption in the United States. Since the 1973 oil embargo, however, this segment of all US energy consumers has probably experienced the least overall percentage reduction in energy consumption per unit output. There are many reasons for this, but a primary cause is the tremendous cost of new equipment required in existing plants to upgrade processes to more state-of-the-art heating and cooling techniques. New plants, however, can incorporate high-efficiency systems during their initial design phase and realize a significant reduction in traditional energy requirements per unit of production.

Because of high Btu per unit output and relatively low temperature requirements, the food processing industry and breweries in particular are appropriate candidates for solar process heat generation and other designed-in energy-conserving strategies. A typical 3 million barrel a year brewery in the US consumes 750,000 MBtu annually for low temperature heating (60°-100°C) and refrigeration (756 Btu per 12 oz [US fluid,0.355 liter] can of beer).

This paper outlines the conceptual design of an energy-efficient brewery of 3 million bbl/yr capacity to be built in San Bernadino,California (34° 15' N latitude). Highlights of the design include 30,000 square feet (2,787 square meters) of solar collectors with a net generation of 10,800 MBtu annually for beer pasteurization, energy-efficient heat pump refrigeration, gas turbine cogeneration and direct sun heat recovery skylights. None of the energy-efficient strategies will compromise proprietary production techniques or capacities of output. Anticipated energy savings of over 30% of conventional brewery requirements will result in annual fuel savings of 230,000 MBtu.

KEYWORDS

Industrial plant design; process solar; breweries; cogeneration; skylights.

THE PLANT

The first phase of this project will result in a plant of 46,450 square meters with
final expansion to 139,350 square meters and a final capacity of 10,000,000 barrels* per year. It will be served by rail and highway as shown in the site plan
Fig. 1. The initial plant must supply continuous refrigeration for 45.42 million
liters (12 million US gallons) of liquid and heating (cooking) for 175,000 liters
of liquid per hour. The plant will incorporate five mechanical bottling and can-
ning lines with a total capacity of 5,650 units per minute. Cans and bottles are
then pasteurized (heated to 60°C held for 20 minutes then cooled to 27°C) at the
same rate. A return bottle handling facility will wash and sterilize 10,000 bottles per hour. Warehouse facilities, with computer controled pallet conveyors,
will handle the continuous movement and storage of 300,000 cases. Malt and grain
silos will vacuum-distribute to mash and wort cookers and spent grains will be
pumped out at the rate of 4.16 truck-loads per hour.

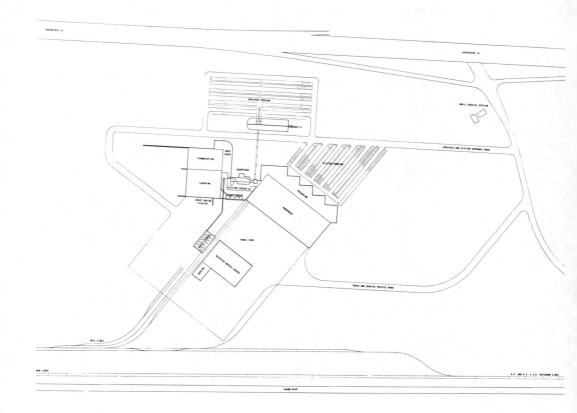

Fig. 1 Brewery site plan.

*One barrel = 117.34 liters (31 US gallons).

This tremendous heating and cooling load is primarily provided by steam while electricity is used for mechanical operations, pumps and lighting.

COGENERATION

Because very large quantities of steam (for heating and cooling) and electricity are constantly required, breweries are excellent candidates for cogeneration systems. Figure 2 shows typical steam requirements, per batch, for beer production. Part of Fig. 3 shows a double cogeneration system incorporating gas-fired turbines to generate electricity with waste heat boilers to generate steam. This high-pressure steam is superheated and used in steam turbines to run refrigeration compressors. The resultant low-pressure exhaust steam serves the other plant process steam heating requirements. This technique of efficient three-stage fuel use (gas turbine electrical generation, steam turbine refrigeration and process steam use) will reduce purchased fuels requirements by as much as 35%. Because over 78% of conventional brewery energy use is boiler fuels, it is likely that this system will also result in electrical generation beyond plant needs and even further savings can therefore be realized by selling excess electricity back to the utility.

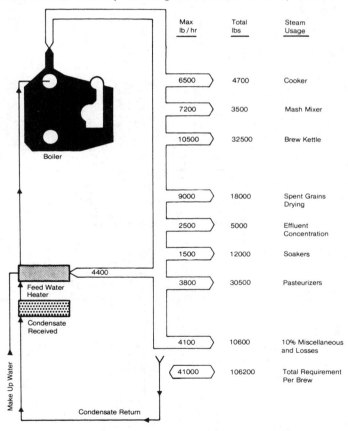

Max lb / hr	Total lbs	Steam Usage
6500	4700	Cooker
7200	3500	Mash Mixer
10500	32500	Brew Kettle
9000	18000	Spent Grains Drying
2500	5000	Effluent Concentration
1500	12000	Soakers
3800	30500	Pasteurizers
4100	10600	10% Miscellaneous and Losses
41000	106200	Total Requirement Per Brew

Boiler

4400

Feed Water Heater

Condensate Received

Make Up Water

Condensate Return

Fig. 2   Steam usage per batch (660 barrels, 4 hour cycle)*

*Source: *The Practical Brewer*, Master Brewers Association of the Americas, 1977

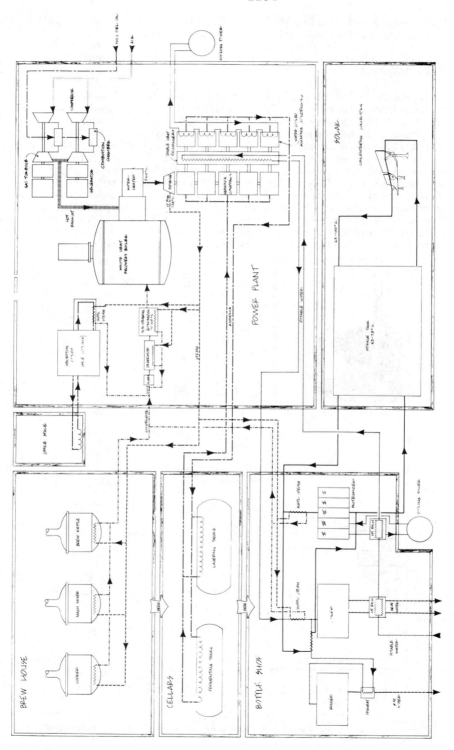

Fig. 3   Energy cogeneration, distribution, conservation.

SOLAR

Sixty degrees centigrade to 90°C bottle soaking and pasturization requirements are ideal temperatures for solar water-heating. Two thousand seven hundred and eighty-seven square meters of concentrating-type collectors at 34° north latitude will produce a net generation of 10,800 MBtu yearly, based on annual radiation and including collector efficiency and line and storage losses. This represents approximately 7% of the total thermal energy requirements for bottle washing and pasteurization. Although the initial investment of $1.3 million for this system produces a simple payback of 15.0 years, as energy costs spiral upward this cost will look increasingly attractive.* Additionally, various state and US federal solar tax incentives will reduce this payback considerably. Figures 4 and 5 show the collector locations on warehouse roofs with additional space available on the fermenting and lagering buildings. Figure 3 shows the integration of the solar system with the pasteurizers and soakers. *Based on $8 per MBtu for No. 2 fuel oil.

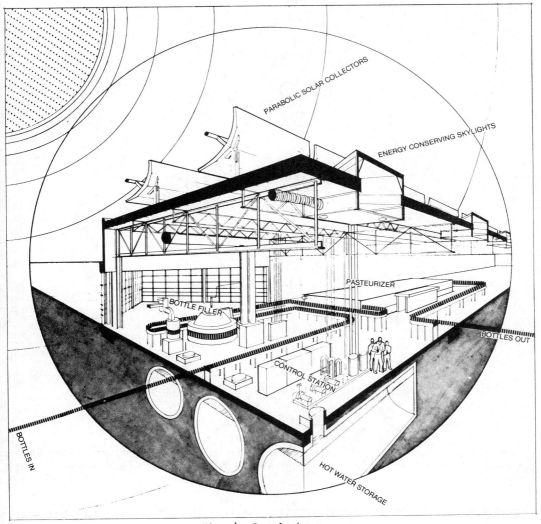

Fig. 4 Bottle house.

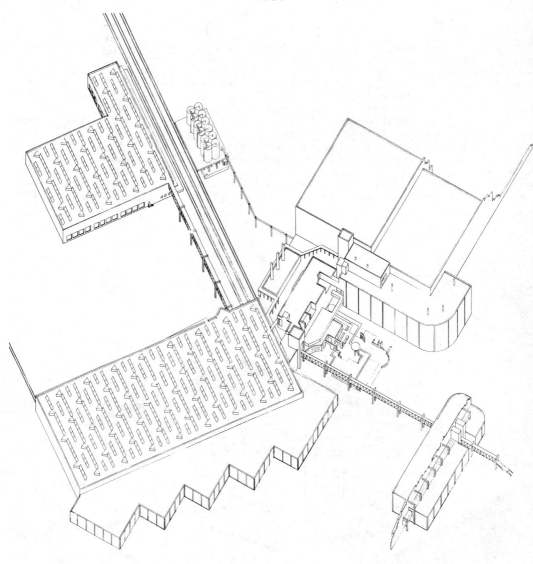

Fig. 5 Axonometric Showing collector and skylights locations.

HEAT RECOVERY SKYLIGHTS

An integrated direct sun skylight has been designed to provide natural light at a constant 540 to 650 lumens/square meters (50-60 fc) during daylight hours.*

---

*See ICEUM-II Conference Proceedings pgs. 1222 through 1228 "Direct Sun Skylights" Newcomb and Anderson, ANCO Engineers, Inc.

Indoor light distribution boxes, capture solar heat gain, fixture heat, and return air from interior spaces, utilizing the collected heat for heating in the
winter and exhausting it in the summer. Figure 6 illustrates the operation of
this system. Annual lighting and cooling costs will be reduced by over 2,900 MWh
as a result of these skylights.

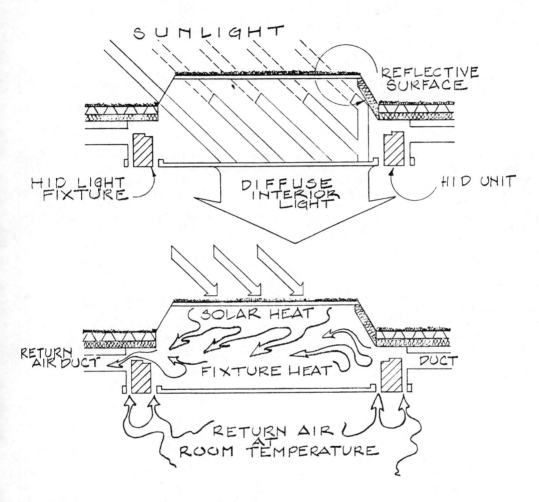

Fig. 6 Energy-efficient skylights.

OTHER STRATEGIES

There are many other energy conserving strategies that can be utilized in breweries, among them are:

- regenerative pasteurizers;

- using steam jackets to eliminate direct steam injection
  and loss of return condensate;

- recycling of water among soakers;
- cycling water between pasteurizers and rinsers;
- using rinse water for cooling tower makeup water;
- utilizing heat exchangers to precool ventilation air in lagering and fermentation cellars;
- various control systems to limit electrical peak power demand;
- by-product recovery and many others.

All of these can be designed in or retrofitted for a significant impact on overall plant energy consumption.

CONCLUSIONS

An ever increasing number of effective energy conserving strategies are being developed with direct application to industrial processes. Because their relative energy consumption is so high, measures providing only one or two percent reduction in a large plant's energy use can have a significant impact. Energy conserving strategies factored into new plant design are often as inexpensive as conventional systems and even with slight increases in front end costs the savings will quickly offset costs as fuel scarcities increase and prices escalate.

The architectural design of this project was developed by NSV, Newcomb, Stanislaw, and Voorhies Associates.

EMPIRICAL ENERGY REQUIREMENTS FOR SEVERAL ETHANOL-FROM-GRAIN OPERATIONS

R. A. Plant* and R. A. Herendeen**

*Environmental Studies Program, University of Lund, Gerdagatan 13,
S-223 62 Lund, Sweden
**Energy Research Group, Office of Vice Chancellor for Research,
University of Illinois at Urbana-Champaign, Urbana, IL 61801, USA

ABSTRACT

We have measured on site energy requirements for three sizes of ethanol-from-grain operations in the midwestern United States. Including the "energy cost of energy," the requirement ranges from 86 thousand to 36 thousand BTU per gallon of ethanol (20-8 MJ/ℓ). The smallest plant requires the most energy, and the largest plant requires the least. Indirect energy requirements (to produce capital equipment, enzymes, yeast, etc.) do not change this conclusion. These values are in the range expected from design studies.

KEYWORDS
Ethanol; Process Energy Analysis; Gasohol

The question of net energy balance for ethanol (EtOH) from grain continues to be addressed (Chambers and co-workers, 1979; DaSilva and co-workers, 1978; OTA, 1979; Scheller, 1979; U.S. National Fuels Alcohol Comm., 1981). Quoted energy require-ments for fermentation and distilling diminish with time. On one hand this is reasonable since the older figures often referred to beverage-grade ethanol produc-tion, which requires additional energy to purify the produce to a degree unnecessary for fuel ethanol. On the other hand, some of the recent claims represent such dramatic reduction in process energy that one wishes empirical verification. For example, in 1979, 40 to 50 thousand BTU per gallon of ethanol (9-12 MJ/ℓ) was con-sidered good (Chambers and co-workers, 1979; Scheller, 1979). Today there are claims based on design studies of 10 thousand BTU per gallon, or 2.3 MJ/ℓ (Harding, 1981).

There is, then, a need to determine if in actual practice energy requirements are diminishing as claimed. We have measured the energy used by three sizes of ethanol-from-corn operations in the midwestern United States:

1. Single farm size 1×10^4 gal. ethanol/year   ($45M^3$/yr)
2. Farmer consortium size 25×10^4 gal. ethanol/year   (1140 M^3/yr)
3. Larger gasohol plant 250×10^4 gal. ethanol/year   (11,400 M^3/yr)

In the first two cases (both in Illinois) the product is hydrous ethanol (180-190 proof) intended for farm equipment use in unblended form. The last case (in

Arkansas) produces a mixture of anhydrous ethanol and gasoline (gasoline being a process feedstock) intended for further blending with gasoline to produce gasohol (90% gasoline, 10% ethanol) for road vehicle use.

We stress that these different size plants use different types and sophistication of technology. The single farm operation is a "pot boiler," in which fermentation and distillation occur in the same tank; distillation involves boiling the entire batch and condensing the vapors. There is only one automatic valve in the entire operation. The farmer-consortium operation employs more complex distillation technology in which the mash is pumped into a three column (stipper-rectifier-condensing) distillation section. The gasohol plant is largely computer controlled, and achieves anhydrous ethanol by use of gasoline as a dehydrating agent and careful control of many process variables.

In spite of this, we feel justified in comparing energy use of the three operations. This is intentionally an empirical study, intended to obtain real data instead of theoretical. There are various hypothetical reasons why (say) the larger technology could not be applied to the smaller operation, and some why it could.[1] This is, to be sure, interesting, but for the purposes here it is peripheral.

All of the measurements were made on operating units that are available commercially. The producers of the first two have requested confidentiality; the last is manufactured by the ACR Process Corporation of Urbana, Illinois. We took data of their technology in place at White Flame Fuels, Van Buren, Arkansas.

Before listing results, we review the magnitudes of the energy requirements for ethanol-from-grain. A nominal yield is about 2.5 gallons of ethanol per bushel of corn, or 0.54 liter/kg. (The three operations here fall 8-15% short of this, however). Assuming the nominal yield, the energy requirements (including the "energy cost of energy") for production of ethanol[2] are:

Agricultural inputs[3]...................... $50-70 \times 10^3$ BTU/gal $(12-16$ MJ/ℓ)

Process Energy (fermentation, distillation). $40-80 \times 10^3$ BTU/gal $(9-18$ MJ/ℓ)

Depreciation of Plant Equipment .... $\underline{5-10 \times 10^3}$ BTU/gal $(1-2$ MJ/ℓ)

TOTAL.......$95-160 \times 10^3$ BTU/gal $(22-36$ MJ/ℓ)

In this work we concentrate on ("direct") process energy, while analyzing the other inputs only in sufficient detail to ensure that they do not affect our conclusions. For comparison, the combustion energy of ethanol is $75-80 \times 10^3$ BTU/gal $(17-19$ MJ/ℓ). Detailed comparison of input and output energy involves attention to "credits" for the fermentation by-product, for miles-per-gallon obtained (or for other indicators of end-use efficiency), and so on. This is covered in Chambers and co-workers (1979), Scheller (1979), and TRW (1980).

---

[1]For example, it is unlikely that the more extensive heat exchangers for heat recuperation in the farmer consortium operation could be employed economically in the single-farm size.

[2]Results are expressed per gallon and liter of pure ethanol even though that may not be the actual output.

[3]The range given for agricultural inputs is for the 10 main corn producing states. A much greater range exists if one accounts for heavily irrigated corn. For example, in 1974 Arizona corn production averaged 625 kBTU/bushel, equivalent to 250 kBTU per gallon ethanol or 5.8×10^7 J/ℓ (USDA, 1976).

Our measurements were made in late spring and early fall of 1980, on runs of one to three days.   Energy use was measured by us, yield and proof by us and by the operating personnel.   Table 1 summarizes the results.   More details on data collection and analysis are in Dovring and co-workers (1980).

The energy requirements do show an apparent economy of size;   the larger operation uses less energy per gallon.   The lowest energy, about 36×10^3 BTU/gal (8.4 MJ/ℓ) of EtOH for the ACR process, agrees with the claims of that firm (Chambers and co-workers, 1979).   The larger operations use less energy for heat, but more as electricity for running pumps and control systems, which are largely absent from the single farm operation.

In all cases yields are less than the industry goal, 2.5 gal/bu and there is an obvious sensitivity to this input variable.   In addition, we must comment on indirect energy requirements such as capital equipment, chemicals (yeast, enzymes, etc.).   In Dovring and co-workers (1980) we have calculated most of these, and have verified that they are relatively small.   More importantly, any inequality in the indirect energy requirements is in the same direction as the direct energy requirements, so that the ordering is unchanged.   In saying this, we have not considered different agricultural energy inputs, which can be very significant if one allows for use of irrigated grain (Dovring, 1980).   We have, however, included the energy consequence of different transportation distances between grain supplier and distillery, and between distillery and grain by-product.   The larger the ethanol operation, the greater these distances.

The uncertainties indicated in Table 1 are based on 1) instrumental uncertainty, 2) a limited amount of statistical analysis of data, and 3) in the end, some guesswork.   They do not adequately reflect some other sources of error, such as operator inexperience.

In Fig. 1 we display the results from Table 1, as well as those from an operation in Colorado (Jantzen and McKinnon, 1980).   The latter is the same size as our farmer consortium, but uses about 1/2 as much energy per gallon.   One reason for this may be more extensive use of heat recovery in the Colorado operation and the use of inexperienced help in the comparable Illinois operation (by the latter's own account).   During data collection we saw some signs of inexperience, including one instance in which a grain-water batch was simultaneously heated and cooled.   (The energy consequence was small since this was only one-fourth of a day's output;   we corrected our results nonetheless.)   But at this point we cannot comment knowledgeably on the large differences in direct energy inputs.[4]

Note from Table 1 that the three operations use either propane or natural gas. The producers of all three claim the potential for substitution to less premium fuels.   The maker of the farm-size operation is enthusiastic about the potential for burning wood, stover, or hay;   while the larger operations are claimed to be suitable for coal.   (The Colorado operation, during the published test results, was using diesel oil.)   We again take an empirical approach to this;   we have not yet seen such conversion.   One possible contradiction occurs for the farm-size still, which can be left untended the better part of an 8-hour distillation run. This neglect is possible with easily manageable gas from a tank or main as fuel, but seems much less likely with a hay or wood burner.   On the other hand, the grain product and stover could be digested to produce methane.   A study of this

---

[4]We have discussed the discrepancy with Thomas McKinnon, and he stands by his results (19 November, 1980).

TABLE 1   Comparison of Three Ethanol-From-Grain Operations: Energy Requirements
Figures in parentheses are energies, including "energy cost of energy"
EtOH = 100 percent ethanol

	SINGLE FARM[5]	FARMER CONSORTIUM[6]	LARGE GASOHOL PLANT[7]
Proof of product	∼180	190	200 (5-15% gasoline)
Output	35 gal EtOH/day (132 l/day)	600 gal EtOH/day (2261 l/day)	6785 gal EtOH/day (25570 l/day)
Yield	2.14 gal EtOH/bu. ±5% (0.46 l/kg)	2.08 gal/bu. ±4% (0.45 l/kg)	2.32 gal/bu. ±5% (0.50 l/kg)
Non-electric energy cons.			
Cooking	9100 BTU (9.6 MJ) ±6%	19800 BTU (20.9 MJ) ±13%	20300 BTU (21.4 MJ) ±9%
Distilling	77600 BTU (81.9 MJ) ±6%	36300 BTU (38.3 MJ) ±6%	
Electricity consumption	1810 BTU (1.9 MJ) ±3%	20000 BTU (21.1 MJ) ±4%	15500 BTU (16.3 MJ) ±3%
Total energy use	88500 BTU/gal ±4% (20.5 MJ/ℓ)	76800 BTU/gal ±4% (17.8 MJ/ℓ)	35800 BTU/gal ±5% (8.3 MJ/ℓ)

[5]The feed by-product is neither dried nor centrifuged. The unit is a "pot-boiler," with a small amount of insulation and some recycling of hot water. Operation is eight hours per day, and the column diameter is four inches.

[6]The feed by-product is centrifuged to 60% water content. Column diameter is 15", and the unit is intended to run continuously. There is extensive insulation and fairly extensive recycling of hot water.

[7]Output is a mixture of anhydrous alcohol and gasoline. The solid by-product is dried in a separate drier, and an evaporator is normally used to reduce the liquids from the stillage (though it was not in operation for these measurements). There is extensive use of heat recovery systems. Column diameter is 42", and the unit is designed for continuous operation.

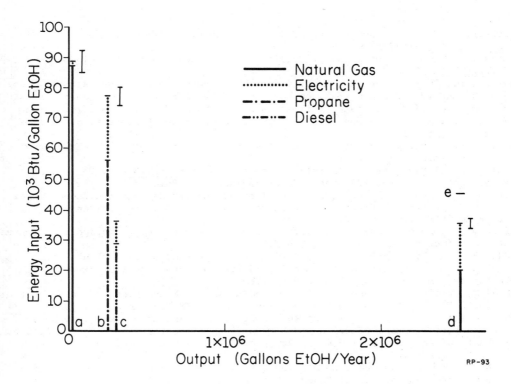

Fig. 1.   Direct energy inputs for four ethanol operations. This
          includes all energy inputs to fermentation and distilling,
          including the "energy cost of energy." It does not include
          agricultural energy requirements for feedstock grain, or
          energy requirements of chemicals and capital equipment at
          the distillery.   Shown is the energy to produce one gallon
          of EtOH, though the actual product may be a mixture of
          ethanol and water, or ethanol and gasoline; see Table 1.
          Error bars are indicated.   a) Single-farm unit, b) Farmer
          consortium unit, c) Schroder plant (Colorado), as discussed
          in Jantzen and McKinnon, d) Gasohol plant.   The line at
          "e" represents the manufacturer's estimate if the grain
          by-product had been dried.

possibility is underway at the University of Illinois (Rodda and Steinberg, 1980).
However, this discussion leads back to design data and speculation, which we hoped
to avoid in this report of actual results from three operating ethanol producers.

REFERENCES

Chambers, R. S., R. A. Herendeen, J. J. Joyce and P. S. Penner (1979).
  Science, 206, 789-795.
DaSilva, J.G., G. E. Serra, J. R. Moreira, J. C. Concalves and J. Goldemberg (1978).
  Science, 201, 903.
Dovring, F., R. A. Herendeen, R. L. Plant, M. A. Ross (1980. "Fuel Alcohol From
  Grain: Energy and dollar balances of small ethanol distilleries and their
  economies of size and scale." Illinois Agricultural Economics Staff Paper 80
  E-151, Dec. 1980; ERG Doc. 313, Energy Research Group, University of Illinois,
  Urbana, IL, 61801.
Dovring, F. (1980). Export or Burn? American Grain and the Energy Equations.
  Illinois Business Review, Vol. 37, No. 4.
Harding, J. (1981). Ethanol's Balance Sheet. Soft Energy Notes, Vol. 3, No. 6.
  Friends of the Earth, 124 Spear, San Francisco, CA, 94105.
Jantzen, D., and T. McKinnon (980). Preliminary Energy Balance and Economics of
  a Farm-Scale Ethanol Plant. Solar Energy Research Institute, Golden, CO.
  Stock No. SERI/RR-624-669R.
Office of Technology Assessment (1979). Gasohol, a Technical Memorandum. Congress
  of the United States, Washington, D.C.
Rodda, E., and M. Steinberg (1980). Energy Analysis of an Agricultural Alcohol
  Fuel System. Presented at the meeting of the American Institute of Chemical
  Engineers, Chicago, IL., 20 November, 1980.
Scheller, W. (1979). Gasohol, Ethanol and Energy. Manuscript presented at the
  National Gasohol Commission Meeting, San Antonio, TX, 2-5 December, 1979.
TRW, Inc. (1980). Energy Balances in the Production and End-Uses of Alcohols
  Derived from Biomass. Stock No. O-BID-33-250, U.S. Government Printing Office.
U.S. Dept. of Agriculture (1976). Energy and U.S. Agriculture: 1974 Data Base.
  Stock No. FEA/D-761459.
U.S. National Alcohols Fuels Commission (1981). Fuel Alcohol, An Energy Alterna-
  tive for the 1980's. Final Report, Washington, D.C.

ENERGY RECOVERY FROM LOW-GRADE FUELS AND WASTES

M. Rasmussen

Vølund Miljøteknik A/S, Denmark

Denmark has very limited resources of fuels for energy
production. The recent years have given us hope of collecting some gas and oil from the North See, but still not
nearly enough to be self-supporting with energy. In the
recent years an intensive work has been carried out to
develop technologies which shall utilize the energy content in low-grade fuels and even waste products.

Most of the newly designed plants are all based on a
technology which has been in use before the oil boom swept
the country in the fifties, but the technology has been
improved by utilizing the development in the technique.

Straw is one of the fuels which in Denmark is utilized for
mainly heat production.

Already during the Second World War, the Danish farmers to
a great extent utilized the straw, in specially designed
boilers utilizing the underfire system. The combustion
rate was closely controlled by regulation of the combustion
air rate in relation to the water temperature.

Within the last five years, a new generation of those
straw boilers have been brought on the market, but still
based on the same principles. The disadvantage of those
boilers are the frequency of feeding the boiler, every 3 -
6 hours, and the rather low efficiency.

Tests have been carried out showing efficiencies as low as
40 - 45%.

The feeding problem has been solved with some automatically feeding system, which only demands manually operation every 24 hours, but to solve the problem of efficiency some other combustion technology has to be used,
and the straw has to pass through a cutter before combustion.

The combustion can take place in two ways.

For the smaller plants, a special type stoker feeds the
cut straw to a combustion grade where adequate combustion
air is added. The feeding rate and the combustion air is
controlled according to the necessary heat demands and can
also, at a very low rate, be regulated on/off. In this
way, the plant can be made fully automatically operated,
thus obtaining a higher efficiency in the range of 68 -
70%. Plants of this type can be built in the capacity
range of 0.2 - 5 Gcal/h per unit.

To further improve the efficiency, a new technology has
been developed, utilizing combustion of the cut straw in
suspension. This type of plant can be operated fully
automatically and controlled, obtaining efficiencies up to
78 - 80%.

In 1980, a plant of this type was put into operation, as an E.E.C. demonstration project, in the town Svendborg in Denmark.

The plant can be divided into four main sections:

- Preparation and storage
- Transport and firing equipment
- Boiler section
- Flue gas treatment section.

The straw is received in compressed max. bales, which after reception pass through a cutter.

By means of a pneumatically transport system, the straw is delivered to a storage silo with a capacity equal to 3 x 24 hours operation. The storage silo at Svendborg has a capacity of 3,200 m^3.

From the silo a mechanical transport system feeds the straw into the burners, mounted directly on the radiation part of the boiler. By means of the combustion air, the straw is blown into the combustion chamber and is mainly burned in suspension. Some bigger particles will be burned out on a fixed grate in the bottom of the boiler.

The flue gases are cooled partly in the radiation part and in two smoketube convection parts. The energy produces hot water for utilization in the district heating scheme of the town.

The plant is capable of burning 3 tons/h and produces 7.4 Gcal/h. The saving in fuel oil is app. 1,000 kg oil per hour.

The plant has been in operation for app. 6 months and so far operated according to expectations with regard to the thermal efficiency. Due to a very wet autumn and winter, some problems have occured in the cutting and transport system, which presently are undergoing some modifications.

Due to the demand for storage room, this type of energy plants do have an upper limit.

It is recommended to build plants in the range of:

5 - 15 Gcal/h per plant and unit.

One of the first fuels ever used was wood. To use wood as fuel today would be considered a luxus, but to utilize the energy content in the great amounts of waste wood, has been more and more common in Scandinavia.

For this purpose, different types of technologies have been developed and more than 300 plants are today in operation all over Scandinavia utilizing wood waste.

A wood waste energy recovery system can be divided up into:

- Storage and feeding system
- Firing equipment
- Furnace and boiler system
- Clinker discharge system
- Fluegas cleaning system.

Each part of the plant will be designed according to:

- Type of wood waste
- Energy demand
- Emission control standard.

Two types of storage are used. For dry wood a cylindrical silo, with height 2 - 3 times the diameter, and with a rotating screw for feeding to the firing equipment.

For wood waste with high humidity, squared silos with a low height, and with a scraber arrangement, as feeding system has to be used.

For capacities in the range of

2 - 5 Gcal/h per unit

the most frequent used firing technology is the stoker firing, specially adapted for wood firing. To prevent that backfire through the stoker ignites the fuel in the sto-rage, all plants are equipped with a water quenching system. The furnace shall be designed with due respect to the humidity content in the wood waste. High humidity content demands a high zone of refractory lining, whereas wood waste with low humidity can be incinerated with very low zone of refractory lining.

The necessary combustion air is introduced through the grates and as overfire air in the combustion zone. This is to ensure complete burn-up of the flue gases before they are cooled down in the boiler. The boiler can be made as hot water, pressurized hot water and steam boiler, and the total system can be regulated continuously according to the energy demand, and be automatically operated. The efficiency is in the range of 70 - 75%.

All plants are equipped with flue gas cleaning system and most frequently in the form of multicylones. The precipitated fly ash consists primarily of carbon, and is being recycled to the stoker system.

For capacities greater than 5 Gcal/h, plants are specially designed for each job.

In northern Sweden, a special waste problem in the forests have been the start to a new technology.

The fuel on this plant consists of

- Bark
- Bark waste

and - branches from the forest.

The water content goes as high as 75% in this waste, which is produced in amounts of 800 - 1,000 tons/day.

In 1975, S.C.A. ordered a steam raising plant to utilize this waste. The technology chosen was the VOLUND forward pushing grate, which for more than 50 years have proved superior to burn garbage.

During 1975 - 76, the plant was designed and constructed with start-up in January 1977.

The plant can incinerate 34 tons wood waste per hour and produces 70 tons steam of 64 bar, superheated to 495°C.

The plant can shortly be described as follows:

From an open storage area, the waste is fed into the furnace with an overhead travelling crane. The feeding chute acts as a sealing between the furnace room and the surroundings.

The grate system consists of 5 grates. The first grate acts as feeding grate, whereas the rest act as drying, ignition, and combustion grates.

The furnace is completely refractory lined, to ensure complete control of the combustion process, and with the boiler attached to the furnace.

The flue gas cleaning takes place in two electro-static precipitators.

The plant has been in operation since start-up in 1977 and until 1st January, 1981 incinerated 1.6 Mill. m^3 bark. The energy produced equals a saving of app. 53,000 tons oil, representing a value of 64 Mill. d.kr..

The total investment including buildings amounts to 54 Mill. d.kr.

S.C.A. considers today this investment as one of the very attractive ones made.

In order to control the waste production it is necessary to make use of a technology which is flexible in respect of variations in the composition of the fuel.

The technology used here allows us, in addition to the burning of bark, to burn domestic waste which in turn lead us into this type of alternative energy.

Incineration of domestic waste with heat recovery has been practised for 50 years as in 1931 the first incinerator with heat recovery was commissioned. This plant was in operation till 1972.

Waste is by nature inhomogeneous and has a high content of humidity and incombustible matters.

A specialized technology with the main stress on the grate system and furnace construction shall be applied.

The grates shall have close contact to the wastes and ensure an even good contact for the combustion air.

The furnace shall be large enough to ensure a complete burn-up of the flue gases from the incineration. This is the only way to ensure destruction of environmental pollutants. Further, as a consequence of this design the damages of corrosion are minimized.

In order to illustrate the application for heat production the following two cases shall be mentioned:

Videbæk District Heating Corporation has seen the possibilities in utilizing the energy from the domestic wastes for heat production for district heating.

With an amount of waste of

$$8 - 10,000 \text{ tonnes/year}$$

they have installed a waste-heating plant for the district heating with a capacity of

$$2 \text{ tonnes waste/hour.}$$

The plant can be operated according to the load on the district heating network from 100% down to 50%.

Through variations in the number of hours during the day the plant is operating, i.e. 8, 16, or 24 hours, it is possible to make the heat production match the demand of heat at the network.

The investment is around 10 million Danish kroner. The following budget can be made:

Wages (6 persons)	690,000
Administration, ½ day + 20%	51,000
Maintenance, machinery, and building	185,000
Sundries	167,000
Miscellaneous	112,500
Cost of operation	1,206,000
Cost of capital	
(20 years, 15% interest)	1,597,610
Total cost	2,803,610
Income of operation	
14,300 Gcal of 215 D.kr.	3,074,500
Cost of treating 10,000 tonnes	
of municipal waste	- 270,890

Corresponding to an income of 27 Danish kroner per tonne of waste.

The other case in Copenhagen: On the large-scale plant, I/S Amagerforbrænding, waste from approximately 550,000 inhabitants is received and turned into heat.

The plant operates continously and the furnaces achieve more than 7,500 hours of operation each (or 85%). On availability corresponding with power stations and other heating plants.

The annual heat production corresponds to an amount of approx. 40,000 tonnes oil.

Further, the clinkers are utilized after a simple mechanical separation as:

20 - 30 %	fine clinkers for slabstone production.
30 - 40 %	coarse clinkers for road foundation.
10 - 15 %	scrap iron for remelting.

The remainder, which shall be deposited, is less than 1% of the original volume.

Also here it is possibe in the same way to make a budget which from the start of the plant will show an equilibrium between the expenditures and the incomes.

At the end, we would like to present a technology which is under development, with the aim of converting the energy content in wet biomass (e.g. animal manure and catch crops) into low grade heat energy suitable for space and water heating. The system consists of a drier, a burner, and a gas scrubber. The heat from combustion of already dried biomass is used for drying of wet biomass on the way to combustion and the heat of the humid flue gases from the drier is recovered by means of the gas scrubber.

Predictions of the performance show that the thermal efficiency could be around 65% and that the maximum acceptable
humidity in the biomass could be 80% without the use of
auxiliary fuel. The maximum temperature on the water side
is estimated to app. 80 - 90°C.

Preliminary, pilot tests have been carried out and full-
scale tests will be started within soon.

The equipment is planned to be made in the capacity range
of 20,000-500,000 kcal/h for use in farm houses and in
connection with a district heating net.

These were some examples of the waste products which are
being utilized in Scandinavia. Further, it shall be mentioned that more than 65% of all industrial and domestic
refuse is incinerated with full energy recovery, both on
large scale plants treating refuse from 600,000 inhabitants to small scale plants treating refuse from 8-10,000
inhabitants.

It is estimated that 6-8% of the demand for heating will
be produced from different types of waste products.

FINITE TIME CONSTRAINTS AND AVAILABILITY

M. H. Rubin*[1], B. Andresen* and R. S. Berry**

*Physics Laboratory II, University of Copenhagen, Universitetsparken 5,
2100 Copenhagen, Denmark
**Department of Chemistry, University of Chicago, Chicago,
Illinois 60637, USA
[1]On leave from Department of Physics, University of Maryland
Baltimore County, Catonsville, Maryland 21228, USA

ABSTRACT

The concept of availability is extended to processes constrained to operate at
nonzero rates or in finite times. The effects of the time constraints are
explored for a model system in which extraction of work competes with internal
relaxation. The goal is the establishment of the finite-time availability as
a standard of performance more useful than the traditional availability based
on reversible processes.

I. INTRODUCTION

Gibbs[1] showed that the availability $A = U + p_o V - T_o S - \Sigma \mu_{oi} N_i$ is the function
whose changes give the maximum work extractable from a system going from an
initial state to equilibrium with an environment whose intensive variables are
the pressure p_o, the temperature T_o and the chemical potentials μ_{oi}. Availability and its dimensionless counterpart of effectiveness are guides to how we can
modify existing processes to extract as much use as possible from process energy
that would otherwise be discarded. If one foresaw that fuel costs would be high
but very uncertain, one might even wish to use availability as a surrogate for
cost, especially if one were comparing alternative processes.

The evaluation of availability is normally done[2] with the initial and final
states of the process given by the definition of the process, including its constraints on the variation of temperature, pressure and other thermodynamic
variables. In practice, the Gibbs availability cannot be completely captured
because of the familiar difficulties with reversible processes. Strictly, the
work that can be captured from a system that begins in a given state 1 and approaches equilibrium in state 2 with its environments at p_o, T_o, ..., is

$$W_{1 \to 2} \leq A_{1 \to 2} = A(1) - A(2)$$

It would be useful to extend the concept of availability to allow us to evaluate
the maximum work that could be extracted from a system whose initial state is
specified, which approaches equilibrium with its environment whose intensive
variables are at p_o, T_o, ..., and for which the allowable processes are subject
to a constraint on time or rate. This quantity would not only be a more reali-

stic measure of the performance that a process might achieve; it also would allow us to optimize rates of operation in order to match the availability of the system with the uses to which that availability could be put.

It is our purpose here to introduce an extension of the definition of availability to include processes constrained to operate in finite time or at finite rates. In the following section, we give the definition itself, a brief description of the generalization, and some conclusions that will be derived and discussed in detail elsewhere. The third section, which is the main body of this paper, treats a model system in detail. The system we have chosen is a set of reservoirs connected to heat engines and to each other by finite heat conductances that satisfy Fourier's law of heat conduction, $\dot{Q} = \kappa(T_2-T_1)$. This system is simple enough that we can get a clear picture of its behavior from algebraic analysis alone and quantitative solutions from straight forward computer solutions. Moreover it is a system that serves as a prototype for many more complex systems that can be described by perturbations of our model system.

II. FINITE-TIME AVAILABILITY

To extend the definition of availability from the conventional one given above, we turn to the statement of the work done by a system in the form of an equality, rather than an inequality. We use the form of this equality formulated by Tolman and Fine[3]:

$$W = -\Delta A - T_o \int_0^{t_f} \dot{S}_{tot} \, dt \tag{1}$$

where

$$\Delta A = A(t_f) - A(0) \tag{2}$$

and t_f is the duration of the process.

Next we suppose that there is a set of constraints $g(x_1,\ldots,t) = 0$ that we must accept, and some control variables $u(t) \varepsilon U$ as well. The problem we wish to solve is to find the maximum of W by suitable choice of the control variables, subject to the constraints. This maximum is the quantity we <u>define</u> as the <u>finite-time availability</u> ΔA:

$$-\Delta A = W_{max}$$

$$= \max_{u \varepsilon U} \left(A(0) - A(t_f) - T_o \int_0^{t_f} \dot{S}_{tot} \, dt \right) \tag{3}$$

subject to $g(x_1,\ldots,t) = 0$.

The quantity ΔA has several properties of interest:

1) In the limit of reversible processes, the rate of entropy production $\dot{S}_{tot}(t)$ is identically zero, so the finite-time availability is the same as the Gibbs availability, $\Delta A \to \Delta A$.

2) If the end points of the process are both fixed, then $A(t_f)$ and $A(0)$ are determined, and the evaluation of ΔA is equivalent to minimization of the entropy production for the fixed (finite) time t_f.

3) In general, the constraints and process definitions need not specify the state of the system at the final time t_f. In particular, the values of the intensive variables of the system at time t_f need not be the values p_o, T_o, μ_{oi} of these variables in the environment. If this is the case, ΔA is not

determined by the statement of the problem, and the problem of finding ΔA includes the problem of finding the state at t_f as well as that of finding the entropy production from 0 to t_f, which together maximize W.

4) In general, any value found for ΔA with the end point specified will be less than or equal to a value determined with the same initial state and all other constraints but without the specification of the final state.

We wish to broaden the class of processes we can describe with the tools of thermodynamics. The introduction of irreversible, finite-time phenomena so broadens this class that we are forced to give more specification of our processes than one is required to make in most of the literature of traditional reversible thermodynamics. It is no longer sufficient to say that a branch is isothermal, isobaric or adiabatic; we must also say what are the significant relaxation processes, entropy generating processes or rate-limiting phenomena.

With this further specification, we must provide more than the traditional constitutive parameters of heat capacity, thermal expansion and compressibility. We need the parameters such as heat conductances, friction coefficients or chemical rate coefficients that quantify the temporal behavior of the system.

III. MODEL

In order to illustrate the finite time availability we consider the problem of extracting the maximum work from a relaxing system. The system is composed of two finite size reservoirs which are connected by a thermal conductor. To extract work from this system we connect them to the environment through two endoreversible heat engines (see Fig. 1). All heat links are assumed to satisfy Fourier's law of heat conduction,

$$\dot{Q} = \kappa (T_i - T_j). \tag{4}$$

Our objective is to extract as much work as possible from this system during the time period 0 to t_f. The heat capacities C_i and the conductance κ are fixed, whereas the other conductances may vary within the bound

$$0 \leq \kappa_{1h}, \ \kappa_{1l}, \ \kappa_{2h}, \ \kappa_{21} \leq \kappa_{max}. \tag{5}$$

We thus have the κ_i and the heat rates $\dot{Q}_i$ (which in turn determine the intermediate temperatures T_i) at our disposal to achieve our goal. Obviously, if the two reservoirs are initially at different temperatures, there will be competition between availability loss through κ and work production by the two heat engines.

Of course, this problem is of most interest when the time constant of the system is comparable to that of the extraction process. If the system time constant is very short then the system will quickly equilibrate and the problem will reduce to a one-reservoir problem with some average initial temperature. In the opposite extreme the link between the reservoirs is irrelevant and the problem reduces to two single reservoir problems.

Although our model is cast in the language of heat transfer and mechanical work, it is actually much more general, describing all thermodynamic processes in which desired product formation competes with internal relaxation, with all losses being linear in rate. If one is concerned with chemical reactions like

$$R_1 + R_2 \rightleftarrows P + \Delta E,$$

$$\text{e.g. } 2 H_2 + O_2 \rightleftarrows 2 H_2O + 572 \text{ J}, \tag{6}$$

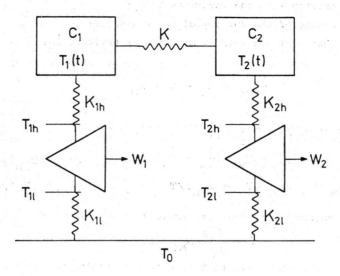

Fig. 1. Model system consisting of two finite size reservoirs at temperatures $T_1(t)$ and $T_2(t)$ and the environment at temperature T_o. The reservoirs produce work through their respective engines in competition with internal relaxation through the conductance κ.

imagine the heat reservoirs to be supplies of reactants at chemical potential μ_i and the environment to be the product reservoir. Physically the reaction can then be carried out reversibly on an electrode surface with the resistances κ_i signifying diffusion to and from the surface or surface potentials. Then eq. (4) is simply replaced by the diffusion equation

$$\frac{dN}{dt} = \kappa(c_i - c_j). \tag{7}$$

The internal relaxation through κ will be stray reactions or autocatalyzed reaction where the product cannot be collected. More complicated, or multistep, reactions can be pictured by putting two systems back to back, so that W is only an imaginary energy transfer from reactants to products, it only appears for calculational purposes. Set up exactly as described above the optimization maximizes energy output from the "battery", but by changing objective functions it is just as easy to maximize chemical output.

The analysis of thermodynamic processes in terms of availability usually starts by dividing the (closed) total system into two interacting parts: The environment characterized by constant intensive variables, and the system of interest. In our case it is more convenient to divide the total system into three parts: The environment, the interacting system, and the machines that are used to perform the useful work. This is done in order to emphasize the generic nature of our analysis.

The determination of W by eq. (1) is based on $\dot{S}_{tot}$ which, with our three-way division, consists of

1) the flow of entropy into the environment, $\dot{S}_o$;

2) the rate of change of the entropy of the system, $\dot{S}$;

3) the rate of change of the entropy of the work devices, $\dot{S}_M$;

so that

$$\dot{S}_{tot} = \dot{S}_o + \dot{S} + \dot{S}_M. \tag{8}$$

Specifically

$$\dot{S}_o = \dot{Q}_o/T_o = \sum_i \kappa_{il} (T_{il} - T_o)/T_o$$

$$\dot{S} = \kappa(T_1 - T_2)(\frac{1}{T_2} - \frac{1}{T_1}) - \sum_i \kappa_{ih} (T_i - T_{ih})/T_i \tag{9}$$

$$\dot{S}_M = \sum_i \kappa_{ih} (T_i - T_{ih})/T_{ih} - \sum_i \kappa_{il} (T_{il} - T_o)/T_{il}$$

since the heat engines themselves are assumed reversible

$$0 = \int_0^{t_f} \dot{S}_{Mi} \, dt. \tag{10}$$

Finally the heat reservoirs change temperature according to the energy conservation equations

$$C_1 \dot{T}_1 = -\kappa_{1h} (T_1 - T_{1h}) + \kappa (T_2 - T_1)$$

$$C_2 \dot{T}_2 = -\kappa_{2h} (T_2 - T_{2h}) - \kappa (T_2 - T_1). \tag{11}$$

We thus wish to optimize eq. (1) with the constraints eqs. (5), (10) and (11) and apply the method of optimal control. Details of the calculation will be presented in a forth-coming publication.

IV. RESULTS

Before doing serious calculations we observe that the lower temperatures T_{il} only appear in $\dot{S}_{tot}$ and in eq. (8) and not mixed with other variables. This implies that for optimal operation the T_{il} are kept constant. Their specific values are determined through the constraints in eq. (10). One should not be disturbed that these constants are larger than the equilibrium temperature T_o, since the final state of the system will not reach T_o in the time t_f.

The existence of a constant of the motion greatly simplifies the analysis. It is

$$\sum_i \psi_i T_i = \text{const.} = C_1 + C_2 - (C_1 T_{1f} + C_2 T_{2f})/T_o, \tag{12}$$

where ψ_i is the variable conjugate to T_i through eqs. (11), and subscript f denotes the value at the final time t_f.

The constants of motion T_{il} and $\sum \psi_i T_i$ are what are called local constants of motion in classical mechanics. That is they depend on the particular path the system evolves along. This is contrasted with general constants of the motion which arise in classical mechanics because of some symmetry of the system, such as rotational symmetry which leads to the angular momentum being conserved for all possible paths.

It turns out that this constant leads to a relation between the way the two reservoirs should be pumped.

In the special case that $C_1 = C_2$ and $\kappa_1 = \kappa_2 = \kappa$ for sufficiently large t_f this leads to the intuitive solution of pumping the system until $T_1 = T_2$ and then maintaining this condition. However, in general the pumping procedure is more complicated and the optimal solution does not lead to a configuration with $T_1 = T_2$.

In order to picture the optimal path it is convenient to introduce two new variables,

$$B \equiv T_2/T_1 \qquad (13)$$

and an angle θ defined through eq. (12) which is related in a complicated way to the rates at which the engines drain the reservoirs.

The analysis of the optimal trajectory is in terms of the time evolution of θ and B. One remarkable result of this analysis is that the possible final states of the system must lie on a curve

$$B_f = (C_1/C_2) \tan^2 \theta_f \qquad (14)$$

which is shown in heavy line in Fig. 2.

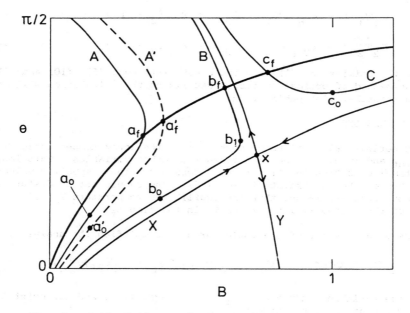

Fig. 2. Optimal time evolution paths for the model system. All final states lie on the heavy curve, and the infinite-time paths are labeled X and Y. B and θ are defined in eq. (13) and below.

Possible time evolution trajectories determined from the optimal control equations, appear in lighter line. They all cross the curve of final states which should be reached in time t_f. This illustrates the usual problem with optimal control solutions that one needs to solve the time evolution both forward (from the given initial state) and backward (to ensure that the trajectory crosses F at t_f) in time. Thus added process time cannot be taken full advantage of simply by continuing the original trajectory, the optimal solution requires a completely new path. The dashed trajectory A' with $t_f' > t_f$ deviates from A right from the beginning a_o. That the two temperatures T_1 and T_2 do not necessarily approach equilibrium is seen from B, where the temperature ratio does diminish from b_o to b_1 but finally is increased from b_1 to b_f. It is even more striking with C, where the reservoirs are initially in equilibrium but develop a strong asymmetry. This effect is caused by their different heat capacities, so that they cool at different rates. At t_f the intermediate temperatures obey the same equation

$$(T_{i1}/T_{ih})_f = \sqrt{T_o/T_{if}} \tag{15}$$

as a single optimized endoreversible engine[4], which is to say that at the end each reservoir is drained as quickly as possible since the interaction κ has negligible effect on this time scale; the engines are essentially decoupled.

The trajectories labelled by X and Y pass through the point x at which both θ and B become time independent. Along the paths X the system requires infinite time to reach x. Thus as t_f increases the optimal trajectory approached one of the curves X and moves towards the point x before turning to the line of final states. It should be noticed that taking t_f to infinity does not lead to a reversible process because of the internal relaxation.

ACKNOWLEDGEMENT

This research was supported by the United State Department of Energy under Contract De-AC02-80ER10646. Bjarne Andresen and R. Stephen Berry would also like to acknowledge the assistance of a NATO Travel Grant.

REFERENCES

1. Gibbs, J.W. (1948). The Collected Works of J. Willard Gibbs (Yale University Press, New Haven, CT) Vol. 1.
2. Hatsopoulos, G.N., and J.H. Keenan (1965). Principles of General Thermodynamics (John Wiley & Sons, New York, N.Y.).
3. Tolman, R.C., and P.C. Fine (1948). Rev. Mod. Phys. 20, 51.
4. Curzon, E.L., and B. Ahlborn (1975). Am. J. Phys. 43, 22.

THE SCIENCE OF ENERGETICS IN THE EXERGY CRISIS
OR
HOW IS THERMODYNAMICS MADE REALLY USEFUL?

T. S. Sørensen

*Fysisk-Kemisk Institut, Technical University of Denmark,
DK2800 Lyngby, Denmark*

ABSTRACT

A general energetics treating energy transformations in all macroscopic physical and chemical disciplines is proposed and some links to more classical expositions are pointed out. The absolute exergy measure is discussed and compared to relative measures of exergy (available energy). The importance of dimensionless exergy economic numbers is emphasized in connection with the thermodynamic analysis of a combined heat and electric power plant. The importance of "dissipative structures" in industrial processes is finally touched upon.

KEYWORDS

Energetics ; rational thermodynamics ; energy ; exergy ; entropy ; heat and power plant ; dimensionless exergy economics ; dissipative structures.

INTRODUCTION

The practically working engineer generally remembers thermodynamics as a "difficult" discipline which some queer teachers tried to knock into his head without much success. At best, some rudimentary knowledge about Carnot cycles and phase equilibria rests with him. The engineer is not alone with that feeling. Most educated persons who have somehow encountered the laws of thermodynamics during their academic carreer have some vague feeling of the importance of the subject, but they are unable to use thermodynamics as a framework for practical thinking.

From time to time I have in the past decade made research into the structure of thermodynamics (Sørensen,1971,1973,1976,1977a,1977b,1977c 1978a,1978b,1978c) as an outgrow of teaching experience from basic and advanced courses in thermodynamics for engineering students at the Technical University of Denmark. I have little by little arrived at the conclusion that thermodynamics as scientific discipline needs to be reformulated to cope with future demands. The historical evolution of thermodynamics caused it to center too narrowly around thermal phenomena, although it is possible to propose a general science of energetics concerned with the various forms of energy in macroscopic sys-

tems. Furthermore, terms as "heat" and "work" are taken over from daily language and from mechanics, respectively. In energetics, "heat"
is not a necessary concept and "work" is given a much broader and more
fundamental significance than in usual thermodynamics. Analytical mechanics is just one special case of energetics as are the thermodynamics of "heat engines", chemical thermodynamics, electrochemistry and
surface or colloid science. In my opinion, the use of statistical
mechanical concepts at an introductory level will only confuse the
students. What is needed is a completely general science of energy
transformations kept at the phenomenological level. Such a science
will then be the point of departure of any statistical interpretation.
I have found strong support for my line of reasoning from previous
energetic theories of Le Chatelier (1894,1928) and of the Danish
thermodynamicist J.N. Brønsted (1940a,1940b,1941,1955). Especially
the last author has strongly criticised the conventional formulations
of thermodynamics.

It is impossible here to give you more than a basic collection of
formulae and of ideas. The reader should be warned that even if the
formulae of energetics may look simple, it is by no means a simple
or trivial affair to apply energetics to practical industrial problems
However, it is not in my might to make Nature simpler than it is, and
engineers as well as city planners and economists (and who not ?)
should understand Nature when they act.

POTENTIALS AND QUANTITIES

In energetics, the processes of Nature are viewed as "motion" of certain quantities (K) between conjugate potentials (P).Work is the
basic concept making different processes commensurable just as money
in economics. The work absorbed when an infinitesimal amount of quantity $\delta K'$ is transferred from potential P_1 to potential P_2 is

$$\delta W = (P_2 - P_1)\delta K' = \Delta P \ \delta K' \tag{1}$$

The work absorbed when an infinitesimal amount of nonconserved quantity $\delta K''$ is created at P is

$$\delta W = P \ \delta K'' \tag{2}$$

Finally, a finite amount of quantity may be moved across an infinitesimal potential difference δP and the work absorbed is then

$$\delta W = K \ \delta P \tag{3}$$

The three types of processes are visualized in Fig. 1 and TABLE 1
surveys the different basic processes with conjugate potentials and
quantities. It is indicated in TABLE 1 whether a quantity is conserved (C) or not conserved (NC). The energetic content of different
macroscopic disciplines appear by combination of the various basic
processes using the work principle (conservation of work or energy)

$$\Sigma P \ \delta K \equiv \Sigma P \delta K' + \Sigma P \delta K'' + \Sigma K \delta P = 0 \tag{4}$$

The summation goes over different quantities and over different positions in physical space. Summations are replaced by integrations in
the case of continuously distributed systems. Products are used for
scalar potentials and quantities, dot products for vectorial and doub-

le dot products for tensorial sets. Examples of combinations are:

A1A2A3A4E1E2   Classical and analytical mechanics. Nonviscous hydrody-
                 namics and elasticity theory.

A1A4aB   Restricted classical thermodynamics of thermal engines.

A1A4aBD   Chemical thermodynamics.

A1A4aBCD   Electrochemistry.

A1A3A4aA5BCDF   Colloid- and interface science (+ E1E2: Interfacial
                 dynamics).

A1A4BDG   Magnetochemistry.

BCE3F   Classical electronic circuitry.

Equally important as the laws of energetics are also the <u>constitutive relations</u> , that is how potentials change when quantities are moved, created or destroyed.

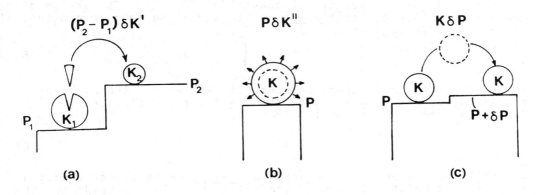

Fig. 1

TABLE 1   Pairs of Conjugate Potentials/Quantities

Label	Basic Process	Potential	Quantity	C/NC
A	Mechanical			
A1	Fundamental form	Force	Distance	C
A2	Angular process	Torque vector	Torsion vector	C
A3	Gravitational	Gravitation potential	Heavy mass	C(NC)
A4	Elastic	Stress tensor	Strain tensor	C
A4a	Spatial(isotropic)	Normal tension(-p)	Volume (V)	C
A5	Interfacial	Interfacial tension	Interfacial area	NC
A6	Nuclear	c^2	Mass	NC
B	Thermal	Absolute temp. (T)	Entropy(S)	C: revers. NC:irrevers.
C	Electric	Electric potential	Charge	C
D	Chemical	Chemical potential(μ_i)	Mol (n_i)	NC

TABLE 1   (Continued)

E	Inertial			
E1	Mechanical	Velocity vector($\underset{\sim}{v}$)	Linear momentum($\underset{\sim}{p}$)	C
E2	Angular	Angular velocity vector	Angular momentum vector	C
E3	Self induction	Electric current	Flux linkage(B-flux)	NC
F	Dielectric	Electric field	Dielectric polarisation	NC
G	Magnetic	Magnetic field	Magnetisation	NC

ENERGY

An energetic system is in any given moment characterised by a "vector" of potentials and a "vector" of quantities ($\underset{\sim}{P}$, $\underset{\sim}{K}$). The energy of the system is the reversible work of construction from zero potentials (or standard potentials):

$$\delta E = \Sigma P \, \delta K \quad ; \quad E = \int_{\underset{\sim}{0}}^{\underset{\sim}{K}} \Sigma P \, dK \tag{5}$$

Reversible means no entropy produced ($\delta S'' = 0$). The existence of a function like E means that the constitutive relations in Nature have to be reciprocal:

$$\frac{\partial P_i}{\partial K_j} = \frac{\partial P_j}{\partial K_i} \qquad (i \neq j) \tag{6}$$

We shall now introduce the concept of degree of an energy function. E is said to be of degree m if the following relation holds for the process of magnification of the system n times:

$$E(n \underset{\sim}{K}) = n^m E(\underset{\sim}{K}) \tag{7}$$

Then, E is a homogeneous function of m'th degree in the variables K . The integration in (5) can be carried out immediately by means of Euler's theorem:

$$E = \frac{1}{m} \Sigma P \cdot K \tag{8}$$

Linear constitutive relations between potentials and quantities lead to E's of degree m = 2 (kinetic energy, elastic energy, energy of capacitor). Nonlinear relations lead in general to no degree (for example relativistic kinetic energy). The important exceptions are phase systems in internal equilibrium as described by J.W.Gibbs(1875 -1878) where m = 1 and where consequently the Gibbs-Duhem equation determines the internal equilibrium

$$\Sigma K \, \delta P = 0 \quad (\text{only } m = 1) \tag{9}$$

The "Gibbsian" energy is usually called internal energy (U). The energies of degree two may be classified into kinetic energy (KE) and potential energy (PE) such as energy stored in gravitation or electric fields or as elastic potential energy. Energy conservation requires that we have

$$\delta U + \delta KE + \delta PE = 0 \tag{10}$$

The sign of the time derivatives $\dot{U}$, $\dot{KE}$ and $\dot{PE}$ yields useful information on the nature of the processes involved. Further subdivision of energies can be made according to TABLE 1.

EXERGY

Exergy is the work which can reversibly be extracted from a system ($\underset{\sim}{P}$, $\underset{\sim}{K}$). Thus we have

$$Ex = -\int_{\underset{\sim}{K}}^{\underset{\sim}{K}_o} \Sigma P dK = E - E_o \qquad (11)$$

When all work has been extracted, the system is in internal equilibrium in the socalled <u>energetic</u> <u>zero</u> <u>point</u> ($\underset{\sim}{P}_o$, $\underset{\sim}{K}_o$). E_o is the energy of the system in that point.

Exergy is "energy" in the usual sense of "ability to do work". Whereas real energy (E) is additive on combination of systems, exergy is not since E_o is generally lower for a combined system than the sum of E_o's for the systems taken separately. Thus for two systems I and II we have

$$Ex(I+II) \geq Ex(I) + Ex(II) \qquad (12)$$

Since holism in philosophy means that there is more in a totality than the sum of the constituent parts, (12) might be called the <u>holistic</u> <u>principle of energetics</u> (Sørensen,1973, 1976). This simple and much overlooked principle is the energetic basis of creativity in Nature and in technology every time subsystems in internal equilibrium are brought together.

ENERGY QUALITY

The somewhat vague concept of "quality" of a given form of energy can be given a precise meaning in terms of the relative content of exergy in the energy form concerned. Thus, the <u>quality</u> <u>coefficient</u> is defined as

$$q = Ex/E = 1 - (E_o/E) \qquad (13)$$

It should be stressed here that it is not only thermal energy which may be of inferior quality. The appearance of residual energy which cannot be exploited is a common phenomenon. Consider for example a system of masses M_i (i = 1,2...n) with velocities $\underset{\sim}{v}_i$ in a given inertial frame. After the moving round of momentum between the masses the energetic zero is reached, where all masses move with the same speed $\underset{\sim}{v}_o$. Due to momentum conservation in the total system we have

$$M_{total}\underset{\sim}{v}_o = \Sigma M_i \cdot \underset{\sim}{v}_i \qquad (14)$$

Thus, $\underset{\sim}{v}_o$ is also the center of mass velocity in the original system. Performing the exergy integration (11) we obtain for the <u>kinetic</u> <u>exergy</u>

$$KEx = \frac{1}{2}\Sigma M_i v_i^2 - \frac{1}{2}M_{total}v_o^2 \qquad (15)$$

so only the KE relative to the center of mass is useful. This is a well known fact in collision studies of particles. Only if we choose M_1 to be much greater than all the other M's (momentum reservoir) and choose $v_1 = 0$, the KE becomes 100% useful. Said Archimede: "Give me a place to stand and I shall move the world !"

Quality coefficients as defined in (13) are average qualities . Often marginal quality is more useful:

$$q_{marginal} = \delta Ex / \delta E \tag{16}$$

An example is the usual Carnot factor for the work to be gained when "heat" (i.e. energy or enthalpy) is extracted from a hot reservoir in a cycle transporting entropy reversibly from a high to a low temperature.

ENERGETIC ZERO AND ENTROPY DEATH

The energetic zero point is reached when a system is brought to equilibrium reversibly. If irreversible processes occur in the system as they do in the real world then another equilibrium point is reached. Clausius spoke about the "heat death" of the universe, but we shall be less ambitious and speak about the entropy death of a given isolated and closed system. Since entropy is the sum of a thermal contribution (disorder in momentum distribution of molecules) and a configurational contribution (disorder in position of molecules), viz.

$$S = S_{thermal} + S_{configuration} \tag{17}$$

maximal entropy does not mean maximal "heat" in the common language sense (i.e. maximal temperature). Therefore, entropy death is more precise than heat death.

At any precise moment we can freeze the irreversible processes and examine the system for exergy content. Not surprisingly we find that irreversible processes decrease the exergy of a system. It is possible to derive in full generality (Sørensen,1977a) a relation between produced entropy and lost exergy:

$$-dEx = T_o dS'' \geq 0 \tag{18}$$

as an expression of the second law of energetics. The conversion temperature T_o is the temperature in the energetic zero point in the given moment[o] and not the temperature or the many different temperatures in the system in which the irreversible processes go on. In general T_o will change all the way towards entropy death. This is visualised in Fig. 2 in the case of heat conduction (i.e. entropy transport with creation of entropy) between two bodies with equal and constant heat capacities (Sørensen, 1977a).

CONDITIONAL EXERGY

Conditional exergy functions can be derived relative to constant surroundings. The most commonly used exergy functions were explored already by Gibbs. With modern names and symbols they are: Enthalpy H, Helmholtz' free energy F and Gibbs' free energy G, viz.

$$H = U[p] = U + pV \quad ; \quad F = U[T] = U - TS \; ; \; G=U[p,T] =U+pV-TS \tag{19}$$

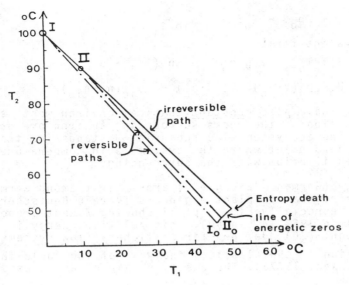

Fig. 2

Apart from causing a change in independent variables such Legendre
transformations of the internal energy have the effect that they correct for exergy-less transports of quantities between equal potentials
In Brønsted's terminology such transports are called neutral transports. Therefore, $H-H_o$ is the exergy function when p is buffered,
$F-F_o$ if T is buffered and $G-G_o$ if as well p and T are buffered by
contact with volume and entropy reservoirs (presso- and thermostats).
There is a certain difference, however, between between the set $U-U_o$
and $H-H_o$ on one hand and the set $F-F_o$ and $G-G_o$ on the other which is
manifest when irreversible processes occur. For the first set dU and
dH are zero in isolated systems whereas dU_o and dH_o are positive.For
the second set dF_o and dG_o are zero whereas -dF and -dG are equal to
TdS". Therefore, U and H are energylike whereas F and G are (neg)entro-
py-like.

In analogy to Gibbs' free energy an available energy can be defined
for systems in a given environment (T_o,p_o) even when the systems studied have temperatures and pressures different from T_o and p_o :

$$\Phi = U\left[p_o,T_o\right] = U + p_oV - T_oS \tag{20}$$

This function has minimum for $T=T_o$ and $p=p_o$ and internal chemical
equilibrium. Therefore $\Phi - \Phi_o$ is a conditional exergy function
since it is also the conditional work function. The function Φ was
proposed already in the previous century by Maxwell and Gouy (Jouguet
1909) and was more recently discussed by Landau and Lifschitz(1958).
A generalisation has been discussed by R.B.Evans who used the name
essergy (Evans 1969) for the function

$$Es = U + p_o V - T_o S - \Sigma \mu_{io} n_i \tag{21}$$

with the equivalent forms

$$Es = S(T-T_o) - V(p-p_o) + \Sigma n_i (\mu_i - \mu_{io}) \tag{22}$$

$$Es = U-U_o + p_o(V-V_o) - T_o(S-S_o) - \Sigma \mu_{io}(n_i-n_{io}) \tag{23}$$

In this case we have also constant chemical environ ment such as the average composition of the rocks surrounding an iron ore or the sea water in a marine ecosystem. The subscript o stands for the conditional energetic zero point where the system is in temperature, pressure and chemical equilibrium with the surroundings.

The name exergy (German: "Exergie") arose first among German power station and refrigeration plant engineers (Verein Deutscher Ingenieure 1965, see especially Fig. 5 p. 11 showing Z.Rant proposing the word "Exergie" in 1953 in Lindau). This relative exergy is a variant of $\Phi - \Phi_o$ in eqn. (20) defined for stationary flow systems as a kind of generalisation of the classical analysis of the Joule-Thomson experiment (Sørensen, 1977a). The exergy per kg of a process stream was defined as

$$ex = h - h_o - T_o(s - s_o) \tag{24}$$

where h and s are specific enthalpy and entropy, respectively.Because of the stationarity such exergy calculations should be classified as marginal accounts. An exergetic analysis of crude oil destillation using this exergy concept has been performed recently by Fratzsher and Michalek (1978).

In all the mentioned cases the loss of exergy is given by $T_o dS"$ in analogy to (18). However, dEx(conditional) has contributions from as well import as export tems as from the exergy losses. Conditional exergy may of course be considered to be a special case of our general exergy definition (11). The very large and "potential fixing" reservoirs have just to be incorporated into the total energetic universe.

It is fundamental of course that the exergy (absolute or conditional) is never negative:

$$Ex \geq 0 \tag{25}$$

This requirement gives rise to energetic stability criteria. As an example we calculate the thermal exergy per kg of a process stream using (24):

$$ex(thermal) = c_p(T-T_o) - c_p T_o \ln \frac{T}{T_o} \simeq c_p \frac{(T-T_o)^2}{T_o}$$

Thus, the specific heat capacity c_p has to be positive. Similarly,in a system of moving masses the kinetic exergy will only fulfil the requirement (25) when all the masses are positive.

EXERGY ECONOMICS: ENTHALPY OR ELECTRICITY ?

On Fig.3 is shown a simplified flow sheet of a conventional steam
cycle electricity plant. Fuel and air enter the combustion chamber C
at a rate of $\dot{M}_f$. The (lower) specific heat of combustion is h_c. T_{fl}
is the temperature of the flue gas. The other symbols should be self
explanatory from inspection of the figure. In the middle, an ideali-
sed cycle with one kg steam and/or water in a T-s diagram is inserted
Even a schematic plot of a real modern steam power plant is much more
involved than shown on Fig.3. Economising measures are: Reheating of
take-off steam of intermediate pressure from the turbine(T),preheat-
ing of air and of feed water by the flue gas. The optimisation of the
modern power station has been possible using the exergy concept of
eqn.(24), see for example Baehr(1966). For the present purpose we
shall be content with a simpler analysis, since our topic is rather
an exergy economic version of the "Butter oder Kanonen" problem:
Should the power station produce electricity only or less electricity and some enthalpy for house heating ?

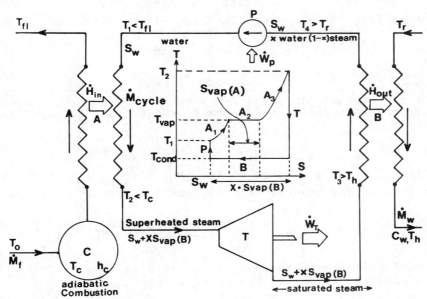

Fig.3

In a simple steam power plant the main exergy loss is situated in the
combustion chamber C (31%).This loss is inavoidable in thermal energy conversion, but it could be avoided by direct conversion from
chemical to electrical exergy in a fuel cell.The next largest loss
is in the heat exchanger A (26%).The third largest drain of exergy
is in the turbine T (7% of the chemical exergy),but we shall neglect
this loss and assume the steam expansion in T to be isentropic.The
enthalpy flux into the steam cycle is given by $\dot{H}_{in} = \phi_c \dot{M}_f h_c$ where ϕ_c
is the energetic combustion efficiency. The net shaft power deliver-
ed is $\dot{W}_{net} = \dot{W}_T - \dot{W}_P$ and since energy is conserved we have:

$$\dot{H}_{out} = \dot{H}_{in} - \dot{W}_{net} \qquad (26)$$

The entropy transferred through the power station per unit time is $\dot{S}' = \dot{H}_{out}/T_{cond}$. The electric power delivered at the consumers plugs is $\dot{W}_{el} = \phi_{el}\dot{W}_{net}$ where ϕ_{el} corrects for generator and transmission losses. Thus, for the overall energetic efficiency we have

$$\phi_E \equiv \dot{W}_{el}/(\phi_{el}\phi_c\dot{M}_f h_c) = 1 - (\dot{S}'/\phi_c\dot{M}_f h_c)T_{cond} \qquad (27)$$

If h_{vap} is the specific heat of vaporisation at T_{cond} and x the weight fraction of water condensed in B we also have:

$$\dot{S}' = xh_{vap}\dot{M}_{cycle}/T_{cond} \qquad (28)$$

Now considering the enthalpy balance for A and using the relation

$$h_{vap}(T_{vap}) = h_{vap} + (c_{vap} - c_w)(T_{vap} - T_{cond}) \qquad (29)$$

we obtain from (28) for the dimensionless entropy flux

$$\hat{S} \equiv T_o\dot{S}'/\phi_c h_c \dot{M}_f = (x/\tau_{cond})\cdot B^{-1} \qquad (30)$$

$$B \equiv 1 + \hat{c}_w(\tau_1 - \tau_{cond}) + \hat{c}_{vap}(\tau_2 - \tau_{cond}) \qquad (31)$$

All the τ's are absolute temperatures made dimensionless by dividing by the environment temperature T_o. The dimensionless heat capacities are given by:

$$\hat{c}_w = T_o c_w/h_{vap} ; \quad \hat{c}_{vap} = T_o c_{vap}/h_{vap} \qquad (32)$$

By inspection of the T-s diagram we see that $x \cdot s_{vap}(B)$ should equal the sum of the entropy changes in A_1, A_2 and A_3. From this we obtain:

$$A \equiv x/\tau_{cond} = \tau_{vap}^{-1} + \hat{c}_w\ln(\frac{\tau_{vap}}{\tau_{cond}}) + (\hat{c}_{vap} - \hat{c}_w)(1 - \frac{\tau_{cond}}{\tau_{vap}}) \qquad (33)$$

Finally, considering the enthalpy balance in B we obtain for the dimensionless cooling ratio K of the power plant:

$$K \equiv \frac{\dot{M}_w c_w T_o}{\phi_c h_c \dot{M}_f} = \frac{\tau_{cond}}{\tau_h - \tau_r}\cdot\hat{S} = \frac{\tau_{cond}}{\tau_h - \tau_r}\cdot(A/B) \qquad (34)$$

From (27),(30) and (33) we have for the overall energetic efficiency

$$\phi_E = 1 - A\tau_{cond}(B_o - \{\hat{c}_{vap} + \hat{c}_w\}\Delta\tau)^{-1} = 1 - (\tau_h - \tau_r)K \qquad (35)$$

where a temperature difference $\Delta\tau$ has been assumed in the two ends of the heat exchanger A and where

$$B_o = 1 + \hat{c}_w(\tau_{f1} - \tau_{cond}) + \hat{c}_{vap}(\tau_c - \tau_{cond}) \qquad (36)$$

In (35) ϕ_E is basically the Carnot factor $1 - (T_{cond}/T_{vap})$.

The $\hat{c}$-terms correct for the fact that not all entropy is fetched from T_{vap} and the $\Delta\tau$-term for the exergy loss in the boiler (A).

Consider now the profit function

$$\dot{P} = e\dot{W} + \phi_h \cdot h \cdot \dot{H}_{out} - d\dot{M}_w \tag{37}$$

where e is the consumers electricity price per J, h is the enthalpy price per J, ϕ_h accounts for heat transmission efficiency and d is a price per kg cooling water reflecting capital and maintenance costs of the heat transmission tubes and pumping costs in excess of the necessary pumping costs for pure electricity production. Fuel costs are not taken into account since we want only to compare el + enthalpy production with pure electricity production. Eqn. (37) is readily transformed to dimensionless form

$$\Pi \equiv \dot{P}/(\Phi_{el}\Phi_c e\dot{M}_f h_c) = 1 + \frac{A}{B}(\ h' - 1 - d'\{\tau_h - \tau_r\}^{-1}\ )\cdot\tau_h \tag{38}$$

where we have put $\tau_h \simeq \tau_{cond}$ and defined

$$h' \equiv \phi_h h / \phi_{el} e \qquad ; \qquad d' \equiv d/(c_w T_o \cdot \phi_{el} e) \tag{39}$$

In the case of a pure electricity plant, the profit function is also given by (38) with $h' = d' = 0$, $\tau_r = 1$ and with

$$\tau_h = \tau_{el} \qquad ; \qquad 1 < \tau_{el} < \tau_h \text{(heat+power)} \tag{40}$$

The temperature τ_{el} is determined as an optimum temperature determined primarily by the strongly increasing pumping costs when $\tau_h \to \tau_r = 1$ where we have $K \to \infty$ according to (34). Now, the profit difference is

$$\Delta\Pi = \Pi - \Pi_{el} = \frac{A}{B} \cdot (\ \tau_{el} - \tau_h + h'\tau_h - \frac{d'\tau_h}{\tau_h - \tau_r}\ ) \tag{41}$$

The price h will be strongly dependent on the temperature τ_h. For the purpose of house heating one is selling temperature just as much as enthalpy. We shall require the hot water to be 60°C when leaving the plant. When T_o = 283 K this means τ_h = 1.18. The return water temperature is taken as normal room temperature τ_r = 1.035 (20°C). Let us say that the optimum temperature for a pure electricity plant is τ_{el} = 1.05 (24°C). For heat and power production to be profitable we should therefore have according to (41):

$$h' > 0.11 + 6.9d' \tag{42}$$

The present example should convince the reader of the necessity of performing cost-benefit analysis in terms of dimensionless exergy-economic variables. This is the only way to make conclusions independent of time and place and plant dimensions!

FINAL REMARKS

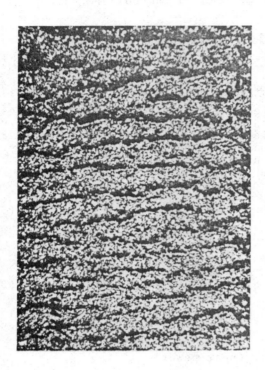

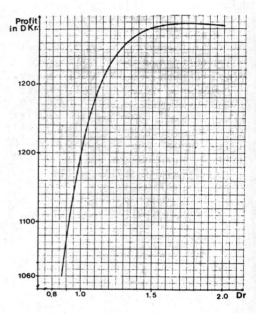

Fig.4   Fig.5

We have only been able here to scratch the surface of the problems
and opportunities of a rational energetics and a more scientific
exergy economics. Taking the energetics first, I would like to say
that the discipline centers around five laws. The first is the conservation of work or energy. The second is the law of production of entropy or consumption of exergy with the "creative" holistic principle
as an important addendum. The third law is the law of vanishing entropy at T= 0 which is strictly speaking an important constitutive relation rather than a fundamental law of phenomenological energetics.
The fourth law concerns the symmetry of phenomenological coefficients
in the flux-force relationships of linear, irreversible thermodynamics
as demonstrated by Lars Onsager(1931). Linear irreversible thermodynamics is a rather complex science in itself (Prigogine 1947; de Groot
and Mazur 1962) but of limited scope in comparison to a real non-equilibrium thermodynamics of nonlinear nature. In my opinion this branch
of thermodynamics should also be reformulated in terms of "forces"
referring to exergy consumption instead of entropy production. Mathe-
matically, the change is trivial, but conceptually and pedagogically
it is far easier to envisage the exergy as the "driving force" in
energetic systems than the entropy production, and the "forces" conjugate to the flux of Brønsted-quantities become of more familiar
dimensions to practically thinking engineers.

The "fifth law" could be described as the law of instabilities which has arised as a result of recent research in nonlinear,irreversible thermodynamics, notably by Ilya Prigogine and his school (Glansdorff and Prigogine 1971 ; Nicolis and Prigogine 1977). It shows up,that it happens very often in nonlinear, dissipative systems that stationary states which are stable in the linear region of irreversible thermodynamics are no longer stable in the nonlinear region above certain critical thresholds of the parameters. We shall give here the necessary thermodynamic or energetic criterion for a system to be unstable in terms of the second variation of exergy $\delta^2 Ex$

$$Ex = Ex_0 + \delta Ex + \tfrac{1}{2}\delta^2 Ex + \ldots \ldots \tag{43}$$

where the expansion goes from the stationary state to be investigated for stability. At equilibrium $\delta Ex = 0$, but not in the case of a dissipative stationary state. Nevertheless, one can always show that

$$\delta^2 Ex > 0 \tag{44}$$

by means of the energetic stability criteria (positive heat capacities, masses, compressibilities and so on). Now if

$$\partial \delta^2 Ex / \partial t < 0 \tag{45}$$

the stability of the stationary state can be guaranteed, since the second variation of exergy is then a socalled Lyapounov function. Thus, it is a necessary,but not sufficient, condition for instability that (45) is violated by some perturbations $\delta \underset{\sim}{K}$ of our stationary, dissipative energetic system $(\underset{\sim}{P}_0, \underset{\sim}{K}_0)$. The criterium stated here can be found by a slight reformulation of the criteria given in the monograph of Glansdorff and Prigogine(1971) for $\delta^2 S$ (or $\delta^2 Z$).

The existence of instabilities in energetic systems "far from equilibrium is of much more than academic significance. Hundreds of instabilities of industrial importance could be mentioned ranging from surface tension driven instabilities in destillation processes and heat and mass transfer equipment (Sawistowski 1971 ; Linde,Schwartz and Wilke 1979 ; Sørensen 1979) to osmotic instabilities in the process of hydration of Portland cement paste (Double,Hellawell and Perry 1978 ; Sørensen 1981). To remain inside the sphere of power station technology, Fig.4 shows the emergence of socalled "ripple magnetite" inside a tube in the steam generator of a power station.The wavelength is typically around 0.1 mm . Such ripple magnetite has in the past been a problem in some German and Danish power stations with boilers of the Benson type (Heimsch and coworkers 1978, Høstgaard-Jensen 1980) because of the increased friction loss in the tubes. Even if the mechanism of formation of ripple magnetite is poorly understood today we have here a beautiful illustration of the importance of dissipative structures arisen by fluctuations and instabilities. Characteristically, ripple magnetite is formed only above certain thresholds for the mass stream in the tubes.

Turning finally to the topic of exergy economics, I would like to pinpoint the fact that it is not enough to minimise primary exergy consumption in industrial processes, i.e. consumption of oil, gas and electricity. As an illustration we consider Fig.5 which stems from a recent optimisation study concerning the extraction of apple juice in

the Danish firm Rynkeby A/S (Østerberg and Sørensen, 1980,1981). The
juice is extracted in a counter current diffuser. The figure shows
a profit function analogous to the one studied for heat and power
stations. The abscissa variable Dr is the draft which is (roughly)
the volume ratio between the stream of water and the stream of apple
slices. The optimum draft is around 1.75. The value is determined by
the cost of superheated steam used in the steam evaporator, where the
juice leaving the diffuser is concentrated to 70 weight percent for
temporary storage. If the draft is further increased, the exit juice
is weaker and more steam (primary exergy) has to be used in the evaporator. From Fig. 5 we see, however, that it is much more serious
to use a draft which is too low, even if this saves primary exergy.
The reason is the substantial loss of raw material (soluble matter in
apples) in the diffuser if the draft is too low. Thus, raw material
conservation cannot be separated from an exergy economic analysis.The
apple price which enters the profit function reflects the exergy content in the apples and the exergy used in producing apples, for example transportation costs and primary exergy consumption of fertilizer
production, apple-tree shuttering machines and the households of the
apple farmers. We have here a basic weakness in all exergy economics
which will not be removed before the science of economics has become
reformulated in terms of real, physical variables, if that ever happens !

REFERENCES

Baehr, H. D. (1966). Thermodynamik , 2nd ed. Springer, Berlin-N.Y.
Brønsted, J. N. (1940a). The fundamental principles of energetics.
    Phil.Mag. 29 ,Ser.7, 449 ;(1940b). The derivation of the equilib-
    rium conditions in physical chemistry on the basis of the work
    principle. J. Phys. Chem. 44 , 699 ; (1941). On the concept of
    heat. Kgl. danske Videnskabernes Selskab math.-fys. Meddelelser
    19 no. 8 ; Principles and Problems in Energetics. Interscience,
    New York(1955).
de Groot, S.R. and Mazur, P. (1962). Non-Equilibrium Thermodynamics
    North-Holland, Amsterdam.
Double, D.D., A. Hellawell, and S.J. Perry (1978).The hydration of
    Portland cement. Proc. Roy. Soc. Lond. A 359 , 435.
Evans, R. B. (1969). A Proof that Essergy is the only Consistent
    Measure of Potential Work , Ph.D. Thesis, Dartmouth College.Uni-
    versity Microfilms, Ann Arbor, Michigan.
Fratscher, W. and K. Michalek (1978). Energetische und exergetische
    Analyse einer Rohöldestillationsanlage. Hungarian Journal of In-
    dustrial Chemistry, Veszprém 6 , 163.
Gibbs, J. W. (1875-1878).On the equilibrium of heterogeneous substan-
    ces. Trans. Connecticut Academy III 108-248 (Oct.1875-May1876);
    343-524 (May 1877- July 1878).
Glansdorff, P. and I. Prigogine (1971) Thermodynamics of Structure,
    Stability and Fluctuations. Wiley, London-N.Y.
Heimsch, R. and coworkers (1978). Beobachtungen über den Einfluss von
    Massenstrom, Geschwindigkeit und mechanischer Beanspruchung auf
    das Schichtwachstum in Heisswasser. VGB Kraftwerkstechnik 58 no.2
    117-126.
Høstgaard-Jensen, P. I/S Nordkraft, Aalborg, Denmark. Personal com-
    munication (1980).
Jouguet, E. (1909) Etude thermodynamique des machines thermiques.
    Ed. Doin, Paris.
Landau, L. and E.M. Lifschitz (1958) Statistical Physics. Addison-
    Wesley, Reading, Mass.

Le Chatelier, H. (1894) . Les principes fondamentaux de l'énergetique
    et leur application aux phénomènes chimiques. J.de Physique thé-
    orique et appliquée 23 289 ; 352; (1928). Les pricipes fondamen-
    taux de l'energetique. Revue des Questions Scientifiques 94,363.
Linde, H., P. Schwartz and H. Wilke (1979) . In T.S.Sørensen (Ed.),
    Dynamics and Instability of Fluid Interfaces. Springer, Berlin-
    Heidelberg-N.Y. pp. 75- 119.
Nicolis, G. and I. Prigogine (1977). Self-Organization in Nonequilib-
    rium Systems. Wiley, New York.
Onsager, L. (1931) Reciprocal relations in irreversible processes.
    Phys. Rev. 37 405 ; 38 2265.
Prigogine, I. (1947) Etude thermodynamique des phénomènes irreversib-
    les, Dunod/Paris,Desoer/Liège.
Sawistowski, H. (1971). In C. Hanson (Ed.), Recent Advances in Liquid-
    Liquid Extraction. Pergamon, Oxford. Chap. 9.
Sørensen, T. S. (1971) The energetics of J.N.Brønsted and analytical
    mechanics. (In Danish). Dansk Kemi 52 138 ; (1973) Studies on the
    Statics, Dynamics and Kinetics of Physico-Chemical Systems (In
    Danish).Ph.D. Thesis, Technical University of Denmark. Chap.I&II;
    (1976) Brønstedian Energetics, Classical Thermodynamics and the
    Exergy. Acta Chemica Scandinavica A 30 555 ; (1977a) Exergy Loss,
    Dissipation and Entropy Production.Acta Chem. Scand. A 31 347 ;
    (1977b) The Gibbs-Duhem equation and equilibrium of matter in ex-
    ternal fields and temperature gradients. Acta Chem. Scand. A 31
    437 ; (1977c) Rational thermodynamics and mechanics. Acta Chem.
    Scand. A 31 892 ; Generalised Gibbs-Duhem equations and quasi-
    thermostatic methods. Acta Chem. Scand. A 32 (1978a)277 ;(1978b)
    General Energetics (In Danish).Lecture notes, Technical Universi-
    ty of Denmark ; (1978c) Irreversible Thermodynamics of Transport
    Processes (In Danish). Lecture notes, Technical University of
    Denmark; (1979). Instabilities induced by mass transfer, low sur-
    face tension and gravity at isothemal and deformable fluid inter-
    faces.In T.S.Sørensen (Ed.), Dynamics and Instabilities of Fluid
    Interfaces, Springer, Berlin-N.Y. pp. 1-74 ; (1980) Marangoni In-
    stability at a Spherical Interface. J.Chemical Society Faraday
    Transactions II , 76 1170 ; (1981) A theory of osmotic instabi-
    lities of a moving semipermeable membrane: Preliminary model for
    the initial stages of silicate garden formation and of Portland
    cement hydration. J. Colloid and Interface Sci. 79 192.
VDI-Fachgruppe Energietechnik (1965) Energie und Exergie , Verlag des
    Vereins Deutscher Ingenieure, Düsseldorf.
Østerberg, N. O. and T.S. Sørensen (1980) Optimisation of apple juice
    extraction using the DdS-diffuser principle. In Proceedings of
    the 5th International Congress in Scandinavia on Chemical Engin-
    eering pp. 278- 292 ; (1981) Apple juice extraction in a counter-
    current diffuser. J. Food Technology (24 pp., in press).

ACKNOWLEDGEMENTS

The late Prof. Jørgen Koefoed and his widely spanning field of interests is commemorated by this paper. Until his death 1980 we have in
the past decade had numerous stimulating discussions. I am also grate-
ful to Prof. Hartmut Linde (Berlin) and to Prof. Albert Sanfeld (Brussels) for useful discussions and comments.

PRODUCTION WITH MINIMAL ENERGY USE

M. Splinter and W. Willeboer

*University of Technology, Dept. of Industrial Engineering (Bedrijfskunde),
P.O. Box 513, 5600 MB Eindhoven, The Netherlands*

ABSTRACT

Energy is an essential factor of production. In the long run, it is expected that
the developments in the energy situation will change the structure of production
systems.
Our research is aimed at learning about effects of these structural changes:
"Production with minimal energy use". In this case, it concerns an existing
coating machine for photocopying paper where most of the (direct) energy is used
for drying of the paper. An energy analysis has been made to show the indirect
energy use of the production system.
On the basis of an energy model for paper drying, it is possible to simulate that
part of the system to minimal energy use, without changing the product quality
or quantity. Experimentally, the results of the simulation are tested.
Only at this phase of our research it is possible to use the acquired knowledge
for other processes and production systems and to compare it with possible
conflicting goals in the fields of economics, environment etc.

KEYWORDS

Energy; energy analysis; energy model; production system; energy conservation;
simulation; drying.

INTRODUCTION

It seems very likely that the world will be plagued by "the energy problem" for
a long time to come, in spite of people who try to "solve" it. We start from the
principle that it is more meaningful to describe the present energy situation as
a move in an inevitable development rather than in terms of problem and solution.
When the prices of the most important energy resources keep rising, because of
physical scarcity and the related rise in production costs, not only energy will
become more expensive, but after some time, the other factors of production too.
This means that after energy conservation, one has to consider the conservation
of materials, labour and other factors of production. The structure of the production system will not be changed; only waste will be minimised.
When, after all, energy prices continu to rise, structural changes in the

production system will be necessary for keeping production costs at an acceptable
level even for survival of the system. Our research is aimed at the multidisciplinary consequences of the former description for the energy situation. These consequences can best be analysed when the theoretical result of the expected development is known: "Production with minimal energy use".
An existing, functioning production system, of which paper drying is energetically
the main process, is chosen as a starting point for our research. An energy
analysis was made, so that the influence of the indirect energy use is known.
The main question is how this system is functioning, with the same product quality
and quantity as before, when it uses a minimum of indirect and direct energy.
When comparing the last situation with the former, original one, it will be
possible to draw conclusions on aspects as management, economics, environment,
employment, etc.
This kind of knowledge will play an important role in the future in fields like
energy conservation and scientific engineering education.

RESEARCHPROGRAM

As a result of the energy analysis of the real production system it was decided
to concentrate on the subsystem of paper drying.
- Analysis.
  This part of the program is concerned with a quite fundamental analysis of
  drying and the introduction of a measuring system.

- Energy model.
  With the results of the former analysis a model is composed to simulate the
  subsystem numerically. This model is compared with the real system.

- Simulation.
  By changing the value of the relevant variables, the influence on energy use is
  simulated.
At this moment we face the problem of adapting the production system to conditions
found by simulation without disturbing production.

- Assessment.
  The consequences of every change in the production system will be analysed and
  evaluated. Notably the economic consequences of production with minimal energy
  use will obtain our attention.
Apart of this research we started with student research in other production
systems where drying is an important process.

ANALYSIS

Up to now, the research applies to one of the two drying sections of a paper
coating machine. This section consists of six directly heated drying boxes; an
outline of one of these boxes is given in figure 1. The whole section is
controlled by one gas control valve.
Measurements on the drying section during production runs have led to the heat
balance which is given in table I. From this heat balance it is seen that just
a small part of the generated heat is used for evaporation of the water; most
of the heat is lost in the flue gases.

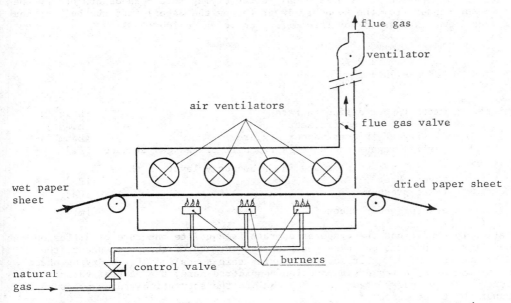

Fig. 1. Schematic presentation of a drying box as used in the paper coating
machine.

Table I: Heat balance of the drying section (in %).

Heat in:		Heat out:	
in the wet paper	3	in the dried paper	6
natural gas (heating value)	80	in the flue gas (total)	81
in the air used	17	(of which 32 is latent heat)	
		heat losses of the equipment	13
	100		100

N.B. The heat of vaporization of the water removed from the paper amounts to 15%.

In this drying process two phenomena are essential, i.e.
- the evaporation on and in the paper sheet and the diffusion of the vapour into
  the surrounding gas mixture
- the transmission of heat to the paper sheet
Since the process is stationary, the heat transmission to the paper sheet is in
equilibrium with the heat of vaporization plus the increase of the heat content of
the paper sheet.
In formula:

$$H_{p\ in} + Q = H_{p\ out} + H_{vap}$$

in which: Q = transmitted heat $[kJ.s^{-1}]$
   $H_{p\ in}$, $H_{p\ out}$ = heat content of the paper sheet $[kJ.s^{-1}]$
   H_{vap} = heat of vaporization $[kJ.s^{-1}]$

The rate of evaporation – which is equal to the drying rate – is determined by the diffusion of vapour from the boundary layer (along the paper) into the bulk of the gas mixture. The expression for this diffusion is as follows:

$$D = K.\rho.A.\ln \frac{1-P_m/P_t}{1-P_p/P_t}$$

in which:
D	= water vapour diffusion rate	$[kg.s^{-1}]$
K	= mass transfer coefficient	$[m.s^{-1}]$
ρ	= density of the gas/vapour mixture	$[kg.m^{-3}]$
P_m	= partial vapour pressure in the bulk of the gas mixture	$[Pa]$
P_p	= partial vapour pressure in the boundary layer on the paper sheet	$[Pa]$
P_t	= total pressure	$[Pa]$
A	= area of paper sheet	$[m^2]$

In stationary conditions the evaporation rate is equal to the rate of diffusion. As long as the vaporization takes place on the surface of the paper sheet, the sheet behaves as a free water surface. During this so-called "first drying phase", which is dealt with in the drying section considered here, the partial vapour pressure in the boundary layer (P_p) is equal to the saturation vapour pressure at the paper sheet temperature.

The heat transmission to the paper sheet is performed by:
- convection
- radiation from the inner walls of the drying box
- radiation from the burners, the flames, and the flue gas

The basic expressions for these three contributions are as follows:

- convection: $Q_c = \alpha .A.\Delta\Theta$

in which:
Q_c	= transmitted heat	$[kJ.s^{-1}]$
α	= heat transfer coefficient	$[kJ.m^{-2}.s^{-1}.^0C^{-1}]$
$\Delta\Theta$	= mean temperature difference between the flue gas and the paper sheet	$[^0C]$

- radiation from the inner walls of the drying box:
$Q_{rw} = \sigma.RF_w. A . (T_w^4 - T_p^4)$

in which:
Q_{rw}	= transmitted heat	$[kJ.s^{-1}]$
σ	= constant of Boltzmann = $5.77 * 10^{-11}$	$[kJ.s^{-1}.m^{-2}.K^{-4}]$
RF_w	= complex radiation factor	$[-]$
T_w	= inner wall temperature	$[K]$
T_p	= average paper sheet temperature	$[K]$

- radiation from the burners, the flames and the flue gas:
$Q_{rb} = RF_b. M_g.A. (T_i^4 - T_p^4)$

in which:
Q_{rb}	= transmitted heat	$[kJ.s^{-1}]$
RF_b	= radiation factor	$[kJ.kg^{-1}.m^{-2}.K^{-4}]$
T_i	= initial gas temperature after burning	$[K]$
M_g	= gas mixture mass flow	$[kg.s^{-1}]$

In the present production situation, the contribution of each of these phenomena
to the total heat transmission is:

- convection	30-35%
- radiation from the walls	25-30%
- radiation from burners, flames and gas	35-40%

ENERGY MODEL

In order to simulate the drying process under consideration, a computer model has
been developed. With this program (written in FORTRAN V) the stationary behaviour
of the process can be calculated as a function of the process variables and the
geometrical data. The basis of the program consists of the expressions for the
heat transmission to the paper sheet and the diffusion of vapour from the sheet.
These expressions are given in a general form in the preceding chapter. With the
help of the measurement data of the production machine, these expressions have
been adapted to this particular case and the parameters have been determined.
The calculations are made separately for each drying box. As mentioned before,
the stationary conditions of the process are determined by the equilibrium
between the heat transfered to the paper sheet at one side, and the heat of
vaporization of the removed water plus the increase of the heat content of the
paper sheet at the other side. Due to the fact that the heat transfer and the
evaporation mutually influence each other, the equilibrium conditions have to be
calculated with help of an iteration procedure. The main parameter in this
iteration is the mean paper sheet temperature within the drying box. Some test
results of the simulation model are given in figure 2.

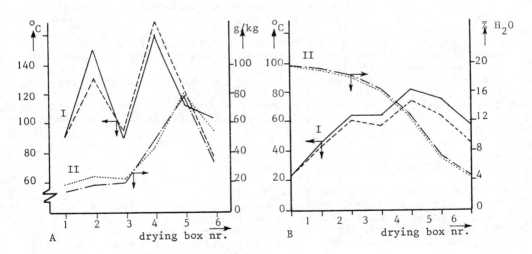

Fig. 2. Test results of the simulation model:
    A: Temperature (I) and humidity (II) of the flue gas.
    B: Temperature (I) and moisture content (II) of the paper sheet.
    ━━·━━ Measurement data.
    ········ Results of simulation.

Based upon the simulation program mentioned above, a "variation program" has been written. The aim of this program is to determine the influence of variations of the process parameters on the specific energy use (i.e. natural gas consumption). The program starts with a case that can be considered as a reference base (for instance, the situation with measured conditions), and calculates the end moisture content of the paper sheet for this case. After that for each variation a new simulation is made, in which the consumption of natural gas is determined so that the end—moisture content is the same as in the basis case. This last calculation is performed with the help of iteration.

SIMULATION

Figure 3 gives some preliminary results of simulations performed with the variation program. The variations shown in figures 3B en 3C concern environmental conditions, which cannot be controlled in the existing installation. However, these results will be interesting when thinking about recycling the flue gases etc. The results shown in figures 3A en 3D concern variations of parameters which can be controlled in the coating machine. It is seen from these figures that the air flow as well as the production speed have great influence on the specific natural gas consumption. Especially the influence of the air flow is interesting, because this means that in the existing installation an important reduction of the natural gas consumption is possible without reducing the production speed. The first test runs have proved the relationship shown in figure 3A.

CONCLUSIONS

- The analysis of the drying process has given an insight in the efficiency of the process, the contribution of convection and radiation to the total heat transfer, the distribution of the air flow and the natural gas flow over the different drying boxes, etc.
- A tool has been developed for simulating the behaviour of the drying process under different circumstances and conditions, and with varying geometrical data and production speed.
- Simulation results show quantitatively the possibilities of conserving energy in the existing equipment.
- With respect to the design of new dryers, or the modification of existing ones, the simulation model can be used to determine the design criteria with respect to the energy use.

CONTINUATION

The following steps of the research will be:
- Development of an economic model to study the relations between different costs and geometrical – and other – data from the dryer.
- Determination of the indirect energy use of the process, in relationship with the geometry of the construction, the materials used, the production speed etc.
- Comparison of the energy optimum (direct and indirect energy use) and the economic optimum; analysis of the factors that influence the discrepancy between these two optima.

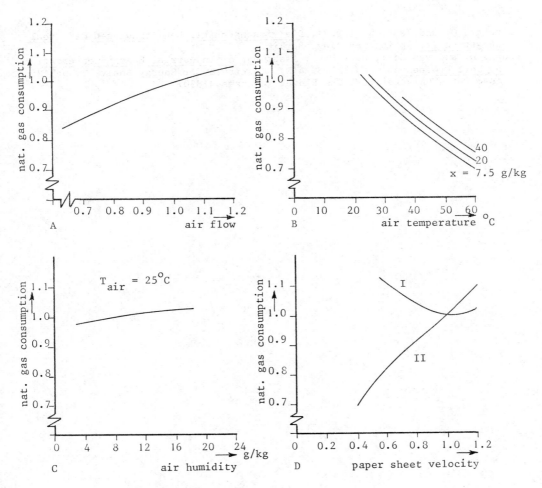

Fig. 3. Results of drying process simulation:
Specific natural gas consumption* as a function of:
A: air flow.
B: air temperature (for several air humidities).
C: air humidity.
D: paper sheet velocity (production speed).
    I : with constant air flow.
    II: with the air flow proportional to the gas consumption.

---

* Reference base = 1,0

ACKNOWLEDGEMENT

The authors would like to express their gratitude to Océ-Nederland B.V. for putting the research facilities at their disposal.

REFERENCES

Krischer, O. and Kast, W. (1978). Die wissenschaftlichen Grundlagen der Trocknungstechnik. Springer-Verlag, Berlin.
Phung, D. and van Gool, W. (1980). Assessment of Industrial Energy Conservation by Unit Processes. ORAU/IEA-80-4(M), Institute for Energy Analysis, Oak Ridge Associated Universities, Oak Ridge. Tennessee37830.

ENERGY MANAGEMENT - A METHODOLOGY FOR PROJECT EVALUATION

T. J. Stenlake

Energy Studies Unit, University of Strathclyde, 100 Montrose Street, Glasgow, UK

ABSTRACT

The object of this paper is to dispel the myth that the only way to do something about higher energy prices is to affect the thermo-dynamic efficiencies of the activities within the organisation.

Techniques of operational research are applied to the whole system and case studies are used to demonstrate the effectiveness of improving the energy related managerial decisions. A methodology for an initial study, applicable to most systems, is developed, providing an ordered approach and indicating areas of greatest potential return.

Aspects of wide applicability such as energy producurement, including purchasing, discounts and delivery options, storage and handling are illustrated. An inventory model for coal storage which takes into account deterioration in the bing, price rises and interest rates, and which saved 5 per cent of the energy bill for a chemical works, is presented and its general application elsewhere is discussed. The value of resource mix models is demonstrated by showing the output of a deterministic simulation in which a linear programming model is used as the core descriptor of a large industrial procurement system. Lagging, maintenance and repair models complete the examination of energy supply.

Demand analysis can be useful too. Load spreading and shedding can alleviate maximum demand and capacity constraints of electrical or gas supply. Space heating savings are easy to spot but are not just a matter of cost-effective insulation, but also of the way in which the building is being used.

Orders of magnitude of current expenditure are used initially to guide the practitioner to areas of greatest potential return. These areas are examined using extreme value analysis to indicate and assess those aspects which would be worthy of a more thorough investigation.

KEYWORDS

Energy management; services management; inventory policy; project evaluation methodology; coal; deterministic simulation linear programming; steam line management.

INTRODUCTION

Boiler houses, water treatment plants, and service distribution systems have often been considered by management to be an unglamourous aspect of the organisation worthy of rare and scant attention. The role of the service manager, if one exists, is likened to that of a forgotten hero. Over the past few years, with higher energy and water costs, there has been a significant shift in the distribution of costs for most organisations, with a consequent increase in the importance of economic service provision.

The Myth: The view that energy consumption is determined by the thermo-dynamic efficiency of the relevant process is widely held.   Fundamental, undisputed laws of physics are quoted as the basis of this opinion. The myth is that the process constraint of the thermo-dynamic efficiency relationship is binding, when in practice this is rarely the case.

Systems are mixtures of processes. The result of operating the myth in a system is a suboptimisation of the use of resources. To operate each process entirely within a system at its optimum individual thermo-dynamic efficiency does not ensure system maximum efficiency.

The myth is that energy management is about saving energy.   Energy, though, is not alone amongst scarce resources, so saving energy doesn't always make sense.

Energy: Energy, like labour, is a basic parameter of the production function. All organisations are consumers of energy, since energy is a necessary input to any activity. There is universal scope for energy management. Energy may be substituted for the other factors of production (materials, capital, labour, etc.). The best production mix is not constant and depends on the demand for goods and services, the natural and man-made entropy of the factors of production, and the set of production functions.

Dispelling the Myth: The success of an organisation is to be measured in the degree to which its goals are achieved. For example, consider an organisation whose goal is to produce a service at minimum cost.   The object of the energy management of that concern should be to further the overall objectives and in doing so should only consider improving the thermo-dynamic efficiencies where this is consistent with the overall objectives. Energy management is about management as well as about engineering.

This paper will demonstrate the value of operational research techniques in this area. Good energy management is about when to save energy, and when to waste energy, and when to do both.

The practitioner of energy management often has to justify any lenthy study and so has to present a case for this to his sponsor. Even where this is not a requirement, it is at any rate prudent to have an ordered approach so that effort is expended in areas of greatest potential return. A methodology for this initial reconnaissance stage is useful and appropriate.

The very first act should be to establish the nature of the overall goals of the organisation and to determine whether, or to what degree, these objectives are being satisfied. Are these goals of a satisfying or optimising type?

To get a feel for the role of energy in the organisation it is useful to spend a few hours gathering data to enable a sketch plan of the system and a table of orders of magnitude to be drawn up. Precision at this stage is not essential. What are the flows of materials?   What are the processes?   What is the annual expenditure

on oil, coal, electricity, water treatment, power plant labour, etc.. What is
the stocking level of these items? How much capital and labour is tied up in
energy procurement? What does a tonne of oil, coal, ice, cost?

These rough and ready numbers will give an idea of where effort might best be
spent. Clearly the largest potential reductions are likely to be in the areas of
most expenditure.

A Case Study: The writer was recently involved in a 3 month project, the purpose
of which was to examine energy procurement for a large chemical works in the United
Kingdom. This works was part of a well managed multi-national corporation.

There were about 50 workshops on site, housing batch-process manufacturing facilities for dyestuffs and organic chemicals. Energy represented about 25% of all
costs and in 1979 energy expenditure was at circa $7million. The author was able
to identify energy cost savings of around 15% which could be achieved by good
housekeeping and without the need to introduce new technologies or re-equipment.
This study is illustrative of the hitherto little recognised value of operational
research techniques in energy management.

Process energy demands were for steam at 3 pressures and electricity. Energy
was purchased in the form of coal, oil or electricity, and converted in a small
power station to produce the process energy requirements. There were 3 coal fired
and 2 oil fired boilers, and 2 steam turbines to produce electricity. There was
also a system of reducing valves to enable the mix of steam pressures delivered to
be altered. Electricity could be generated, bought in, or sold back to the
national grid. Demands were stochastic.

Table 1 is a crude representation of the energy procurement system; that is the
supply of energy to the processes within the works.

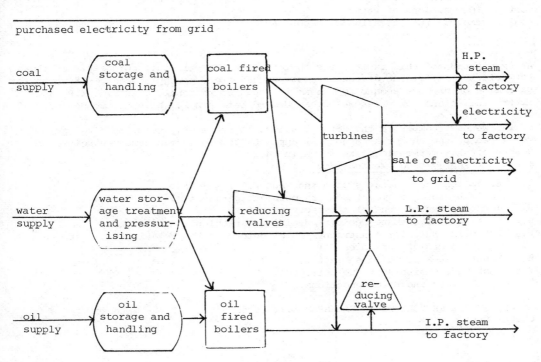

Water was chemically treated in a treatment plant to remove solid impurities and to prevent scaling and corrosion in the boilers and turbines. The water feedpumps took water from the treatment plant and pressurised it so that it could be fed into the boilers.

Oil was held in storage tanks and pumped to boilers as required. Coal could be tipped onto outdoor stockpiles and consequently hauled by contractor to receiving bays, or delivered directly to receiving bays. There were various handling systems to move coal from the receiving bays to the boiler hoppers.

Table 2 shows the orders of magnitude of some of the important parameters of this system.

TABLE 2   Orders of Magnitude

Coal purchases	(49621 tonnes)	$5.75 million	p.a.
Oil purchases	(2899.5 tonnes)	$575,000	p.a.
Water purchases		$150,000	p.a.
Electricity purchases		$900,000	p.a.
Chemical purchases for water treatment		$ 15,000	p.a.
Labour		$450,000	p.a.
Average factory demand for H.P.steam		5,000 lbs/hour	
Average factory demand for I.P. steam		30,000 lbs/hour	
Average factory demand for L.P. steam		100,000 lbs/hour	
Electricity purchased		14270000 k.w.h.	p.a.
Electricity generated		34322000 k.w.h.	p.a.
Electricity sold		5000 k.w.h.	
Price paid for one k.w.h. of imported electricity		6.3 cents	
Price received for one k.w.h. of exported electricity		2.1 cents	
Cost of one tonne of coal		$118	
Cost of one tonne of oil		$200	
Cost of 1000 gallons of water		£0.77	

Ash is removed free of charge by a contractor

An examination of this table shows that coal is the principal purchase. It shows the relative importance of the various purchases and energy demands. If we try to economise on process demand for steam it would be rational to examine I.P. and L.P. (intermediate and low pressure) steam demand before H.P. (high) steam demand.

So far we have completed the first 2 stages in the energy study methodology, developed by the author as a general approach to energy management studies. The stages of this methodology are shown in table 3.

TABLE 3   Energy Study Methodology

1. Establish what is the system and what are its goals.
2. Collect orders of magnitude of important parameters.
3. Identify system control and decisions.
4. Re-examine system to identify decision variables.
5. Establish rules or constraints which determine the relationships between the decision variables.
6. Model system or valid subsystem.
7. Test Model using historic data.
8. Conclusions and Recommendations.

AT ALL STAGES RECYCLE WHERE NECESSARY ...... do not treat the methodology as a check-list.

The identification of system control is a crucial step in any analysis. The crude
system identification and order of magnitude stages can be rushed, but this step is
critical and the thoroughness of decision identification will be reflected in the
conclusions which will be drawn. Decisions are not always at all obvious.

Manually operated decisions were found in the chemical works study. Decisions as
to which boiler, turbine, reducing valve or pump to operate. Decisions as to stock
levels, handling systems to use, delivery and discount options to take, and machine
settings to select. These are all conscious decisions taken by people at the time
of action. But there were also automatic and built-in decisions. Automatic control
equipment on boilers and turbines, which responded in a pre-programmed or designed
manner; and built-in decision, determined by the nature of the plant itself.

The importance of identifying all the decisions is that any improvements that are to
be made in the energy management of the system will be achieved by improving the
decision making within that system ... by making better decisions.

When the control decisions have been established, each must be examined in turn to
determine those parameters which affect the outcomes of each decision. In the case
study, some such parameters were those parameters determining the efficiencies and
capacities of the boilers, turbines and other entities of the power station.

Consider the decision as to the setting of a flow control on a steam reducing valve.
See table 5.

TABLE 5   Steam Reducing Valve System

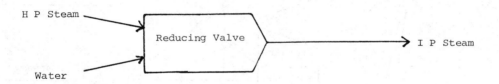

The decision is to decide the output flow from the valve and it is
enacted by the setting of a gate valve. The parameters $x_1, x_2 ... x_5$ are known as
the decision variables and determine the outcome of the decision.

TABLE 6   Decision Variables of Steam Reducing Valve System

x_1 upper capacity limit of reducing valve
x_2 lower capacity limit of reducing valve
x_3 rate of flow of I.P. steam output
x_4 rate of flow of H.P. steam input
x_5 rate of flow of water input

The decision variables limit the decision area within which a decision may be made.
We now proceed to relate these variables to each other by producing the set of
relationships that exist between decision variables. This requires a more detailed
examination of the system. For the reducing valve decision we have a conservation
of flow relationship, a steam water input mix relationship, and two capacity
constraints. These relationships must hold true for any decision that is made,
and therefore restrict the set of possible decisions .

The overall objective of the chemical works study was to minimise the cost of energy procurement, subject to meeting process energy demands. Optimisation of decision making over the whole system is the requirement. Where there are parallel system decisions, individual decision optimisation in isolation of the rest of the system cannot be assumed to lead to global optimisation. Only where there is a series decision is it possible for this decision to be optimised in insolation without compromising global system optimisation.

In the chemical works it was possible in this way to remove immediate consideration of the coal handling and purchase decision, the water treatment decision and the ash handling decision. It was considered that the outcomes of these decisions ny the nature of the particular system, did not affect the optimisation of the operation of the boilers, water pumps, reducing valves, turbines and electricity trading.

This reduced system, shown in table 6, involved 42 decision variables and 51 equations or inequalities relating the decision variables to each other and to the overall objective function.

TABLE 6   Reduced Procurement System

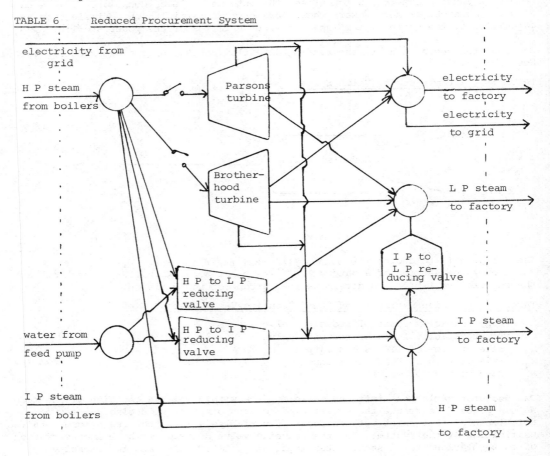

A linear programming model (aside to describe nature of L.P.s) of the type shown in table 7 was developed.

TABLE 7   Linear Programming Model

minimise cost:

subject to 50 constraints : capacity constraints, cost constraints, demands, conservation of flow constraints, flow relationships.

This model was designed to operate over a period of one day which was considered to be a suitable time period over which decisions might be varied. There would be no difficulty in changing the time span of the model.   Process energy demands were entered as input to the model and the output was the minimum possible operating cost with optimal values for the 42 decision variables.   Non linearity of relationships was not a problem.   Zero-one variables, such as those describing the on/off state of turbines were set at one value and then the other for successive runs of the model and the results for the run with minimum cost selected.   It was surprising to find relatively linear efficiency curves for the boilers and turbines.   In practice they tended to operate at much the same output levels over which it was reasonable to make assumptions about linearity. Should this not be the case elsewhere, the problem could be overcome by using non-linear programming, or by successive linear approximation in repeated sweeps of the L.P. model.

The model took about 2½ minutes of c.p.u. time on an I.B.M. mainframe 360 series computer using a simplex algorithm.

The one-day L.P. model was useful for day to day decision making. Its solution involved such a large number of calculations and comparisons that it could be shown that optimal decision making would not be possible, even in this relatively simple system, without a computer aid.

The value of the model was still restricted though.   Historic daily energy demands and actual expenditures were available.   It was possible to repeatedly run the model to simulate optimal decision making over a year, allowing for actual breakdowns in equipment, and using actual historic demands as input.   By comparing with actual incurred costs, the value of improving decision making by using the model was determined.   A shortened version of this deterministic simulation L.P. model was produced and tested on historical data.

It was now possible not only to know the optimal decisions for the existing system, but to get answers to what-if situations.   The client drew up a list of such questions and conclusions and recommendations were produced based on these questionsand the results from the model.   Table 8 shows some of these questions which it was possible to answer in this way.   In addition the model was available as a package on the firm's computer as an aid to future decision making.

TABLE 8.   Results from Power Procurement Model

1. Optimal mix of resources.
2. Benefit of additional reducing valve capacity.
3. Steam or electric water pump decision rules. In the form of a rule of thumb.
4. Cost of outage for various pieces of equipment.
5. Marginal costs of steam and electricity.
6. Effect of freeing the system from demand control.
7. Effect of changes of maximum demand of import electricity.
8. Value of altering the efficiencies of boilers and turbines.
9. Total outage cost.

Models do not have to be complex, computer solved, algorithms. Consider the
following system for the delivery, handling and storage of coal for small coal
power station.   See table 9.

TABLE 9.   A Coal Handling System

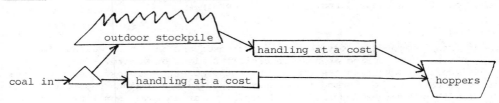

Coal could be discharged directly into hoppers for immediate use, or alternatively
could be stored in a stockpile for future use.   The reason for the stockpile was
to prevent shortage in the event of a strike by the supplier. There were
different delivery and handling costs associated with different coal routes.
Operators in the power station maintained that the coal via outdoor stockpiles
was of poorer quality, although from the same source.   Coal price was determined
by an annual contract which was negotiated 3 months before the expiry of the last
contract.   Historically coal prices have risen by about 23% annually over the past
5 years.

The client had reduced the stock level, having analysed the record of reliability
of the supplier and in line with a general corporate policy of reducing stock
levels to increase group liquidity.   The policy was to set a reorder level of
stock, and buy in coal to maintain this safety stock.

In establishing this policy, they had not considered the whole system and consequently had sub-optimised (or not optimised!).   The author was able to show that
because the price rise could be predicted, and because there was plenty of unused
land available on site, that the current decision was wrong.

The trade-off of holding cost against shortage cost was made, but allowing for an
increase in value of stocks at each price rise.

It was not easy to estimate shortage cost in this case.   However it could be
assumed that the current safety stock level, which had proved sufficient to cover
the event of non-supply over a decade, could be used to given an upper bound on
shortage cost.

Table 10 shows the coal inventory decision trade-off :

TABLE 10:

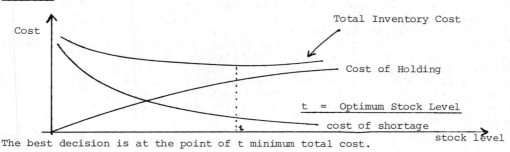

The best decision is at the point of t minimum total cost.

Holding cost was calculated to include handling costs, loss in value due to deterioration, cost of finance, difference in delivery cost over direct delivery, and change in value at price change point.

The solution, shown in Table 11, was to buy in before the price rise. The optimal stocking level at the time of the study was 90 weeks. There was not sufficient land for a 90 week policy, and also taking into account the risk involved in restricting policy options, a recommendation of 26 weeks stock level was made, to be bought in prior to the price rise, followed by a reduction of stock to the current level after the price rise. This involved borrowing about $2.25 million, the cost of which, of course, was included in the holding cost calculation. This new policy was to save the client $500,000 by reducing operating costs on that site alone in the first 9 months of operation. The bought in stock provided security for the finance.

The reasoning behind the saving is quite simple. Assume price-rise of 20% and cost of money also 20%. Buy 12 months sotck prior to the price rise. Assume continuous steady consumption of stock, hence average holding time is 6 months. Interest for 6 months is 10%, therefore a saving of 10% of annual cost after paying for finance. The sum is modified to take into account uncertainty, capacities, handling, deterioration etc.

A decision graph was produced, see table 11, to show action to be taken for different predicted price rises and interest rates. A further refinement would be to allow for uncertainty in the predicted price rise and interest rates.

TABLE 11.   Recommended Coal Inventory Policy

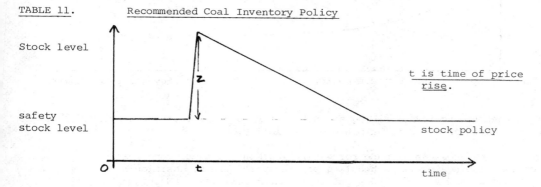

Stock level

t is time of price rise.

safety stock level

stock policy

time

Case Study : A steam Distribution System

A power plant produces steam for distribution through pipework to process locations elsewhere. Losses arise in the distribution network due to radiation through the walls of the pipework and because of leaks. Leaks usually start in a small way and gradually get larger, until a point comes when they should be repaired. The larger the leak the higher the cost of repair, but to repair all the leaks as they occur would mean having a repair team idle for a lot of the time waiting for leaks to occur. How many repair men are needed? Pipes can be lagged to reduce radiation losses. What is the best lagging thickness?

The principal difficulty here is measurement. Flow meters are notoriously inaccurate and errors of around 20% are not uncommon. Working in the area of residuals is generally not productive. Furtunately in this case there was an

holiday on which there was an almost complete process shutdown, and so the total
distribution loss could be established.

Radiation losses were determined by measuring surface temperature, using infra-red
temperature measuring equipment, which can be hired. The actual radiation losses
for given ambient conditions could then be calculated, and an estimate of actual
ambient conditions was obtained from a nearby airport. The effects of additional
increments of lagging thickness in terms of cost saving was found. This was
traded off against the costs of providing these various additional lagging thicknesses. The recommendation was to lag to the thickness which displayed the
greatest saving in cost.

The steam leakage loss is the total loss determined from the shutdown less the
radiation loss. Careful monitoring of leaks as they occurred enabled the comparison between repair policy and cost of leaks to be made. See table 12.

TABLE 12   Leak Repair Decision

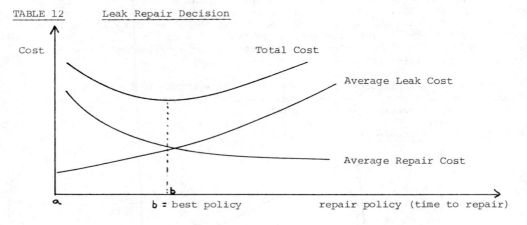

a

b = best policy   repair policy (time to repair)

Note that the policy of saving as much energy as possible, here consistent with
policy position a, is not a minimum cost policy.

Similar analyses can be made for water treatment systems, repair and maintenance
programs for equipment, scheduling of transport fleets, positioning of plant and
warehouses, and vehicle and plant replacement policies. Such analyses can
produce fuel and money savings without the need for better equipment or technology.

In this paper I have concentrated on the area of energy supply because it is here
that a general methodology can usefully be applied. Demand management can benefit
from operational research techniques such as scheduling, optimum resource mix,
game and decision theory. However, in this area applications are more specific
to the particular process or system and it is not easy to use a general approach.

Step changes in maximum demand utility tariffs are often met. These tariff jumps
are used to reflect the fact that it is the peak demand that determines the utility
company's plant investment and capacity. It has been found that by operating
either or both a load-shedding policy, or demand scheduling, that consumer costs
can be held to a lower tariff band. This saving just involves planning.

Space heating often attracts attention and the benefits of economic insulation
of buildings are well known. Buildings display the need for access, which
allows movement of air and consequent heat loss. A general purpose building
may not be suited to its particular use in this respect and by either redesigning

entrances, or scheduling access, savings which can be surprising can be made. Draughts and vents cause about 50% of total heat loss in industrial buildings. Double doors, for example, can have an investment payback time of less than a year.

CONCLUSION.

Orders of magnitude of current expenditure are used initially to guide the practitioner to areas of potential saving. This information can be used to give an estimate of extreme values of potential savings as a result of improving decision making in specific areas of the system under examination.

The methodology of the analysis is to identify the system and its goals, the decisions within it, and those parameters which determine the outcomes of those decisions. Models are produced to describe the decision making process of selected systems, bearing in mind the dangers of sub-optimisation. These models are tested against historical data and also for reasonableness. At any stage of the methodology, there may be feedback to an earlier stage. When the model is tested and accepted by the client as reasonable, conclusions are drawn.

Case studies have been used to illustrate the view that operational research and good housekeeping is as important as new technology and thermo-dynamic efficiency, as a tool for energy management.

DEVELOPMENT TRENDS IN THE FIELD OF WASTE AND RESIDUE INCINERATION

K. J. Thomé-Kozmiensky

*Technische Universität Berlin, Institut für Technischen Umweltschutz,
Fachgebiet Abfallwirtschaft, Kaiserin-Augusta-Allee 5,
D-1000 Berlin 21, Federal Republic of Germany*

Keywords

Recycling, Incineration, Gasification, Degasification, Municipal solid waste, Residues trade and industry, Emmissions

Introduction

Recycling methods are dependent for their success on a number of different parameters:

- t e c h n o l o g i c a l: state of art, availability, compatibility with other processes, operational reliability, etc.

- e c o l o g i c a l: resource conservation, avoidance of the adverse effects of other treatment and disposal methods, environmental impact of the recycling process itself, etc.

- e c o n o m i c: absolute costs and costs as compared with other methods and processes, prices and expected price trends of the recycled products and of the competitive primary raw materials and forms of energy, marketability of the products, structure of the markets, etc.

Waste and residue incineration, which can look back on fifty years of industrial development and which thus constitutes the oldest processing technique for waste treatment, doubtless displays the most sophisticated state of development of all recycling methods. Nevertheless, the development potential of this process, which is effective both as a waste treatment method and for energy recovery and even material recycling, is not yet exhausted. Thanks to the extensive experience gained from the operation of hundreds of incinerators for the treatment of both municipal waste and residues from trade and industry, the drawbacks of this process are also known. Efforts to eliminate these disadvantages and further develop the relevant process technology are being made by engineers and scientists in the operating organizations, the plants themselves, in industry and scientific and research institutes.

In connection with the assumed development trends, let us examine the following postulates:

1. The number of plants incinerating municipal waste and residues from trade and industry will continue to increase in both absolute and relative terms.

2. Existing design consepts will be improved by advances in control technology, process engineering and plant construction.

3. Pollutant emissions will be further reduced.

4. Special processes will be developed to cope with the different characteristics of the various wastes and residues.

5. The construction of large-scaled and centralized plants will be limited to the treatment of hazardous wastes.

1. The number of plants incinerating municipal waste and residues from trade and industry will continue to increase in both absolute and relative terms.

The factors which point to the likelihood of this prediction's proving correct are: dwindling landfilling capacity, the necessary of developing alternative disposal methods and the inability of existing incineration plants to develop alternatives to energy recycling.

While waste generation has increased over the past years and will continue to grow - albeit at a somewhat slower rate - in the foreseeable future, the amount of land available for landfilling operations is continually decreasing. By the year 1990 the landfilling space required for the disposal of the waste generated by a million inhabitants of the Federal Republic of Germany will be exhausted. This applies in particular measure (75 %) to the following Federal State: Baden-Württemberg (33 %), Bavaria (28 %) and Lower Saxony (14 %). An additional problem is the disposal capacity lost through the closure of uncontrolled landfills. In view of this situation, at least 70 disposal contractors are confronted with the necessity of evolving alternative techniques.

The acquisition of new sites is just as problematical for landfilling operations as for other waste treatment plants. It is therefore imperative that the available landfilling space be used as economically as possible. All methods for the volume reduction of waste serve to increase the working life expectancy of landfills.

Today greater demands are made as regards the characteristics of prospective landfill sites than it was the case only a few years ago. Abandoned open-work mining operations, the geological nature of which is such as to allow contamination of the ground and /or surface water, are no longer considered acceptable as landfill sites.

With growing recognition of the necessity of treating landfill emissions-, leachate-, and gasprocess engineering is being increasingly applied in landfilling operations. This means that

landfilling - once all low-cost solution to the problems of
waste disposal - is becoming an expensive waste treatment method.
At present 41 municipal waste incinerators are in operation in
the Federal Republic of Germany. Of these, 2 are more than 15
years old, 6 between 10 and 15 years old and 14 between 5 and
10 years old. This means that in the 8 plants that are more 10
years old the mechanical equipment is likely to require partial
or complete renewal in the foreseeable future. Moreover, owing
to new technological developments in recent years, which have
been designed to improve operational reliability, energy recovery
rates and emission control, some of the 14 plants between 5 and
10 years old will also be requiring renewal of equipment in the
near future.
These renewal measures will be limited exclusively to incineration plants. At present, incineration is still the only waste
treatment method by which energy recycling, i.e. the recovery of
energy from waste, can be realized on a full commercial scale.
New refinement processes for the conversion of waste into storable fuels, the so-called gasification and degasification processes, are still in the development stage and have not yet proved
suitable for fullscale commercial realization.

Since in recent years the high hopes placed in these new processes have tended to defer necessary decisions on the application
of waste treatment processes - a situation which to some extent
still persists today - it is worth briefly examining the state
of art in this field.

The superiority of gasification and degasification processes as
compared with incineration is usually upheld on the basis of the
following arguments:

- Gasification and degasification are less sensitive than incineration to variations in waste composition. High-calorific
  wastes in particular can be treated more easily.

- Process products - gas, oil and char - are storable and transportable. This enables higher-quality products to be marketed
  than is possible in the case of incineration.

- Particularly in the case of degasification processes, smaller
  quantities of gas are generated, which means that there is
  less expenditure for gas scrubbing and the subsequent reprocessing of gas-scrubbing effluent. As a result, specific investment costs are reduced, which in turn means that small-
  scale plants can also be constructed economically.

- High-temperature gasification enables greater volume reduction
  than incineration, which permits an increased saving of landfilling space.

In recent years extensive R & D work has been done on the gasification and degasification of wastes. To verify the above-mentioned arguments, it would seem useful to strike an intermediate
balance, though so far this can only be based essentially on the
results obtained from small-scale plants - only for two gasification processes are larger-scale plants under trial operation:

- With the exception of the Andco-Torrax process, all gasification and degasification processes for the treatment of municipal waste are preceded by a size reduction process, and occa-
sionally by further pre-treatment of the waste. This additional process stage, not required for incineration, is necessary
to ensure reactor density. Entry of coarse waste particles into the reactor is generally prevented by specially constructed charging gates. For the treatment of high-calorific waste,
tires, oil, sludges and plastics, specially developed processes are available in which particular care must be taken as
regards product refinement. Since here potential customers are
obliged to make exacting demands in respect of product quality, plants developed for the treatment of municipal waste will
offer no advantages over incineration plants. Nor can highly
problematical (hazardous) wastes, such as acid sludges from
the re-refining industry, be treated in such plants without
alterations being made to improve the gas cleaning systems and
eliminate corrosion problems. Long-term operational tests are
also required before the co-treatment of waste and sewage
sludge can be applied on a commercial scale. Particular attention must be paid here to effects of heavy metals as regards
corrosion problems and their distribution in the products.

So far it has not proved possible to confirm the claim of lesser sensitivity of gasification and degasification processes
to variations in waste composition, as compared with incineration processes. On the contrary, more extensive pre-treatment
of the waste is necessary for such processes.

- As regards product utilization, a fundamental distinction must
be made between the products obtained from the degasification
of municipal waste and those obtained form the degasification
of homogeneous residues from trade and industry. Of the products generated by the degasification of municipal waste, only
that part of the gas stream not required for process maintenance can be utilized. However, in order to be able to utilize
this product, the consumer will be obliged to adapt his requirements to the properties of this gas. Mixing this gas with
towngas is inadvisable, as this would necessitate additional
processing measures which would be uneconomical in view of the
relatively small quantities of gas involved. The amount of
oils and tars generated are too small to warrant considering
their industrial utilization. The solid residues, greater in
quantity than those produces by incineration in view of additional presence of fixed barbon, must be dumped as their quality is insufficient to permit re-utilization.

- On the other hand, the prospects for the utilization of products from the degasification of high-quality commercial and
industrial wastes are better, though here too further R & D
work is necessary.

- At present,then, we are not in a position to verify the claims
made regarding better utilization of thes process products.

- The markedly smaller quantities of gas generated during degasification as compared with incineration are the root of the
most difficult problems involved in this technique. Both the

new variety of pollutants typical of the degasification process and the problems caused by oils/tars and effluent during
gas cleaning call for new developments which must be successfully implemented before waste degasification can be accepted
as an alternative technology.

Where as during degasification slightly greater, during medium-
temperature gasification approximately the same, quantities of
residues are produced as in incineration, high-temperature gasification causes a greater reduction in the volume of solid residues. However, this advantage can only be considered of real
value, if no profitable use can be made of the waste incineration residues. These are, however, utilized today in various
ways.

So far it has not proved possible to verify the claims made regarding the competitiveness, let alone superiority of these new thermal waste treatment techniques as compared with incineration. If
verifications of these claims should fail to materialize in the
forseeable future, the 60 pyrolysis plants envisaged by Simon for
1990/95 are likely to be constructed, in the majority of cases,
as incineration plants.

2. Existing design concepts will be improved by advances in control technology, process engineering and plant construction

Municipal waste and municipal-type commercial waste differ from
conventional fuels in particular with respect to their heterogeneous composition as regards particle size and heating value.
These factors, together with seasonal fluctuations in quality,
present problems with regard to control of the incineration process. Incinerators must therefore be operated using air quantities which greatly exceed stoichiometric levels. This, in turn,
necessitates the construction of very large combustion chambers
and additional downstream equipment which reduce efficiency.

Design optimization involves the following objectives:

- increased waste throughput
- increased energy recovery rate
- increased plant availability
- prevention or reduction of pollutants.

In order to realize these objectives, various developments can
be envisaged, and we shall now look at each of these briefly:

Basic research and experiments on automation are being carried
out and preliminary results are available. In Japan extensive research is being done into automatic combustion control with the
aim of minimizing NO_x emissions. At present a large-scale waste
incineration plant is under construction at which automatic combustion control technology will be tested and further developed
on a full commercial scale.

Another promising technique for optimization and control of the
combustion process is the recirculation of part of the flue gas
stream. A corresponding research project is being carried out in
the Institute for Environmental Protection Technology at the Tech-

nical University of Berlin in conjunction the municipal waste incineration plant at Rotterdam. As fluegas recirculation enables temperatures, volume flows and concentrations of the gas components involved in incineration to be altered, depending on the channeling of the recirculated gases, the waste can be predried, the combustion chamber temperature lowered, chemically latent energy released and controllability improved particularly during partial load operation. Tests have been delayed by finincial difficulties besetting the AVR in Rotterdam, which means that results are not expected fo be available until next year.

Utilization of the thermal energy from waste incineration is being increasingly improved. In other words, the energy is being used not only for electrictiy generation, but also for supplying district heating networts, drying sewage sludge, heating gardening nurseries, manufacturing distilled water, etc. Design measures, such as the optimization of combustion chamber configuration and heating surface arrangement, redesign of grates and alterations in the combustion air supply are serving to increase availability and efficiency.

3. Pollutant emissions will be further reduced

Even assuming that measures designed to reduce the generation of pollutants continue to be successfully implemented, waste incineration will still produce emissions whose reduction is dependent on the development of suitable technologies. This applies particularly to the exhaust gases and the solid residues.

Generally speaking, the installation of electrostatic precipitators for cleaning the exhaust gases from municipal waste incinerators has yielded satisfactory results as regards dust removal.

Where the hazardous nature of the particulate emissions makes it advisable, more efficient measures can be taken, e.g. the installation of special fabric and ceramic filters.

On the other hand, the development of devices for the removal of gaseous pollutants is a research field which is still comperatively in its infancy. The relatively stringent air pollution control regulations (TA Luft) in Germany - as compared with other countries - , together with the recognition of the fact that by shifting the exhaust gas problem to another demium-water or soil - emissions problems are merely given a different emphasis and, in some cases, made more complex, have given rise to integral approaches to this problem. Research is being done into wet as well as semi-dry processes. Within the next few years the technology on which these processes are based is likely to be perfected.

During waste incineration, various solid residues are generated: ash and slag from the combustion chamber, fly ash from the boiler and dust filter and, in future, the residues from the pollutant gas cleaning. Since waste incinerator slag is relatively insusceptible to leaching, and exhibits a favourable grain size distribution and good compactability characteristics, it is being increasingly used in the Netherlands, northern Germany and Switzerland for the construction of traffic noise reduction barriers. A similar application is envisaged for Denmark, though certain legislative regulations remain to be met here. At present, all the solid redidues generated by a waste incinerator are usually composidet. This mixture fo slag and fly ash contrains a considerably high concentration of soluble salts. Tests performed in Bavaria have shown that slag which contains no fly ash exhibits only small quantities of water-soluble components, these being found mainly in the fly ash. If, then the fly ash, which accounts for 5 - 10 % by weight of the solid residues, is kept separate from the slag, the latter can be reutilized or dumped even in gravel pits without risk. However, since the dumping of the fly ash requires special precautionary measures, tests are being carried out at the Ingolstadt municipal waste incinerator to solidify the fly ash with the aim of absorbing the pollutants.

On the basis of the positive results obtained to date, the constuction of a pilot plant with a fly ash thoughput capacity of 2,5 t/h is planned. The work done in this field indicates new possibilities of further reducing the environmental impact from waste incineration.

In the USA municipal waste contains up to 0,7 % of aluminium and 0,3 % of other non-ferous metals, which are found in concentrated form in the incinerator slag. Tests are being performed to process these materials with a view to their recovery.

In Japan tests are underway to extract the heavy metals from the effluent of gascleaning units by means of organic solvents and subsequently to precipitate them out in the form of salts or subject them to electrolytic separation. In Sweden the exhaust gases from a municipal incinerator are cleaned in a filter containing caustic lime. The lime is subsequently spread on land used for agricultural and forestry purposes. Success and possible side-effects of this method are still being tested.

Although at this stage it is impossible to make predictions about the prospects of success of a number of these techniques, nevertheless a tendency is perceptible towards highly individual approaches to the problem of emission reduction with the aims of pollution control and resource recovery.

4. Special processes will be developed to cope with the different characteristics of the various wastes and residues

Whereas it was long considered desirable to incinerate as many materials as possible together, there is today growing recognition of the fact that plants which take into account the proparties of the particular type of waste involved demonstrate better control characteristics both with regard to efficiency and environmental impact. This became apparent initially through the construction of special plants for the incineration of hazardous wastes. The co-combustion of municipal waste and sewage sludge also appears to be stagnating in favour of incinerator systems which take into account the characteristics of the particular fuels used. This does not necessarily rule out the development of combined incinerator plants with mutual energy utilization, such as are in operation for the co-combustion of municipal waste and sewage sludge in Bologne and for the incineration of hazardous wastes in Vienna.

Consequently further tests are underway to separate a fuel fraction from municipal waste (RDF oder BRAM), the USA being the leader in this field. For example, a process developed in the USA
has been adapted to German waste composition and is being tried
out on a pilot scale in the Ruhr district. Both here and in other
pilot and experimental plants efforts are being made to determine
whether this waste processing can be profitably applied at justi-
fiable expense. A point requiring particular attention here is
the fact that after separation of the combustible fraction and
the scrap metal a considerable residue must be landfilled, unless
some further use can be found for it, e.g. by composting or additional sorting processes.

5. The construction of large-scale and centralized plants will be limited fo the treatment of hazardous wastes

So far most of the incineration plants in the Federal Republic
of Germany have been located in densely populated areas. However,
these incinerate only slightly more than 25 % of the total amount
of waste generated, compared with approximately 70 % in Switzerland and Japan, for instance. Although the specific waste treatment costs decrease with increasing plant size, it will only be
possible to push the percentage of waste treated by incineration
clearly above the 30 % mark if small-scale incineration plants,
which are competitive in respect of efficiency and emission levels, can be successfully constructed. This will only be possible
by designing plants according to the modular construction principle. The feasibility of this design principle has already been
demonstrated by the construction of small-scale industrial facilities. With rising prices for energy, there is a growing tendency on the part of industry to meet at least a part of in-plant
energy requirements by incineration of internally generated industrial residues.

Experiences gained from the incineration of industrial residues
in small and medium-scale plants could prove of great benefit for
these developments. However, it will be necessary to develop new
techniques, such as those using the incineration of wood wastes
and cable residues, for example.

A fact which should be borne in mind by those who advocate the
exclusive construction of large-scale plants is that by far the
greater part of waste disposal costs arise in the area of collection, transfer and transport, where as the actual waste treatment
costs are comparatively low. In addition, energy is consumed and
emissions caused by the transportation of the waste. However, in
planning small and medium-scale waste treatment plants, energy
utilization must be considered as an important factor from the
start. Potential customers for the energy are public utilities
and commercial enterprises.

The greatest development potential for waste incineration lies
in the field of small and medium-scale facilities, provided that
emission regulations can be met.

INDUSTRIAL ENERGY AS AN INVESTMENT ALLOCATION PROBLEM

B. de Vries, D. Dijk and E. Nieuwlaar

Vrije Studierichting, Chemische Laboratoria, Rijks Universiteit Groningen, The Netherlands

ABSTRACT

A fruitful approach to the diverse ways of increasing energy productivity in industry is to make a distinction between basic processes, auxiliary processes and energy services. Each class consists of different types of unit-equipment, for which production functions derived from engineering and economic data are estimated. We discuss ways to construct these production functions from available information and consider the role of a lower, thermodynamic bound on the exergy input. Next, the optimized conservation vs. additional capital curve is calculated for a set of three different unit-processes: evaporation, compression and steam generation. Sensitivity of the results for changes in important parameters is discussed.

KEYWORDS

Industrial energy conservation; unit-processes; production function; second-law efficiency; investment allocation.

ENGINEERING AND ECONOMIC PRODUCTION FUNCTIONS

In micro-economic production theory most research focuses on statistical examination of existing plant operation to establish input substitution and effects of scale and technical change. Although the statistical treatment of the data and the mathematical formulation of the resulting empirical production functions have become highly sophisticated, this kind of research yields only quite limited insight because the engineering principles underlying the process are not explicitly taken into account.

Many of the industrial processes are governed by a small number of relatively simple physical-chemical laws. This enables one to consider the ex ante construction of production functions. Chenery has been the first to set forth this approach in a systematic way. (1949, 1953) His concept of an industrial process is that input materials "receive" factor services. The factor services are energy supply, energy transformation and control. Accordingly, the core of the process description is the material transformation function, connecting output quantity with input quality, output quality and energy requirements by way of design laws. These may be either analytical or, for more complex processes, empirical. In combination with an energy supply function, accounting for the energy conversions taking place within the system boundary, the material transformation function is called engineering production function.

To make the transition to the economic production function one needs the input functions. These connect the process variables of the engineering production function to economic quantities, that is quantities to which a price can be attached, like weight, capacity, manhours etc. Often it is here that problems arise. Only for the most simple or simplified processes it is possible to characterize equipment ("capital goods") in terms of physical parameters such that on multiplication with prices one gets the capital cost. In practice capital cost has to be correlated from empirical data to the relevant physical parameters, as a way to bypass the tedious task of calculating inputs for the multitude of possible design, material and construction configurations.

If an economic production function can be constructed along these lines, at least partly ex ante from engineering principles, its merits are evident. Even if confined to the most important process variables, it contains most of the wealth of ways of performing industrial processes. In the literature examples of such ex ante construction of production functions are given by Chenery (1949, 1953), Smith (1974), Pearl (1975) and others. An outstanding feature of these production functions is that continuous isoquants can be drawn due to the existence of well-defined substitution variables, like insulation thickness for heat transfer, pipe diameter for fluid flow and conductor diameter for electric power transmission. For each isoquant drawn, a set of process, design and material parameters has to be specified. Usually, innovation can be related to changes in these parameters.

Most production functions derived from engineering laws are non-homogeneous. Only if the process is reduced to its most elementary form one may get a production function of the simple Cobb-Douglas type:

$$Q = a \ K^{\alpha} \ \Phi^{\beta} \qquad\qquad [1]$$

assuming only two production factors, capital and exergy. Here Q represents the size or scale of the process, i.e. the design capacity of the equipment. Using a total cost function of the form

$$C = \sum_i \ p_i \ X_i = p_k \ K + p_{\varphi} \ \Phi \qquad\qquad [2]$$

it can easily be shown that for a given ratio of exergy price p_{φ} and capital price p_k all points of minimum cost operation are on a straight line

$$K = (\alpha / \beta) \ p \ \Phi \qquad\qquad [3]$$

with $p = p_{\varphi} / p_k$. Increase of output under minimum cost conditions must satisfy

$$\frac{C}{C_r} = \frac{K}{K_r} = \frac{\Phi}{\Phi_r} = \left(\frac{Q}{Q_r}\right)^{1/(\alpha + \beta)} \qquad\qquad [4]$$

as is clear from combining the above three equations (r refers to a reference installation). In the K - Φ plane the points of minimum cost operation are along a straight line.

In engineering literature there is much information on the factor (capital, energy) input as a function of equipment size. Is it possible to derive substitution data from these factor scaling equations? For exergy one can divide the input to perform a given process into two terms:
- a lower thermodynamic bound, representing the minimum exergy requirement for reversible operation and the irreversibilities characteristic for the rate and type of process considered, and
- a sum of terms that are exponential in size.

In equational form:

$$\Phi = \Phi_0 + \sum_i \alpha_i Q^{y_i}$$ [5]

In relation to well-defined reference unit-equipment, this yields

$$\frac{\varphi - \varphi_0}{\varphi_r - \varphi_0} = \left(\frac{1 - \varepsilon}{1 - \varepsilon_r}\right)\frac{\varepsilon_r}{\varepsilon} = \frac{\sum_i \alpha_i Q^{y-1}}{\sum_i \alpha_i}$$ [6]

with $\varepsilon = \varphi_0/\varphi$ either the second-law efficiency when φ is exergy or the first-law efficiency when φ denotes enthalpy input. Q_r is taken unity, which implies that $\Phi - \Phi_r = \sum_i \alpha_i$. A similar equation can be written for other factor inputs.

Suppose now that from detailed engineering studies the α_i and y_i can be established for both exergy and capital, then it is possible to construct a "pure scaling" curve which can be approximated by

$$K = K_r \left(\frac{\Phi}{\Phi_r}\right)^{y_k/y_\varphi}$$ [7]

In reality, however, the Φ, K vs. Q relations contain not only scaling but also substitution. This can be illustrated by an example. Let an energy conserving investment ΔK be defined, which is size-independent ($y_i = 0$) and results in an exergy input reduction which is linear in size ($y_i = 1$). Neglecting different α-values this would imply that for increasing size the extra capital cost per unit product ($\sim 1/Q$) falls whereas the amount of exergy saved per unit product remains constant. Thus, only above a given size the investment will be done.

This type of scale-induced substitution[1] can, at least theoretically, be separated from "pure scaling" as defined above. Further on we do a calculation for steam boilers, with satisfying results. Although in this case the method appears feasible, more complex equipment will not yield reliable estimates in view of the difficulty of assessing α- and y-values. In these cases substitution data have to be derived from explicit engineering analysis of substitution. This will be discussed further on. First, we take a short side-step to discuss the above introduced lower thermodynamic bound.

THERMODYNAMICS: THE LOWER BOUND ON EXERGY

From a thermodynamic point of view energy inputs and outputs are to be evaluated in terms of exergy (availability, available work), that is, all flows are aggregated by way of their specific (i.e. per unit of product) exergy. As is well known the use of specific enthalpy of combustion in evaluating fuel flows gives only a slight deviation from the exergetic evaluation. For the sensible heat of mass flows the difference may be considerable.

From thermodynamics it is known that there is a lower bound on the exergy requirement to perform a process, reached in the limit of a reversible process - infinitely slow and at zero output, by definition. Let us call this lower bound Φ_0. For a chemical reaction it equals the change in Gibbs free energy between products and reactants, ΔG. More in general it is for steady state steady flow systems equal to

[1] In their study on industrial energy conservation Van Gool and Phung (1979) interpret Q as "complexity" and assume y_φ to be negative. Then, eqn. [7] can be considered as the engineering basis for a Cobb-Douglas type of production function. Complexity in this sense is meant to be a set of interdependent substitution variables of a technical nature.

the sum of all in- and outgoing exergy flows, $\sum_i (H_i - T_0 S_i)$ with T_0 the temperature of the surroundings.
The so-called effectiveness or second-law efficiency is defined as

$$\varepsilon = \frac{\Phi_0}{\Phi_{in}} = \frac{\Phi_0}{\Phi_{out}} \; n_{ex} \qquad\qquad [8]$$

with n_{ex} the exergetic effiency and is a measure of the degree to which the ideal, reversible process is approached.

To establish the values of Φ_{out}, Φ_{in} and the lower bound Φ_0 one has to use proper accounting rules as to system boundary and time horizon and material and fuel properties. Once this is agreed upon, the difference between Φ_{in} and Φ_0 can be identified with losses due to dissipational processes such as heat, mass and charge transfer across finite potential differences. It depends on the actual process characteristics and can be written as:

$$\Phi_{in} - \Phi_0 = T_0 \, \Delta S_{irr} \qquad\qquad [9]$$

with ΔS_{irr} the irreversible entropy increase for system and environment in producing one unit of output.

Any process can be thought of as a series of unit operations; the way to reduce the overall exergy loss is by either eliminating or reducing ΔS_{irr} for one or more unit processes. Elimination involves new process technologies with different design and unit process configuration. The three processes to separate uranium isotopes, diffusion, ultracentrifuge and laser, may serve as an example. Reduction of the exergy loss can be achieved by way of substitution, scaling up and innovation. Heat transfer, fluid flow and electric power transmission are among the unit processes for which the exergy loss is governed by simple equations, at least in first approximation. They are well known as respectively Fourier's law, Newton's law and Ohm's law. Processes involving mass transfer and chemical reactions are more complicated.

Several problems and possibilities as to the proper choice of Φ_{out}, Φ_{in} and Φ_0 arise. One may wish to omit certain input and/or output flows from an exergetic evaluation as a result of specific supply or demand conditions, like having no possibility to use hot water or conversely getting it free from an adjacent plant; explicit rules are required to establish a lower thermodynamic bound. If only dissipational processes are involved, Φ_0 will be zero. However, sometimes the exergy content of (part of) the products is not recovered within the given system boundary and time horizon; the process now is not an exergy store but an exergy sink. This yields a positive value of Φ_0 that is the basis for the distinction between gross and net energy requirement as made in the IFIAS report (1974) on energy analysis. On the other hand, Φ_0 may become negative for processes like paper manufacturing and petroleum refining, due to the omission of the input materials exergy content.

Figure 1 serves as an example to illustrate the role of the lower bound on exergy input for large-scale power generation. To our knowledge, Cootner and Löf (1965) are the only ones who tried to establish the marginal capital cost of increasing steam pressure and temperature for power generation of a simple Rankine cycle. This resulted in a supply curve of thermal efficiency for the year 1957, with steam pressure and temperature as the substitution variables. Trenkler (1976) observes that the construction of a general cost vs. efficiency function is impossible, due to lacking capital cost data. He does present, however, a set of

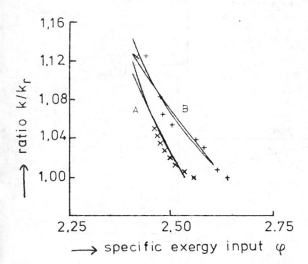

relative capital cost vs. efficiency data for varying steam pressures and temperatures and for varying number and conditions of reheat and feed-water heating cycles for a 300 MWe coal-fired power-plant without flue gas desulfurisation. Figure 1 shows the isoquants which result from least-square fits of these data to the form

$$\frac{k}{k_r} = \left(\frac{\varphi - \varphi_0}{\varphi_r - \varphi_0}\right)^s \qquad [10]$$

with $\varphi - \varphi_0$ the substitutable part of the specific exergy input. As has been discussed previously , this limit is from the second law point of view equal to 1. But as long as heat is an intermediate step in power generation, it is more appropriate to take a higher value. It turns out that for $\varphi_0 = 1$ correlation coefficients are 0.92 - 0.93; for higher values of φ_0 based on adiabatic combustion of coal (1.3 - 1.55) or on material constraints (2.25) the correlation improves continuously up to values of $\varphi_0 = 2.4$. For a more systematic treatment of the role of a lower bound on

Figure 1 Specific exergy input vs. normalized specific capital input for large-scale electric power generation. The data are from Cootner (1965), curve A (s=-0.36 for $\varphi_0=1$, s=-0.18 for $\varphi_0=2.25$, $r^2=0.93$) and from Trenkler (1976), curve B (s=-0.79 for $\varphi_0=1$, $r^2=0.93$, s= -0.14 for $\varphi_0=2.25$, $r^2=0.95$).

curve-fitting results we refer to the work of Berry and Meshkov (1979). From the case studied here the conclusion must be that the best possible fit is not necessarily consistent with a meaningful choice of a lower bound on thermodynamic or technical grounds. For the form of the production function we have chosen (eqn. 10), the elasticity of substitution σ is critically dependent on the lower bound; any calculation of σ is meaningless unless reference is made to a lower bound assumption.

ENERGY CONSERVATION IN INDUSTRIAL PROCESSES

One of the problems in assessing the energy conservation potential and its cost in industrial processes is the large diversity and complexity of energy use patterns. To deal with this fact we divide the process - related capital goods into three classes: basic, auxiliary and energy equipment (see Fig. 2). To each "unit-process" such as separation, melting, transport, power generation etc. corresponds "unit-equipment". For each one a production function is estimated, which represents all possible substitutions of non-energy production factors for energy and includes estimates of the effects of innovation and feedstock, product and fuel quality parameters. In following this approach we are much indebted to the analysis by Van Gool, Phung a.o. (1979), which can be considered as the first systematic approach for estimating energy conservation by way of unit-processes.

Among the investments that increase the energy productivity of the basic unit-equipment are: increase of the number of trays in a distillation column, increase of the number of stages in an evaporator, more furnace-insulation etc. They are hard to separate from innovative changes like improved column design, better catalysts etc. Increased energy-productivity for functions like transport, cold storage, effluent treatment etc. can be realized by additional investment in auxiliary unit-equipment such as compressors, electric motors etc. A third way to conserve

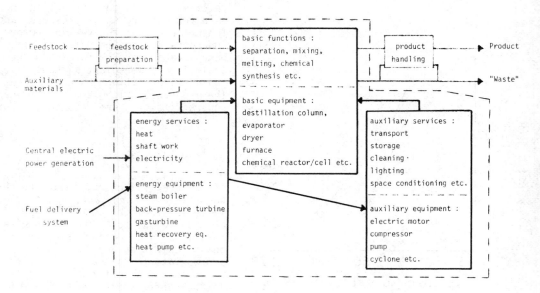

Figure **1**   Schematic representation of industrial processes and the role of energy services.

energy is to reduce energy losses in the energy sypply system, either within (steam boiler, heat pump etc.) or outside (central electric power generation, oil refinery) the industrial system under consideration. As usual in a production function approach, it is hard to isolate the substitution and scale effects from technical progress and external parameters like feedstock quality, environmental constraints, etc. In integrating unit-equipment within or between classes, additional energy conservation can be realized; then the analysis, however, becomes increasingly more complex. Important examples are heat recovery equipment and application of heat pumps and bottoming power cycles. It becomes more and more evident that integrating energy and material flows offers fascinating perspectives for the longer term. It is still unclear to what extent system reliability and complexity will be a more stringent constraint than economic factors.

First we will outline a way to construct an aggregate cost-energy function for a set of unit-processes competing for a finite amount of capital available for energy conservation. The unit-equipment considered here are an evaporator, an electrically driven compressor and a steam boiler. It will be clear from Fig. 2 that only a very limited part of the energy conservation potential is considered; future work aims at extending the system boundaries of the unit equipment and including energy conserving investments of a more integrating nature.

In this section we present a working model for calculating the optimal allocation of investments over several unit processes. This is equivalent to minimizing the over-all process energy requirement given a fixed amount of capital available for energy conserving investments.

As our starting point we take a Cobb-Douglass type of production function as given by eqn. [1] for each unit-process. Using specific quantities, $x_i = X_i/Q$, eqn [1] changes with respect to a specified reference state (k_r, φ_r) at the same output level into

$$-\frac{\varphi}{\varphi_r} = \left(\frac{k}{k_r}\right)^{-\alpha/\beta} \qquad [11]$$

For convenience we put $-\frac{\alpha}{\beta} = s$. It can be easily verified that low absolute values of s indicate that the slope of the isoquant will be rather steep and therefore much capital has to be invested in order to realize only moderate energy savings. The reverse holds equally good for high absolute values of s. Since we are interested in energy savings we find it convenient to introduce the following fractions

$$\alpha = \frac{p \, \Delta\varphi}{\varphi_r} \quad \text{and} \quad \beta = \frac{\Delta k}{k_r} \qquad [12]$$

where $\Delta\varphi = \varphi_r - \varphi$, $\Delta k = k - k_r$ and p the energy price. Now, let us suppose that the production of a fixed amount of a certain good involves n independent unit processes. We then have n equations of the form

$$\alpha_i = 1 - (\beta_i + 1)^{s_i} \qquad i = 1 \ldots n \qquad [13]$$

We further suppose that for each unit process s can be determined from a fit of the empirical data (i.e. α_i and β_i) to eqn. [13].Once we have thus established a continuous relation between α_i and β_i our problem can be formulated as follows: find the maximum of $\Delta C_\varphi = \Sigma_i \varphi_{ri} \, \alpha_i$ under the constraint that the sum of all extra investments, ΔK,does not exceed a fixed percentage of the capital stock $K_r = \Sigma k_{ri}$. Now we can solve our problem, using the method of Lagrange, by seeking the maximum in

$$L(\beta_i) = \Sigma_i \, p_i \varphi_{ri} \left[1-(\beta_i + 1)\right]^{s_i} + \lambda \left[\Delta K - \Sigma_i \beta_i \, k_{ri} \right] \qquad [14]$$

Differentiating [14] with respect to β_i and equating the result to zero, we get

$$\lambda = - \frac{\varphi_{ri} \, s_i}{k_{ri}} (\beta_i + 1)^{s_i - 1} \qquad [15]$$

which holds for all values of i between 1 and n.
From [15] it follows that we can express any β_i in terms of every β_j; thus:

$$\frac{\varphi_{ri} \, s_i}{k_{ri}} (\beta_i + 1)^{s_i - 1} = \frac{\varphi_{rj} \, s_j}{k_{rj}} (\beta_j + 1)^{s_j - 1} \qquad [16]$$

From this we have,

$$\beta_i = A_{ji} (\beta_j + 1)^{s_i^*} - 1 \qquad [17]$$

where $A_{ji} = \left[\frac{\varphi_{rj} \, s_j \, k_{oi} \, p_j}{\varphi_{ri} \, s_i \, k_{oj} \, p_i} \right]^{\frac{1}{s_i-1}}$ and $s_i^* = \frac{s_j - 1}{s_i - 1}$ $\qquad [18]$

In passing we note that only relative energy prices matter in the calculation of the elements A_{ji}. Recalling that $\beta_i = \frac{\Delta k_i}{k_{oi}}$ and defining $K_0 + \Delta K = K$ we have after summation the following implicit eqn.

$$K = \Sigma_i \, k_{oi} \, A_{ji} (\beta_j + 1)^{s_i*} \qquad [19]$$

which can be solved for β_j. Substitution of the result in eqn. [17] gives the other β_i's. In using this procedure one has to keep in mind however that none of the β_i's may, as a result of the optimisation, become negative for this would imply disinvestment in the corresponding unit-process. A negative value of some particular β_i simply indicates that the i-th unit-process cannot succesfully compete with the other unit-processes for the energy conservation funds. However at larger values of ΔK the i-th unit-process may as yet come into the picture. Once the β_i's for some value of ΔK are known we can calculate the corresponding α_i's from eqn. [13] and determine ΔC_φ, i.e. the money worth of the total energy saved.

As stated before we consider by way of example a set of unit-processes consisting of an evaporator, a steam boiler and electrically driven compressors. For evapora-tors the main energy conservation measures are increasing the number of effects and increasing the number of heat exchangers. The values we have used are taken from Van Gool a.o. (1979) and are mainly based on a detailed engineering analysis. For steam boilers we constructed s-values from manufacturer's data on cost and efficiency-improvement of economisers, air preheat, spray tower and flue gas condensor. The results vary from -2 for a 15 t/h fire-tube boiler to -0.7 for a 45 t/h boiler.

We also established a s-value from the previously discussed scale-and-substitution curve, using the data of Fichtner (1967) for natural-gas-fired steam boilers. For the "pure scaling" we assume that boiler energy losses are for one-sixth part size-dependent radiation losses and for five-sixth flue-gas heat losses which are pro-portional with size; K is assumed to scale completely. The resulting curves show a large difference between the "pure scaling" and the scale-plus-substitution curve; the latter one can be interpreted as minimum total cost design for the given size. Thus calculated values for s vary from -1.6 (y = 0.6) to -0.5 (y = 0.8), which is in good agreement with the values found above. Part of the capital for substitution, it should be noted, has been used for an increase of steam pressure; correcting for this would yield slightly lower s-values. For electrically driven compressors the most important improvement is the use of more stages and inter-cooling to approach more closely isothermal operation. Besides, one can improve on electric motor performance.

Table 1 summarizes the values of s, k_r, φ_r and p[a] we used in our calculations.

	evaporator	boiler	compressor
s	-0.67	-1.00	-0.25
k_r	DFL 1.3 x 10^6	DFL 0.6 x 10^6	DFL 0.7 x 10^6
φ_r	11.00 MW	3.52 MW	1.64 MW
p	1.19	1.00	5.28

TABLE 1

[a] all prices are normalized to the natural gas price .

The results of the allocation procedure for different values of ΔK (up to 50% of K_o) are shown in Fig. 3. From these results we draw the following conclusions :
- both the s-value and the reference situation of the pieces of unit-equipment determine the extent to which energy can be conserved at a given cost
- the optimisation procedure yields only significant changes in the aggregate capital-energy relation when the s-values and the characteristics of the average equipment can be estimated with a fair accuracy.

REFERENCES

Chenery, H.B. (1949). Engineering production functions. Quarterly Jounal of
    Economics, 63, 507
Chenery, H.B. (1953). Process and production functions from engineering data.
    In: W.W. Leontief et al., Studies in the structure of the american economy.
Cootner, P. and G. Löf (1965). Water demand for steam electric generation.
    Resources for the Future.
Fichtner (Ingenieurgesellschaft). Survey of steam production costs - using as fuel
    hard coal, heavy fuel oil and natural gas (1966).
Gool, W.van , D. Phung a.o. (1979). Assessment of industrial energy conservation
    by unit processes. Oak Ridge, Institute for Energy Analysis, Report ORAU/
    IEA-79-17 (R).
IFIAS-report nr. 6 (1974). Energy Analysis. Stockholm.
Meshkov, N. and R.S. Berry. Can thermodynamics say anything useful about the
    economics of production? ICEUM-II Conference: Changing energy use futures,
    volume I, p. 374. Pergamon Press (1979).
Pearl, D.J. and J.L. Enos (1975). Engineering production functions and technological
    progress. The journal of industrial economics, 24, 55.
Smith, V. (1966). Investment and Production (Harvard Press). See also: J. Marsden,
    Engineering foundations of production functions. Journal of Economic Theory,
    9, 124-140 (1974).
Trenkler, H. (1976). Grenzen der Möglichkeiten besserer Energienutzung aus
    betriebswirtschaftlicher Sicht; H. Haas und H. Maghon, Mögliche Wirkungs-
    gradverbesserung bei Dampf und Gasturbinen. Both in: Technische Mitteilungen,
    69. Jahrgang, Heft 9/10. sept./okt.

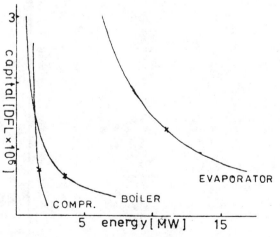

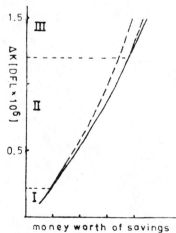

Figure 3 a   shows the isoquants for three types
of unit equipment. In each curve the reference
situation is indicated by a cross-mark.

Figure 3 b shows a plot of   K·vs. the money-worth
of energy savings. The solid line represents the
optimal investment curve. The dashed lines show
the energy savings when all the available capital
is invested in the evaporator only (left one) or
in both the evaporator and the steam boiler (right
one). In the optimal investment scheme up to   K=0.2
investment takes place in the evaporator only (I) ;
up to   K=1.2 investment takes place in both evaporator and steam boiler (II). Finally, also the
compressor comes into the picture (III).